T0178504

Physics of Semiconductor Devices

Massimo Rudan

Physics of Semiconductor Devices

 Springer

Massimo Rudan
University of Bologna
Bologna
Italy

ISBN 978-1-4939-4699-0   ISBN 978-1-4939-1151-6 (eBook)
DOI 10.1007/978-1-4939-1151-6
Springer New York Heidelberg Dordrecht London

Printed on acid-free paper

Springer is part of Springer Science+Business Media (www.springer.com)

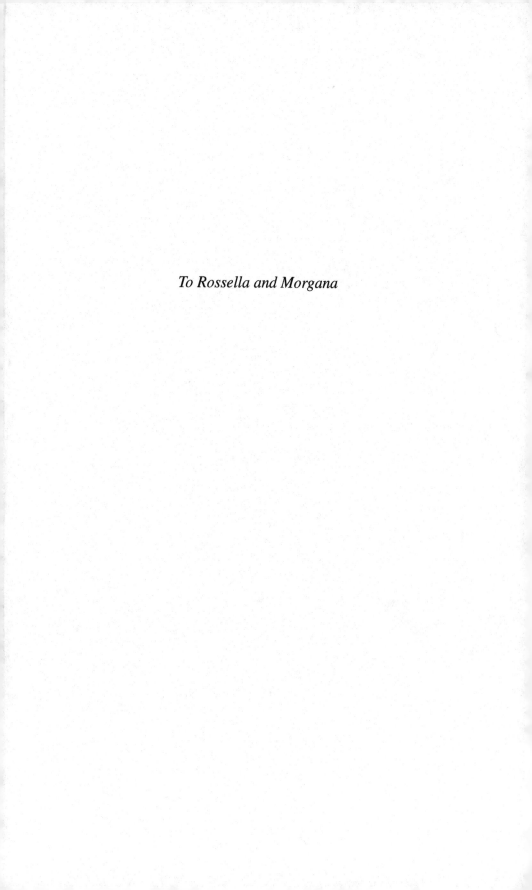

To Rossella and Morgana

Preface

This volume originates from the lectures on Solid-State Electronics and Microelectronics that I have been giving since 1978 at the School of Engineering of the University of Bologna. Its scope is to provide the reader with a book that, starting from the elementary principles of classical mechanics and electromagnetism, introduces the concepts of quantum mechanics and solid-state theory, and describes the basic physics of semiconductors including the hierarchy of transport models, ending up with the standard mathematical model of semiconductor devices and the analysis of the behavior of basic devices. The ambition of the work has been to write a book, self contained as far as possible, that would be useful for both students and researchers; to this purpose, a strong effort has been made to elucidate physical concepts, mathematical derivations, and approximation levels, without being verbose.

The book is divided into eight parts. Part I deals with analytical mechanics and electromagnetism; purposedly, the material is not given in the form of a resumé: quantum-mechanics and solid-state theory's concepts are so richly intertwined with the classical ones that presenting the latter in an abridged form may make the reading unwieldy and the connections more difficult to establish. Part II provides the introductory concepts of statistical mechanics and quantum mechanics, followed by the description of the general methods of quantum mechanics. The problem of bridging the classical concepts with the quantum ones is first tackled using the historical perspective, covering the years from 1900 to 1926. The type of statistical description necessary for describing the experiments, and the connection with the limiting case of the same experiments involving massive bodies, is related to the properties of the doubly-stochastic matrices. Part III illustrates a number of applications of the Schrödinger equation: elementary cases, solutions by factorization, and time-dependent perturbation theory. Part IV analyzes the properties of systems of particles, with special attention to those made of identical particles, and the methods for separating the equations. The concepts above are applied in Part V to the analysis of periodic structures, with emphasis to crystals of the cubic type and to silicon in particular, which, since the late 1960s, has been and still is the most important material for the fabrication of integrated circuits. Part VI illustrates the single-electron dynamics in a periodic structure and derives the semiclassical Boltzmann Transport

Equation; from the latter, the hydrodynamic and drift-diffusion models of semiconductor devices are obtained using the moments expansion. The drift-diffusion model is used in Part VII to work out analytically the electrical characteristics for the basic devices of the bipolar and MOS type. Finally, Part VIII presents a collection of items which, although important *per se*, are not in the book's mainstream: some of the fabrication-process steps of integrated circuits (thermal diffusion, thermal oxidation, layer deposition, epitaxy), and methods for measuring the semiconductor parameters.

In the preparation of the book I have been helped by many colleagues. I wish to thank, in particular, Giorgio Baccarani, Carlo Jacoboni, and Rossella Brunetti, who gave me important suggestions about the matter's distribution in the book, read the manuscript and, with their observations, helped me to clarify and improve the text; I wish also to thank, for reading the manuscript and giving me their comments, Giovanni Betti Beneventi, Fabrizio Buscemi, Gaetano D'Emma, Antonio Gnudi, Elena Gnani, Enrico Piccinini, Susanna Reggiani, Paolo Spadini.

Last, but not least, I wish to thank the students, undergraduate, graduate, and postdocs, who for decades have accompanied my teaching and research activity with stimulating curiosity. Many comments, exercises, and complements of this book are the direct result of questions and comments that came from them.

Bologna   Massimo Rudan
September 2014

Contents

Acronyms

Abbreviations

BJT   Bipolar junction transistor. A transistor whose operation is obtained by a suitable arrangement of two *p-n* junctions. The term "bipolar" is used because both electrons and holes are involved in the device functioning.

BTE   Boltzmann transport equation. The equation expressing the continuity of the distribution function in the phase space.

CVD   Chemical vapor deposition. A deposition process in which the material to be deposited is the product of a chemical reaction that takes place on the surface of the substrate or in its vicinity.

DD   Drift-diffusion. The term indicates a transport model for semiconductors made, for each band, of the balance equation for the carrier number and average velocity. Such equations contain the electric field and the magnetic induction; as a consequence, their solution must be calculated consistently with that of the Maxwell equations. Compare with the HD model.

HD   HydroDynamic. The term indicates a transport model for semiconductors made, for each band, of the balance equation for the carrier number, average velocity, average kinetic energy, and average flux of the kinetic energy. Such equations contain the electric field and the magnetic induction; as a consequence, their solution must be calculated consistently with that of the Maxwell equations. Compare with the DD model.

IC   Integrated circuit. Also called chip or microchip. An assembly of electronic circuits on the same plate of semiconductor material. The idea was proposed in the early 1950s, and demonstrated in 1958; it provided an enormous improvement, both in cost and performance, with respect to the manual assembly of circuits using discrete components.

IGFET   Insulated-gate field-effect transistor. A device architecture demonstrated in the early 1930s. Its first implementation (1960) using a thermally-oxidized silicon layer gave rise to the MOSFET architecture.

LASER   Light amplification by stimulated emission of radiation.

LOCOS   Local oxidation. The technological process consisting in depositing and patterning a layer of silicon nitride over the areas where the substrate's oxidation must be prevented.

MBE   Molecular beam epitaxy. A low-temperature epitaxial process based on evaporation.

MIS   Metal insulator semiconductor. Structure made of the superposition of a metal contact, an insulator, and a semiconductor.

MOS   Metal oxide semiconductor. Structure made of the superposition of a metal contact, an oxide that acts as an insulator, and a semiconductor.

MOSFET   Metal-oxide-semiconductor, field-effect transistor. A transistor whose active region is an MOS structure. In last-generation devices the insulator may be deposited instead of being obtained by oxidizing the semiconductor underneath. The MOSFET has been for decades, and still is, the fundamental device of the integrated-circuit architecture.

PDE   Partial-differential equation.

PVD   Physical vapor deposition. A deposition process in which the material to be deposited does not react chemically with other substances.

SGOI   Silicon-germanium-on-insulator. A technology analogous to SOI. SGOI increases the speed of the transistors by straining the material under the gate, this making the electron mobility higher.

SOI   Silicon on insulator. A technology introduced in 1998 for semiconductor manufacturing, in which the standard silicon substrate is replaced with a layered structure of the silicon-insulator-silicon type. SOI reduces the parasitic capacitances and the short-channel effect in MOS transistors.

SOS   Silicon on sapphire. A technological process that consists in growing a thin layer of silicon on a wafer made of sapphire (Al_2O_3).

VLSI   Very-large-scale integration. The process of creating an integrated circuit by placing a very large number of transistors in a single chip.

List of Tables

Part I
A Review of Analytical Mechanics and Electromagnetism

Chapter 1
Analytical Mechanics

1.1 Introduction

The differential equations that govern the dynamical problems can be derived from variational principles, upon which the theories of Euler and Lagrange, and those of Hamilton and Jacobi, are based. Apart from being the source of the greatest intellectual enjoyment, these theories have the definite advantage of generality. Their concepts can in fact be extended to cases where the Newtonian equation of dynamics does not apply. Among such cases there are the equations governing the electromagnetic field and those related to the quantum description of the particles' motion.

The invariance property of the Lagrange equations with respect to a change of coordinates gives origin to the concept of generalized coordinates and conjugate momenta; in turn, the introduction of the Hamilton equations provides a picture in which the intrinsic symmetry of the roles of coordinates and momenta becomes apparent. Basing on the concept of conjugate coordinates, the time evolution of a particle or of a system of particles is described in the phase space instead of the coordinate space. This chapter and the next one illustrate the basic principles of Analytical Mechanics. Their purpose is to introduce a number of concepts that are not only useful *per se*, but also constitute a basis for the concepts of Quantum Mechanics that are introduced in later chapters. A third chapter devoted to Analytical Mechanics describes a number of important examples, that will be applied to later developments illustrated in the book.

As the velocity of particles within a semiconductor device is small with respect to that of light, the non-relativistic form of the mechanical laws is sufficient for the purposes of this book. The relativistic form is used only in a few paragraphs belonging to the chapter devoted to examples, to the purpose of describing a specific type of collision between particles. This chapter starts with the description of the Lagrangian function and the Lagrange equations, that are derived as a consequence of the variational calculus, followed by the derivation of the Hamiltonian function and Hamilton equations. Next, the Hamilton–Jacobi equation is derived after discussing the time–energy conjugacy. The chapter continues with the definition of the Poisson

© Springer Science+Business Media New York 2015
M. Rudan, *Physics of Semiconductor Devices,*
DOI 10.1007/978-1-4939-1151-6_1

brackets and the derivation of some properties of theirs, and concludes with the description of the phase space and state space.

1.2 Variational Calculus

Consider a real function $w(\xi)$ defined in the interval $\xi \in [a, b]$ and differentiable in its interior at least twice. The first two derivatives will be indicated with $\dot{w}$ e $\ddot{w}$. Now, define the integral

$$G[w] = \int_a^b g(w, \dot{w}, \xi)\, d\xi, \qquad (1.1)$$

where the form of the function $g(w, \dot{w}, \xi)$ is prescribed. If (1.1) is calculated for any function w fulfilling the requisites stated above, with a and b fixed, the result is some real number G whose value depends on the choice of w. By this procedure, (1.1) establishes a correspondence $G[w]$ between a set of functions and a set of numbers. Such a correspondence is called *functional*.

It is interesting to extend to the case of functionals some concepts and procedures that apply to functions proper; among these it is important the concept of extremum. In fact, one defines the *extremum function* of a functional by a method similar to that used for defining the extremum point of a function: some w is an extremum of G if a variation dw in (1.1) produces a variation dG that is infinitesimal of an order higher than that of dw. The procedure by which the extremum functions are calculated is called *variational calculus*.

To proceed it is necessary to define the variation dw. For this one lets $\delta w = \alpha \eta$, with $\eta(\xi)$ an arbitrary function defined in $[a, b]$ and differentiable in its interior, and α a real parameter. The function δw thus defined is the finite variation of w. The sum $w + \delta w$ tends to w in the limit $\alpha \to 0$. As a consequence, such a limit provides the infinitesimal variation dw. For simplicity it is convenient to restrict the choice of η to the case $\eta(a) = \eta(b) = 0$, so that $w + \delta w$ coincides with w at the integration boundaries for any value of α.

Now, replacing w with $w + \alpha \eta$ in (1.1) makes G a function of α, whose derivative is

$$\frac{dG}{d\alpha} = \int_a^b \left[\frac{\partial g}{\partial(w + \alpha \eta)} \eta + \frac{\partial g}{\partial(\dot{w} + \alpha \dot{\eta})} \dot{\eta} \right] d\xi. \qquad (1.2)$$

According to the definition given above, if w is an extremum function of G then it must be $\lim_{\alpha \to 0} dG/d\alpha = 0$; in this case, in fact, the first-order term in the power expansion of G with respect to α vanishes, and the variation of G becomes second order in α or higher. In conclusion, one proceeds by imposing that the right hand side of (1.2) vanishes for $\alpha = 0$. Then, integrating by parts the second term in brackets yields

$$\int_a^b \frac{\partial g}{\partial w} \eta\, d\xi + \left[\frac{\partial g}{\partial \dot{w}} \eta \right]_a^b = \int_a^b \left(\frac{d}{d\xi} \frac{\partial g}{\partial \dot{w}} \right) \eta\, d\xi \qquad (1.3)$$

where, in turn, the integrated part vanishes because $\eta(a) = \eta(b) = 0$. This makes the two integrals in (1.3) equal to each other. On the other hand, such an equality must hold for any choice of η due to the arbitrariness of the latter. It follows that the integrands must be equal to each other, namely,

$$\frac{d}{d\xi}\frac{\partial g}{\partial \dot{w}} = \frac{\partial g}{\partial w}. \tag{1.4}$$

The relation (1.4) thus found is a second-order differential equation in the unknown w, whose explicit form is easily calculated:

$$\frac{\partial^2 g}{\partial \dot{w}^2}\ddot{w} + \frac{\partial^2 g}{\partial w \partial \dot{w}}\dot{w} + \frac{\partial^2 g}{\partial \xi \partial \dot{w}} = \frac{\partial g}{\partial w}. \tag{1.5}$$

Its solution provides the extremum function w sought. To actually find a solution one must associate to (1.4) suitable boundary conditions, e.g., $w(a) = w_a$, $\dot{w}(a) = \dot{w}_a$, or $w(a) = w_a$, $w(b) = w_b$, and so on. As g does not contain $\ddot{w}$, (1.4) is linear with respect to $\ddot{w}$. It is also worth noting that, consistently with what happens in the case of functions proper, the above calculation does not provide in itself any information about w being a minimum or maximum of G. Such an information must be sought through additional calculations.

The analysis above is easily extended to the case where g depends on several functions w_1, w_2,... and the corresponding derivatives. Introducing the vectors $\mathbf{w}(\xi) = (w_1, w_2, \ldots, w_n)$, $\dot{\mathbf{w}}(\xi) = (\dot{w}_1, \dot{w}_2, \ldots, \dot{w}_n)$ one finds that the set of n extremum functions $w_i(\xi)$ of functional

$$G[\mathbf{w}] = \int_a^b g(\mathbf{w}, \dot{\mathbf{w}}, \xi)\,d\xi \tag{1.6}$$

is the solution of the set of differential equations

$$\frac{d}{d\xi}\frac{\partial g}{\partial \dot{w}_i} = \frac{\partial g}{\partial w_i}, \qquad i = 1, \ldots, n, \tag{1.7}$$

supplemented with suitable boundary conditions. Equations (1.7) are called *Euler equations* of the functional G.

Each equation (1.7) is homogeneous with respect to the derivatives of g and does not contain g itself. As a consequence, the differential equations (1.7) are invariant when g is replaced with $A\,g + B$, where $A, B \neq 0$ are constants. As the boundary conditions of w_i are not affected by that, the solutions w_i are invariant under the transformation. Moreover, it can be shown that the solutions are invariant under a more general transformation. In fact, consider an arbitrary function $h = h(\mathbf{w}, \xi)$ and let $g' = g + dh/d\xi$, this transforming (1.6) into

$$G'[\mathbf{w}] = A\int_a^b g(\mathbf{w}, \dot{\mathbf{w}}, \xi)\,d\xi + h(\mathbf{w}_b, \xi_b) - h(\mathbf{w}_a, \xi_a). \tag{1.8}$$

When each w_i is replaced with $w_i + dw_i$, the terms involving h do not vary because the variations vanish at the boundaries of the integration domain. Thus, the variation of

G' equals that of the integral, namely, it is of a higher order than dw_i. In conclusion, the extremum functions of G are also extremum functions of G'. This means that the solutions $w_i(\xi)$ are invariant under addition to g of the total derivative of an arbitrary function that depends on $\mathbf{w}$ and ξ only. This reasoning does not apply if h depends also on the derivatives $\dot{\mathbf{w}}$, because in general the derivatives of the variations do not vanish at the boundaries.

1.3 Lagrangian Function

In many cases the solution of a physical problem is achieved by solving a set of second-order differential equations of the form $\ddot{w}_i = \ddot{w}_i(\mathbf{w}, \dot{\mathbf{w}}, \xi)$. For instance, for non-relativistic velocities the law of motion of a particle of constant mass m is Newton's law $\mathbf{F} = m\mathbf{a}$ which, in a Cartesian frame, takes the form

$$m\ddot{x}_i = F_i(\mathbf{r}, \dot{\mathbf{r}}, t), \qquad i = 1, 2, 3. \tag{1.9}$$

In (1.9), $\mathbf{r}(t) = (x_1, x_2, x_3)$ is the particle's position vector[1] at t. In the following the particle's velocity will be indicated with $\mathbf{u} = \dot{\mathbf{r}}$.

Equations (1.9) and (1.7) have the same form, as is easily found by observing that t is the analogue of ξ and x_i is that of w_i. As a consequence, one may argue that (1.9) could be deduced as Euler equations of a suitable functional. This problem is in fact the inverse of that solved in Sect. 1.2: there, the starting point is the function g, whose derivatives provide the coefficients of the differential equations (1.7); here, the coefficients of the differential equation are given, while the function g is to be found. For the inverse problem the existence of a solution is not guaranteed in general; if a solution exists, finding the function g may be complicated because the process requires an integration. In other terms, the direct problem involves only the somewhat "mechanical" process of calculating derivatives, whereas the inverse problem involves the integration which is, so to speak, an art.

When dealing with the dynamics of a particle or of a system of particles, the function g, if it exists, is called *Lagrangian function* and is indicated with L. The equations corresponding to (1.7) are called *Lagrange equations*. The expression of the Lagrangian function depends on the form of the force F_i in (1.9). Some examples are given in the following. It is important to note that by "system of particles" it is meant a collection of particles that interact with each other. If there were no interactions it would be possible to tackle the dynamics of each particle separately; in other terms, each particle would constitute a system in itself, described by a smaller number of degrees of freedom.

---

[1] The units in (1.9) are: $[m] = \text{kg}$, $[\mathbf{r}] = \text{m}$, $[\dot{\mathbf{r}}] = \text{m s}^{-1}$, $[\ddot{x}_i] = \text{m s}^{-2}$, $[F_i] = \text{N}$, where "N" stands for Newton.

1.3.1 Force Deriving from a Potential Energy

Consider the case of a force deriving from a potential energy, namely $\mathbf{F} = -\mathrm{grad}\,V$ with $V = V(\mathbf{r}, t)$, so that (1.9) becomes

$$m\dot{u}_i = -\frac{\partial V}{\partial x_i}. \tag{1.10}$$

Using the replacements $\xi \leftarrow t$, $w_i \leftarrow x_i$, $g \leftarrow L$ and equating (1.7) and (1.10) side by side yields

$$\frac{\partial L}{\partial x_i} = -\frac{\partial V}{\partial x_i}, \qquad \frac{\mathrm{d}}{\mathrm{d}t}\frac{\partial L}{\partial u_i} = \frac{\mathrm{d}}{\mathrm{d}t}(mu_i), \qquad i = 1, 2, 3. \tag{1.11}$$

The first of (1.11) shows that the sum $T = L + V$ does not depend on the coordinates x_i. Inserting $L = T - V$ into the second of (1.11) and taking $i = 1$ shows that the difference $\Phi = \partial T / \partial u_1 - mu_1$ does not depend on time either, so it depends on the u_i components at most. Integrating Φ with respect to u_1 yields $T = mu_1^2/2 + T_1(u_2, u_3, t)$, with T_1 yet undetermined. Differentiating this expression of T with respect to u_2, and comparing it with the second of (1.11) specified for $i = 2$, yields $T = m(u_1^2 + u_2^2)/2 + T_2(u_3, t)$, with T_2 undetermined. Repeating the procedure for $i = 3$ finally provides $T = m(u_1^2 + u_2^2 + u_3^2)/2 + T_0(t)$, with T_0 an undetermined function of time only. The latter, in turn, can be viewed as the time derivative of another function h. Remembering the invariance property discussed at the end of Sect. 1.2 with reference to (1.8), one lets $T_0 = 0$. In conclusion, indicating with u the modulus of $\mathbf{u}$ it is $T = mu^2/2$, and the Lagrangian function reads

$$L = \frac{1}{2}mu^2 - V. \tag{1.12}$$

The derivation of (1.12) may appear lengthy. However, the procedure is useful because it is applicable to forces of a more complicated form.

1.3.2 Electromagnetic Force

Consider a charged particle subjected to an electromagnetic field and let m, e be its mass and charge, respectively. The particle's velocity $\mathbf{u}$ is assumed to be non-relativistic. The electromagnetic field acts on the particle with the *Lorentz force* (Sect. 4.11) [2]

$$\mathbf{F} = e(\mathbf{E} + \mathbf{u} \wedge \mathbf{B}), \tag{1.13}$$

---

[2] The units in (1.13) are: $[\mathbf{F}] = \mathrm{N}$, $[e] = \mathrm{C}$, $[\mathbf{E}] = \mathrm{V\,m^{-1}}$, $[\mathbf{u}] = \mathrm{m\,s^{-1}}$, $[\mathbf{B}] = \mathrm{V\,s\,m^{-2}} = \mathrm{Wb\,m^{-2}} = \mathrm{T}$, where "N", "C", "V", "Wb", and "T" stand for Newton, Coulomb, Volt, Weber, and Tesla, respectively. The coefficients in (1.13) differ from those of [4] because of the different units adopted there. In turn, the units in (1.14) are: $[\varphi] = \mathrm{V}$, $[\mathbf{A}] = \mathrm{V\,s\,m^{-1}} = \mathrm{Wb\,m^{-1}}$.

where the electric field $\mathbf{E}$ and the magnetic induction $\mathbf{B}$ are in turn expressed through the scalar potential $\varphi = \varphi(\mathbf{r}, t)$ and the vector potential $\mathbf{A} = \mathbf{A}(\mathbf{r}, t)$ as (Sect. 4.5)

$$\mathbf{E} = -\text{grad}\varphi - \frac{\partial \mathbf{A}}{\partial t}, \qquad \mathbf{B} = \text{rot}\mathbf{A}. \tag{1.14}$$

Letting $i = 1$ in (1.9) one finds from (1.13) $m\dot{u}_1 = e(E_1 + u_2 B_3 - u_3 B_2)$. Using for E_1, B_3, B_2 the expressions extracted from (1.14) yields

$$m\dot{u}_1 + e\left(\frac{\partial A_1}{\partial t} + u_2 \frac{\partial A_1}{\partial x_2} + u_3 \frac{\partial A_1}{\partial x_3}\right) = e\left(-\frac{\partial \varphi}{\partial x_1} + u_2 \frac{\partial A_2}{\partial x_1} + u_3 \frac{\partial A_3}{\partial x_1}\right). \tag{1.15}$$

Now, using $u_i = \dot{x}_i$ transforms the term in parentheses at the left hand side of (1.15) into $dA_1/dt - u_1 \, \partial A_1/\partial x_1$, which gives (1.15) the more compact form

$$\frac{d}{dt}(mu_1 + eA_1) = \frac{\partial}{\partial x_1}(e\mathbf{u} \cdot \mathbf{A} - e\varphi). \tag{1.16}$$

Similar expressions are found for $i = 2, 3$. Comparing with (1.7) in the same manner as in Sect. 1.3.1 yields

$$\frac{\partial L}{\partial x_i} = \frac{\partial}{\partial x_i}(e\mathbf{u} \cdot \mathbf{A} - e\varphi), \qquad \frac{d}{dt}\frac{\partial L}{\partial u_i} = \frac{d}{dt}(mu_i + eA_i), \qquad i = 1, 2, 3. \tag{1.17}$$

Note that (1.17) reduce to (1.11) when $\mathbf{A} = 0$, with $e\varphi = V$. The first of (1.17) shows that the sum $T = L + e\varphi - e\mathbf{u} \cdot \mathbf{A}$ does not depend on the coordinates x_i. Inserting $L = T - e\varphi + e\mathbf{u} \cdot \mathbf{A}$ into the second of (1.17) transforms the latter into $d(\partial T/\partial u_i)/dt = d(mu_i)/dt$. Like in Sect. 1.3.2 the procedure eventually yields $T = mu^2/2$. In conclusion, the Lagrangian function of a particle subjected to the Lorentz force (1.13) is

$$L = \frac{1}{2}mu^2 - e\varphi + e\mathbf{u} \cdot \mathbf{A}. \tag{1.18}$$

It is shown in Sect. 4.5 that the $\mathbf{E}$ and $\mathbf{B}$ fields are invariant under the *gauge transformation*

$$\varphi \leftarrow \varphi - \frac{\partial h}{\partial t}, \qquad \mathbf{A} \leftarrow \mathbf{A} + \text{grad}h, \tag{1.19}$$

where $h(\mathbf{r}, t)$ is an arbitrary function. Using (1.19) in (1.18) transforms the terms containing the potentials as

$$-e\varphi + e\mathbf{u} \cdot \mathbf{A} \leftarrow -e\varphi + e\mathbf{u} \cdot \mathbf{A} + e\frac{dh}{dt}, \tag{1.20}$$

namely, the transformed Lagrangian function differs from the original one by the total derivative of an arbitrary function that depends on position and time only. As a consequence, the solutions $x_i(t)$ are invariant under the gauge transformation (1.19). This is easily understood by observing that the invariance of the $\mathbf{E}$ and $\mathbf{B}$ fields makes the Lorentz force (1.13) invariant as well. As a consequence, the particle's dynamics is not influenced by the gauge transformation.

1.3.3 Work

The elementary work exerted by a force $\mathbf{F}$ acting on a particle of mass m during the time dt is $\mathbf{F} \cdot d\mathbf{r}$, where $\mathbf{r}$ is the particle's position at t in a Cartesian frame and $d\mathbf{r} = \mathbf{u}dt$ the elementary displacement. Let $P = \mathbf{r}(t = a)$, $Q = \mathbf{r}(t = b)$ be the boundaries of the particle's trajectory. The work exerted from P to Q is found by integrating $\mathbf{F} \cdot d\mathbf{r}$ over the trajectory, namely,

$$\int_P^Q \mathbf{F} \cdot d\mathbf{r} = m \int_a^b \dot{\mathbf{u}} \cdot \mathbf{u} dt = \frac{1}{2}m \int_a^b \frac{du^2}{dt} dt = T(b) - T(a), \quad (1.21)$$

where the relation $T = mu^2/2$ has been used. The exerted work is then equal to the variation of T, which is the same quantity that appears in (1.12, 1.18) and is called *kinetic energy* of the particle.

If a system having n degrees of freedom is considered instead of a single particle, the work exerted by the forces is defined as the sum of terms of the form (1.21). As a consequence, the kinetic energy of the system is the sum of the kinetic energies of the individual particles. The expression of the system's kinetic energy in Cartesian coordinates is

$$T = \sum_{i=1}^n \frac{1}{2} m_i u_i^2 = \sum_{i=1}^n \frac{1}{2} m_i \dot{x}_i^2, \quad (1.22)$$

that is, a positive-definite quadratic form in the velocities. The masses in (1.22) take the same value when they are referred to the same particle. When other types of coordinates are used, the kinetic energy is still a second-degree function of the velocities, however the function's coefficients may depend on the coordinates (an example is given in Sect. 2.8).

When a force deriving from a potential energy $V = V(\mathbf{r}, t)$ is considered, like that of Sect. 1.3.1, the integrand of (1.21) becomes $-\text{grad} V \cdot d\mathbf{r}$. To calculate the integral it is necessary to account for the explicit dependence of V on t by using mutually consistent values of $\mathbf{r}$ and t; in other terms, the integral in (1.21) can actually be calculated only after determining the function $\mathbf{r}(t)$. An exception occurs when V has no explicit dependence on time; in this case one finds

$$\int_P^Q \mathbf{F} \cdot d\mathbf{r} = -\int_P^Q \text{grad} V \cdot d\mathbf{r} = -\int_P^Q dV = V(P) - V(Q), \quad (1.23)$$

namely, to calculate the integral it suffices to know the boundaries of the trajectory. Moreover, when $V = V(\mathbf{r})$ the Lagrangian function (1.12) does not depend explicitly on time either. It is shown in Sect. 1.6 that in this case also the sum $T + V$ of the kinetic and potential energies is independent of time. A dynamical property that does not depend on time is called *constant of motion*. A force field that makes $T + V$ a constant of motion is called *conservative*.

When a force of the form $\mathbf{F} = e\,(\mathbf{E} + \mathbf{u} \wedge \mathbf{B})$ is considered, like that of Sect. 1.3.2, the scalar multiplication by $d\mathbf{r} = \mathbf{u}\,dt$ shows that the second term of the force does not contribute to the work because $\mathbf{u} \wedge \mathbf{B} \cdot \mathbf{u} = 0$ (Sect. A.7). Remembering the first of (1.14), the integral corresponding to that of (1.23) reads

$$\int_P^Q \mathbf{F} \cdot d\mathbf{r} = -e \int_P^Q \left(\text{grad}\varphi + \frac{\partial \mathbf{A}}{\partial t} \right) \cdot d\mathbf{r}. \qquad (1.24)$$

If the electromagnetic field is independent of time, the calculation is the same as in (1.23) and the exerted work is $e\varphi(P) - e\varphi(Q)$.

1.3.4 Hamilton Principle—Synchronous Trajectories

From the analysis of Sect. 1.2 it follows that the solutions $x_i(t)$ of the motion equations (1.9) are the extremum functions of the functional

$$S[\mathbf{r}] = \int_a^b L(\mathbf{r}, \dot{\mathbf{r}}, t) \, dt. \qquad (1.25)$$

On the other hand, $\mathbf{r}(t)$ describes the particle's trajectory. The latter is also called *natural trajectory* to distinguish it from the $\mathbf{r} + \delta\mathbf{r}$ trajectories that are obtained through a variation. In summary, the natural trajectory of the particle is the extremum function of (1.25). This statement is called *Hamilton principle*.

The integration boundaries in (1.25) determine a time interval $b - a$ that measures the motion's duration between the initial and final position of the particle, $\mathbf{r}(a)$ and $\mathbf{r}(b)$ respectively. The duration is the same also for the $\mathbf{r} + \delta\mathbf{r}$ trajectories. In fact, remembering the derivation of Sect. 1.2, the variation $\delta\mathbf{r}$ vanishes at the integration boundaries, so that any trajectory obtained through a variation has the same initial and final positions as the natural one at the same instants a and b. Moreover, any position $\mathbf{r} + \delta\mathbf{r}$ between $\mathbf{r}(a)$ and $\mathbf{r}(b)$ is considered at the same instant as the position $\mathbf{r}$ of the natural trajectory. For this reason the $\mathbf{r} + \delta\mathbf{r}$ trajectories of the functional (1.25) are called *synchronous*.

1.4 Generalized Coordinates

The extremum functions are calculated as shown in Sect. 1.3 also when a system of N particles, instead of a single particle, is considered. The starting point is still (1.9), where a new index is introduced to distinguish the masses. The number of coordinates that describe the motion of all particles in the system is not necessary equal to $3\,N$; in fact, a number of constraints may exist that limit the relative positions of the particles. As a consequence, letting $n \leq 3\,N$ indicate the number of degrees of freedom of the system, the set $x_1(t), \ldots, x_n(t)$ suffices to determine the positions of the particles at time t.

Depending on the problem in hand it may be more convenient to use a new set of coordinates $q_1(t), \ldots, q_n(t)$ instead of the Cartesian set $x_1(t), \ldots, x_n(t)$. For this, it is necessary to extend the calculation of the extremum functions to the case where the new set is used. Let the relation between the old and new coordinates be

$$
\begin{cases}
q_1 & = & q_1(x_1, \ldots, x_n, t) \\
& \vdots & \\
q_n & = & q_n(x_1, \ldots, x_n, t)
\end{cases}
\qquad
\begin{cases}
x_1 & = & x_1(q_1, \ldots, q_n, t) \\
& \vdots & \\
x_n & = & x_n(q_1, \ldots, q_n, t)
\end{cases}
\tag{1.26}
$$

The coordinates q_i, whose units are not necessarily a length, are called *generalized coordinates*. Their time derivatives $\dot{q}_i = dq_i/dt$ are called *generalized velocities*. The explicit dependence on time in (1.26) is present if a relative motion of the two frames exists: e.g., the relations $q_1 = x_1 - v_0 t$, $q_2 = x_2$, $q_3 = x_3$ transform the $x_1 x_2 x_3$ set into the $q_1 q_2 q_3$ set, that moves with respect to the former one with the velocity v_0 along the first axis.

Differentiating q_i twice with respect to time and using the first of (1.26) provides a relation of the form $\ddot{q}_i = \ddot{q}_i(x_1, \dot{x}_1, \ddot{x}_1, \ldots, x_n, \dot{x}_n, \ddot{x}_n, t)$. The above, after eliminating the second derivatives $\ddot{x}_1, \ldots, \ddot{x}_n$ through (1.9), becomes $\ddot{q}_i = \ddot{q}_i(x_1, \dot{x}_1, \ldots, x_n, \dot{x}_n, t)$. Finally, replacing $x_1, \dot{x}_1, \ldots, x_n, \dot{x}_n$ extracted from the second of (1.26) yields

$$
\ddot{q}_i = \ddot{q}_i(\mathbf{q}, \dot{\mathbf{q}}, t), \qquad i = 1, \ldots, n,
\tag{1.27}
$$

where $\mathbf{q}$ indicates the set $q_1, \ldots, q_n$, and $\dot{\mathbf{q}}$ indicates the corresponding derivatives. Equations (1.27) have the same form as (1.9), hence they must be deducible as the extremum functions of a functional. To show this, one starts from (1.25) by writing the Lagrangian function in the new set of coordinates. A rather lengthy calculation based on the chain-differentiation rule yields

$$
\frac{d}{dt} \frac{\partial L}{\partial \dot{q}_i} = \frac{\partial L}{\partial q_i}, \qquad i = 1, \ldots, n,
\tag{1.28}
$$

that is, the Lagrange equations written in the q_i coordinates. Specifically, (1.28) turn out to be the Lagrange equations of the functional

$$
S[\mathbf{q}] = \int_a^b L(\mathbf{q}, \dot{\mathbf{q}}, t) \, dt.
\tag{1.29}
$$

This result is very important because it shows that the Lagrange equations are invariant under a change of coordinates of the type (1.26).

The solution of (1.28) provides the time evolution of the coordinates q_i describing the particles' motion. As (1.28) are n second-order equations, to determine their solution it is necessary to specify at $t = a$ the values of the n functions q_i and of the correspondent derivatives $\dot{q}_i$, namely, a total of $2n$ constants. The function

$p_i = \partial L / \partial \dot{q}_i$ is called *generalized momentum* or *conjugate momentum* of q_i. From this definition and from (1.28) it follows

$$p_i = \frac{\partial L}{\partial \dot{q}_i}, \qquad \dot{p}_i = \frac{\partial L}{\partial q_i}, \qquad i = 1, \ldots, n. \tag{1.30}$$

The derivative $\dot{p}_i$ is called *generalized force*. Due to the definitions (1.30), the generalized momentum and force depend on the same coordinates as the Lagrangian function, namely, $p_i = p_i(\mathbf{q}, \dot{\mathbf{q}}, t)$, $\dot{p}_i = \dot{p}_i(\mathbf{q}, \dot{\mathbf{q}}, t)$.

1.5 Hamiltonian Function

From (1.30) one derives the following expression of the total derivative with respect to time of the Lagrangian function:

$$\frac{\mathrm{d}L}{\mathrm{d}t} = \frac{\partial L}{\partial t} + \sum_{i=1}^{n} \left(\frac{\partial L}{\partial q_i} \dot{q}_i + \frac{\partial L}{\partial \dot{q}_i} \ddot{q}_i \right) = \frac{\partial L}{\partial t} + \sum_{i=1}^{n} (\dot{p}_i \dot{q}_i + p_i \ddot{q}_i). \tag{1.31}$$

The quantity in parentheses in (1.31) is the time derivative of $p_i \dot{q}_i$, so that

$$\frac{\partial L}{\partial t} = -\frac{\mathrm{d}H}{\mathrm{d}t}, \qquad H = \sum_{i=1}^{n} p_i \dot{q}_i - L. \tag{1.32}$$

The quantity H defined by (1.32) is called *Hamiltonian function*. Remembering the derivation of the Lagrangian function one observes that L, H, and $p_i \dot{q}_i$ have the units of an energy. In turn, the product energy × time is called *action*. In particular, the functional (1.29) is called *action integral* [42, Chap. 8]. From the above observation it follows that $q_i \, p_i$ has the units of an action in all coordinate sets.

By way of example one takes the single-particle Lagrangian functions (1.12) and (1.18), where the Cartesian coordinates are used. The momentum conjugate to x_i is given, respectively, by

$$L = \frac{1}{2} m u^2 - V \rightarrow p_i = m u_i, \quad L = \frac{1}{2} m u^2 - e\varphi + e\mathbf{u} \cdot \mathbf{A} \rightarrow p_i = m u_i + e A_i. \tag{1.33}$$

The expression of H is found from (1.32) after introducing the vector $\mathbf{p} = (p_1, p_2, p_3)$ and indicating its modulus with p. For the case $L = m u^2 / 2 - V$ one finds

$$H = \frac{1}{2} m u^2 + V = \frac{1}{2m} p^2 + V, \tag{1.34}$$

while the case $L = m u^2 / 2 - e\varphi + e\mathbf{u} \cdot \mathbf{A}$ yields

$$H = \frac{1}{2} m u^2 + e\varphi = \frac{1}{2m} |\mathbf{p} - e\mathbf{A}|^2 + e\varphi. \tag{1.35}$$

Still using the Cartesian coordinates, (1.34) is readily extended to the case of a system of particles having n degrees of freedom. The force acting on the ith degree of freedom at time t is given by a generalization of (1.10),

$$m_i \dot{u}_i = -\frac{\partial V}{\partial x_i}, \tag{1.36}$$

where the time derivative is calculated at t and the $x_1, \ldots, x_n$ coordinates appearing in V are calculated at t as well. For the sake of simplicity the coordinate index i is also used to distinguish the masses in (1.36). It is implied that the same value of m_i must be applied to the indices associated with the same particle. The Lagrangian function is calculated in the same manner as in Sect. 1.3 and reads

$$L = \sum_{i=1}^{n} \frac{1}{2} m_i u_i^2 - V(\mathbf{r}, t), \qquad p_i = m_i u_i, \tag{1.37}$$

whence

$$H = \sum_{i=1}^{n} \frac{1}{2} m_i u_i^2 + V(\mathbf{r}, t) = \sum_{i=1}^{n} \frac{1}{2m_i} p_i^2 + V(\mathbf{r}, t). \tag{1.38}$$

Comparing the two forms of H shown in (1.38), one notes that the second of (1.37) is exploited to express the Hamiltonian function in terms of the $\mathbf{r}, \mathbf{p}$ sets instead of the $\mathbf{r}, \mathbf{u}$ sets. This procedure is generalized in Sect. 1.6.

1.6 Hamilton Equations

As the Lagrangian function depends on $\mathbf{q}$, $\dot{\mathbf{q}}$, and t, the generalized momentum p_i defined by (1.30) depends on the same variables at most. It is useful to consider also the inverse relations, where the generalized velocities $\dot{q}_i$ are expressed in terms of $\mathbf{q}$, $\mathbf{p}$, and t. The two sets of relations are

$$
\begin{cases}
p_1 = p_1(q_1, \dot{q}_1, \ldots, q_n, \dot{q}_n, t) \\
\vdots \\
p_n = p_n(q_1, \dot{q}_1, \ldots, q_n, \dot{q}_n, t)
\end{cases}
\qquad
\begin{cases}
\dot{q}_1 = \dot{q}_1(q_1, p_1, \ldots, q_n, p_n, t) \\
\vdots \\
\dot{q}_n = \dot{q}_n(q_1, p_1, \ldots, q_n, p_n, t)
\end{cases}
\tag{1.39}
$$

A simple example is given by the two cases of (1.33). Letting $q_i = x_i$, $\dot{q}_i = u_i$, the first case gives (1.39) the form $p_i = m \dot{q}_i$ and $\dot{q}_i = p_i / m$, while the second one gives (1.39) the form $p_i = m \dot{q}_i + e A_i(\mathbf{q}, t)$ and $\dot{q}_i = [p_i - e A_i(\mathbf{q}, t)]/m$.

Introducing the second of (1.39) into the definition (1.32) of the Hamiltonian function expresses the latter in terms of $\mathbf{q}, \mathbf{p}$, and t. The derivatives of the Hamiltonian function with respect to the new variables q_i, p_i are very significant. In fact, for any index r one finds

$$\frac{\partial H}{\partial q_r} = \sum_{i=1}^{n} p_i \frac{\partial \dot{q}_i}{\partial q_r} - \left(\frac{\partial L}{\partial q_r} + \sum_{i=1}^{n} \frac{\partial L}{\partial \dot{q}_i} \frac{\partial \dot{q}_i}{\partial q_r} \right) = -\frac{\partial L}{\partial q_r} = -\dot{p}_r. \qquad (1.40)$$

The two sums in (1.40) cancel each other thanks to the first of (1.30), while the last equality is due to the second of (1.30). The derivative with respect to p_r is found by the same token,

$$\frac{\partial H}{\partial p_r} = \left(\dot{q}_r + \sum_{i=1}^{n} p_i \frac{\partial \dot{q}_i}{\partial p_r} \right) - \sum_{i=1}^{n} \frac{\partial L}{\partial \dot{q}_i} \frac{\partial \dot{q}_i}{\partial p_r} = \dot{q}_r. \qquad (1.41)$$

The results of (1.40, 1.41) are condensed in the *Hamilton equations*

$$\dot{q}_i = \frac{\partial H}{\partial p_i}, \qquad \dot{p}_i = -\frac{\partial H}{\partial q_i}, \qquad i = 1, \ldots, n, \qquad (1.42)$$

that provide a set of $2n$ differential equations of the first order in the $2n$ independent unknowns $q_1, \ldots, q_n, p_1, \ldots, p_n$. It is important to note that from (1.42) one readily derives the following:

$$\frac{\partial \dot{q}_i}{\partial q_i} + \frac{\partial \dot{p}_i}{\partial p_i} = \frac{\partial^2 H}{\partial q_i \partial p_i} - \frac{\partial^2 H}{\partial p_i \partial q_i} = 0. \qquad (1.43)$$

The Hamilton equations (1.42) provide the time evolution of the generalized coordinates q_i; as a consequence, they are equivalent to the Lagrange equations (1.28). Another way of obtaining the Hamilton equations is to derive them as the extremum equations of a suitable functional. This is shown in Sect. 1.7.

In contrast to the Lagrange equations (1.28), that are n second-order differential equations, the Hamilton equations (1.42) are $2n$, first-order differential equations. To determine the solution of the latter it is necessary to prescribe the values of the $2n$ unknowns $q_1, \ldots, q_n, p_1, \ldots, p_n$ at the initial time $t = a$, that is, $2n$ constants. Therefore, the number of constants to be prescribed is the same as for the Lagrange equations. The independent functions $q_1, \ldots, q_n, p_1, \ldots, p_n$ are called *canonical coordinates*. For each index i the functions q_i, p_i are called *conjugate coordinates*.

Thanks to (1.42) the total derivative of H reads

$$\frac{dH}{dt} = \frac{\partial H}{\partial t} + \sum_{i=1}^{n} \left(\frac{\partial H}{\partial p_i} \dot{p}_i + \frac{\partial H}{\partial q_i} \dot{q}_i \right) = \frac{\partial H}{\partial t} = -\frac{\partial L}{\partial t}, \qquad (1.44)$$

where the last equality derives from the first of (1.32). If the Lagrangian function does not depend explicitly on time it follows $dH/dt = 0$, namely, H is a constant of motion. Its value is fixed by the values of the canonical coordinates at the initial time $t = a$. From (1.44) it also follows that $dH/dt = 0$ is equivalent to $\partial H/\partial t = 0$. In other terms, the Hamiltonian function is a constant of motion if it does not depend explicitly on time, and vice versa.

If the Lagrangian function does not depend on one of the coordinates, say, q_r, the latter is called *cyclic* or *ignorable*. From the second of (1.30) it follows that, if q_r is cyclic, its conjugate momentum p_r is a constant of motion. Moreover, due to the second of (1.42) it is $\partial H/\partial q_r = 0$, namely, the Hamiltonian function does not depend on q_r either.

1.7 Time–Energy Conjugacy—Hamilton–Jacobi Equation

Equations (1.42) can also be derived as the extremum equations of a functional. To show this it suffices to replace the Lagrangian function taken from the second of (1.32) into the functional (1.29), this yielding

$$S = \int_a^b \left(\sum_{i=1}^n p_i \dot{q}_i - H \right) dt. \tag{1.45}$$

Using in (1.45) the expressions of the generalized velocities given by the second of (1.39), the integrand becomes a function of q_i, p_i, and t. Then, the extremum equations are found by introducing the variations in the coordinates, that become $q_i + \alpha_i \eta_i$. Like in the case of (1.1) it is assumed that η_i vanishes at a and b. Similarly, the conjugate momenta become $p_i + \beta_i \zeta_i$. Differentiating (1.45) with respect to α_i or β_i yields, respectively,

$$\frac{\partial S}{\partial \alpha_i} = \int_a^b \left[(p_i + \beta_i \zeta_i) \dot{\eta}_i - \frac{\partial H}{\partial (q_i + \alpha_i \eta_i)} \eta_i \right] dt, \tag{1.46}$$

$$\frac{\partial S}{\partial \beta_i} = \int_a^b \left[(\dot{q}_i + \alpha_i \dot{\eta}_i) \zeta_i - \frac{\partial H}{\partial (p_i + \beta_i \zeta_i)} \zeta_i \right] dt. \tag{1.47}$$

Letting $\alpha_1 = \ldots = \beta_n = 0$ in (1.46, 1.47), integrating by parts the term containing $\dot{\eta}_i$, and using the condition $\eta_i(a) = \eta_i(b) = 0$ provides

$$\left(\frac{\partial S}{\partial \alpha_i} \right)_0 = -\int_a^b \left(\dot{p}_i + \frac{\partial H}{\partial q_i} \right) \eta_i \, dt, \quad \left(\frac{\partial S}{\partial \beta_i} \right)_0 = \int_a^b \left(\dot{q}_i - \frac{\partial H}{\partial p_i} \right) \zeta_i \, dt. \tag{1.48}$$

As in Sect. 1.2 the equations for the extremum functions are found by letting $(\partial S/\partial \alpha_i)_0 = 0$, $(\partial S/\partial \beta_i)_0 = 0$. Such equations coincide with (1.42). It is worth observing that, as no integration by part is necessary for obtaining the second of (1.48), the derivation of (1.42) does not require any prescription for the boundary conditions of ζ_i. On the other hand, considering that in the Hamilton equations q_i and p_i are independent variables, one can add the prescription $\zeta_i(a) = \zeta_i(b) = 0$. Although the latter is not necessary here, it becomes useful in the treatment of the canonical transformations, as shown in Sect. 2.2.

In the coordinate transformations discussed so far, time was left unchanged. This aspect is not essential: in fact, within the coordinate transformation one can replace t with another parameter that depends single-valuedly on t. This parameter, say, $\theta(t)$, is equally suitable for describing the evolution of the particles' system; proceeding in this way transforms (1.45) into

$$S = \int_{\theta(a)}^{\theta(b)} \left(\sum_{i=1}^{n} p_i \dot{q}_i \frac{dt}{d\theta} - H \frac{dt}{d\theta} \right) d\theta = \int_{\theta(a)}^{\theta(b)} \left(\sum_{i=1}^{n} p_i q_i' - H t' \right) d\theta, \quad (1.49)$$

where the primes indicate the derivatives with respect to θ. Now, letting $q_{n+1} = t$, $p_{n+1} = -H$, the Lagrangian function is recast in the more compact form $L = \sum_{i=1}^{n+1} p_i q_i'$. Remembering the definition (1.30) of the conjugate momenta, it follows that the latter becomes $p_i = \partial L / \partial q_i'$. In conclusion, the negative Hamiltonian function is the momentum conjugate to θ. This result is not due to any particular choice of the relation $\theta(t)$, hence it holds also for the identical transformation $\theta = t$; in other terms, $-H$ is the momentum conjugate to t.

If the upper limit b in the action integral S (Eq. (1.29)) is considered as a variable, the Langrangian is found to be the total time derivative of S. Letting $b \leftarrow t$, from the (1.45) form of S one derives its total differential

$$dS = \sum_{i=1}^{n} p_i dq_i - H dt. \quad (1.50)$$

As a consequence it is $p_i = \partial S / \partial q_i$, $H = -\partial S / \partial t$. Remembering that H depends on the generalized coordinates and momenta, and on time, one may abridge the above findings into the relation

$$\frac{\partial S}{\partial t} + H \left(q_1, \ldots, q_n, \frac{\partial S}{\partial q_1}, \ldots, \frac{\partial S}{\partial q_n}, t \right) = 0, \qquad p_i = \frac{\partial S}{\partial q_i}, \quad (1.51)$$

that is, a partial-differential equation in the unknown function S. The former is called *Hamilton–Jacobi equation*, while the latter in this context is called *Hamilton's principal function*. As (1.51) is a first-order equation in the $n + 1$ variables $q_1, \ldots, q_n, t$, the solution S contains $n + 1$ integration constants. One of them is an additive constant on S, as is easily found by observing that (1.51) contains the derivatives of S, not S itself. For this reason the additive constant is irrelevant and can be set to zero, so that the integration constants reduce to n. It is shown in Sect. 2.2 that (1.51) provides the time evolution of the generalized coordinates q_i. As a consequence it is equivalent to the Lagrange equations (1.28) and to the Hamilton equations (1.42) for describing the system's dynamics.

1.8 Poisson Brackets

Let ρ, σ be arbitrary functions of the canonical coordinates, differentiable with respect to the latter. The *Poisson bracket* of ρ and σ is defined as the function[3]

$$[\rho, \sigma] = \sum_{i=1}^{n} \left(\frac{\partial \rho}{\partial q_i} \frac{\partial \sigma}{\partial p_i} - \frac{\partial \rho}{\partial p_i} \frac{\partial \sigma}{\partial q_i} \right). \tag{1.52}$$

From (1.52) it follows $[\rho, \sigma] = -[\sigma, \rho]$, $[\rho, \rho] = 0$. Also, due to (1.42) it is

$$\frac{d\rho}{dt} = \frac{\partial \rho}{\partial t} + [\rho, H]. \tag{1.53}$$

Letting $\rho = H$ shows that (1.44) is a special case of (1.53). If ρ is a constant of motion, then

$$\frac{\partial \rho}{\partial t} + [\rho, H] = 0. \tag{1.54}$$

If ρ does not depend explicitly on time, (1.53) yields

$$\frac{d\rho}{dt} = [\rho, H] \tag{1.55}$$

where, in turn, the right hand side is equal to zero if ρ is a constant of motion, while it is different from zero in the other case. Special cases of the Poisson bracket are

$$[q_i, q_j] = 0, \qquad [p_i, p_j] = 0, \qquad [q_i, p_j] = \delta_{ij}, \tag{1.56}$$

with δ_{ij} the Kronecker symbol (A.18). Other interesting expressions are found by introducing the $2\,n$-dimensional vectors $\mathbf{s}$, $\mathbf{e}$ defined as

$$\mathbf{s} = \begin{bmatrix} q_1 \\ \vdots \\ q_n \\ p_1 \\ \vdots \\ p_n \end{bmatrix}, \qquad \mathbf{e} = \begin{bmatrix} \partial H/\partial p_1 \\ \vdots \\ \partial H/\partial p_n \\ -\partial H/\partial q_1 \\ \vdots \\ -\partial H/\partial q_n \end{bmatrix}. \tag{1.57}$$

Using the definitions (1.57) one finds

$$\dot{\mathbf{s}} = \mathbf{e}, \qquad \mathrm{div}_{\mathbf{s}}\,\dot{\mathbf{s}} = \sum_{i=1}^{n} \left(\frac{\partial \dot{q}_i}{\partial q_i} + \frac{\partial \dot{p}_i}{\partial p_i} \right) = 0, \tag{1.58}$$

---

[3] The definition and symbol (1.52) of the Poisson bracket conform to those of [42, Sect. 9–5]. In [67, Sect. 42], instead, the definition has the opposite sign and the symbol $\{\rho, \sigma\}$ is used. In [110, Sect. 11] the definition is the same as that adopted here, while the symbol $\{\rho, \sigma\}$ is used.

the first of which expresses the Hamilton equations (1.42) in vector form, while the second one derives from (1.43). The symbol div_s indicates the divergence with respect to all the variables that form vector s (Sect. A.3). Now, taking an arbitrary function ρ like that used in (1.53) and calculating the divergence of the product $\rho\,\dot{s}$ yields, thanks to (1.58) and to (A.16, A.12),

$$\text{div}_s(\rho\,\dot{s}) = \rho\,\text{div}_s\,\dot{s} + \dot{s} \cdot \text{grad}_s \rho = \sum_{i=1}^{n} \left(\frac{\partial\rho}{\partial q_i} \dot{q}_i + \frac{\partial\rho}{\partial p_i} \dot{p}_i \right) = [\rho, H]. \qquad (1.59)$$

1.9   Phase Space and State Space

Given a system of particles having n degrees of freedom it is often convenient to describe its dynamical properties by means of a geometrical picture. To this purpose one introduces a $2n$-dimensional space whose coordinates are $q_1, \ldots, q_n, p_1, \ldots, p_n$. This space has no definite metrical structure, one simply assumes that q_i and p_i are plotted as Cartesian coordinates of an Euclidean space [65, Chap. 6-5]. Following Gibbs, the space thus defined is often called *phase space*. However, it is convenient to better specify the terminology by using that of Ehrenfest, in which the term γ-*space* is used for this space (the citations of Gibbs and Ehrenfest are in [110, Sect. 17]). At some instant t the whole set of canonical coordinates $q_1, \ldots, q_n, p_1, \ldots, p_n$ corresponds to a point of the γ-space. Such a point is called *phase point*. In turn, the *state* of a mechanical system at some instant t is defined as the set of its canonical coordinates at that instant. It follows that the phase point represents the dynamical state of the system at t. As time evolves, the phase points representing the state at different instants provide a curve of the γ-space called *phase trajectory*.

A generalization of the γ-space is obtained by adding the time t as a $(2n + 1)$th coordinate. The $(2n + 1)$-dimensional space thus obtained is called *state space* [65, Chap. 6–5]. The curve of the state space describing the system's dynamics is called *state trajectory*. Consider two systems governed by the same Hamiltonian function and differing only in the initial conditions. The latter are represented by two distinct points of the $2n$-dimensional section of the state space corresponding to $t = 0$. The subsequent time evolution of the two systems provides two state trajectories that never cross each other. In fact, if a crossing occurred at, say, $t = \bar{t}$, the canonical coordinates of the two Hamiltonian functions would be identical there, this making the initial conditions of the subsequent motion identical as well. As a consequence, the two state trajectories would coincide for $t \geq \bar{t}$. However, the same reasoning holds when considering the motion backward in time ($t \leq \bar{t}$). Thus, the two trajectories should coincide at all times, this contradicting the hypothesis that the initial conditions at $t = 0$ were different.

A similar reasoning about the crossing of trajectories is possible in the γ-space. The conclusion is that the phase trajectories do not cross each other if the Hamiltonian function does not depend explicitly on time. Instead, they may cross each other if the Hamiltonian function depends on time; the crossing, however, occurs at different times (in other terms, the set of canonical coordinates of the first system calculated at $t = t_1$ may coincide with the set of canonical coordinates of the second system calculated at $t = t_2$ only in the case $t_2 \neq t_1$).

Despite of the larger number of dimensions, the adoption of the state space is convenient for the geometrical representation of the system's dynamics, because a trajectory is uniquely specified by the initial point and no crossing of trajectories occurs. With the provision stated above, this applies also to the γ-space. In contrast, consider a geometrical picture of the Lagrangian type, in which the generalized coordinates $q_1, \ldots, q_n$ only are used. The latter may be considered as the Cartesian coordinates of an n-dimensional Euclidean space called *configuration space*. To specify a trajectory in such a space it is necessary to prescribe the position $q_1, \ldots, q_n$ and velocity $\dot{q}_1, \ldots, \dot{q}_n$ of the system at $t = 0$. If one considers two or more systems differing only in the initial conditions, the motion of each system could start from every point of the configuration space and in every direction. As a consequence, it would be impossible to obtain an ordered picture of the trajectories, which will inevitably cross each other.

As mentioned above, the γ-space for a system having n degrees of freedom is a $2n$-dimensional space whose coordinates are $q_1, \ldots, q_n, p_1, \ldots, p_n$. It is sometimes convenient to use a different type of phase space whose dimension is twice the number of degrees of freedom possessed by each of the system's particles. To specifiy this issue, consider the case of a system made of N point-like particles, with no constraints. In this case each particle (say, the jth one) has three degrees of freedom and its dynamical state at the time t is determined by the six canonical coordinates $\bar{q}_{1j}, \bar{q}_{2j}, \bar{q}_{3j}, \bar{p}_{1j}, \bar{p}_{2j}, \bar{p}_{3j}$. Together, the latter identify a point X_j of a six-dimensional phase space called μ-*space*[4]. At the time t the system as a whole is represented in the μ-space by the set of N points $X_1, \ldots, X_N$.

1.10 Complements

1.10.1 Higher-Order Variational Calculus

The variational calculus described in Sect. 1.2 can be extended to cases were the function g in (1.1) depends on derivatives of a higher order than the first. Consider for instance the functional

$$G[w] = \int_a^b g(w, \dot{w}, \ddot{w}, \xi) \, d\xi. \tag{1.60}$$

Following the procedure of Sect. 1.2 and assuming that the derivative $\dot{\eta}$ vanishes at a and b, yields the following differential equation for the extremum functions of (1.60):

$$-\frac{d^2}{d\xi^2} \frac{\partial g}{\partial \ddot{w}} + \frac{d}{d\xi} \frac{\partial g}{\partial \dot{w}} = \frac{\partial g}{\partial w}. \tag{1.61}$$

---

[4] The letter "μ" stands for "molecule", whereas the letter "γ" in the term "γ-space" stands for "gas".

1.10.2   *Lagrangian Invariance and Gauge Invariance*

It is shown in Sect. 1.2 that the extremum functions $w_i(\xi)$ are invariant under addition to g of the total derivative of an arbitrary function h that depends on $\mathbf{w}$ and ξ only (refer to Eq. (1.8)). Then, it is mentioned in Sect. 1.3.2 that the $\mathbf{E}$ and $\mathbf{B}$ fields are invariant under the gauge transformation (1.19), where $h(\mathbf{r}, t)$ is an arbitrary function. These two properties have in fact the same origin, namely, the description based upon a Lagrangian function. In fact, as shown in Sect. 4.2, a Lagrangian description is possible also in the case of a system having a continuous distribution of the degrees of freedom like, for instance, the electromagnetic field.

1.10.3   *Variational Calculus with Constraints*

In several problems it is required that the function w, introduced in Sect. 1.2 as the extremum function of functional (1.1), be able to fulfill one or more constraints. By way of example consider the constraint

$$G_0 = \int_a^b g_0(w, \dot{w}, \xi)\, d\xi, \tag{1.62}$$

where the function g_0 and the number G_0 are prescribed. A typical case where (1.62) occurs is that of finding the maximum area bounded by a perimeter of given length (*Dido's problem*). For this reason, extremum problems having a constraint like (1.62) are called *isoperimetric* even when they have no relation with geometry [115, Par. 4-1].

To tackle the problem one extends the definition of the variation of w by letting $\delta w = \alpha_1 \eta_1 + \alpha_2 \eta_2$, where $\eta_1(\xi)$, $\eta_2(\xi)$ are arbitrary functions that are differentiable in the interior of $[a, b]$ and fulfill the conditions $\eta_1(a) = \eta_1(b) = 0$, $\eta_2(a) = \eta_2(b) = 0$.

If w is an extremum function of G that fulfills (1.62), replacing w with $w + \delta w$ transforms (1.1, 1.62) to a pair of functions of the α_1, α_2 parameters, namely,

$$G = G(\alpha_1, \alpha_2), \qquad G_0(\alpha_1, \alpha_2) = G_0(0, 0) = \text{const.} \tag{1.63}$$

The first of (1.63) has an extremum at $\alpha_1 = \alpha_2 = 0$, while the second one establishes a relation between α_1 and α_2. The problem is thus reduced to that of calculating a constrained extremum, and is solved by the method of the *Lagrange multipliers*.

For this, one considers the function $G_\lambda = G(\alpha_1, \alpha_2) + \lambda\, G_0(\alpha_1, \alpha_2)$, with λ an indeterminate parameter, and calculates the free extremum of G_λ by letting

$$\left(\frac{\partial G_\lambda}{\partial \alpha_1}\right)_0 = 0, \qquad \left(\frac{\partial G_\lambda}{\partial \alpha_2}\right)_0 = 0, \tag{1.64}$$

where index 0 stands for $\alpha_1 = \alpha_2 = 0$. The rest of the calculation is the same as in Sect. 1.2; the two relations (1.64) turn out to be equivalent to each other and provide the same Euler equation. More specifically, from the definition of G and G_0 as integrals of g and g_0 one finds that the Euler equation of this case is obtained from that of Sect. 1.2 by replacing g with $g_\lambda = g + \lambda g_0$:

$$\frac{d}{d\xi} \frac{\partial g_\lambda}{\partial \dot{w}} = \frac{\partial g_\lambda}{\partial w}. \tag{1.65}$$

As (1.65) is a second-order equation, its solution w contains two integration constants. The λ multiplier is an additional indeterminate constant. The three constants are found from the constraint (1.62) and from the two relations provided by the boundary or initial conditions of w.

1.10.4 An Interesting Example of Extremum Equation

Consider the Hamilton–Jacobi Eq. (1.51) for a single particle of mass m. Using the Cartesian coordinates and a Hamiltonian function of the form

$$H = \frac{p^2}{2m} + V(x_1, x_2, x_3, t), \quad p^2 = p_1^2 + p_2^2 + p_3^2, \tag{1.66}$$

the Hamilton–Jacobi equation reads

$$\frac{\partial S}{\partial t} + \frac{|\mathrm{grad}\, S|^2}{2m} + V(x_1, x_2, x_3, t) = 0, \quad p_i = \frac{\partial S}{\partial q_i}. \tag{1.67}$$

If V is independent of time, then $H = E$ and the separation $S = W - Et$ (Sect. 2.4) yields $\partial S / \partial t = -E$, $\mathrm{grad}\, S = \mathrm{grad}\, W = \mathbf{p}$. It follows

$$\frac{|\mathrm{grad}\, W|^2}{2m} + V(x_1, x_2, x_3) = E. \tag{1.68}$$

Both Hamilton's principal (S) and characteristic (W) functions have the dimensions of an action and are defined apart from an additive constant. Also, the form of $|\mathrm{grad}\, W|$ is uniquely defined by that of $V - E$. In turn, E is prescribed by the initial conditions of the particle's motion.

Consider now the case where $E \geq V$ within a closed domain Ω whose boundary is $\partial \Omega$. As $\mathrm{grad}\, W$ is real, the motion of the particle is confined within Ω, and $\mathrm{grad}\, W$ vanishes at the boundary $\partial \Omega$. The Hamilton–Jacobi equation for W (1.68) is recast in a different form by introducing an auxiliary function w such that

$$w = w_0 \exp(W/\mu), \tag{1.69}$$

with μ a constant having the dimensions of an action. The other constant w_0 is used for prescribing the dimensions of w. Apart from this, the choice of w_0 is arbitrary due to the arbitrariness of the additive constant of W. Taking the gradient of (1.69) yields $\mu\, \mathrm{grad}\, w = w\, \mathrm{grad}\, W$, with $w \neq 0$ due to the definition. As $\mathrm{grad}\, W$ vanishes at the boundary, grad$w$ vanishes there as well. As a consequence, w is constant over the boundary. Inserting (1.69) into (1.68) yields

$$\frac{\mu^2}{2m} \frac{|\text{grad}w|^2}{w^2} + V(x_1, x_2, x_3) = E, \tag{1.70}$$

which determines $|\text{grad}w/w|$ as a function of $V - E$. Rearranging the above and observing that $\text{div}(w\text{grad}w) = w\nabla^2 w + |\text{grad}w|^2$ (Sect. A.1) provides

$$\frac{\mu^2}{2m} \left[\text{div}(w\text{grad}w) - w\nabla^2 w \right] + (V - E)w^2 = 0. \tag{1.71}$$

Integrating (1.71) over Ω and remembering that grad$w$ vanishes at the boundary,

$$\int_\Omega w \left[-\frac{\mu^2}{2m} \nabla^2 w + (V - E)w \right] d\Omega = 0. \tag{1.72}$$

The term in brackets of (1.72) does not necessarily vanish. In fact, the form of w is such that only the integral as a whole vanishes. On the other hand, by imposing that the term in brackets vanishes, and replacing μ with the reduced Planck constant $\hbar$, yields

$$-\frac{\hbar^2}{2m} \nabla^2 w + (V - E)w = 0, \tag{1.73}$$

namely, the Schrödinger equation independent of time (7.45). This result shows that the Schrödinger equation derives from a stronger constraint than that prescribed by the Hamilton–Jacobi equation.

An immediate consequence of replacing the integral relation (1.72) with the differential Eq. (1.73) is that the domain of w is not limited any more by the condition $E \geq V$, but may extend to infinity.

Another consequence is that, if the boundary conditions are such that w vanishes over the boundary (which, as said above, may also be placed at infinity), then (1.73) is solvable only for specific values of E, that form its spectrum of eigenvalues. Moreover it can be demonstrated, basing on the form of the Schrodinger equation, that the condition $E \geq V_{\text{min}}$ must be fulfilled (Sect. 8.2.3).

It is interesting to note another relation between the Schrödinger and the Hamilton–Jacobi equations. For the sake of simplicity one takes the one-dimensional case of the Hamilton–Jacobi equation expressed in terms of w (1.70):

$$\frac{\mu^2}{2m} (w')^2 + V(x)w^2 = E w^2, \tag{1.74}$$

where the prime indicates the derivative with respect to x. The left hand side of the equation may be considered the generating function $g = g(w, w', x)$ of a functional G, defined over an interval of the x axis that may extend to infinity:

$$G[w] = \int_a^b \left[\frac{\mu^2}{2m} (w')^2 + V w^2 \right] dx. \tag{1.75}$$

One then seeks the extremum function w of G that fulfills the constraint

$$G_0[w] = \int_a^b w^2 \, dx = 1. \tag{1.76}$$

The problem is solved by the method of Sect. 1.10.3, namely, by letting $g_0 = w^2$, $g_E = g - E\, g_0$, and applying the Euler equation to g_E:

$$\frac{d}{dx} \frac{\partial g_E}{\partial w'} = \frac{d}{dx} \frac{\mu^2}{m} w' = \frac{\mu^2}{m} w'', \qquad \frac{\partial g_E}{\partial w} = 2\,(V - E)\,w, \tag{1.77}$$

showing that the Schrödinger equation is actually the Euler equation of the functional G subjected to the constraint G_0, with the eigenvalue E provided by the Lagrange multiplier. This result holds also in the higher-dimensional cases, and is in fact the method originally used by Schrödinger to determine the time-independent equation [94, Eqs. (23, 24)].

1.10.5 Constant-Energy Surfaces

Consider the γ-space for a system having n degrees of freedom (Sect. 1.9). If the system is conservative, the relation $H(q_1, \ldots, q_n, p_1, \ldots, p_n) = E$ introduces a constraint among the canonical coordinates. Due to this, at each instant of time the latter must belong to the $(2n - 1)$-dimensional surface $H = E$ of the phase space, that is called *constant-energy surface*. As E is prescribed by the initial conditions, the phase point of a conservative system always belongs to the same constant-energy surface.

For a system having one degree of freedom the relation describing the constant-energy surface reduces to $H(q, p) = E$, that describes a curve in the $q\,p$ plane. The corresponding state trajectory is a curve of the three-dimensional $q\,p\,t$ space.

Problems

1.1 In the xy plane find the geodesic $y = y(x)$ through the points $A \equiv (a, y_a)$, $B \equiv (b, y_b)$, $A \neq B$.

1.2 Given the Hamiltonian function $H = p^2/(2\,m) + (c/2)\,x^2$, $m, c > 0$ (that describes the *linear harmonic oscillator*, Sect. 3.3), find the constant-energy curves of the xp plane corresponding to different values of the total energy E.

1.3 Given the Hamiltonian function of the harmonic oscillator of the general form $H = p^2/(2\,m) + (c/s)\,|x|^s$, $m, c, s > 0$, find the constant-energy curves of the xp plane corresponding to a fixed total energy E and to different values of parameter s.

Chapter 2
Coordinate Transformations and Invariance Properties

2.1 Introduction

An important generalization of the subject of coordinate transformation is that of canonical transformation, which leads to the concept of generating function and, through it, to the definition of the principal function and characteristic function of Hamilton. The principal function is found to coincide with the solution of the Hamilton–Jacobi equation introduced in the previous chapter, this showing the equivalence of the approach based on the variational principles with that based on the canonical transformations. Connected with the Hamiltonian formalism is also the important concept of separability. Still based on Hamilton's principal function is the concept of phase velocity applied to a mechanical system, that brings about an analogy with the electromagnetic theory. The aspects mentioned above give another indication about the generality that is reached by the formalism of Analytical Mechanics illustrated in this part of the book.

Another fundamental issue is that of the invariance properties of the mechanical systems. It is shown that, basing only on the observation of symmetries possessed by the Lagrangian function or other functions connected to it, one derives the existence of invariance properties of the system. Among these are the constants of motion, namely, the dynamical properties that are constant in time and are therefore known from the motion's initial condition.

Of special relevance among the constants of motion are the total energy, the total momentum, and the total angular momentum of the system. The conservation of the total energy is related to the uniformity of time, that of the total momentum is related to the uniformity of space and, finally, that of the total angular momentum is related to the isotropy of space. Besides the theoretical interest connected to it, the knowledge of a constant of motion is important also for practical purposes: by introducing a known constraint among the canonical coordinates, it is of use in the separation procedure.

© Springer Science+Business Media New York 2015
M. Rudan, *Physics of Semiconductor Devices,*
DOI 10.1007/978-1-4939-1151-6_2

   This chapter completes the illustration of the basic principles of Analytical Mechanics started in the previous one. The purpose of the two chapters is to introduce a number of concepts that are not only useful *per se*, but also constitute a basis for the concepts of Quantum Mechanics that are introduced in later chapters. The first subject is that of the canonical transformations, followed by the definition and properties of the Hamilton characteristic function and of the phase velocity. Then, the invariance properties that derive from the symmetries of the Lagrangian function are discussed. The chapter continues with a short description of the Maupertuis principle and of the expression of the angular momentum in spherical coordinates. The last paragraphs deal with the linear motion and the action-angle variables.

2.2   Canonical Transformations

Section 1.4 introduced the generalized coordinates $q_1, \ldots, q_n$, that are defined by the first of (1.26) starting from a set of Cartesian coordinates $x_1, \ldots, x_n$. From this, one defines the generalized velocities $\dot{q}_i$ and, from the second of (1.26), calculates the Lagrangian function in the new variables $L(\mathbf{q}, \dot{\mathbf{q}}, t)$. The conjugate momenta p_i are derived from the latter using the first of (1.30) and, finally, the new Hamiltonian function is determined from the second of (1.32). From this, the Hamilton equations (1.42) in the new coordinates are deduced. The process depicted here is a series of replacement, elimination, and differentiation steps.

   Relations like (1.26), that transform a set of coordinates into another one, are called *point transformations*. It has been observed in Sect. 1.4 that the canonical coordinates $q_1, \ldots, q_n, p_1, \ldots, p_n$ are mutually independent. It follows that the point transformations are not the most general coordinate transformations, because they act on the $q_1, \ldots, q_n$ only. The most general transformations act simultaneously on the generalized coordinate and momenta, hence they have the form

$$
\left\{
\begin{aligned}
\tilde{q}_1 &= \tilde{q}_1(q_1, \ldots, q_n, p_1, \ldots, p_n, t) \\
&\vdots \\
\tilde{q}_n &= \tilde{q}_n(q_1, \ldots, q_n, p_1, \ldots, p_n, t) \\
\tilde{p}_1 &= \tilde{p}_1(q_1, \ldots, q_n, p_1, \ldots, p_n, t) \\
&\vdots \\
\tilde{p}_n &= \tilde{p}_n(q_1, \ldots, q_n, p_1, \ldots, p_n, t)
\end{aligned}
\right.
\tag{2.1}
$$

where q_i, p_i indicate the old canonical coordinates and $\tilde{q}_i, \tilde{p}_i$ indicate the new ones. If H is the Hamiltonian function in the old coordinates, introducing into H the inverse transformations of (2.1) yields a function $\tilde{H}$ that depends on the new coordinates and on time.

   For an arbitrary choice of the form of (2.1) it is not possible in general to deduce from $\tilde{H}$ the Hamilton equations in the new coordinates. In fact it is necessary to limit the choice of the transformation (2.1) to the cases where the resulting $\tilde{H}$ is a Hamiltonian function proper, namely, it is such that

$$\frac{d\tilde{q}_i}{dt} = \frac{\partial \tilde{H}}{\partial \tilde{p}_i}, \qquad \frac{d\tilde{p}_i}{dt} = -\frac{\partial \tilde{H}}{\partial \tilde{q}_i}, \qquad i = 1, \dots, n \qquad (2.2)$$

are fulfilled. The transformations (2.1) that make (2.2) to hold are called *canonical transformations*. The procedure by which the Hamilton equations in the old coordinates are found has been illustrated in Sect. 1.4 and is based on the derivation of the extremum equation of the action integral (1.45). To obtain (2.2) the same calculation must be repeated based on the action integral defined in the new coordinates. It follows that for two sets of coordinates q_i, p_i and $\tilde{q}_i$, $\tilde{p}_i$ connected by a canonical transformation, the following must hold simultaneously:

$$S = \int_a^b \left(\sum_{i=1}^n p_i \frac{dq_i}{dt} - H \right) dt, \qquad \tilde{S} = \int_a^b \left(\sum_{i=1}^n \tilde{p}_i \frac{d\tilde{q}_i}{dt} - \tilde{H} \right) dt. \qquad (2.3)$$

The difference between the two integrals in (2.3) can be set equal to an arbitrary constant because the calculation uses only the variations of S or $\tilde{S}$. As the limits of the two integrals are fixed to the same values a and b, the constant can in turn be written as the integral between a and b of the total time derivative of an arbitrary function K. In this way the relation between the two integrands in (2.3) reads

$$\sum_{i=1}^n p_i \frac{dq_i}{dt} - H = \sum_{i=1}^n \tilde{p}_i \frac{d\tilde{q}_i}{dt} - \tilde{H} + \frac{dK}{dt}. \qquad (2.4)$$

It is worth reminding that in the derivation of the Hamilton equations in Sect. 1.6 it is assumed that all variations of generalized coordinates and momenta vanish at the integration limits. Here this applies to both the old and new sets of coordinates. As a consequence, K can be made to depend on all $4n$ coordinates q_i, p_i, $\tilde{q}_i$, $\tilde{p}_i$, and on time t. Due to the $2n$ relations (2.1) that define the canonical transformation, only $2n$ coordinates are independent. Thus, K can be made to depend on $2n$ coordinates chosen among the $4n$ available ones, and on time. The most interesting cases are those where K has one of the following forms [42, Sect. 9-1]:

$$K_1 = K_1(\mathbf{q}, \tilde{\mathbf{q}}, t), \quad K_2 = K_2(\mathbf{q}, \tilde{\mathbf{p}}, t), \quad K_3 = K_3(\mathbf{p}, \tilde{\mathbf{q}}, t), \quad K_4 = K_4(\mathbf{p}, \tilde{\mathbf{p}}, t). \qquad (2.5)$$

By way of example, select the first form: replacing K_1 into (2.4), calculating dK_1/dt, and multiplying both sides by dt yields

$$\sum_{i=1}^n \left(\frac{\partial K_1}{\partial q_i} - p_i \right) dq_i + \sum_{i=1}^n \left(\frac{\partial K_1}{\partial \tilde{q}_i} + \tilde{p}_i \right) d\tilde{q}_i + \left(\frac{\partial K_1}{\partial t} + H - \tilde{H} \right) dt = 0,$$

$$(2.6)$$

where the left hand side is a total differential in the independent variables q_i, $\tilde{q}_i$, and t. To fulfill (2.6) the parentheses must vanish independently from each other, whence

$$p_i = \frac{\partial K_1}{\partial q_i}, \qquad \tilde{p}_i = -\frac{\partial K_1}{\partial \tilde{q}_i}, \qquad \tilde{H} = H + \frac{\partial K_1}{\partial t}, \qquad i = 1, \ldots, n. \qquad (2.7)$$

As K_1 is prescribed, the first two equations in (2.7) provide $2n$ relations involving the $4n$ coordinates q_i, p_i, $\tilde{q}_i$, $\tilde{p}_i$, that constitute the canonical transformation sought. Using the latter one expresses the right hand side of the third of (2.7) in terms of $\tilde{q}_i$, $\tilde{p}_i$, this yielding the new Hamiltonian function $\tilde{H}(\tilde{\mathbf{q}}, \tilde{\mathbf{p}}, t)$. The procedure is the same for the other functions listed in (2.5), that can all be defined starting from K_1. In fact, letting

$$K_2(\mathbf{q}, \tilde{\mathbf{p}}, t) = K_1(\mathbf{q}, \tilde{\mathbf{q}}, t) + \sum_{i=1}^{n} \tilde{p}_i \tilde{q}_i, \qquad (2.8)$$

and applying the same procedure used to determine (2.7), yields

$$p_i = \frac{\partial K_2}{\partial q_i}, \qquad \tilde{q}_i = \frac{\partial K_2}{\partial \tilde{p}_i}, \qquad \tilde{H} = H + \frac{\partial K_2}{\partial t}, \qquad i = 1, \ldots, n. \qquad (2.9)$$

In (2.8) the independent variables are q_i, $\tilde{p}_i$, so that the coordinates $\tilde{q}_i$ are expressed through them. Similarly, when K_3 is used one lets

$$K_3(\mathbf{p}, \tilde{\mathbf{q}}, t) = K_1(\mathbf{q}, \tilde{\mathbf{q}}, t) - \sum_{i=1}^{n} p_i q_i, \qquad (2.10)$$

to find

$$q_i = -\frac{\partial K_3}{\partial p_i}, \qquad \tilde{p}_i = -\frac{\partial K_3}{\partial \tilde{q}_i}, \qquad \tilde{H} = H + \frac{\partial K_3}{\partial t}, \qquad i = 1, \ldots, n. \qquad (2.11)$$

Finally, in the case of K_4 one lets

$$K_4(\mathbf{p}, \tilde{\mathbf{p}}, t) = K_1(\mathbf{q}, \tilde{\mathbf{q}}, t) + \sum_{i=1}^{n} \tilde{p}_i \tilde{q}_i - \sum_{i=1}^{n} p_i q_i, \qquad (2.12)$$

whence

$$q_i = -\frac{\partial K_4}{\partial p_i}, \qquad \tilde{q}_i = \frac{\partial K_4}{\partial \tilde{p}_i}, \qquad \tilde{H} = H + \frac{\partial K_4}{\partial t}, \qquad i = 1, \ldots, n. \qquad (2.13)$$

Regardless of the choice of K, the relation between the old and new Hamiltonian function is always of the form $\tilde{H} = H + \partial K / \partial t$. As the canonical transformation is completely determined when K is prescribed, the latter is called *generating function of the canonical transformation*. Two interesting examples are those produced by the generating functions $F_1 = \sum_{i=1}^{n} q_i \tilde{q}_i$ and $F_2 = \sum_{i=1}^{n} q_i \tilde{p}_i$. Applying (2.7) to

F_1 yields $\tilde{q}_i = p_i$ and $\tilde{p}_i = -q_i$. As a consequence, the effect of the transformation generated by F_1 is that of exchanging the roles of the generalized coordinates and momenta. This result shows that the distinction between coordinates and momenta is not fundamental, namely, these two groups of variables globally constitute a set of $2n$ independent coordinates. Applying (2.9) to F_2 provides the identical transformation $\tilde{q}_i = q_i$, $\tilde{p}_i = p_i$. A generalization of this example is found using $F_2 = \sum_{i=1}^n z_i(\mathbf{q}, t)\tilde{p}_i$, where z_i are arbitrary functions. The new coordinates are in this case $\tilde{q}_i = z_i(\mathbf{q}, t)$ which, as indicated at the beginning of this section, are point transformations. This example shows that all point transformations are canonical.

2.3 An Application of the Canonical Transformation

The discussion of Sect. 2.2 has shown that a canonical transformation based on an arbitrary generating function K brings a Hamiltonian function $H(\mathbf{q}, \mathbf{p}, t)$ into a new one $\tilde{H}(\tilde{\mathbf{q}}, \tilde{\mathbf{p}}, t)$. One may then exploit the arbitrariness of K to obtain the form of $\tilde{H}$ that is most convenient for solving the problem in hand. For instance, remembering the definition of cyclic coordinate given in Sect. 1.42, one may seek a transformation such that the new canonical coordinates $\tilde{q}_i$, $\tilde{p}_i$ are all cyclic. In this case, thanks to (2.2), it is $d\tilde{q}_i/dt = \partial\tilde{H}/\partial\tilde{p}_i = 0$, $d\tilde{p}_i/dt = -\partial\tilde{H}/\partial\tilde{q}_i = 0$, namely, each new canonical coordinate is a constant of motion.

The simplest way to obtain this result is to set the new Hamiltonian function equal to zero. Remembering from Sect. 2.2 that in every canonical transformation the relation between the old and new Hamiltonian function is $\tilde{H} = H + \partial K/\partial t$, one finds in this case the relation $\partial K/\partial t + H = 0$. To proceed it is convenient to choose a generating function of the $K_2 = K_2(\mathbf{q}, \tilde{\mathbf{p}}, t)$ type in which, as noted above, the new momenta $\tilde{p}_i$ are constants of motion. Given that the aim is to obtain the relation $\tilde{H} = 0$, the generating function of this problem is the particular function of the coordinates q_i, $\tilde{p}_i$, and t, that fulfills the equation $\partial K_2/\partial t + H = 0$. In other terms, the generating function becomes the problem's unknown. A comparison with (1.51) shows that the equation to be solved is that of Hamilton–Jacobi, and that K_2 coincides with Hamilton's principal function S.

As mentioned in Sect. 1.7, (1.51) is a first-order, partial differential equation in the unknown S and in the $n+1$ variables $q_1, \dots, q_n, t$. As one of the $n+1$ integration constants can be set to zero, the actual number of integration constants is n. This seems contradictory, because the Hamilton–Jacobi equation is expected to be equivalent to those of Hamilton or Lagrange for the description of the system's motion. As a consequence, the number of constants involved should be $2n$. The contradiction is easily removed by observing that n more constants appear in S, to be identified with the new momenta $\tilde{p}_1, \dots, \tilde{p}_n$: remembering the canonical transformations (2.9) to be used in connection with the generating function of the K_2 type one finds

$$p_i = \frac{\partial S}{\partial q_i}, \qquad \tilde{q}_i = \frac{\partial S}{\partial \tilde{p}_i}, \qquad i = 1, \dots, n. \qquad (2.14)$$

Calculating the first of (2.14) at the initial time $t = a$ yields a set of n algebraic equations in the n unknowns $\tilde{p}_1, \ldots, \tilde{p}_n$. In fact, at $t = a$ the old canonical coordinates q_i, p_i are known because they are the problem's initial conditions. The solution of such algebraic equations yields the first set of motion's constants $\tilde{p}_1, \ldots, \tilde{p}_n$. Then, one considers the second of (2.14) at $t = a$, whose right hand sides, at this point of the procedure, are known. As a consequence, the second of (2.14) yield the new generalized coordinates $\tilde{q}_1, \ldots, \tilde{q}_n$, that are the second set of motion's constants.

It is worth observing that the procedure depicted above provides also the time evolution of the old canonical coordinates. In fact, after all constants have been calculated, Eqs. (2.14) form $2n$ relations in the $2n+1$ variables $q_1, \ldots, q_n, p_1, \ldots, p_n, t$. From them one extracts the relations $q_1 = q_1(t), \ldots, p_n = p_n(t)$. This shows that the Hamilton–Jacobi picture is equivalent to those based on the Hamilton or Lagrange equations for the solution of the mechanical problem.

2.4   Separation—Hamilton's Characteristic Function

The Hamilton–Jacobi equation (1.51) can be recast in a more symmetric form by letting $q_{n+1} = t$ and incorporating $\partial S/\partial t = \partial S/\partial q_{n+1}$ into the other term:

$$C\left(q_1, \ldots, q_{n+1}, \frac{\partial S}{\partial q_1}, \ldots, \frac{\partial S}{\partial q_{n+1}}\right) = 0. \qquad (2.15)$$

Solving (2.15) becomes simpler if one of the coordinates, say q_i, and the corresponding momentum $p_i = \partial S/\partial q_i$ appear in (2.15) only through a relation $c_i = c_i(q_i, \partial S/\partial q_i)$ that does not contain any other coordinate, nor derivatives with respect to them, nor time. In this case q_i is called *separable coordinate* and the solution of (2.15) can be written as $S = S_i + W_i$, where S_i depends only on q_i and W_i depends on the other coordinates and time [67, Sect. 48]. Replacing this expression of S into (2.15) and extracting c_i yields a relation of the form $C_i = c_i$ with

$$C_i = C_i\left(q_1, \ldots, q_{i-1}, q_{i+1}, \ldots, q_{n+1}, \frac{\partial W_i}{\partial q_1}, \ldots, \frac{\partial W_i}{\partial q_{i-1}}, \frac{\partial W_i}{\partial q_{i+1}}, \ldots, \frac{\partial W_i}{\partial q_{n+1}}\right),$$

$$c_i = c_i\left(q_i, \frac{\partial S_i}{\partial q_i}\right). \qquad (2.16)$$

The equality $C_i = c_i$ must hold for any value of the coordinates. As this is possible only if the two sides are constant, $C_i = c_i$ separates and yields the pair

$$c_i\left(q_i, \frac{\partial S_i}{\partial q_i}\right) = c_{i0}, \qquad (2.17)$$

$$C\left(q_1, \ldots, q_{i-1}, q_{i+1}, \ldots, q_{n+1}, \frac{\partial W_i}{\partial q_1}, \ldots, \frac{\partial W_i}{\partial q_{i-1}}, \frac{\partial W_i}{\partial q_{i+1}}, \ldots, \frac{\partial W_i}{\partial q_{n+1}}, c_{i0}\right) = 0,$$

where C does not contain q_i nor the corresponding derivative. The solution of the first of (2.17) provides $S_i(q_i)$. The latter contains two constants, namely, c_{i0} and the integration constant. As noted earlier, the latter can be set to zero because an additive constant on S is irrelevant. In conclusion, c_{i0} is the only constant that remains after this step. In turn, the solution of the second of (2.17), which is an n-variable differential equation, contains n more constant, one of which is additive and can be disregarded. It follows that the total number of integration constants in the set (2.17) is still n.

If all coordinates are separable one has $S = \sum_{i=1}^{n} S_i(q_i)$ and the problem is solved by n individual integrations (an example is given in Sect. 3.10). In this case one says that the Hamilton–Jacobi equation is *completely separable*. A special case of separable coordinate is that of the cyclic coordinate. If q_i is cyclic, in fact, (2.17) reduces to $\partial S_i / \partial q_i = c_{i0}$, whence $S_i = c_{i0}q_i$ and $S = c_{i0}q_i + W_i$. If the cyclic coordinate is $q_{n+1} = t$, the above becomes

$$\frac{\partial S_{n+1}}{\partial t} = -E, \qquad S_{n+1} = -Et, \qquad (2.18)$$

where the symbol E is used for the constant $c_{n+1,0}$. It is worth noting that the units of E are always those of an energy regardless of the choice of the generalized coordinates q_i. Comparing (2.18) with (1.51) yields $H = E = \text{cost}$, consistently with the hypothesis that H does not depend on t. Using the symbol W instead of W_{n+1} provides the pair

$$H\left(q_1, \ldots, q_n, \frac{\partial W}{\partial q_1}, \ldots, \frac{\partial W}{\partial q_n}\right) = E, \qquad S = W - Et, \qquad (2.19)$$

that holds when H is a constant of motion. The first of (2.19) is a differential equation in the generalized coordinates only, called *time-independent Hamilton–Jacobi equation*. The unknown function W is called *Hamilton's characteristic function*.

2.5 Phase Velocity

The dynamics of a mechanical system can be obtained from Hamilton's principal function $S(\mathbf{q}, \tilde{\mathbf{p}}, t)$ as shown in Sect. 2.3. After S has been determined it is possible to build up an interesting geometrical construction, that is shown below. The indication of the constants $\tilde{p}_i$ is omitted for the sake of conciseness.

To begin, fix the time t and let $S(\mathbf{q}, t) = S_0$, where S_0 is some constant. This relation describes a surface belonging to the configuration space $q_1, \ldots, q_n$. Now change the time by dt: the corresponding variation in S is obtained from (1.50) and reads $dS = \sum_{i=1}^{n} p_i dq_i - H dt = \mathbf{p} \cdot d\mathbf{q} - H dt$. In this relation each component $p_i = \partial S / \partial q_i$ is calculated in terms of the coordinates $q_1, \ldots, q_n$ at the instant t, hence the vector $\mathbf{p} = \text{grad}_q S$ is a function of $\mathbf{q}$ calculated at that instant. If $\mathbf{q}$ belongs to the surface $S = S_0$, then $\mathbf{p}$ is normal to the surface at $\mathbf{q}$. Now let $S' = S + dS = S_0$, where S_0 is the same constant as before. The relation $S'(\mathbf{q}, t) = S_0$ provides the new surface into which $S = S_0$ evolves in the interval dt. As both $S = S_0$, $S + dS = S_0$ hold, it must be $dS = 0$, namely, $\mathbf{p} \cdot d\mathbf{q} = H dt$.

When H has no explicit dependence on t, thanks to (2.19) the relation $\mathbf{p}\cdot d\mathbf{q} = H dt$ becomes $\mathbf{p}\cdot d\mathbf{q} = E dt$, with $\mathbf{p} = \operatorname{grad}_q W$. In this case, letting φ be the angle between the vectors $\mathbf{p}$ and $d\mathbf{q}$ (whose moduli are indicated with p, dq), and excluding the points where $\mathbf{p} = 0$, one obtains

$$\cos \varphi \, \frac{dq}{dt} = \frac{E}{p}. \tag{2.20}$$

The product $\cos \varphi \, dq$ in (2.20) is the projection of $d\mathbf{q}$ over the direction of $\mathbf{p}$, hence it provides the variation of $\mathbf{q}$ in the direction normal to the surface $S = S_0$. When the Cartesian coordinates are used, the product $\cos \varphi \, dq$ is a length and the left hand side of (2.20) is a velocity, that provides the displacement of the point $\mathbf{q}$ during the time interval dt and in the direction normal to the surface $S = S_0$.

As shown above, the vector $\mathbf{p}$ is normal to the $S = S_0$ surface at each point of the latter. Consider for simplicity the case of a single particle of mass m in the conservative case, and use the Cartesian coordinates; from $\mathbf{p} = m \dot{\mathbf{q}}$ one finds that at each instant the surface $S = S_0$ is normal to the particle's trajectory. This makes the surface $S = S_0$ the analogue of the constant-phase surface found in the wave theory (e.g., Sect. 5.9). For this reason, $\cos \varphi \, dq/dt$ is called *phase velocity*.

Due to its definition, the phase velocity depends on the position $\mathbf{q}$ and is the velocity with which each point of the $S = S_0$ surface moves. It is worth adding that the phase velocity does not coincide with the actual velocity of any of the system's particles. To show this it suffices to consider the single particle's case with $\mathbf{p} = m \dot{\mathbf{q}}$: from (2.20) one finds that the modulus of the phase velocity is in fact inversely proportional to that of the particle.

2.6   Invariance Properties

A number of dynamical properties of the system of particles under consideration can be inferred directly from the form of the Lagrangian or Hamiltonian function, without the need of solving the motion's equations. An example is the conservation of the momentum conjugate to a cyclic coordinate (Sect. 1.4). Other properties are discussed below.

2.6.1   Time Reversal

It is found by inspection that the expression (1.22) of the system's kinetic energy in Cartesian coordinates is invariant when t is replaced with $-t$ (time reversal). A rather lengthy calculation based on the chain-differentiation rule shows that this property still holds after a coordinate transformation.

In some cases the whole Lagrangian function is invariant under time reversal. This makes the Lagrange equations (1.28) invariant as well. Assume that (1.28) are solved starting from a given initial condition at $t = a$ to the final instant $t = b$. Then, replace t with $t' = -t$ and solve the Lagrange equations again, using $q_i(t = b)$ and $-\dot{q}_i(t = b)$ as initial conditions. Letting $q'_i = dq_i/dt' = -\dot{q}_i$, (1.28) become

$$\frac{d}{d(-t')} \frac{\partial L}{\partial(-q'_i)} = \frac{d}{dt'} \frac{\partial L}{\partial q'_i} = \frac{\partial L}{\partial q_i}, \qquad i = 1, \ldots, n \qquad (2.21)$$

where, due to the hypothesis of invariance, the Lagrangian function is the same as that used to describe the motion from $t = a$ to $t = b$. It follows that the trajectories of the second motion are equal to those of the first one. Moreover, the initial velocities of the second motion, calculated at $t' = -b$, are opposite to those of the first motion at the same point. Due to the arbitrariness of b, at each point of a trajectory the velocity described by the time t' is opposite to that described by t. A motion having this property is called *reversible*.

Taking the examples of Sect. 1.3 and remembering the form (1.12, 1.18) of the corresponding Lagrangian functions one finds that, in the first example, the motion is reversible if the potential energy V is invariant under time reversal, namely, $V(-t) = V(t)$, while in the second example the motion is reversible if $\varphi(-t) = \varphi(t)$ and $\mathbf{A}(-t) = -\mathbf{A}(t)$.

2.6.2 Translation of Time

Consider the case where the Hamiltonian function is invariant with respect to translations of the origin of time. The invariance holds also for an infinitesimal translation dt, hence it is $dH/dt = 0$. In other terms H is a constant of motion. When this happens, as illustrated in Sect. 1.4, the Lagrangian and Hamiltonian functions have no explicit dependence on time, and vice versa.

2.6.3 Translation of the Coordinates

Another interesting case occurs when the Lagrangian function is invariant with respect to translations of the coordinates' origin. By way of example consider an N-particle system with no constraints, whence $n = 3N$, and use the Cartesian coordinates x_{js}. Here the first index is associated with the particles and the second one with the axes. Then, choose an infinitesimal translation dh_1 in the direction of the first axis and, similarly, infinitesimal translations dh_2 and dh_3 in the other two directions. Thus, each coordinate x_{j1}, $j = 1, \ldots, N$ within the Lagrangian function is replaced by $x_{j1} + dh_1$, and so on. The translational invariance then yields

$$dL = dh_1 \sum_{j=1}^{N} \frac{\partial L}{\partial x_{j1}} + dh_2 \sum_{j=1}^{N} \frac{\partial L}{\partial x_{j2}} + dh_3 \sum_{j=1}^{N} \frac{\partial L}{\partial x_{j3}} = 0. \qquad (2.22)$$

Each sum in (2.22) vanishes independently of the others due to the arbitrariness of the translations. Taking the sum multiplying dh_1 and using (1.28) yields

$$\sum_{j=1}^{N} \frac{\partial L}{\partial x_{j1}} = \sum_{j=1}^{N} \frac{d}{dt} \frac{\partial L}{\partial \dot{x}_{j1}} = \frac{d}{dt} \sum_{j=1}^{N} p_{j1} = \frac{dP_1}{dt} = 0, \qquad (2.23)$$

where $P_1 = \sum_{j=1}^{N} p_{j1}$ is the first component of the *total momentum*

$$\mathbf{P} = \sum_{j=1}^{N} \mathbf{p}_j \qquad (2.24)$$

of the system of particles. The other two components are treated in the same manner. In conclusion, if the Lagrangian function is invariant with respect to translations of the coordinates' origin, then the total momentum of the system is a constant of motion.

The above reasoning applies independently to each axis. As a consequence, if the Lagrangian function is such that the sum $\sum_{j=1}^{N} \partial L/\partial x_{j1}$ vanishes, while the analogous sums associated with the other two axes do not vanish, then P_1 is a constant of motion, while P_2, P_3 are not.

An important example of a Langrangian function, that is invariant with respect to translations of the coordinates' origin, is found when the force $\mathbf{F}_i$ acting on the ith particle derives from a potential energy V that depends only on the relative distances $r_{jk} = |\mathbf{r}_j - \mathbf{r}_k|$ among the particles, $k \neq j$. An example is given in Sect. 3.7.

2.6.4 Rotation of the Coordinates

Consider the case where the Lagrangian function is invariant with respect to rotations of the coordinates around an axis that crosses the origin. Like in Sect. 2.6.3 a system of N particles with no constraints is assumed, and the Cartesian coordinates are used. Let π be the plane that contains the origin and is normal to the rotation axis. It is convenient to use on π a polar reference (Sect. B.2) in which the rotation is defined over π by the angle φ. In turn, let ϑ_j be the angle between the rotation axis and the position vector $\mathbf{r}_j = (x_{j1}, x_{j2}, x_{j3})$ of the jth particle. The meaning of the angles is the same as in Fig. B.1, where the axes x, y define plane π; then, ϑ and $\mathbf{r}_j$ are represented by ϑ and $\mathbf{r}$ of the figure, respectively. If an infinitesimal rotation $d\varphi$ is considered, the position vector $\mathbf{r}_j$ undergoes a variation $d\mathbf{r}_j$ parallel to π and of magnitude $|d\mathbf{r}_j| = r_j \sin \vartheta_j \, d\varphi$. To specify the direction of $d\mathbf{r}_j$ one takes the unit vector $\mathbf{a}$ of the rotation axis and associates to the rotation the vector $\mathbf{a} \, d\varphi$ such that

$$d\mathbf{r}_j = \mathbf{a} \, d\varphi \wedge \mathbf{r}_j. \qquad (2.25)$$

The corresponding variations $d\dot{\mathbf{r}}_j$ of the velocities are found by differentiating (2.25) with respect to time. The variation of the Lagrangian function is

$$dL = \sum_{j=1}^{N} \sum_{s=1}^{3} \left(\frac{\partial L}{\partial x_{js}} dx_{js} + \frac{\partial L}{\partial \dot{x}_{js}} d\dot{x}_{js} \right), \qquad (2.26)$$

where the variations of the components are found from (2.25) and read $dx_{js} = d\varphi \left(\mathbf{a} \wedge \mathbf{r}_j \right)_s$, $d\dot{x}_{js} = d\varphi \left(\mathbf{a} \wedge \dot{\mathbf{r}}_j \right)_s$. Replacing the latter in (2.26) and using (1.28) yields

$$dL = d\varphi \sum_{j=1}^{N} \left(\mathbf{a} \wedge \mathbf{r}_j \cdot \dot{\mathbf{p}}_j + \mathbf{a} \wedge \dot{\mathbf{r}}_j \cdot \mathbf{p}_j \right). \tag{2.27}$$

Due to the rotational invariance, (2.27) vanishes for any $d\varphi$. Letting the sum to vanish after exchanging the scalar and vector products, and remembering that $\mathbf{a}$ is constant, one finds

$$\mathbf{a} \cdot \sum_{j=1}^{N} \left(\mathbf{r}_j \wedge \dot{\mathbf{p}}_j + \dot{\mathbf{r}}_j \wedge \mathbf{p}_j \right) = \mathbf{a} \cdot \frac{d}{dt} \sum_{j=1}^{N} \mathbf{r}_j \wedge \mathbf{p}_j = \mathbf{a} \cdot \frac{d\mathbf{M}}{dt} = \frac{d}{dt} \left(\mathbf{M} \cdot \mathbf{a} \right) = 0, \tag{2.28}$$

where

$$\mathbf{M} = \sum_{j=1}^{N} \mathbf{r}_j \wedge \mathbf{p}_j \tag{2.29}$$

is the *total angular momentum* of the system of particles. In conclusion, if the Lagrangian function is invariant with respect to rotations of the coordinates around an axis that crosses the origin, then the projection of the system's total angular momentum $\mathbf{M}$ over the rotation axis is a constant of motion.

2.7   Maupertuis Principle

Besides the Hamilton principle described in Sect. 1.3.4, other variational principles exist. Among them is the *Maupertuis*, or *least action*, principle, that applies to a particle subjected to conservative forces. Let $V = V(x_1, x_2, x_3)$ be the potential energy and $E = $ const the total energy, and let A and B indicate the two points of the (x_1, x_2, x_3) space that limit the trajectory of the particle. The Maupertuis principle states that the natural trajectory between A and B is the one that minimizes the functional

$$G = \int_{AB} \sqrt{E - V} \, ds, \qquad ds^2 = dx_1^2 + dx_2^2 + dx_3^2, \tag{2.30}$$

where the integral is carried out along the trajectory. The form of (2.30) explains the term "least action": in fact, the relation $p^2/(2m) = m \, u^2/2 = E - V$ shows that the integrand $\sqrt{E - V}$ is proportional to the particle's momentum p; as a multiplicative constant is irrelevant for calculating the extremum functions, the minimization of G is equivalent to that of the action $\int_{AB} p \, ds$.

To calculate the extremum functions of (2.30) it is convenient to parametrize the coordinates in the form $x_i = x_i(\xi)$, where the parameter ξ takes the same limiting values, say, $\xi = a$ at A and $\xi = b$ at B, for the natural and all the virtual trajectories. Letting $\dot{x}_i = dx_i/d\xi$ one finds $(ds/d\xi)^2 = \dot{x}_1^2 + \dot{x}_2^2 + \dot{x}_3^2$ which, remembering (1.7), yields the extremum condition for (2.30):

$$\delta \int_a^b \theta \, d\xi = 0, \qquad \theta = \sqrt{E - V} \sqrt{\dot{x}_1^2 + \dot{x}_2^2 + \dot{x}_3^2}, \qquad \frac{d}{d\xi} \frac{\partial \theta}{\partial \dot{x}_i} = \frac{\partial \theta}{\partial x_i}. \qquad (2.31)$$

The following relations are useful to work out the last of (2.31): $ds/dt = u = \sqrt{(2/m)(E - V)}$, $dx_i = \dot{x}_i \, d\xi$, $dx_i/dt = u_i$. One finds

$$\frac{\partial \theta}{\partial \dot{x}_i} = \sqrt{E - V} \frac{\dot{x}_i}{ds/d\xi} = \sqrt{\frac{m}{2}} u \frac{\dot{x}_i \, d\xi}{ds} = \sqrt{\frac{m}{2}} \frac{dx_i}{dt} = \sqrt{\frac{m}{2}} u_i, \qquad (2.32)$$

$$\frac{\partial \theta}{\partial x_i} = \frac{ds}{d\xi} \frac{-\partial V/\partial x_i}{2\sqrt{E - V}} = \frac{ds}{d\xi} \frac{F_i}{2\sqrt{m/2}u} = \frac{dt}{d\xi} \frac{F_i}{\sqrt{2m}}, \qquad (2.33)$$

with F_i the ith component of the force. The last of (2.31) then yields

$$\frac{d}{d\xi} \left(\sqrt{\frac{m}{2}} u_i \right) = \frac{dt}{d\xi} \frac{F_i}{\sqrt{2m}}, \qquad F_i = m \frac{du_i}{d\xi} \frac{d\xi}{dt} = m \frac{du_i}{dt}. \qquad (2.34)$$

In conclusion, the equation that provides the extremum condition for functional G is equivalent to Newton's second law $\mathbf{F} = m\mathbf{a}$.

2.8   Spherical Coordinates—Angular Momentum

Consider a single particle of mass m and use the transformation from the Cartesian (x, y, z) to the spherical (r, ϑ, φ) coordinates shown in Sect. B.1. The kinetic energy is given by (B.7), namely,

$$T = \frac{m}{2} (\dot{r}^2 + r^2\dot{\vartheta}^2 + r^2\dot{\varphi}^2 \sin^2 \vartheta). \qquad (2.35)$$

If the force acting onto the particle is derivable from a potential energy $V = V(x, y, z, t)$, the Lagrangian function in the spherical reference is $L = T - V(r, \vartheta, \varphi, t)$, where T is given by (2.35). The momenta conjugate to the spherical coordinates are

$$\begin{cases} p_r &= \partial L/\partial \dot{r} &= m\dot{r} \\ p_\vartheta &= \partial L/\partial \dot{\vartheta} &= mr^2 \dot{\vartheta} \\ p_\varphi &= \partial L/\partial \dot{\varphi} &= mr^2 \dot{\varphi} \sin^2 \vartheta \end{cases} \qquad (2.36)$$

Using (2.36), the kinetic energy is recast as

$$T = \frac{1}{2m} \left(p_r^2 + \frac{p_\vartheta^2}{r^2} + \frac{p_\varphi^2}{r^2 \sin^2 \vartheta} \right). \tag{2.37}$$

The components of the momentum $\mathbf{p}$ derived from the Lagrangian function written in the Cartesian coordinates are $m\dot{x}$, $m\dot{y}$, $m\dot{z}$. It follows that the components of the angular momentum $\mathbf{M} = \mathbf{r} \wedge \mathbf{p}$ written in the Cartesian and spherical references are

$$\begin{cases} M_x &=& m\,(y\dot{z} - z\dot{y}) &=& -mr^2\,(\dot{\vartheta}\,\sin\varphi + \dot{\varphi}\,\sin\vartheta\,\cos\vartheta\,\cos\varphi) \\ M_y &=& m\,(z\dot{x} - x\dot{z}) &=& mr^2\,(\dot{\vartheta}\,\cos\varphi - \dot{\varphi}\,\sin\vartheta\,\cos\vartheta\,\sin\varphi) \\ M_z &=& m\,(x\dot{y} - y\dot{x}) &=& mr^2\,\dot{\varphi}\,\sin^2\vartheta \end{cases} \tag{2.38}$$

The square modulus of the angular momentum in spherical coordinates reads

$$M^2 = m^2 r^4 \left(\dot{\vartheta}^2 + \dot{\varphi}^2 \sin^2 \vartheta \right) = p_\vartheta^2 + \frac{p_\varphi^2}{\sin^2 \vartheta}, \tag{2.39}$$

where the last equality is due to (2.36). From (2.37, 2.39) one finds

$$T = \frac{1}{2m} \left(p_r^2 + \frac{M^2}{r^2} \right). \tag{2.40}$$

If M is a constant of motion, (2.40) shows that the kinetic energy depends on r and $\dot{r}$ only. Comparing (2.36) with (2.38) one also notices that

$$p_\varphi = M_z, \tag{2.41}$$

namely, the component along the z axis of the angular momentum turns out to be the momentum conjugate to the φ coordinate. The latter describes the rotations along the same axis. In contrast, the other two components of $\mathbf{M}$ are not conjugate momenta. This result is due to the asymmetry of the relations (B.1) that connect the Cartesian to the spherical coordinates, and does not ascribe any privileged role to the z axis. In fact, by exchanging the Cartesian axes one makes p_φ to coincide with M_x or M_y.

Another example refers to a particle of mass m and charge e subjected to an electromagnetic field. Remembering (1.33) one has $L = (1/2)\,mu^2 - eU + e\mathbf{u} \cdot \mathbf{A}$, where the scalar potential is indicated with U to avoid confusion with the φ coordinate. It follows

$$L = \frac{1}{2}\,m\,\left(\dot{r}^2 + r^2\,\dot{\vartheta}^2 + r^2\,\dot{\varphi}^2\,\sin^2\vartheta \right) - e\,U + e\mathbf{u} \cdot \mathbf{A}, \tag{2.42}$$

where the components of $\mathbf{u} = \dot{\mathbf{r}}$ are given by (B.6), and U, A depend on the coordinates and time. Indicating the components of $\mathbf{A}$ with A_x, A_y, A_z, the momenta read

$$\begin{cases} p_r = \partial L / \partial \dot{r} = m\dot{r} + e\,A_x\,\sin\vartheta\,\cos\varphi + e\,A_y\,\sin\vartheta\,\sin\varphi + e\,A_z\,\cos\vartheta\,\cos\varphi \\ p_\vartheta = \partial L / \partial \dot{\vartheta} = m\,r^2\,\dot{\vartheta} + e\,A_x\,r\,\cos\vartheta\,\cos\varphi + e\,A_y\,r\,\cos\vartheta\,\sin\varphi - e\,A_z\,r\,\sin\vartheta \\ p_\varphi = \partial L / \partial \dot{\varphi} = m\,r^2\,\dot{\varphi}\,\sin^2\vartheta - e\,A_x\,r\,\sin\vartheta\,\sin\varphi + e\,A_y\,r\,\sin\vartheta\,\cos\varphi \end{cases}$$
$$\tag{2.43}$$

Thanks to (B.1, B.6), the third of (2.43) can be written as

$$p_\varphi = m\,(x\,\dot{y} - y\,\dot{x}) + e\,(x\,A_y - y\,A_x) = x\,(m\,\dot{y} + e\,A_y) - y\,(m\,\dot{x} + e\,A_x),$$
$$(2.44)$$

that coincides with the component of the angular momentum $\mathbf{M} = \mathbf{r} \wedge \mathbf{p} = \mathbf{r} \wedge (m\,\mathbf{u} + e\mathbf{A})$ along the z axis. This result shows that (2.41) holds also when the force acting onto the particle derives from an electromagnetic field.

2.9   Linear Motion

The *linear motion* is the motion of a system having only one degree of freedom. Using the Cartesian coordinate x, and assuming the case where the force acting onto the particle derives from a potential energy $V(x)$, gives the Hamiltonian function (1.32) the form $H = p^2/(2\,m) + V(x)$. As shown in Sect. 1.4, a Hamiltonian function of this type is a constant of motion whence, remembering that here it is $p = m\,\dot{x}$,

$$\frac{1}{2}\,m\,\dot{x}^2 + V(x) = E = \text{const.} \tag{2.45}$$

The constant E is called *total energy*. Its value is given by the initial conditions $x_0 = x(t = a)$, $\dot{x}_0 = \dot{x}(t = a)$. As the kinetic energy $m\,\dot{x}^2/2$ can not be negative, the motion is possible only in the intervals of the x axis such that $V(x) \leq E$. In particular, the velocity $\dot{x}$ vanishes at the points where $V = E$. Instead, the intervals where $V > E$ can not be reached by the particle. Equation (2.45) is separable and provides a relation of the form $t = t(x)$,

$$t = a \pm \sqrt{\frac{m}{2}} \int_{x_0}^{x} \frac{\mathrm{d}\xi}{\sqrt{E - V(\xi)}}. \tag{2.46}$$

By way of example consider a situation like that shown in Fig. 2.1, where it is assumed that to the right of x_C the potential energy V keeps decreasing as $x \to \infty$. If the initial position of the particle is $x_0 = x_C$, there the velocity vanishes and the particle is subjected to a positive force $F = -\mathrm{d}V/\mathrm{d}x > 0$. As a consequence, the particle's motion will always be oriented to the right starting from x_C. Such a motion is called *unlimited*. If the initial position is $x_0 > x_C$ and the initial velocity is negative, the particle moves to the left until it reaches the position x_C, where it bounces back. The subsequent motion is the same as described above.

A different situation arises when the initial position of the particle belongs to an interval limited by two zeros of the function $E - V(x)$ like, e.g., x_A and x_B in Fig. 2.1. The motion is confined between x_A and x_B and, for this reason, is called *limited*. The particle bounces back and forth endlessly under the effect of a force that does not depend on time. As a consequence, the time necessary to complete a cycle $x_A \to x_B \to x_A$ is the same for each cycle. In other terms, the motion is periodic in time. Also, from (2.46) it is found by inspection that the time spent by the particle

Fig. 2.1 Example of potential energy discussed in Sect. 2.9

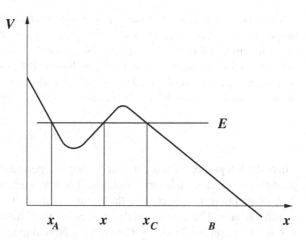

in the $x_A \to x_B$ part of the cycle is the same as that spent in the $x_B \to x_A$ part. The period of the oscillation is then found to be

$$T = 2\sqrt{\frac{m}{2}} \int_{x_A}^{x_B} \frac{\mathrm{d}x}{\sqrt{E - V(x)}}. \qquad (2.47)$$

Note that the period depends on the total energy E. However there are exceptions, as the example of Sect. 3.3 shows.

2.10 Action-Angle Variables

Consider a linear, conservative motion of constant energy E (Sect. 2.9) and let q, p be two canonical coordinates describing it. The following hold

$$H(q, p) = E, \qquad p = p(q, E). \qquad (2.48)$$

The second of (2.48) is derived from the first one by solving for the momentum, and provides the phase trajectory (Sect. 1.9) starting from the initial conditions q_0, p_0 of the motion. As shown below, in several mechanical systems of interest the phase trajectory has some special characteristic that is worth examining.

Consider, first, the situation where the phase trajectory is closed: in this case, after a time T has elapsed from $t = 0$, the canonical coordinates take again the values q_0, p_0. As a consequence, for $t > T$ the motion repeats itself, and so on. It follows that both q and p are periodic functions of time with the same period T. As discussed in Sect. 2.9, this type of periodic motion, in which both q and p are bounded, is typically found when the initial position q_0 lies between two zeros of $E - V$, and is of the oscillatory type. It is also indicated with the astronomical term *libration*.

A second important situation occurs when p is a periodic function of q. In this type of motion q is unbounded. However, when q increases by a period the configuration of the mechanical system remains practically unchanged. In fact, in this type of motion the canonical coordinate q is always an angle of rotation: the motion is still periodic and is referred to as *rotation*. Note that the same mechanical system may give rise to a libration or a rotation, depending on the motion's initial conditions: a typical example is that of the simple pendulum where q is identified with the angle of deflection [42, Chap. 10.6]. The *action variable* is defined as

$$J(E) = \oint p(q, E)\, dq, \tag{2.49}$$

where the integral is carried out over a complete period of libration or rotation, depending on the case under investigation. The name given to J stems from the fact that, as mentioned in Sect. 1.5, the product $q\, p$ has the units of an action in all coordinate sets. The action variable is a constant of motion because it depends on E only. Inverting $J(E)$ yields $H = H(J)$. Now one applies a canonical transformation generated by a Hamilton characteristic function of the form $W = W(q, J)$. Remembering the procedure depicted in Sect. 2.4, W is the solution of

$$H\left(q, \frac{\partial W}{\partial q}\right) = E. \tag{2.50}$$

Applying (2.14) one finds the generalized coordinate $w = \partial W / \partial J$, called *angle variable*, conjugate to J. The pair J, w constitutes the set of canonical coordinates called *action-angle variables*. Finally, the Hamilton equations in the new coordinates read

$$\dot{w} = \frac{\partial H}{\partial J} = \text{const}, \qquad \dot{J} = -\frac{\partial H}{\partial w} = 0. \tag{2.51}$$

The time evolution of the action-angle variables is then $w = \dot{w} t + w_0$, $J = \text{const}$. From the first of (2.51) it also follows that the units of $\dot{w}$ are those of a frequency. The usefulness of the action-angle variables becomes apparent when one calculates the change Δw over a complete libration or rotation cycle of q:

$$\Delta w = \oint dw = \oint \frac{\partial w}{\partial q}\, dq = \oint \frac{\partial^2 W}{\partial q\, \partial J}\, dq = \frac{d}{dJ} \oint \frac{\partial W}{\partial q}\, dq = 1, \tag{2.52}$$

where the last equality derives from combining $p = \partial W / \partial q$ with (2.49). On the other hand, if T is the time necessary for completing a cycle of q, then it is $\Delta w = w(T) - w(0) = \dot{w} T$, whence $\dot{w} = 1/T$. Thus, the frequency $\nu = \dot{w}$ is that associated with the periodic motion of q. In conclusion, the action-angle variables provide a straightforward method to determine the frequency of a periodic motion without the need of solving the motion equation. The method is applicable also to conservative systems having more than one degree of freedom, provided there exists at least one set of coordinates in which the Hamilton–Jacobi equation is completely separable [42, Chap. 10.7].

2.11 Complements

2.11.1 Infinitesimal Canonical Transformations

Consider a system with n degrees of freedom whose Hamiltonian function is $H(q_1, \ldots, q_n, p_1, \ldots, p_n, t)$. Remembering (1.42) one finds that the canonical coordinates at $t + dt$ are expressed, in terms of the same coordinates at t, by the relations

$$q_i + dq_i = q_i + \frac{\partial H}{\partial p_i} dt, \qquad p_i + dp_i = p_i - \frac{\partial H}{\partial q_i} dt. \tag{2.53}$$

Letting $\tilde{q}_i = q_i + dq_i$, $\tilde{p}_i = p_i + dp_i$ gives (2.53) the same form as (2.1), namely, that of a coordinate transformation. It is interesting to check whether such a transformation is canonical. For this, one notes that the transformation (2.53) differs by infinitesimal quantities from the identical transformation $\tilde{q}_i = q_i$, $\tilde{p}_i = p_i$; as a consequence one expects the generating function of (2.53), if it exists, to differ by an infinitesimal function from $F_2 = \sum_{i=1}^{n} q_i \tilde{p}_i$ which, as shown in Sect. 2.2, generates the identical transformation. One then lets

$$K_2 = \sum_{i=1}^{n} q_i \tilde{p}_i + \epsilon G(\mathbf{q}, \tilde{\mathbf{p}}, t), \tag{2.54}$$

where ϵ is an infinitesimal quantity. From the first two equations in (2.9) it follows

$$p_i = \frac{\partial K_2}{\partial q_i} = \tilde{p}_i + \epsilon \frac{\partial G}{\partial q_i}, \qquad \tilde{q}_i = \frac{\partial K_2}{\partial \tilde{p}_i} = q_i + \epsilon \frac{\partial G}{\partial \tilde{p}_i}. \tag{2.55}$$

In the last term of (2.55) one may replace $\tilde{p}_i$ with p_i on account of the fact that the difference between $\epsilon \, \partial G / \partial \tilde{p}_i$ and $\epsilon \, \partial G / \partial p_i$ is infinitesimal of a higher order. Then, letting $\epsilon = dt$, $G(\mathbf{q}, \mathbf{p}, t) = H(\mathbf{q}, \mathbf{p}, t)$, and $K_2 = \sum_{i=1}^{n} q_i \tilde{p}_i + H(\mathbf{q}, \mathbf{p}, t) dt$, makes (2.55) identical to (2.53). Note that this replacement transforms the third of (2.9) into $\tilde{H} = H + (\partial H / \partial t) dt$, as should be (compare with (1.44)).

The above reasoning shows that the $H dt$ term in (2.54) generates a canonical transformation that produces the variations of the canonical coordinates in the time interval dt. Such a transformation is called *infinitesimal canonical transformation*. On the other hand, as the application of more than one canonical transformation is still canonical, the evolution of the coordinates q_i, p_i during a finite interval of time can be thought of as produced by a succession of infinitesimal canonical transformations generated by the Hamiltonian function. In other terms, the Hamiltonian function generates the motion of the system.

2.11.2 Constants of Motion

It has been shown in Sect. 2.11.1 that a succession of infinitesimal canonical transformation generated by the Hamiltonian function determines the time evolution of the canonical coordinates q_i, p_i. If such a succession starts with the initial conditions q_{i0}, p_{i0}, at some later time t the transformations Eqs. (2.53) take the form

$$q_i = q_i(\mathbf{q}_0, \mathbf{p}_0, t), \qquad p_i = p_i(\mathbf{q}_0, \mathbf{p}_0, t). \qquad (2.56)$$

The relations (2.56) are nothing else than the solution of the mechanical problem; in fact, they express the canonical coordinates at time t, given the initial conditions. From another viewpoint, they show that the solution of the problem contains $2n$ constants. They are not necessarily constants of motion, in fact, their values at $t > 0$ are in general different from those at $t = 0$. If the system has extra properties (like, e.g., the invariance properties discussed in Sect. 2.6), it also has one or more constants of motion. The latter keep the value that they possessed at $t = 0$, so they are expressible as combinations of the canonical coordinates at $t = 0$; by way of example, the total energy E of a single particle subjected to a conservative force reads $E = (p_{01}^2 + p_{02}^2 + p_{03}^2)/(2\,m) + V(x_{10}, x_{20}, x_{30})$.

For a system having n degrees of freedom the total number of independent combinations of the initial conditions can not exceed the number of the initial conditions themselves. As a consequence, for such a system the maximum number of independent constants of motion is $2n$.

Problems

2.1 Given the Hamiltonian function $H = p^2/(2\,m) + (c/2)x^2$, $m, c > 0$ (that describes the *linear harmonic oscillator*, Sect. 3.3), find the oscillation frequency ν using the action-angle variables.

Chapter 3
Applications of the Concepts of Analytical Mechanics

3.1 Introduction

This chapter provides a number of important examples of application of the principles of Analytical Mechanics. The examples are chosen with reference to the applications to Quantum Mechanics shown in later chapters. The first sections treat the problems of the square well, linear harmonic oscillator, and central motion. The subsequent sections deal with the two-particle interaction: first, the description of the collision is given, along with the calculation of the energy exchange involved in it, with no reference to the form of the potential energy; this is followed by the treatment of the collision when the potential energy is of the repulsive-Coulomb type. The chapter continues with the treatment of a system of strongly-bound particles: the diagonalization of its Hamiltonian function shows that the motion of the particles is a superposition of harmonic oscillations. Finally, the motion of a particle subjected to a periodic potential energy is analyzed, including the case where a weak perturbation is superimposed to the periodic part. A number of complements are also given, that include the treatment of the collision with a potential energy of the attractive-Coulomb type, and that of the collision of two relativistic particles.

3.2 Particle in a Square Well

As a first example of linear motion consider the case of a potential energy V of the form shown in Fig. 3.1. Such a potential energy is called *square well* and is to be understood as the limiting case of a potential-energy well whose sides have a finite, though very large, slope. It follows that the force $F = -\mathrm{d}V/\mathrm{d}x$ is everywhere equal to zero with the exception of the two points $-x_M$ and $+x_M$, where it tends to $+\infty$ and $-\infty$, respectively. From the discussion of Sect. 2.9 it follows that the case $E < 0$

© Springer Science+Business Media New York 2015
M. Rudan, *Physics of Semiconductor Devices*,
DOI 10.1007/978-1-4939-1151-6_3

Fig. 3.1 The example of the
square well analyzed in
Sect. 3.2. Only the case
$0 \le E \le V_0$ is shown

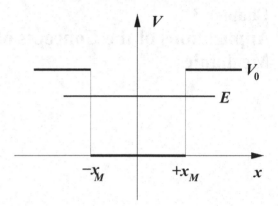

is forbidden. The motion of the particle is finite for $0 \le E \le V_0$, while it is infinite
for $E > V_0$.

Considering the $0 \le E \le V_0$ case first, the motion is confined within the well
and the velocity of the particle is constant in the interval $-x_M < x < +x_M$, where
the Hamiltonian function yields $m\,\dot{x}^2/2 = E$. If the particle's motion is oriented to
the right, the velocity is $\dot{x} = \sqrt{2\,E/m}$. When the particle reaches the position x_M
its velocity reverses instantly to become $\dot{x} = -\sqrt{2\,E/m}$. The motion continues at a
constant velocity until the particle reaches the position $-x_M$ where is reverses again,
and so on. As the spatial interval corresponding to a full cycle is $4\,x_M$, the oscillation
period is $T = \sqrt{8\,m/E}\,x_M$.

To treat the $E > V_0$ case assume that the particle is initially at a position $x < -x_M$
with a motion oriented to the right. The Hamiltonian function outside the well yields
$m\,\dot{x}^2/2 + V_0 = E$. The constant velocity is $\dot{x} = \sqrt{2\,(E - V_0)/m}$ until the particle
reaches the position $-x_M$. There the velocity increases abruptly to $\dot{x} = \sqrt{2\,E/m}$ and
keeps this value until the particle reaches the other edge of the well, $+x_M$. There,
the velocity decreases abruptly back to the initial value $\dot{x} = \sqrt{2\,(E - V_0)/m}$, and
the particle continues its motion at a constant velocity in the positive direction.

3.3 Linear Harmonic Oscillator

An important example of linear motion is found when the force derives from a
potential energy of the form $V = c\,x^2/2$, with $c > 0$. The force acting on the
particle turns out to be $F = -\mathrm{d}V/\mathrm{d}x = -c\,x$, namely, it is linear with respect to x,
vanishes at $x = 0$, and has a modulus that increases as the particle departs from the
origin. Also, due to the positiveness of c, the force is always directed towards the
origin. A force of this type is also called *linear elastic force*, and c is called *elastic
constant* (Fig. 3.2).

From the discussion of Sect. 2.9 it follows that the case $E < 0$ is forbidden. The
motion of the particle is always finite because for any $E \ge 0$ it is confined between
the two zeros $x_M = \pm\sqrt{2\,E/c}$ of the equation $V = E$. The Hamiltonian function
reads

Fig. 3.2 The example of the linear harmonic oscillator analyzed in Sect. 3.3

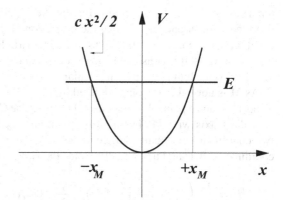

$$H = \frac{1}{2m} p^2 + \frac{1}{2} cx^2 = E = \text{const},$$ (3.1)

yielding the motion's equation $\dot{p} = m\ddot{x} = -\partial H/\partial x = -cx$ whose solution is

$$x(t) = x_M \cos(\omega t + \alpha_0), \quad \dot{x}(t) = -\omega x_M \sin(\omega t + \alpha_0), \quad \omega = \sqrt{c/m}. \quad (3.2)$$

Due to the form of (3.2), a particle whose motion is derived from the Hamiltonian function (3.1) is called *linear harmonic oscillator*. The maximum elongation $x_M > 0$ and the initial phase α_0 are readily expressed in terms of the initial conditions $x_0 = x(t = 0)$, $\dot{x}_0 = \dot{x}(t = 0)$. In fact, letting $t = 0$ in (3.2) yields $x_M^2 = x_0^2 + \dot{x}_0^2/\omega^2$ and $\tan \alpha_0 = -\dot{x}_0/(\omega x_0)$. The total energy in terms of the initial conditions reads $E = m\dot{x}_0^2/2 + cx_0^2/2$ and, finally, the oscillation's period is $T = 2\pi/\omega$. Note that T depends on the two parameters m, c appearing in the Hamiltonian function, but not on the total energy (in other terms, for a given pair m, c the oscillation period does not depend on the initial conditions). As mentioned in Sect. 2.9, this is an exceptional case.

3.4 Central Motion

Consider the case of a particle of mass m acted upon by a force that derives from a potential energy of the form $V = V(r)$, where $r = \sqrt{x^2 + y^2 + z^2}$ is the modulus of the position vector $\mathbf{r}$ of the particle. The force

$$\mathbf{F} = -\text{grad}\, V = -\frac{dV}{dr} \frac{\mathbf{r}}{r},$$ (3.3)

depends on r only and is oriented along $\mathbf{r}$. For this reason it is called *central force*. In turn, the point whence the force originates (in this case, the origin of the reference) is called *center of force*. The corresponding Lagrangian function

$$L = \frac{1}{2} m\dot{r}^2 - V(r), \quad \mathbf{p} = m\dot{\mathbf{r}}, \quad \dot{r} = |\dot{\mathbf{r}}|$$ (3.4)

turns out to be invariant under any rotation (Sect. 2.6.4). Remembering (2.28, 2.29), this type of invariance means that the projection of the angular momentum $\mathbf{M}$ onto any direction is conserved. It follows that, for a particle acted upon by a central force, the vector $\mathbf{M}$ itself is conserved. On the other hand it is $\mathbf{M} = \mathbf{r} \wedge m\dot{\mathbf{r}}$, so the constant angular momentum is fixed by the initial conditions of the motion.

As $\mathbf{M}$ is normal to the plane defined by $\mathbf{r}$ and $\dot{\mathbf{r}}$, the trajectory of the particle lies always on such a plane. It is then useful to select the Cartesian reference by aligning, e.g., the z axis with $\mathbf{M}$. In this way, the trajectory belongs to the x, y plane and two coordinates eventually suffice to describe the motion. Turning to the spherical coordinates (B.1) and using (2.40) yields the Hamiltonian function

$$H = \frac{1}{2m} \left(p_r^2 + \frac{M^2}{r^2} \right) + V(r) = \frac{p_r^2}{2m} + V_e(r), \quad V_e = V + \frac{M^2}{2mr^2}, \quad (3.5)$$

with $M^2 = p_\vartheta^2 + p_\varphi^2/\sin^2\vartheta$, $\mathbf{M} = $ const, and $p_r = m\dot{r}$, $p_\vartheta = mr^2\dot{\vartheta}$, $p_\varphi = M_z = mr^2\dot{\varphi}\sin\vartheta$. However, the z axis has been aligned with $\mathbf{M}$, which is equivalent to letting $\vartheta = \pi/2$. It turns out $M_z = M$, and $p_r = m\dot{r}$, $p_\vartheta = 0$, $p_\varphi = M = mr^2\dot{\varphi}$, so that

$$H = \frac{1}{2m} \left(p_r^2 + \frac{p_\varphi^2}{r^2} \right) + V(r) = \frac{p_r^2}{2m} + V_e(r), \quad V_e = V + \frac{p_\varphi^2}{2mr^2}. \quad (3.6)$$

As the total energy is conserved it is $H = E$, where E is known from the initial conditions. The intervals of r where the motion can actually occur are those in which $E \geq V_e$. Letting $r_0 = r(t = 0)$, the time evolution of the radial part is found from $p_r^2 = m^2(dr/dt)^2 = 2m(E - V_e)$, namely

$$t(r) = \pm\sqrt{\frac{m}{2}} \int_{r_0}^{r} \frac{d\xi}{\sqrt{E - V_e(\xi)}}. \quad (3.7)$$

From $p_\varphi = mr^2\dot{\varphi} = $ const it follows that φ depends monotonically on time, and also that $dt = (mr^2/p_\varphi)\,d\varphi$. Combining the latter with (3.7) written in differential form, $dt = \pm\sqrt{m/2}\,[E - V_e(r)]^{-1/2}\,dr$, yields the equation for the trajectory,

$$\varphi(r) = \varphi_0 \pm \frac{p_\varphi}{\sqrt{2m}} \int_{r_0}^{r} \frac{d\xi}{\xi^2\sqrt{E - V_e(\xi)}}, \quad (3.8)$$

with $\varphi_0 = \varphi(t = 0)$. Finally, elimination of r from $t(r)$ and $\varphi(r)$ provides the time evolution of φ. It is convenient to let the initial time $t = 0$ correspond to an extremum of the possible values of r. In this way the sign of t and $\varphi - \varphi_0$ changes at $r = r_0$. By this choice the trajectory is symmetric with respect to the line drawn from the origin to the point of coordinates r_0, φ_0, and the evolution of the particle's motion over each half of the trajectory is symmetric with respect to time.

3.5   Two-Particle Collision

Consider a system made of two particles whose masses are m_1, m_2. The system is isolated, namely, the particles are not subjected to any forces apart those due to the mutual interaction. As a consequence, the Lagrangian function is invariant under coordinate translations, and the Hamiltonian function is invariant under time translations. Thus, as shown in Sect. 2.6, the total momentum and total energy of the system are conserved.

The type of motion that is considered is such that the distance between the particles is initially so large as to make the interaction negligible. The interaction becomes significant when the particles come closer to each other; when they move apart, the interaction becomes negligible again. This type of interaction is called *collision*. The values of the dynamical quantities that hold when the interaction is negligible are indicated as *asymptotic values*. The labels *a* and *b* will be used to mark the asymptotic values before and after the interaction, respectively.

It is worth specifying that it is assumed that the collision does not change the internal state of the particles (for this reason it is more appropriately termed *elastic collision*, [67, Sect. 17]). When the distance is sufficiently large, the particles can be considered as isolated: they move at a constant velocity and the total energy of the system is purely kinetic, $E_a = T_a$ and $E_b = T_b$. On the other hand, due to the invariance under time translation the total energy of the system is conserved, $E_b = E_a$. In conclusion it is $T_b = T_a$, namely, in an elastic collision the asymptotic kinetic energy of the system is conserved.

An analysis of the collision based only on the asymptotic values is incomplete because it does not take into account the details of the interaction between the two particles. However it provides a number of useful results, so it is worth pursuing. Letting $\mathbf{r}_1$ and $\mathbf{r}_2$ be the positions of the particles in a reference O, the position of the center of mass and the relative position of the particles are

$$\mathbf{R} = \frac{m_1\,\mathbf{r}_1 + m_2\,\mathbf{r}_2}{m_1 + m_2}, \qquad \mathbf{r} = \mathbf{r}_1 - \mathbf{r}_2. \tag{3.9}$$

The corresponding velocities are $\mathbf{v}_1 = \dot{\mathbf{r}}_1$, $\mathbf{v}_2 = \dot{\mathbf{r}}_2$, and $\mathbf{v} = \dot{\mathbf{r}}$. The relations between the velocities are obtained by differentiating (3.9) with respect to time. Solving for $\mathbf{v}_1$, $\mathbf{v}_2$ yields

$$\mathbf{v}_1 = \dot{\mathbf{R}} + \frac{m_2}{m_1 + m_2}\,\mathbf{v}, \qquad \mathbf{v}_2 = \dot{\mathbf{R}} - \frac{m_1}{m_1 + m_2}\,\mathbf{v}. \tag{3.10}$$

Letting $\dot{R} = |\dot{\mathbf{R}}|$, the system's kinetic energy before the interaction is

$$T_a = \frac{1}{2}\,m_1\,v_{1a}^2 + \frac{1}{2}\,m_2\,v_{2a}^2 = \frac{1}{2}\,(m_1 + m_2)\,\dot{R}_a^2 + \frac{1}{2}\,m\,v_a^2, \tag{3.11}$$

where $m = m_1 m_2/(m_1 + m_2)$ is called *reduced mass*. The expression of the kinetic energy after the interaction is obtained from (3.11) by replacing *a* with *b*.

The total momentum before the collision is $\mathbf{P}_a = m_1\,\mathbf{v}_{1a} + m_2\,\mathbf{v}_{2a} = (m_1 + m_2)\,\dot{\mathbf{R}}_a$. The conservation of $\mathbf{P}$ due to the invariance under coordinate translations yields $\mathbf{P}_b = \mathbf{P}_a$, whence $\dot{\mathbf{R}}_b = \dot{\mathbf{R}}_a$. Using (3.11) in combination with the conservation rules $\dot{\mathbf{R}}_b = \dot{\mathbf{R}}_a$ and $T_b = T_a$ yields $v_b = v_a$, namely, the asymptotic modulus of the relative velocity is conserved.

The analysis is now repeated in a new reference B in which the particles' positions are defined as

$$\mathbf{s}_1 = \mathbf{r}_1 - \mathbf{R} = \frac{m_2}{m_1 + m_2}\,(\mathbf{r}_1 - \mathbf{r}_2), \quad \mathbf{s}_2 = \mathbf{r}_2 - \mathbf{R} = \frac{m_1}{m_1 + m_2}\,(\mathbf{r}_2 - \mathbf{r}_1). \quad (3.12)$$

By construction, the origin of B coincides with the system's center of mass. The relative position in B is the same as in O, in fact

$$\mathbf{s} = \mathbf{s}_1 - \mathbf{s}_2 = \mathbf{r}_1 - \mathbf{r}_2 = \mathbf{r}. \quad (3.13)$$

From (3.12, 3.13) one finds

$$m_1\,\mathbf{s}_1 = -m_2\,\mathbf{s}_2, \quad \mathbf{s}_1 = \frac{m_2}{m_1 + m_2}\,\mathbf{s}, \quad \mathbf{s}_2 = -\frac{m_1}{m_1 + m_2}\,\mathbf{s}. \quad (3.14)$$

The velocities in reference B are $\mathbf{u}_1 = \dot{\mathbf{s}}_1$, $\mathbf{u}_2 = \dot{\mathbf{s}}_2$, and $\mathbf{u} = \dot{\mathbf{s}}$. The relations among the latter are found by differentiating (3.12, 3.13) and read

$$\mathbf{u}_1 = \mathbf{v}_1 - \dot{\mathbf{R}}, \quad \mathbf{u}_2 = \mathbf{v}_2 - \dot{\mathbf{R}}, \quad \mathbf{u} = \mathbf{v}, \quad (3.15)$$

$$\mathbf{u}_1 = \frac{m_2}{m_1 + m_2}\,\mathbf{u}, \quad \mathbf{u}_2 = -\frac{m_1}{m_1 + m_2}\,\mathbf{u}, \quad (3.16)$$

which in turn yield

$$\mathbf{v}_1 = \dot{\mathbf{R}} + \frac{m_2}{m_1 + m_2}\,\mathbf{u}, \quad \mathbf{v}_2 = \dot{\mathbf{R}} - \frac{m_1}{m_1 + m_2}\,\mathbf{u}. \quad (3.17)$$

Thanks to (3.16) the system's kinetic energy before and after the interaction, in reference B, is

$$K_a = \frac{1}{2}\,m_1 u_{1a}^2 + \frac{1}{2}\,m_2 u_{2a}^2 = \frac{1}{2}\,m u_a^2, \quad K_b = \frac{1}{2}\,m u_b^2. \quad (3.18)$$

The conservation of the kinetic energy, $K_b = K_a$, yields $u_b = u_a$. Using the third of (3.15) then yields

$$u_b = u_a = v_b = v_a, \quad (3.19)$$

that is, the asymptotic modulus of the relative velocity is conserved and has the same value in the two references. Moreover, (3.16) show that it is also $u_{1b} = u_{1a}$ and $u_{2b} = u_{2a}$, namely, in reference B the asymptotic kinetic energy is conserved for each particle separately.

Fig. 3.3 Graphic
representation of the vector
relation (3.20)

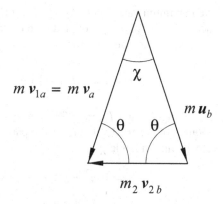

$$m\,\mathbf{v}_{1a} = m\,\mathbf{v}_a$$

3.6   Energy Exchange in the Two-Particle Collision

To complete the asymptotic analysis of the two-particle collision it is useful to choose
for O a reference such that $\mathbf{v}_{2a} = 0$. In this case (3.10) yield $\mathbf{v}_a = \mathbf{v}_{1a}$, whence the
total momentum reads $(m_1 + m_2)\,\dot{\mathbf{R}}_a = m_1 \mathbf{v}_{1a} = m_1 \mathbf{v}_a$. Remembering that $\dot{\mathbf{R}}_b = \dot{\mathbf{R}}_a$
one finds $\dot{\mathbf{R}}_b = m_1\,\mathbf{v}_a/(m_1 + m_2)$. Using the latter relation in the second of (3.17)
specified with the b label yields, after multiplying both sides by m_2,

$$m_2 \mathbf{v}_{2b} = m\mathbf{v}_a - m\mathbf{u}_b. \qquad (3.20)$$

The triangle formed by the vectors $m_2 \mathbf{v}_{2b}$, $m\mathbf{v}_a$, and $m\mathbf{u}_b$ is isosceles because $\mathbf{v}_a$ and
$\mathbf{u}_b$ have the same modulus (Fig. 3.3). Letting χ, θ be the angle between $m\,\mathbf{v}_a$ and $m\,\mathbf{u}_b$
and, respectively, the common value of the other two angles, a scalar multiplication
of (3.20) by $\mathbf{v}_a$ yields $m_2 v_{2b} \cos\theta = mv_a - mu_b \cos\chi = mv_a (1 - \cos\chi)$. Using
$2\theta + \chi = \pi$ and $v_a = v_{1a}$ transforms the latter into

$$m_2 v_{2b} = mv_{1a}\,\frac{1 - \cos\chi}{\cos\left[(\pi - \chi)/2\right]} = 2mv_{1a}\,\sin\left(\chi/2\right). \qquad (3.21)$$

This relation allows one to calculate, in reference O where the particle of mass m_2
is initially at rest, the modulus of the final velocity of this particle in terms of the
initial velocity of the other particle and of angle χ. As only v_{1a} is prescribed, the two
quantities v_{2b} and χ can not be determined separately. The reason for this (already
mentioned in Sect. 3.5) is that (3.21) is deduced using the motion's asymptotic
conditions without considering the interaction's details. In fact, the calculation is
based only on the momentum and total-energy conservation and on the hypothesis
that the collision is elastic. From (3.21) one derives the relation between the kinetic
energies $T_{1a} = (1/2)\,m_1 v_{1a}^2$ and $T_{2b} = (1/2)\,m_2 v_{2b}^2$,

$$T_{2b}\,(\chi) = \frac{4\,m_1\,m_2}{(m_1 + m_2)^2}\,T_{1a}\,\sin^2\left(\chi/2\right). \qquad (3.22)$$

As in reference O the particle of mass m_2 is initially at rest, T_{2b} is the variation of
the kinetic energy of this particle due to to the collision, expressed in terms of χ.

The maximum variation is $T_{2b}(\chi = \pm\pi)$. The conservation relation for the kinetic energy $T_{1b} + T_{2b} = T_{1a}$ coupled with (3.22) yields the kinetic energy of the particle of mass m_1 after the collision,

$$T_{1b} = T_{1a} - T_{2b} = \left[1 - \frac{4\,m_1\,m_2}{(m_1 + m_2)^2}\sin^2{(\chi/2)}\right]T_{1a}. \qquad (3.23)$$

Expressing T_{1a} and T_{1a} in (3.23) in terms of the corresponding velocities yields the modulus of the final velocity of the particle of mass m_1 as a function of its initial velocity and of angle χ,

$$v_{1b} = [(m_1^2 + m_2^2 + 2m_1m_2\cos\chi)^{1/2}/(m_1 + m_2)]\,v_{1a}. \qquad (3.24)$$

Although expressions (3.21–3.24) are compact, the use of angle χ is inconvenient. It is preferable to use the angle, say ψ, between vectors $\mathbf{v}_{1b}$ and $\mathbf{v}_{1a} = \mathbf{v}_a$ that belong to the same reference O (Fig. 3.7).[1] A scalar multiplication by $\mathbf{v}_{1a}$ of the conservation relation for momentum, $m_1\mathbf{v}_{1b} = m_1\mathbf{v}_{1a} - m_2\mathbf{v}_{2b}$, followed by the replacement of the expressions of v_{2b} and v_{1b} extracted from (3.21, 3.24), eventually yields $\cos\psi = (m_1 + m_2\cos\chi)/(m_1^2 + m_2^2 + 2m_1m_2\cos\chi)^{1/2}$. Squaring both sides of the latter provides the relation between χ and ψ,

$$\tan\psi = \frac{\sin\chi}{m_1/m_2 + \cos\chi}. \qquad (3.25)$$

Using (3.25) one expresses (3.21–3.24) in terms of the deflection ψ (in reference O) of the particle of mass m_1. If $m_1 > m_2$, then $\psi < \pi/2$, while it is $\psi = \chi/2$ if $m_1 = m_2$. When $m_1 < m_2$ and $\chi = \arccos{(-m_1/m_2)}$, then $\psi = \pi/2$; finally, if $m_1 \ll m_2$ it is $\psi \simeq \chi$ and, from (3.21–3.24), it follows $v_{2b} \simeq 0$, $T_{2b} \simeq 0$, $v_{1b} \simeq v_{1a}$, $T_{1b} \simeq T_{1a}$. In other terms, when $m_1 \ll m_2$ the particle of mass m_2 remains at rest; the other particle is deflected, but its kinetic energy is left unchanged.

In reference O, the angle between the final velocity of the particle of mass m_2, initially at rest, and the initial velocity of the other particle has been defined above as $\theta = (\pi - \chi)/2$. Replacing the latter in (3.25) provides the relation between ψ and θ,

$$\tan\psi = \frac{\sin{(2\theta)}}{m_1/m_2 - \cos{(2\theta)}}. \qquad (3.26)$$

---

[1] Note that the angle γ between two momenta $\mathbf{p}' = m'\mathbf{v}'$ and $\mathbf{p}'' = m''\mathbf{v}''$ is the same as that between the corresponding velocities because the masses cancel out in the angle's definition $\gamma = \arccos{[\mathbf{p}' \cdot \mathbf{p}''/(p'p'')]}$. In contrast, a relation like (3.20) involving a triad of vectors holds for the momenta but not (with the exception of the trivial cases of equal masses) for the corresponding velocities.

3.7 Central Motion in the Two-Particle Interaction

Consider an isolated two-particle system where the force acting on each particle derives from a potential energy $V = V(\mathbf{r}_1, \mathbf{r}_2)$. Using the symbols defined in Sect. 3.5 yields the Lagrangian function $L = m_1 v_1^2/2 + m_2 v_2^2/2 - V(\mathbf{r}_1, \mathbf{r}_2)$. Now assume that the potential energy V depends on the position of the two particles only through the modulus of their distance, $r = |\mathbf{r}_1 - \mathbf{r}_2|$. In this case it is convenient to use the coordinates and velocities relative to the center of mass, (3.12–3.17), to find

$$L = \frac{1}{2}(m_1 + m_2)\dot{R}^2 + \frac{1}{2}m\dot{s}^2 - V(s), \qquad \dot{s} = |\dot{\mathbf{s}}| = |\mathbf{u}|. \tag{3.27}$$

As discussed in Sects. 2.6.3 and 3.5 the total momentum is conserved, whence $\dot{R}$ is constant. Another way of proving this property is noting that the components of $\mathbf{R}$ are cyclic (Sect. 1.6). The first term at the right hand side of (3.27), being a constant, does not influence the subsequent calculations. The remaining terms, in turn, are identical to those of (3.4). This shows that, when in a two-particle system the potential energy depends only on the relative distance, adopting suitable coordinates makes the problem identical to that of the central motion. One can then exploit the results of Sect. 3.4. Once the time evolution of $\mathbf{s}$ is found, the description of the motion of the individual particles is recovered from (3.12–3.17), where the constant $\dot{R}$ is determined by the initial conditions.

The total energy of the two-particle system is conserved and, in the new reference, it reads

$$\frac{1}{2}m\dot{s}^2 + V(s) = E_B, \qquad E_B = E - \frac{1}{2}(m_1 + m_2)\dot{R}^2. \tag{3.28}$$

The total angular momentum is constant as well. Starting from the original reference and using (3.12–3.17) yields

$$\mathbf{M} = \mathbf{r}_1 \wedge m_1 \mathbf{v}_1 + \mathbf{r}_2 \wedge m_2 \mathbf{v}_2 = (m_1 + m_2)\mathbf{R} \wedge \dot{\mathbf{R}} + m\mathbf{s} \wedge \mathbf{u}. \tag{3.29}$$

The constancy of $\dot{\mathbf{R}}$ yields $\mathbf{R} = \mathbf{R}_0 + \dot{\mathbf{R}}t$, with $\mathbf{R}_0$ the initial value of $\mathbf{R}$, whence $(\mathbf{R}_0 + \dot{\mathbf{R}}t) \wedge \dot{\mathbf{R}} = \mathbf{R}_0 \wedge \dot{\mathbf{R}}$. Thus, the first term at the right hand side of (3.29) is constant, which makes $\mathbf{M}_B = m\mathbf{s} \wedge \mathbf{u}$ a constant as well. The latter vector is parallel to $\mathbf{M}$ because the motion is confined to a fixed plane (Sect. 3.4). Then, aligning the z axis with $\mathbf{M}$, turning to polar coordinates over the x, y plane ($s_x = s\cos\varphi$, $s_y = s\sin\varphi$), and using (3.8), one finds

$$\varphi(s) = \varphi_0 \pm \frac{M_B}{\sqrt{2m}} \int_{s_0}^{s} \frac{d\xi}{\xi^2\sqrt{E_B - V_e(\xi)}}, \tag{3.30}$$

with $V_e(s) = V(s) + M_B^2/(2ms^2)$. It is important to note that the factor M_B in (3.30) is the scalar coefficient of $\mathbf{M}_B = M_B\mathbf{k}$, with $\mathbf{k}$ the unit vector of the z axis. As a consequence, M_B may have sign. As observed in Sect. 2.9, the admissible values of s are those belonging to the interval such that $E_B \geq V_e(s)$. If two or more disjoint

Fig. 3.4 Graphic
representation of the
trajectory (3.35) for different
values of the angular
momentum. The curves have
been obtained by setting the
parameters' values to $s_0 = 1$,
$\varphi_0 = 0$, $\lambda = 0.5$, and
$\mu = 0.01, \dots, 0.6$ (the units
are arbitrary)

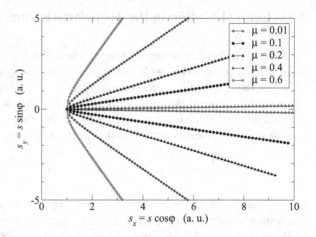

intervals exist that have this property, the actual interval of the motion is determined
by the initial conditions. The motion is limited or unlimited, depending on the extent
of this interval.

The analysis can not be pursued further unless the form of the potential energy V
is specified. This is done in Sect. 3.8 with reference to the Coulomb case.

3.8 Coulomb Field

An important example is that of a potential energy of the form $V \propto 1/r$, that occurs
for the gravitational and for the electrostatic force. In the latter case the term *Coulomb
potential energy* is used for V, that reads

$$V(s) = \frac{\kappa \, Z_1 \, Z_2 \, q^2}{4\pi \, \varepsilon_0 \, s}, \qquad s > 0, \tag{3.31}$$

with $q > 0$ the elementary electric charge, $Z_1 \, q$ and $Z_2 \, q$ the absolute value of the net
charge of the first and second particle, respectively, ε_0 the vacuum permittivity and,
finally, $\kappa = 1 \, (-1)$ in the repulsive (attractive) case. The form of V fixes the additive
constant of the energy so that $V(\infty) = 0$. The repulsive case only is considered here,
whence V_e is strictly positive and $E_B \geq V_e > 0$. Defining the lengths

$$\lambda = \frac{Z_1 \, Z_2 \, q^2}{8\pi \, \varepsilon_0 \, E_B} > 0, \qquad \mu = \frac{M_B}{\sqrt{2m \, E_B}} \tag{3.32}$$

yields $V_e/E_B = 2\lambda/s + \mu^2/s^2$. The zeros of $E_B - V_e = E_B \, (s^2 - 2\lambda s - \mu^2)/s^2$ are

$$s_A = \lambda - \sqrt{\lambda^2 + \mu^2}, \qquad s_B = \lambda + \sqrt{\lambda^2 + \mu^2}, \tag{3.33}$$

where s_A is negative and must be discarded as s is strictly positive. The only acceptable zero is then $s_B \geq 2\lambda > 0$, that corresponds to the position where the radial

velocity $\dot{s} = \pm\sqrt{2(E_B - V)/m}$ reverses, and must therefore be identified with s_0 (Sect. 3.7). The definitions (3.32, 3.33) are now inserted into the expression (3.30) of the particle's trajectory. To calculate the integral it is convenient to use a new variable w such that $(s_0 - \lambda)/(\xi - \lambda) = (\mu^2 - s_0^2 w^2)/(\mu^2 + s_0^2 w^2)$. The range of w corresponding to $s_0 \le \xi \le s$ is

$$0 \le w \le \frac{|\mu|}{s_0}\sqrt{\frac{s - s_0}{s + s_0 - 2\lambda}}, \qquad s \ge s_0 \ge 2\lambda > 0. \tag{3.34}$$

From (3.30) the trajectory in the s, φ reference is thus found to be

$$\varphi(s) = \varphi_0 \pm 2 \arctan\left(\frac{\mu}{s_0}\sqrt{\frac{s - s_0}{s + s_0 - 2\lambda}}\right). \tag{3.35}$$

Next, the trajectory in the Cartesian reference s_x, s_y is found by replacing (3.35) into $s_x = s \cos\varphi$, $s_y = s \sin\varphi$ and eliminating s from the pair $s_x(s)$, $s_y(s)$ thus found. A graphic example is given in Fig. 3.4. It is worth observing that in the derivation of (3.35) the factor $|\mu|$ appears twice, in such a way as to compensate for the sign of μ. The result then holds irrespective of the actual sign of $M_B = \sqrt{2m\,E_B}\,\mu$. It still holds for $M_B = 0$, that yields $\varphi(s) = \varphi_0$; such a case corresponds to a straight line crossing the origin of the s, φ reference: along this line the modulus s of the relative position decreases until it reaches s_0, then it increases from this point on.

When $M_B \ne 0$ the angles corresponding to the asymptotic conditions of the motion are found by letting $s \to \infty$, namely, $\varphi_a = \varphi_0 - 2\arctan(\mu/s_0)$ and $\varphi_b = \varphi_0 + 2\arctan(\mu/s_0)$. The total deflection is then $\varphi_b - \varphi_a$ which, in each curve of Fig. 3.4, is the angle between the two asymptotic directions. Now one combines the definition of angle χ given in Sect. 3.6 with the equality $\mathbf{u} = \mathbf{v}$ taken from the last of (3.15); with the aid of Fig. 3.5 one finds

$$\chi = \pi - (\varphi_b - \varphi_a) = \pi - 4\arctan\left(\frac{\mu}{s_0}\right). \tag{3.36}$$

The definitions (3.32, 3.33) show that (3.36) eventually provides the relation $\chi = \chi(E_B, M_B)$. In contrast with the approach of Sect. 3.6, where the asymptotic conditions only were considered, here the analysis has been brought to the end by considering a specific type of interaction.

When μ ranges from $-\infty$ to $+\infty$ at a fixed E_B, the definitions (3.32, 3.33) make the ratio $\mu/s_0 = \mu/s_A$ to range from -1 to $+1$. If $\mu/s_0 = 1\,(-1)$, then $\chi = 0\,(2\pi)$, namely, no deflection between $\mathbf{u}_a$ and $\mathbf{u}_b$ occurs. If $\mu/s_0 = 0$, then $\chi = \pi$, namely, the motion's direction reverses at s_0 as noted above. From $\mathbf{u} = \mathbf{v}$ one finds that χ is also the angle between $\mathbf{v}_a = \mathbf{v}_{1a} - \mathbf{v}_{2a}$ and $\mathbf{v}_b = \mathbf{v}_{1b} - \mathbf{v}_{2b}$.

3.9 System of Particles near an Equilibrium Point

Consider a system of N particles, not necessarily identical to each other, subjected to conservative forces. The mass and instantaneous position of the jth particle are indicated with m_j and $\mathbf{R}_j = (X_{j1}, X_{j2}, X_{j3})$, respectively. It is assumed that there

Fig. 3.5 Graphic
representation of (3.36)

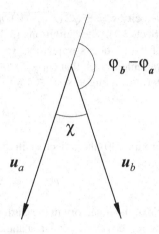

are no constraints, so that the number of degrees of freedom of the system is $3N$.
The Hamiltonian function reads

$$H_a = T_a + V_a = \sum_{j=1}^{N} \frac{P_j^2}{2 m_j} + V_a(X_{11}, X_{12}, \dots), \qquad (3.37)$$

with $P_j^2 = m_j^2 (\dot{X}_{j1}^2 + \dot{X}_{j2}^2 + \dot{X}_{j3}^2)$. The force acting on the jth particle along the kth
axis is $F_{jk} = -\partial V_a / \partial X_{jk}$. The Hamilton equations (Sect. 1.6) read

$$\dot{X}_{jk} = \frac{\partial H_a}{\partial P_{jk}} = \frac{P_{jk}}{m_j}, \qquad \dot{P}_{jk} = -\frac{\partial H_a}{\partial X_{jk}} = -\frac{\partial V_a}{\partial X_{jk}} = F_{jk}. \qquad (3.38)$$

They show that the relation $F_{jk} = m_j \ddot{X}_{jk}$, which yields the dynamics of the jth
particle along the kth axis, involves the positions of all particles in the system due to
the coupling of the latter.

Define the $3N$-dimensional vector $\mathbf{R} = (X_{11}, \dots, X_{N3})$ that describes the instantaneous position of the system in the configuration space, and let $\mathbf{R}_0$ be a position
where the potential energy V_a has a minimum, namely, $(\partial V_a / \partial X_{jk})_{\mathbf{R}_0} = 0$ for all
j, k. Such a position is called *equilibrium point* of the system. To proceed, assume
that the instantaneous displacement $\mathbf{R} - \mathbf{R}_0$ with respect to the equilibrium point
is small. In this case one approximates V with a second-order Taylor expansion
around $\mathbf{R}_0$. To simplify the notation new symbols are adopted, namely, $s_1 = X_{11}$,
$s_2 = X_{12}, \dots, s_{3j+k-3} = X_{jk}, \dots$, and $h_n = s_n - s_{n0}$, with $n = 1, 2, \dots, 3N$ and
s_{n0} the equilibrium position. Remembering that the first derivatives of V_a vanish at
$\mathbf{R}_0$ one finds

$$V_a \simeq V_{a0} + \frac{1}{2} \sum_{k=1}^{3N} h_k \sum_{n=1}^{3N} c_{kn} h_n, \qquad c_{kn} = \left(\frac{\partial^2 V_a}{\partial h_k \partial h_n} \right)_{\mathbf{R}_0}. \qquad (3.39)$$

In (3.39) it is $V_{a0} = V_a(\mathbf{R}_0)$, and the terms c_{kn} are called *elastic coefficients*. As
the approximate form of the potential energy is quadratic in the displacements, each

component of the force is a linear combination of the latter,

$$F_r = -\frac{\partial V_a}{\partial s_r} = -\frac{\partial V_a}{\partial h_r} = -\sum_{n=1}^{3N} c_{rn} h_n, \qquad r = 3j + k - 3. \tag{3.40}$$

To recast the kinetic energy in terms of the new symbols it is necessary to indicate the masses with μ_n, $n = 1, \ldots, 3N$, where $\mu_{3j-2} = \mu_{3j-1} = \mu_{3j} = m_j$, $j = 1, \ldots, N$. Observing that $\dot{X}_{jk} = \dot{s}_{3j+k-3} = \dot{h}_{3j+k-3}$, one finds a quadratic form in the derivatives of the displacements,

$$T_a = \frac{1}{2} \sum_{j=1}^{N} \sum_{k=1}^{3} \mu_{3j+k-3} \dot{X}_{jk}^2 = \frac{1}{2} \sum_{j=1}^{N} \sum_{k=1}^{3} \mu_{3j+k-3} \dot{h}_{3j+k-3}^2 = \frac{1}{2} \sum_{n=1}^{3N} \mu_n \dot{h}_n^2. \tag{3.41}$$

The relations obtained so far are readily recast in matrix form. First, one defines the *mass matrix* $\mathbf{M}$ as the real, $3N \times 3N$ diagonal matrix whose entries are $[\mathbf{M}]_{kn} = \mu_n \delta_{kn} > 0$, with δ_{kn} the Kronecker symbol (A.18). By construction, the mass matrix is symmetric and positive definite; the entries of its inverse are $[\mathbf{M}^{-1}]_{kn} = \delta_{kn}/\mu_n$. Then, one defines the *elastic matrix* $\mathbf{C}$ as the real, $3N \times 3N$ matrix whose entries are $[\mathbf{C}]_{kn} = c_{kn}$. The entries of the elastic matrix are the second derivatives of the potential energy V_a; as the order of the derivation is irrelevant, the matrix is symmetric. Also, the derivatives are calculated in a minimum of V_a; from the first of (3.39) it follows that the quadratic form at the right hand side equals $V_a - V_{a0}$ which, by construction, is positive. It follows that the elastic matrix is positive definite, namely, for any choice of the displacements (excluding the case where all displacements are zero) the quadratic form generated by the matrix is positive. Finally, let $\mathbf{h}$ be the column vector of entries $h_1, h_2, \ldots$, and $\mathbf{h}^T$ its transpose. Combining (3.37, 3.39, 3.41) expresses the Hamiltonian function in terms of the sum of two quadratic forms,

$$H_a - V_{a0} = \frac{1}{2} \dot{\mathbf{h}}^T \mathbf{M} \dot{\mathbf{h}} + \frac{1}{2} \mathbf{h}^T \mathbf{C} \mathbf{h}. \tag{3.42}$$

3.10 Diagonalization of the Hamiltonian Function

Thanks to the properties of the matrices $\mathbf{M}$ and $\mathbf{C}$, the right hand side of (3.42) can be set in diagonal form. To this purpose one considers the eigenvalue equation

$$\mathbf{C} \mathbf{g}_\sigma = \lambda_\sigma \mathbf{M} \mathbf{g}_\sigma, \qquad \sigma = 1, \ldots, 3N, \tag{3.43}$$

where the eigenvalues λ_σ are real because $\mathbf{C}$ and $\mathbf{M}$ are real and symmetric. As all coefficients of (3.43) are real, the eigenvectors $\mathbf{g}_\sigma$ are real as well. Also, due to the positive definiteness of $\mathbf{C}$ and $\mathbf{M}$, the eigenvalues are positive and the eigenvectors are linearly independent. They can also be selected in order to fulfill the property of being orthonormal with respect to $\mathbf{M}$, namely, $\mathbf{g}_\sigma^T \mathbf{M} \mathbf{g}_\tau = \delta_{\sigma\tau}$.

Each of the $3N$ eigenvectors $\mathbf{g}_\sigma$ has $3N$ entries. Thus, the set of eigenvectors can be arranged to form a $3N \times 3N$ real matrix $\mathbf{G}$, whose σth column is the σth eigenvector. The inverse matrix $\mathbf{G}^{-1}$ exists because, by construction, the columns of $\mathbf{G}$ are linearly independent. The orthonormality relation between the eigenvectors can now be expressed in matrix form as

$$\mathbf{G}^T \mathbf{M} \mathbf{G} = \mathbf{I}, \tag{3.44}$$

with $\mathbf{I}$ the identity matrix. Equation (3.44) is the basic ingredient for the diagonalization of (3.42). From it one preliminarily derives four more relations,

$$\mathbf{G}^T \mathbf{M} = \mathbf{G}^{-1}, \qquad \mathbf{G} \mathbf{G}^T \mathbf{M} = \mathbf{I}, \qquad \mathbf{M} \mathbf{G} = \left(\mathbf{G}^T \right)^{-1}, \qquad \mathbf{M} \mathbf{G} \mathbf{G}^T = \mathbf{I}. \tag{3.45}$$

The first of (3.45) is obtained by right multiplying (3.44) by $\mathbf{G}^{-1}$ and using $\mathbf{G} \mathbf{G}^{-1} = \mathbf{I}$. Left multiplying by $\mathbf{G}$ the first of (3.45) yields the second one. The third of (3.45) is obtained by left multiplying (3.44) by $(\mathbf{G}^T)^{-1}$. Finally, right multiplying by $\mathbf{G}^T$ the third of (3.45) yields the fourth one. To complete the transformation of the equations into a matrix form one defines the eigenvalue matrix $\mathbf{L}$ as the real, $3N \times 3N$ diagonal matrix whose entries are $[\mathbf{L}]_{\sigma\tau} = \lambda_\tau \delta_{\sigma\tau} > 0$. The set of $3N$ eigenvalue Eqs. (3.43) then takes one of the two equivalent forms

$$\mathbf{C} \mathbf{G} = \mathbf{M} \mathbf{G} \mathbf{L}, \qquad \mathbf{G}^T \mathbf{C} \mathbf{G} = \mathbf{L}. \tag{3.46}$$

The first of (3.46) is the analogue of (3.43), while the second form is obtained from the first one by left multiplying by $\mathbf{G}^T$ and using (3.44). The diagonalization of (3.42) is now accomplished by inserting the second and fourth of (3.45) into the potential-energy term of (3.42) to obtain

$$\mathbf{h}^T \mathbf{C} \mathbf{h} = \mathbf{h}^T (\mathbf{M} \mathbf{G} \mathbf{G}^T) \mathbf{C} (\mathbf{G} \mathbf{G}^T \mathbf{M}) \mathbf{h} = (\mathbf{h}^T \mathbf{M} \mathbf{G}) (\mathbf{G}^T \mathbf{C} \mathbf{G}) (\mathbf{G}^T \mathbf{M} \mathbf{h}), \tag{3.47}$$

where the associative law has been used. At the right hand side of (3.47), the term in the central parenthesis is replaced with $\mathbf{L}$ due to the second of (3.46). The term in the last parenthesis is a column vector for which the short-hand notation $\mathbf{b} = \mathbf{G}^T \mathbf{M} \mathbf{h}$ is introduced. Note that $\mathbf{b}$ depends on time because $\mathbf{h}$ does. The first of (3.45) shows that $\mathbf{h} = \mathbf{G} \mathbf{b}$, whence $\mathbf{h}^T = \mathbf{b}^T \mathbf{G}^T$. Finally, using (3.44), transforms the term in the first parenthesis at the right hand side of (3.47) into $\mathbf{h}^T \mathbf{M} \mathbf{G} = \mathbf{b}^T \mathbf{G}^T \mathbf{M} \mathbf{G} = \mathbf{b}^T$. In conclusion, the potential-energy term of (3.42) is recast in terms of $\mathbf{b}$ as $\mathbf{h}^T \mathbf{C} \mathbf{h} = \mathbf{b}^T \mathbf{L} \mathbf{b}$, which is the diagonal form sought. By a similar procedure one finds for the kinetic-energy term $\dot{\mathbf{h}}^T \mathbf{M} \dot{\mathbf{h}} = \dot{\mathbf{b}}^T \mathbf{G}^T \mathbf{M} \mathbf{G} \dot{\mathbf{b}} = \dot{\mathbf{b}}^T \dot{\mathbf{b}}$.

The terms $\dot{\mathbf{b}}^T \dot{\mathbf{b}}$ and $\mathbf{b}^T \mathbf{L} \mathbf{b}$ have the same units. As a consequence, the units of $\mathbf{L}$ are the inverse of a time squared. Remembering that the entries of $\mathbf{L}$ are positive, one introduces the new symbol $\omega_\sigma^2 = \lambda_\sigma$ for the eigenvalues, $\omega_\sigma > 0$. In conclusion, the diagonal form of (3.42) reads

$$H_a - V_{a0} = \sum_{\sigma=1}^{3N} H_\sigma, \qquad H_\sigma = \frac{1}{2} \dot{b}_\sigma^2 + \frac{1}{2} \omega_\sigma^2 b_\sigma^2. \tag{3.48}$$

Apart from the constant V_{a0}, the Hamiltonian function H_a is given by a sum of terms, each associated with a single degree of freedom. A comparison with (3.1) shows that the individual summands H_σ are identical to the Hamiltonian function of a linear harmonic oscillator with $m = 1$. As a consequence, the two canonical variables of the σth degree of freedom are $q_\sigma = b_\sigma$, $p_\sigma = \dot{b}_\sigma$, and the time evolution of b_σ is the same as that in (3.2),

$$b_\sigma(t) = \beta_\sigma \cos(\omega_\sigma t + \varphi_\sigma) = \frac{1}{2}\left[\tilde{\beta}_\sigma \exp(-i\omega_\sigma t) + \tilde{\beta}_\sigma^* \exp(i\omega_\sigma t)\right]. \quad (3.49)$$

The constants β_σ, φ_σ depend on the initial conditions $b_\sigma(0)$, $\dot{b}_\sigma(0)$. The complex coefficients are related to the above constants by $\tilde{\beta}_\sigma = \beta_\sigma \exp(-i\varphi_\sigma)$. In turn, the initial conditions are derived from those of the displacements, $\mathbf{b}(0) = \mathbf{G}^{-1}\mathbf{h}(0)$, $\dot{\mathbf{b}}(0) = \mathbf{G}^{-1}\dot{\mathbf{h}}(0)$.

The $3N$ functions $b_\sigma(t)$ are called *normal coordinates* or *principal coordinates*. Once the normal coordinates have been found, the displacements of the particles are determined from $\mathbf{h} = \mathbf{G}\,\mathbf{b}$. It follows that such displacements are superpositions of oscillatory functions. Despite the complicacy of the system, the approximation of truncating the potential energy to the second order makes the Hamiltonian function completely separable in the normal coordinates. The problem then becomes a generalization of that of the linear harmonic oscillator (Sect. 3.3), and the frequencies of the oscillators are determined by combining the system parameters, specifically, the particle masses and elastic constants. The Hamiltonian function associated with each degree of freedom is a constant of motion, $H_\sigma = E_\sigma$, whose value is prescribed by the initial conditions. The total energy of the system is also a constant and is given by

$$E = V_{a0} + \sum_{\sigma=1}^{3N} E_\sigma, \qquad E_\sigma = \frac{1}{2}\dot{b}_\sigma^2(0) + \frac{1}{2}\omega_\sigma^2 b_\sigma^2(0). \quad (3.50)$$

The oscillation of the normal coordinate of index σ is also called *mode* of the vibrating system.

3.11 Periodic Potential Energy

An interesting application of the action-angle variables introduced in Sect. 2.10 is found in the case of a conservative motion where the potential energy V is periodic. For simplicity a linear motion is considered (Sect. 2.9), whence $V(x + a) = V(x)$, with $a > 0$ the spatial period. Letting E be the total energy and m the mass of the particle, an unlimited motion is assumed, namely, $E > V$; it follows that the momentum $p = \sqrt{2m[E - V(x)]}$ is a spatially-periodic function of period a whence, according to the definition of Sect. 2.10, the motion is a rotation. For any position g, the time τ necessary for the particle to move from g to $g + a$ is found from (2.47), where the positive sign is provisionally chosen:

$$\tau = \sqrt{\frac{m}{2}} \int_g^{g+a} \frac{dx}{\sqrt{E - V(x)}} > 0. \tag{3.51}$$

The integral in (3.51) is independent of g due to the periodicity of V. As a consequence, for any g the position of the particle grows by a during the time τ. The action variable is found from (2.49):

$$J(E) = \int_g^{g+a} p \, dx = \sqrt{2m} \int_g^{g+a} \sqrt{E - V(x)} \, dx = \text{const.} \tag{3.52}$$

In turn, the derivative of the angle variable is found from (2.51). It reads $\dot{w} = \partial H / \partial J = 1/(dJ/dE) = \text{const}$, with H the Hamiltonian function. The second form of $\dot{w}$ holds because H does not depend on w, and $H = E$. Using (3.52) and comparing with (3.51) one finds

$$\frac{1}{\dot{w}} = \frac{dJ}{dE} = \int_g^{g+a} \frac{m}{[2m (E - V(x))]^{1/2}} \, dx = \tau. \tag{3.53}$$

As expected, $1/\tau$ is the rotation frequency. In conclusion, the time evolution of the action-angle variables is given by $w = t/\tau + w_0$, $J = \text{const}$. Note that the relation (3.52) between E and J holds when the positive sign is chosen in (2.47); if the above calculations are repeated after choosing the negative sign, one finds that $-J$ is associated to the same E. As a consequence, E is an even function of J.

Another observation is that the action-angle variables can be scaled by letting, e.g., $w J = (a w)(J/a)$. In this way the property that the product of two canonically-conjugate variables is dimensionally an action is still fulfilled. A comparison with (3.52) shows that, thanks to this choice of the scaling factor, $P = J/a$ is the average momentum over a period, while $X = a w$ is a length. The Hamilton equations and the time evolution of the new variables are then

$$\dot{X} = \frac{\partial H}{\partial P}, \quad \dot{P} = -\frac{\partial H}{\partial X} = 0, \quad X = \frac{a}{\tau} t + X_0, \quad P = P_0 = \text{const}, \tag{3.54}$$

where a/τ is the average velocity of the particle over the spatial period, and $X_0 = X(0)$, $P_0 = P(0)$. In conclusion, in the new canonical variables no force is acting ($\dot{P} = 0$), and the motion of the new position X is uniform in time. However, the relation between E and $P = J/a$, given by (3.52), is not quadratic as it would be in free space.[2]

In many cases it is of interest to investigate the particle's dynamics when a perturbation δH is superimposed to the periodic potential energy V. It is assumed that δH depends on x only, and that E is the same as in the unperturbed case (the latter assumption is not essential). The Hamiltonian function of the perturbed case is then

---

[2] Compare with comments made in Sect. 19.6.1.

written as the sum of the unperturbed one and of the perturbation; also in this case an unlimited motion is assumed, specifically, $E > V$ and $E > V + \delta H$. Still using the positive sign for the momentum, one finds

$$H(x, p) = \frac{p^2}{2m} + V(x) + \delta H(x) = E, \qquad p(x, E) = \sqrt{2m(E - V - \delta H)}.$$

(3.55)

As in the unperturbed case one defines the average momentum over a,

$$\tilde{P}(g, E) = \frac{\sqrt{2m}}{a} \int_g^{g+a} \sqrt{E - V - \delta H}\, dx,$$

(3.56)

which depends also on g because δH is not periodic. Differentiating (3.56) with respect to E and comparing with (3.51) shows that

$$\frac{\partial \tilde{P}}{\partial E} = \frac{\tilde{\tau}}{a}, \qquad \tilde{\tau}(g) = \int_g^{g+a} \frac{m}{[2m(E - V(x) - \delta H)]^{1/2}}\, dx,$$

(3.57)

with $\tilde{\tau}$ the time necessary for the particle to move from g to $g + a$ in the perturbed case. Using $H = E$ in the above yields

$$\frac{\partial H}{\partial \tilde{P}} = \frac{a}{\tilde{\tau}} = \frac{(g + a) - g}{\tilde{\tau}}.$$

(3.58)

So far no hypothesis has been made about the perturbation. Now one assumes that δH is weak and varies little over the period a. The first hypothesis implies $|\delta H| \ll E - V$ so that, to first order, $[2m(E - V - \delta H)]^{1/2} \simeq [2m(E - V)]^{1/2} - m[2m(E - V)]^{-1/2}\delta H$. Using $P = J/a$ and (3.52), the average momentum (3.56) becomes

$$\tilde{P}(g, E) \simeq P(E) - \frac{1}{a} \int_g^{g+a} \frac{m\,\delta H}{[2m(E - V)]^{1/2}}\, dx.$$

(3.59)

In turn, the hypothesis that the perturbation varies little over the period a implies that in the interval $[g, g + a]$ one can approximate $\delta H(x)$ with $\delta H(g)$, which transforms (3.59), due to (3.53), into

$$\tilde{P}(g, E) \simeq P(E) - \frac{\tau}{a}\delta H(g).$$

(3.60)

If the procedure leading to (3.60) is repeated in the interval $[g + a, g + 2a]$ and the result is subtracted from (3.60), the following is found:

$$\frac{\tilde{P}(g + a, E) - \tilde{P}(g, E)}{\tau} = -\frac{\delta H(g + a) - \delta H(g)}{a}.$$

(3.61)

The above shows that the perturbed momentum $\tilde{P}$ varies between g and $g + a$ due to the corresponding variation in δH. Fixing the time origin at the position g and letting $\tau \simeq \tilde{\tau}$ in the denominator transforms (3.61) into

$$\frac{\tilde{P}(\tilde{\tau}, E) - \tilde{P}(0, E)}{\tilde{\tau}} \simeq -\frac{\delta H(g + a) - \delta H(g)}{a}.$$

(3.62)

The relations (3.58, 3.62) are worth discussing. If one considers g as a position coordinate and $\tilde{P}$ as the momentum conjugate to it, (3.58, 3.62) become a pair of Hamilton equations where some derivatives are replaced with difference quotients. Specifically, (3.62) shows that the average momentum varies so that its "coarse-grained" variation with respect to time, $\Delta \tilde{P}/\Delta \tilde{\tau}$, is the negative coarse-grained variation of the Hamiltonian function with respect to space, $-\Delta H/\Delta g = -\Delta \delta H/\Delta g$. In turn, (3.58) shows that the coarse-grained variation of position with respect to time, $\Delta g/\Delta \tilde{\tau}$, is the derivative of the Hamiltonian function with respect to the average momentum. In conclusion, (3.58, 3.62) are useful when one is not interested in the details of the particle's motion within each spatial period, but wants to investigate on a larger scale how the perturbation influences the average properties of the motion.

3.12 Energy-Momentum Relation in a Periodic Potential Energy

It has been observed, with reference to the non-perturbed case, that the relation (3.52) between the total energy and the average momentum is not quadratic. In the perturbed case, as shown by (3.56), the momentum depends on both the total energy and the coarse-grained position. To investigate this case it is then necessary to fix g and consider the dependence of $\tilde{P}$ on E only. To proceed one takes a small interval of $\tilde{P}$ around a given value, say, $\tilde{P}_s$, corresponding to a total energy E_s, and approximates the $E(\tilde{P})$ relation with a second-order Taylor expansion around $\tilde{P}_s$,

$$E \simeq E_s + \left(\frac{dE}{d\tilde{P}}\right)_s (\tilde{P} - \tilde{P}_s) + \frac{1}{2} \left(\frac{d^2E}{d\tilde{P}^2}\right)_s (\tilde{P} - \tilde{P}_s)^2. \qquad (3.63)$$

Although in general the $\tilde{P}(E)$ relation (3.56) can not be inverted analytically, one can calculate the derivatives that appear in (3.63). The latter are worked out in the unperturbed case $\delta H = 0$ for simplicity. Using (3.53), the first derivative is found to be $(dE/d\tilde{P})_s \simeq (dE/dP)_s = a/\tau_s$, with $\tau_s = \tau(E_s)$. For the second derivative,

$$\frac{d^2E}{d\tilde{P}^2} \simeq \frac{d^2E}{dP^2} = \frac{d(a/\tau)}{dP} = -\frac{a}{\tau^2} \frac{d\tau}{dE} \frac{dE}{dP} = -\frac{a^3}{\tau^3} \frac{d^2P}{dE^2}. \qquad (3.64)$$

On the other hand, using (3.53) again, it is

$$\frac{d^2P}{dE^2} = \frac{d(\tau/a)}{dE} = -\frac{m^2}{a} \int_g^{g+a} K^3 \, dx, \qquad K = [2m \, (E - V(x))]^{-1/2}. \qquad (3.65)$$

Combining (3.64) with (3.65) and defining the dimensionless parameter

$$r_s = \int_{g/a}^{g/a+1} K^3 \, d(x/a) \times \left[\int_{g/a}^{g/a+1} K \, d(x/a)\right]^{-3}, \qquad (3.66)$$

transforms (3.63) into

$$E \simeq E_s + \frac{a}{\tau_s}(\tilde{P} - \tilde{P}_s) + \frac{r_s}{2m}(\tilde{P} - \tilde{P}_s)^2, \tag{3.67}$$

where the coefficients, thanks to the neglect of the perturbation, do not depend on g. The linear term is readily eliminated by shifting the origin of the average momentum; in fact, letting $\tilde{P} - \tilde{P}_s = \tilde{p} - (m/r_s)(a/\tau_s)$ yields

$$E - E(0) = \frac{r_s}{2m}\tilde{p}^2, \qquad E(0) = E_s - \frac{1}{2}\frac{m}{r_s}\left(\frac{a}{\tau_s}\right)^2. \tag{3.68}$$

In conclusion, in a small interval of $\tilde{p}$ the relation between energy and average momentum of a particle of mass m subjected to a periodic potential has the same form as that of a free particle of mass m/r_s. In other terms, the ratio m/r_s acts as an *effective mass* within the frame of the coarse-grained dynamics.

A bound for r_s is obtained from Hölder's inequality (C.110). Letting $|F| = K$, $G = 1$, $b = 3$, $x_1 = g/a$, $x_2 = g/a + 1$ in (C.110), and using the definition (3.66), yields $r_s \geq 1$, whence $m/r_s \leq m$: the effective mass can never exceed the true mass. The equality between the two masses is found in the limiting case $E - V_M \gg V_M - V - \delta H$, with V_M the maximum of V. In fact, (3.66) yields $r \simeq 1$ and, from (3.56), it is $\tilde{P} \simeq \sqrt{2mE}$. As expected, this limiting case yields the dynamics of a free particle.

3.13 Complements

3.13.1 Comments on the Linear Harmonic Oscillator

The paramount importance of the example of the linear harmonic oscillator, shown in Sect. 3.3, is due to the fact that in several physical systems the position of a particle at any instant happens to depart little from a point where the potential energy V has a minimum. As a consequence, the potential energy can be approximated with a second-order expansion around the minimum, that yields a positive-definite quadratic form for the potential energy and a linear form for the force. The theory depicted in this section is then applicable to many physical systems, as shown by the examples of Sects. 3.9 and 5.6. The approximation of the potential energy with a second-order expansion, like that discussed in Sects. 3.9, 3.10, is called *harmonic approximation*. The terms beyond the second order in the expansion are called *anharmonic*.

3.13.2 Degrees of Freedom and Coordinate Separation

With reference to the analysis of the central motion carried out in Sect. 3.4, it is worth noting that the constancy of **M** reduces the number of degrees of freedom of the problem from three to two. Also, the form (3.6) of the Hamiltonian function is such as to provide a relation containing only r and the corresponding momentum p_r. Thus, the coordinate r is separable according to the definition of Sect. 2.4. This allows one to independently find the time evolution (3.7) of r by solving an ordinary differential equation of the first order. Then one finds (3.8), that is, the trajectory $\varphi(r)$, through another equation of the same type. Finally, combining (3.7) with (3.8) yields the time evolution of the remaining coordinate φ.

It has been noted in Sect. 3.4 that, thanks to the constancy of the angular momentum, the adoption of spherical coordinates allows one to separate the radial coordinate r. This simplifies the problem, whose solution is in fact reduced to the successive solution of the evolution equations for r and φ. The same problem, instead, is not separable in the Cartesian coordinates. In other terms, separability may hold in some coordinate reference, but does not hold in general in an arbitrarily-chosen reference.

Another example of separability is that illustrated in Sects. 3.9, 3.10. In general the Hamiltonian function is not separable in the Cartesian coordinates, whereas it is completely separable in the normal coordinates, no matter how large the number of the degrees of freedom is. Moreover, after the separation has been accomplished, one finds that all the equations related to the single degrees of freedom (the second relation in (3.48)) have the same form. In fact, they differ only in the numerical value of the angular frequency ω_σ. As a consequence, the expression of the solution is the same for all. Also, as the energy H_σ of each degree of freedom is independently conserved, no exchange of energy among the normal coordinates occurs: therefore, the distribution of energy among the normal coordinates that is present at $t = 0$ is maintained forever. This result is baffling because, for instance, it seems to prevent the condition of thermal equilibrium from being established; actually it is due to the fact that the system under investigation is isolated: if it were put in contact with a thermal reservoir, the exchanges of energy occurring with the reservoir would eventually bring the energy distribution of the system to the condition of thermal equilibrium.

Still with reference to the system discussed in Sects. 3.9, 3.10 it is important to underline the formal analogy between the modes of a mechanical, vibrating system and those of the electromagnetic field *in vacuo* described in Sect. 5.6. In both cases the energy of each mode is that of a linear harmonic oscillator of unit mass (Eq. (3.48) and, respectively, (5.40)).

3.13.3 Comments on the Normal Coordinates

It has been shown in Sect. 3.9 that the elastic matrix $\mathbf{C}$ is positive definite. One may argue that in some cases the matrix is positive semi-definite. Consider, for instance, the case where the potential energy depends on the relative distance of the particles, $V_a = V_a(\mathbf{R}_1 - \mathbf{R}_2, \mathbf{R}_1 - \mathbf{R}_3, \dots)$. For any set of positions $\mathbf{R}_1, \mathbf{R}_2, \dots$, a uniform displacement $\mathbf{R}_\delta$ of all particles, that tranforms each $\mathbf{R}_j$ into $\mathbf{R}_j + \mathbf{R}_\delta$, leaves V_a unchanged. As a consequence, if the positions prior to the displacement correspond

Fig. 3.6 Graphic
representation of (3.69)

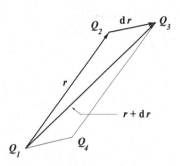

to the equilibrium point $\mathbf{R}_{01}, \mathbf{R}_{02}, \ldots$, it is $V_a(\mathbf{R}_{01} + \mathbf{R}_\delta, \ldots) = V_a(\mathbf{R}_{01}, \ldots) = V_{a0}$. In such a case all terms beyond the zero-order term in the Taylor expansion of V_a around the equilibrium point vanish, which implies that the elastic matrix $\mathbf{C}$ is positive semi-definite. In the case examined in Sect. 3.9 the eigenvalues are real and positive; here, instead, they are real and non-negative. Remembering (3.48), one finds that the Hamiltonian function of the degree of freedom corresponding to the null eigenvalue reads $H = \dot{b}_\sigma^2/2$, whence $\ddot{b}_\sigma = 0$, $b_\sigma = b_\sigma(0) + a\,t$, with a a constant.

The problem tackled in Sect. 3.10 is that of diagonalizing the right hand side of (3.42). The diagonalization of a quadratic form entails a linear transformation over the original vector ($\mathbf{h}$ in this case) using a matrix formed by eigenvectors. One may observe that, in (3.42), the kinetic energy $\dot{\mathbf{h}}^T \mathbf{M} \dot{\mathbf{h}}/2$ is already diagonal in the original vector, while the potential energy $\mathbf{h}^T \mathbf{C} \mathbf{h}/2$ is not. If the diagonalization were carried out using the matrix formed by the eigenvalues of $\mathbf{C}$ alone, the outcome of the process would be that of making the potential energy diagonal while making the kinetic energy non-diagonal (both in the transformed vector). The problem is solved by using the eigenvalue Eq. (3.43), that involves both matrices $\mathbf{M}$ and $\mathbf{C}$ in the diagonalization process. In fact, as shown in Sect. 3.10, in the transformed vector $\mathbf{b}$ the potential energy becomes diagonal, and the kinetic energy remains diagonal.

One may observe that, given the solutions of the eigenvalue Eq. (3.43), the process of diagonalizing (3.42) is straightforward. The real difficulty lies in solving (3.43). When the number of degrees of freedom is large, the solution of (3.43) must be tackled by numerical methods and may become quite cumbersome. In practical applications the elastic matrix $\mathbf{C}$ exhibits some structural properties, like symmetry or periodicity (e.g., Sect. 17.7.1), that are exploited to ease the problem of solving (3.43).

3.13.4 Areal Velocity in the Central-Motion Problem

Consider the central-motion problem discussed in Sect. 3.4. In the elementary time-interval dt the position vector changes from $\mathbf{r}$ to $\mathbf{r} + d\mathbf{r}$. The area dA of the triangle whose sides are $\mathbf{r}$, $\mathbf{r} + d\mathbf{r}$, and $d\mathbf{r}$ is half the area of the parallelogram $Q_1 Q_2 Q_3 Q_4$

Fig. 3.7 Definition of the
angles used in Sects. 3.6 and
3.13.5

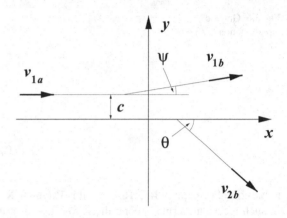

whose consecutive sides are, e.g., $\mathbf{r}$ and $d\mathbf{r}$ (Fig. 3.6). Thus,

$$dA = \frac{1}{2}|\mathbf{r} \wedge d\mathbf{r}| = \frac{1}{2}|\mathbf{r} \wedge \dot{\mathbf{r}} dt| = \frac{|\mathbf{M}|}{2m}dt, \qquad \frac{dA}{dt} = \frac{|\mathbf{M}|}{2m}, \qquad (3.69)$$

with $\mathbf{M}$ the angular momentum. The derivative dA/dt is called *areal velocity*. The
derivation of (3.69) is based purely on definitions, hence it holds in general. For a
central motion the angular momentum $\mathbf{M}$ is constant, whence the areal velocity is
constant as well: the area swept out by the position vector $\mathbf{r}$ in a given time interval
is proportional to the interval itself (*Kepler's second law*). If the particle's trajectory
is closed, the time T taken by $\mathbf{r}$ to complete a revolution and the area A enclosed by
the orbit are related by

$$A = \int_0^T \frac{dA}{dt} dt = \frac{|\mathbf{M}|}{2m}T, \qquad T = \frac{2mA}{|\mathbf{M}|}. \qquad (3.70)$$

3.13.5 Initial Conditions in the Central-Motion Problem

The theory of the central motion for a two-particle system has been worked out in
Sects. 3.7, 3.8 without specifying the initial conditions. To complete the analysis it
is convenient to use the same prescription as in Sect. 3.6, namely, to select an O
reference where the particle of mass m_2 is initially at rest ($\mathbf{v}_{2a} = 0$). Moreover, here
reference O is chosen in such a way as to make the initial position of the particle of
mass m_2 to coincide with the origin ($\mathbf{r}_{2a} = 0$), and the initial velocity $\mathbf{v}_a = \mathbf{v}_{1a}$ of
the particle of mass m_1 to be parallel to the x axis (Fig. 3.7), so that $\mathbf{v}_{1a} = (\mathbf{v}_{1a} \cdot \mathbf{i})\mathbf{i}$.
From (3.11) one finds $E = E_a = T_a = T_{1a} = m_1 v_{1a}^2/2$ and, from Sect. 3.5,
$(m_1 + m_2)\dot{\mathbf{R}}_a = m_1 \mathbf{v}_{1a}$. Using $\mathbf{r}_{1a} = x_{1a}\mathbf{i} + y_{1a}\mathbf{j}$ and (3.28) then yields

$$E_B = \frac{1}{2}m v_{1a}^2, \qquad \mathbf{M} = \mathbf{r}_{1a} \wedge m_1 \mathbf{v}_{1a} = -m_1 y_{1a}(\mathbf{v}_{1a} \cdot \mathbf{i})\mathbf{k}, \qquad (3.71)$$

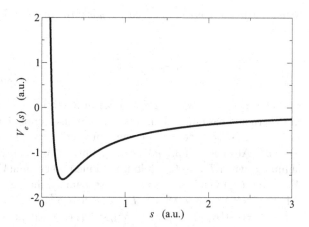

Fig. 3.8 Dependence of V_e on the distance s from the center of force, as given by (3.74) in arbitrary units

with $\mathbf{i}, \mathbf{j}, \mathbf{k}$ the unit vectors of the x, y, z axes and m the reduced mass. On the other hand, (3.29) shows that $\mathbf{M} = (m_1 + m_2)\, \mathbf{R}_a \wedge \dot{\mathbf{R}} + \mathbf{M}_B$ whence, writing $\mathbf{R}_a$, $\dot{\mathbf{R}}$ in terms of $\mathbf{r}_{1a}$, $\mathbf{v}_{1a}$ and equating the two expressions of $\mathbf{M}$ provides

$$M_B = -y_{1a}\, m\, \mathbf{v}_{1a} \cdot \mathbf{i}. \tag{3.72}$$

Replacing (3.72) and the first of (3.71) in (3.32, 3.33) yields

$$\mu = \mp y_{1a}, \qquad s_0 = \lambda + \sqrt{\lambda^2 + c^2}, \qquad c = |y_{1a}|. \tag{3.73}$$

The distance c between the x axis and the direction of $\mathbf{v}_{1a}$ (Fig. 3.7) is called *impact parameter*. The outcome of the calculation demonstrates the usefulness of choosing reference O as described above. In fact, for a given form of the potential energy V, the angle χ defined in Sect. 3.6 becomes a function of two easily-specified quantities: kinetic energy ($E = m_1 v_{1a}^2/2$ or $E_B = m\, v_{1a}^2/2$) and impact parameter c (compare, e.g., with (3.36)). Once χ is determined, the final kinetic energies T_{1b}, T_{2b} and the angles ψ, θ are recovered from (3.22, 3.23) and (3.25, 3.26), respectively. Another property of reference O is that θ turns out to be the angle between the x axis and the final direction of the particle of mass m_2.

3.13.6 The Coulomb Field in the Attractive Case

To treat the attractive case one lets $\kappa = -1$ in (3.31). The trajectory lies in the x, y plane; in polar coordinates it is still given by (3.30), with

$$V_e(s) = \frac{M_B^2}{2ms^2} - \frac{Z_1\, Z_2\, q^2}{4\pi\, \varepsilon_0\, s}, \qquad s > 0. \tag{3.74}$$

In this case V_e becomes negative for some values of s. As a consequence, E_B may also be negative, provided the condition $E_B \geq V_e$ is fulfilled. Then, it is not possible to use the definitions (3.32) because E_B is positive there. The following will be used instead,

$$\alpha = \frac{Z_1 Z_2 q^2}{8\pi \, \varepsilon_0} > 0, \qquad \beta = \frac{M_B}{\sqrt{2m}}, \tag{3.75}$$

so that $V_e(s) = (\beta/s)^2 - 2\alpha/s$. Like in Sect. 3.8 it is assumed that M_B differs from zero and has either sign. It is found by inspection that V_e has only one zero at $s = s_c = \beta^2/(2\alpha)$ and only one minimum at $s = 2s_c$, with $\min(V_e) = V_e(2s_c) = -\alpha^2/\beta^2$. Also, it is $\lim_{s\to 0} V_e = \infty$, $\lim_{s\to\infty} V_e = 0$ (Fig. 3.8). The motion is unlimited when $E_B \geq 0$, while it is limited when $\min(V_e) \leq E_B < 0$. The case $E_B = \min(V_e)$ yields $s = 2s_c = \text{const}$, namely, the trajectory is a circumference. When $\min(V_e) = -\alpha^2/\beta^2 < E_B < 0$ it is $\alpha^2 > \beta^2 |E_B|$. Then, the difference $E_B - V_e = -(|E_B| s^2 - 2\alpha s + \beta^2)/s^2$ has two real, positive zeros given by

$$s_0 = \frac{\alpha - \sqrt{\alpha^2 - \beta^2 |E_B|}}{|E_B|}, \qquad s_1 = \frac{\alpha + \sqrt{\alpha^2 - \beta^2 |E_B|}}{|E_B|}, \qquad s_0 < s_1. \tag{3.76}$$

Using the zeros one finds $s^2 \sqrt{E_B - V_e} = \sqrt{|E_B|} \, s \sqrt{(s - s_0)(s_1 - s)}$, that is replaced within (3.30) after letting $s \leftarrow \xi$. The upper limit of the integral belongs to the interval $s_0 \leq s \leq s_1$. To calculate the integral it is convenient to use a new variable w such that $2 s_0 s_1/\xi = (s_1 - s_0) w + s_1 + s_0$. The range of w corresponding to the condition $s_0 \leq \xi \leq s$ is

$$\frac{2 s_0 s_1 - (s_0 + s_1) s}{(s_1 - s_0) s} \leq w \leq 1, \qquad w(s_1) = -1. \tag{3.77}$$

From (3.30) the trajectory in the s, φ reference is thus found to be

$$\varphi(s) = \varphi_0 \pm \frac{M_B}{|M_B|} \arccos \left[\frac{2 s_0 s_1 - (s_0 + s_1) s}{(s_1 - s_0) s} \right]. \tag{3.78}$$

As noted in Sect. 3.4, the trajectory is symmetric with respect to φ_0. When (3.78) is inverted, the $\pm M_B/|M_B|$ factor is irrelevant because the cosine is an even function of the argument. Thus,

$$\frac{1}{s} = \frac{s_1 + s_0}{2 s_0 s_1} \left[1 + \frac{s_1 - s_0}{s_1 + s_0} \cos(\varphi - \varphi_0) \right]. \tag{3.79}$$

When $\varphi = \varphi_0$ it is $s = s_0$; when $\varphi = \varphi_0 + \pi$ it is $s = s_1$. The $s(\varphi)$ relation (3.79) is the equation of an ellipse of eccentricity $e = (s_1 - s_0)/(s_1 + s_0)$, where the center of force $s = 0$ is one of the foci. The distance between the foci is $s_1 - s_0$. With the aid of Fig. 3.9 one finds that the semimajor and semiminor axes are obtained, respectively, from $a = (s_1 + s_0)/2$, $b^2 = a^2 - (s_1 - s_0)^2/4$ whence, using (3.76),

$$a = \frac{\alpha}{|E_B|}, \qquad b = \frac{|\beta|}{|E_B|} = \frac{|M_B|}{\sqrt{2m |E_B|}}. \tag{3.80}$$

Fig. 3.9 The elliptical trajectory described by (3.79) with $\varphi_0 = 0$

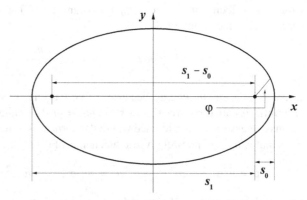

As the particle's trajectory is two-dimensional, the problem has four constants of motion (Sect. 2.11.2); the total energy E_B and the angular momentum M_B are two of such constants. As shown by (3.80), the semimajor axis of the elliptical trajectory depends only on E_B; the area of the ellipse in terms of the constants of motion is $A = \pi\, a\, b = (\pi\, \alpha/\sqrt{2m})\,|M_B|\,|E_B|^{-3/2}$. The position vector **s** completes a full orbit in a period T given by (3.70); combining the latter with the expression of A yields

$$T = \frac{\pi\,\alpha\,\sqrt{2m}}{|E_B|^{3/2}} = \pi\,\sqrt{\frac{2m}{\alpha}}\,a^{3/2}, \tag{3.81}$$

namely, the period depends on the total energy, but not on the angular momentum. Thus, the period is still given by (3.81) in the limiting case $M_B \to 0$, which makes the trajectory to shrink into a segment of length a crossing the origin of the s, φ reference, and the position vector to oscillate along this segment (compare with problem 3.2). The second form of (3.81) shows that $T^2 \propto a^3$ (*Kepler's third law*).

3.13.7 Dynamic Relations of Special Relativity

The dynamic relations considered in this book refer almost invariably to situations where the particle velocity is small with respect to that of light. For this reason it is sufficient to use the non-relativistic relations. The only exception where the velocities of the particles involved do not allow for such an approximation is considered in Sects. 3.13.8 and 7.4.3. For this reason a set of relations of the Special-Relativity Theory are given here, that apply to the case of a free particle. The first of them is the relation between velocity **u** and momentum **p**,

$$\mathbf{p} = \frac{m_0\,\mathbf{u}}{\sqrt{1 - u^2/c^2}}, \qquad u = |\mathbf{u}|, \tag{3.82}$$

with c the velocity of light and m_0 a constant mass. The second relation involves energy and velocity and reads

$$E = m c^2, \qquad m = \frac{m_0}{\sqrt{1 - u^2/c^2}}, \tag{3.83}$$

where E is a kinetic energy because a free particle is considered. In the above, $m = m(u)$ is called *relativistic mass* and $m_0 = m(0)$ is called *rest mass*. The latter is the mass measured in a reference where the particle is at rest, and is the value of the mass that is used in non-relativistic mechanics. From (3.82, 3.83) it follows

$$\mathbf{p} = m \, \mathbf{u}, \qquad m^2 c^2 - m^2 u^2 = m_0^2 c^2, \qquad m^2 c^2 = E^2/c^2, \tag{3.84}$$

whence the elimination of u provides the relation between E and the modulus of $\mathbf{p}$:

$$E^2/c^2 - p^2 = m_0^2 c_2, \qquad p = \sqrt{E^2/c^2 - m_0^2 c^2}. \tag{3.85}$$

In general the $p = p(E)$ relation is non linear. However, for particles with $m_0 = 0$ the expressions (3.85) simplify to the linear relation $p = E/c$. An example of this is found in the theory of the electromagnetic field, where the same momentum-energy relation is derived from the Maxwell equations (Sect. 5.7). It is worth observing that if a particle has $m_0 = 0$ and $p \neq 0$, its velocity is necessarily equal to c; in fact, the first of (3.84) yields $\lim_{m_0 \to 0} \mathbf{p} = 0$ when $u < c$. Another limiting case is found when $u/c \ll 1$. In fact, the second of (3.83) simplifies to

$$m \simeq \frac{m_0}{1 - u^2/(2 \, c^2)} \simeq m_0 \left(1 + \frac{u^2}{2 \, c^2} \right). \tag{3.86}$$

Inserting the last form into the first of (3.83) yields

$$E \simeq m_0 \, c^2 + \frac{1}{2} \, m_0 \, u^2. \tag{3.87}$$

The constant $m_0 \, c^2$ is called *rest energy*. The limiting case $u/c \ll 1$ then renders for $E - m_0 \, c^2$ the non-relativistic expression of the kinetic energy.

3.13.8   Collision of Relativistic Particles

This section illustrates the collision between two relativistic particles that constitute an isolated system. The same approach of Sect. 3.5 is used here, namely, the asymptotic values only are considered. Also, the case where the particles' trajectories belong to the same plane, specifically, the x, y plane, is investigated. The initial conditions are the same as in Sect. 3.6: the asymptotic motion of the first particle before the collision is parallel to the x axis, while the second particle is initially at rest. Finally, it is assumed that the rest mass of the first particle is zero, so that the momentum-energy relation of this particle is $p = E/c$ as shown in Sect. 3.13.7, while the rest mass of the second particle is $m_0 \neq 0$.

The collision is treated by combining the conservation laws of energy and momentum of the two-particle system. Let E_a, E_b be the asymptotic energies of the first particle before and after the collision, respectively. As for the second particle, which is initially at rest, the energy before the collision is its rest energy $m_0 c^2$, while that after the collision is $m c^2$ (Sect. 3.13.7). The conservation of energy then reads

$$E_a + m_0 c^2 = E_b + mc^2, \tag{3.88}$$

while the conservation of momentum reads, respectively for the x and y components,

$$\frac{E_a}{c} = \frac{E_b}{c} \cos \psi + mu \cos \theta, \qquad 0 = \frac{E_b}{c} \sin \psi - mu \sin \theta. \tag{3.89}$$

The angles ψ and θ in (3.89) are the same as in Fig. 3.7. Extracting $m c^2$ from (3.88) and squaring both sides yields

$$(E_a - E_b)^2 + 2 m_0 c^2 (E_a - E_b) = m^2 c^4 - m_0^2 c^4 = m^2 u^2 c^2, \tag{3.90}$$

where the last equality derives from the second of (3.84). Then, the momentum-conservation relations (3.89) are used to eliminate θ by squaring and adding up the results, to find

$$m^2 u^2 c^2 = E_a^2 + E_b^2 - 2 E_a E_b \cos \psi = (E_a - E_b)^2 + 4 E_a E_b \sin^2 (\psi/2). \tag{3.91}$$

Eliminating $m^2 u^2 c^2 - (E_a - E_b)^2$ between (3.90) and (3.91) yields

$$\frac{1}{E_b} - \frac{1}{E_a} = \frac{2}{m_0 c^2} \sin^2 \left(\frac{\psi}{2} \right), \tag{3.92}$$

that provides the asymptotic energy after the collision of the first particle, as a function of the asymptotic energy before the collision, the deflection angle of the same particle, and the rest energy of the second particle. Equation (3.92) is used in Sect. 7.4.3 for the explanation of the Compton effect.

The non-relativistic analogue of the above procedure is illustrated in Sect. 3.6. It is interesting to note that the calculation carried out here seems rather less involved than the non-relativistic one. This surprising fact is actually due to the special choice of the first particle, whose rest energy is zero. In this case, in fact, the relation between momentum and energy becomes linear. That of the second particle, which is non linear, is eliminated from the equations. On the contrary, in the non-relativistic case treated in Sect. 3.6 the energy-momentum relations are non linear for both particles, this making the calculation more laborious.

3.13.9 *Energy Conservation in Charged-Particles' Interaction*

The two-particle interaction considered in Sect. 3.8 involves charged particles. As the particles' velocity during the interaction is not constant, the particles radiate (Sect. 5.11.2) and, consequently, lose energy. This phenomenon is not considered in the analysis carried out in Sect. 3.8, where the total energy of the two-particle system is assumed constant. The assumption is justified on the basis that the radiated power is relatively small. This subject is further discussed in Sect. 5.11.3.

Problems

3.1 Given the Hamiltonian function of the one-dimensional harmonic oscillator of the general form $H = p^2/(2\,m) + (c/s)\,|x|^s$, $m, c, s > 0$, find the oscillator's period.

3.2 Given the Hamiltonian function of the one-dimensional harmonic oscillator of the general form $H = p^2/(2\,m) - (k/s)\,|x|^{-s} = E < 0$, $m, k, s > 0$, find the oscillator's period.

3.3 Consider the collision between two particles in the repulsive Coulomb case. Calculate the relation $T_{1b}(T_{1a}, c)$, with c the impact parameter (hint: follow the discussion of Sect. 3.13.5 and use (3.23, 3.36), (3.32, 3.33), and (3.73)).

Chapter 4
Electromagnetism

4.1 Introduction

This chapter outlines the basic principles of the electromagnetic theory *in vacuo*. First, the extension of the Lagrangian formalism to functions that depend on more than one variable is tackled: this yields useful tools for the analysis of continuous media. Next, the Maxwell equations are introduced along with the derivation of the electric and magnetic potentials, and the concept of gauge transformation is illustrated. The second part of the chapter is devoted to the Helmholtz and wave equations, both in a finite and infinite domain. The chapter finally introduces the Lorentz force, that connects the electromagnetic field with the particles' dynamics. The complements discuss some invariance properties of the Euler equations, derive the wave equations for the electric and magnetic field, and clarify some issues related to the boundary conditions in the application of the Green method to the boundary-value problem.

4.2 Extension of the Lagrangian Formalism

In Sect. 1.2 the derivation of the extremum functions has been carried out with reference to a functional $G[w]$ of the form (1.1). Such a functional contains one unknown function w that, in turn, depends on one independent variable ξ. The result has been extended to the case where the functional depends on several unknown functions $w_1, w_2, \ldots$, each dependent on one variable only (compare with (1.6)). The extension to more than one independent variable is shown here.

To proceed it suffices to consider a single unknown function w that depends on two independent variables ξ, σ and is differentiable at least twice with respect to each. The first and second derivatives of w are indicated with w_ξ, w_σ, $w_{\xi\xi}$, $w_{\sigma\sigma}$, and $w_{\xi\sigma}$. Letting Ω be the domain over which w is defined, and g the generating function, the functional reads

$$G[w] = \int_\Omega g(w, w_\xi, w_\sigma, \xi, \sigma) \, \mathrm{d}\Omega. \qquad (4.1)$$

© Springer Science+Business Media New York 2015
M. Rudan, *Physics of Semiconductor Devices*,
DOI 10.1007/978-1-4939-1151-6_4

Then, let $\delta w = \alpha\,\eta$, with $\eta(\xi, \sigma)$ an arbitrary function defined in Ω and differentiable in its interior, and α a real parameter. Like in the case of one independent variable the choice is restricted to those functions η that vanish at the boundary of Ω, so that w and $w + \delta w$ coincide along the boundary for any value of α. If w is an extremum function of G, the extremum condition if found by replacing w with $w + \alpha\,\eta$ and letting $(\mathrm{d}G/\mathrm{d}\alpha)_0 = 0$, where suffix 0 indicates that the derivative is calculated at $\alpha = 0$ (compare with Sect. 1.2). Exchanging the integral with the derivative in (4.1) yields

$$\left(\frac{\mathrm{d}G}{\mathrm{d}\alpha}\right)_0 = \int_\Omega \left(\frac{\partial g}{\partial w}\,\eta + \frac{\partial g}{\partial w_\xi}\,\eta_\xi + \frac{\partial g}{\partial w_\sigma}\,\eta_\sigma\right)\,\mathrm{d}\Omega = 0. \tag{4.2}$$

The second and third term of the integrand in (4.2) are recast in compact form by defining vector $\mathbf{u} = (\partial g/\partial w_\xi, \partial g/\partial w_\sigma)$ and using the second identity in (A.16), so that the sum of the two terms reads $\mathbf{u} \cdot \mathrm{grad}\eta = \mathrm{div}(\eta\mathbf{u}) - \eta\,\mathrm{div}\mathbf{u}$. Integrating over Ω and using the divergence theorem (A.23) yields

$$\int_\Omega \mathbf{u} \cdot \mathrm{grad}\eta\,\mathrm{d}\Omega = \int_\Sigma \eta\,\mathbf{u} \cdot \mathbf{n}\,\mathrm{d}\Sigma - \int_\Omega \eta\,\mathrm{div}\mathbf{u}\,\mathrm{d}\Omega, \tag{4.3}$$

where Σ is the boundary of Ω and $\mathbf{n}$ the unit vector normal to $\mathrm{d}\Sigma$, oriented in the outward direction with respect to Σ. The first term at the right hand side of (4.3) is equal to zero because η vanishes over Σ.

It is important to clarify the symbols that will be used to denote the derivatives. In fact, to calculate $\mathrm{div}\mathbf{u}$ one needs, first, to differentiate $\partial g/\partial w_\xi$ with respect to ξ considering also the implicit ξ-dependence within w, w_ξ, and w_σ; then, one differentiates in a similar manner $\partial g/\partial w_\sigma$ with respect to σ. The two derivatives are summed up to form $\mathrm{div}\mathbf{u}$. For this type of differentiation the symbols $\mathrm{d}/\mathrm{d}\xi$ and $\mathrm{d}/\mathrm{d}\sigma$ are used, even if the functions in hand depend on two independent variables instead of one. The symbols $\partial/\partial\xi$ and $\partial/\partial\sigma$ are instead reserved to the derivatives with respect to the explicit dependence on ξ or σ only. With this provision, inserting (4.3) into (4.2) yields the extremum condition

$$\int_\Omega \left(\frac{\partial g}{\partial w} - \frac{\mathrm{d}}{\mathrm{d}\xi}\frac{\partial g}{\partial w_\xi} - \frac{\mathrm{d}}{\mathrm{d}\sigma}\frac{\partial g}{\partial w_\sigma}\right)\eta\,\mathrm{d}\Omega = 0. \tag{4.4}$$

As (4.4) holds for any η, the term in parentheses must vanish. In conclusion, the extremum condition is

$$\frac{\mathrm{d}}{\mathrm{d}\xi}\frac{\partial g}{\partial w_\xi} + \frac{\mathrm{d}}{\mathrm{d}\sigma}\frac{\partial g}{\partial w_\sigma} = \frac{\partial g}{\partial w}, \tag{4.5}$$

namely, a second-order partial-differential equation in the unknown function w, that must be supplemented with suitable boundary conditions. The equation is linear with respect to the second derivatives of w because g does not depend on such derivatives.

The result is readily extended to the case where g depends on several functions $w_1, w_2, \ldots, w_l$ and the corresponding derivatives. Defining the vectors $\mathbf{w}(\xi, \sigma) =$

$(w_1, \ldots, w_l)$, $\mathbf{w}_\xi = (\partial w_1/\partial \xi, \ldots, \partial w_l/\partial \xi)$, $\mathbf{w}_\sigma = (\partial w_1/\partial \sigma, \ldots, \partial w_l/\partial \sigma)$, the set of the l extremum functions w_i of functional

$$G[\mathbf{w}] = \int_\Omega g(\mathbf{w}, \mathbf{w}_\xi, \mathbf{w}_\sigma, \xi, \sigma) \, d\Omega \tag{4.6}$$

is found by solving the equations

$$\frac{d}{d\xi} \frac{\partial g}{\partial(\partial w_i/\partial \xi)} + \frac{d}{d\sigma} \frac{\partial g}{\partial(\partial w_i/\partial \sigma)} = \frac{\partial g}{\partial w_i}, \qquad i = 1, \ldots, l, \tag{4.7}$$

supplemented with the suitable boundary conditions. It follows that (4.7) are the Euler equations of G. Finally, the case where the independent variables are more than two is a direct extension of (4.7). For instance, for m variables $\xi_1, \ldots, \xi_m$ one finds

$$\sum_{j=1}^m \frac{d}{d\xi_j} \frac{\partial g}{\partial(\partial w_i/\partial \xi_j)} = \frac{\partial g}{\partial w_i}, \qquad i = 1, \ldots, l. \tag{4.8}$$

If g is replaced with $g' = g + \text{div}\mathbf{h}$, where $\mathbf{h}$ is an arbitrary vector of length m whose entries depend on $\mathbf{w}$ and $\xi_1, \ldots, \xi_m$, but not on the derivatives of $\mathbf{w}$, then (4.8) is still fulfilled. The replacement, in fact, adds the same term to both sides. For instance, the term added to the left hand side is

$$\sum_{j=1}^m \frac{d}{d\xi_j} \frac{\partial}{\partial(\partial w_i/\partial \xi_j)} \sum_{r=1}^m \left(\frac{\partial h_r}{\partial \xi_r} + \sum_{s=1}^l \frac{\partial h_r}{\partial w_s} \frac{\partial w_s}{\partial \xi_r} \right), \qquad i = 1, \ldots, l, \tag{4.9}$$

where the sum over r is the explicit expression of $\text{div}\mathbf{h}$. Remembering that $\mathbf{h}$ does not depend on the derivatives of w_i one recasts (4.9) as

$$\sum_{j=1}^m \frac{d}{d\xi_j} \sum_{r=1}^m \sum_{s=1}^l \frac{\partial h_r}{\partial w_s} \frac{\partial(\partial w_s/\partial \xi_r)}{\partial(\partial w_i/\partial \xi_j)} = \sum_{j=1}^m \frac{\partial}{\partial \xi_j} \frac{\partial h_j}{\partial w_i}, \qquad i = 1, \ldots, l, \tag{4.10}$$

where the equality is due to the relation $\partial(\partial w_s/\partial \xi_r)/\partial(\partial w_i/\partial \xi_j) = \delta_{is}\delta_{jr}$, with $\delta_{is(jr)}$ the Kronecker symbol (A.18). Inverting the order of the derivatives at the right hand side of (4.10) yields $\partial \, \text{div}\mathbf{h}/\partial w_i$, that coincides with the term added to the right hand side of (4.8). Finally, (4.8) is recast in compact form by defining a vector $\mathbf{u}_i$ and a scalar s_i as

$$\mathbf{u}_i = \left[\frac{\partial g}{\partial(\partial w_i/\partial \xi_1)}, \ldots, \frac{\partial g}{\partial(\partial w_i/\partial \xi_m)} \right], \qquad s_i = \frac{\partial g}{\partial w_i} \tag{4.11}$$

to find

$$\text{div}_\xi \mathbf{u}_i = s_i, \qquad i = 1, \ldots, l. \tag{4.12}$$

If w_i depends on one variable only, say ξ, (4.8, 4.12) reduce to (1.7). Using the language of the Lagrangian theory, the comparison between the one-dimensional and multi-dimensional case shows that in both cases the functions w_i play the role of generalized coordinates; in turn, the scalar parameter ξ of (1.7) becomes the vector $(\xi_1, \ldots, \xi_m)$ of (4.8) and, finally, each generalized velocity $\dot{w}_i$ becomes the set $\partial w_i/\partial \xi_1, \ldots, \partial w_i/\partial \xi_m$.

4.3   Lagrangian Function for the Wave Equation

It has been shown in Sect. 1.3 that the relations $\ddot{w}_i = \ddot{w}_i(\mathbf{w}, \dot{\mathbf{w}}, \xi)$, $i = 1, \ldots, n$, describing the motion of a system of particles with n degrees of freedom, are the Euler equations of a suitable functional. Then, the analysis of Sect. 4.2 has shown that, when the unknown functions $w_1, \ldots, w_l$ depend on more than one variable, the Euler equations are the second-order partial-differential equations (4.8). The form (4.8) is typical of the problems involving continuous media (e.g., elasticity field, electromagnetic field). Following the same reasoning as in Sect. 1.3 it is possible to construct the Lagrangian function whence the partial-differential equation derives. This is done here with reference to the important case of the *wave equation*[1]

$$\nabla^2 w - \frac{1}{u^2} \frac{\partial^2 w}{\partial t^2} = s, \tag{4.13}$$

where $u = $ const is a velocity and, for the sake of simplicity, s is assumed to depend on x and t, but not on w or its derivatives. It is worth noting that, when a differential equation other than Newton's law is considered, the corresponding Lagrangian function is not necessarily an energy. For this reason it will provisionally be indicated with L_e instead of L. To proceed one considers the one-dimensional form of (4.13), $\partial^2 w/\partial x^2 - (1/u^2)\,\partial^2 w/\partial t^2 = s$ and replaces ξ, σ, g with x, t, L_e, respectively. Then, one makes the one-dimensional form identical to (4.5) by letting

$$\frac{\partial^2 L_e}{\partial w_x^2} = 1, \qquad \frac{\partial^2 L_e}{\partial w_t^2} = -\frac{1}{u^2}, \qquad \frac{\partial L_e}{\partial w} = s, \tag{4.14}$$

with $w_x = \partial w/\partial x$ and $w_t = \partial w/\partial t$. The second derivatives of $L_e(w, w_x, w_t, x, t)$ with respect to the combinations of the arguments not appearing in the first two equations of (4.14) are set to zero. The third of (4.14) provides $L_e = s\,w + c$, with c independent of w. Replacing $L_e = sw + c$ into the first two equations in (4.14), and integrating the first one with respect to w_x, yields $\partial c/\partial w_x = w_x + a_{01}$, with a_{01} independent of w_x. Similarly, from the second equation in (4.14), $\partial c/\partial w_t = -w_t/u^2 + a_{02}$, with a_{02} independent of w_t. Also, remembering that c is independent of w, one finds that a_{01} and a_{02} do not depend on w either. Considering that all the second derivatives of L_e not appearing in (4.14) are equal to zero shows that a_{01} depends on t at most, while a_{02} depends on x at most. Integrating $\partial c/\partial w_x = w_x + a_{01}$ and $\partial c/\partial w_t = -w_t/u^2 + a_{02}$ one finds

$$c = \frac{1}{2} w_x^2 + a_{01}(t)\,w_x + a_{11}, \qquad c = -\frac{1}{2u^2} w_t^2 + a_{02}(x)\,w_t + a_{12}, \tag{4.15}$$

where a_{11} does not depend on w or w_x, while a_{12} does not depend on w or w_t. Also, a_{11} can not depend on both t and w_t due to $\partial^2 L_e/(\partial t\,\partial w_t) = 0$; similarly, a_{12} can not depend on both x and w_x due to $\partial^2 L_e/(\partial x\,\partial w_x) = 0$. On the other hand, as both

---

[1] Also called *D'Alembert equation* in the homogeneous case.

(4.15) hold, a_{11} must coincide (apart from an additive constant) with the first two terms at the right hand side of the second equation in (4.15), and a_{12} must coincide with the first two terms at the right hand side of the first equation. In conclusion,

$$c = \frac{1}{2} w_x^2 - \frac{1}{2u^2} w_t^2 + a_{01}(t) w_x + a_{02}(x) w_t, \tag{4.16}$$

with $a_{01}(t)$, $a_{02}(x)$ arbitrary functions. The last two terms in (4.16) are equal to $d(a_{01} w)/dx + d(a_{02} w)/dt$, namely, they form the divergence of a vector. As shown in Sect. 4.2 such a vector is arbitrary, so it can be eliminated by letting $a_{01} = 0$, $a_{02} = 0$. The relation $L_e = sw + c$ then yields

$$L_e = \frac{1}{2} w_x^2 - \frac{1}{2u^2} w_t^2 + sw. \tag{4.17}$$

The generalization to the three-dimensional case (4.13) is immediate,

$$L_e = \frac{1}{2} |\text{grad}w|^2 - \frac{1}{2u^2} \left(\frac{\partial w}{\partial t} \right)^2 + sw. \tag{4.18}$$

with $|\text{grad}w|^2 = w_x^2 + w_y^2 + w_z^2$.

4.4 Maxwell Equations

The *Maxwell equations*, that describe the electromagnetic field, lend themselves to an interesting application of the results of Sect. 4.3. The first group of Maxwell equations reads

$$\text{div}\mathbf{D} = \rho, \qquad \text{rot}\mathbf{H} - \frac{\partial \mathbf{D}}{\partial t} = \mathbf{J}, \tag{4.19}$$

where $\mathbf{D}$ is the electric displacement and $\mathbf{H}$ the magnetic field.[2] The sources of the electromagnetic field are the charge density ρ and the current density $\mathbf{J}$. When point-like charges are considered, they read

$$\rho_c = \sum_j e_j \delta \left(\mathbf{r} - \mathbf{s}_j(t) \right), \qquad \mathbf{J}_c = \sum_j e_j \delta \left(\mathbf{r} - \mathbf{s}_j(t) \right) \mathbf{u}_j(t), \tag{4.20}$$

where index c is used to distinguish the case of point-like charges from that of a continuous charge distribution. In (4.20), e_j is the value of the jth charge, $\mathbf{s}_j$ and

---

[2] The units in (4.19, 4.23, 4.24) are: $[\mathbf{D}] = \text{C m}^{-2}$, $[\rho] = \text{C m}^{-3}$, $[\mathbf{H}] = \text{A m}^{-1}$, $[\mathbf{J}] = \text{C s}^{-1} \text{m}^{-2} = \text{A m}^{-2}$, $[\mathbf{B}] = \text{V s m}^{-2} = \text{Wb m}^{-2} = \text{T}$, $[\mathbf{E}] = \text{V m}^{-1}$, where "C", "A", "V", "Wb", and "T" stand for Coulomb, Ampere, Volt, Weber, and Tesla, respectively. The coefficients in (4.19, 4.23, 4.24) differ from those of [4] because of the different units adopted there. In turn, the units in (4.25) are $[\varepsilon_0] = \text{C V}^{-1} \text{m}^{-1} = \text{F m}^{-1}$, $[\mu_0] = \text{s}^2 \text{F}^{-1} \text{m}^{-1} = \text{H m}^{-1}$, where "F" and "H" stand for Farad and Henry, respectively, and those in (4.26) are $[\varphi] = \text{V}$, $[\mathbf{A}] = \text{V s m}^{-1} = \text{Wb m}^{-1}$.

$\mathbf{u}_j$ its position and velocity at time t, respectively, and $\mathbf{r}$ the independent positional variable. If the spatial scale of the problem is such that one can replace the point-like charges with a continuous distribution, one applies the same procedure as in Sect. 23.2. The number of charges belonging to a cell of volume ΔV centered at $\mathbf{r}$ is $\int_{\Delta V} \rho_c \, d^3 s' = \sum'_j e_j$, where the prime indicates that the sum is limited to the charges that belong to Δ at time t. Then one defines $\rho(\mathbf{r}, t) = \sum'_j e_j / \Delta V$. The continuous distribution of the current density is obtained in a similar manner,

$$\mathbf{J} = \frac{1}{\Delta V} \int_{\Delta V} \mathbf{J}_c \, d^3 s' = \frac{1}{\Delta V} \sum_j' e_j \mathbf{u}_j = \rho \, \mathbf{v}, \qquad \mathbf{v} = \frac{\sum_j' e_j \mathbf{u}_j}{\sum_j' e_j}, \qquad (4.21)$$

with $\mathbf{v}(\mathbf{r}, t)$ the average velocity of the charges. If all charges are equal, $e_1 = e_2 = \ldots = e$, then $\rho = e N$, with $N(\mathbf{r}, t)$ the concentration, and $\mathbf{J} = e N \mathbf{v} = e \mathbf{F}$, with $\mathbf{F}(\mathbf{r}, t)$ the flux density (compare with the definitions of Sect. 23.2). If the charges are different from each other it is convenient to distribute the sum $\sum_j$ over the groups made of equal charges. In this case the charge density and current density read

$$\rho = \rho_1 + \rho_2 + \ldots, \qquad \mathbf{J} = \rho_1 \mathbf{v}_1 + \rho_2 \mathbf{v}_2 + \ldots, \qquad (4.22)$$

where ρ_1, $\mathbf{v}_1$ are the charge density and average velocity of the charges of the first group, and so on. Taking the divergence of the second equation in (4.19) and using the third identity in (A.35) yields the *continuity equation*

$$\frac{\partial \rho}{\partial t} + \text{div} \mathbf{J} = 0. \qquad (4.23)$$

Apart from the different units of the functions involved, the form of (4.23) is the same as that of (23.3). The meaning of (4.23) is that of conservation of the electric charge. The second group of Maxwell equations is

$$\text{div} \mathbf{B} = 0, \qquad \text{rot} \mathbf{E} + \frac{\partial \mathbf{B}}{\partial t} = 0, \qquad (4.24)$$

where $\mathbf{B}$ and $\mathbf{E}$ are the magnetic induction and the electric field, respectively. Here the Maxwell equations are considered *in vacuo*, so that the following hold

$$\mathbf{D} = \varepsilon_0 \mathbf{E}, \qquad \mathbf{B} = \mu_0 \mathbf{H}, \qquad \frac{1}{\sqrt{\varepsilon_0 \mu_0}} = c, \qquad (4.25)$$

with $\varepsilon_0 \simeq 8.854 \times 10^{-12}$ F m^{-1} and $\mu_0 \simeq 1.256 \times 10^{-6}$ H m^{-1} the vacuum permittivity and permeability, respectively, and $c \simeq 2.998 \times 10^8$ m s^{-1} the speed of light *in vacuo*.

4.5 Potentials and *Gauge* Transformations

Thanks to (4.25), the electromagnetic field *in vacuo* is determined by two suitably-chosen vectors—typically, **E** and **B**—out of the four ones appearing in (4.25). This amounts to using six scalar functions of position and time. However, the number of scalar functions is reduced by observing that, while (4.19) provide relations between the electromagnetic field and its sources, (4.24) provide relations among the field vectors themselves; as a consequence, (4.24) reduce the number of independent vectors. In fact, using the properties illustrated in Sect. A.9, one finds that from divB = 0 one derives **B** = rotA, where **A** is called *vector potential* or *magnetic potential*. In turn, the vector potential transforms the second of (4.24) into rot(**E** + $\partial \mathbf{A}/\partial t$) = 0; using again the results of Sect. A.9 shows that the term in parentheses is the gradient of a scalar function, that is customarily indicated with $-\varphi$. Such a function[3] is called *scalar potential* or *electric potential*. In summary,

$$\mathbf{B} = \text{rot}\mathbf{A}, \qquad \mathbf{E} = -\text{grad}\varphi - \frac{\partial \mathbf{A}}{\partial t}, \tag{4.26}$$

showing that for determining the electromagnetic field *in vacuo* it suffices to know four scalar functions, namely, φ and the three components of **A**. To proceed, one replaces (4.26) into (4.19) and uses the third relation in (4.25), to find

$$\nabla^2\varphi + \frac{\partial}{\partial t}\text{div}\mathbf{A} = -\frac{\rho}{\varepsilon_0}, \qquad -\text{rot rot}\mathbf{A} - \frac{1}{c^2}\frac{\partial^2\mathbf{A}}{\partial t^2} = -\mu_0\mathbf{J} + \frac{1}{c^2}\text{grad}\frac{\partial\varphi}{\partial t}. \tag{4.27}$$

Thanks to the first identity in (A.36) the second equation in (4.27) becomes

$$\nabla^2\mathbf{A} - \frac{1}{c^2}\frac{\partial^2\mathbf{A}}{\partial t^2} = -\mu_0\mathbf{J} + \text{grad}\theta, \qquad \theta = \text{div}\mathbf{A} + \frac{1}{c^2}\frac{\partial\varphi}{\partial t} \tag{4.28}$$

while, using the definition (4.28) of θ, one transforms the first equation in (4.27) into

$$\nabla^2\varphi - \frac{1}{c^2}\frac{\partial^2\varphi}{\partial t^2} = -\frac{\rho}{\varepsilon_0} - \frac{\partial\theta}{\partial t}. \tag{4.29}$$

In conclusion, (4.29) and the first equation in (4.28) are a set of four scalar differential equations whose unknowns are φ and the components of **A**. Such equations are coupled because θ contains all unknowns; however, they become decoupled after suitable transformations, shown below.

To proceed, one observes that only the derivatives of the potentials, not the potential themselves, appear in (4.26); as a consequence, while the fields **E**, **B** are uniquely defined by the potentials, the opposite is not true. For instance, replacing **A**

---

[3] The minus sign in the definition of φ is used for consistency with the definition of the gravitational potential, where the force is opposite to the direction along which the potential grows.

with $\mathbf{A}' = \mathbf{A} + \mathrm{grad}\,f$, where $f(\mathbf{r}, t)$ is any differentiable scalar function, and using the second identity in (A.35), yields $\mathbf{B}' = \mathrm{rot}\mathbf{A}' = \mathbf{B}$, namely, $\mathbf{B}$ is invariant with respect to such a replacement. If, at the same time, one replaces φ with a yet undetermined function φ', (4.26) yields $\mathbf{E}' = -\mathrm{grad}(\varphi' + \partial f/\partial t) - \partial \mathbf{A}/\partial t$. It follows that by choosing $\varphi' = \varphi - \partial f/\partial t$ one obtains $\mathbf{E}' = \mathbf{E}$. The transformation $(\varphi, \mathbf{A}) \to (\varphi', \mathbf{A}')$ defined by

$$\varphi' = \varphi - \frac{\partial f}{\partial t}, \qquad \mathbf{A}' = \mathbf{A} + \mathrm{grad}\,f. \tag{4.30}$$

is called *gauge transformation*. As shown above, $\mathbf{E}$ and $\mathbf{B}$ are invariant with respect to such a transformation. One also finds that (4.29) and the first equation in (4.28) are invariant with respect to the transformation: all terms involving f cancel each other, so that the equations in the primed unknowns are identical to the original ones. However, the solutions φ', $\mathbf{A}'$ are different from φ, $\mathbf{A}$ because, due to (4.30), their initial and boundary conditions are not necessarily the same. The difference between the primed and unprimed solutions is unimportant because the fields, as shown above, are invariant under the transformation. Using (4.30) in the second equation of (4.28) shows that θ transforms as

$$\theta' = \mathrm{div}\mathbf{A}' + \frac{1}{c^2} \frac{\partial \varphi'}{\partial t} = \theta + \nabla^2 f - \frac{1}{c^2} \frac{\partial^2 f}{\partial t^2}. \tag{4.31}$$

The arbitrariness of f may be exploited to give θ' a convenient form. For instance one may choose f such that $\theta' = (1/c^2)\,\partial \varphi/\partial t$, which is equivalent to letting

$$\mathrm{div}\mathbf{A}' = 0, \tag{4.32}$$

called *Coulomb gauge*. The latter yields

$$\nabla^2 \varphi' = -\frac{\rho}{\varepsilon_0}, \qquad \nabla^2 \mathbf{A}' - \frac{1}{c^2} \frac{\partial^2 \mathbf{A}'}{\partial t^2} = -\mu_0 \mathbf{J} + \frac{1}{c^2} \frac{\partial}{\partial t} \mathrm{grad}\varphi', \tag{4.33}$$

the first of which (the *Poisson equation*) is decoupled from the second one. After solving the Poisson equation, the last term at the right hand side of the second equation is not an unknown any more, this showing that the equations resulting from the Coulomb gauge are indeed decoupled. Another possibility is choosing f such that $\theta' = 0$, which is equivalent to letting

$$\mathrm{div}\mathbf{A}' = -\frac{1}{c^2} \frac{\partial \varphi'}{\partial t}, \tag{4.34}$$

called *Lorentz gauge*. This transformation yields

$$\nabla^2 \varphi' - \frac{1}{c^2} \frac{\partial^2 \varphi'}{\partial t^2} = -\frac{\rho}{\varepsilon_0}, \qquad \nabla^2 \mathbf{A}' - \frac{1}{c^2} \frac{\partial^2 \mathbf{A}'}{\partial t^2} = -\mu_0 \mathbf{J}. \tag{4.35}$$

that are decoupled and have the form of the wave Eq. (4.13). Another interesting application of the gauge transformation is shown in Sect. 5.11.4.

4.6 Lagrangian Density for the Maxwell Equations

To apply the Lagrangian formalism to the Maxwell equations it is useful to use the expressions (4.26) of the fields in terms of the potentials. It follows that the functions playing the role of generalized coordinates and generalized velocities are φ, A_i, and, respectively, $\partial \varphi / \partial x_k$, $\partial A_i / \partial x_k$, $\partial A_i / \partial t$, with $i, k = 1, 2, 3$, $k \neq i$. The Lagrangian density, whose units are J m^{-3}, then reads

$$L_e = \frac{\varepsilon_0}{2} E^2 - \frac{1}{2\mu_0} B^2 - \rho\,\varphi + \mathbf{J} \cdot \mathbf{A}, \tag{4.36}$$

with

$$E^2 = \left(\frac{\partial \varphi}{\partial x_1} + \frac{\partial A_1}{\partial t}\right)^2 + \left(\frac{\partial \varphi}{\partial x_2} + \frac{\partial A_2}{\partial t}\right)^2 + \left(\frac{\partial \varphi}{\partial x_3} + \frac{\partial A_3}{\partial t}\right)^2 \tag{4.37}$$

and

$$B^2 = \left(\frac{\partial A_3}{\partial x_2} - \frac{\partial A_2}{\partial x_3}\right)^2 + \left(\frac{\partial A_1}{\partial x_3} - \frac{\partial A_3}{\partial x_1}\right)^2 + \left(\frac{\partial A_2}{\partial x_1} - \frac{\partial A_1}{\partial x_2}\right)^2, \tag{4.38}$$

To show that (4.36) is in fact the Lagrangian function of the Maxwell equations one starts with the generalized coordinate φ, to find $\partial L_e / \partial \varphi = -\rho$. Then, considering the kth component,

$$\frac{\partial L_e / \varepsilon_0}{\partial(\partial \varphi / \partial x_k)} = \frac{\partial E^2 / 2}{\partial(\partial \varphi / \partial x_k)} = \frac{\partial E^2 / 2}{\partial(\partial A_k / \partial t)} = \frac{\partial \varphi}{\partial x_k} + \frac{\partial A_k}{\partial t} = -E_k = -\frac{D_k}{\varepsilon_0}. \tag{4.39}$$

Using (4.8) after replacing g with L_e, ξ_j with x_j, and w_i with φ yields div$\mathbf{D} = \rho$, namely, the first equation in (4.19). Turning now to another generalized coordinate, say, A_1, one finds $\partial L_e / \partial A_1 = J_1$. As L_e depends on the spatial derivatives of A_1 only through B^2, (4.38) and the first of (4.26) yield

$$\frac{\partial B^2 / 2}{\partial(\partial A_1 / \partial x_3)} = \frac{\partial A_1}{\partial x_3} - \frac{\partial A_3}{\partial x_1} = B_2, \qquad \frac{\partial B^2 / 2}{\partial(\partial A_1 / \partial x_2)} = \frac{\partial A_1}{\partial x_2} - \frac{\partial A_2}{\partial x_1} = -B_3. \tag{4.40}$$

In contrast, L_e depends on the time derivative of A_1 only through E^2, as shown by (4.39). To use (4.8) one replaces g with L_e and w_i with A_1, then takes the derivative with respect to x_3 in the first relation in (4.40), the derivative with respect to x_2 in the second relation, and the derivative with respect to t of the last term in (4.39). In summary this yields

$$\frac{1}{\mu_0}\left(\frac{\partial B_3}{\partial x_2} - \frac{\partial B_2}{\partial x_3}\right) - \frac{\partial D_1}{\partial t} = J_1, \tag{4.41}$$

namely, the first component of the second equation in (4.19).

Fig. 4.1 The domain V used
for the solution of the
Helmholtz equation (4.43).
The three possible positions
of point $\mathbf{r}$ are shown: external
to V, internal to V, or on the
boundary S

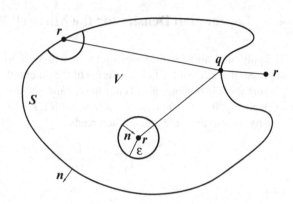

4.7   Helmholtz Equation

Consider the wave equations (4.35) and assume that the charge density ρ and current
density $\mathbf{J}$ are given as functions of position and time. In the following, the apex
in φ and $\mathbf{A}$ will be dropped for the sake of conciseness. The four scalar equations
(4.35) are linear with respect to the unknowns and have the same structure; also,
their coefficients and unknowns are all real. The solution of (4.35) will be tackled
in this section and in the following ones, basing upon the Fourier transform whose
general properties are depicted in Sect. C.2. This solution procedure involves the use
of complex functions. The starting assumption is that the condition for the existence
of the Fourier transform with respect to time holds (such a condition is found by
replacing x with t in (C.19)). Then one obtains

$$\nabla^2 \mathcal{F}_t \varphi + \frac{\omega^2}{c^2}\, \mathcal{F}_t \varphi = -\frac{1}{\varepsilon_0}\, \mathcal{F}_t \rho, \qquad \nabla^2 \mathcal{F}_t \mathbf{A} + \frac{\omega^2}{c^2}\, \mathcal{F}_t \mathbf{A} = -\mu_0\, \mathcal{F}_t \mathbf{J}. \qquad (4.42)$$

Indicating with f the transform of φ or A_i, and with b the transform of $-\rho/\varepsilon_0$ or
$-\mu_0 J_i$, $i = 1, 2, 3$, and letting $k^2 = \omega^2/c^2$, each scalar equation in (4.42) has the
form of the *Helmholtz equation*

$$\nabla^2 f + k^2 f = b. \qquad (4.43)$$

The solution of (4.43) is sought within a finite domain V (Fig. 4.1), for a given
set of boundary conditions defined over the boundary S of V, and for a given right
hand side b defined within V and over S. Let $\mathbf{r} = (x, y, z)$ be a point external to V,
$\mathbf{q} = (\xi, \eta, \zeta)$ a point internal to V, and

$$\mathbf{g} = \mathbf{r} - \mathbf{q}, \qquad g = \left[(x - \xi)^2 + (y - \eta)^2 + (z - \zeta)^2\right]^{1/2} \qquad (4.44)$$

where, by construction, it is $g > 0$. In the following calculation, $\mathbf{r}$ is kept fixed while
$\mathbf{q}$ varies. As a consequence, the derivatives of g are calculated with respect to ξ, η,
and ζ. It is easily shown that in a three-dimensional space the auxiliary function

$$G(g) = \frac{1}{g}\, \exp(-i\,k\,g), \qquad k \text{ real}, \qquad (4.45)$$

fulfills the homogeneous Helmholtz equation $\nabla^2 G + k^2 G = 0$. Using the procedure that leads to the second Green theorem (Sect. A.5, Eq. (A.25)) yields the integral relation

$$\int_S \left(G \frac{\partial f}{\partial n} - f \frac{\partial G}{\partial n} \right) dS = \int_V G b \, dV, \tag{4.46}$$

where the unit vector $\mathbf{n}$ over S is oriented in the outward direction and, by construction, point $\mathbf{r}$ is external to V.

4.8 Helmholtz Equation in a Finite Domain

The relation (4.46) would not be applicable if $\mathbf{r}$ were internal to V, because G diverges for $g \to 0$ and, as a consequence, is not differentiable in $\mathbf{q} = \mathbf{r}$. On the other hand, in many applications $\mathbf{r}$ happens to be internal to V. In such cases one must exclude from the integral a suitable portion of volume V; this is achieved by considering a spherical domain of radius ε centered on $\mathbf{r}$ and internal to V (Fig. 4.1). Letting V_ε, S_ε be, respectively, the volume and surface of such a sphere, and considering the new volume $V' = V - V_\varepsilon$, having $S' = S \cup S_\varepsilon$ as boundary, makes (4.46) applicable to V', to yield

$$\int_S \left(G \frac{\partial f}{\partial n} - f \frac{\partial G}{\partial n} \right) dS + \int_{S_\varepsilon} (\ldots) \, dS_\varepsilon = \int_V G b \, dV - \int_{V_\varepsilon} G b \, dV_\varepsilon, \tag{4.47}$$

where the dots indicate that the integrand is the same as in the first integral at the left hand side. Over S_ε it is $G = (1/\varepsilon) \exp(-ik\,\varepsilon)$, with the unit vector $\mathbf{n}$ pointing from the surface towards the center of the sphere, namely, opposite to the direction along which ε increases. It follows $\partial G / \partial n = -\partial G / \partial \varepsilon = (ik + 1/\varepsilon)\, G$. Letting $[f]$ and $[\partial f / \partial n]$ be the average values of f and, respectively, $\partial f / \partial n$ over S_ε, and observing that G and $\partial G / \partial \varepsilon$ are constant there, yields

$$\int_{S_\varepsilon} \left(G \frac{\partial f}{\partial n} - f \frac{\partial G}{\partial n} \right) dS_\varepsilon = 4\pi \, \exp(-i\,k\,\varepsilon) \left(\varepsilon \left[\frac{\partial f}{\partial n} \right] - (1 + i\,k\varepsilon)[f] \right). \tag{4.48}$$

As for the integral $I = \int_{V_\varepsilon} G b \, dV_\varepsilon$ it is useful to adopt the spherical coordinates (B.1) after shifting the origin to the center of the sphere. In the new reference it is $\mathbf{r} = 0$, so that the radial coordinate coincides with g. It follows

$$I = \int_0^\varepsilon \int_0^\pi \int_0^{2\pi} g \sin \vartheta \exp(-i\,k\,g)\, b(g, \vartheta, \varphi) \, dg \, d\vartheta \, d\varphi. \tag{4.49}$$

Taking the absolute value of I and observing that g and $\sin \vartheta$ are positive yields $|I| \leq 2\pi\, \varepsilon^2 \sup_{V_\varepsilon} |b|$. To proceed, one assumes that f and b are sufficiently smooth as to fulfill the conditions

$$\lim_{\varepsilon \to 0} \varepsilon\, [f] = 0, \qquad \lim_{\varepsilon \to 0} \varepsilon \left[\frac{\partial f}{\partial n} \right] = 0, \qquad \lim_{\varepsilon \to 0} \varepsilon^2 \sup_{V_\varepsilon} |b| = 0. \tag{4.50}$$

Thanks to (4.50) one restores the original volume V by taking the limit $\varepsilon \to 0$. Observing that $\lim_{\varepsilon \to 0} [f] = f(\mathbf{r})$, one finds

$$4\pi\, f(\mathbf{r}) = \int_S \left(G\, \frac{\partial f}{\partial n} - f\, \frac{\partial G}{\partial n} \right) \mathrm{d}S - \int_V G b\, \mathrm{d}V, \qquad (4.51)$$

that renders $f(\mathbf{r})$ as a function of b, of the boundary values of f and $\partial f/\partial n$, and of the auxiliary function G. It is easily found that, if $\mathbf{r}$ were on the boundary S instead of being internal to V, the left hand side of (4.51) would be replaced by $2\pi\, f(\mathbf{p})$. Similarly, if $\mathbf{r}$ were external to V, the left hand side would be zero. In conclusion one generalizes (4.51) to

$$\omega_r\, f(\mathbf{r}) = \int_S \left(G\, \frac{\partial f}{\partial n} - f\, \frac{\partial G}{\partial n} \right) \mathrm{d}S - \int_V G b\, \mathrm{d}V, \qquad (4.52)$$

where ω_r is the solid angle under which the surface S is seen from $\mathbf{r}$ considering the orientation of the unit vector $\mathbf{n}$. Namely, $\omega_r = 0$, $\omega_r = 2\pi$, or $\omega_r = 4\pi$ when $\mathbf{r}$ is external to V, on the boundary of V, or internal to V, respectively.

Letting $k = 0$ in (4.45), namely, taking $G = 1/g$, makes the results of this section applicable to the Poisson equation $\nabla^2 f = b$. It must be noted that (4.52) should be considered as an integral relation for f, not as the solution of the differential equation whence it derives. In fact, for actually calculating (4.52) it is necessary to prescribe both f and $\partial f/\partial n$ over the boundary. This is an overspecification of the problem: in fact, the theory of boundary-value problems shows that the solution of an equation of the form (4.43) is found by specifying over the boundary either the unknown function only (*Dirichlet boundary condition*), or its normal derivative only (*Neumann boundary condition*). To find a solution starting from (4.52) it is necessary to carry out more steps, by which either f or $\partial f/\partial n$ is eliminated from the integral at the right hand side [53, Sect. 1.8–1.10]. In contrast, when the solution is sought in a domain whose boundary extends to infinity, and the contribution of the boundary conditions vanish as shown in Sect. 4.9, the limiting case of (4.52) provides a solution proper. More comments about this issue are made in Sect. 4.12.3.

4.9   Solution of the Helmholtz Equation in an Infinite Domain

The procedure shown in Sect. 4.8 is readily extended to the case $V \to \infty$. Here one may replace S with a spherical surface of radius $R \to \infty$, centered on $\mathbf{r}$; this makes the calculation of the integral over S similar to that over the sphere of radius ε outlined in Sect. 4.8, the only difference being that the unit vector $\mathbf{n}$ now points in the direction where R increases. Shifting the origin to $\mathbf{r}$ and observing that $\omega_r = 4\pi$ yields

$$\int_S \left(G\, \frac{\partial f}{\partial n} - f\, \frac{\partial G}{\partial n} \right) \mathrm{d}S = 4\pi\, \exp(-\mathrm{i}\, k\, R) \left(R \left[\frac{\partial f}{\partial n} \right] + (1 + \mathrm{i}\, k\, R)[f] \right),$$

$$(4.53)$$

where the averages are calculated over S. To proceed one assumes that the following relations hold,

$$\lim_{R\to\infty} [f] = 0, \qquad \lim_{R\to\infty} R \left(\left[\frac{\partial f}{\partial n} \right] + ik\,[f] \right) = 0, \tag{4.54}$$

that are called *Sommerfeld asymptotic conditions*. Due to (4.54) the surface integral (4.53) vanishes. Shifting the origin back from $\mathbf{r}$ to the initial position, the solution of the Helmholtz equation (4.43) over an infinite domain finally reads

$$f(\mathbf{r}) = -\frac{1}{4\pi} \int_\infty b(\mathbf{q})\, \frac{\exp\left(-ik\,|\mathbf{r} - \mathbf{q}|\right)}{|\mathbf{r} - \mathbf{q}|}\, d^3 q, \tag{4.55}$$

where $\int_\infty$ indicates the integral over the whole three-dimensional $\mathbf{q}$ space. The $k = 0$ case yields the solution of the Poisson equation $\nabla^2 f = b$ in an infinite domain,

$$f(\mathbf{r}) = -\frac{1}{4\pi} \int_\infty b(\mathbf{q})\, \frac{1}{|\mathbf{r} - \mathbf{q}|}\, d^3 q. \tag{4.56}$$

4.10 Solution of the Wave Equation in an Infinite Domain

The solutions of the Helmholtz equation found in Sects. 4.8, 4.9 allow one to calculate that of the wave equation. In fact, it is worth reminding that the Helmholtz Eq. (4.43) was deduced in Sec. 4.7 by Fourier transforming the wave equation (4.35) and *i)* letting f indicate the transform of the scalar potential φ or of any component A_i of the vector potential, *ii)* letting b indicate the transform of $-\rho/\varepsilon_0$ or $-\mu_0 J_i$, $i = 1, 2, 3$. As a consequence, f and b depend on the angular frequency ω besides the spatial coordinates. From the definition $k^2 = \omega^2/c^2$ one may also assume that both k and ω have the same sign, so that $k = \omega/c$. Considering for simplicity the case $V \to \infty$, applying (C.17) to antitransform (4.56), and interchanging the order of integrals yields

$$\mathcal{F}^{-1} f = -\frac{1}{4\pi} \int_\infty \frac{1}{g} \left[\frac{1}{\sqrt{2\pi}} \int_{-\infty}^{+\infty} b(\mathbf{q}, \omega) \exp\left[i\,\omega\,(t - g/c)\right] d\omega \right] d^3 q, \tag{4.57}$$

with $g = |\mathbf{r} - \mathbf{q}|$. Now, denote with a the antitransform of b, $a(\mathbf{q}, t) = \mathcal{F}^{-1} b$. It follows that the function between brackets in (4.57) coincides with $a(\mathbf{q}, t - g/c)$. As remarked above, when f represents φ, then a stands for $-\rho/\varepsilon_0$; similarly, when f represents a component of $\mathbf{A}$, then a stands for the corresponding component of $-\mu_0 \mathbf{J}$. In conclusion,

$$\varphi(\mathbf{r}, t) = \frac{1}{4\pi\,\varepsilon_0} \int_\infty \frac{\rho(\mathbf{q}, t - |\mathbf{r} - \mathbf{q}|/c)}{|\mathbf{r} - \mathbf{q}|}\, d^3 q, \tag{4.58}$$

$$\mathbf{A}(\mathbf{r}, t) = \frac{\mu_0}{4\pi} \int_\infty \frac{\mathbf{J}(\mathbf{q}, t - |\mathbf{r} - \mathbf{q}|/c)}{|\mathbf{r} - \mathbf{q}|}\, d^3 q, \tag{4.59}$$

that express the potentials in terms of the field sources ρ and $\mathbf{J}$, when the asymptotic behavior of the potentials fulfills the Sommerfeld conditions (4.54). The functions rendered by the antitransforms are real, as should be. Note that $|\mathbf{r} - \mathbf{q}|/c > 0$ is the time necessary for a signal propagating with velocity c to cross the distance $|\mathbf{r} - \mathbf{q}|$. As $t - |\mathbf{r} - \mathbf{q}|/c < t$, the above expressions of φ and $\mathbf{A}$ are called *retarded potentials*[4].

4.11 Lorentz Force

It has been assumed so far that the sources of the electromagnetic field, namely, charge density and current density, are prescribed functions of position and time. This is not necessarily so, because the charges are in turn acted upon by the electromagnetic field, so that their dynamics is influenced by it. Consider a test charge of value e immersed in an electromagnetic field described by the vectors $\mathbf{E}$, $\mathbf{B}$ generated by other charges. The force acting upon the test charge is the *Lorentz force* [4, Vol. I, Sect. 44]

$$\mathbf{F} = e\,(\mathbf{E} + \mathbf{u} \wedge \mathbf{B}), \tag{4.60}$$

where $\mathbf{u}$ is the velocity of the test charge and $\mathbf{E}$, $\mathbf{B}$ are independent of $\mathbf{u}$. The expression of the Lorentz force does not derive from assumptions separate from Maxwell's equations; in fact, it follows from Maxwell's equations and Special Relativity [109, 39]. The extension of (4.60) to the case of a number of point-like charges follows the same line as in Sect. 4.4: considering the charges belonging to a cell of volume ΔV centered at $\mathbf{r}$, one writes (4.60) for the jth charge and takes the sum over j, to find

$$\mathbf{f} = \frac{\sum'_j \mathbf{F}_j}{\Delta V} = \rho\,(\mathbf{E} + \mathbf{v} \wedge \mathbf{B}), \tag{4.61}$$

where ρ, $\mathbf{v}$ are defined in (4.21) and $\mathbf{f}$ is the force density ($[\mathbf{f}] = \mathrm{N}\,\mathrm{m}^{-3}$). The fields in (4.61) are calculated in $\mathbf{r}$ and t.

Consider a small time interval δt during which the charge contained within ΔV is displaced by $\delta \mathbf{r} = \mathbf{v}\,\delta t$. The work per unit volume exchanged between the charge and the electromagnetic field due to such a displacement is

$$\delta w = \mathbf{f} \cdot \delta \mathbf{r} = \rho\,(\mathbf{E} + \mathbf{v} \wedge \mathbf{B}) \cdot \mathbf{v}\,\delta t = \mathbf{E} \cdot \mathbf{J}\,\delta t, \qquad [w] = \mathrm{J}\,\mathrm{m}^{-3}, \tag{4.62}$$

---

[4] Expressions of φ and $\mathbf{A}$ obtained from (4.58, 4.59) after replacing $t - |\mathbf{r} - \mathbf{q}|/c$ with $t + |\mathbf{r} - \mathbf{q}|/c$ are also solutions of the wave equations (4.35). This is due to the fact that the Helmholtz equation (4.43) can also be solved by using G^* instead of G, which in turn reflects the time reversibility of the wave equation. However, the form with $t - |\mathbf{r} - \mathbf{q}|/c$ better represents the idea that an electromagnetic perturbation, that is present in $\mathbf{r}$ at the time t, is produced by a source acting in $\mathbf{q}$ at a time prior to t.

where (A.32) and (4.21) have been used. When the scalar product is positive, the charge acquires kinetic energy from the field, and vice versa. Letting $\delta t \to 0$ yields

$$\frac{\partial w}{\partial t} = \mathbf{E} \cdot \mathbf{J}, \tag{4.63}$$

where the symbol of partial derivative is used because (4.63) is calculated with $\mathbf{r}$ fixed.

4.12 Complements

4.12.1 Invariance of the Euler Equations

It has been shown in Sect. 4.2 that the Euler equations (4.8) are still fulfilled if the generating function g is replaced with $g' = g + \mathrm{div}\mathbf{h}$, where $\mathbf{h}$ is an arbitrary vector of length m whose entries depend on $\mathbf{w}$ and $\xi_1, \dots, \xi_m$, but not on the derivatives of $\mathbf{w}$. This property is a generalization of that illustrated in Sect. 1.2 with reference to a system of particles, where it was shown that the solutions $w_i(\xi)$ are invariant under addition to g of the total derivative of an arbitrary function that depends on $\mathbf{w}$ and ξ only.

4.12.2 Wave Equations for the E and B Fields

The Maxwell equations can be rearranged in the form of wave equations for the electric and magnetic fields. To this purpose, one takes the rotational of both sides of the second equation in (4.24). Using the first identity in (A.36) and the relation $\mathbf{D} = \varepsilon_0 \mathbf{E}$ provides $-\partial \mathrm{rot}\mathbf{B}/\partial t = \mathrm{rot}\,\mathrm{rot}\,\mathbf{E} = \mathrm{graddiv}(\mathbf{D}/\varepsilon_0) - \nabla^2\mathbf{E}$. Replacing $\mathrm{div}\mathbf{D}$ and $\mathrm{rot}\mathbf{H} = \mathrm{rot}\mathbf{B}/\mu_0$ from (4.19) and using $\varepsilon_0 \mu_0 = 1/c^2$ then yields

$$\nabla^2\mathbf{E} - \frac{1}{c^2}\frac{\partial^2\mathbf{E}}{\partial t^2} = \frac{1}{\varepsilon_0}\mathrm{grad}\rho + \mu_0\frac{\partial\mathbf{J}}{\partial t}. \tag{4.64}$$

Similarly, one takes the rotational of both sides of the second equation in (4.19). Using the relation $\mathbf{B} = \mu_0 \mathbf{H}$ provides $\varepsilon_0\,\partial \mathrm{rot}\mathbf{E}/\partial t + \mathrm{rot}\mathbf{J} = \mathrm{rot}\,\mathrm{rot}\,\mathbf{H} = \mathrm{graddiv}(\mathbf{B}/\mu_0) - \nabla^2\mathbf{H}$. Replacing $\mathrm{div}\mathbf{B}$ and $\mathrm{rot}\mathbf{E}$ from (4.24) yields

$$\nabla^2\mathbf{H} - \frac{1}{c^2}\frac{\partial^2\mathbf{H}}{\partial t^2} = -\mathrm{rot}\mathbf{J}. \tag{4.65}$$

4.12.3 Comments on the Boundary-Value Problem

Considering relation (4.52) derived in Sect. 4.8, one notes that the right hand side is made of the difference between two terms; the first one depends on the boundary values of f, $\partial f/\partial n$, but not on b, while the second one depends only on the values of b within V and over the boundary. In these considerations it does not matter whether point $\mathbf{r}$ is external to V, on the boundary of V, or internal to it. In latter case the two terms at the right hand side of (4.52) balance each other.

If b is replaced with a different function $\tilde{b}$, and thereby the value of the second integral changes, it is possible to modify the boundary values in such a way as to balance the variation of the second integral with that of the first one; as a consequence, $f(\mathbf{r})$ is left unchanged. A possible choice for the modified b is $\tilde{b} = 0$; by this choice one eliminates the data of the differential equation and suitably modifies the boundary values, leaving the solution unaffected. An observer placed at $\mathbf{r}$ would be unable to detect that the data have disappeared. The same process can also be carried out in reverse, namely, by eliminating the boundary values and suitably changing the data.

An example is given in Prob. 4.4 with reference to a one-dimensional Poisson equation where the original charge density differs from zero in a finite interval $[a, b]$. The charge density is removed and the boundary values at a are modified so that the electric potential φ is unaffected for $x \geq b$. Obviously φ changes for $a < x < b$ because both the charge density and boundary conditions are different, and also for $x \leq a$ because the boundary conditions are different.

Problems

4.1 Solve the one-dimensional Poisson equation $\mathrm{d}^2\varphi/\mathrm{d}x^2 = -\rho(x)/\varepsilon_0$, with ρ given, using the integration by parts to avoid a double integral. The solution is prescribed at $x = a$ while the first derivative is prescribed at $x = c$.

4.2 Let $c = a$ in the solution of Prob. 4.1 and assume that the charge density ρ differs from zero only in a finite interval $a \leq x \leq b$. Find the expression of φ for $x > b$ when both the solution and the first derivative are prescribed at $x = a$.

4.3 In Prob. 4.2 replace the charge density ρ with a different one, say, $\tilde{\rho}$. Discuss the conditions that leave the solution unchanged.

4.4 In Prob. 4.2 remove the charge density ρ and modify the boundary conditions at a so that the solution for $x > b$ is left unchanged.

4.5 Using the results of Probs. 4.2 and 4.3, and assuming that both M_0 and M_1 are different from zero, replace the ratio ρ/ε_0 with $\mu\, \delta(x-h)$ and find the parameters μ, h that leave M_0, M_1 unchanged. Noting that h does not necessarily belong to the interval $[a, b]$, discuss the outcome for different positions of h with respect to a.

Chapter 5
Applications of the Concepts of Electromagnetism

5.1 Introduction

This chapter provides a number of important applications of the concepts of Electromagnetism. The solution of the wave equation found in Chap. 4 is used to calculate the potentials generated by a point-like charge; this result is exploited later to analyze the decay of atoms in the frame of the classical model, due to the radiated power. Next, the continuity equations for the energy and momentum of the electromagnetic field are found. As an application, the energy and momentum of the electromagnetic field are calculated in terms of modes in a finite domain, showing that the energy of each mode has the same expression as that of a linear harmonic oscillator. The analysis is extended also to an infinite domain. The chapter is concluded by the derivation of the eikonal equation, leading to the approximation of Geometrical Optics, followed by the demonstration that the eikonal equation is generated by a variational principle, namely, the Fermat principle. The complements show the derivation of the fields generated by a point-like charge and the power radiated by it. It is found that the planetary model of the atom is inconsistent with electromagnetism because it contradicts the atom's stability. Finally, a number of analogies are outlined and commented between Mechanics and Geometrical Optics, based on the comparison between the Maupertuis and Fermat principles. The course of reasoning deriving from the comparison hints at the possibility that mechanical laws more general than Newton's law exist.

5.2 Potentials Generated by a Point-Like Charge

The calculation of φ and $\mathbf{A}$ based upon (4.58, 4.59) has the inconvenience that, as $\mathbf{q}$ varies over the space, it is necessary to consider the sources ρ and $\mathbf{J}$ at different time instants. This may be avoided by recasting a in the form

© Springer Science+Business Media New York 2015
M. Rudan, *Physics of Semiconductor Devices*,
DOI 10.1007/978-1-4939-1151-6_5

$$a(\mathbf{q}, t - |\mathbf{r} - \mathbf{q}|/c) = \int_{-\infty}^{+\infty} a(\mathbf{q}, t') \delta(t' - t + |\mathbf{r} - \mathbf{q}|/c) \, dt', \qquad (5.1)$$

and interchanging the integration over $\mathbf{q}$ in (4.58, 4.59) with that over t'. This procedure is particularly useful when the source of the field is a single point-like charge. Remembering (4.20), one replaces ρ and $\mathbf{J}$ with

$$\rho_c(\mathbf{q}, t') = e\, \delta\left(\mathbf{q} - \mathbf{s}(t')\right), \qquad \mathbf{J}_c(\mathbf{q}, t') = e\, \delta\left(\mathbf{q} - \mathbf{s}(t')\right) \mathbf{u}(t'), \qquad (5.2)$$

where e is the value of the point-like charge, $\mathbf{s} = \mathbf{s}(t')$ its trajectory, and $\mathbf{u}(t') = d\mathbf{s}/dt'$ its velocity. First, the integration over space fixes $\mathbf{q}$ at $\mathbf{s}' = \mathbf{s}(t')$, this yielding

$$\varphi(\mathbf{r}, t) = \frac{e}{4\pi\,\varepsilon_0} \int_{-\infty}^{+\infty} \frac{\delta[\beta(t')]}{|\mathbf{r} - \mathbf{s}'|} \, dt', \quad \mathbf{A}(\mathbf{r}, t) = \frac{e\,\mu_0}{4\pi} \int_{-\infty}^{+\infty} \frac{\delta[\beta(t')]\,\mathbf{u}(t')}{|\mathbf{r} - \mathbf{s}'|} \, dt', \qquad (5.3)$$

with $\beta(t') = t' - t + |\mathbf{r} - \mathbf{s}'|/c$. Next, the integration over t' fixes the latter to the value that makes the argument of δ to vanish. Such a value is the solution of

$$|\mathbf{r} - \mathbf{s}(t')| = c\,(t - t'), \qquad (5.4)$$

where t, $\mathbf{r}$, and the function $\mathbf{s}(t')$ are prescribed. As $|\mathbf{u}| < c$ it can be shown that the solution of (5.4) exists and is unique [68, Sect. 63]. Observing that the argument of δ in (5.3) is a function of t', to complete the calculation one must follow the procedure depicted in Sect. C.5, which involves the derivative

$$\frac{d\beta}{dt'} = 1 + \frac{1}{c} \frac{d|\mathbf{r} - \mathbf{s}'|}{dt'} = 1 + \frac{d[(\mathbf{r} - \mathbf{s}') \cdot (\mathbf{r} - \mathbf{s}')]}{2c\,|\mathbf{r} - \mathbf{s}'|\,dt'} = 1 - \frac{\mathbf{r} - \mathbf{s}'}{|\mathbf{r} - \mathbf{s}'|} \frac{\mathbf{u}}{c}. \qquad (5.5)$$

Then, letting $t' = \tau$ be the solution of (5.4) and $\dot{\beta} = (d\beta/dt')_{t'=\tau}$, one applies (C.57) to (5.3). The use of the absolute value is not necessary here, in fact one has $[(\mathbf{r} - \mathbf{s}')/|\mathbf{r} - \mathbf{s}'|] \cdot (\mathbf{u}/c) \le u/c < 1$, whence $|\dot{\beta}| = \dot{\beta}$. In conclusion one finds

$$\varphi(\mathbf{r}, t) = \frac{e/(4\pi\,\varepsilon_0)}{|\mathbf{r} - \mathbf{s}(\tau)| - (\mathbf{r} - \mathbf{s}(\tau)) \cdot \mathbf{u}(\tau)/c}, \qquad (5.6)$$

$$\mathbf{A}(\mathbf{r}, t) = \frac{e\,\mu_0/(4\pi)\,\mathbf{u}(\tau)}{|\mathbf{r} - \mathbf{s}(\tau)| - (\mathbf{r} - \mathbf{s}(\tau)) \cdot \mathbf{u}(\tau)/c}, \qquad (5.7)$$

that provide the potentials generated in $\mathbf{r}$ and at time t by a point-like charge that follows the trajectory $\mathbf{s} = \mathbf{s}(\tau)$. The relation between t and τ is given by $t = \tau + |\mathbf{r} - \mathbf{s}(\tau)|/c$, showing that $t - \tau$ is the time necessary for the electromagnetic perturbation produced by the point-like charge at $\mathbf{s}$ to reach the position $\mathbf{r}$. The expressions (5.6, 5.7) are called *Liénard and Wiechert potentials*. In the case $\mathbf{u} = 0$ they become

$$\varphi(\mathbf{r}) = \frac{e}{4\pi\,\varepsilon_0\,|\mathbf{r} - \mathbf{s}|}, \qquad \mathbf{A}(\mathbf{r}) = 0, \qquad (5.8)$$

$\mathbf{s} = \text{const}$, the first of which is the Coulomb potential. The fields $\mathbf{E}$, $\mathbf{B}$ generated by a point-like charge are obtained from (5.6, 5.7) using (4.26). The calculation is outlined in Sect. 5.11.1.

5.3 Energy Continuity—Poynting Vector

The right hand side of (4.63) is recast in terms of the fields by replacing $\mathbf{J}$ with the left hand side of the second equation in (4.19); using $\mathbf{D} = \varepsilon_0\,\mathbf{E}$,

$$\frac{\partial w}{\partial t} = \mathbf{E} \cdot \left(\text{rot}\mathbf{H} - \varepsilon_0 \frac{\partial \mathbf{E}}{\partial t} \right) = \mathbf{E} \cdot \text{rot}\mathbf{H} - \frac{\varepsilon_0}{2} \frac{\partial E^2}{\partial t}. \tag{5.9}$$

The above expression is given a more symmetric form by exploiting the first equation in (4.19). In fact, a scalar multiplication of the latter by $\mathbf{H}$ along with the relation $\mathbf{B} = \mu_0\,\mathbf{H}$ provides $0 = \mathbf{H} \cdot \text{rot}\mathbf{E} + \mu_0\,\partial(H^2/2)/\partial t$ which, subtracted from (5.9), finally yields

$$\frac{\partial w}{\partial t} = \mathbf{E} \cdot \text{rot}\mathbf{H} - \mathbf{H} \cdot \text{rot}\mathbf{E} - \frac{\partial w_{\text{em}}}{\partial t}, \qquad w_{\text{em}} = \frac{1}{2}\left(\varepsilon_0\,E^2 + \mu_0\,H^2 \right). \tag{5.10}$$

Then, using the second identity in (A.36) transforms (5.10) into

$$\frac{\partial}{\partial t}(w + w_{\text{em}}) + \text{div}\mathbf{S} = 0, \qquad \mathbf{S} = \mathbf{E} \wedge \mathbf{H}. \tag{5.11}$$

As $w + w_{\text{em}}$ is an energy density, $\mathbf{S}$ (called *Poynting vector*) is an energy-flux density ($[\mathbf{S}] = \text{J m}^{-2}\,\text{s}^{-1}$). To give w_{em} and $\mathbf{S}$ a physical meaning one notes that (5.11) has the form of a continuity equation (compare, e.g., with (23.3) and (4.23)) where two interacting systems are involved, namely, the charges and the electromagnetic field. Integrating (5.11) over a volume V yields

$$\frac{d}{dt}(W + W_{\text{em}}) = -\int_{\Sigma} \mathbf{S} \cdot \mathbf{n}\,d\Sigma, \qquad W = \int_V w\,dV, \qquad W_{\text{em}} = \int_V w_{\text{em}}\,dV, \tag{5.12}$$

where Σ is the boundary of V, $\mathbf{n}$ is the unit vector normal to $d\Sigma$ oriented in the outward direction with respect to V, and W, W_{em} are energies. If V is let expand to occupy all space, the surface integral in (5.12) vanishes because the fields $\mathbf{E}$, $\mathbf{H}$ vanish at infinity; it follows that for an infinite domain the sum $W + W_{\text{em}}$ is conserved in time, so that $dW_{\text{em}}/dt = -dW/dt$. Observing that W is the kinetic energy of the charges, and that the latter exchange energy with the electromagnetic field, gives W_{em} the meaning of energy of the electromagnetic field; as a consequence, w_{em} is the energy density of the electromagnetic field, and the sum $W + W_{\text{em}}$ is the constant energy of the two interacting systems.

When V is finite, the surface integral in (5.12) may be different from zero, hence the sum $W + W_{\text{em}}$ is not necessarily conserved. This allows one to give the surface integral the meaning of energy per unit time that crosses the boundary Σ, carried by the electromagnetic field. In this reasoning it is implied that, when V is finite, it is chosen in such a way that no charge is on the boundary at time t. Otherwise the kinetic energy of the charges crossing Σ during dt should also be accounted for.

5.4 Momentum Continuity

The procedure used in Sect. 5.3 to derive the continuity equation for the charge energy can be replicated to obtain the continuity equation for the charge momentum per unit volume, $\mathbf{m}$. Remembering (4.61) one finds the relation $\mathbf{f} = \dot{\mathbf{m}} = \sum'_j \dot{\mathbf{p}}_j / \Delta V$, with $\mathbf{p}_j$ the momentum of the jth charge contained within ΔV. Using (4.19) along with $\mathbf{J} = \rho \mathbf{v}$ yields

$$\dot{\mathbf{m}} = \rho \, \mathbf{E} + \mathbf{J} \wedge \mathbf{B} = \mathbf{E} \operatorname{div} \mathbf{D} + \left(\operatorname{rot} \mathbf{H} - \frac{\partial \mathbf{D}}{\partial t} \right) \wedge \mathbf{B}. \tag{5.13}$$

Adding $\mathbf{D} \wedge \partial \mathbf{B}/\partial t$ to both sides of (5.13), using $\partial \mathbf{B}/\partial t = -\operatorname{rot} \mathbf{E}$, and rearranging:

$$\dot{\mathbf{m}} + \frac{\partial \mathbf{D}}{\partial t} \wedge \mathbf{B} + \mathbf{D} \wedge \frac{\partial \mathbf{B}}{\partial t} = \mathbf{E} \operatorname{div} \mathbf{D} + (\operatorname{rot} \mathbf{H}) \wedge \mathbf{B} + (\operatorname{rot} \mathbf{E}) \wedge \mathbf{D}. \tag{5.14}$$

Poynting vector's definition (5.11) transforms the left hand side of (5.14) into $\dot{\mathbf{m}} + \varepsilon_0 \mu_0 \, \partial(\mathbf{E} \wedge \mathbf{H})/\partial t = \partial(\mathbf{m} + \mathbf{S}/c^2)/\partial t$. In turn, the kth component of $\mathbf{E} \operatorname{div} \mathbf{D} + (\operatorname{rot} \mathbf{E}) \wedge \mathbf{D}$ can be recast as

$$\varepsilon_0 \, (E_k \operatorname{div} \mathbf{E} + \mathbf{E} \cdot \operatorname{grad} E_k) - \frac{\varepsilon_0}{2} \frac{\partial E^2}{\partial x_k} = \varepsilon_0 \operatorname{div}(E_k \, \mathbf{E}) - \frac{\varepsilon_0}{2} \frac{\partial E^2}{\partial x_k}. \tag{5.15}$$

Remembering that $\operatorname{div} \mathbf{B} = 0$, the kth component of the term $(\operatorname{rot} \mathbf{H}) \wedge \mathbf{B} = \mathbf{H} \operatorname{div} \mathbf{B} + (\operatorname{rot} \mathbf{H}) \wedge \mathbf{B}$ is treated in the same manner. Adding up the contributions of the electric and magnetic parts and using the definition (5.10) of w_{em} yields

$$\frac{\partial}{\partial t} \left(m_k + \frac{1}{c^2} S_k \right) + \operatorname{div} \mathbf{T}_k = 0, \qquad \mathbf{T}_k = w_{\text{em}} \, \mathbf{i}_k - \varepsilon_0 \, E_k \, \mathbf{E} - \mu_0 \, H_k \mathbf{H}, \tag{5.16}$$

with $\mathbf{i}_k$ the unit vector of the kth axis. As $m_k + S_k/c^2$ is a momentum density, $\mathbf{T}_k$ is a momentum-flux density ($[\mathbf{T}_k] = \text{J m}^{-3}$). Following the same reasoning as in Sect. 5.3 one integrates (5.16) over a volume V, to find

$$\frac{\mathrm{d}}{\mathrm{d}t} \int_V \left(m_k + \frac{1}{c^2} S_k \right) \mathrm{d}V = - \int_\Sigma \mathbf{T}_k \cdot \mathbf{n} \, \mathrm{d}\Sigma, \tag{5.17}$$

where Σ and $\mathbf{n}$ are defined as in (5.12). If V is let expand to occupy all space, the surface integral in (5.17) vanishes because the fields $\mathbf{E}, \mathbf{H}$ vanish at infinity; it follows that for an infinite domain the sum

$$\int_V \left(\mathbf{m} + \frac{1}{c^2} \mathbf{S} \right) \mathrm{d}V = \mathbf{p} + \int_V \frac{1}{c^2} \mathbf{S} \, \mathrm{d}V, \qquad \mathbf{p} = \int_V \mathbf{m} \, \mathrm{d}V \tag{5.18}$$

is conserved in time. As $\mathbf{p}$ is the momentum of the charges, $\int_V \mathbf{S}/c^2 \, \mathrm{d}^3 r$ takes the meaning of momentum of the electromagnetic field within V. As a consequence, $\mathbf{S}/c^2$ takes the meaning of momentum per unit volume of the electromagnetic field.

Fig. 5.1 The domain used
for the expansion of the
vector potential into a Fourier
series (Sect. 5.5)

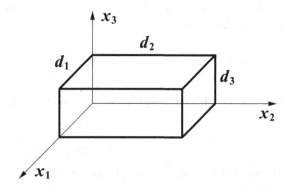

When V is finite, the surface integral in (5.17) may be different from zero, hence
the sum (5.18) is not necessarily conserved. This allows one to give the surface
integral in (5.17) the meaning of momentum per unit time that crosses the boundary
Σ, carried by the electromagnetic field. In this reasoning it is implied that, when V is
finite, it is chosen in such a way that no charge is on the boundary at time t. Otherwise
the momentum of the charges crossing Σ during dt should also be accounted for.

5.5   Modes of the Electromagnetic Field

The expressions of the energy and momentum of the electromagnetic field, worked
out in Sects. 5.3 and 5.4, take a particularly interesting form when a spatial region
free of charges is considered. In fact, if one lets $\rho = 0$, $\mathbf{J} = 0$, equations (4.33) or
(4.35) that provide the potentials become homogeneous. To proceed one takes a finite
region of volume V; the calculation will be extended in Sect. 5.8 to the case of an
infinite domain. As the shape of V is not essential for the considerations illustrated
here, it is chosen as that of a box whose sides d_1, d_2, d_3 are aligned with the coordinate
axes and start from the origin (Fig. 5.1). The volume of the box is $V = d_1 d_2 d_3$.

The calculation is based on (4.33), that are the equations for the potentials deriving
from the Coulomb gauge (4.32). Letting $\rho = 0$, $\mathbf{J} = 0$ and dropping the primes yields

$$\nabla^2\varphi = 0, \qquad \nabla^2\mathbf{A} - \frac{1}{c^2}\frac{\partial^2\mathbf{A}}{\partial t^2} = \frac{1}{c^2}\frac{\partial}{\partial t}\mathrm{grad}\varphi, \tag{5.19}$$

the first of which is a Laplace equation. It is shown in Sect. 5.11.4 that a gauge
transformation exists such that $\varphi = 0$ here. The system (5.19) then reduces to the
linear, homogeneous wave equation for the vector potential $\mathbf{A} = \mathbf{A}(\mathbf{r}, t)$,

$$\nabla^2\mathbf{A} - \frac{1}{c^2}\frac{\partial^2\mathbf{A}}{\partial t^2} = 0. \tag{5.20}$$

As the vector potential is defined within a finite volume and has a finite module as well, one can expand it into the Fourier series

$$\mathbf{A} = \sum_{\mathbf{k}} \mathbf{a}_{\mathbf{k}} \exp{(\mathrm{i}\,\mathbf{k}\cdot\mathbf{r})}, \qquad \mathbf{a}_{\mathbf{k}} = \frac{1}{V} \int_V \mathbf{A} \exp{(-\mathrm{i}\,\mathbf{k}\cdot\mathbf{r})} \mathrm{d}V, \qquad (5.21)$$

where $\mathbf{a}_{\mathbf{k}} = \mathbf{a}(\mathbf{k}, t)$ is complex and the *wave vector* $\mathbf{k}$ is given by

$$\mathbf{k} = n_1 \frac{2\pi}{d_1} \mathbf{i}_1 + n_2 \frac{2\pi}{d_2} \mathbf{i}_2 + n_3 \frac{2\pi}{d_3} \mathbf{i}_3, \qquad n_i = 0, \pm 1, \pm 2, \ldots \qquad (5.22)$$

The symbol $\sum_{\mathbf{k}}$ indicates a triple sum over all integers n_1, n_2, n_3. The definition of $\mathbf{a}_{\mathbf{k}}$ yields

$$\mathbf{a}_{-\mathbf{k}} = \mathbf{a}_{\mathbf{k}}^*, \qquad \mathbf{a}_0 = \frac{1}{V} \int_V \mathbf{A} \, \mathrm{d}V, \qquad (5.23)$$

with $\mathbf{a}_0$ real. Applying Coulomb's gauge $\mathrm{div}\mathbf{A} = 0$ to the expansion (5.21) provides

$$\mathrm{div}\mathbf{A} = \sum_{\mathbf{k}} \sum_{m=1}^{3} a_{\mathbf{k}m} \, \mathrm{i}\, k_m \exp{(\mathrm{i}\,\mathbf{k}\cdot\mathbf{r})} = \mathrm{i} \sum_{\mathbf{k}} \mathbf{a}_{\mathbf{k}} \cdot \mathbf{k} \exp{(\mathrm{i}\,\mathbf{k}\cdot\mathbf{r})} = 0, \qquad (5.24)$$

that is, a linear combination of functions of $\mathbf{r}$. As such functions are linearly independent from each other, (5.24) vanishes only if the coefficients vanish, so it is $\mathbf{a}_{\mathbf{k}} \cdot \mathbf{k} = 0$. Replacing $\mathbf{k}$ with $-\mathbf{k}$ and using (5.23) shows that $\mathbf{a}_{-\mathbf{k}} \cdot \mathbf{k} = \mathbf{a}_{\mathbf{k}}^* \cdot \mathbf{k} = 0$. In conclusion, $\mathbf{a}_{\mathbf{k}}$ has no components in the direction of $\mathbf{k}$, namely, it has only two independent (complex) components that lie on the plane normal to $\mathbf{k}$: letting $\mathbf{e}_1, \mathbf{e}_2$ be unit vectors belonging to such a plane and normal to each other, one has

$$\mathbf{a}_{\mathbf{k}} = \mathbf{a}_{\mathbf{k}} \cdot \mathbf{e}_1 \, \mathbf{e}_1 + \mathbf{a}_{\mathbf{k}} \cdot \mathbf{e}_2 \, \mathbf{e}_2. \qquad (5.25)$$

Clearly the reasoning above does not apply to $\mathbf{a}_0$; however, it is shown below that eventually this term does not contribute to the fields. The Fourier series (5.21) is now inserted into the wave equation (5.20), whose two summands become

$$\nabla^2 \mathbf{A} = \sum_{\mathbf{k}} \mathbf{a}_{\mathbf{k}} \sum_{m=1}^{3} (\mathrm{i}\, k_m)^2 \exp{(\mathrm{i}\,\mathbf{k}\cdot\mathbf{r})} = - \sum_{\mathbf{k}} \mathbf{a}_{\mathbf{k}} \, k^2 \exp{(\mathrm{i}k\cdot\mathbf{r})}, \qquad (5.26)$$

$$-\frac{1}{c^2} \frac{\partial^2 \mathbf{A}}{\partial t^2} = -\frac{1}{c^2} \sum_{\mathbf{k}} \ddot{\mathbf{a}}_{\mathbf{k}} \exp{(\mathrm{i}\,\mathbf{k}\cdot\mathbf{r})}. \qquad (5.27)$$

Adding up yields $\sum_{\mathbf{k}} \left(\ddot{\mathbf{a}}_{\mathbf{k}} + c^2 \, k^2 \, \mathbf{a}_{\mathbf{k}} \right) \exp{(\mathrm{i}\,\mathbf{k}\cdot\mathbf{r})} = 0$ whence, using the same reasoning as that used for discussing (5.24),

$$\ddot{\mathbf{a}}_{\mathbf{k}} + \omega^2 \, \mathbf{a}_{\mathbf{k}} = 0, \qquad \omega(\mathbf{k}) = c\, k \geq 0, \qquad \omega(-\mathbf{k}) = \omega(\mathbf{k}). \qquad (5.28)$$

The case $\mathbf{k} = 0$ yields $\ddot{\mathbf{a}}_0 = 0$ whence $\mathbf{a}_0(t) = \mathbf{a}_0(t = 0) + \dot{\mathbf{a}}_0(t = 0)\,t$. The constant $\dot{\mathbf{a}}_0(t = 0)$ must be set to zero to prevent $\mathbf{a}_0$ from diverging. When $\mathbf{k} \neq 0$ the solution of (5.28) is readily found to be $\mathbf{a_k}(t) = \mathbf{c_k} \exp(-\mathrm{i}\omega t) + \mathbf{c'_k} \exp(\mathrm{i}\omega t)$, where the complex vectors $\mathbf{c_k}, \mathbf{c'_k}$ depend on $\mathbf{k}$ only and lie on the plane normal to it. Using the first relation in (5.23) yields $\mathbf{c'_k} = \mathbf{c}^*_{-\mathbf{k}}$ and, finally,

$$\mathbf{a_k} = \mathbf{s_k} + \mathbf{s}^*_{-\mathbf{k}}, \qquad \mathbf{s_k}(t) = \mathbf{c_k} \exp(-\mathrm{i}\omega t), \qquad \mathbf{k} \neq 0. \tag{5.29}$$

Thanks to (5.29) one reconstructs the vector potential $\mathbf{A}$ in a form that shows its dependence on space and time explicitly. To this purpose one notes that the sum (5.21) contains all possible combinations of indices n_1, n_2, n_3, so that a summand corresponding to $\mathbf{k}$ is paired with another summand corresponding to $-\mathbf{k}$. One can then rearrange (5.21) as $\mathbf{A} = (1/2) \sum_{\mathbf{k}} \left[\mathbf{a_k} \exp(\mathrm{i}\mathbf{k} \cdot \mathbf{r}) + \mathbf{a_{-k}} \exp(-\mathrm{i}\mathbf{k} \cdot \mathbf{r}) \right]$, where the factor $1/2$ is introduced to eliminate a double counting. Using (5.29), and remembering from (5.28) that $\omega(\mathbf{k})$ is even, renders $\mathbf{A}$ as a sum of real terms,

$$\mathbf{A} = \sum_{\mathbf{k}} \Re \left\{ \mathbf{c_k} \exp[\mathrm{i}(\mathbf{k} \cdot \mathbf{r} - \omega t)] + \mathbf{c}^*_{-\mathbf{k}} \exp[\mathrm{i}(\mathbf{k} \cdot \mathbf{r} + \omega t)] \right\}, \tag{5.30}$$

The summands of (5.30) corresponding to $\mathbf{k}$ and $-\mathbf{k}$ describe two plane and monochromatic waves that propagate in the $\mathbf{k}$ and $-\mathbf{k}$ direction, respectively. The two waves together form a *mode* of the electromagnetic field, whose angular frequency is $\omega = ck$. The summands corresponding to $\mathbf{k} = 0$ yield the real constant $\mathbf{c}_0 + \mathbf{c}^*_0 = \mathbf{a}_0$. Finally, the $\mathbf{E}$ and $\mathbf{B}$ fields are found by introducing the expansion of $\mathbf{A}$ into (4.26) after letting $\varphi = 0$. For this calculation it is convenient to use the form of the expansion bearing the factor $1/2$ introduced above: from the definition (5.29) of $\mathbf{s_k}$ and the first identity in (A.35) one finds

$$\mathbf{E} = -\frac{\partial \mathbf{A}}{\partial t} = \frac{1}{2} \sum_{\mathbf{k}} \mathrm{i}\omega \left[\left(\mathbf{s_k} - \mathbf{s}^*_{-\mathbf{k}} \right) \exp(\mathrm{i}\mathbf{k} \cdot \mathbf{r}) + \left(\mathbf{s_{-k}} - \mathbf{s}^*_{\mathbf{k}} \right) \exp(-\mathrm{i}\mathbf{k} \cdot \mathbf{r}) \right], \tag{5.31}$$

$$\mathbf{B} = \mathrm{rot}\mathbf{A} = \frac{1}{2} \sum_{\mathbf{k}} \mathrm{i}\mathbf{k} \wedge \left[\left(\mathbf{s_k} + \mathbf{s}^*_{-\mathbf{k}} \right) \exp(\mathrm{i}\mathbf{k} \cdot \mathbf{r}) - \left(\mathbf{s_{-k}} + \mathbf{s}^*_{\mathbf{k}} \right) \exp(-\mathrm{i}\mathbf{k} \cdot \mathbf{r}) \right]. \tag{5.32}$$

As anticipated, the constant term $\mathbf{a}_0$ does not contribute to the fields. Also, due to the second relation in (5.29), the vectors $\mathbf{s_k}, \mathbf{s}^*_{-\mathbf{k}}, \mathbf{s_{-k}}$, and $\mathbf{s}^*_{\mathbf{k}}$ lie over the plane normal to $\mathbf{k}$. Due to (5.31, 5.32) the $\mathbf{E}$ and $\mathbf{B}$ fields lie on the same plane as well, namely, they have no component in the propagation direction. For this reason they are called *transversal*.

5.6   Energy of the Electromagnetic Field in Terms of Modes

The expressions of the $\mathbf{E}, \mathbf{B}$ fields within a finite volume V free of charges have been calculated in Sect. 5.5 as superpositions of modes, each of them associated with a wave vector $\mathbf{k}$ and an angular frequency $\omega = ck$. Basing upon such expressions one is able to determine the electromagnetic energy within V in terms of modes. To this purpose one calculates from (5.31, 5.32) the squares $E^2 = \mathbf{E} \cdot \mathbf{E}$ and $B^2 = \mathbf{B} \cdot \mathbf{B}$, inserts the resulting expression into the second relation of (5.10) to obtain the energy per unit volume and, finally, integrates the latter over V (last relation in (5.12)). Letting $\mathbf{I_k}$ be the quantity enclosed within brackets in (5.31), it is $E^2 = -(1/4) \sum_{\mathbf{k}} \sum_{\mathbf{k'}} \omega \omega' \, \mathbf{I_k} \cdot \mathbf{I_{k'}}$, where $\omega' = ck'$. The integration over V avails itself of the integrals (C.121), to yield

$$-\frac{1}{4} \sum_{k'} \omega \omega' \int_V \mathbf{I_k} \cdot \mathbf{I_{k'}} \, dV = V \omega^2 \left(\mathbf{s_k} - \mathbf{s_{-k}^*} \right) \cdot \left(\mathbf{s_k^*} - \mathbf{s_{-k}} \right), \tag{5.33}$$

so that the part of the electromagnetic energy deriving from $\mathbf{E}$ reads

$$\int_V \frac{\varepsilon_0}{2} E^2 \, dV = \frac{\varepsilon_0}{2} V \sum_{k} \omega^2 \left(\mathbf{s_k} - \mathbf{s_{-k}^*} \right) \cdot \left(\mathbf{s_k^*} - \mathbf{s_{-k}} \right). \tag{5.34}$$

By the same token one lets $\mathbf{Y_k}$ be the quantity enclosed within brackets in (5.32), whence $B^2 = -(1/4) \sum_{\mathbf{k}} \sum_{\mathbf{k'}} (\mathbf{k} \wedge \mathbf{Y_k}) \cdot (\mathbf{k'} \wedge \mathbf{Y_{k'}})$ and

$$-\frac{1}{4} \sum_{k'} \int_V (\mathbf{k} \wedge \mathbf{Y_k}) \cdot (\mathbf{k'} \wedge \mathbf{Y_{k'}}) \, dV = V \left[\mathbf{k} \wedge \left(\mathbf{s_k} + \mathbf{s_{-k}^*} \right) \right] \cdot \left[\mathbf{k} \wedge \left(\mathbf{s_{-k}} + \mathbf{s_k^*} \right) \right]. \tag{5.35}$$

The expression at the right hand side of (5.35) simplifies because, due to (5.29), $\mathbf{k}$ is normal to the plane where $\mathbf{s_k}$, $\mathbf{s_{-k}^*}$, $\mathbf{s_{-k}}$, and $\mathbf{s_k^*}$ lie, so that $[\mathbf{k} \wedge (\mathbf{s_k} + \mathbf{s_{-k}^*})] \cdot [\mathbf{k} \wedge (\mathbf{s_{-k}} + \mathbf{s_k^*})] = k^2 (\mathbf{s_k} + \mathbf{s_{-k}^*}) \cdot (\mathbf{s_{-k}} + \mathbf{s_k^*})$. Using the relation $k^2 = \omega^2/c^2 = \varepsilon_0 \mu_0 \omega^2$ yields the part of the electromagnetic energy deriving from $\mathbf{H} = \mathbf{B}/\mu_0$,

$$\int_V \frac{1}{2 \mu_0} B^2 \, dV = \frac{\varepsilon_0}{2} V \sum_{k} \omega^2 (\mathbf{s_k} + \mathbf{s_{-k}^*}) \cdot (\mathbf{s_{-k}} + \mathbf{s_k^*}). \tag{5.36}$$

Adding up (5.34) and (5.36) one finally obtains

$$W_{\text{em}} = \varepsilon_0 V \sum_{k} \omega^2 \left(\mathbf{s_k} \cdot \mathbf{s_k^*} + \mathbf{s_{-k}} \cdot \mathbf{s_{-k}^*} \right) = 2 \varepsilon_0 V \sum_{k} \omega^2 \mathbf{s_k} \cdot \mathbf{s_k^*}. \tag{5.37}$$

This result shows that the energy of the electromagnetic field within V is the sum of individual contributions, each associated to a wave vector $\mathbf{k}$ through the complex vector $\mathbf{s_k}$. As the latter lies on the plane normal to $\mathbf{k}$, it is expressed in terms of two scalar components as $\mathbf{s_k} = s_{k1} \mathbf{e_1} + s_{k2} \mathbf{e_2}$. Such components are related to the polarization of the electromagnetic field over the plane [9, Sect. 1.4.2]. These considerations allow one to count the number of indices that are involved in the representation (5.37) of W_{em}: in fact, the set of $\mathbf{k}$ vectors is described by the triple infinity of indices n_1, n_2, $n_3 \in \mathbf{Z}$ that appear in (5.22), while the two scalar components require another index $\sigma = 1, 2$. The $\mathbf{s_k}$ vectors describe the electromagnetic field through (5.31) and (5.32), hence one may think of each scalar component $s_{k\sigma}$ as a degree of freedom of the field; the counting outlined above shows that the number of degrees of freedom is $2 \times \mathbf{Z}^3$. In turn, each degree of freedom is made of a real and an imaginary part, $s_{\mathbf{k}\sigma} = R_{\mathbf{k}\sigma} + i\, I_{\mathbf{k}\sigma}$, this yielding

$$W_{\mathrm{em}} = \sum_{\mathbf{k}\sigma} W_{\mathbf{k}\sigma}, \qquad W_{\mathbf{k}\sigma} = 2\,\varepsilon_0\, V\, \omega^2 \left(R_{\mathbf{k}\sigma}^2 + I_{\mathbf{k}\sigma}^2 \right). \tag{5.38}$$

As $\omega = ck$, the mode with $\mathbf{k} = 0$ does not contribute to the energy. In (5.38) it is $R_{\mathbf{k}\sigma} = |c_{\mathbf{k}\sigma}|\cos[i\,\omega\,(t_0 - t)]$, $I_{\mathbf{k}\sigma} = |c_{\mathbf{k}\sigma}|\sin[i\,\omega\,(t_0 - t)]$, where the polar form $|c_{\mathbf{k}\sigma}|\exp(i\,\omega\,t_0)$ has been used for $c_{\mathbf{k}\sigma}$. One notes that each summand in (5.38) is related to a single degree of freedom and has a form similar to the Hamiltonian function of the linear harmonic oscillator discussed in Sect. 3.3. To further pursue the analogy one defines the new pair

$$q_{\mathbf{k}\sigma}(t) = 2\sqrt{\varepsilon_0\, V}\, R_{\mathbf{k}\sigma}, \qquad p_{\mathbf{k}\sigma}(t) = 2\,\omega\sqrt{\varepsilon_0\, V}\, I_{\mathbf{k}\sigma}, \tag{5.39}$$

whence

$$W_{\mathbf{k}\sigma} = \frac{1}{2}\left(p_{\mathbf{k}\sigma}^2 + \omega^2\, q_{\mathbf{k}\sigma}^2\right), \qquad \frac{\partial W_{\mathbf{k}\sigma}}{\partial p_{\mathbf{k}\sigma}} = p_{\mathbf{k}\sigma}, \qquad \frac{\partial W_{\mathbf{k}\sigma}}{\partial q_{\mathbf{k}\sigma}} = \omega^2\, q_{\mathbf{k}\sigma}. \tag{5.40}$$

On the other hand, the time dependence of $R_{\mathbf{k}\sigma}$, $I_{\mathbf{k}\sigma}$ is such that

$$\dot{q}_{\mathbf{k}\sigma} = p_{\mathbf{k}\sigma} = \frac{\partial W_{\mathbf{k}\sigma}}{\partial p_{\mathbf{k}\sigma}}, \qquad \dot{p}_{\mathbf{k}\sigma} = -\omega^2\, q_{\mathbf{k}\sigma} = -\frac{\partial W_{\mathbf{k}\sigma}}{\partial q_{\mathbf{k}\sigma}}. \tag{5.41}$$

Comparing (5.41) with (1.42) shows that $q_{\mathbf{k}\sigma}$, $p_{\mathbf{k}\sigma}$ are canonically-conjugate variables and $W_{\mathbf{k}\sigma}$ is the Hamiltonian function of the degree of freedom associated to $\mathbf{k}\sigma$. Then, comparing (5.40) with (3.1, 3.2) shows that $W_{\mathbf{k}\sigma}$ is indeed the Hamiltonian function of a linear harmonic oscillator of unit mass.

The energy associated to each degree of freedom is constant in time. In fact, from the second relation in (5.29) one derives $W_{\mathbf{k}\sigma} = 2\,\varepsilon_0\, V\, \omega^2\, \mathbf{c}_{\mathbf{k}\sigma} \cdot \mathbf{c}_{\mathbf{k}\sigma}^*$. The same result can be obtained from the properties of the linear harmonic oscillator (Sect. 3.3). It follows that the total energy W_{em} is conserved. As shown in Sect. 5.11.4 this is due to the periodicity of the Poynting vector (5.11): in fact, the electromagnetic energies the cross per unit time two opposite faces of the boundary of V are the negative of each other.

5.7 Momentum of the Electromagnetic Field in Terms of Modes

It has been shown in Sect. 5.4 that the momentum per unit volume of the electromagnetic field is $\mathbf{S}/c^2 = \mathbf{E} \wedge \mathbf{B}/(\mu_0\, c^2) = \varepsilon_0 \mathbf{E} \wedge \mathbf{B}$. Using the symbols defined in Sect. 5.6 one finds $\varepsilon_0 \mathbf{E} \wedge \mathbf{B} = -(\varepsilon_0/4) \sum_{\mathbf{k}} \sum_{\mathbf{k}'} \omega\, \mathbf{I}_{\mathbf{k}} \wedge (\mathbf{k}' \wedge \mathbf{Y}_{\mathbf{k}'})$ and

$$-\frac{\varepsilon_0}{4} \sum_{\mathbf{k}'} \int_V \omega\, \mathbf{I}_{\mathbf{k}} \wedge (\mathbf{k}' \wedge \mathbf{Y}_{\mathbf{k}'})\, \mathrm{d}V = \frac{\varepsilon_0}{2}\, \omega\, V\, (\mathbf{Z}_{\mathbf{k}} + \mathbf{Z}_{-\mathbf{k}}), \tag{5.42}$$

with $\mathbf{Z_k} = (\mathbf{s_k} - \mathbf{s}^*_{-\mathbf{k}}) \wedge [\mathbf{k} \wedge (\mathbf{s_{-k}} + \mathbf{s}^*_{\mathbf{k}})]$. The expression of $\mathbf{Z_k}$ simplifies because, due to (5.29), $\mathbf{k}$ is normal to the plane where $\mathbf{s_k}$, $\mathbf{s}^*_{-\mathbf{k}}$, $\mathbf{s_{-k}}$, and $\mathbf{s}^*_{\mathbf{k}}$ lie, so that $\mathbf{Z_k} = \mathbf{k}(\mathbf{s_k} - \mathbf{s}^*_{-\mathbf{k}}) \cdot (\mathbf{s_{-k}} + \mathbf{s}^*_{\mathbf{k}})$ and $\mathbf{Z_k} + \mathbf{Z_{-k}} = 2\,\mathbf{k}\,\mathbf{s_k} \cdot \mathbf{s}^*_{\mathbf{k}} + 2\,(-\mathbf{k})\,\mathbf{s_{-k}} \cdot \mathbf{s}^*_{-\mathbf{k}}$. In conclusion, observing that $\omega\,\mathbf{k} = (\omega^2/c)\,\mathbf{k}/k$,

$$\int_V \frac{\mathbf{S}}{c^2}\,\mathrm{d}V = 2\,\varepsilon_0\,V \sum_\mathbf{k} \omega\,\mathbf{s_k} \cdot \mathbf{s}^*_\mathbf{k}\,\mathbf{k} = \sum_{\mathbf{k}\sigma} \frac{1}{c}\,W_{\mathbf{k}\sigma}\,\frac{\mathbf{k}}{k} = \frac{1}{2c} \sum_{\mathbf{k}\sigma} \left(p^2_{\mathbf{k}\sigma} + \omega^2 q^2_{\mathbf{k}\sigma}\right) \frac{\mathbf{k}}{k},$$

(5.43)

where the last two equalities derive from (5.37, 5.38, 5.40). One notes from (5.43) that the momentum of the electromagnetic field is the sum of individual momenta, each related to a single degree of freedom. The modulus of the individual momentum is equal to the energy $W_{\mathbf{k}\sigma}$ pertaining to the same degree of freedom divided by c. The same relation between momentum and energy has been derived in Sect. 3.13.7 with reference to the dynamic relations of Special Relativity. Each summand in (5.43) is constant in time, so the electromagnetic momentum is conserved; as noted in Sects. 5.6, 5.11.4, this is due to the periodicity of the Poynting vector.

5.8 Modes of the Electromagnetic Field in an Infinite Domain

The treatment of Sects. 5.5, 5.6, 5.7 is extended to the case of an infinite domain by means of the Fourier transform (Sect. C.2)[1]

$$\mathbf{A} = \iiint_{-\infty}^{+\infty} \mathbf{b_k}\,\frac{\exp(\mathrm{i}\,\mathbf{k} \cdot \mathbf{r})}{(2\,\pi)^{3/2}}\,\mathrm{d}^3 k, \qquad \mathbf{b_k} = \iiint_{-\infty}^{+\infty} \mathbf{A}\,\frac{\exp(-\mathrm{i}\,\mathbf{k} \cdot \mathbf{r})}{(2\,\pi)^{3/2}}\,\mathrm{d}^3 r. \quad (5.44)$$

where $\mathbf{b_k} = \mathbf{b}(\mathbf{k}, t)$ is complex, with $\mathbf{b_{-k}} = \mathbf{b}^*_\mathbf{k}$, and the components of the wave vector $\mathbf{k}$ are continuous. Relations of the same form as (5.29) hold for $\mathbf{b_k}$, yielding

$$\mathbf{b_k} = \tilde{\mathbf{s}}_\mathbf{k} + \tilde{\mathbf{s}}^*_{-\mathbf{k}}, \qquad \tilde{\mathbf{s}}_\mathbf{k}(t) = \mathbf{d_k}\,\exp(-\mathrm{i}\,\omega\,t), \qquad \mathbf{k} \neq 0. \quad (5.45)$$

where the complex vector $\mathbf{d_k}$ depend on $\mathbf{k}$ only and lies on the plane normal to it. Relations similar to (5.30, 5.31, 5.32) hold as well, where $\mathbf{c_k}$ and $\mathbf{s_k}$ are replaced with $\mathbf{d_k}$ and $\tilde{\mathbf{s}}_\mathbf{k}$, respectively, and the sum is suitably replaced with an integral over $\mathbf{k}$. To determine the energy of the electromagnetic field one must integrate over the whole space the energy density w_{em}. Using (C.56) and the second relation in (5.45) one finds

$$W_{\mathrm{em}} = \iiint_{-\infty}^{+\infty} w_{\mathrm{em}}\,\mathrm{d}^3 r = 2\,\varepsilon_0 \iiint_{-\infty}^{+\infty} \omega^2\,\mathbf{d_k} \cdot \mathbf{d}^*_\mathbf{k}\,\mathrm{d}^3 k. \quad (5.46)$$

---

[1] For the existence of (5.44) it is implied that the three-dimensional equivalent of condition (C.19) holds.

It is sometimes useful to consider the frequency distribution of the integrand at the right hand side of (5.46). For this one converts the integral into spherical coordinates k, ϑ, γ and uses the relation $k = \omega/c = 2\pi \nu/c$ to obtain $\mathbf{d_k} = \mathbf{d}(\nu, \vartheta, \gamma)$; then, from (B.3),

$$W_{em} = \int_{-\infty}^{+\infty} U_{em}(\nu)\, d\nu, \quad U_{em} = \frac{2\varepsilon_0}{c^2} (2\pi \nu)^4 \int_0^\pi \int_0^{2\pi} |\mathbf{d}|^2 \sin\vartheta \, d\vartheta \, d\gamma,$$

(5.47)

where U_{em} (whose units are J s) is called *spectral energy* of the electromagnetic field. By a similar procedure one finds the total momentum, that reads

$$\iiint_{-\infty}^{+\infty} \frac{\mathbf{S}}{c^2}\, d^3r = 2\varepsilon_0 \iiint_{-\infty}^{+\infty} \omega |\mathbf{d}|^2 \mathbf{k}\, d^3k.$$

(5.48)

5.9 Eikonal Equation

Consider the case of a monochromatic electromagnetic field with angular frequency ω. For the calculation in hand it is convenient to consider the Maxwell equations in complex form; specifically, $\mathrm{rot}\mathbf{H} = \partial \mathbf{D}/\partial t$ yields, in vacuo,

$$\Re \left[(\mathrm{rot}\mathbf{H}_c + i\omega\varepsilon_0 \mathbf{E}_c) \exp(-i\omega t) \right] = 0,$$

(5.49)

while $\mathrm{rot}\mathbf{E} = -\partial \mathbf{B}/\partial t$ yields

$$\Re \left[(\mathrm{rot}\mathbf{E}_c - i\omega\mu_0 \mathbf{H}_c) \exp(-i\omega t) \right] = 0.$$

(5.50)

The solution of (5.49, 5.50) has the form $\mathbf{E}_c = \mathbf{E}_{c0} \exp(i\mathbf{k}\cdot\mathbf{r})$, $\mathbf{H}_c = \mathbf{H}_{c0} \exp(i\mathbf{k}\cdot\mathbf{r})$, with $\mathbf{E}_{c0}, \mathbf{H}_{c0} = $ const, i.e., a planar wave propagating along the direction of $\mathbf{k}$. In a non-uniform medium it is $\varepsilon = \varepsilon(\mathbf{r})$, $\mu = \mu(\mathbf{r})$, and the form of the solution differs from the planar wave. The latter can tentatively be generalized as

$$\mathbf{E}_c = \mathbf{E}_{c0}(\mathbf{r}) \exp[i k S(\mathbf{r})], \quad \mathbf{H}_c = \mathbf{H}_{c0}(\mathbf{r}) \exp[i k S(\mathbf{r})],$$

(5.51)

with $k = \omega\sqrt{\varepsilon_0\mu_0} = \omega/c$. Function $k S$ is called *eikonal* ($[S] = $ m). Replacing (5.51) into (5.49, 5.50) and using the first identity in (A.35) yields

$$\mathrm{grad}S \wedge \mathbf{H}_{c0} + c\varepsilon \mathbf{E}_{c0} = -\frac{c}{i\omega} \mathrm{rot}\mathbf{H}_{c0},$$

(5.52)

$$\mathrm{grad}S \wedge \mathbf{E}_{c0} - c\mu \mathbf{H}_{c0} = -\frac{c}{i\omega} \mathrm{rot}\mathbf{E}_{c0}.$$

(5.53)

Now it is assumed that ω is large enough to make the right hand side of (5.52, 5.53) negligible; in this case $\mathrm{grad}S$, $\mathbf{E}_{c0}$, $\mathbf{H}_{c0}$ become normal to each other. Vector multiplying (5.52) by $\mathrm{grad}S$ and using (5.53) then yields

$$\mathrm{grad}S \wedge (\mathrm{grad}S \wedge \mathbf{H}_{c0}) + \frac{\varepsilon\mu}{\varepsilon_0\mu_0} \mathbf{H}_{c0} = 0.$$

(5.54)

Remembering that $c = 1/\sqrt{\varepsilon_0 \mu_0}$ one defines the *phase velocity, refraction index,* and *wavelength* of the medium as

$$u_f(\mathbf{r}) = \frac{1}{\sqrt{\varepsilon \mu}}, \qquad n(\mathbf{r}) = \frac{c}{u_f}, \qquad \lambda(\mathbf{r}) = \frac{u_f}{\nu}, \tag{5.55}$$

respectively, so that $\varepsilon \mu/(\varepsilon_0 \mu_0) = n^2$. Using the first identity in (A.33) and remembering that $\operatorname{grad} S \cdot \mathbf{H}_{c0} = 0$ transforms (5.54) into $(|\operatorname{grad} S|^2 - n^2) \mathbf{H}_{c0} = 0$. As $\mathbf{H}_{c0} \neq 0$ it follows

$$|\operatorname{grad} S|^2 = n^2, \qquad n = n(\mathbf{r}), \tag{5.56}$$

that is, a partial-differential equation, called *eikonal equation,* in the unknown S. The equation has been derived in the hypothesis that $\omega = 2\pi \nu$ is large, hence $\lambda = u_f/\nu$ is small; it is shown below that this condition is also described by stating that $\mathbf{E}_{c0}(\mathbf{r})$, $\mathbf{H}_{c0}(\mathbf{r})$, and $S(\mathbf{r})$ vary little over a distance of the order of λ.

The form of (5.51) is such that $S(\mathbf{r}) = \text{const}$ defines the constant-phase surface (the same concept has been encountered in Sect. 2.5 for the case of a system of particles). It follows that the normal direction at each point $\mathbf{r}$ of the surface is that of $\operatorname{grad} S$. Let $\mathbf{t} = d\mathbf{r}/ds$ be the unit vector parallel to $\operatorname{grad} S$ in the direction of increasing S. A *ray* is defined as the envelope of the $\mathbf{t}$ vectors, taken starting from a point A in a given direction. The description of rays obtained through the approximation of the eikonal equation is called *Geometrical Optics.*

The eikonal equation (5.56) can be given a different form by observing that from the definition of $\mathbf{t}$ it follows $\operatorname{grad} S = n\mathbf{t}$ and $\mathbf{t} \cdot \operatorname{grad} S = dS/ds = n$, whence

$$\operatorname{grad} n = \operatorname{grad} \frac{dS}{ds} = \frac{d\operatorname{grad} S}{ds} = \frac{d(nt)}{ds} = \frac{d}{ds}\left(n\frac{d\mathbf{r}}{ds}\right). \tag{5.57}$$

This form of the eikonal equation is more often used. It shows that the equation is of the second order in the unknown function $\mathbf{r}(s)$, where $\mathbf{r}$ is the point of the ray corresponding to the curvilinear abscissa s along the ray itself. The equation's coefficient and data are given by the refraction index n. As the equation is of the second order, two boundary conditions are necessary to completely define the solution; for instance, the value of $\mathbf{r}(s = 0)$ corresponding to the initial point A, and the direction $\mathbf{t} = d\mathbf{r}/ds$ of the ray at the same point. Remembering that $d\mathbf{t}/ds = \mathbf{n}/\rho_c$, where ρ_c is the *curvature radius* of the ray at $\mathbf{r}$, and $\mathbf{n}$ the *principal normal unit vector,* the eikonal equation may also be recast as $\operatorname{grad} n = (dn/ds)\mathbf{t} + (n/\rho_c)\mathbf{n}$. Using the curvature radius one can specify in a quantitative manner the approximation upon which the eikonal equation is based; in fact, for the approximation to hold it is necessary that the electromagnetic wave can be considered planar, namely, that its amplitude and direction do not significantly change over a distance of the order of λ. This happens if at each point $\mathbf{r}$ along the ray it is $\rho_c \gg \lambda$.

5.10   Fermat Principle

It is worth investigating whether the eikonal equation (5.57) worked out in Sect. 5.9 is derivable from a variational principle. In fact it is shown below that the *Fermat* (or *least time*) principle holds, stating that, if A and B are two different points belonging to a ray, the natural ray (that is, the actual path followed by the radiation between the given points) is the one that minimizes the time $\int_{AB} \mathrm{d}t$. The principle thus reads

$$\delta \int_{AB} \mathrm{d}t = 0, \qquad (5.58)$$

where the integral is carried out along the trajectory. The analysis is similar to that carried out in Sect. 2.7 with reference to the Maupertuis principle.

Using the relations (5.55) and observing that $\mathrm{d}t = \mathrm{d}s/u_f = n\,\mathrm{d}s/c$ transforms (5.58) into $\delta \int_{AB} n\,\mathrm{d}s = 0$. Introducing a parametric description $\mathbf{r} = \mathbf{r}(\xi)$ of the ray, with $\xi = a$ when $\mathbf{r} = A$ and $\xi = b$ when $\mathbf{r} = B$, yields

$$\int_{AB} n\,\mathrm{d}s = \int_a^b g\,\mathrm{d}\xi, \qquad g = n\frac{\mathrm{d}s}{\mathrm{d}\xi} = n(x_1, x_2, x_3)\sqrt{\dot{x}_1^2 + \dot{x}_2^2 + \dot{x}_3^2}, \qquad (5.59)$$

$$\frac{\partial g}{\partial \dot{x}_i} = n\frac{2\dot{x}_i}{2\,\mathrm{d}s/\mathrm{d}\xi} = n\frac{\mathrm{d}x_i}{\mathrm{d}s}, \qquad \frac{\partial g}{\partial x_i} = \frac{\partial n}{\partial x_i}\frac{\mathrm{d}s}{\mathrm{d}\xi}. \qquad (5.60)$$

Remembering (1.7), the Euler equation for the ith coordinate reads

$$\frac{\mathrm{d}}{\mathrm{d}\xi}\left(n\frac{\mathrm{d}x_i}{\mathrm{d}s}\right) = \frac{\partial n}{\partial x_i}\frac{\mathrm{d}s}{\mathrm{d}\xi}, \qquad i = 1, 2, 3, \qquad (5.61)$$

whence

$$\frac{\mathrm{d}}{\mathrm{d}s}\left(n\frac{\mathrm{d}x_i}{\mathrm{d}s}\right) = \frac{\partial n}{\partial x_i}. \qquad (5.62)$$

As (5.62) is the ith component of the eikonal equation (5.57), such an equation is indeed derivable from the variational principle (5.58). Some comments about the formal analogy between the Maupertuis and Fermat principles are made in Sect. 5.11.6.

5.11   Complements

5.11.1   Fields Generated by a Point-Like Charge

The Liénard and Wiechert expressions (5.6, 5.7) provide the potentials generated in $\mathbf{r}$ at time t by a point-like charge that follows a trajectory $\mathbf{s}$. More specifically, if $\mathbf{s} = \mathbf{s}(\tau)$ is the position occupied by the charge at the instant τ, and $\mathbf{r}$ is the position where the potentials produced by the charge are detected at time $t > \tau$, the relation (5.4) holds, namely, $|\mathbf{r} - \mathbf{s}(\tau)| = c\,(t - \tau)$, that links the spatial coordinates with the time instants. Letting

$$\mathbf{g} = \mathbf{r} - \mathbf{s}(\tau), \qquad g = |\mathbf{g}|, \qquad \mathbf{u} = \frac{d\mathbf{s}}{d\tau}, \qquad \dot{\mathbf{u}} = \frac{d\mathbf{u}}{d\tau}, \qquad (5.63)$$

the fields $\mathbf{E}$, $\mathbf{B}$ are determined by applying (4.26) to (5.6, 5.7), which amounts to calculating the derivatives with respect to t and the components of $\mathbf{r}$. This is somewhat complicated because (4.26) introduces a relation of the form $\tau = \tau(\mathbf{r}, t)$, so that $\varphi = \varphi(\mathbf{r}, \tau(\mathbf{r}, t))$ and $\mathbf{A} = \mathbf{A}(\mathbf{r}, \tau(\mathbf{r}, t))$. It is therefore convenient to calculate some intermediate steps first. To this purpose, (5.4) is recast in implicit form as

$$\sigma(x_1, x_2, x_3, t, \tau) = \left[\sum_{i=1}^{3} (x_i - s_i(\tau))^2 \right]^{1/2} + c\,(\tau - t) = 0, \qquad (5.64)$$

whence $\mathrm{grad}\,\sigma = \mathbf{g}/g$, $\partial\sigma/\partial t = -c$, $\partial\sigma/\partial\tau = c - \mathbf{u}\cdot\mathbf{g}/g$. The differentiation rule of the implicit functions then yields

$$\frac{\partial\tau}{\partial t} = -\frac{\partial\sigma/\partial t}{\partial\sigma/\partial\tau} = \frac{c}{c - \mathbf{u}\cdot\mathbf{g}/g} \qquad (5.65)$$

Basing on (5.64, 5.65) and following the calculation scheme reported, e.g., in [96, Chap. 6] one obtains

$$\mathbf{E} = \frac{e\,(\partial\tau/\partial t)^3}{4\pi\,\varepsilon_0\,g^3} \left\{ \left(1 - \frac{u^2}{c^2}\right) \left(\mathbf{g} - g\,\frac{\mathbf{u}}{c}\right) + \mathbf{g} \wedge \left[\left(\mathbf{g} - g\,\frac{\mathbf{u}}{c}\right) \wedge \frac{\dot{\mathbf{u}}}{c^2} \right] \right\}, \qquad (5.66)$$

$$\mathbf{B} = \frac{\mathbf{g}}{g} \wedge \frac{\mathbf{E}}{c}. \qquad (5.67)$$

This result shows that $\mathbf{E}$ and $\mathbf{B}$ are the sum of two terms, the first of which decays at infinity like g^{-2}, while the second decays like g^{-1}. The latter term differs from zero only if the charge is accelerated ($\dot{\mathbf{u}} \neq 0$); its contribution is called *radiation field*. Also, $\mathbf{E}$ and $\mathbf{B}$ are orthogonal to each other, while $\mathbf{g}$ is orthogonal to $\mathbf{B}$ but not to $\mathbf{E}$; however, if g is large enough to make the second term in (5.66) dominant, $\mathbf{g}$ becomes orthogonal also to $\mathbf{E}$ and (5.67) yields $B = E/c$. In the case $\mathbf{u} = 0$ the relations (5.66, 5.67) simplify to

$$\mathbf{E} = \frac{e}{4\pi\,\varepsilon_0\,g^2}\,\frac{\mathbf{g}}{g}, \qquad \mathbf{B} = 0, \qquad (5.68)$$

that hold approximately also for $\mathbf{u} = \mathrm{const}$, $u/c \ll 1$.

5.11.2 Power Radiated by a Point-Like Charge

The expressions of the **E** and **B** fields worked out in Sect. 5.11.1 are readily exploited to determine the power radiated by a point-like charge. Remembering the results of Sect. 5.3, it suffices to integrate the Poynting vector over a surface Σ surrounding the charge. Introducing (5.67) into the definition (5.11) of Poynting's vector and using the first identity in (A.33) yields

$$\mathbf{S} = \frac{1}{\mu_0 c} \mathbf{E} \wedge \left(\frac{\mathbf{g}}{g} \wedge \mathbf{E} \right) = \frac{\varepsilon_0 c}{g} \left(E^2 \mathbf{g} - \mathbf{E} \cdot \mathbf{g} \, \mathbf{E} \right). \tag{5.69}'$$

The case $\dot{\mathbf{u}} \neq 0$, $u/c \ll 1$ is considered, which is typical of a bound particle. As the surface Σ can be chosen arbitrarily, it is convenient to select it at a large distance from the charge in order to make the second term in (5.66) dominant and **E** practically normal to **g**. This simplifies (5.66) and (5.69) to

$$\mathbf{S} \simeq \varepsilon_0 c E^2 \frac{\mathbf{g}}{g}, \qquad \mathbf{E} \simeq \frac{e}{4\pi \varepsilon_0 g} \left(\frac{\mathbf{g}}{g} \cdot \frac{\dot{\mathbf{u}}}{c^2} \frac{\mathbf{g}}{g} - \frac{\dot{\mathbf{u}}}{c^2} \right), \tag{5.70}$$

where the first identity in (A.33) has been used. Letting ϑ be the angle between **g** and $\dot{\mathbf{u}}$, one combines the two expressions in (5.70) to find

$$\mathbf{S} \simeq \varepsilon_0 c \, \mathbf{E} \cdot \mathbf{E} \frac{\mathbf{g}}{g} = \frac{1}{4\pi \varepsilon_0} \frac{e^2 \dot{u}^2}{4\pi c^3} \frac{\sin^2 \vartheta}{g^2} \frac{\mathbf{g}}{g}. \tag{5.71}$$

To proceed one chooses for Σ a spherical surface centered at $\mathbf{s}(\tau)$ and shifts the origin to its center. This yields $\mathbf{g} = \mathbf{r}$ at time τ, whence the unit vector normal to Σ becomes $\mathbf{n} = \mathbf{g}/g$. The radiation emitted by the charge reaches Σ at a later time $t = \tau + g/c$; however, thanks to the hypothesis $u/c \ll 1$, during the interval $t - \tau$ the charge moves little with respect to center of the sphere. For this reason, the surface integral can be calculated by keeping the charge fixed in the center of the spherical surface, so that the integral $\int_\Sigma (\sin^2 \vartheta / g^2) \, d\Sigma$ must be evaluated with $g = \text{const}$. Such an integral is easily found to equal $8\pi/3$: first, one turns to spherical coordinates and expresses the volume element as $J \, d\vartheta \, d\varphi \, dg = d\Sigma \, dg$; then, one finds from (B.3) the ratio $d\Sigma / g^2 = \sin \vartheta \, d\theta \, d\varphi$ and replaces it in the integral. In conclusion, combining the above result with (5.12, 5.71),

$$-\frac{d(W + W_{em})}{dt} = \frac{1}{4\pi \varepsilon_0} \frac{e^2 \dot{u}^2}{4\pi c^3} \int_\Sigma \frac{\sin^2 \vartheta}{g^2} \mathbf{n} \cdot \mathbf{n} \, d\Sigma = \frac{2 \, e^2 / 3}{4\pi \varepsilon_0 c^3} \dot{u}^2. \tag{5.72}$$

The expression at the right hand side of (5.72), called *Larmor formula*, gives an approximate expression of the power emitted by a point-like charge, that is applicable when $u/c \ll 1$. As shown by the left hand side, part of the emitted power $(-dW/dt)$ is due to the variation in the charge's mechanical energy, while the other part $(-dW_{em}/dt)$ is due to the variation in the electromagnetic energy within the volume enclosed by Σ.

5.11.3   Decay of Atoms According to the Classical Model

The power radiated by a point-like charge has been determined in Sect. 5.11.2 under the approximations $\dot{\mathbf{u}} \neq 0$, $u/c \ll 1$, typical of a bound particle. The radiated power (5.72) is proportional to the square of the particle's acceleration: this result is of a paramount importance, because it shows that the so-called *planetary model* of the atom is not stable. Considering for instance the simple case of hydrogen, the model describes the atom as a planetary system whose nucleus is fixed in the reference's origin while the electron orbits around it. If no power were emitted the motion's description would be that given, e.g., in Sect. 3.13.6, where the Hamiltonian function is a constant of motion. In other terms, the total energy would be conserved. In fact, in the planetary motion the electron's acceleration, hence the emitted power, differ from zero; the emission produces an energy loss which was not considered in the analysis of Sect. 3.13.6. Some comments about this problem were anticipated in Sect. 3.13.9.

To proceed it is useful to carry out a quantitative estimate of the emitted power. The outcome of it is that in the case of a bound electron the emission is relatively weak, so that one can consider it as a perturbation with respect to the conservative case analyzed in Sect. 3.13.6. The estimate starts from the experimental observation of the emission of electromagnetic radiation by excited atoms; here the datum that matters is the minimum angular frequency ω_0 of the emitted radiation, which is found to be in the range $[10^{15}, 10^{16}]$ rad s^{-1}. The simplest model for describing the unperturbed electron's motion is that of the linear harmonic oscillator [4, Vol. II, Sect. 4]

$$\mathbf{s}(\tau) = \mathbf{s}_0 \cos(\omega_0 \tau), \tag{5.73}$$

with $s_0 = |\mathbf{s}_0|$ the maximum elongation with respect to the origin, where the nucleus is placed. Equation (5.73) may be thought of as describing the projection over the direction of $\mathbf{s}_0$ of the instantaneous position of an electron that follows a circular orbit. The product $e\,\mathbf{s}$ is called *electric dipole moment* of the oscillator. Other experimental results, relative to the measure of the atom's size, show that s_0 is of the order of 10^{-10} m so that, calculating $\mathbf{u} = d\mathbf{s}/d\tau = -\mathbf{s}_0\,\omega_0 \sin(\omega_0\tau)$ from (5.73) and letting $\omega_0 = 5 \times 10^{15}$, one finds $u/c \leq s_0\,\omega_0/c \simeq 2 \times 10^{-3}$. This shows that the approximations of Sect. 5.11.2 are applicable.

It is worth noting that the type of motion (5.73) is energy conserving, hence it must be understood as describing the unperturbed dynamics of the electron. Remembering the discussion of Sect. 3.3 one finds for the total, unperturbed energy the expression $E_u = m\,\omega_0^2\,s_0^2/2$, with $m = 9.11 \times 10^{-31}$ kg the electron mass. To tackle the perturbative calculation it is now necessary to estimate the energy E_r lost by the electron during an oscillation period $2\pi/\omega_0$ and compare it with E_u. From $\dot{\mathbf{u}} = d\mathbf{u}/d\tau = \omega_0^2\,\mathbf{s}$ one obtains the maximum square modulus of the electron's acceleration, $\dot{u}_M^2 = \omega_0^4\,s_0^2$; inserting the latter into (5.72) and using $e = -q = -1.602 \times 10^{-19}$ C for the electron charge provides the upper bounds

$$E_r \leq \frac{2\pi}{\omega_0} \frac{2e^2/3}{4\pi \varepsilon_0 c^3} \omega_0^4 s_0^2 = \frac{e^2 s_0^2 \omega_0^3}{3\varepsilon_0 c^3}, \qquad \frac{E_r}{E_u} \leq \frac{2e^2\omega_0}{\varepsilon_0 m c^3} \simeq 4 \times 10^{-7}. \quad (5.74)$$

This result shows that the energy lost during an oscillation period is indeed small, so that the electron's motion is only slightly perturbed with respect to the periodic case. The equation of motion of the perturbed case can now tentatively be written as

$$m\ddot{\mathbf{s}} + m\omega_0^2 \mathbf{s} = \mathbf{F}_r, \quad (5.75)$$

where $\mathbf{F}_r$ is a yet unknown force that accounts for the emitted power. A scalar multiplication of (5.75) by $\mathbf{u}$ yields $m\,\mathbf{u} \cdot \dot{\mathbf{u}} + m\omega_0^2 \mathbf{s} \cdot \mathbf{u} = \mathrm{d}W/\mathrm{d}\tau = \mathbf{u} \cdot \mathbf{F}_r$, with $W = (m/2)(u^2 + \omega_0^2 s^2)$. One notes that W has the same expression as the total energy E_u of the unperturbed case; however, W is not conserved due to the presence of $\mathbf{F}_r \neq 0$ at the right hand side. In fact, $-\mathrm{d}W/\mathrm{d}\tau = -\mathbf{u} \cdot \mathbf{F}_r > 0$ is the power emitted by the electron, and its time average over $2\pi/\omega_0$,

$$-\langle \mathbf{u} \cdot \mathbf{F}_r \rangle = -\frac{\omega_0}{2\pi} \int_0^{2\pi/\omega_0} \mathbf{u} \cdot \mathbf{F}_r \, \mathrm{d}\tau > 0, \quad (5.76)$$

is the variation in the oscillator's energy during a period; a part of it crosses the surface Σ, while the other part is the variation in the electromagnetic energy within Σ (Sects. 5.3 and 5.11.2). The part that crosses Σ is the time average of (5.72);[2] for the sake of simplicity it is assumed that it is dominant with respect to the other one. The factor $\dot{u}^2$ that appears in (5.72) is worked out by taking the time average of the identity $\mathrm{d}(\mathbf{u} \cdot \dot{\mathbf{u}})/\mathrm{d}\tau = \dot{u}^2 + \mathbf{u} \cdot \ddot{\mathbf{u}}$ and observing that $\langle \mathrm{d}(\mathbf{u} \cdot \dot{\mathbf{u}})/\mathrm{d}\tau \rangle$ is negligibly small, whence $\langle \dot{u}^2 \rangle = -\langle \mathbf{u} \cdot \ddot{\mathbf{u}} \rangle > 0$. In conclusion, defining a time τ_0 such that $e^2/(6\pi \varepsilon_0 c^3) = m\tau_0$ and equating (5.76) to the time average of (5.72) yields $\langle \mathbf{u} \cdot m\tau_0 \ddot{\mathbf{u}} \rangle = \langle \mathbf{u} \cdot \mathbf{F}_r \rangle$. It is found $\tau_0 \simeq 6 \times 10^{-24}$ s.

As a crude approximation one finally converts the equality of the averages just found into an equality of the arguments, whence $\mathbf{F}_r \simeq m\tau_0 \ddot{\mathbf{u}}$. Replacing the latter into (5.75) yields $\ddot{\mathbf{s}} + \omega_0^2 \mathbf{s} = \tau_0 \dddot{\mathbf{u}}$, that is, a linear, homogeneous equation of the third order in $\mathbf{s}$ with constant coefficients. The equation is solved by letting $\mathbf{s} = \mathbf{s}(\tau = 0) \exp(\alpha\tau) \cos(\omega\tau)$, with α, ω undetermined. Using the tentative solution provides the system of characteristic algebraic equations

$$\tau_0 \omega^2 = 3\tau_0 \alpha^2 - 2\alpha, \qquad \alpha^2 + \omega_0^2 = \tau_0 \alpha^3 + (1 - 3\tau_0\alpha)\omega^2, \quad (5.77)$$

whence the elimination of ω^2 yields $8\alpha^2 - 2\alpha/\tau_0 - \omega_0^2 = 8\tau_0\alpha^3$. Thanks to the smallness of τ_0 the latter equation may be solved by successive approximations starting from the zeroth-order solution $\alpha^{(0)} \simeq -\tau_0 \omega_0^2/2$ (this solution is found by

---

[2] Remembering the discussion of Sect. 5.11.2, the use of (5.72) implies that the particle's position departs little from the center of the spherical surface. Thus the radius of Σ must be much larger than the size of the atom.

solving $8\alpha^2 - 2\alpha/\tau_0 - \omega_0^2 = 0$ and using the binomial approximation; the other possible value of $\alpha^{(0)}$ is positive and must be discarded to prevent s from diverging). Replacing $\alpha^{(0)}$ into the first equation in (5.77) yields $\omega^2 = \omega_0^2 (1 + 3\tau_0^2 \omega_0^2/4) \simeq \omega_0^2$. In conclusion, the zeroth-order solution of the differential equation for s reads

$$s(\tau) \simeq s(\tau = 0) \cos(\omega_0 \tau) \exp(-\tau_0 \omega_0^2 \tau/2). \tag{5.78}$$

Basing upon (5.78) one can identify the decay time of the atom with the time necessary for the modulus of s to reduce by a factor $1/e$ with respect to the initial value. The decay time is thus found to be $2/(\tau_0 \omega_0^2) \simeq 13 \times 10^{-9}$ s. As the ratio of the decay time to the period $2\pi/\omega_0$ is about 10^7, the perturbative approach is indeed justified.

As anticipated at the beginning of this section, the planetary model of the atom is not stable. The approximate solution (5.78) of the electron's dynamics shows that according to this model the electron would collapse into the nucleus in a very short time due to the radiation emitted by the electron. This behavior is not observed experimentally: in fact, the experiments show a different pattern in the energy-emission or absorption behavior of the atoms. The latter are able to absorb energy from an external radiation and subsequently release it: an absorption event brings the atom to a higher-energy state called *excited state*; the absorbed energy is then released by radiation in one or more steps (*emissions*) until, eventually, the atom reaches the lowest energy state (*ground state*). However, when the atom is in the ground state and no external perturbations is present, the atom is stable and no emission occurs. In conclusion, the experimental evidence shows that the planetary model is not applicable to the description of atoms.[3]

5.11.4 Comments about the Field's Expansion into Modes

The homogeneous wave equation (5.20) used in Sects. 5.5, 5.8 as a starting point for the derivation of the field's expansion into modes is based on the hypothesis that a gauge transformation exists such that $\varphi = 0$. In turn, (5.20) derives from (5.19), that implies the Coulomb gauge $\operatorname{div}\mathbf{A} = 0$. To show that these conditions are mutually compatible one chooses f in (4.30) such that $\varphi' = 0$, whence $\mathbf{E}' = -\partial\mathbf{A}'/\partial t$ due to the second relation in (4.26). In a charge-free space it is $\operatorname{div}\mathbf{D}' = \varepsilon_0 \operatorname{div}\mathbf{E}' = 0$; it follows $\partial\operatorname{div}\mathbf{A}'/\partial t = 0$, namely, $\operatorname{div}\mathbf{A}'$ does not depend on time. The second equation in (4.19) with $\mathbf{J} = 0$ yields $(1/c^2) \partial\mathbf{E}'/\partial t = \operatorname{rot}\mathbf{B}'$, so that $-(1/c^2) \partial^2\mathbf{A}'/\partial t^2 = \operatorname{rot}\operatorname{rot}\mathbf{A}'$. Now let $\mathbf{A}' = \mathbf{A}'' + \operatorname{grad}g$, where g is an arbitrary function of the coordinates only; the second identity in (A.35) and the first identity in (A.36) then yield $-(1/c^2) \partial^2\mathbf{A}''/\partial t^2 = \operatorname{grad}\operatorname{div}\mathbf{A}'' - \nabla^2\mathbf{A}''$, with $\operatorname{div}\mathbf{A}'' = \operatorname{div}\mathbf{A}' -$

---

[3] In this respect one might argue that the inconsistency between calculation and experiment is due to some flaw in the electromagnetic equations. However, other sets of experiments show that it is not the case.

$\nabla^2 g$. Choosing g such that $\text{div}\mathbf{A}'' = 0$ and dropping the double apex finally yields (5.20) [68, Sect. 46].

The vector potential $\mathbf{A}$ has been expressed in (5.21) as a Fourier series and in (5.44) as a Fourier integral. Such expressions produce a separation of the spatial coordinates from the time coordinate: the former appear only in the terms $\exp(i\,\mathbf{k}\cdot\mathbf{r})$, while the latter appears only in the terms $\mathbf{a_k}$ and, respectively, $\mathbf{b_k}$.

The Fourier series (5.21) applies to the case of a finite domain of the form shown in Fig. 5.1 and prescribes the spatial periodicity of $\mathbf{A}$ at all times. By way of example, let $0 \le x_2 \le d_2, 0 \le x_3 \le d_3$ and consider the point $\mathbf{r}_A = x_2\,\mathbf{i}_2 + x_3\,\mathbf{i}_3$; then, consider a second point $\mathbf{r}_B = d_1\,\mathbf{i}_1 + x_2\,\mathbf{i}_2 + x_3\,\mathbf{i}_3$. By construction, $\mathbf{r}_A$ and $\mathbf{r}_B$ belong to two opposite faces of the domain of Fig. 5.1 and are aligned with each other in the x_1 direction. From (5.22) one obtains for any $\mathbf{k}$

$$\exp(i\,\mathbf{k}\cdot\mathbf{r}_B) = \exp(i\,2\,\pi\,n_1)\,\exp(i\,\mathbf{k}\cdot\mathbf{r}_A) = \exp(i\,\mathbf{k}\cdot\mathbf{r}_A) \qquad (5.79)$$

which, combined with (5.21), yields $\mathbf{A}(\mathbf{r}_B,t) = \mathbf{A}(\mathbf{r}_A,t)$. Clearly an equality of this form is found for any pair of opposite boundary points that are aligned along the coordinate direction normal to the faces where the points lie. On the other hand, such an equality is a homogeneous relation among the boundary values of the solution of the differential equation (5.20), namely, it is a homogeneous boundary condition of the Dirichlet type.

The reasoning based on (5.79) is applicable also to the expressions (5.31, 5.32) to yield $\mathbf{E}(\mathbf{r}_B,t) = \mathbf{E}(\mathbf{r}_A,t)$ and $\mathbf{B}(\mathbf{r}_B,t) = \mathbf{B}(\mathbf{r}_A,t)$, namely, the fields have the same periodicity as $\mathbf{A}$. The Poynting vector $\mathbf{S} = \mathbf{E} \wedge \mathbf{H}$ has this property as well, whence $\mathbf{S}(\mathbf{r}_B,t)\cdot\mathbf{n}_B = -\mathbf{S}(\mathbf{r}_A,t)\cdot\mathbf{n}_A$; in fact, the unit vector $\mathbf{n}$ is oriented in the outward direction with respect to the domain (Sect. 5.3), so that when two opposite faces of V are considered it is $\mathbf{n}_B = -\mathbf{n}_A$. Using (5.12) with $W = 0$ shows that $dW_{\text{em}}/dt = 0$ namely, as noted in Sect. 5.6, the electromagnetic energy within V is conserved. The same reasoning applies to the conservation of the electromagnetic momentum found in Sect. 5.7. As for the initial condition on $\mathbf{A}$, from (5.21) and (5.29) one derives $\mathbf{A}(\mathbf{r},t = 0) = \sum_{\mathbf{k}}(\mathbf{c_k} + \mathbf{c^*_{-k}})\exp(i\,\mathbf{k}\cdot\mathbf{r})$. It follows that the initial condition is provided by the vectors $\mathbf{c_k}$.

5.11.5 Finiteness of the Total Energy

The differential equation (5.20) is linear and homogeneous with respect to the unknown $\mathbf{A}$; when the Fourier series (5.21) is replaced in it, the resulting equation (5.28) is linear and homogeneous with respect to $\mathbf{a_k}$, hence (due to (5.29)) with respect to $\mathbf{s_k}$ and $\mathbf{c_k}$ as well. It follows that the fields (5.31, 5.32) are linear and homogeneous functions of these quantities. The same applies in the case of an infinite domain (Sect. 5.8), in which the fields $\mathbf{E}$, $\mathbf{B}$ are linear and homogeneous functions of $\tilde{\mathbf{s}}_{\mathbf{k}}$ and $\mathbf{d_k}$.

In turn, the energy density w_{em} of the electromagnetic field, given by the second relation in (5.11), is a quadratic and homogeneous function of the fields; this explains why the expressions (5.37) and (5.46) are quadratic and homogeneous functions of s_k, c_k or, respectively, $\tilde{s}_k$, d_k.

When (5.37) is considered, the energy associated to the individual degree of freedom is $W_{k\sigma} = 2\,\varepsilon_0\,V\,\omega^2\,|c_{k\sigma}|^2$; as the sum $\sum_{k\sigma} W_{k\sigma}$ spans all wave vectors, the factor $\omega^2 = c^2\,k^2$ diverges. On the other hand, the energy of the electromagnetic field within a finite region of space is finite; this means that the term $|c_{k\sigma}|^2$ becomes vanishingly small as $|k| \to \infty$, in such a way as to keep the sum $\sum_{k\sigma} W_{k\sigma}$ finite. The same reasoning applies to the term $|d_{k\sigma}|^2$ in (5.46); in this case the finiteness of the total energy W_{em} is due to the fact that the vanishing of the fields at infinity makes the Fourier transform in (5.44) to converge.

5.11.6 Analogies between Mechanics and Geometrical Optics

A number of analogies exist between the Maupertuis principle, discussed in Sect. 2.7, and the Fermat principle discussed in Sect. 5.10. The principles read, respectively,

$$\delta \int_{AB} \sqrt{E - V}\, ds = 0, \qquad \delta \int_{AB} n\, ds = 0, \tag{5.80}$$

and the analogies are:

1. A constant parameter is present, namely, the total energy E on one side, the frequency ν on the other side (in fact, the Fermat principle generates the eikonal equation which, in turn, applies to a monochromatic electromagnetic field, Sect. 5.9).
2. Given the constant parameter, the integrand is uniquely defined by a property of the medium where the physical phenomenon occurs: the potential energy $V(\mathbf{r})$ and the refraction index $n(\mathbf{r})$, respectively.
3. The outcome of the calculation is a curve of the three-dimensional space: the particle's trajectory and the optical ray, respectively. In both cases the initial conditions are the starting position and direction (in the mechanical case the initial velocity is obtained by combining the initial direction with the momentum extracted from $E - V$).

In summary, by a suitable choice of the units, the same concept is applicable to both mechanical and optical problems. In particular it is used for realizing devices able to obtain a trajectory or a ray of the desired form: the control of the ray's shape is achieved by prescribing the refraction index n by means of, e.g., a lens or a system of lenses; similarly, the trajectory of a particle of charge e is controlled by a set of electrodes (*electrostatic lenses*) that prescribe the electric potential $\varphi = V/e$. The design of *electron guns* and of the equipments for *electron-beam lithography* and *ion-beam lithography* is based on this analogy.

It must be emphasized that the Maupertuis principle is derived without approximations: as shown in Sect. 2.7, the principle is equivalent to Newton's law applied to a particle of constant energy. The Fermat principle, instead, is equivalent to the eikonal equation; the latter, in turn, is derived from the Maxwell equations in the hypothesis that at each position along the ray the curvature radius of the ray is much larger than the wavelength. In other terms, the mechanical principle is exact whereas the optical principle entails an approximation. If the exact formulation of electro-magnetism given by the Maxwell equation were used, the analogy discussed here would be lost.

The rather surprising asymmetry outlined above could be fixed by speculating that Newton's law is in fact an approximation deriving from more general laws, possibly similar to the Maxwell equations. In this case one could identify in such laws a parameter analogue of the wavelength, and deduce Newton's as the limiting case in which the parameter is small. It will be shown later that mechanical laws more general than Newton's laws indeed exist: they form the object of Quantum Mechanics.[4]

The analogy between Mechanics and Geometrical Optics discussed here is one of the possible courses of reasoning useful for introducing the quantum-mechanical concepts; however, in this reasoning the analogy should not be pushed too far. In fact, one must observe that the Maupertuis principle given by the first expression in (5.80) provides the non-relativistic form of Newton's law, whereas the Maxwell equations, of which the Fermat principle is an approximation, are intrinsically relativistic. As a consequence, the analogy discussed in this section is useful for generalizing the geometrical properties of the motion, but not the dynamical properties.

Problems

5.1 Solve the eikonal equation (5.57) in a medium whose refraction index depends on one coordinate only, say, $n = n(x_1)$.

5.2 Use the solution of Problem 5.1 to deduce the Descartes law of refraction.

---

[4] Once Quantum Mechanics is introduced, Newtonian Mechanics is distinguished from it by the designation *Classical Mechanics*.

Part II
Introductory Concepts to Statistical and Quantum Mechanics

Chapter 6
Classical Distribution Function and Transport Equation

6.1 Introduction

When a system made of a large number of molecules is considered, the description of the dynamics of each individual member of the system is practically impossible, and it is necessary to resort to the methods of Statistical Mechanics. The chapter introduces the concept of distribution function in the phase space and provides the definition of statistical average (over the phase space and momentum space) of a dynamical variable. The derivation of the equilibrium distribution in the classical case follows, leading to the Maxwell–Boltzmann distribution. The analysis proceeds with the derivation of the continuity equation in the phase space: the collisionless case is treated first, followed by the more general case where the collisions are present, this leading to the Boltzmann Transport Equation. In the Complements, after a discussion about the condition of a vanishing total momentum and angular momentum in the equilibrium case, and the derivation of statistical averages based on the Maxwell-Boltzmann distribution, the Boltzmann H-theorem is introduced. This is followed by an illustration of the apparent paradoxes brought about by Boltzmann's Transport Equation and H-theorem: the violation of the symmetry of the laws of mechanics with respect to time reversal, and the violation of Poincaré's time recurrence. The illustration is carried out basing on Kac's ring model. The chapter is completed by the derivation of the equilibrium limit of the Boltzmann Transport Equation.

6.2 Distribution Function

Consider a system made of N identical particles with no constraints. For the sake of simplicity, point-like particles are assumed, so that the total number of degrees of freedom is $3N$. The dynamics of the jth particle is described by the canonical

© Springer Science+Business Media New York 2015

M. Rudan, *Physics of Semiconductor Devices,*

DOI 10.1007/978-1-4939-1151-6_6

coordinates $q_{1j}, q_{2j}, q_{3j}, p_{1j}, p_{2j}, p_{3j}$, that belong to the 6-dimensional μ-space introduced in Sect. 1.9.

If the number of particles is large, the description of the dynamics of each individual belonging to the system is in fact impossible. For instance, the number density of air at 20°C and 1 atm is about 2.5×10^{19} cm^{-3}. Even if the measurement of the initial conditions were possible, it would not be feasible in practice ([66], Sect. 1). This problem is present also when the number of particles is much lower than in the example above.

The problem is faced by adopting the viewpoint of *statistical mechanics*, whose object is not the description of the dynamical behavior of the individual particles but, rather, that of the distribution of the dynamical properties over the phase space. To this purpose one identifies each point of the μ-space with the pair of vectors $\mathbf{q} = (q_1, q_2, q_3)$, $\mathbf{p} = (p_1, p_2, p_3)$ pertaining to it, and considers the elementary volume $d\omega = d^3q \, d^3p$ of the μ-space centered at $(\mathbf{q}, \mathbf{p})$, with $d^3q = dq_1 \, dq_2 \, dq_3$ and the like for d^3p. Then, the number of particles dN that at time t belong to $d\omega$ is given by

$$dN = f_\mu(\mathbf{q}, \mathbf{p}, t) \, d\omega, \tag{6.1}$$

with f_μ the concentration in the μ-space. The procedure here is similar to that carried out in Sect. 23.2, the difference being that the space considered here is the phase space instead of the configuration space of Sect. 23.2.[1] In both cases, the motion of the particles is described as that of a continuous fluid: in fact, index j is dropped from the canonical coordinates, which do not indicate any more a specific particle, but the center of the elementary cell of the phase space where the concentration f_μ is considered. As in Sect. 23.2, this procedure is legitimate if the cells of volume $d\omega$ into which the phase space is partitioned can be treated as infinitesimal quantities in the scale of the problem that is being investigated, and the number of particles within each cell is large enough to make their average properties significant. The concentration f_μ is also called *distribution function*. By definition it fulfills the normalization condition

$$\int f_\mu(\mathbf{q}, \mathbf{p}, t) \, d\omega = N, \tag{6.2}$$

where the integral is 6-dimensional and extends over the whole μ-space. As the order of integration is immaterial, the calculation can be split into two steps, namely,

$$n(\mathbf{q}, t) = \iiint_{-\infty}^{+\infty} f_\mu(\mathbf{q}, \mathbf{p}, t) \, d^3p, \qquad N = \iiint_{-\infty}^{+\infty} n(\mathbf{q}, t) \, d^3q. \tag{6.3}$$

The function $n(\mathbf{q}, t)$ defined by the first of (6.3) is the concentration of the particles in the configuration space.

---

[1] Note that here the symbol N indicates the number of particles; instead, in Sect. 23.2 the number of particles is indicated with $\mathcal{N}$, whereas N indicates the concentration.

Basing on the distribution function it is possible to define the average of a dynamical function. For the sake of generality the dynamical function is considered to be a vector that depends on all canonical coordinates and time, say, $\mathbf{a} = \mathbf{a}(\mathbf{q}, \mathbf{p}, t)$. Due to the smallness of the cell size one assumes that the dynamical function takes the same value for all particles within the same cell. As a consequence, the product $\mathbf{a} f_\mu \, d\omega$ is the cell value of the function weighed by the number of particles belonging to the cell. The *statistical average* of $\mathbf{a}$ over the phase space is then

$$\mathrm{Av}\,[\mathbf{a}](t) = \frac{1}{N} \int \mathbf{a}(\mathbf{q}, \mathbf{p}, t) \, f_\mu(\mathbf{q}, \mathbf{p}, t) \, d\omega, \qquad (6.4)$$

where the integral is 6-dimensional, while the average over the momentum space is

$$\overline{\mathbf{a}}(\mathbf{q}, t) = \frac{1}{n(\mathbf{q}, t)} \iiint_{-\infty}^{+\infty} \mathbf{a}(\mathbf{q}, \mathbf{p}, t) \, f_\mu(\mathbf{q}, \mathbf{p}, t) \, d^3 p. \qquad (6.5)$$

Using the expression of N given by (6.2), and that of n given by the first of (6.3), shows that the definitions (6.4, 6.5) indeed provide the weighed averages of interest. By way of example, the dynamical function may be identified with the particle velocity $\mathbf{u}$: using the Cartesian coordinates one finds for the average velocity $\mathbf{v}$ in the configuration space the expression

$$\mathbf{v}(\mathbf{r}, t) = \frac{1}{n(\mathbf{r}, t)} \iiint_{-\infty}^{+\infty} \mathbf{u}(\mathbf{r}, \mathbf{p}, t) \, f_\mu(\mathbf{r}, \mathbf{p}, t) \, d^3 p. \qquad (6.6)$$

Similarly, the average Hamiltonian function in the configuration space reads

$$\overline{H}(\mathbf{r}, t) = \frac{1}{n(\mathbf{r}, t)} \iiint_{-\infty}^{+\infty} H(\mathbf{r}, \mathbf{p}, t) \, f_\mu(\mathbf{r}, \mathbf{p}, t) \, d^3 p. \qquad (6.7)$$

6.3 Statistical Equilibrium

This section deals with the properties of a system of particles in a condition of macroscopic equilibrium. Considering that in general the systems that are considered are composed of a large number of particles or molecules, the statistical concepts introduced in Sect. 6.2 will be used. Generally speaking, the condition of statistical equilibrium is fulfilled if the distribution function is independent of time. This condition may be achieved in different ways: for instance, $f_\mu = \mathrm{const}$ fulfills the required condition. A more general definition of a distribution function fulfilling the condition of statistical equilibrium is $f_\mu = f_\mu(c)$, where c is any constant of motion of the system. In case of a conservative system, energy is the most natural constant of motion to be used.

To proceed, consider a conservative system having a total energy E_S, enclosed in a stationary container of volume Ω. Let the walls of the container be such that no energy flows across them. Also, the container is assumed to be sufficiently massive so that it can be regarded as stationary despite the transfer of kinetic energy due to the molecules' collisions with the walls. If any external force acts on the molecules, it is assumed to be independent of time and conservative ([110], Sect. 26). Finally, the total momentum and angular momentum of the system are assumed to vanish; this condition is by no means obvious and requires some reasoning, as detailed in Sect. 6.6.1.

So far it has not been explicitly indicated whether the molecules that form the system are identical to each other or not; in the practical cases it is to be expected that the system under consideration be made of a mixture of different atoms or molecules. As the extension to a mixture is straightforward ([110], Sect. 30), the analysis is limited here to a system made of identical molecules. It should be noted that the molecules are identical, but distinguishable from each other: from the point of view of Classical Mechanics a continuous observation of the trajectory of each molecule is in fact possible, without disturbing its motion. As a consequence, systems that differ by the exchange of two molecules are to be considered as different from each other.

To proceed one assumes that the number of molecules forming the system is N, and that each molecule has R degrees of freedom. The canonical coordinates that describe the motion of a single molecule are then $q_1, \ldots, q_R, p_1, \ldots, p_R$, so that the number of dimensions of the μ-space is $2R$. As anticipated in Sect. 6.2, the description of the precise state of each molecule is impossible in practice; the state of the system will then be specified in a somewhat less precise manner, as detailed below. First, each q axis of the μ-space is divided into equal intervals of size $\Delta q_1, \ldots, \Delta q_R$ and, similarly, each p axis is divided into equal intervals of size $\Delta p_1, \ldots, \Delta p_R$. As a consequence, the μ-space is partitioned into elements, called *cells*, whose volume and units are, respectively,

$$\Delta M = (\Delta q_1 \, \Delta p_1) \ldots (\Delta q_R \, \Delta p_R), \quad [\Delta M] = (\text{J s})^R. \tag{6.8}$$

The partitioning of the μ-space into cells has the advantage, first, that the set of cells is countable. Besides that, the partitioning into cells of finite size has a deeper meaning, that becomes apparent when the theory outlined in this section is extended to the quantum-mechanical case. In fact, due to the Heisenberg uncertainty relation (Sect. 10.6), the precision by which two conjugate variables can simultaneously be known is limited, so that each product $\Delta q_i \, \Delta p_i$ in (6.8) is bounded below, the bound being of the order of the Planck constant.

After the partitioning is accomplished, the state of the system is assigned by specifying the numbers $N_1, N_2, \ldots \geq 0$ of molecules that belong, respectively, to the cell labeled $1, 2, \ldots$; such numbers are subjected to the constraint $N_1 + N_2 + \ldots = N$ which, in view of the calculations that follow, is more conveniently expressed as

$$F_N(N_1, N_2, \ldots) = 0, \quad F_N = N - \sum_i N_i. \tag{6.9}$$

The sum in (6.9) may be thought of as extending to all cells, due to the fact that only in a finite number of cases the cell population N_i differs from zero. Clearly, the description using the numbers $N_1, N_2, \ldots$ is less precise than that given by the molecules' trajectories, and provides a partial specification of the state of each molecule within the limits of the size of ΔM. This means, among other things, that identical molecules belonging to different cells are distinguishable from each other, whereas identical molecules belonging to the same cell are not distinguishable.

As mentioned above, the total energy of the system is E_S, which provides a second constraint to be vested with mathematical form. It is provisionally assumed that the system is *dilute*, namely, that the energy of the interaction among the molecules is negligible. It follows that one can approximately assign to each molecule a total energy that corresponds to its position, momentum, and internal configuration, in other terms, to the cell where the molecule belongs. Letting the energy corresponding to the ith cell be E_i, the constraint on energy reads $N_1 E_1 + N_2 E_2 + \ldots = E_S$, namely,

$$F_E(N_1, N_2, \ldots) = 0, \qquad F_E = E_S - \sum_i N_i E_i. \qquad (6.10)$$

The above reasoning does not apply to concentrated systems, where the interaction energy is strong. However, it can be shown that (6.10) still holds, albeit with a different interpretation of E_i ([110], Sect. 29). Another observation is that, given the constraints to which the system is subjected, the set of numbers $N_1, N_2, \ldots$ may not be unique. It is therefore necessary to extract the set, that actually describes the system, from a larger number of sets made of all possible sets that are compatible with the constraints.

To proceed, let $N_1, N_2, \ldots$ be a set that provides a molecules' distribution compatible with the constraints; such a set is called *accessible state*. It is postulated that no accessible state is privileged with respect to any other; this is in fact the fundamental hypothesis of equal *a priori* probability of the accessible states, upon which Statistical Mechanics is based ([110], Sect. 23). Remembering that the particles are identical to each other, any system obtained from the original distribution by exchanging two molecules is also compatible with the constraints. However, the system resulting from the exchange of molecules belonging to different cells is different from the original one because such molecules are distinguishable; in contrast, an exchange within the same cell does not produce a different system. As the total number of possible exchanges of N molecules of the system as a whole is $N!$, and the number of possible exchanges within the ith cell is $N_i!$, the total number of different systems corresponding to a set $N_1, N_2, \ldots$ is

$$W(N_1, N_2, \ldots) = \frac{N!}{N_1! \, N_2! \ldots} \qquad (6.11)$$

As time evolves, the interactions among the molecules makes the numbers $N_1, N_2, \ldots$ to change, so that, if the system is inspected at successive times, it may be found to belong to different accessible states (as the principles of Classical Mechanics apply here, such an inspection does not perturb the state of the system); in principle, given enough time, the system will go through all accessible states. Now, the dependence of W on $N_1, N_2, \ldots$ is quite strong, so that some accessible states correspond to a large number W of systems, others to a much smaller number.[2] As no accessible state is privileged with respect to any other, the majority of inspections carried out onto the system will provide the accessible state that maximizes W. Such an accessible state corresponds, by definition, to the condition of *statistical equilibrium* of the system under consideration.

6.4 Maxwell-Boltzmann Distribution

The analysis carried out in Sect. 6.3 led to the conclusion that the condition of statistical equilibrium of the system is found by maximizing the expression of W given by (6.11), under the constraints (6.9) and (6.10). The calculation is based upon the Lagrange method, that determines the free maximum of an auxiliary function F embedding the constraints. It is convenient to maximize $\log W$, which is a monotonic function of W, instead of W itself, so that the auxiliary function reads

$$F(N_1, N_2, \ldots, \alpha, \beta) = \log W + \alpha F_N + \beta F_E, \qquad (6.12)$$

where α, β are the *Lagrange multipliers*, respectively related to the total number of molecules and total energy of the system. In a typical system the total number of molecules and the populations of the majority of non empty cells are very large,[3] so that the Stirling approximation (C.97) is applicable; it follows, after neglecting terms of the form $\log \sqrt{2\pi N}$ and $\log \sqrt{2\pi N_i}$,

$$\log W = \log(N!) - \sum_i \log(N_i!) \simeq N \log(N) - N - \sum_i N_i \log(N_i) + \sum_i N_i,$$
$$(6.13)$$

where $-N$ and $\sum_i N_i$ cancel each other due to (6.9). The function to maximize then becomes $F = N \log(N) - \sum_i N_i \log(N_i) + \alpha F_N + \beta F_E$. Here the property of N_i of being very large is again of help, because, on account of the fact that a change of N_i by one unit is negligibly small with respect to N_i itself, in calculating the maximum one treats the integers N_i as continuous variables. Taking the derivative of F with respect to, say, N_r, and equating it to zero, yields $\log N_r + 1 = -\alpha - \beta E_r$. Neglecting the unity at the left hand side eventually yields the *Maxwell-Boltzmann distribution law*

$$N_r = \exp(-\alpha - \beta E_r). \qquad (6.14)$$

---

[2] This is apparent even if the numbers $N_1, N_2, \ldots$ are much smaller than in realistic systems. Let for instance $N = 8$: the combination $N_1 = 8$, $N_2 = N_3 = \ldots = 0$ yields $W = 1$, whereas the combination $N_2 = N_3 = N_4 = N_5 = 2$, $N_1 = N_6 = N_7 = \ldots = 0$ yields $W = 2.520$. It is implied that $8 E_1 = 2(E_2 + E_3 + E_4 + E_5) = E_S$.

[3] The hypothesis that the populations are large is not essential. A more complicate calculation, in which such a hypothesis is not present, leads to the same result [22].

The Lagrange multipliers are then determined from (6.9, 6.10); the first one yields

$$\sum_r \exp(-\alpha - \beta E_r) = N, \qquad N \exp(\alpha) = \sum_r \exp(-\beta E_r) = Z, \qquad (6.15)$$

where Z denotes the *partition function*.[4]. Extracting $\exp(\alpha)$ from (6.15) and replacing it into (6.10) provides the relation

$$\frac{E_S}{N} = \frac{1}{Z} \sum_r E_r \exp(-\beta E_r) = -\frac{\partial}{\partial \beta} \log Z, \qquad (6.16)$$

with β the only unknown. This procedure is able to express β and, consequently, α, in terms of the dynamical properties of the molecules. A different method to determine the parameters, that relates them with macroscopic quantities typical of Thermodynamics, like pressure and temperature, is shown in the following. First, one introduces a constant C such that

$$CN \Delta M = \exp(-\alpha), \qquad (6.17)$$

where ΔM is given by (6.8). After replacing (6.17) in (6.14), the cell's volume ΔM is made smaller and smaller so that, after dropping index r, (6.14) is recast in differential form as

$$dN = CN \exp(-\beta E) dM, \qquad \frac{1}{C} = \int \exp(-\beta E) dM, \qquad (6.18)$$

where the integral is extended over the μ-space and energy E is expressed in terms of the $2R$ coordinates $q_1, \ldots, q_R, p_1, \ldots, p_R$. The nature of the system is now specified as that of a monatomic gas of mass m, described by the Cartesian coordinates x_i and conjugate momenta p_i. Integrating (6.18) over the container's volume Ω yields the number dN_p of atoms belonging to the elementary volume of the momentum space. As the gas is assumed to be dilute, the energy in (6.18) is substantially of the kinetic type,

$$E \simeq \frac{1}{2m} \left(p_1^2 + p_2^2 + p_3^2 \right), \qquad (6.19)$$

so that the integrand is independent of the spatial coordinates. It follows

$$dN_p = CN\Omega \exp(-\beta E) dp_1 dp_2 dp_3. \qquad (6.20)$$

The integral of (6.20) over the momentum space yields N, whence

$$\frac{1}{C\Omega} = \iiint_{-\infty}^{+\infty} \exp(-\beta E) dp_1 dp_2 dp_3. \qquad (6.21)$$

---

[4] Symbol Z comes from the German term *Zustandssumme* ("sum over states") ([95], Chap. II).

The pressure P exerted by the gas is uniform over all the container's walls, hence it can be calculated with reference to any surface element $d\Sigma$ belonging to them. One can then choose a surface element normal to the x_1 axis, placed in such a way that the atoms impinging on it travel in the positive direction. Let p_1 be the component of momentum of one of such atoms before hitting the wall; after the atom is reflected by the wall, the component transforms into $-p_1$, so that the variation in the momentum component along the x_1 direction is $2p_1$. The p_2, p_3 components, instead, are left unchanged. The product of $2\,p_1$ by the number of atoms hitting $d\Sigma$ in the unit time provides the force dF exerted by the gas on the surface element, whence the pressure is obtained as $P = dF/d\Sigma$. Now, consider an elementary cylinder whose base and height are $d\Sigma$ and dx_1 respectively; the number $d\tilde{N}_1$ of atoms that belong to such a cylinder and whose momentum p_1 belongs to dp_1 is obtained as the product of the atoms' concentration dN_p/Ω times the cylinder's volume $d\Sigma\,dx_1 = d\Sigma\,dt\,dx_1/dt = d\Sigma\,(p_1/m)\,dt$, integrated over the other two momenta p_2, p_3:

$$d\tilde{N}_1 = d\Sigma\,\frac{p_1}{m}\,dt\,dp_1\,C\,N\iint_{-\infty}^{+\infty}\exp(-\beta E)\,dp_2\,dp_3. \tag{6.22}$$

As each atom in (6.22) undergoes a change $2p_1$ in the momentum component due to the reflexion at the wall, the force $dF = Pd\Sigma$ is obtained by multiplying (6.22) by $2p_1$, dividing it by dt, and integrating over p_1 between 0 and $+\infty$; in fact, only the atoms that move towards the wall must be accounted for. Eliminating $d\Sigma$ yields

$$\frac{P}{C\,N} = \frac{2}{m}\int_0^{+\infty}\iint_{-\infty}^{+\infty}p_1^2\exp(-\beta E)\,dp_1\,dp_2\,dp_3. \tag{6.23}$$

Due to the form of (6.19), the integrals in (6.21) and (6.23) are products of one-dimensional integrals over p_1, p_2, and p_3. As a consequence, in the ratio between (6.23) and (6.21) the integrals over p_2 and p_3 cancel each other, to yield

$$\frac{P\Omega}{N} = \frac{2}{m}\frac{\int_0^{+\infty}p_1^2\exp[-\beta\,p_1^2/(2\,m)]\,dp_1}{\int_{-\infty}^{+\infty}\exp[-\beta\,p_1^2/(2\,m)]\,dp_1}. \tag{6.24}$$

Letting Y indicate the integral at the denominator of (6.24), the integral at the numerator is found to be $-m\,dY/d\beta$. Using (C.27) one finds

$$P\Omega = -2\frac{d\sqrt{2\pi\,m/\beta}/d\beta}{\sqrt{2\pi\,m/\beta}} = \frac{N}{\beta}. \tag{6.25}$$

The assumption that the gas used in the derivation of (6.25) is dilute makes it possible to consider it as a perfect gas, for which the phenomenological relation $P\Omega = Nk_BT$ holds, with $k_B = 1.38\times10^{-23}$ J/K the *Boltzmann constant* and T the gas temperature. Comparing with (6.25) yields

$$\beta = \frac{1}{k_B\,T}. \tag{6.26}$$

It can be shown that the validity of (6.26) is not limited to the case where the simple derivation shown here applies. Actually, (6.26) is found to hold for any system that follows the Maxwell-Boltzmann distribution law ([110], Sect. 32) and also, as shown in Sects. 15.9.1, 15.9.5, for quantum systems in equilibrium.

6.5 Boltzmann Transport Equation

The expressions worked out in Sect. 6.4 show the important role of the distribution function. It is then necessary to determine the equation fulfilled by it when the system is not in an equilibrium condition. The derivation is made of two steps; in the first one the interactions between molecules are neglected, in the second one they are accounted for. To start with the first step one observes that, due to the neglect of collisions, the only force acting on the molecules is that of an external field. To denote the position in the phase space it is convenient to use the symbol s introduced in Sect. 1.8. Here the symbol has a slightly different meaning, because the space is 6-dimensional instead of being $2n$-dimensional. However, the relations (1.57, 1.58, 1.59) still hold. Applying to the μ-space the same reasoning used in Sect. 23.2 to find the continuity equation (23.3), and considering the case where no particles are generated or destroyed, yields

$$\frac{\partial f_\mu}{\partial t} + \mathrm{div}_s \left(\dot{s}\, f_\mu \right) = 0. \tag{6.27}$$

From (1.58, 1.59) it follows $\mathrm{div}_s(\dot{s}\, f_\mu) = \dot{s} \cdot \mathrm{grad}_s f_\mu$. Replacing the latter into (6.27) and using the Cartesian coordinates yields the *Boltzmann collisionless equation*[5]

$$\frac{\partial f_\mu}{\partial t} + \dot{\mathbf{r}} \cdot \mathrm{grad}_r f_\mu + \dot{\mathbf{p}} \cdot \mathrm{grad}_p f_\mu = 0. \tag{6.28}$$

From the meaning of a continuity equation it follows that $-\mathrm{div}_s \left(\dot{s}\, f_\mu \right)$ is the time variation of f_μ per unit volume $d\omega$ of the μ-space. As f_μ depends on $\mathbf{r}$, $\mathbf{p}$, and t, (6.28) is recast in compact form as $df_\mu/dt = 0$.

To accomplish the second step, namely, adding the effects of collisions, one observes that the latter produce a further time change in f_μ. In principle, one might incorporate such effects into $-\mathrm{div}_s \left(\dot{s}\, f_\mu \right)$; however, it is more convenient to keep the effects of collisions separate from those of the external field. In fact, assuming as before that the system under consideration is dilute, each molecule spends a relatively large fraction of time far enough from the other molecules not to suffer any interaction; in other terms, the time during which a molecule is subjected to the external field is much longer than that involved in a collision. For this reason it is preferable to write the continuity equation, when the collisions are accounted for, as

$$\frac{\partial f_\mu}{\partial t} + \dot{\mathbf{r}} \cdot \mathrm{grad}_r f_\mu + \dot{\mathbf{p}} \cdot \mathrm{grad}_p f_\mu = C, \tag{6.29}$$

---

[5] In plasma physics, (6.28) is also called *Vlasov equation* ([85], Sect. 13.2).

called *Boltzmann Transport Equation*. In (6.29), term C indicates the time variation of f_μ per unit volume $d\omega$ due to collisions, whereas $-\text{div}_s (\dot{s}\, f_\mu)$ keeps the meaning of variation due to the external field. The compact form of (6.29) reads in this case

$$\frac{df_\mu}{dt} = C, \qquad \int C\, d\omega = 0. \qquad (6.30)$$

where the second relation is due to the normalization condition (6.2). In the equilibrium condition the distribution function has no explicit dependence on time $(\partial f_\mu / \partial t = 0)$ and depends on constants of motion only, so that $C = 0$. The condition $C = 0$ does not prevent collisions from happening; in fact, in the equilibrium condition the change in the state of two colliding particles is balanced by simultaneous state changes of other particles, that occur in the same elementary volume $d\omega$, in such a way that the distribution function is left unchanged (*principle of detailed balance*).

In the calculation of C it is assumed that collisions are of the *binary* type, namely, that they involve only two particles at the time because the probability of a simultaneous interaction of more than two particles is negligibly small. This hypothesis, along with the assumption of a short duration of the interactions, greatly simplifies the calculation of C. This issue will not be pursued further here, because it will be developed directly in the quantum case (Sect. 19.3.1). It is worth observing that in a general non-equilibrium condition it is $C \neq 0$; the second relation in (6.30) then indicates that the form of C must be such, that in the integration over the μ-space every elementary contribution to the integral is balanced by an opposite contribution.

When the system under consideration is made of charged particles, the external field that matters is the electromagnetic one; if the particles are identical to each other, (6.29) takes the form

$$\frac{\partial f_\mu}{\partial t} + \mathbf{u} \cdot \text{grad}_r f_\mu + e\,(\mathbf{E} + \mathbf{u} \wedge \mathbf{B}) \cdot \text{grad}_p f_\mu = C, \qquad (6.31)$$

with $\mathbf{E}(\mathbf{r}, t)$ the electric field, $\mathbf{B}(\mathbf{r}, t)$ the magnetic induction, e the common value of the particles' charge, and $\mathbf{u}$ the velocity (Sect. 1.3.2).

6.6 Complements

6.6.1 *Momentum and Angular Momentum at Equilibrium*

In the introductory discussion about statistical equilibrium, carried out in Sect. 6.3, it has been assumed that the total momentum and angular momentum of the system vanish. To discuss this issue, consider the box-shaped container whose cross section is shown in Fig. 6.1, filled with molecules identical to each other, having initial positions near the right wall of the container. If the initial velocities are normal to the wall and equal to each other, as schematically indicated by the arrows, the total momentum $\mathbf{P}$ of the particles at $t = 0$ is different from zero. If the left and right walls are perfectly reflecting and parallel to each other, the particles keep bouncing

Fig. 6.1 Schematic picture used for discussing the issue of the total momentum of identical molecules within a container

back and forth between the two walls, and the total momentum alternates between **P** and −**P**. As the container's mass is large, absorbing the momentum 2 **P** leaves the stationary condition of the container unaltered. On the other hand, as remarked in Sect. 1.3, this picture should not be regarded as describing a "system of particles", because the latter have no mutual interaction. To establish the interaction one must assume that the initial velocities are slightly misaligned, or the walls are not exactly parallel, or both; in this way the molecules will start colliding with each other and, after some time, their velocities will not be parallel any more. If each collision, and reflection at the walls, is energy conserving, the total energy of the system does not change; in contrast, opposite velocity components of different molecules compensate each other in the calculation of the total momentum, so that the latter will vanish after a sufficiently large number of collisions.[6] A similar argument is applicable to the case of the cylindrical container whose cross section is shown in Fig. 6.2, where the initial positions and velocities of the molecules are such that all molecules would move along the square described by the arrows. In this case the total angular momentum with respect to the cylinder's axis is different from zero. A slight misalignment of the initial conditions, or a deviation of the container's wall from the perfectly-cylindrical form, or both, will eventually make the molecules to collide with each other.

6.6.2   Averages Based on the Maxwell-Boltzmann Distribution

In a system made of classical particles in equilibrium at temperature T, each having R degrees of freedom, the average occupation number at energy E_r is given by (6.14). In general, the number of energy levels is large and their separation small, so that one

---

[6] If the walls are perfectly reflecting, and the collisions are elastic (Sect. 3.5), the molecular motions are reversible so that, in both examples of this section, the initial condition is recovered by reversing the direction of time. More comments about this are made in Sect. 6.6.4.

Fig. 6.2 Schematic picture
used for discussing the issue
of the total angular
momentum of identical
molecules within a container

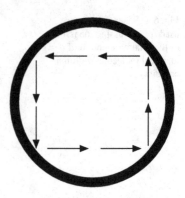

disposes of the index and considers the number of particles belonging to the infinitesimal interval $dq_1 \, dp_1 \ldots dq_R \, dp_R$ centered at $(q_1, p_1, \ldots, q_R, p_R)$. After dropping the index, the energy becomes a function of the position $(q_1, p_1, \ldots, q_R, p_R)$ of the interval's center, $E = E(q_1, p_1, \ldots, q_R, p_R)$; in turn, the Maxwell–Boltzmann distribution (6.14) takes the form $\exp(-\alpha - \beta E)$. Given these premises, and extending to the case of R degrees of freedom the definitions of Sect. 6.2, the statistical average over the Maxwell–Boltzmann distribution of a function $\zeta(q_1, p_1, \ldots, q_R, p_R)$ is

$$\overline{\zeta} = \frac{\int \ldots \int \zeta \, \exp(-\beta E) \, dq_1 \, dp_1 \ldots dq_R \, dp_R}{\int \ldots \int \exp(-\beta E) \, dq_1 \, dp_1 \ldots dq_R \, dp_R}, \tag{6.32}$$

where the factor $\exp(-\alpha)$ has been canceled out. An interesting case occurs when ζ depends on the generalized coordinates through the total energy only ($\zeta = \zeta(E)$) and, in turn, the energy is a positive-definite quadratic form of the coordinates,

$$E = a_1 \, q_1^2 + b_1 \, p_1^2 + \ldots + a_R \, q_R^2 + b_R \, p_R^2, \qquad a_i, b_i > 0. \tag{6.33}$$

Letting $n = 2R$ and using the Herring–Vogt transformation (17.66) yields

$$E = \eta/\beta, \qquad \eta = u_1^2 + \ldots + u_n^2, \tag{6.34}$$

where

$$u_1 = \sqrt{\beta \, a_1} \, q_1, \; u_2 = \sqrt{\beta \, b_1} \, p_1, \ldots, u_{n-1} = \sqrt{\beta \, a_R} \, q_R, \; u_n = \sqrt{\beta \, b_R} \, p_R. \tag{6.35}$$

In turn it is $du_1 \ldots du_n = c \, dq_1 \, dp_1 \ldots dq_R \, dp_R$, with $c = \beta^R \sqrt{a_1 \, b_1 \ldots a_R \, b_R}$. Using the procedure involving the density of states illustrated in Sect. B.5 yields

$$\overline{\zeta} = \frac{\int_0^{+\infty} \zeta(\eta) \, \exp(-\eta) \, b(\eta) \, d\eta}{\int_0^{+\infty} \exp(-\eta) \, b(\eta) \, d\eta} = \frac{\int_0^{+\infty} \zeta(\eta) \, \exp(-\eta) \, \eta^{n/2-1} \, d\eta}{\int_0^{+\infty} \exp(-\eta) \, \eta^{n/2-1} \, d\eta}, \tag{6.36}$$

where the last expression is obtained after canceling out the numerical factors appearing in (B.40). An important case of (6.36) is $\zeta = E = \eta/\beta$, which yields the average energy of the particles. Remembering (C.88, C.89), and using (6.26), one finds

$$\overline{E} = k_B T \frac{\Gamma(n/2 + 1)}{\Gamma(n/2)} = \frac{n}{2} k_B T = R k_B T. \tag{6.37}$$

The physical systems where the energy is a quadratic form of the type (6.33) are made of linear-harmonic oscillators, like those deriving from the diagonalization of the Hamiltonian function of a system of particles near the mechanical-equilibrium point (Sect. 3.10), or from the expression of the energy of the electromagnetic field *in vacuo* in terms of modes (Sect. 5.6). For systems of this type the average energy of the particles equals $k_B T$ times the number R of degrees of freedom of each particle.

Another important system is the dilute one, where the energy is essentially kinetic. In this case the form of the latter is found by letting $a_i \to 0$ in (6.33), so that the energy is made of a sum of R terms instead of $n = 2 R$. Thus, the average energy of the particles is given by an expression of the form (6.37) where $n/2$ is replaced with $R/2$:

$$\overline{E} = k_B T \frac{\Gamma(R/2 + 1)}{\Gamma(R/2)} = R \frac{k_B T}{2}. \tag{6.38}$$

The above shows that in a dilute system the average energy of the particles equals $k_B T/2$ times the number R of degrees of freedom of each particle. The result expressed by (6.37) or (6.38) is called *principle of equipartition of energy*.

6.6.3 Boltzmann's H-Theorem

A measure of the extent by which the condition of a system departs from the equilibrium case is given by Boltzmann's H_B quantity, whose definition is that of statistical average of $\log f_\mu$. The case considered here refers to a spatially-homogeneous system in the absence of external forces, so that $d f_\mu/dt = \partial f_\mu/\partial t$. Remembering (6.4) one finds

$$H_B(t) = \mathrm{Av}\,[\log f_\mu] = \frac{1}{N} \int f_\mu \log f_\mu \, d\omega, \tag{6.39}$$

whose time derivative, using (6.30), reads

$$\frac{d H_B}{dt} = \frac{1}{N} \int \left(\frac{\partial f_\mu}{\partial t} \log f_\mu + \frac{\partial f_\mu}{\partial t} \right) d\omega = \int C \log f_\mu \, d\omega. \tag{6.40}$$

As indicated in Sect. 6.5, the collision term will be worked out directly in the quantum case. However, it is worth anticipating that the analysis of the collision term C leads, both in the classical and quantum cases, to an important conclusion: the time derivative $d H_B/dt$ is negative for any distribution function f_μ different from the equilibrium one, while it vanishes in the equilibrium case. This result is the *Boltzmann*

H-theorem. It implies that if a uniform system is initially set in a condition described by a non-equilibrium distribution function, and the external forces are removed, then the initial distribution can not be stationary: an equilibration process occurs, that brings the distribution to the equilibrium one, and whose duration is dictated by the time constants typical of C. The decrease of H_B with respect to time while the system reaches the equilibrium condition reminds one of the behavior of entropy. In fact, it can be shown that H_B is the entropy apart from a negative multiplicative factor and an additive constant ([113], Sect. 18.3).[7]

6.6.4 Paradoxes — Kac-Ring Model

It is known that the Boltzmann Transport Equation (6.29), and the H-theorem derived from it, bring about two apparent paradoxes: the first one is that the equation contains irreversibility, because any initial distribution function, different from the equilibrium one, evolves towards the equilibrium distribution when the external forces are removed, whereas an opposite evolution never occurs. This outcome is in contrast with the symmetry of the laws of mechanics with respect to time reversal (Sect. 2.6.1). The second paradox is the violation of Poincaré's time recurrence, which states that every finite mechanical system returns to a state arbitrarily close to the initial one after a sufficiently long time (called *Poincaré cycle*); this is forbidden by the H-theorem, that prevents entropy from decreasing back to the initial value.

A thorough discussion about the mathematical origin of the paradoxes can be found, e.g., in ([113], Sect. 18.4); a qualitative insight into the question is given by a simple model, called *Kac's ring model*, also reported in [113] and taken from [60]. In the model, N objects are uniformly distributed over a circle, so that at time $t = 0$ each object is ascribed to a specific arc. The objects have two possible states, say, either "0" or "1". The time variable is discrete so that, when time evolves from $k \Delta t$ to $(k + 1) \Delta t$, $k = 0, 1, 2, \ldots$, each object moves clockwise from the arc it occupied at time $k \Delta t$ to the next arc. A number $n < N$ of markers is present along the circle: specifically, the markers' positions are at the junctions between two neighboring arcs. The objects that cross the position of a marker change the state from "0" to "1" or vice versa; those that do not cross the position of a marker keep their state.

Given the number of objects and markers, the initial state of each object, and the markers' positions along the circle, one wants to investigate the time evolution of the states. Such an evolution is obviously time reversible and fulfills Poincaré's time recurrence; in fact, the set of objects goes back into the initial condition after N time steps if n is even, and after $2 N$ time steps if n is odd.

Providing the time evolution of the individual object's state is in fact a microscopic description of the system; as remarked in Sect. 6.2, such a description

---

[7] Compare with the definition of entropy given in Sect. 15.9.1 which, at first sight, looks different. The equivalence between the two definitions is justified in Sects. 47 and 102 of [110].

Fig. 6.3 Kac-ring model: computer calculation of the time evolution of the number of "0" states in two samples made of $N = 4{,}000$ objects, which at time $t = 0$ were all set to "0". The markers of the two samples are $n = 4$ and $n = 8$, respectively, and the number of time steps is much smaller than N

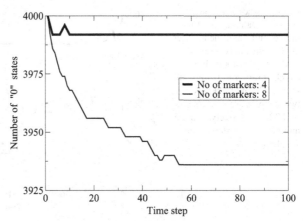

becomes impossible when the number of objects in the system is large. A less detailed, macroscopic description of the Kac ring consists, for instance, in providing the time evolution of the number of "0" states. However, the outcome of the latter analysis seems to indicate that an irreversible process takes place; for instance, Fig. 6.3 shows a computer calculation of the time evolution of the number of "0" states in two samples made of $N = 4000$ objects, which at time $t = 0$ were all set to "0". The markers of the two samples are $n = 4$ and $n = 8$, respectively, and the number of time steps is much smaller than N. Both curves tend to decrease and, after some fluctuations (that depend on the markers' positions), stabilize to a constant value; the same behavior occurs at a larger numbers of markers, although the number of time steps necessary to reach a constant value increases (curve $n = 16$ in Fig. 6.4). A further increase in the number of markers makes the fluctuations more pronounced (curve $n = 32$ in the same figure).

On the other hand, a similar calculation using a number of time steps larger than the number of objects shows that the stabilization at or around a constant value is eventually lost: the system fulfills Poincaré's time recurrence and recovers the initial condition (Fig. 6.5). Such an outcome is not detectable in real many-body systems, because the Poincaré cycle is enormously long with respect to the typical time scales of experiments.[8]

6.6.5 Equilibrium Limit of the Boltzmann Transport Equation

As remarked in Sect. 6.5, in the equilibrium condition the distribution function has no explicit dependence on time ($\partial f_\mu / \partial t = 0$) and depends on constants of motion

---

[8] A crude estimate of the Poincaré cycle yields $\sim \exp(N)$, with N the total number of molecules in the system ([50], Sect. 4.5). In typical situations such a time is longer than the age of the universe.

Fig. 6.4 Kac-ring model: computer calculation of the time evolution of the number of "0" states in two samples made of $N = 4000$ objects, which at time $t = 0$ were all set to "0". The markers of the two samples are $n = 16$ and $n = 32$, respectively, and the number of time steps is much smaller than N

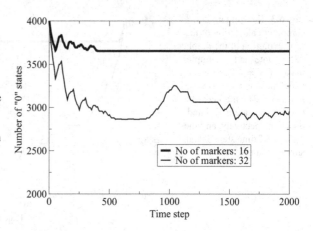

Fig. 6.5 Kac-ring model: computer calculation of the time evolution of the number of "0" states in two samples made of $N = 4000$ objects, which at time $t = 0$ were all set to "0". The markers of the two samples are $n = 16$ and $n = 32$, respectively, and the number of time steps is larger than N

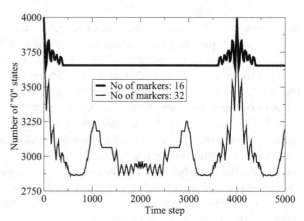

only, so that $C = 0$. From (6.29) it then follows $\dot{\mathbf{r}} \cdot \mathrm{grad}_{\mathbf{r}} f_\mu + \dot{\mathbf{p}} \cdot \mathrm{grad}_{\mathbf{p}} f_\mu = 0$. In case of a conservative system, energy is the most natural constant of motion to be used; in fact it is $H(\mathbf{r}, \mathbf{p}) = E = $ const, with H the Hamiltonian function (Sect. 1.5). From $f_\mu = f_\mu(H)$ one derives $\mathrm{grad}_{\mathbf{r}} f_\mu = (\mathrm{d} f_\mu/\mathrm{d} H) \, \mathrm{grad}_{\mathbf{r}} H$ and $\mathrm{grad}_{\mathbf{p}} f_\mu = (\mathrm{d} f_\mu/\mathrm{d} H) \, \mathrm{grad}_{\mathbf{p}} H$, so that the equilibrium limit of the Boltzmann Transport Equation reads

$$\frac{\mathrm{d} f_\mu}{\mathrm{d} H} \left(\dot{\mathbf{r}} \cdot \mathrm{grad}_{\mathbf{r}} H + \dot{\mathbf{p}} \cdot \mathrm{grad}_{\mathbf{p}} H \right) = 0, \tag{6.41}$$

Apart from the trivial case $f_\mu = $ const, it is $\mathrm{d} f_\mu/\mathrm{d} H \neq 0$. On the other hand, recasting the Hamilton equations (1.42) as $\dot{\mathbf{r}} = \mathrm{grad}_{\mathbf{p}} H$, $\dot{\mathbf{p}} = -\mathrm{grad}_{\mathbf{r}} H$, and replacing them in (6.41), reduces the term in parentheses to $-\dot{\mathbf{r}} \cdot \dot{\mathbf{p}} + \dot{\mathbf{p}} \cdot \dot{\mathbf{r}}$, this showing that the equation is fulfilled identically regardless of the explicit form of $f_\mu(H)$. This result implies that the equilibrium distribution (6.14) can not be extracted solely from the equilibrium limit of (6.29); its derivation requires also the maximization procedure described in Sects. 6.3 and 6.4.

The equilibrium limit of the Boltzmann Transport Equation described in this section applies in general; as a consequence, it includes also cases like that of (6.31), where a magnetic force is involved. This seems to contradict one of the hypotheses used in Sect. 6.3 to derive the equilibrium distribution, namely, that the system is acted upon by conservative forces. To clarify the issue one uses the concepts illustrated in Sects. 1.5, 1.6; first, one notes that the equilibrium limit is achieved by making the external force independent of time, so that the scalar and vector potentials whence $\mathbf{E}$ and $\mathbf{B}$ derive are independent of time as well: $\varphi = \varphi(\mathbf{r})$, $\mathbf{A} = \mathbf{A}(\mathbf{r})$ (the dependence on space is kept to account for the possibility that the system under consideration is not uniform). It follows that the Hamiltonian function, which in this case is given by (1.35), is still a constant of motion; as a consequence, the procedure leading to (6.41) is applicable. One notes in passing that each summand in the resulting identity $-\dot{\mathbf{r}} \cdot \dot{\mathbf{p}} + \dot{\mathbf{p}} \cdot \dot{\mathbf{r}}$ becomes in this case

$$\dot{\mathbf{p}} \cdot \mathbf{u} = e\,(\mathbf{E} + \mathbf{u} \wedge \mathbf{B}) \cdot \mathbf{u} = e\,\mathbf{E} \cdot \mathbf{u}, \qquad (6.42)$$

where the mixed product vanishes due to (A.32).

Problems

6.1 Calculate the average energy like in (6.37) assuming that energy, instead of being a continuous variable, has the discrete form $E_n = nh\nu$, $n = 0, 1, 2, \ldots$, $h\nu = $ const. This is the hypothesis from which Planck deduced the black-body's spectral energy density (Sect. 7.4.1).

Chapter 7
From Classical Mechanics to Quantum Mechanics

7.1 Introduction

The chapter tackles the difficult problem of bridging the concepts of Classical Mechanics and Electromagnetism with those of Quantum Mechanics. The subject, which is fascinating *per se*, is illustrated within a historical perspective, covering the years from 1900, when Planck's solution of the black-body radiation was given, to 1926, when Schrödinger's paper was published.

At the end of the 1800s, the main branches of physics (mechanics, thermodynamics, kinetic theory, optics, electromagnetic theory) had been established firmly. The ability of the physical theories to interpret the experiments was such, that many believed that all the important laws of physics had been discovered: the task of physicists in the future years would be that of clarifying the details and improving the experimental methods. Fortunately, it was not so: the elaboration of the theoretical aspects and the refinement in the experimental techniques showed that the existing physical laws were unable to explain the outcome of some experiments, and could not be adjusted to incorporate the new experimental findings. In some cases, the theoretical formulations themselves led to paradoxes: a famous example is the *Gibbs entropy paradox* [70]. It was then necessary to elaborate new ideas, that eventually produced a consistent body generally referred to as *modern physics*. The elaboration stemming from the investigation of the microscopic particles led to the development of Quantum Mechanics, that stemming from investigations on high-velocity dynamics led to Special Relativity.

The chapter starts with the illustration of the planetary model of the atom, showing that the model is able to justify a number of experimental findings; this is followed by the description of experiments that can not be justified in full by the physical theories existing in the late 1800s: stability of the atoms, spectral lines of excited atoms, photoelectric effect, spectrum of the black-body radiation, Compton effect. The solutions that were proposed to explain such phenomena are then illustrated; they proved to be correct, although at the time they were suggested a comprehensive theory was still lacking. This part is concluded by a heuristic derivation of the time-independent Schrödinger equation, based upon the analogy between the variational principles of Mechanics and Geometrical Optics.

© Springer Science+Business Media New York 2015   129
M. Rudan, *Physics of Semiconductor Devices*,
DOI 10.1007/978-1-4939-1151-6_7

In the final part of the chapter the meaning of the wave function is given: for this, an analysis of the measuring process is carried out first, showing the necessity of describing the statistical distribution of the measured values of dynamical quantities when microscopic particles are dealt with; the connection with the similar situations involving massive bodies is also analyzed in detail. The chapter is concluded with the illustration of the probabilistic interpretation of the wave function.

7.2   Planetary Model of the Atom

Several experiments were carried out in the late 1800s and early 1900s, whose outcome was the determination of a number of fundamental constants of atomic physics; among them, the *electron charge-to-mass ratio* was measured by J. J. Thomson in 1897, and the *electron charge* was measured by R. Millikan in 1909 (Table D.1). A theory of the atom was proposed by E. Rutherford after a series of experiments in 1909–1914, that led to the measurement of the atomic radius r_a. The experiments consisted in measuring the broadening of beams of finely collimated α particles[1] passing through thin metal foils. The latter were typically made of gold sheets with a thickness of a few thousand atomic layers; the dynamics of the interaction between an α particle and the metal foil was treated by Classical Mechanics, using an interaction of the Coulomb type (Sect. 3.8). The outcome of the experiments led Rutherford to conceive the *planetary model* of the atom; this model depicts the atom as made of a small nucleus of atomic charge $Z q$ surrounded by Z electrons, where q is the absolute value of the electron charge and Z indicates the position of the element in the periodic table. The model assumes that, as the α-particles are rather heavy, they are deflected mainly by the nuclei[2] of the foil; the type of deflection implies that the majority of the α particles is deflected only once when crossing the foil, this indicating that the foil's atoms are placed far apart from each other. In fact, the deflection experiments made it possible to estimate[3] the atom's and nucleus' diameters, respectively, as

$$r_a \approx 0.1 \text{ nm}, \qquad r_e \approx 2 \times 10^{-6} \sqrt{Z} \text{ nm}, \qquad Z = 1, 2, \ldots . \tag{7.1}$$

As Z ranges from 1 to about 100, the second relation in (7.1) shows that $r_e \ll r_a$ (compare with Table D.1).

The simplest atom is that of hydrogen. The planetary model depicts it as an electron moving nearby the proton under the effect of a Coulomb potential V, where the energy reference is chosen in such a way as to make $V(\infty) = 0$. Then, the theory

---

[1] These particles are obtained by ionizing helium atoms.

[2] The meaning of term "nucleus" in this context is clarified in Sect. 7.8.1.

[3] The estimate means that for scale lengths equal or larger than those indicated in (7.1), an atom or a nucleus can be considered as geometrical points having no internal structure. The electron's radius can be determined in a similar way using X-ray diffusion.

Fig. 7.1 Classical description
of the electron's orbit for
$E \geq 0$

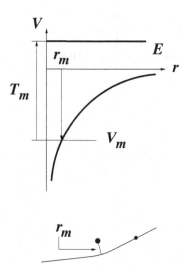

of the Coulomb interaction in the attractive case applies (Sect. 3.13.6). As the proton
is much more massive than the electron, its position can be approximated with that
of the atom's center of mass, and placed in the origin; in summary, it is

$$V(\mathbf{r}) = V(r) = -\frac{q^2}{4\pi\varepsilon_0 r}, \qquad T + V = \frac{1}{2}mu^2 - \frac{q^2}{4\pi\varepsilon_0 r} = E = \text{const}, \quad (7.2)$$

where $\mathbf{r}$ is the electron position, u its velocity's module, T and E its kinetic and
total energies, respectively, and ε_0 the permittivity of vacuum. The planetary model
is extended to more complicated atoms by considering an outer electron moving
nearby a core of net charge q embedding Z protons and $Z - 1$ electrons, or to even
more complicated cases (*hydrogenic-like systems*). Observing that $T = E - V =
E + |V| \geq 0$, and remembering the analysis of Sect. 3.8, one finds that two cases are
possible: the first one, shown in Fig. 7.1, is $E \geq 0$, corresponding to $V \leq 0 \leq E$ and
$r_{\max} = \infty$ (*free electron*). The second case is $E < 0$, corresponding to $V \leq E < 0$
and $r_{\max} < \infty$ (*bound electron*). For the qualitative reasoning to be carried out here,
it is sufficient to consider the simpler case of a bound electron whose trajectory is
circular (Fig. 7.2). In such a case, using the results of Sects. 3.7 and 3.13.6 after letting
$Z_1 = Z_2 = 1$ and replacing s with r, yields $r = 4\pi\varepsilon_0 M_B^2/(m q^2) = \text{const}$, where
$M_B^2 = m^2 r^2 u^2$. Combining these relations with the first one in (7.2) shows that
$F = |\mathbf{F}| = m |\mathbf{a}| = m u^2/r = 2T/r$ whence, using $F = q^2/(4\pi\varepsilon_0 r^2) = -V/r$,
one finds

$$T = -\frac{V}{2}, \qquad E = T + V = \frac{V}{2} = -\frac{q^2}{8\pi\varepsilon_0 r} = \text{const} < 0. \qquad (7.3)$$

It follows $dE/dr = |E|/r > 0$, that is, the total energy is larger at larger orbits (this
is a particular case of the general theory worked out in Prob. 3.2).

Fig. 7.2 Classical
description of the electron's
orbit for $E < 0$. For
simplicity, a circular orbit is
considered

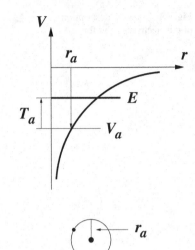

Fig. 7.3 Schematic
description of the potential
energy in a linear monatomic
chain

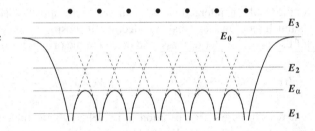

Despite its simplicity, the planetary model is able to explain phenomena like the excitation and ionization of atoms; *excitation* corresponds to the absorption of energy from an external electromagnetic field, such that an initially-bound electron increases its energy from E_1 to E_2, where $E_1 < E_2 < 0$: the electron in the final state is still bound. The inverse process is the emission of energy in form of an electromagnetic radiation, so that the electron's total energy decreases from E_2 to E_1. In turn, *ionization* corresponds to the absorption of energy such that an initially-bound electron becomes free: $E_1 < 0$ and $E_2 \geq 0$. The inverse process is the capture of a free electron by an ionized atom, with an energy emission equal to $E_2 - E_1$.

The above reasoning can also be used to explain the behavior of systems more complicate than single atoms. For instance, consider a finite *linear monatomic chain*, namely, a system made of a finite number of identical atoms placed along a line, at equal mutual distances (Fig. 7.3), which can be thought of as a rudimental version of a crystal. The positions of the atoms are marked by the dots visible in the upper part of the figure. Let the chain be aligned with the x axis; if each nucleus is made to coincide with a local origin of the reference, the distance $r = \sqrt{x^2 + y^2 + z^2}$ in the expression of the potential energy becomes $|x|$; the potential energy pertaining to each nucleus is proportional to $1/|x|$ and is indicated with a dashed line in the figure. An electron placed at some position in the chain is subjected to the sum of

Fig. 7.4 The same structure of Fig. 7.3, where the peaks are replaced with the envelope

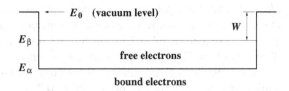

the potential energies; being the latter negative, the sum is lower than the individual contributions: in the figure, it is indicated by the continuous line, which for simplicity is drawn by adding up the contributions of the nearest-neighboring nuclei only. The chain, however, has a finite length; when the leftmost nucleus is considered, the potential energy on its left does not suffer any lowering: this creates the energy step visible in the figure. The same happens on the right side of the righmost nucleus. The shape of the potential energy thus obtained is able to qualitatively explain several features of crystals. For instance, consider the case where the only force acting on an electron inside the crystal derives from the potential energy of Fig. 7.3, namely, it is a conservative force. If the total energy of the electron under consideration is E_1, the electron's position is confined within the potential well dictated by the initial position of its motion. Thus, the electron oscillates within the well like in the example of Prob. 3.2, and its motion can not extend out of it; if all electrons of the crystal are bound, the material is an insulator. This reasoning implies that the situation where all electrons are bound is maintained also under the application of an external voltage; due to this, no electric current ensues.

If the total energy of the electron under consideration is E_2, the electron can move within the whole crystal; finally, if the total energy is E_3, the electron overcomes one or the other of the energy steps and moves into vacuum: for this reason, the minimum energy E_0 necessary for the electron to leave the crystal is called *vacuum level*. If the two ends of the crystal are connected to a voltage generator by suitable contacts, and an external voltage is applied, the material can carry an electric current, whose amplitude depends also on the number of electrons whose energy is sufficiently high.[4] It is worth noting that, although this is prohibited in the frame of Classical Mechanics, the electrons whose energy is of the type E_2 may also contribute to the current; in fact, Quantum Mechanics shows that they have a finite probability to penetrate the energy step and reach the contact. This phenomenon is called *tunnel effect* (Sect. 11.3.1).

To proceed it is convenient to give Fig. 7.3 a simpler appearance: in fact, considering that the interatomic distance is a fraction of a nanometer, the spatial extent in the x direction where each peak is placed is hardly visible in a macroscopic representation; for this reason, it is sufficient to graphically indicate the envelope E_α of the peaks. By the same token, the steps on the sides are described as discontinuities (Fig. 7.4). The electrons with $E < E_\alpha$ or $E \geq E_\alpha$ are called, respectively, *bound*

---

[4] The combination of the number of such electrons with other factors also determines whether the material is a conductor or a semiconductor (Chap. 18).

electrons and *free electrons*.[5] In the equilibrium condition the total energy of the electrons is prescribed; it follows that the majority of the electrons has an energy E lower than a given value E_β (which is not necessarily larger than E_α). Thus, the difference $W = E_0 - E_\beta$ is a measure of the energy that is necessary to extract an electron from the material: among other things, the model explains the existence of a minimum extraction energy of the electrons.[6]

7.3 Experiments Contradicting the Classical Laws

About 1900, experimental evidence was found for a number of phenomena that contradict the calculations based on the known physical laws of the time, that is, the laws of Analytical Mechanics, Electromagnetism, and Statistical Mechanics. A number of such phenomena are listed in this section.

Stability of the Atom
The solution of the electromagnetic equations shows that an accelerated electron radiates a power given by (5.72), namely, $q^2\dot{v}^2/(6\pi\,\varepsilon_0\,c^3)$, with $\dot{v}$ the electron's acceleration. As discussed in Sect. 5.11.2, this is in contradiction with the planetary model of the atom (7.3), in which $T + V = E = \text{const}$: due to the radiated power, the electron should lose energy and, as a consequence, the atom should shrink. The possibility of an extremely slow, non-detectable shrinking must be ruled out: in fact, a perturbative calculation (Sect. 5.11.3) shows that due to the energy loss the atomic radius should decrease from the initial value, say, r, to the value r/e in about 10^{-8} s. This, however, is not observed (Sect. 9.7.2).

Spectral Lines of Excited Atoms
The planetary model explains the emission of electromagnetic radiation by excited atoms. The explanation, however, is qualitative only, because the model does not impose any constraint on the frequency of the emitted waves. In contrast, the experiments show that the waves emitted by, e.g., hydrogen atoms have frequencies v of the form (*Balmer law*, 1885),[7]

$$v_{nm} = v_R \left(\frac{1}{n^2} - \frac{1}{m^2} \right), \qquad v_R \simeq 3.3 \times 10^{15}\ \text{s}^{-1}, \qquad (7.4)$$

---

[5] The description is qualitative; for instance, it does not consider the band structure of the solid (Sect. 17.6).

[6] When the material is a conductor, E_β coincides with the Fermi level (Sect. 15.8.1), and W is called *work function*; in a semiconductor, E_β coindices with the lower edge E_C of the conduction band (Sect. 17.6.5) and the minimum extraction energy (typically indicated with a symbol different from W) is called *electron affinity* (Sect. 22.2).

[7] The ratio $R = v_R/c \simeq 1.1 \times 10^5$ cm^{-1} is called *Rydberg constant*. The formula was generalized in the 1880s to the hydrogenic-like atoms by Rydberg: the expression (7.4) of the frequencies must be multiplied by a constant that depends on the atom under consideration.

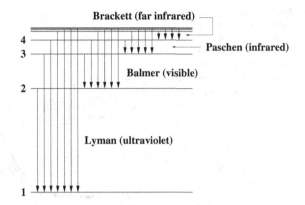

Fig. 7.5 Designation of the lower series of spectral lines (7.4)

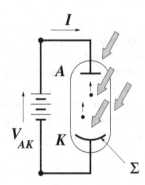

Fig. 7.6 Schematic cross-section of the apparatus used for measuring the photoelectric effect

where n, m are integers, $m > n \geq 1$. The emissions described by (7.4) are also called *spectral lines*. The lower series of spectral lines are shown in Fig. 7.5 along with their designations; the numbers in the figure correspond to n, m in (7.4). Another experimental finding shows that, instead of occurring with a single emission of frequency ν_{nm}, the release of electromagnetic energy by the atom may be accomplished in steps; if that happens, the frequencies associated to the individual steps fulfill a relation called *Ritz emission rule*: considering, e.g., two steps, it reads

$$\nu_{nm} = \nu_{nk} + \nu_{km}, \tag{7.5}$$

with ν_{nk}, ν_{km} the frequencies of the individual steps.

Photoelectric Effect

It is found that an impinging electromagnetic radiation extracts charges from a metal (H. Hertz, 1887) and that these charges are electrons (J. J. Thomson, 1899). The phenomenon is ascribed to the absorption of energy from the radiation: the electron absorbs an energy sufficiently large to be extracted from the metal. An electron thus extracted is also called *photoelectron*. A sketch of the measuring apparatus is given in Fig. 7.6, where two electrodes, anode (A) and cathode (K), are placed inside a vacuum

Fig. 7.7 The $I = I(V_{AK})$ curves, in arbitrary units, obtained from the photoelectric effect at constant frequency of the radiation, with the spectral power used as a parameter

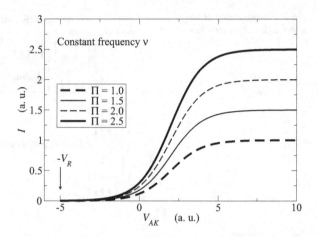

tube in order to prevent interactions between the photoelectrons and the atmosphere. A voltage V_{AK} is applied to the electrodes, such that $V_{AK} > 0$ when the electric potential at the anode is larger than that at the cathode. A monochromatic radiation of a given intensity is made to impinge on the cathode, whose surface is marked with Σ, and the current I flowing in the tube is recorded. Important parameters are the radiation's frequency ν, the spectral intensity of the radiation, $\eta = dE/(d\Sigma\, dt\, d\nu)$, where $d\Sigma$ is the surface element of the cathode, and the spectral power[8]

$$\Pi = \int_\Sigma \eta\, d\Sigma = \frac{dE}{dt\, d\nu}. \tag{7.6}$$

The outcome of the experiment is shown in arbitrary units in Figs. 7.7 and 7.8. The first one shows a set of the $I = I(V_{AK})$ curves at constant ν, with Π a parameter. When V_{AK} is positive and sufficiently high, it is expected that practically all electrons extracted from the cathode be driven to the anode; as a consequence, the slope of the curves should be negligible. Also, when the intensity of the radiation increases, the number of extracted electrons, and the current with it, should also increase. This is in fact confirmed by the curves of Fig. 7.7. When, instead, V_{AK} is negative, only the electrons leaving the cathode with a sufficiently high kinetic energy are able to reach the anode, whereas those whose initial kinetic energy is low are repelled towards the cathode by the electric field imposed by the reverse bias.[9] Considering for simplicity a one-dimensional case, energy conservation yields for an electron traveling from cathode to anode,

$$\frac{1}{2}m\, u_A^2 - \frac{1}{2}m\, u_K^2 = q V_{AK}, \tag{7.7}$$

---

[8] The units of η and Π are $[\eta] = \mathrm{J\ cm^{-2}}$ and $[\Pi] = \mathrm{J}$, respectively.

[9] The concentration of electrons in the vacuum tube is small enough not to influence the electric field; thus, the latter is due only to the value of V_{AK} and to the form of the electrodes.

Fig. 7.8 The $I = I(V_{AK})$ curves, in arbitrary units, obtained from the photoelectric effect at constant spectral power of the radiation, with frequency used as a parameter

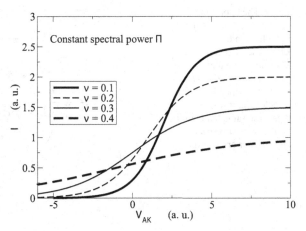

where u_K is the electron's velocity at the cathode and u_A that at the anode. The *blocking voltage* $V_R > 0$ is the value $V_{AK} = -V_R$ such that $u_A = 0$; from (7.7) one obtains the relation

$$\frac{1}{2} m u_K^2 = q V_R, \tag{7.8}$$

which allows one to measure the kinetic energy of the most energetic electrons that are extracted from the cathode at given spectral power and frequency of the radiation.[10] Such electrons are those that inside the cathode have an energy in the vicinity of E_β (Fig. 7.4) and do not suffer energy losses while being extracted. If E_L is the energy that the most energetic electron absorbs from the radiation, its kinetic energy at the cathode is $(1/2) m u_K^2 = E_L - W$, with W the metal's work function, whence

$$q V_R = E_L - W, \tag{7.9}$$

so that the photoelectric effect provides in fact a method for measuring E_L. The classical model predicts that the blocking voltage should increase with Π; this, however, does not happen: as shown in Fig. 7.7, at a given frequency the blocking voltage is the same for all values of Π.

In addition, it is unexpectedly found that both I and V_R depend on the frequency ν (Fig. 7.8). In fact, the comparison between the experimental blocking voltages and (7.9) shows that the energy E_L that the electron absorbs from the electromagnetic field is proportional to the frequency,

$$E_L = h \nu, \tag{7.10}$$

with $h \simeq 6.626 \times 10^{-34}$ J s the *Planck constant*. If $h \nu < W$, no current is measured; this provides a threshold value for the frequency to be used in the experiment.

---

[10] The most energetic electrons succeed in overcoming the effect of the reverse bias and reach the vicinity of the anode; they constantly slow down along the trajectory, to the point that their velocity at the anode vanishes. Then, their motion reverses and they are driven back to the cathode.

Fig. 7.9 The approximation
to a black body consisting in
a small hole in the wall of an
enclosure kept at constant
temperature. If a thermometer
(represented by the *shaded
area*) were suspended within
the enclosure, it would
indicate the same temperature
T as the walls, irrespective of
its position or orientation

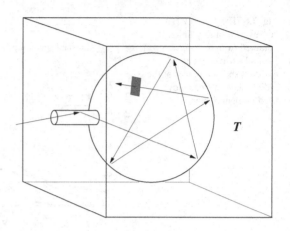

Spectrum of the Black-Body Radiation

Consider a body at temperature T in equilibrium with an electromagnetic field. Due
to the detailed-balance principle, the spectral intensity η_B emitted by the body, that
is, the electromagnetic power emitted by it per unit surface element $d\Sigma$ and unit
frequency $d\nu$, in the direction normal to $d\Sigma$, fulfills the relation

$$\eta_B = \alpha\, \eta, \qquad (7.11)$$

where η is the spectral intensity of the radiation (compare with (7.6)), and $0 \le \alpha \le 1$
the fraction of η absorbed by the body at frequency ν. By Kirchhoff's law (1859),
for any body in thermal equilibrium with radiation it is

$$\frac{\eta_B}{\alpha} = K(\nu, T), \qquad (7.12)$$

where K is a universal function of ν and T [85, Sect. 9–15]. A *black body* is a body
such that $\alpha = 1$ at all frequencies; thus, for a black body at equilibrium with radiation
it is $\eta_B = K$. A good approximation to a black body is a small hole in the wall of an
enclosure kept at constant temperature, like that illustrated in Fig. 7.9: any radiation
entering the hole has a negligible probability of escaping, due to multiple reflections at
the walls; as a consequence, the hole acts like a perfect absorber. Thanks to $\eta_B = K$,
the spectral intensity emitted by any black body has the same characteristics: in
particular, it is not influenced by the form of the enclosure, the material of which
the walls are made, or other bodies present in the enclosure. As a consequence, η_B,
or any other function related to it, can be calculated by considering a convenient
geometry of the problem and assuming that the radiation propagates *in vacuo*.

To proceed, consider a surface element $d\Sigma$ of the black body, and connect a local
Cartesian reference to it such that dx and dy belong to $d\Sigma$; it follows

$$\eta_B = \frac{dE}{dx\, dy\, dt\, d\nu} = c\, \frac{dE}{dx\, dy\, dz\, d\nu} = c\, u, \qquad (7.13)$$

where u is the *spectral energy density* of the black body, that is, the energy per unit volume and frequency. The integral of u over the frequencies yields the energy density; remembering that equilibrium is assumed, one finds[11]

$$w_{em}^{eq}(T) = \int_0^\infty u(v, T) \, dv. \tag{7.14}$$

In turn, the integral of u over the coordinates gives the equilibrium value of the spectral energy, whose general definition is given by (5.47). It is found experimentally (Stefan's law, 1879) that

$$\int_0^\infty \eta_B(v, T) \, dv = \sigma \, T^4, \tag{7.15}$$

where $\sigma = 5.67 \times 10^{-12}$ W cm^{-2} K^{-4} is the Stefan-Boltzmann constant. Combining (7.15) with (7.13) and (7.14) yields $w_{em}^{eq}(T) = \sigma \, T^4/c$.

The spectral energy density u can be calculated as the product of the number of monochromatic components of the electromagnetic field per unit volume and frequency, times the energy of each monochromatic component. The first factor is readily found by taking an enclosure of prismatic form like that of Sect. 15.9.4; the calculation yields $8 \pi v^2/c^3$, which is obtained by dividing both sides of (15.74) by the enclosure's volume V. As for the energy of each monochromatic component, the only assumption possible in the frame of Classical Mechanics is that the energy of the electromagnetic field at equilibrium is distributed over the frequencies according to the Maxwell–Bolzmann distribution (6.14). Assuming that each monochromatic component is equivalent to a one-dimensional linear-harmonic oscillator, the energy to be associated to it is the average energy of a system with one degree of freedom; thus, letting $R = 1$ in (6.37), yields for the average energy the value $k_B T$. The product of the two factors thus found yields for the spectral energy density of the black body the expression

$$u(v, T) = 8 \pi \, \frac{k_B T}{c^3} \, v^2, \tag{7.16}$$

called *Rayleigh-Jeans law*. Experimental results for u as a function of frequency are shown in Fig. 7.10, with temperature a parameter. The comparison with experiments shows that the parabolic behavior of (7.16) approximates the correct form of the curves only at low frequencies; clearly the result expressed by (7.16) can not be correct, because it makes the equilibrium energy density (7.14) to diverge.[12]

---

[11] Compare with the general definition (5.10) of w_{em}, where the assumption of equilibrium is not made.

[12] This unphysical outcome is also called *ultraviolet catastrophe*.

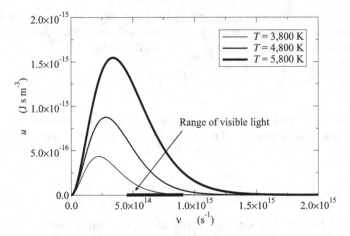

Fig. 7.10 Spectral energy density of the black body at different temperatures. The value $T = 5800$ K corresponds to the surface temperature of the sun

Compton Effect

When X-rays of a given frequency interact with atoms and are scattered with an angle ψ with respect to the direction of incidence, the frequency of the emerging rays is found to depend on ψ. This outcome is in contrast with the prediction of the electromagnetic theory, according to which the frequency of the scattered rays should be equal to that of the impinging ones. The dependence of frequency on the scattering angle is also called *Compton effect*.

The experimental setup for measuring the Compton effect is schematically shown in Fig. 7.11. The gray box in the middle of the figure is a piece of solid material, onto which the radiation impinges from the left (dark arrows); the vertical lines are the intersections of the constant-phase planes with the plane of the figure. The gray arrows on the right represent the part of the radiation that does not interact with the material and exits from it unaltered, while the white arrows indicate some of the directions of the rays scattered by the material. The circumferences are the intersections with the figure's plane of the spherical waves produced by the scattering. The origin of the measuring angle is aligned with the direction of the incoming radiation, so that $\psi = 0$ corresponds to the absence of scattering, $\psi = 2\pi$ to reflection.

7.4   Quantum Hypotheses

In the early 1900s, a number of hypotheses were made to solve the contradictions between the experimental evidence and the calculations based on the physical laws known at that time. The solutions thus found and the new concepts originating from them were eventually combined into a unified and coherent theory, Quantum Mechanics.

Fig. 7.11 Scheme of the experimental setup for measuring the Compton effect

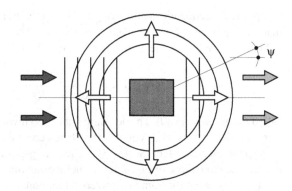

In essence, the contradictions with the physical laws known in the early 1900s were found thanks to the refinement of the experimental techniques. Such refinements were in turn made available by the general advancement of science that had taken place in the preceding decades. Thanks to them, it was possible to start investigating the microscopic world, namely, the dynamics of elementary particles. A parallel improvement took place in the same years in the investigation of the dynamics at high velocities, and led to the concepts of Special Relativity (1905).[13]

7.4.1 Planck's Solution of the Black-Body Problem

To explain the features of the black-body radiation, Planck made in 1900 the hypothesis that a monochromatic electromagnetic energy is absorbed or emitted only in quantities that are integer multiples of a fixed quantity $h\nu$, where h is a suitable constant [82]. The occupation number then becomes

$$P_n = P_0 \exp(-n\beta h \nu), \qquad \beta = 1/(k_B T). \tag{7.17}$$

As a consequence, using the same procedure as in Sect. 6.6.2 after replacing the integrals with sums, yields for the average energy $\overline{n h \nu}$ the expression [14]

$$\overline{n h \nu} = \frac{\sum_{n=0}^{\infty} n h \nu \, P_n}{\sum_{n=0}^{\infty} P_n} = \frac{h \nu}{\exp(\beta h \nu) - 1}. \tag{7.18}$$

In contrast with the constant value $k_B T$ used in the determination of the Rayleigh-Jeans law, here the average energy of each monochromatic component depends on the

---

[13] As the particles' velocities that occur in solid-state physics are low, Special Relativity is not used in this book; the only exception is in the explanation of the Compton effect, illustrated in Sect. 7.4.3.

[14] The detailed calculation leading to (7.18) is shown in Prob. 6.1.

component's frequency. Multiplying (7.18) by the number $8\pi\,\nu^2/c^3$ of monochromatic components of the electromagnetic field per unit volume and frequency, found in Sect. 7.3, yields for the spectral energy density of the black body the expression

$$u(\nu, T) = 8\pi\frac{h\,\nu^3/c^3}{\exp[h\,\nu/(k_B\,T)] - 1},\qquad(7.19)$$

called *Planck law* (1900). The derivation of (7.19) involves one undetermined parameter, h. If the latter is made equal to the Planck constant introduced in the description of the photoelectric effect (Sect. 7.3), the resulting expression fits perfectly the experimental data like those of Fig. 7.10. Remembering that the spectral energy density of a black body in equilibrium is a universal function, it follows that h does not depend on the specific experiment, namely, it is a universal constant.

The low-frequency limit of (7.19), $h\,\nu \ll k_B\,T$, is independent of h and renders the Rayleigh–Jeans law (7.16).

7.4.2 Einstein's Solution of the Photoelectric Effect

In 1905, Einstein proposed the following explanation of the photoelectric effect: the transport of electromagnetic energy is *quantized*; specifically, a monochromatic electromagnetic wave of frequency ν is made of the flux of identical objects, called *photons*, each carrying the energy $h\,\nu$. In the interaction with a photon, an electron may absorb an energy up to $h\,\nu$. If the absorbed energy is exactly $h\nu$, the photon is annihilated [34].[15] This theory provides a correct explanation of the photoelectric effect: with reference to Fig. 7.7, the photoelectric current increases as the spectral power Π increases at constant ν, because the number of photons is larger: as a consequence, the number of photoelectrons is larger as well. In turn, with reference to Fig. 7.8, the blocking voltage V_R increases as ν increases at constant Π, because the photons are more energetic; however, they are fewer, which explains why the curves intersect each other: the spectral power, in fact, can be written as $\Pi = dE/(dt\,d\nu) = h\,\nu\,[dN/(dt\,d\nu)]$, where the quantity in brackets is the number of photons per unit time and frequency; as a consequence, the constraint $\Pi = $ const of the experiment of Fig. 7.8 makes the quantity in brackets to decrease when the photon energy $h\,\nu$ increases.

7.4.3 Explanation of the Compton Effect

The concept of photon, introduced in Sect. 7.4.2, explains the Compton effect by describing the interaction of the electron with the electromagnetic field as the collision

---

[15] Einstein's hypothesis is more general than Planck's: the latter, in fact, assumes that energy is quantized only in the absorption or emission events.

between the electron and a photon [19]. As the photon's velocity is c, its rest mass is zero (Sect. 3.13.7); in turn, the modulus of the photon's momentum is $p = E/c$, which is consistent with classical electromagnetism (compare with (5.43)).

The analysis of the electron-phonon collision is worked out assuming that the system made of the two particles under consideration is isolated; thus, the calculation is based upon the energy- and momentum-conservation equations, and the results of Sect. 3.13.8 hold. The dynamical quantities for the photon are given by

$$E = h\nu, \qquad p = \frac{E}{c} = \frac{h\nu}{c} = \frac{h}{\lambda}, \tag{7.20}$$

the second of which derives from (5.55) expressed *in vacuo*. Defining the *reduced Planck constant* $\hbar = h/(2\pi) \simeq 1.055 \times 10^{-34}$ J s, and using the modulus k of the wave vector, (7.20) becomes

$$E = \hbar\, 2\pi\, \nu = \hbar\,\omega, \qquad p = \frac{\hbar}{\lambda/(2\pi)} = \hbar\, k. \tag{7.21}$$

The second relation of (7.21) in vector form reads

$$\mathbf{p} = \hbar\, \mathbf{k}. \tag{7.22}$$

Here the useful outcome of the analysis of Sect. 3.13.8 is (3.92), that relates the photon's energies prior and after the collision (E_a and E_b, respectively) with the deflection angle ψ (Fig. 3.7). Using $E = c\, h/\lambda$ in (3.92) yields

$$\lambda_b - \lambda_a = 2\,\lambda_0 \sin^2\left(\frac{\psi}{2}\right), \qquad \lambda_0 = \frac{h}{m_0\, c}, \tag{7.23}$$

with $\lambda_0 \simeq 2.43 \times 10^{-12}$ m the *Compton wavelength* (1923). The frequency corresponding to it is $\nu_0 = c/\lambda_0 \simeq 1.2 \times 10^{20}$ Hz. The maximum difference in wavelength corresponds to the case of reflection, $\max(\lambda_b - \lambda_a) = 2\,\lambda_0$. Even in this case, the smallness of λ_0 makes the effect difficult to measure; in practice, the shift in wavelength is detectable only for sufficiently small values of λ_a, typically in the range of 10^{-10} m corresponding to the X-ray frequencies ($\nu \sim 10^{18}$ s^{-1}). Due to the large energy of the photon, the energy transferred to the electron brings the latter into a high-velocity regime; this, in turn, imposes the use of the relativistic expressions for describing the electron's dynamics.

7.4.4 Bohr's Hypothesis

The description of the monochromatic components of the electromagnetic field as a flow of identical photons with energy $h\,\nu$ lends itself to the explanation of the Balmer law (7.4). Such an explanation (*Bohr's hypothesis*, 1913) is based on the idea that, if

ν_{nm} is the frequency of the emitted radiation, the corresponding energy of the emitted photon is $h\,\nu_{nm}$; multiplying (7.4) by h and remembering that $m > n$ then yields

$$h\,\nu_{nm} = h\,\nu_R\left(\frac{1}{n^2} - \frac{1}{m^2}\right) = \left(-\frac{h\,\nu_R}{m^2}\right) - \left(-\frac{h\,\nu_R}{n^2}\right). \tag{7.24}$$

As the left hand side is the energy of the emitted photon, the terms on right hand side can be recast as

$$E_m = -\frac{h\,\nu_R}{m^2}, \qquad E_n = -\frac{h\,\nu_R}{n^2}, \qquad E_n < E_m < 0; \tag{7.25}$$

then, if E_m (E_n) is interpreted as the atom's energy before (after) emitting the photon, Balmer's law becomes the expression of energy conservation. From this, the emission rule of Ritz is easily explained; in fact, (7.5) is equivalent to

$$E_m - E_n = (E_m - E_k) + (E_k - E_n). \tag{7.26}$$

Bohr's hypothesis is expressed more precisely by the following statements:

1. The energy variations of the atom are due to the electrons of the outer shell, that exchange energy with the electromagnetic field.
2. The total energy of a non-radiative state is quantized, namely, it is associated to an integer index: $E_n = -h\,\nu_R/n^2$, $n = 1, 2, \ldots$; the values of energy thus identified are called *energy levels*. The lowest level corresponds to $n = 1$ and is called *ground level* or *ground state*.
3. The total energy can vary only between the quantized levels by exchanging with the electromagnetic field a photon of energy $\nu_{nm} = (E_m - E_n)/h$.

It is interesting to note that, by combining Bohr's hypothesis with the planetary model of the atom, the quantization of the other dynamical quantities follows from that of energy; again, the case of a circular orbit is considered. By way of example, using $E_n = -h\,\nu_R/n^2$ in the second relation of (7.3) provides the quantization of the orbit's radius:

$$r = r_n = -\frac{q^2}{8\,\pi\,\varepsilon_0\,E_n} = \frac{q^2}{8\,\pi\,\varepsilon_0}\,\frac{n^2}{h\,\nu_R}. \tag{7.27}$$

The smallest radius r_1 corresponds to the ground state $n = 1$; taking ν_R from (7.4) and the other constants from Table D.1 one finds $r_1 \simeq 0.05$ nm; despite the simplicity of the model, r_1 is fairly close to the experimental value r_a given in (7.1).

In turn, the velocity is quantized by combining (7.3) to obtain $T = -V/2 = -E$; replacing the expressions of T and E then yields

$$\frac{1}{2}\,m\,u^2 = \frac{h\,\nu_R}{n^2}, \qquad u = u_n = \sqrt{\frac{2\,h\,\nu_R}{m\,n^2}}. \tag{7.28}$$

The largest velocity is found from (7.28) by letting $n = 1$ and using the minimum value for the mass, that is, the rest mass $m = m_0$. It turns out $u_1 \simeq 7 \times 10^{-3}\,c$;

as a consequence, the velocity of a bound electron belonging to the outer shell of the atom can be considered non relativistic. Thanks to this result, from now on the electron's mass will be identified with the rest mass. Finally, for the angular momentum $M = r\,p = r\,m\,u$ one finds

$$M = M_n = \frac{q^2 n^2}{8\pi\,\varepsilon_0\,h\,v_R}\,m\,\sqrt{\frac{2\,h\,v_R}{m\,n^2}} = \frac{1}{2\pi}\left[\frac{q^2}{\varepsilon_0}\,\sqrt{\frac{m}{8\,h\,v_R}}\right] n. \tag{7.29}$$

The quantity in brackets in (7.29) has the same units as M, namely, an action (Sect. 1.5) and, replacing the constants, it turns out[16] to be equal to h. Using the reduced Planck constant it follows

$$M_n = n\,\hbar. \tag{7.30}$$

The Bohr hypothesis provides a coherent description of some atomic properties; yet it does not explain, for instance, the fact that the electron belonging to an orbit of energy $E_n = -h\,v_R/n^2$ does not radiate, in contrast to what is predicted by the electromagnetic theory (compare with the discussion in Sect. 7.3). Another phenomenon not explained by the hypothesis is the fact that only the ground state of the atom is stable, whereas the excited states are unstable and tend to decay to the ground state.

7.4.5 De Broglie's Hypothesis

The explanation of the Compton effect (Sect. 7.4.3) involves a description of the photon's dynamics in which the latter is treated like a particle having energy and momentum. Such mechanical properties are obtained from the wave properties of a monochromatic component of the electromagnetic field: the relations involved are (7.20) (or (7.21)), by which the photon energy is related to the frequency, and its momentum to the wave vector. It is worth specifying that such relations are applied to the asymptotic part of the motion, namely, when the photon behaves like a free particle. In 1924, de Broglie postulated that analogous relations should hold for the free motion of a real particle: in this case, the fundamental dynamic properties are energy and momentum, to which a frequency and a wavelength (or a wave vector) are associated by relations identical to (7.20), (7.21),[17]

$$\omega = 2\pi\,\nu = 2\pi\,\frac{E}{h} = \frac{E}{\hbar}, \qquad k = \frac{2\pi}{\lambda} = \frac{2\pi}{h/p} = \frac{p}{\hbar}, \qquad \mathbf{k} = \frac{\mathbf{p}}{\hbar}. \tag{7.31}$$

The usefulness of associating, e.g., a wavelength to a particle's motion lies in the possibility of qualitatively justifying the quantization of the mechanical properties

---

[16] This result shows that the physical constants appearing in (7.29) are not independent from each other. Among them, v_R is considered the dependent one, while q, $m = m_0$, ε_0, and h are considered fundamental.

[17] The wavelength associated to the particle's momentum is called *de Broglie's wavelength*.

illustrated in Sect. 7.4.4. For this, consider the case of the circular orbit of the planetary motion, and associate a wavelength to the particle's momentum, $\lambda = h/p$. Such an association violates the prescription that (7.31) apply only to a free motion; however, if the orbit's radius is very large, such that $\lambda \ll r$, the orbit may be considered as locally linear and the concept of wavelength is applicable. Replacing $\lambda = h/p$ in (7.30) yields

$$2\pi r = n\lambda, \tag{7.32}$$

namely, the quantization of the mechanical properties implies that the orbit's length is an integer multiple of the wavelength associated to the particle. This outcome suggests that the formal description of quantization should be sought in the field of eigenvalue equations.

De Broglie also postulated that a function $\psi = \psi(\mathbf{r}, t)$, containing the parameters $\omega, \mathbf{k}$ defined in (7.31), and called *wave function*, is associated to the particle's motion. Its meaning is provisionally left indefinite; as for its form, it is sensible to associate to the free motion, which is the simplest one, the simplest wave function, that is, the planar monochromatic wave. The latter is conveniently expressed in complex form as

$$\psi = A \exp[i(\mathbf{k} \cdot \mathbf{r} - \omega t)], \tag{7.33}$$

where $A \neq 0$ is a complex constant, not specified. Due to (7.31), the constant wave vector $\mathbf{k}$ identifies the momentum of the particle, and the angular frequency ω identifies its total energy, which in a free motion coincides with the kinetic energy. It is worth pointing out that, despite its form, the wave function is not of electromagnetic nature; in fact, remembering that in a free motion it is $H = p^2/(2m) = E$, with H the Hamiltonian function, it follows

$$\hbar\omega = \frac{1}{2m}\hbar^2 k^2, \qquad \omega(\mathbf{k}) = \frac{\hbar}{2m}k^2, \tag{7.34}$$

which is different from the electromagnetic relation $\omega = ck$. By the same token it would not be correct to identify the particle's velocity with the phase velocity u_f derived from the electromagnetic definition; in fact, one has

$$u_f = \frac{\omega}{k} = \frac{E/\hbar}{p/\hbar} = \frac{p^2/(2m)}{p} = \frac{p}{2m}. \tag{7.35}$$

The proper definition of velocity is that deriving from Hamilton's Eqs. (1.42); its ith component reads in this case

$$u_i = \dot{x}_i = \frac{\partial H}{\partial p_i} = \frac{1}{\hbar}\frac{\partial H}{\partial k_i} = \frac{\partial \omega}{\partial k_i} = \frac{\hbar k_i}{m} = \frac{p_i}{m}. \tag{7.36}$$

The concepts introduced so far must now be extended to motions of a more general type. A sensible generalization is that of the conservative motion of a particle subjected to the force deriving from a potential energy $V(\mathbf{r})$. In this case the association described by (7.31) works only partially, because in a conservative motion the total energy is a constant, whereas momentum is generally not so. As a consequence, letting $\omega = E/\hbar$ yields for the wave function the form

$$\psi = w(\mathbf{r}) \exp(-i\omega t), \tag{7.37}$$

which is still monochromatic but, in general, not planar. Its *spatial part* $w(\mathbf{r})$ reduces to $A \exp(i\mathbf{k} \cdot \mathbf{r})$ for the free motion. The function of the form (7.37) is postulated to be the wave function associated with the motion of a particle at constant energy $E = \hbar\omega$. While the time dependence of ψ is prescribed, its space dependence must be worked out, likely by solving a suitable equation involving the potential energy V, the particle's mass and, possibly, other parameters.

7.5 Heuristic Derivation of the Schrödinger Equation

The concept of wave function introduced in Sect. 7.4.5 has been extended from the case of a free motion, where the wave function is fully prescribed apart from the multiplicative constant A, to the case of a conservative motion (7.37), where only the time dependence of the wave function is known. It is then necessary to work out a general method for determining the spatial part $w(\mathbf{r})$. The starting point is the observation that w is able at most to provide information about the particle's trajectory, not about the particle's dynamics along the trajectory. One of the methods used in Classical Mechanics to determine the trajectories is based on the Maupertuis principle (Sect. 2.7); moreover, from the discussion carried out in Sect. 5.11.6 it turns out that the analogy between the Maupertuis principle and the Fermat principle of Geometrical Optics (compare with (5.80)) provides the basis for a generalization of the mechanical laws. The first of the two principles applies to a particle (or system of particles) subjected to a conservative force field prescribed by a potential energy $V(\mathbf{r})$, with E a given constant; the second one applies to a monochromatic ray ($\nu = $ const) propagating in a medium whose properties are prescribed by the refraction index $n(\mathbf{r})$. The latter is related to frequency and wavelength by $n = c/(\lambda\nu)$ (compare with (5.55)); as a consequence, (5.80) can be rewritten as

$$\delta \int_{AB} \sqrt{E - V}\, ds = 0, \qquad \delta \int_{AB} \frac{1}{\lambda}\, ds = 0. \tag{7.38}$$

Considering that the variational principles hold apart from a multiplicative constant, the two expressions in (7.38) transform into each other by letting

$$\sqrt{E - V} = \frac{\alpha}{\lambda}, \tag{7.39}$$

where α is a constant that must not depend on the form of V or λ, nor on other parameters of the problem. For this reason, α is left unchanged also after removing the Geometrical-Optics approximation; when this happens, the Fermat principle is replaced with the Maxwell equations or, equivalently, with the wave equations for

the electric field (4.64) and magnetic field (4.65). For simplicity, the latter equations
are solved in a uniform medium with no charges in it, on account of the fact that α is
not influenced by the medium's properties. Also, considering that in the uniform case
(4.64) and (4.65) have the same structure, and that the function w under investigation
is scalar, the analysis is limited to any scalar component C of $\mathbf{E}$ or $\mathbf{H}$; such a
component fulfills the equation

$$\nabla^2 C - \frac{1}{u_f^2} \frac{\partial^2 C}{\partial t^2} = 0, \tag{7.40}$$

with $u_f = $ const the medium's phase velocity. Solving (7.40) by separation with
$C(\mathbf{r}, t) = \eta(\mathbf{r}) \theta(t)$ yields

$$\theta \nabla^2 \eta = \frac{1}{u_f^2} \ddot{\theta} \eta, \qquad \frac{\nabla^2 \eta}{\eta} = \frac{1}{u_f^2} \frac{\ddot{\theta}}{\theta} = -k^2, \tag{7.41}$$

where the separation constant $-k^2$ must be negative to prevent θ from diverging. As
a consequence, k is real and can be assumed to be positive. The solution for the time
factor is $\theta = \cos(\omega t + \varphi)$, where the phase φ depends on the initial conditions, and
$\omega = 2\pi \nu = u_f k > 0$. It follows $k = 2\pi \nu/u_f = 2\pi/\lambda$, whence the spatial part of
(7.40) reads

$$\nabla^2 \eta + \frac{(2\pi)^2}{\lambda^2} \eta = 0, \tag{7.42}$$

namely, a Helmholtz equation (Sect. 4.7). By analogy, the equation for the spatial
part w of the wave function is assumed to be

$$\nabla^2 w + \frac{(2\pi)^2}{\alpha^2} E w = 0, \tag{7.43}$$

which is obtained from (7.42) by replacing η with w and using (7.39) with $V = 0$. The
value of α is determined by expressing E in (7.43) in terms of the de Broglie wavelength; using the symbol λ_{dB} for the latter to avoid confusion with the electromagnetic
counterpart, one finds $E = p^2/(2m) = h^2/(2m \lambda_{dB}^2)$, namely,

$$\nabla^2 w + \frac{(2\pi)^2}{\alpha^2} \frac{h^2}{2m} \frac{1}{\lambda_{dB}^2} w = 0. \tag{7.44}$$

Equation (7.44) becomes identical to (7.42) by letting $\alpha^2 = h^2/(2m)$ whence, using
the reduced Planck constant $\hbar$, (7.43) becomes $\nabla^2 w + (2m E/\hbar^2) w = 0$. Such
a differential equation holds in a uniform medium; hence, the dynamical property
involved is the kinetic energy of the particle. The extension to the case of a non-
uniform medium is then obtained by using the general form $E - V$ of the kinetic
energy in terms of the coordinates; in conclusion, the equation for the spatial part of
the wave function in a conservative case is

$$\nabla^2 w + \frac{2m}{\hbar^2} (E - V) w = 0, \qquad -\frac{\hbar^2}{2m} \nabla^2 w + V w = E w. \tag{7.45}$$

The above is the *time-independent Schrödinger equation*. It is a homogenous equation, with E the eigenvalue and w the eigenfunction.[18] Although the derivation based on the analogy between mechanical and optical principles is easy to follow, it must be remarked that the step leading from (7.44) to (7.45) is not rigorous; in the electromagnetic case, in fact, Eq. (7.42) for the spatial part holds only in a uniform medium; when the latter is non uniform, instead, the right hand side of (7.42) is different from zero, even in a charge-free case, because it contains the gradient of the refraction index. As shown in Sect. 1.10.4, the actual method used in 1926 by Schrödinger for deriving (7.45) consists in seeking the constrained extremum of a functional generated by the Hamilton–Jacobi equation; in such a procedure, the hypothesis of a uniform medium is not necessary.

It is also worth noting that in the analogy between mechanical and optical principles the spatial part of the wave function, and also the wave function as a whole, is the analogue of a component of the electromagnetic field. From this standpoint, the analogue of the field's intensity is the wave function's square modulus. In the monochromatic case, the latter reads $|\psi|^2 = |w|^2$. This reasoning is useful in the discussion about the physical meaning of ψ.

7.6 Measurement

To make the wave function a useful tool for the description of the particles' dynamics it is necessary to connect the value taken by ψ, at a specific position and time, with some physical property of the particle (or system of particles) under investigation. To make such a connection it is in turn necessary to measure the property of interest; otherwise, the hypotheses illustrated in the previous sections would be relegated to a purely abstract level. In other terms, the meaning of the wave function can be given only by discussing the measuring process in some detail. The analysis is carried out below, following the line of [69]; in particular, a general formalism is sought which applies to both the macroscopic and microscopic bodies; the specific features brought about by the different size of the objects that are being measured are made evident by suitable examples.

The measurement of a dynamical variable A pertaining to a physical body is performed by making the body to interact with a measuring apparatus and recording the reading shown by the latter. For simplicity it is assumed that there is a finite number of possible outcomes of the measurement, say, $A_1, \ldots, A_M$. The extension to the case of a continuous, infinitely extended set of outcomes can be incorporated into the theory at the cost of a more awkward notation. Letting A_i be the outcome of the measurement of A, consider the case where the body is later subjected to the measurement of another dynamical variable B. Assume that the outcome of such a measurement is B_j, out of the possible outcomes $B_1, \ldots, B_N$. Next, the body is subjected to the measurement of a third variable C, this yielding the value C_k, and so on.

---

[18] The structure of (7.45) is illustrated in detail in Chap. 8.

As in general the dynamical variables depend on time, it is necessary to specify the time of each measurement. The most convenient choice is to assume that the time interval between a measurement and the next one is negligibly small, namely, that the measurement of B takes place immediately after that of A, similarly for that of C, and so on. The duration of each measurement is considered negligible as well. A special consequence of this choice is the following: if the measurement of A yielded A_i, and the measurement is repeated (namely, $B = A$), the outcome of the second measurement is again A_i.

Consider now the case where, after finding the numbers A_i, B_j, and C_k from the measurements of A, B, and C, respectively, the three variables are measured again, in any order. The experiments show that the results depend on the size of the body being measured. For a massive body the three numbers A_i, B_j, and C_k are always found. One concludes that the dynamical state of a massive body is not influenced by the interaction with the measuring apparatus or, more precisely, that if such an influence exists, it is so small that it can not be detected. As a consequence one may also say that the values of the dynamical variables are properties of the body that exist prior, during, and after each measurement.

The situation is different for a microscopic body. By way of example, consider the case of a measurement of B followed by a measurement of A, the first one yielding B_n, the second one yielding A_i. If the measurement of B is carried out again after that of A, the result is still one of the possible outcomes $B_1, \ldots, B_N$, but it is not necessarily equal to B_n. In other terms, the individual outcome turns out to be unpredictable. For a microscopic body one concludes that the interaction with the measuring apparatus is not negligible. It is worth observing that the apparatus able to measure the dynamical variable A may also be conceived in such a way as to block all outcomes that are different from a specific one, say, A_i. In such a case the apparatus is termed *filter*. Using the concept of filter one may build up the statistical distribution of the outcomes, for instance by repeating a large number of times the experiment in which the measurement of B is carried out after filtering A_i. The statistics is built up by recording the fraction of cases in which the measurement of B carried out on an A_i-filtered body yields the result B_j, $j = 1, \ldots, N$.

7.6.1 Probabilities

The fraction of measurements of the type described above, namely, of those that yield B_j after a measurement of A that has yielded A_i, will be indicated with the symbol $P(A_i \to B_j)$. Obviously the following hold:

$$0 \le P(A_i \to B_j) \le 1, \qquad \sum_{j=1}^{N} P(A_i \to B_j) = 1. \tag{7.46}$$

The first relation in (7.46) is due to the definition of $P(A_i \to B_j)$, the second one to the fact that the set of values $B_1, \ldots, B_N$ encompasses all the possible outcomes of the measurement of B. It follows that $P(A_i \rightarrow B_j)$ is the probability that a measurement of the dynamical variable B, made on a particle that prior to the measurement is in the state A_i of the dynamical variable A, yields the value B_j. The possible combinations of $P(A_i \rightarrow B_j)$ are conveniently arranged in the form of an M-row$\times N$-column matrix:

$$
\mathbf{P}_{AB} = \begin{bmatrix}
P(A_1 \rightarrow B_1) & \cdots & P(A_1 \rightarrow B_N) \\
\vdots & & \vdots \\
P(A_i \rightarrow B_1) & \cdots & P(A_i \rightarrow B_N) \\
\vdots & & \vdots \\
P(A_M \rightarrow B_1) & \cdots & P(A_M \rightarrow B_N)
\end{bmatrix}. \tag{7.47}
$$

Due to (7.46), each row of $\mathbf{P}_{AB}$ adds up to unity. As the number of rows is M, the sum of all entries of matrix (7.47) is M. The same reasoning can also be made when the measurement of B is carried out prior to that of A. In this case the following N-row$\times M$-column matrix is obtained:

$$
\mathbf{P}_{BA} = \begin{bmatrix}
P(B_1 \rightarrow A_1) & \cdots & P(B_1 \rightarrow A_M) \\
\vdots & & \vdots \\
P(B_j \rightarrow A_1) & \cdots & P(B_j \rightarrow A_M) \\
\vdots & & \vdots \\
P(B_N \rightarrow A_1) & \cdots & P(B_N \rightarrow A_M)
\end{bmatrix}, \tag{7.48}
$$

with

$$
0 \leq P(B_j \rightarrow A_i) \leq 1, \qquad \sum_{i=1}^{M} P(B_j \rightarrow A_i) = 1. \tag{7.49}
$$

As the number of rows in (7.49) is N, the sum of all entries of matrix $\mathbf{P}_{BA}$ is N. It can be proven that it must be

$$
P(B_j \rightarrow A_i) = P(A_i \rightarrow B_j) \tag{7.50}
$$

for any pair of indices ij. In fact, if (7.50) did not hold, thermodynamic equilibrium would not be possible [69, Ch. V-21]. Equality (7.50) makes $\mathbf{P}_{BA}$ the transpose of $\mathbf{P}_{AB}$. As a consequence, the sum of all entries of the two matrices must be the same, namely, $N = M$. In other terms the outcomes of the measurements have the same multiplicity, and the matrices (7.47), (7.48) are square matrices of order $N = M$. Combining (7.50) with the second of (7.49) yields

$$
\sum_{i=1}^{M} P(A_i \rightarrow B_j) = \sum_{i=1}^{M} P(B_j \rightarrow A_i) = 1, \tag{7.51}
$$

showing that in the matrices (7.47), (7.48) not only each row, but also each column adds up to unity. A square matrix where all entries are non negative and all rows and columns add up to unity is called *doubly stochastic matrix*. Some properties of this type of matrices are illustrated in [76, Ch. II-1.4] and in Sect. A.11.

Note that (7.50) does not imply any symmetry of $\mathbf{P}_{AB}$. In fact, symmetry would hold if $P(A_j \rightarrow B_i) = P(A_i \rightarrow B_j)$. If the filtered state is A_i and the measurement of the dynamical variable A is repeated, the result is A_i again. In other terms,

$$P(A_i \rightarrow A_i) = 1, \qquad P(A_i \rightarrow A_k) = 0, \quad k \neq i. \tag{7.52}$$

This result can be recast in a more compact form as $\mathbf{P}_{AA} = \mathbf{I}$, with $\mathbf{I}$ the identity matrix.

7.6.2 Massive Bodies

It is useful to consider the special case where the measurement of B does not change the outcome of a previous measurement of A, and viceversa. In other terms, assume that the measurement of A has yielded A_i and the subsequent measurement of B has yielded B_j; then, another measure of A yields A_i again, a later measure of B yields B_j again, and so on. It follows that in $\mathbf{P}'_{AB}$ it is $P'(A_i \rightarrow B_j) = 1$, while all remaining entries in the ith row and jth column are equal to zero. This situation is typical of the bodies that are sufficiently massive, such that the interference suffered during the measurement of a dynamical variable is not detectable. For the sake of clarity an apex is used here to distinguish the probabilities from those of the general case where the body's mass can take any value. Considering a 4×4 matrix by way of example, a possible form of the matrix would be

$$\mathbf{P}'_{AB} = \begin{bmatrix} 0 & 1 & 0 & 0 \\ 0 & 0 & 1 & 0 \\ 1 & 0 & 0 & 0 \\ 0 & 0 & 0 & 1 \end{bmatrix}, \tag{7.53}$$

that is, one of the 4! possible permutation matrices of order 4. Clearly all the other permutation matrices of order 4 different from (7.53) are equally possible. The meaning of a matrix like (7.53) is that the successive measurements of A and B yield either the pair A_1, B_2, or the pair A_2, B_3, or A_3, B_1, or A_4, B_4. Matrix (7.53) may be thought of as a limiting case: starting from a microscopic body described by a 4×4, doubly stochastic matrix whose entries are in general different from zero, the size of the body is increased by adding one atom at a time, and the set of measurements of A and B is repeated at each time. As the reasoning that prescribes the doubly-stochastic nature of the matrix holds at each step, the successive matrices must tend to the limit of a permutation matrix. Which of the 4! permutation matrices will be reached by this process depends on the initial preparation of the experiments. One may wonder why a matrix like

$$\mathbf{P}'_{AB} = \begin{bmatrix} 0 & 1 & 0 & 0 \\ 0 & 1 & 0 & 0 \\ 1 & 0 & 0 & 0 \\ 0 & 0 & 0 & 1 \end{bmatrix}, \tag{7.54}$$

should not be reached. In fact, such a matrix is not acceptable because it is not doubly stochastic: its transpose implies that the outcomes B_1 and B_2 are simultaneously associated to A_2 with certainty, which is obviously impossible not only for a massive body, but for any type of body. This reasoning is associated to another argument, based on the theorem mentioned in Sect. A.11, stating that a doubly-stochastic matrix is a convex combination of permutation matrices. Letting $\theta_1, \dots, \theta_M$ be the combination's coefficients as those used in (A.42), in the process of transforming the microscopic body into a macroscopic one all the coefficients but one vanish, and the non-vanishing one tends to unity. As a consequence, out of the original combination, only one permutation matrix is left.

7.6.3   Need of a Description of Probabilities

The non-negligible influence of the measuring apparatus on the dynamical state of a microscopic body makes it impossible to simultaneously measure the dynamical variables that constitute the initial conditions of the motion. As a consequence, the possibility of using the Hamiltonian theory for describing the dynamics is lost. As outlined in the above sections, the distinctive mark of experiments carried out on microscopic objects is the statistical distribution of the outcomes; thus, a theory that adopts the wave function as the basic tool must identify the connection between the wave function and such a statistical distribution. The theory must also contain the description of the massive bodies as a limiting case.

7.7   Born's Interpretation of the Wave Function

Basing on the optical analogy and the examination of experiments, the *probabilistic interpretation* of the wave function introduced by Born states that the integral

$$\int_\tau |\psi(\mathbf{r}, t)|^2 \, \mathrm{d}^3 r \tag{7.55}$$

is proportional to the probability that a measuring process finds the particle within the volume τ at the time t.[19] Note that the function used in (7.55) is the square modulus of ψ, namely, as noted in Sect. 7.5, the counterpart of the field's intensity in the optical analogy. Also, considering that by definition the integral of (7.55) is dimensionless, the units[20] of ψ are $m^{-3/2}$.

When $\tau \to \infty$ the integral in (7.55) may, or may not, converge. In the first case, ψ is said to be *normalizable*, and a suitable constant σ can be found such that the integral of $|\sigma \psi|^2$ over the entire space equals unity. The new wave function provides a probability proper,

$$\int_\tau |\sigma \psi|^2 \, d^3r \leq 1, \qquad \sigma^{-2} = \int_\infty |\psi|^2 \, d^3r. \qquad (7.56)$$

In the second case ψ is not normalizable:[21] a typical example is the wave function of a free particle, $\psi = A \exp[i(\mathbf{k} \cdot \mathbf{r} - \omega t)]$; however, it is still possible to define a probability ratio

$$\int_{\tau_1} |\psi|^2 \, d^3r \left(\int_{\tau_2} |\psi|^2 \, d^3r \right)^{-1}, \qquad (7.57)$$

where both volumes τ_1 and τ_2 are finite. Relation (7.57) gives the ratio between the probability of finding the particle within τ_1 and that of finding it within τ_2.

Consider a particle whose wave function at time t differs from zero within some volume τ, and assume that a process of measuring the particle's position is initiated at t and completed at some later time t'; let the outcome of the experiment be an improved information about the particle's location, namely, at t' the wave function differs from zero in a smaller volume $\tau' \subset \tau$. This event is also called *contraction* of the wave function.

7.8 Complements

7.8.1 Core Electrons

Throughout this book the term "nucleus" is used to indicate the system made of protons, neutrons, and *core electrons*, namely, those electrons that do not belong to the outer shell of the atom and therefore do not participate in the chemical bonds. In solid-state materials, core electrons are negligibly perturbed by the environment, in contrast to the electrons that belong to the outer shell (*valence electrons*).

---

[19] From this interpretation it follows that $|\psi|^2 \, d^3r$ is proportional to an infinitesimal probability, and $|\psi|^2$ to a probability density.

[20] A more detailed discussion about the units of the wave function is carried out in Sect. 9.7.1.

[21] This issue is further discussed in Sect. 8.2.

Chapter 8
Time-Independent Schrödinger Equation

8.1 Introduction

The properties of the time-independent Schrödinger equation are introduced step by step, starting from a short discussion about its boundary conditions. Considering that the equation is seldom amenable to analytical solutions, two simple cases are examined first: that of a free particle and that of a particle in a box. The determination of the lower energy bound follows, introducing more general issues that build up the mathematical frame of the theory: norm of a function, scalar product of functions, Hermitean operators, eigenfunctions and eigenvalues of operators, orthogonal functions, and completeness of a set of functions. The chapter is concluded with the important examples of the Hamiltonian operator and momentum operator. The complements provide examples of Hermitean operators, a collection of operators' definitions and properties, examples of commuting operators, and a further discussion about the free-particle case.

8.2 Properties of the Time-Independent Schrödinger Equation

A number of properties of the time-independent Schrödinger equation are discussed in this section. The form (7.45) holds only when the force is conservative, so it is not the most general one. However, as will be shown later, many of its properties still hold in more complicated cases. Equation (7.45) is a linear, homogeneous partial-differential equation of the second order, with the zero-order coefficient depending on $\mathbf{r}$. As shown in Prob. 8.1, it is a very general form of linear, second-order equation. The boundary conditions are specified on a case-by-case basis depending on the problem under consideration. More details about the boundary conditions are discussed below. One notes that:

1. The coefficients of (7.45) are real. As a consequence, the solutions are real. In same cases, however, it is convenient to express them in complex form. An example is given in Sect. 8.2.1.

© Springer Science+Business Media New York 2015
M. Rudan, *Physics of Semiconductor Devices*,
DOI 10.1007/978-1-4939-1151-6_8

2. The equation is linear and homogeneous and, as shown below, its boundary conditions are homogeneous as well. It follows that its solution is defined apart from a multiplicative constant. The function $w = 0$ is a solution of (7.45), however it has no physical meaning and is not considered.

3. As the equation is of the second order, its solution w and first derivatives $\partial w/\partial x_i$ are continuous. These requirements are discussed from the physical standpoint in Sect. 9.4. The second derivatives may or may not be continuous, depending on the form of the potential energy V.

4. The solution of (7.45) may contain terms that diverge as $|\mathbf{r}| \to \infty$. In this case such terms must be discarded because they are not compatible with the physical meaning of w (examples are given in Sect. 8.2.1).

Given the above premises, to discuss the boundary conditions of (7.45) it is convenient to distinguish a few cases:

A. The domain Ω of w is finite; in other terms, some information about the problem in hand is available, from which it follows that w vanishes identically outside a finite domain Ω. The continuity of w (see point 3 above) then implies that w vanishes over the boundary of Ω, hence the boundary conditions are homogeneous. After discarding possible diverging terms from the solution, the integral $\int_\Omega |w|^2 \, d\Omega$ is finite (the use of the absolute value is due to the possibility that w is expressed in complex form, see point 1 above).

B. The domain of w is infinite in all directions, but the form of w is such that $\int_\Omega |w|^2 \, d\Omega$ is finite. When this happens, w necessarily vanishes as $|\mathbf{r}| \to \infty$. Thus, the boundary conditions are homogeneous also in this case.[1]

C. The domain of w is infinite, and the form of w is such that $\int_\Omega |w|^2 \, d\Omega$ diverges. This is not due to the fact that $|w|^2$ diverges (in fact, divergent terms in w must be discarded beforehand), but to the fact that w, e.g., asymptotically tends to a constant different from zero, or oscillates (an example of asymptotically-oscillating behavior is given in Sect. 8.2.1). These situations must be tackled separately; one finds that w is still defined apart from a multiplicative constant.

As remarked above, the time-independent Schrödinger equation is a second-order differential equation of a very general form. For this reason, an analytical solution can seldom be obtained, and in the majority of cases it is necessary to resort to numerical-solution methods. The typical situations where the problem can be tackled analytically are those where the equation is separable (compare with Sect. 10.3), so that it can be split into one-dimensional equations. Even when this occurs, the analytical solution can be found only for some forms of the potential energy. The rest of this chapter provides examples that are solvable analytically.

---

[1] It may happen that the domain is infinite in some direction and finite in the others. For instance, one may consider the case where w vanishes identically for $x \geq 0$ and differs from zero for $x < 0$. Such situations are easily found to be a combination of cases A and B illustrated here.

8.2.1 Schrödinger Equation for a Free Particle

The equation for a free particle is obtained by letting $V = \text{const}$ in (7.45). Without loss of generality one may let $V = 0$, this yielding $\nabla^2 w = -(2\,m\,E/\hbar^2)\,w$. As the above can be solved by separating the variables, it is sufficient to consider here only the one-dimensional form

$$\frac{\mathrm{d}^2 w}{\mathrm{d}x^2} = -\frac{2\,m\,E}{\hbar^2}\,w. \tag{8.1}$$

The case $E < 0$ must be discarded as it gives rise to divergent solutions, which are not acceptable from the physical standpoint. The case $E = 0$ yields $w = a_1 x + a_2$, where a_1 must be set to zero to prevent w from diverging. As a consequence, the value $E = 0$ yields $w = a_2 = \text{const}$, that is one of the possibilities anticipated at point C of Sect. 8.2. The integral of $|w|^2$ diverges. Finally, the case $E > 0$ yields

$$w = c_1 \exp(\mathrm{i}\,k\,x) + c_2 \exp(-\mathrm{i}\,k\,x), \quad k = \sqrt{2\,m\,E/\hbar^2} = p/\hbar > 0, \tag{8.2}$$

where c_1, c_2 are constants to be determined. Thus, the value $E > 0$ yields the asymptotically-oscillating behavior that has also been anticipated at point C of Sect. 8.2. The integral of $|w|^2$ diverges. One notes that w is written in terms of two complex functions; it could equally well be expressed in terms of the real functions $\cos(k\,x)$ and $\sin(k\,x)$. The time-dependent, monochromatic wave function $\psi = w \exp(-\mathrm{i}\,\omega\,t)$ corresponding to (8.2) reads

$$\psi = c_1 \exp[\mathrm{i}(k\,x - \omega\,t)] + c_2 \exp[-\mathrm{i}(k\,x + \omega\,t)], \quad \omega = E/\hbar. \tag{8.3}$$

The relations $k = p/\hbar$, $\omega = E/\hbar$ stem from the analogy described in Sect. 7.4.5. The total energy E and momentum's modulus p are fully determined; this outcome is the same as that found for the motion of a free particle in Classical Mechanics: for a free particle the kinetic energy equals the total energy; if the latter is prescribed, the momentum's modulus is prescribed as well due to $E = p^2/(2\,m)$. The direction of the motion, instead, is not determined because both the forward and backward motions, corresponding to the positive and negative square root of p^2 respectively, are possible solutions. To ascertain the motion's direction it is necessary to acquire more information; specifically, one should prescribe the initial conditions which, in turn, would provide the momentum's sign.

The quantum situation is similar, because the time-dependent wave function (8.2) is a superposition of a planar wave $c_1 \exp[\mathrm{i}(k\,x - \omega\,t)]$ whose front moves in the positive direction, and of a planar wave $c_2 \exp[-\mathrm{i}(k\,x + \omega\,t)]$ whose front moves in the negative direction. Here to ascertain the motion's direction one must acquire the information about the coefficients in (8.2): the forward motion corresponds to $c_1 \neq 0, c_2 = 0$, the backward motion to $c_1 = 0, c_2 \neq 0$. Obviously (8.2) in itself does not provide any information about the coefficients, because such an expression is the general solution of (8.1) obtained as a combination of the two linearly-independent, particular solutions $\exp(\mathrm{i}\,k\,x)$ and $\exp(-\mathrm{i}\,k\,x)$; so, without further information about c_1 and c_2, both the forward and backward motions are possible.

Another similarity between the classical and quantum cases is that no constraint is imposed on the total energy, apart from the prescription $E \geq 0$. From this viewpoint one concludes that (8.1) is an eigenvalue equation with a continuous distribution of eigenvalues in the interval $E \geq 0$.

8.2.2 Schrödinger Equation for a Particle in a Box

Considering again the one-dimensional case of (7.45),

$$\frac{d^2 w}{dx^2} = -\frac{2\,m}{\hbar^2}(E - V)\,w, \qquad V = V(x), \tag{8.4}$$

let $V = \text{const} = 0$ for $x \in [0, a]$ and $V = V_0 > 0$ elsewhere. The form of the potential energy is that of a square well whose counterpart of Classical Mechanics is illustrated in Sect. 3.2. Here, however, the limit $V_0 \to \infty$ is considered for the sake of simplicity. This limiting case is what is referred to with the term *box*. As shown in Sect. 11.5, here w vanishes identically outside the interval $[0, a]$: this is one of the possibilities that were anticipated in Sect. 8.2 (point A). The continuity of w then yields $w(0) = w(a) = 0$. It is easily found that if $E \leq 0$ the only solution of (8.4) is $w = 0$, which is not considered because it has no physical meaning. When $E > 0$, the solution reads

$$w = c_1 \exp(i\,k\,x) + c_2 \exp(-i\,k\,x), \qquad k = \sqrt{2\,m\,E/\hbar^2} > 0. \tag{8.5}$$

Letting $w(0) = 0$ yields $c_1 + c_2 = 0$ and $w = 2\,i\,c_1 \sin(k\,x)$. Then, $w(a) = 0$ yields $k\,a = n\,\pi$ with n an integer whence, using the relation $k = k_n = n\,\pi/a$ within those of E and w,

$$E = E_n = \frac{\hbar^2\,\pi^2}{2\,m\,a^2}\,n^2, \qquad w = w_n = 2\,i\,c_1 \sin\left(\frac{n\,\pi}{a}\,x\right). \tag{8.6}$$

This result shows that (8.4) is an eigenvalue equation with a discrete distribution of eigenvalues, given by the first relation in (8.6). For this reason, the energy is said to be *quantized*. To each index n it corresponds one and only one eigenvalue E_n, and one and only one eigenfunction w_n; as a consequence, this case provides a one-to-one-correspondence between eigenvalues and eigenfunctions.[2] Not every integer should be used in (8.6) though; in fact, $n = 0$ must be discarded because the corresponding eigenfunction vanishes identically. Also, the negative indices are to be excluded because $E_{-n} = E_n$ and $|w_{-n}|^2 = |w_n|^2$, so they do not add information with respect to the positive ones. In conclusion, the indices to be used are $n = 1, 2, \ldots$

---

[2] The one-to-one correspondence does not occur in general. Examples of the Schrödinger equation are easily given (Sect. 9.6) where to each eigenvalue there corresponds more than one—even infinite— eigenfunctions.

Fig. 8.1 The first eigenfunctions of the Schrödinger equation in the case of a particle in a box

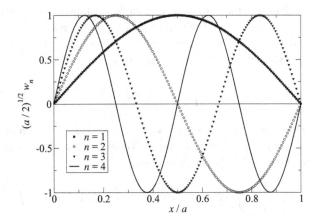

As expected, each eigenfunction contains a multiplicative constant; here the integral of $|w|^2$ converges, so the constant can be exploited to normalize the eigenfunction by letting $\int_0^a |w_n|^2 \, dx = 1$. One finds

$$\int_0^a |w_n|^2 \, dx = 4\,|c_1|^2 \int_0^a \sin^2\left(\frac{n\pi}{a}x\right)\,dx = \frac{4\,|c_1|^2 a}{n\pi} \int_0^{n\pi} \sin^2(y)\,dy. \quad (8.7)$$

Integrating by parts shows that the last integral equals $n\pi/2$, whence the normalization condition yields $4\,|c_1|^2 = 2/a$. Choosing $2c_1 = -j\sqrt{2/a}$ provides the eigenfunctions

$$w_n = \sqrt{\frac{2}{a}}\,\sin\left(\frac{n\pi}{a}x\right). \quad (8.8)$$

The first eigenfunctions are shown in Fig. 8.1. Remembering that $w = 0$ outside the interval $[0, a]$, one notes that dw/dx is discontinuous at $x = 0$ and $x = a$. This apparently contradicts the continuity property of the first derivative mentioned in Sect. 8.2, point 3. However, in the case considered here the limit $V_0 \to \infty$ has introduced a discontinuity of the second kind into the potential energy; for this reason, the property mentioned above does not apply.

8.2.3 Lower Energy Bound in the Schrödinger Equation

In the example of Sect. 8.2.1, where the free particle is considered, the lower bound for the particle's total energy is $E \geq V_{min}$, with V_{min} the minimum[3] of the potential energy; in contrast, in the example of the particle in a box illustrated in Sect. 8.2.2, the lower bound is $E > V_{min}$. A more general analysis of the lower bound for the total energy in the Schrödinger equation is carried out here.

---

[3] Such a minimum is set to zero in the example of Sect. 8.2.1.

Consider the time-independent Schrödinger equation in a conservative case, (7.45), and let Ω be the domain of w (which may extend to infinity), with Σ the boundary of Ω. Recasting (7.45) as $-\nabla^2 w = 2\,m\,(E - V)\,w/\hbar^2$ and integrating it over Ω after multiplying both sides by w^* yields

$$-\int_\Omega w^* \nabla^2 w \, d\Omega = \frac{2m}{\hbar^2} \int_\Omega (E - V)\,|w|^2\,d\Omega. \qquad (8.9)$$

It is implied that w is a physically meaningful solution of (7.45), whence w does not vanish identically within Ω. Thanks to the identity (A.17) and the divergence theorem (A.23) the above becomes

$$\frac{2\,m}{\hbar^2} \int_\Omega (E - V)\,|w|^2\,d\Omega = \int_\Omega |\mathrm{grad}w|^2\,d\Omega - \int_\Sigma w^* \frac{\partial w}{\partial n}\,d\Sigma, \qquad (8.10)$$

with $\partial w/\partial n$ the derivative of w in the direction normal to Σ. Consider now the case where w vanishes over Σ; as w^* vanishes as well, the boundary integral in (8.10) is equal to zero. In contrast, the other integral at the right hand side of (8.10) is strictly positive: in fact, as w vanishes at the boundary while it is different from zero inside the domain, its gradient does not vanish identically in Ω. It follows

$$\int_\Omega (E - V)\,|w|^2\,d\Omega > 0, \qquad E > \frac{\int_\Omega V\,|w|^2\,d\Omega}{\int_\Omega |w|^2\,d\Omega} \ge V_{min}, \qquad (8.11)$$

where the last inequality stems from the fact that $|w|^2$ is strictly positive. In conclusion, when V is such that w vanishes at the boundary, then the strict inequality $E > V_{min}$ holds. When V does not vanish at the boundary, the reasoning leading to (8.11) does not apply and the lower bound for E must be sought by a direct examination of the solutions. An example of this examination is that of the free-particle case shown in Sect. 8.2.1.

8.3   Norm of a Function—Scalar Product

The functions f, g,... that are considered in this section are *square-integrable* complex functions, namely, they have the property that the integrals

$$||f||^2 = \int_{\Omega'} |f|^2\,d\Omega', \qquad ||g||^2 = \int_{\Omega''} |g|^2\,d\Omega'', \ldots \qquad (8.12)$$

converge. In (8.12), Ω' is the domain of f, Ω'' that of g, and so on. The variables in the domains Ω', Ω'',... are real. The positive numbers $||f||$ and $||g||$ are the *norm* of f and g, respectively. If f, g are square integrable over the same domain Ω, a linear combination $\lambda\,f + \mu\,g$, with λ, μ arbitrary complex constants, is also square integrable over Ω ([78], Chap. V.2).

If a square-integrable function f is defined apart from a multiplicative constant, for instance because it solves a linear, homogeneous differential equation with homogeneous boundary conditions, it is often convenient to choose the constant such that the norm equals unity. This is accomplished by letting $\varphi = c f$ and $||\varphi|| = 1$, whence $|c|^2 = 1/||\varphi||^2$.

Consider two square-integrable functions f and g defined over the same domain Ω; their *scalar product* is defined as

$$\langle g|f \rangle = \int_\Omega g^* f \, d\Omega. \tag{8.13}$$

From (8.13) it follows

$$\langle f|g \rangle = \int_\Omega f^* g \, d\Omega = \left(\int_\Omega f g^* \, d\Omega \right)^* = \langle g|f \rangle^*. \tag{8.14}$$

It is implied that f, g are regular enough to make the integral in (8.13) to exist; in fact, this is proved by observing that for square-integrable functions the *Schwarz inequality* holds, analogous to that found in the case of vectors (Sect. A.2): if f and g are square integrable, then

$$|\langle g|f \rangle| \le ||f|| \times ||g||, \tag{8.15}$$

where the equality holds if and only if f is proportional to g (compare with (A.5)). In turn, to prove (8.15) one observes that $\sigma = f + \mu g$, where μ is an arbitrary constant, is also square integrable. Then [47],

$$||\sigma||^2 = ||f||^2 + |\mu|^2 ||g||^2 + \mu \langle f|g \rangle + \mu^* \langle g|f \rangle \ge 0. \tag{8.16}$$

The relation (8.15) is obvious if $f = 0$ or $g = 0$. Let $g \ne 0$ and choose $\mu = -\langle g|f \rangle/||g||^2$. Replacing in (8.16) yields (8.15). For the equality to hold it must be $\sigma = 0$, which implies that f and g are proportional to each other; conversely, from $f = c g$ the equality follows.

The symbol $\langle g|f \rangle$ for the scalar product is called *Dirac's notation*.[4] If $\langle g|f \rangle = 0$, the functions f, g are called *orthogonal*. For any complex constants b, b_1, b_2 the following hold:

$$\langle g|b f \rangle = b \langle g|f \rangle, \qquad \langle g|b_1 f_1 + b_2 f_2 \rangle = b_1 \langle g|f_1 \rangle + b_2 \langle g|f_2 \rangle, \tag{8.17}$$

$$\langle b g|f \rangle = b^* \langle g|f \rangle, \qquad \langle b_1 g_1 + b_2 g_2|f \rangle = b_1^* \langle g_1|f \rangle + b_2^* \langle g_2|f \rangle, \tag{8.18}$$

namely, the scalar product is distributive and bilinear. The properties defined here are the counterpart of those defined in Sect. A.1 for vectors.

---

[4] The two terms $\langle g|$ and $|f \rangle$ of the scalar product $\langle g|f \rangle$ are called *bra vector* and *ket vector*, respectively.

8.3.1   Adjoint Operators and Hermitean Operators

A function appearing within a scalar product may result from the application of a linear operator, say, $\mathcal{A}$, onto another function.[5] For instance, if $s = \mathcal{A}f$, then from (8.13, 8.14) it follows

$$\langle g|s \rangle = \int_\Omega g^* \mathcal{A}f \, d\Omega, \qquad \langle s|g \rangle = \int_\Omega (\mathcal{A}f)^* g \, d\Omega = \langle g|s \rangle^*. \tag{8.19}$$

Given an operator $\mathcal{A}$ it is possible to find another operator, typically indicated with $\mathcal{A}^\dagger$, having the property that, for any pair f, g of square-integrable functions,

$$\int_\Omega \left(\mathcal{A}^\dagger g\right)^* f \, d\Omega = \int_\Omega g^* \mathcal{A}f \, d\Omega \tag{8.20}$$

or, in Dirac's notation, $\langle \mathcal{A}^\dagger g|f \rangle = \langle g|\mathcal{A}f \rangle$. Operator $\mathcal{A}^\dagger$ is called the *adjoint*[6] of $\mathcal{A}$. In general it is $\mathcal{A}^\dagger \neq \mathcal{A}$; however, for some operators it happens that $\mathcal{A}^\dagger = \mathcal{A}$. In this case, $\mathcal{A}$ is called Hermitean. Thus, for Hermitean operators the following holds:

$$\langle g|\mathcal{A}f \rangle = \langle \mathcal{A}g|f \rangle = \langle g|\mathcal{A}|f \rangle. \tag{8.21}$$

The notation on the right of (8.21) indicates that one can consider the operator as applied onto f or g. Examples of Hermitean operators are given in Sect. 8.6.1. It is found by inspection that, for any operator $\mathcal{C}$, the operators $\mathcal{S} = \mathcal{C} + \mathcal{C}^\dagger$ and $\mathcal{D} = -\mathrm{i}\,(\mathcal{C} - \mathcal{C}^\dagger)$ are Hermitean.

The following property is of use: a linear combination of Hermitean operator with real coefficients is Hermitean; considering, e.g., two Hermitean operators $\mathcal{A}, \mathcal{B}$ and two real numbers λ, μ, one finds

$$\int_\Omega g^* \, (\lambda \, \mathcal{A} + \mu \, \mathcal{B})\, f \, d\Omega = \int_\Omega [(\lambda \, \mathcal{A} + \mu \, \mathcal{B})\, g]^* \, f \, d\Omega. \tag{8.22}$$

8.4   Eigenvalues and Eigenfunctions of an Operator

A linear operator $\mathcal{A}$ may be used to generate a homogeneous equation (*eigenvalue equation*) in the unknown v, having the form

$$\mathcal{A}v = A\,v, \tag{8.23}$$

---

[5] In this context the term *operator* has the following meaning: if an operation brings each function f of a given function space into correspondence with one and only one function s of the same space, one says that this is obtained through the action of a given operator $\mathcal{A}$ onto f and writes $s = \mathcal{A}f$. A *linear* operator is such that $\mathcal{A}\,(c_1\, f_1 + c_2\, f_2) = c_1\, \mathcal{A}f_1 + c_2\, \mathcal{A}f_2$ for any pair of functions f_1, f_2 and of complex constants c_1, c_2 ([78], Chap. II.11).

[6] The adjoint operator is the counterpart of the conjugate-transpose matrix in vector algebra.

with A a parameter. Clearly (8.23) admits the solution $v = 0$ which, however, is of no interest; it is more important to find whether specific values of A exist (*eigenvalues*), such that (8.23) admits non-vanishing solutions (*eigenfunctions*). In general (8.23) must be supplemented with suitable boundary or regularity conditions on v.

The set of the eigenvalues of an operator found from (8.23) is the operator's *spectrum*. It may happen that the eigenvalues are distinguished by an index, or a set of indices, that take only discrete values; in this case the spectrum is called *discrete*. If, instead, the eigenvalues are distinguished by an index, or a set of indices, that vary continuously, the spectrum is *continuous*. Finally, it is *mixed* if a combination of discrete and continuous indices occurs.

An eigenvalue is *simple* if there is one and only one eigenfuction corresponding to it, while it is *degenerate of order s* if there are s linearly independent eigenfuctions corresponding to it. The order of degeneracy may also be infinite. By way of example, the Schrödinger equation for a free particle in one dimension discussed in Sect. 8.2.1 has a continuous spectrum of eigenvalues $E = \hbar^2 k^2/(2\,m)$ of index k, namely, $E = E_k$. Each eigenvalue is degenerate of order 2 because to each E there correspond two linearly-independent eigenfunctions $\exp{(i\,k\,x)}$, $\exp{(-i\,k\,x)}$, with $k = \sqrt{2\,m\,E}/\hbar$. Instead, the Schrödinger equation for a particle in a box discussed in Sect. 8.2.2 has a discrete spectrum of eigenvalues E_n given by the first relation in (8.6). Each eigenvalue is simple as already indicated in Sect. 8.2.2.

Let $v^{(1)}, \ldots, v^{(s)}$ be the linearly-independent eigenfunctions belonging to an eigenvalue A degenerate of order s; then a linear combination of such eigenfunctions is also an eigenfunction belonging to A. In fact, letting $\alpha_1, \ldots, \alpha_s$ be the coefficients of the linear combination, from $\mathcal{A}v^{(k)} = A\,v^{(k)}$ it follows

$$\mathcal{A}\sum_{k=1}^{s}\alpha_k\,v^{(k)} = \sum_{k=1}^{s}\alpha_k\mathcal{A}v^{(k)} = \sum_{k=1}^{s}\alpha_k\,A\,v^{(k)} = A\sum_{k=1}^{s}\alpha_k\,v^{(k)}. \tag{8.24}$$

8.4.1 Eigenvalues of Hermitean Operators

A fundamental property of the Hermitean operators is that their eigenvalues are real. Consider, first, the case where the eigenfunctions are square integrable, so that $\langle v|v\rangle$ is different from zero and finite. To proceed one considers the discrete spectrum, where the eigenvalues are A_n. Here n indicates a single index or also a set of indices. If the eigenvalue is simple, let v_n be the eigenfunction belonging to A_n; if it is degenerate, the same symbol v_n is used here to indicate any eigenfunction belonging to A_n. Then, two operations are performed: in the first one, the eigenvalue equation $\mathcal{A}v_n = A_n\,v_n$ is scalarly multiplied by v_n on the left, while in the second one the conjugate equation $(\mathcal{A}v_n)^* = A_n^*\,v_n^*$ is scalarly multiplied by v_n on the right. The operations yield, respectively,

$$\langle v_n|\mathcal{A}v_n\rangle = A_n\,\langle v_n|v_n\rangle, \qquad \langle\mathcal{A}v_n|v_n\rangle = A_n^*\,\langle v_n|v_n\rangle. \tag{8.25}$$

The left hand sides in (8.25) are equal to each other due to the hermiticity of $\mathcal{A}$; as a consequence, $A_n^* = A_n$, that is, A_n is real.

Another fundamental property of the Hermitean operators is that two eigenfunctions belonging to different eigenvalues are orthogonal to each other. Still considering the discrete spectrum, let A_m, A_n be two different eigenvalues and let v_m (v_n) be an eigenfunction belonging to A_m (A_n). The two eigenvalues are real as demonstrated earlier. Then, the eigenvalue equation $\mathcal{A}v_n = A_n v_n$ is scalarly multiplied by v_m on the left, while the conjugate equation for the other eigenvalue, $(\mathcal{A}v_m)^* = A_m v_m^*$, is scalarly multiplied by v_n on the right. The operations yield, respectively,

$$\langle v_m | \mathcal{A}v_n \rangle = A_n \langle v_m | v_n \rangle, \qquad \langle \mathcal{A}v_m | v_n \rangle = A_m \langle v_m | v_n \rangle. \tag{8.26}$$

The left hand sides in (8.26) are equal to each other due to the hermiticity of $\mathcal{A}$; as a consequence, $(A_m - A_n) \langle v_m | v_n \rangle = 0$. But $A_n \neq A_m$, so it is $\langle v_m | v_n \rangle = 0$.

8.4.2   Gram–Schmidt Orthogonalization

When two eigenfunctions belonging to a degenerate eigenvalue are considered, the reasoning that proves their orthogonality through (8.26) is not applicable because $A_n = A_m$. In fact, linearly-independent eigenfunctions of an operator $\mathcal{A}$ belonging to the same eigenvalue are not mutually orthogonal in general. However, it is possible to form mutually-orthogonal linear combinations of the eigenfunctions. As shown by (8.24), such linear combinations are also eigenfunctions, so their norm is different from zero. The procedure (*Gram–Schmidt orthogonalization*) is described here with reference to the case of the nth eigenfunction of a discrete spectrum, with a degeneracy of order s. Let the non-orthogonal eigenfunctions be $v_n^{(1)}, \ldots, v_n^{(s)}$, and let $u_n^{(1)}, \ldots, u_n^{(s)}$ be the linear combinations to be found. Then one prescribes $u_n^{(1)} = v_n^{(1)}$, $u_n^{(2)} = v_n^{(2)} + a_{21} u_n^{(1)}$ where a_{21} is such that $\langle u_n^{(1)} | u_n^{(2)} \rangle = 0$; thus

$$\langle u_n^{(1)} | v_n^{(2)} \rangle + a_{21} \langle u_n^{(1)} | u_n^{(1)} \rangle = 0, \qquad a_{21} = -\frac{\langle u_n^{(1)} | v_n^{(2)} \rangle}{\langle u_n^{(1)} | u_n^{(1)} \rangle}. \tag{8.27}$$

The next function is found be letting $u_n^{(3)} = v_n^{(3)} + a_{31} u_n^{(1)} + a_{32} u_n^{(2)}$, with $\langle u_n^{(1)} | u_n^{(3)} \rangle = 0$, $\langle u_n^{(2)} | u_n^{(3)} \rangle = 0$, whence

$$\langle u_n^{(1)} | v_n^{(3)} \rangle + a_{31} \langle u_n^{(1)} | u_n^{(1)} \rangle = 0, \qquad a_{31} = -\frac{\langle u_n^{(1)} | v_n^{(3)} \rangle}{\langle u_n^{(1)} | u_n^{(1)} \rangle}, \tag{8.28}$$

$$\langle u_n^{(2)} | v_n^{(3)} \rangle + a_{32} \langle u_n^{(2)} | u_n^{(2)} \rangle = 0, \qquad a_{32} = -\frac{\langle u_n^{(2)} | v_n^{(3)} \rangle}{\langle u_n^{(2)} | u_n^{(2)} \rangle}. \tag{8.29}$$

Similarly, the kth linear combination is built up recursively from the combinations of indices $1, \ldots, k - 1$:

$$u_n^{(k)} = v_n^{(k)} + \sum_{i=1}^{k-1} a_{ki} u_n^{(i)}, \qquad a_{ki} = -\frac{\langle u_n^{(i)} | v_n^{(k)} \rangle}{\langle u_n^{(i)} | u_n^{(i)} \rangle}. \tag{8.30}$$

The denominators in (8.30) are different from zero because they are the squared norms of the previously-defined combinations.

8.4.3 Completeness

As discussed in Sect. 8.2.1, the eigenfunctions of the Schrödinger equation for a free particle, for a given $k = \sqrt{2\,m\,E}/\hbar$ and apart from a multiplicative constant, are $w_{+k} = \exp(i\,k\,x)$ and $w_{-k} = \exp(-i\,k\,x)$. They may be written equivalently as $w(x, k) = \exp(i\,k\,x)$, with $k = \pm\sqrt{2\,m\,E}/\hbar$. Taking the multiplicative constant equal to $1/\sqrt{2\,\pi}$, and considering a function f that fulfills the condition (C.19) for the Fourier representation, one applies (C.16) and (C.17) to find

$$f(x) = \int_{-\infty}^{+\infty} \frac{\exp(i\,k\,x)}{\sqrt{2\,\pi}} c(k)\,dk, \qquad c(k) = \int_{-\infty}^{+\infty} \frac{\exp(-i\,k\,x)}{\sqrt{2\,\pi}} f(x)\,dx. \quad (8.31)$$

Using the definition (8.13) of scalar product one recasts (8.31) as

$$f(x) = \int_{-\infty}^{+\infty} c(k)\,w(x, k)\,dk, \qquad c(k) = \langle w | f \rangle. \quad (8.32)$$

In general the shorter notation $w_k(x)$, c_k is used instead of $w(x, k)$, $c(k)$. A set of functions like $w_k(x)$ that allows for the representation of f given by the first relation in (8.32) is said to be *complete*. Each member of the set is identified by the value of the continuous parameter k ranging from $-\infty$ to $+\infty$. To each k it corresponds a *coefficient* of the expansion, whose value is given by the second relation in (8.32).

Expressions (8.31) and (8.32) hold true because they provide the Fourier transform or antitransform of a function that fulfills (C.19). On the other hand, $w_k(x)$ is also the set of eigenfunctions of the free particle. In conclusion, the eigenfunctions of the Schrödinger equation for a free particle form a complete set.

The same conclusion is readily found for the eigenfunctions of the Schrödinger equation for a particle in a box. To show this, one considers a function $f(x)$ defined in an interval $[-\alpha/2, +\alpha/2]$ and fulfilling $\int_{-\alpha/2}^{+\alpha/2} |f(x)|\,dx < \infty$. In this case the expansion into a Fourier series holds:

$$f(x) = \frac{1}{2} a_0 + \sum_{n=1}^{\infty} \left[a_n \cos(2\,\pi\,n\,x/\alpha) + b_n \sin(2\,\pi\,n\,x/\alpha) \right], \quad (8.33)$$

with $a_0/2 = \bar{f} = (1/\alpha) \int_{-\alpha/2}^{+\alpha/2} f(x)\,dx$ the average of f over the interval, and

$$\begin{Bmatrix} a_n \\ b_n \end{Bmatrix} = \frac{2}{\alpha} \int_{-\alpha/2}^{+\alpha/2} \begin{Bmatrix} \cos \\ \sin \end{Bmatrix} \left(\frac{2\,\pi\,n\,x}{\alpha} \right) f(x)\,dx, \qquad n = 1, 2, \ldots \quad (8.34)$$

Equality (8.33) indicates convergence in the mean, namely, using $g = f - \bar{f}$ for the sake of simplicity, (8.33) is equivalent to

$$\lim_{N \to \infty} \int_{-\alpha/2}^{+\alpha/2} \left\{ g - \sum_{n=1}^{N} \left[a_n \cos(2\,\pi\,n\,x/\alpha) + b_n \sin(2\,\pi\,n\,x/\alpha) \right] \right\}^2 dx = 0.$$

$$(8.35)$$

Defining the auxiliary functions

$$\chi_n = \sqrt{2/\alpha} \, \cos{(2\pi n x/\alpha)}, \qquad \sigma_n = \sqrt{2/\alpha} \, \sin{(2\pi n x/\alpha)}, \qquad (8.36)$$

a more compact notation is obtained, namely, $f = \bar{f} + \sum_{n=1}^{\infty} (\langle \chi_n | f \rangle \chi_n + \langle \sigma_n | f \rangle \sigma_n)$ or, observing that $\langle \sigma_n | \text{const} \rangle = \langle \chi_n | \text{const} \rangle = 0$,

$$g = \sum_{n=1}^{\infty} (\langle \chi_n | g \rangle \chi_n + \langle \sigma_n | g \rangle \sigma_n). \qquad (8.37)$$

The norm of the auxiliary functions (8.36) is unity, $\langle \chi_n | \chi_n \rangle = \langle \sigma_n | \sigma_n \rangle = 1$ for $n = 1, 2, \ldots$, and all auxiliary functions are mutually orthogonal: $\langle \chi_m | \chi_n \rangle = \langle \sigma_m | \sigma_n \rangle = 0$ for $n, m = 0, 1, 2, \ldots$, $m \neq n$, and $\langle \sigma_m | \chi_n \rangle = 0$ for $n, m = 0, 1, 2, \ldots$ A set whose functions have a norm equal to unity and are mutually orthogonal is called *orthonormal*. Next, (8.37) shows that the set $\chi_n, \sigma_n, n = 0, 1, 2, \ldots$ is complete in $[-\alpha/2, +\alpha/2]$ with respect to any g for which the expansion is allowed. Letting $c_{2n-1} = \langle \chi_n | g \rangle$, $c_{2n} = \langle \sigma_n | g \rangle$, $w_{2n-1} = \chi_n$, $w_{2n} = \sigma_n$, (8.37) takes the even more compact form

$$g = \sum_{m=1}^{\infty} c_m w_m, \qquad c_m = \langle w_m | g \rangle. \qquad (8.38)$$

From the properties of the Fourier series it follows that the set of the σ_n functions alone is complete with respect to any function that is odd in $[-\alpha/2, +\alpha/2]$, hence it is complete with respect to any function over the half interval $[0, +\alpha/2]$. On the other hand, letting $a = \alpha/2$ and comparing with (8.8) shows that σ_n (apart from the normalization coefficient) is the eigenfunction of the Schrödinger equation for a particle in a box. In conclusion, the set of eigenfunctions of this equation is complete within $[0, a]$.

One notes the striking resemblance of the first relation in (8.38) with the vector-algebra expression of a vector in terms of its components c_m. The similarity is completed by the second relation in (8.38), that provides each component as the projection of g over w_m. The latter plays the same role as the unit vector in algebra, the difference being that the unit vectors here are functions and that their number is infinite. A further generalization of the same concept is given by (8.32), where the summation index k is continuous.

Expansions like (8.32) or (8.38) hold because $w_k(x)$ and $w_m(x)$ are complete sets, whose completeness is demonstrated in the theory of Fourier's integral or series; such a theory is readily extended to the three-dimensional case, showing that also the three-dimensional counterparts of $w_k(x)$ or $w_m(x)$ form complete sets (in this case the indices k or m are actually groups of indices, see, e.g., (9.5)). One may wonder whether other complete sets of functions exist, different from those considered in this section; the answer is positive: in fact, completeness is possessed by many other sets of functions,[7] and those of interest in Quantum Mechanics are made of the eigenfunctions of equations like (8.23). A number of examples will be discussed later.

8.4.4 Parseval Theorem

Consider the expansion of a complex function f with respect to a complete and orthonormal set of functions w_n,

$$f = \sum_n c_n w_n, \qquad c_n = \langle w_n | f \rangle, \qquad \langle w_n | w_m \rangle = \delta_{nm}, \tag{8.39}$$

where the last relation on the right expresses the set's orthonormality. As before, m indicates a single index or a group of indices. The squared norm of f reads

$$||f||^2 = \int_\Omega |f|^2 \, d\Omega = \left\langle \sum_n c_n w_n \middle| \sum_m c_m w_m \right\rangle. \tag{8.40}$$

Applying (8.17, 8.18) yields

$$||f||^2 = \sum_n c_n^* \sum_m c_m \langle w_n | w_m \rangle = \sum_n c_n^* \sum_m c_m \delta_{nm} = \sum_n |c_n|^2, \tag{8.41}$$

namely, the norm of the function equals the norm of the vector whose components are the expansion's coefficients (*Parseval theorem*). The result applies irrespective of the set that has been chosen for expanding f. The procedure leading to (8.41) must be repeated for the continuous spectrum, where the expansion reads

$$f = \int_\alpha c_\alpha w_\alpha \, d\alpha, \qquad c_\alpha = \langle w_\alpha | f \rangle. \tag{8.42}$$

Here a difficulty seems to arise, related to expressing the counterpart of the third relation in (8.39). Considering for the sake of simplicity the case where a single index is present, the scalar product $\langle w_\alpha | w_\beta \rangle$ must differ from zero only for $\beta = \alpha$, while it must vanish for $\beta \neq \alpha$ no matter how small the difference $\alpha - \beta$ is. In other terms, for a given value of α such a scalar product vanishes for any β apart from a null set. At the same time, it must provide a finite value when used as a factor within an integral. An example taken from the case of a free particle shows that the requirements listed above are mutually compatible. In fact, remembering the analysis of Sect. 8.4.3, the scalar product corresponding to the indices α and β reads

$$\langle w_\alpha | w_\beta \rangle = \frac{1}{2\pi} \int_{-\infty}^{+\infty} \exp\left[i\left(\beta - \alpha\right)x\right] dx = \delta(\alpha - \beta), \tag{8.43}$$

---

[7] The completeness of a set of eigenfunctions must be proven on a case-by-case basis.

where the last equality is taken from (C.43). As mentioned in Sect. C.4, such an equality can be used only within an integral. In conclusion,[8]

$$\int_\Omega |f|^2 \, d\Omega = \langle f|f \rangle = \int_{-\infty}^{+\infty} c_\alpha^* \, d\alpha \int_{-\infty}^{+\infty} c_\beta \, \delta(\alpha - \beta) \, d\beta = \int_{-\infty}^{+\infty} |c_\alpha|^2 \, d\alpha.$$

$$(8.44)$$

One notes that (8.44) generalizes a theorem of Fourier's analysis that states that the norm of a function equals that of its transform.

8.5 Hamiltonian Operator and Momentum Operator

As mentioned in Sect. 7.5, the form (7.45) of the time-independent Schrödinger equation holds only when the force is conservative. It is readily recast in the more compact form (8.23) by defining the *Hamiltonian operator*

$$\mathcal{H} = -\frac{\hbar^2}{2\,m} \nabla^2 + V,$$

$$(8.45)$$

that is, a linear, real operator that gives (7.45) the form

$$\mathcal{H}w = E\,w.$$

$$(8.46)$$

The term used to denote $\mathcal{H}$ stems from the formal similarity of (8.46) with the classical expression $H(\mathbf{p}, \mathbf{q}) = E$ of a particle's total energy in a conservative field, where $H = T + V$ is the Hamiltonian function (Sect. 1.5). By this similarity, the classical kinetic energy $T = p^2/(2m)$ corresponds to the kinetic operator $\mathcal{T} = -\hbar^2/(2\,m)\,\nabla^2$; such a correspondence reads

$$T = \frac{1}{2\,m}\left(p_1^2 + p_2^2 + p_3^2\right) \quad\Longleftrightarrow\quad \mathcal{T} = -\frac{\hbar^2}{2\,m}\left(\frac{\partial^2}{\partial x_1^2} + \frac{\partial^2}{\partial x_2^2} + \frac{\partial^2}{\partial x_3^2}\right). \quad (8.47)$$

The units of $\mathcal{T}$ are those of an energy, hence $\hbar^2\,\nabla^2$ has the units of a momentum squared. One notes that to transform T into $\mathcal{T}$ one must replace each component of momentum by a first-order operator as follows:

$$p_i \Leftarrow \hat{p}_i = -\mathrm{i}\,\hbar\,\frac{\partial}{\partial x_i},$$

$$(8.48)$$

where $\hat{p}_i$ is called *momentum operator*. The correspondence (8.47) would still hold if the minus sign in (8.48) were omitted. However, the minus sign is essential for

---

[8] The relation (8.44) is given here with reference to the specific example of the free particle's eigenfunctions. For other cases of continuous spectrum the relation $\langle w_\alpha | w_\beta \rangle = \delta(\alpha - \beta)$ is proven on a case-by-case basis.

a correct description of the particle's motion.[9] From the results of Sect. 8.6.1 one finds that the momentum operator and its three-dimensional form $\hat{\mathbf{p}} = -i\hbar$ grad are Hermitean for square-integrable functions. Their units are those of a momentum. The Hamiltonian operator (8.45) is a real-coefficient, linear combination of ∇^2 and V; combining (8.22) with the findings of Sect. 8.6 shows that (8.45) is Hermitean for square-integrable functions.

The one-dimensional form of the momentum operator yields the eigenvalue equation

$$-i\hbar \frac{dv}{dx} = \tilde{p}\, v, \tag{8.49}$$

where $\tilde{p}$ has the units of a momentum. The solution of (8.49) is $v = \text{const} \times \exp(i\,\tilde{p}/\hbar)$, where $\tilde{p}$ must be real to prevent the solution from diverging. Letting $\text{const} = 1/\sqrt{2\pi}$, $k = \tilde{p}/\hbar$ yields $v = v_k(x) = \exp(i\,k\,x)/\sqrt{2\pi}$, showing that the eigenfunctions of the momentum operator form a complete set (compare with (8.31)) and are mutually orthogonal (compare with (8.43)). As $|v_k(x)|^2 = 1/(2\pi)$, the eigenfunctions are not square integrable; the spectrum is continuous because the eigenvalue $\hbar\,k$ can be any real number.

8.6 Complements

8.6.1 Examples of Hermitean Operators

A real function V, depending on the spatial coordinates over the domain Ω, and possibly on other variables $\alpha, \beta, \ldots$, may be thought of as a purely multiplicative operator. Such an operator is Hermitean; in fact,

$$\int_\Omega g^* V f \, d\Omega = \int_\Omega V g^* f \, d\Omega = \int_\Omega (Vg)^* f \, d\Omega. \tag{8.50}$$

In contrast, an imaginary function $W = iV$, with V real, is not Hermitean because

$$\langle g|W f\rangle = -\langle W g|f\rangle. \tag{8.51}$$

Any operator that fulfills a relation similar to (8.51) is called *anti-Hermitean* or *skew-Hermitean*.

As a second example consider a one-dimensional case defined over a domain Ω belonging to the x axis. It is easily shown that the operator $(i\, d/dx)$ is Hermitean: in

---

[9] Consider for instance the calculation of the expectation value of the momentum of a free particle based on (10.18). If the minus sign were omitted in (8.48), the direction of momentum would be opposite to that of the propagation of the wave front associated to it.

fact, integrating by parts and observing that the integrated part vanishes because f and g are square integrable, yields

$$\int_\Omega g^* \, \mathrm{i} \, \frac{\mathrm{d}f}{\mathrm{d}x} \, \mathrm{d}\Omega = \left[g^* \, \mathrm{i} \, f \right]_\Omega - \int_\Omega \mathrm{i} \, \frac{\mathrm{d}g^*}{\mathrm{d}x} \, f \, \mathrm{d}\Omega = \int_\Omega \left(\mathrm{i} \, \frac{\mathrm{d}g}{\mathrm{d}x} \right)^* f \, \mathrm{d}\Omega. \qquad (8.52)$$

By the same token one shows that the operator $\mathrm{d}/\mathrm{d}x$ is skew-Hermitean. The three-dimensional generalization of $(\mathrm{i}\,\mathrm{d}/\mathrm{d}x)$ is $(\mathrm{i}\,\mathrm{grad})$. Applying the latter onto the product $g^* \, f$ yields $g^* \, \mathrm{i}\,\mathrm{grad} f - (\mathrm{i}\,\mathrm{grad} g)^* \, f$. Integrating over Ω with Σ the boundary of Ω and $\mathbf{n}$ the unit vector normal to it, yields

$$\int_\Omega g^* \, \mathrm{i}\,\mathrm{grad} f \, \mathrm{d}\Omega - \int_\Omega (\mathrm{i}\,\mathrm{grad} g)^* \, f \, \mathrm{d}\Omega = \mathrm{i} \int_\Sigma g^* \, f \, \mathbf{n} \, \mathrm{d}\Sigma. \qquad (8.53)$$

The form of the right hand side of (8.53) is due to (A.25). As f, g vanish over the boundary, it follows $\langle g | \mathrm{i}\,\mathrm{grad} f \rangle = \langle \mathrm{i}\,\mathrm{grad} g | f \rangle$, namely, $(\mathrm{i}\,\mathrm{grad})$ is Hermitean.

Another important example, still in the one-dimensional case, is that of the operator $\mathrm{d}^2/\mathrm{d}x^2$. Integrating by parts twice shows that the operator is Hermitean. Its three-dimensional generalization in Cartesian coordinates is ∇^2. Using the second Green theorem (A.25) and remembering that f, g vanish over the boundary provides $\langle g | \nabla^2 f \rangle = \langle \nabla^2 g | f \rangle$, that is, ∇^2 is Hermitean.

8.6.2 A Collection of Operators' Definitions and Properties

A number of definitions and properties of operator algebra are illustrated in this section. The *identity operator* $\mathcal{I}$ is such that $\mathcal{I} f = f$ for all f; the *null operator* $\mathcal{O}$ is such that $\mathcal{O} f = 0$ for all f. The product of two operators, $\mathcal{A}\mathcal{B}$, is an operator whose action on a function is defined as follows: $s = \mathcal{A}\mathcal{B} f$ is equivalent to $g = \mathcal{B} f$, $s = \mathcal{A} g$; in other terms, the operators $\mathcal{A}$ and $\mathcal{B}$ act in a specific order. In general, $\mathcal{B}\mathcal{A} \neq \mathcal{A}\mathcal{B}$. The operators $\mathcal{A}\mathcal{A}$, $\mathcal{A}\mathcal{A}\mathcal{A}, \ldots$ are indicated with $\mathcal{A}^2, \mathcal{A}^3, \ldots$

An operator $\mathcal{A}$ may or may not have an *inverse*, $\mathcal{A}^{-1}$. If the inverse exists, it is unique and has the property $\mathcal{A}^{-1}\mathcal{A} f = f$ for all f. Left multiplying the above by $\mathcal{A}$ and letting $g = \mathcal{A} f$ yields $\mathcal{A}\mathcal{A}^{-1} g = g$ for all g. The two relations just found can be recast as

$$\mathcal{A}^{-1}\mathcal{A} = \mathcal{A}\mathcal{A}^{-1} = \mathcal{I}. \qquad (8.54)$$

From (8.54) it follows $(\mathcal{A}^{-1})^{-1} = \mathcal{A}$. If $\mathcal{A}$ and $\mathcal{B}$ have an inverse, letting $\mathcal{C} = \mathcal{B}^{-1}\mathcal{A}^{-1}$ one finds, for all f and using the associative property, the two relations $\mathcal{B}\mathcal{C} f = \mathcal{B}\mathcal{B}^{-1}\mathcal{A}^{-1} f = \mathcal{A}^{-1} f$ and $\mathcal{A}\mathcal{B}\mathcal{C} f = \mathcal{A}\mathcal{A}^{-1} f = f$, namely, $\mathcal{A}\mathcal{B}\mathcal{C} = \mathcal{I}$; in conclusion,

$$(\mathcal{A}\mathcal{B})^{-1} = \mathcal{B}^{-1}\mathcal{A}^{-1}. \qquad (8.55)$$

From (8.55) one defines the inverse powers of A as

$$A^{-2} = (A^2)^{-1} = (AA)^{-1} = A^{-1}A^{-1}, \tag{8.56}$$

and so on. Let $Av = \lambda v$ be the eigenvalue equation of A. Successive left multiplications by A yield

$$A^2 v = \lambda^2 v, \quad A^3 v = \lambda^3 v, \quad \dots \tag{8.57}$$

As a consequence, an operator of the polynomial form

$$P_n(A) = c_0 A^n + c_1 A^{n-1} + c_2 A^{n-2} + \dots + c_n \tag{8.58}$$

fulfills the eigenvalue equation

$$P_n(A) v = P_n(\lambda) v, \quad P_n(\lambda) = c_0 \lambda^n + \dots + c_n. \tag{8.59}$$

By definition, an eigenfunction can not vanish identically. If A has an inverse, left-multiplying the eigenvalue equation $Av = \lambda v$ by A^{-1} yields $v = \lambda A^{-1} v \neq 0$, whence $\lambda \neq 0$. Dividing the latter by λ and iterating the procedure shows that

$$A^{-2} v = \lambda^{-2} v, \quad A^{-3} v = \lambda^{-3} v, \quad \dots \tag{8.60}$$

An operator may be defined by a series expansion, if the latter converges:

$$C = \sigma(A) = \sum_{k=-\infty}^{+\infty} c_k A^k. \tag{8.61}$$

By way of example,

$$C = \exp(A) = I + A + \frac{1}{2!} A^2 + \frac{1}{3!} A^3 + \dots \tag{8.62}$$

Given an operator A, its adjoint $A^\dagger$ is defined as in Sect. 8.3.1. Letting $C = A^\dagger$, applying the definition of adjoint operator to C, and taking the conjugate of both sides shows that $(A^\dagger)^\dagger = A$. From the definition of adjoint operator it also follows

$$(AB)^\dagger = B^\dagger A^\dagger. \tag{8.63}$$

An operator is *unitary* if its inverse is identical to its adjoint for all f:

$$A^{-1} f = A^\dagger f. \tag{8.64}$$

Left multiplying (8.64) by A, and left multiplying the result by $A^\dagger$, yields for a unitary operator

$$AA^\dagger = A^\dagger A = I. \tag{8.65}$$

The application of a unitary operator to a function f leaves the norm of the latter unchanged. In fact, using definition (8.12), namely, $||f||^2 = \langle f| f \rangle$, and letting $g = \mathcal{A}f$ with $\mathcal{A}$ unitary, yields

$$||g||^2 = \int_\Omega (\mathcal{A}f)^* \mathcal{A}f \, d\Omega = \int_\Omega (\mathcal{A}^\dagger \mathcal{A}f)^* f \, d\Omega = \int_\Omega f^* f \, d\Omega = ||f||^2, \quad (8.66)$$

where the second equality holds due to the definition of adjoint operator, and the third one holds because $\mathcal{A}$ is unitary. The inverse also holds true: if the application of $\mathcal{A}$ leaves the function's norm unchanged, that is, if $||\mathcal{A}f|| = ||f||$ for all f, then

$$\int_\Omega (\mathcal{A}^\dagger \mathcal{A}f - f)^* f \, d\Omega = 0. \quad (8.67)$$

As a consequence, the quantity in parenthesis must vanish, whence the operator is unitary. The product of two unitary operators is unitary:

$$(\mathcal{A}\mathcal{B})^{-1} = \mathcal{B}^{-1}\mathcal{A}^{-1} = \mathcal{B}^\dagger \mathcal{A}^\dagger = (\mathcal{A}\mathcal{B})^\dagger, \quad (8.68)$$

where the second equality holds because $\mathcal{A}$ and $\mathcal{B}$ are unitary. The eigenvalues of a unitary operator have the form $\exp(i\, v)$, with v a real number. Let an eigenvalue equation be $\mathcal{A}v = \lambda v$, with $\mathcal{A}$ unitary. The following hold,

$$\int_\Omega |\mathcal{A}v|^2 \, d\Omega = |\lambda|^2 \int_\Omega |v|^2 \, d\Omega, \qquad \int_\Omega |\mathcal{A}v|^2 \, d\Omega = \int_\Omega |v|^2 \, d\Omega, \quad (8.69)$$

the first one because of the eigenvalue equation, the second one because $\mathcal{A}$ is unitary. As an eigenfunction can not vanish identically, it follows $|\lambda|^2 = 1$ whence $\lambda = \exp(i\, v)$. It is also seen by inspection that, if the eigenvalues of an operator have the form $\exp(i\, v)$, with v a real number, then the operator is unitary.

It has been anticipated above that in general it is $\mathcal{B}\mathcal{A} \neq \mathcal{A}\mathcal{B}$. Two operators $\mathcal{A}, \mathcal{B}$ are said to *commute* if

$$\mathcal{B}\mathcal{A}f = \mathcal{A}\mathcal{B}f \quad (8.70)$$

for all f. The *commutator* of $\mathcal{A}, \mathcal{B}$ is the operator $\mathcal{C}$ such that

$$i\mathcal{C}f = (\mathcal{A}\mathcal{B} - \mathcal{B}\mathcal{A}) f \quad (8.71)$$

for all f. The definition (8.71) is such that, if both $\mathcal{A}$ and $\mathcal{B}$ are Hermitean, then $\mathcal{C}$ is Hermitean as well. The commutator of two commuting operators is the null operator. A very important example of non-commuting operators is the pair $q, -i\, d/dq$, where q is any dynamical variable. One finds

$$i\mathcal{C}f = -i q \frac{df}{dq} + i \frac{d(q\, f)}{dq} = i f, \quad (8.72)$$

namely, the commutator is in this case the identity operator $\mathcal{I}$.

8.6.3 Examples of Commuting Operators

Operators that contain only spatial coordinates commute; similarly, operators that contain only momentum operators commute. The operators $\mathcal{A}$, $\mathcal{B}$, $\mathcal{C}$ defined in (10.4) commute because they act on different coordinates; note that the definition of $\mathcal{A}$ is such that it may contain both x and $\hat{p}_x = -i\hbar\,\partial/\partial x$, and so on.

As an example of operators containing only momentum operators one may consider the Hamiltonian operator $-(\hbar^2/2\,m)\,\nabla^2$ of a free particle discussed in Sect. 8.2.1 and the momentum operator $-i\hbar\,\nabla$ itself (Sect. 8.5). As for a free particle they commute, a measurement of momentum is compatible in that case with a measurement of energy (Sect. 8.6.4). Considering a one-dimensional problem, the energy is $E = p^2/(2\,m)$, where the modulus of momentum is given by $p = \hbar\,k$; for a free particle, both energy and momentum are conserved. The eigenfunctions are const $\times \exp(\pm i\,p\,x/\hbar)$ for both operators.

Remembering (8.72) one concludes that two operators do not commute if one of them contains one coordinate q, and the other one contains the operator $-i\hbar\,\partial/\partial q$ associated to the momentum conjugate to q.

8.6.4 Momentum and Energy of a Free Particle

The eigenfunctions of the momentum operator are the same as those of the Schrödinger equation for a free particle. More specifically, given the sign of $\tilde{p}$, the solution of (8.49) concides with either one or the other of the two linearly-independent solutions of (8.1). This outcome is coherent with the conclusions reached in Sect. 8.2.1 about the free particle's motion. For a free particle whose momentum is prescribed, the energy is purely kinetic and is prescribed as well, whence the solution of (8.49) must be compatible with that of (8.1). However, prescribing the momentum, both in modulus and direction, for a free particle, provides the additional information that allows one to eliminate one of the two summands from the linear combination (8.2) by setting either c_1 or c_2 to zero. For a given eigenvalue $\tilde{p}$, (8.49) has only one solution (apart from the multiplicative constant) because it is a first-order equation; in contrast, for a given eigenvalue E, the second-order equation (8.1) has two independent solutions and its general solution is a linear combination of them.

In a broader sense the momentum operator $\hat{p}_x = -i\hbar\,d/dx$ is Hermitean also for functions of the form $v_k(x) = \exp(i\,k\,x)/\sqrt{2\pi}$, which are not square integrable. In fact, remembering (C.43) one finds

$$\langle v_{k'}|\hat{p}_x v_k\rangle = -\frac{i\hbar}{2\pi}\int_{-\infty}^{+\infty}\exp(-i\,k'\,x)\frac{d}{dx}\exp(i\,k\,x)\,dx = \hbar\,k\,\delta(k'-k). \quad (8.73)$$

Similarly it is $\langle \hat{p}_x v_{k'}|v_k\rangle = \hbar\,k'\,\delta(k'-k)$. As mentioned in Sect. C.4, the two equalities just found can be used only within an integral over k or k'. In that case, however, they yield the same result $\hbar\,k$. By the same token one shows that

$$\langle v_{k'}|\hat{p}_x^2 v_k\rangle = \hbar^2\,k^2\,\delta(k'-k), \qquad \langle \hat{p}_x^2 v_{k'}|v_k\rangle = \hbar^2\,(k')^2\,\delta(k'-k), \qquad (8.74)$$

hence the Laplacian operator is Hermitean in a broader sense for non-square-integrable functions of the form $v_k(x) = \exp(\mathrm{i}\,k\,x)/\sqrt{2\pi}$.

Problems

8.1 The one-dimensional, time-independent Schrödinger equation is a homogeneous equation of the form

$$w'' + q\,w = 0, \qquad q = q(x), \tag{8.75}$$

where primes indicate derivatives. In turn, the most general, linear equation of the second order with a non-vanishing coefficient of the highest derivative is

$$f'' + a\,f' + b\,f = c, \qquad a = a(x), \quad b = b(x), \quad c = c(x). \tag{8.76}$$

Assume that a is differentiable. Show that if the solution of (8.75) is known, then the solution of (8.76) is obtained from the former by simple integrations.

Chapter 9
Time-Dependent Schrödinger Equation

9.1 Introduction

The time-dependent Schrödinger equation is derived from the superposition principle, in the conservative case first, then in the general case. The derivation of the continuity equation follows, leading to the concept of wave packet and density of probability flux. Then, the wave packet for a free particle is investigated in detail, and the concept of group velocity is introduced. The first complement deals with an application of the semiclassical approximation; through it one explains why an electron belonging to a stationary state emits no power, namely, why the radiative decay predicted by the classical model does not occur. The polar form of the time-dependent Schrödinger equation is then shown, that brings about an interesting similarity with the Hamilton–Jacobi equation of Classical Mechanics. The last complement deals with the Hamiltonian operator of a particle subjected to an electromagnetic field, and shows the effect of a gauge transformation on the wave function.

9.2 Superposition Principle

Following De Broglie's line of reasoning (Sect. 7.4.5) one associates the monochromatic wave function $w(\mathbf{r}) \exp(-\mathrm{i}\,\omega\,t)$ to the motion of a particle with definite and constant energy $E = \hbar\,\omega$. The analogy with the electromagnetic case then suggests that a more general type of wave function—still related to the conservative case—can be expressed as a superposition, that is, a linear combination with constant coefficients, of monochromatic wave functions. This possibility is one of the postulates of De Broglie's theory, and is referred to as *Superposition Principle*. To vest it with a mathematical form one must distinguish among the different types of spectrum; for the discrete spectrum, indicating with c_n the complex coefficients of the linear combination, the general wave function reads

$$\psi(\mathbf{r}, t) = \sum_n c_n\, w_n\, \exp(-\mathrm{i}\,\omega_n\, t), \qquad (9.1)$$

© Springer Science+Business Media New York 2015
M. Rudan, *Physics of Semiconductor Devices*,
DOI 10.1007/978-1-4939-1151-6_9

with E_n, w_n the eigenvalues and eigenfunctions of the time-independent Schrödinger equation $\mathcal{H}w_n = E_n w_n$, and $\omega_n = E_n/\hbar$. As usual, n stands for a single index or a set of indices. The form of (9.1) is such that the spatial coordinates are separated from the time coordinate; fixing the latter by letting, say, $t = 0$, and remembering that the set of eigenfunctions w_n is complete, yields

$$\psi_{t=0} = \psi(\mathbf{r}, 0) = \sum_n c_n w_n, \qquad c_n = \langle w_n | \psi_{t=0} \rangle. \tag{9.2}$$

The above shows that the coefficients c_n are uniquely determined by the initial condition $\psi_{t=0}$. On the other hand, once the coefficients c_n are known, the whole time evolution of ψ is determined, because the angular frequencies appearing in the time-dependent terms $\exp(-i\omega_n t)$ are also known. In other terms, ψ is determined by the initial condition and by the time-independent Hamiltonian operator whence E_n, w_n derive.

An important aspect of (9.1) is that it allows one to construct a wave function of a given form; for such a construction, in fact, it suffices to determine the coefficients by means of the second relation in (9.2). In particular it is possible to obtain a wave function that is square integrable at all times, even if the eigenfunctions w_n are not square integrable themselves. Thanks to this property the wave function (9.1) is localized in space at each instant of time, hence it is suitable for describing the motion of the particle associated to it. Due to the analogy with the electromagnetic case, were the interference of monochromatic waves provides the localization of the field's intensity, a wave function of the form (9.1) is called *wave packet*. Remembering that the wave function provides the probability density $|\psi|^2$ used to identify the position of the particle, one can assume that the wave packet's normalization holds:

$$\int_\Omega |\psi|^2 \, \mathrm{d}^3 r = \sum_n |c_n|^2 = 1, \tag{9.3}$$

where the second equality derives from Parseval's theorem (8.41). From (9.3) it follows that the coefficients are subjected to the constraint $0 \le |c_n|^2 \le 1$.

As all possible energies E_n appear in the expression (9.1) of ψ, the wave packet does not describe a motion with a definite energy. Now, assume that an energy measurement is carried out on the particle, and let $t = t_E$ be the instant at which the measurement is completed. During the measurement the Hamiltonian operator of (8.46) does not hold because the particle is interacting with the measuring apparatus, hence the forces acting on it are different from those whence the potential energy V of (8.45) derives. Instead, for $t > t_E$ the original Schrödinger equation (8.46) is restored, so the expression of ψ is again given by a linear combination of monochromatic waves; however, the coefficients of the combination are expected to be different from those that existed prior to the measurement, due to the perturbation produced by the latter. In particular, the form of ψ for $t > t_E$ must be compatible with the fact that the energy measurement has found a specific value of the energy; this is possible only if the coefficients are set to zero, with the exception of the one corresponding to the energy that is the outcome of the measurement. The latter must be one of the eigenvalues of (8.46) due to the compatibility requirement; if it is, say, E_m, then the form of the wave function for $t > t_E$ is

$$\psi(\mathbf{r}, t) = w_m \exp[-i E_m (t - t_E)/\hbar], \tag{9.4}$$

where the only non-vanishing coefficient, c_m, has provisionally been set to unity.

The reasoning leading to (9.4) can be interpreted as follows: the interaction with the measuring apparatus filters out from (9.1) the term corresponding to E_m; as a consequence, the coefficients c_n whose values were previously set by the original ψ are modified by the measurement and become $c_n = \delta_{nm}$. If the filtered eigenfunction w_m is square integrable, then (9.3) holds, whence $\sum_n |c_n|^2 = |c_m|^2 = 1$, $c_m = \exp(i\Phi)$. As the constant phase Φ does not carry any information, it can be set to zero to yield $c_m = 1$. If w_m is not square integrable, the arbitrariness of the multiplicative constant still allows one to set $c_m = 1$.

As the energy measurement forces the particle to belong to a definite energy state (in the example above, E_m), for $t > t_E$ the particle's wave function keeps the monochromatic form (9.4). If, at a later time, a second energy measurement is carried out, the only possible outcome is E_m; as a consequence, after the second measurement is completed, the form of the wave function is still (9.4), whence $|c_m|^2 = 1$. One notes that the condition $|c_m|^2 = 1$ is associated with the certainty that the outcome of the energy measurement is E_m whereas, when the general superposition (9.1) holds, the coefficients fulfill the relations $\sum_n |c_n|^2 = 1, 0 \leq |c_n|^2 \leq 1$, and the measurement's outcome may be any of the eigenvalues. It is then sensible to interpret $|c_n|^2$ as the *probability* that a measurement of energy finds the result E_n. This interpretation, that has been worked out here with reference to energy, is extended to the other dynamical variables (Sect. 10.2).

When the spectrum is continuous the description of the wave packet is

$$\psi(\mathbf{r}, t) = \iiint_{-\infty}^{+\infty} c_\mathbf{k} w_\mathbf{k} \exp(-i \omega_\mathbf{k} t) \, d^3 k, \tag{9.5}$$

with $E_\mathbf{k}$, $w_\mathbf{k}$ the eigenvalues and eigenfunctions of $\mathcal{H} w_\mathbf{k} = E_\mathbf{k} w_\mathbf{k}$, and $\omega_\mathbf{k} = E_\mathbf{k}/\hbar$. Such symbols stand for $E_\mathbf{k} = E(\mathbf{k})$, $w_\mathbf{k} = w(\mathbf{r}, \mathbf{k})$, and so on, with $\mathbf{k}$ a three-dimensional vector whose components are continuous. The relations corresponding to (9.2, 9.3) are

$$\psi_{t=0} = \psi(\mathbf{r}, 0) = \iiint_{-\infty}^{+\infty} c_\mathbf{k} w_\mathbf{k} \, d^3 k, \qquad c_\mathbf{k} = \langle w_\mathbf{k} | \psi_{t=0} \rangle, \tag{9.6}$$

$$\int_\Omega |\psi|^2 \, d^3 r = \iiint_{-\infty}^{+\infty} |c_\mathbf{k}|^2 \, d^3 k = 1. \tag{9.7}$$

The expression of $\psi_{t=0}$ in (9.6) lends itself to providing an example of a wave function that is square integrable, while the eigenfunctions that build up the superposition are not. Consider, in fact, the relation (C.82) and let $c_k = \sigma \exp(-\sigma^2 k^2/2)$,

$w_k = \exp(\mathrm{i}\,k\,x)/\sqrt{2\pi}$ in it, with σ a length; in this way (C.82) becomes the one-dimensional case of (9.6) and yields $\psi_{t=0} = \exp[-x^2/(2\sigma^2)]$, showing that a square-integrable function like the Gaussian one can be expressed as a combination of the non-square-integrable spatial parts of the plane waves.

The extraction of the probabilities in the continuous-spectrum case accounts for the fact the $E(\mathbf{k})$ varies continuously with $\mathbf{k}$. To this purpose one takes the elementary volume d^3k centered on some $\mathbf{k}$ and considers the product $|c_{\mathbf{k}}|^2\,\mathrm{d}^3k$. Such a product is given the meaning of infinitesimal probability that the outcome of an energy measurement belongs to the range of $E(\mathbf{k})$ values whose domain is d^3k (more comments are made in Sect. 9.7.1).

9.3 Time-Dependent Schrödinger Equation

The Superposition Principle illustrated in Sect. 9.2 prescribes the form of the wave packet in the conservative case. Considering for simplicity a discrete set of eigenfunctions, the time derivative of ψ reads

$$\frac{\partial \psi}{\partial t} = \sum_n c_n\, w_n\, \frac{E_n}{\mathrm{i}\,\hbar}\, \exp(-\mathrm{i}E_n\, t/\hbar). \qquad (9.8)$$

Using the time-independent Schrödinger equation $\mathcal{H}w_n = E_n w_n$ transforms the above into

$$\mathrm{i}\,\hbar\, \frac{\partial \psi}{\partial t} = \sum_n c_n\, \mathcal{H}w_n\, \exp(-\mathrm{i}\,E_n\, t/\hbar), \qquad \mathrm{i}\,\hbar\, \frac{\partial \psi}{\partial t} = \mathcal{H}\psi. \qquad (9.9)$$

The second relation in (9.9) is a linear, homogeneous partial-differential equation, of the second order with respect to the spatial coordinates and of the first order with respect to time, whose solution is the wave function ψ. It is called *time-dependent Schrödinger equation*; as its coefficients are complex, so is ψ. To solve the equation it is necessary to prescribe the initial condition $\psi(\mathbf{r}, t = 0)$ and the boundary conditions. For the latter the same discussion as in Sect. 8.2 applies, because the spatial behavior of ψ is prescribed by the Hamiltonian operator.

The reasoning leading to the second relation in (9.9) is based on the Superposition Principle, namely, once the form of ψ is given, the equation fulfilled by ψ is readily extracted. Such a reasoning is not applicable in the non-conservative cases, because the time-independent equation $\mathcal{H}w_n = E_n w_n$ does not hold then, so the eigenvalues and eigenfunctions upon which the superposition is based are not available. However, another line of reasoning shows that the time-dependent Schrödinger equation holds also for the non-conservative situations [69]. Although in such cases the wave function ψ is not expressible as a superposition of monochromatic waves, it can still be expanded using an orthonormal set. If the set is discrete, $v_n = v_n(\mathbf{r})$, the expansion reads

$$\psi(\mathbf{r}, t) = \sum_n b_n\, v_n(\mathbf{r}), \qquad b_n(t) = \langle v_n|\psi\rangle, \qquad (9.10)$$

whereas for a continuous set $v_{\mathbf{k}} = v(\mathbf{r}, \mathbf{k})$ one finds

$$\psi(\mathbf{r}, t) = \iiint_{-\infty}^{+\infty} b_{\mathbf{k}} \, v_{\mathbf{k}} \, d^3 k, \qquad b_{\mathbf{k}}(t) = \langle v_{\mathbf{k}} | \psi \rangle. \tag{9.11}$$

9.4 Continuity Equation and Norm Conservation

Remembering that the square modulus of the wave function provides the localization of the particle, it is of interest to investigate the time evolution of $|\psi|^2$, starting from the time derivative of ψ given by the time-dependent Schrödinger equation (9.9). Here it is assumed that the wave function is normalized to unity and that the Hamiltonian operator is real, $\mathcal{H}^* = \mathcal{H} = -\hbar^2/(2\,m)\,\nabla^2 + V$; a case where the operator is complex is examined in Sect. 9.5. Taking the time derivative of $|\psi|^2$ yields

$$\frac{\partial |\psi|^2}{\partial t} = \psi^* \frac{\partial \psi}{\partial t} + \psi \frac{\partial \psi^*}{\partial t} = \psi^* \frac{\mathcal{H}\psi}{i\hbar} - \psi \frac{\mathcal{H}\psi^*}{i\hbar}, \tag{9.12}$$

with $\psi^* \mathcal{H}\psi - \psi \mathcal{H}\psi^* = -\hbar^2/(2m)\,(\psi^* \nabla^2 \psi - \psi \nabla^2 \psi^*)$. Identity (A.17) then yields

$$\frac{\partial |\psi|^2}{\partial t} + \mathrm{div}\mathbf{J}_\psi = 0, \qquad \mathbf{J}_\psi = \frac{i\hbar}{2\,m}\,(\psi\,\mathrm{grad}\,\psi^* - \psi^*\,\mathrm{grad}\,\psi). \tag{9.13}$$

The first relation in (9.13) has the form of a continuity equation (compare with (23.3) and (4.23)). As $|\psi|^2$ is the probability density, $\mathbf{J}_\psi$ takes the meaning of *density of the probability flux*;[1] it is a real quantity because the term in parentheses in the second relation of (9.13) is imaginary.

Relations (9.13) provide a physical explanation of the continuity requirements that were discussed from the mathematical standpoint in Sect. 8.2. Such requirements, namely, the continuity of the wave function and of its first derivatives in space, were introduced in Sect. 8.2 with reference to the solutions of the time-independent Schrödinger equation; however, they hold also for the time-dependent one because the spatial behavior of ψ is prescribed by the Hamiltonian operator. Their physical explanation is that they provide the spatial continuity of the probability density and of the probability-flux density. Integrating (9.13) over a volume Ω' whose surface is Σ' yields

$$\frac{d}{dt} \int_{\Omega'} |\psi|^2 \, d\Omega' = - \int_{\Sigma'} \mathbf{J}_\psi \cdot \mathbf{n} \, d\Sigma', \tag{9.14}$$

with $\mathbf{n}$ the unit vector normal to Σ', oriented in the outward direction. The integral at the left hand side of (9.14) is the probability of localizing the particle within Ω',

---

[1] Remembering that $[|\psi|^2] = \mathrm{m}^{-3}$, one finds $[\mathbf{J}_\psi] = \mathrm{m}^{-2}\,\mathrm{t}^{-1}$.

that at the right hand side is the probability flux across Σ' in the outward direction; as a consequence, the meaning of (9.14) is that the time variation of the localization probability within Ω' is the negative probability flux across the surface. If $\Omega' \to \infty$ the surface integral vanishes because ψ is square integrable and, as expected,

$$\frac{d}{dt} \int_\infty |\psi|^2 \, d\Omega = 0. \tag{9.15}$$

The above is another form of the normalization condition and is also termed *norm-conservation condition*. Note that the integral in (9.15) does not depend on time although ψ does.

The density of the probability flux can be given a different form that uses the momentum operator $\hat{\mathbf{p}} = -i\hbar \, \mathrm{grad}$ introduced in Sect. 8.5; one finds

$$\mathbf{J}_\psi = \frac{1}{2m} \left[\psi \, (\hat{\mathbf{p}}\psi)^* + \psi^* \, \hat{\mathbf{p}}\psi \right] = \frac{1}{m} \, \Re \left(\psi^* \, \hat{\mathbf{p}}\psi \right). \tag{9.16}$$

Although this form is used less frequently than (9.13), it makes the analogy with the classical flux density much more intelligible.

When the wave function is of the monochromatic type (9.4), the time-dependent factors cancel each other in (9.13), to yield

$$\mathrm{div}\mathbf{J}_\psi = 0, \qquad \mathbf{J}_\psi = \frac{i\hbar}{2m} \, (w \, \mathrm{grad}w^* - w^* \, \mathrm{grad}w). \tag{9.17}$$

If w is real, then $\mathbf{J}_\psi = 0$.

9.5   Hamiltonian Operator of a Charged Particle

The Hamiltonian function of a particle of mass m and charge e, subjected to an electromagnetic field, is given by (1.35), namely,

$$H = \sum_{i=1}^{3} \frac{1}{2m} \, (p_i - e \, A_i)^2 + e \, \varphi, \tag{9.18}$$

where the scalar potential φ and the components of the vector potential A_i may depend on the spatial coordinates and time. To find the Hamiltonian operator corresponding to H one could apply the same procedure as in (8.48), that consists in replacing p_i with $\hat{p}_i = -i\hbar \, \partial/\partial x_i$. In this case, however, a difficulty arises if A_i depends on the coordinates; in fact, the two equivalent expansions of $(p_i - e \, A_i)^2$, namely, $p_i^2 + e^2 \, A_i^2 - 2 \, p_i \, e \, A_i$ and $p_i^2 + e^2 \, A_i^2 - 2 \, e \, A_i \, p_i$ yield two different operators: the first of them contains the summand $\partial(A_i \, \psi)/\partial x_i$, the other one contains $A_i \, \partial\psi/\partial x_i$, and neither one is Hermitean. If, instead, one keeps the order of the factors in the expansion, namely, $(p_i - e A_i)^2 = p_i^2 + e^2 A_i^2 - e A_i p_i - p_i e A_i$, the resulting Hamiltonian operator reads $\mathcal{H} = \mathcal{H}_R + i \mathcal{H}_I$, with

$$
\mathcal{H}_R = -\frac{\hbar^2}{2m}\nabla^2 + e\varphi + \frac{e^2}{2m}\mathbf{A}\cdot\mathbf{A}, \qquad \mathcal{H}_I = \frac{\hbar e}{2m}\sum_{i=1}^{3}\left(A_i\frac{\partial}{\partial x_i} + \frac{\partial}{\partial x_i}A_i\right),
$$

$$(9.19)$$

and is Hermitean (compare with Sect. 10.2). The particle dynamics is determined by the time-dependent Schrödinger equation $i\hbar\,\partial\psi/\partial t = \mathcal{H}\psi$. The continuity equation fulfilled by ψ is found following the same reasoning as in Sect. 9.4, starting from

$$
\frac{\partial|\psi|^2}{\partial t} = \psi^*\frac{(\mathcal{H}_R + i\mathcal{H}_I)\psi}{i\hbar} - \psi\frac{(\mathcal{H}_R - i\mathcal{H}_I)\psi^*}{i\hbar}. \qquad (9.20)
$$

The terms related to $\mathcal{H}_R$ yield $-\mathrm{div}\Re(\psi^*\,\hat{\mathbf{p}}\,\psi)/m$ as in Sect. 9.4. Those related to $\mathcal{H}_I$ yield $\mathrm{div}(e\,\mathbf{A}\,|\psi|^2)/m$. In conclusion, the continuity equation for the wave function of a charged particle reads

$$
\frac{\partial|\psi|^2}{\partial t} + \mathrm{div}\mathbf{J}_\psi = 0, \qquad \mathbf{J}_\psi = \frac{1}{m}\Re\left[\psi^*\left(\hat{\mathbf{p}} - e\mathbf{A}\right)\psi\right]. \qquad (9.21)
$$

It is worth noting that the transformation from the Hamiltonian function (9.18) to the Hamiltonian operator (9.19) produced by replacing p_i with $\hat{p}_i$ is limited to the dynamics of the particle; the electromagnetic field, instead, is still treated through the scalar and vector potentials, and no transformation similar to that used for the particle is carried out. The resulting Hamiltonian operator (9.19) must then be considered as approximate; the term *semiclassical approximation* is in fact used to indicate the approach based on (9.19), where the electromagnetic field is treated classically whereas Quantum Mechanics is used in the description of the particle's dynamics. The procedure by which the quantum concepts are extended to the electromagnetic field is described in Sect. 12.3.

The semiclassical approximation is useful in several instances, among which there is the calculation of the stationary states of atoms. As shown in Sect. 9.7.2, it explains why the radiative decay predicted in Sect. 5.11.3 using the classical (planetary) model does not actually occur.

9.6   Approximate Form of the Wave Packet for a Free Particle

The energy spectrum of a free particle is continuous, and the wave packet is given by (9.5), with

$$
w_\mathbf{k}(\mathbf{r}) = \frac{1}{(2\pi)^{3/2}}\exp(i\,\mathbf{k}\cdot\mathbf{r}), \qquad E_\mathbf{k} = \frac{\hbar^2}{2m}\left(k_1^2 + k_2^2 + k_3^2\right) = \hbar\,\omega_\mathbf{k}. \qquad (9.22)
$$

As $-\infty < k_i < +\infty$, here the order of degeneracy of $E_\mathbf{k}$ is infinite. Now, remembering that the wave packet is normalized to unity, it follows that $|\psi(\mathbf{r},t)|^2\,\mathrm{d}^3r$ is the infinitesimal probability that at time t the particle is localized within d^3r ; also, from the analysis carried out in Sect. 9.2, the product $|c_{\mathbf{k}}|^2 d^3k$ is the infinitesimal probability that the outcome of an energy measurement belongs to the range of $E(\mathbf{k})$ values whose domain is d^3k. Note that $c_{\mathbf{k}}$ does not depend on time.

Considering the example of a Gaussian wave packet given at the end of Sect. 9.2, one can assume that $|\psi(\mathbf{r}, t)|^2$ is localized in the $\mathbf{r}$ space and $|c_{\mathbf{k}}|^2$ is localized in the $\mathbf{k}$ space. This means that $|\psi(\mathbf{r}, t)|^2$ and $|c_{\mathbf{k}}|^2$ become vanishingly small when $\mathbf{r}$ departs from its average value $\mathbf{r}_0(t)$ and respectively, $\mathbf{k}$ departs from its average value $\mathbf{k}_0$. Such average values are given by[2]

$$\mathbf{r}_0(t) = \iiint_{-\infty}^{+\infty} \mathbf{r}\,|\psi(\mathbf{r}, t)|^2\, d^3r, \qquad \mathbf{k}_0 = \iiint_{-\infty}^{+\infty} \mathbf{k}\,|c_{\mathbf{k}}|^2\, d^3k. \qquad (9.23)$$

An approximate expression of the wave packet is obtained by observing that, due to the normalization condition (9.7), the main contribution to the second integral in (9.7) is given by the values of $\mathbf{k}$ that are in the vicinity of $\mathbf{k}_0$. From the identity $k_i^2 = (k_{0i} - k_{0i} + k_i)^2 = k_{0i}^2 + 2 k_{0i} (k_i - k_{0i}) + (k_i - k_{0i})^2$ it then follows

$$\omega_{\mathbf{k}} = \frac{\hbar}{2m} k_0^2 + \frac{\hbar}{m} \mathbf{k}_0 \cdot (\mathbf{k} - \mathbf{k}_0) + \frac{\hbar}{2m} |\mathbf{k} - \mathbf{k}_0|^2 \qquad (9.24)$$

where, for $\mathbf{k}$ close to $\mathbf{k}_0$, one neglects the quadratic term to find

$$\omega_{\mathbf{k}} \simeq \omega_0 + \mathbf{u} \cdot (\mathbf{k} - \mathbf{k}_0), \qquad \omega_0 = \frac{\hbar}{2m} k_0^2, \qquad \mathbf{u} = \frac{\hbar}{m} \mathbf{k}_0 = (\mathrm{grad}_{\mathbf{k}} \omega_{\mathbf{k}})_{\mathbf{k}_0}, \qquad (9.25)$$

with $\mathbf{u}$ the *group velocity*. The neglect of terms of order higher than the first used to simplify (9.24) could not be applied to $c_{\mathbf{k}}$; in fact, $c_{\mathbf{k}}$ has typically a peak for $\mathbf{k} = \mathbf{k}_0$, so its first derivatives would vanish there. Using (9.25) and letting $\Phi_0 = \mathbf{k}_0 \cdot \mathbf{r} - \omega_0 t$ transform (9.5) into $\psi(\mathbf{r}, t) \simeq \exp(i \Phi_0)\, A\,(\mathbf{r} - \mathbf{u}\,t\,; \mathbf{k}_0)$, where the *envelope function* A is defined as

$$A\,(\mathbf{r} - \mathbf{u}\,t\,; \mathbf{k}_0) = \iiint_{-\infty}^{+\infty} \frac{c_{\mathbf{k}}}{(2\pi)^{3/2}} \exp\left[i\,(\mathbf{r} - \mathbf{u}\,t) \cdot (\mathbf{k} - \mathbf{k}_0)\right] d^3k. \qquad (9.26)$$

Within the limit of validity of (9.25), the envelope function contains the whole information about the particle's localization: $|\psi|^2 = |A|^2$. Also, the dependence of A on $\mathbf{r}$ and t is such that, for any two pairs $(\mathbf{r}_1, t_1)$, $(\mathbf{r}_2, t_2)$ fulfilling $\mathbf{r}_2 - \mathbf{u}\,t_2 = \mathbf{r}_1 - \mathbf{u}\,t_1$ the form of $A(\mathbf{r}_2, t_2)$ is the same as that of $A(\mathbf{r}_1, t_1)$. In other terms, A moves without distortion in the direction of $\mathbf{u}$, and its velocity is $(\mathbf{r}_2 - \mathbf{r}_1)/(t_2 - t_1) = \mathbf{u}$. As time evolves, the approximation leading to (9.25) becomes less and less accurate; taking by way of example $t_1 = 0$ as the initial time, and observing that the summands in (9.25) belong to the phase $-i\,\omega_{\mathbf{k}}\, t$, the approximation holds as long as $\hbar\, |\mathbf{k} - \mathbf{k}_0|^2\, t/(2m) \ll 2\pi$.

---

[2] Definitions (9.23) provide the correct weighed average of $\mathbf{r}$ and $\mathbf{k}$ thanks to the normalization condition (9.7). A more exhaustive treatment is carried out in Sect. 10.5.

9.7 Complements

9.7.1 About the Units of the Wave Function

Consider the wave function ψ associated to a single particle. When the wave function is square integrable and normalized to unity, its units are $[\psi] = m^{-3/2}$ due to $\int |\psi|^2 \, d^3r = 1$. Then, if the eigenvalues of the Hamiltonian operator are discrete, the second equality in (9.3) shows that $|c_n|^2$, c_n are dimensionless and, finally, (9.2) shows that w_n has the same units as ψ. If the eigenvalues are continuous, the second equality in (9.7) shows that $|c_k|^2$ has the dimensions of a volume of the real space, so that $[c_k] = m^{3/2}$ and, finally, (9.6) shows that w_k has the same units as ψ.

There are situations, however, where units different from those illustrated above are ascribed to the eigenfunctions. One example is that of the eigenfunctions of the form $w_k = \exp(i \mathbf{k} \cdot \mathbf{r})/(2\pi)^{3/2}$, worked out in Sect. 8.2.1, which are dimensionless; another example is that of eigenfunctions of the form (10.7), whose units are $[\delta(\mathbf{r} - \mathbf{r}_0)] = m^{-3}$. When such eigenfunctions occur, the units of the expansion's coefficients must be modified in order to keep the correct units of ψ (compare with the calculation shown in Sect. 14.6).

The considerations carried out here apply to the single-particle case. When the wave function describes a system of two or more particles, its units change accordingly (Sect. 15.2).

9.7.2 An Application of the Semiclassical Approximation

As indicated in Sect. 9.5 for the case of a particle subjected to an electromagnetic field, the semiclassical approximation consists in using the Hamiltonian operator (9.19), which is derived from the Hamiltonian function (9.18) by replacing p_i with $\hat{p}_i$. The electromagnetic field, instead, is still treated through the scalar and vector potentials. Experiments show that the approximation is applicable in several cases of interest. For instance, consider again the problem of the electromagnetic field generated by a single electron, discussed in classical terms in Sect. 5.11.2. If ψ is the wave function (assumed to be square integrable) associated to the electron in the quantum description, the electron's localization is given by $|\psi|^2$. It is then sensible to describe the charge density and the current density produced by the electron as

$$\rho = -q \, |\psi|^2, \qquad \mathbf{J} = -q \, \mathbf{J}_\psi, \tag{9.27}$$

where q is the elementary charge and $\mathbf{J}_\psi$ is defined in (9.13). If the electron is in a stationary state, namely, $\psi(\mathbf{r}, t) = w(\mathbf{r}) \exp(-i\omega t)$, then ρ and $\mathbf{J}$ are independent of time. From (4.58, 4.59) it follows that the potentials φ, $\mathbf{A}$ are independent of time as well. As a consequence, the distribution of charge and current density associated to the electron's motion is stationary (compare with (9.17)), which also yields that the acceleration $\dot{u}$ vanishes. From Larmor's formula (5.72) one finally finds that in this situation the electron emits no power; thus, the radiative decay predicted in Sect. 5.11.3 using the classical model does not occur.

9.7.3 Polar Form of the Schrödinger Equation

The time-dependent Schrödinger equation (9.9) is easily split into two real equations by considering the real and imaginary part of ψ. However, in this section the wave function will rather be written in polar form, $\psi = \alpha(\mathbf{r}, t) \exp[i\,\beta(\mathbf{r}, t)]$, $\alpha \geq 0$, which reminds one of that used to derive the eikonal equation (5.51) of Geometrical Optics. Despite the fact that the resulting relations are non linear, the outcome of this procedure is interesting. Considering a Hamiltonian operator of the type $\mathcal{H} = -\hbar^2\,\nabla^2/(2\,m) + V$, replacing the polar expression of ψ in (9.9), and separating the real and imaginary parts, yields two coupled, real equations; the first of them reads

$$\frac{\partial \alpha}{\partial t} = -\frac{\hbar}{2\,m}\left(\alpha\,\nabla^2\beta + 2\mathrm{grad}\alpha \cdot \mathrm{grad}\beta\right),\tag{9.28}$$

where the units of α^2 are those of the inverse of a volume. As for the second equation one finds

$$\frac{\partial(\hbar\,\beta)}{\partial t} + \frac{1}{2\,m}|\mathrm{grad}(\hbar\beta)|^2 + V + Q = 0,\qquad Q = -\frac{\hbar^2}{2\,m}\frac{\nabla^2\alpha}{\alpha},\tag{9.29}$$

where the units of $\mathrm{grad}(\hbar\,\beta)$ are those of a momentum. Using the short-hand notation

$$P = \alpha^2 = |\psi|^2,\quad S = \hbar\,\beta,\quad \mathbf{v}_e = \frac{\mathrm{grad}\,S}{m},\quad H_Q = \frac{1}{2}m v_e^2 + Q + V,\tag{9.30}$$

and multiplying (9.28) by 2α, transforms (9.28, 9.29) into

$$\frac{\partial P}{\partial t} + \mathrm{div}(P\,\mathbf{v}_e) = 0,\qquad \frac{\partial S}{\partial t} + H_Q = 0.\tag{9.31}$$

The wave function is assumed to be square integrable, so that $\int_\Omega P\,\mathrm{d}^3r = 1$. It is easily found that the first of (9.31) is the continuity equation (9.13): from the expression (9.16) of the current density one finds in fact

$$\mathbf{J}_\psi = \frac{1}{m}\,\Re\left(\psi^*\,\hat{\mathbf{p}}\psi\right) = \frac{\hbar}{m}\,\Re\left(\alpha^2\,\mathrm{grad}\beta - i\,\alpha\,\mathrm{grad}\alpha\right) = P\,\mathbf{v}_e.\tag{9.32}$$

The two differential equations (9.31), whose unknowns are α, β, are coupled with each other. The second of them is similar to the Hamilton–Jacobi equation (1.51) of Classical Mechanics, and becomes equal to it in the limit $\hbar \to 0$, that makes Q to vanish and S to become the Hamilton principal function (Sect. 1.7). Note that the limit $Q \to 0$ decouples (9.31) from each other. In the time-independent case (9.31) reduce to $\mathrm{div}(P\,\mathbf{v}_e) = 0$ and $m\,v_e^2/2 + Q + V = E$, coherently with the fact that in this case Hamilton's principal function becomes $S = W - Et$, with W the (time-independent) Hamilton characteristic function. Although $\mathbf{v}_e$ plays the role of an average velocity in the continuity equation of (9.31), determining the expectation value needs a further averaging: in fact, taking definition of (10.13) expectation value and observing that the normalization makes the integral of $\mathrm{grad}\alpha^2$ to vanish, yields

$$m \langle \mathbf{v}_e \rangle = \int_\Omega \psi^* \, \hat{\mathbf{p}} \psi \, \mathrm{d}^3 r = \int_\Omega \alpha^2 \, \nabla(\hbar \, \beta) \, \mathrm{d}^3 r = m \int_\Omega \alpha^2 \, \mathbf{v}_e \, \mathrm{d}^3 r. \qquad (9.33)$$

The last relation in (9.30) seems to suggest that Q is a sort of potential energy to be added to V. In fact this is not true, as demonstrated by the calculation of the expectation value of the kinetic energy T,

$$\langle T \rangle = -\frac{\hbar^2}{2m} \int_\Omega \psi^* \nabla^2 \psi \, \mathrm{d}^3 r = \int_\Omega \psi^* \left(\frac{1}{2} m \, v_e^2 + Q \right) \psi \, \mathrm{d}^3 r, \qquad (9.34)$$

showing that Q enters the expectation value of T, not V. To better investigate the meaning of Q it is useful to consider alternative expressions of $\langle Q \rangle$, like

$$\langle Q \rangle = \frac{\hbar^2}{2m} \int_\Omega |\nabla \alpha|^2 \, \mathrm{d}^3 r = \frac{1}{2m} \left(\langle \hat{\mathbf{p}} \cdot \hat{\mathbf{p}} \rangle - \langle p_e^2 \rangle \right), \qquad (9.35)$$

where $\hat{\mathbf{p}} = -i \hbar \, \mathrm{grad}$ and $\mathbf{p}_e = m \, \mathbf{v}_e$. The derivation of (9.34, 9.35) follows the same pattern as that of (9.28, 9.29). The first form of (9.35) shows that $\langle Q \rangle$ is positive definite irrespective of the shape of α. The second one is the analogue of the definition of dispersion around the average: the analogy with the treatment used in statistical mechanics (compare with (19.79)) suggests that $p_e^2/(2m)$ provides the analogue of the convective part of the kinetic energy, while Q provides the analogue of the thermal part of it [87].

It is interesting to note that the analogy between the Schrödinger equation and a set of a continuity and a Hamilton–Jacobi-like equations had been noted by de Broglie, who introduced the concept of *pilot wave* in [24]. This cost him severe criticism by Pauli at the Fifth Solvay Conference in 1927. He resumed the idea more than 20 years later, stimulated by the papers by Bohm introducing the concept of *quantum potential*, see, e.g., [8]. The most recent paper by de Broglie on the subject is [25], published when the author was 79 years old.

9.7.4 Effect of a Gauge Transformation on the Wave Function

The Hamiltonian function (1.35) of a particle of mass m and charge e, subjected to an electromagnetic field, has been derived in Sect. 1.5 and reads $H = \sum_{i=1}^{3} (p_i - e A_i)^2/(2m) + e \varphi$, with p_i the ith component of momentum in Cartesian coordinates, φ the electric potential, and A_i the ith component of the magnetic potential. If a gauge transformation (Sect. 4.5) is carried out, leading to the new potentials

$$\varphi \leftarrow \varphi' = \varphi - \frac{\partial \vartheta}{\partial t}, \qquad \mathbf{A} \leftarrow \mathbf{A}' = \mathbf{A} + \mathrm{grad}\vartheta, \qquad (9.36)$$

the resulting Hamiltonian function H' differs from the original one. In other terms, the Hamiltonian function is not gauge invariant. However, the Lorentz force $e\,(\mathbf{E} + \dot{\mathbf{r}} \wedge \mathbf{B})$ is invariant, whence the dynamics of the particle is not affected by a gauge transformation.

Also in the quantum case it turns out that the Hamiltonian operator is not gauge invariant, $\mathcal{H}' \neq \mathcal{H}$. As consequence, the solution of the Schrödinger equation is not gauge invariant either: $\psi' \neq \psi$. However, the particle's dynamics cannot be affected by a gauge transformation because the Lorentz force is invariant. It follows that, if the initial condition is the same, $|\psi'|^2_{t=0} = |\psi|^2_{t=0}$, then it is $|\psi'|^2 = |\psi|^2$ at all times; for this to happen, there must exist a real function σ such that

$$\psi' = \psi \exp(\mathrm{i}\,\sigma), \qquad \sigma = \sigma(\mathbf{r}, t). \tag{9.37}$$

From the gauge invariance of $|\psi|^2$ at all times it follows $\partial|\psi'|^2/\partial t = \partial|\psi|^2/\partial t$ whence, from the continuity equation (9.21), one obtains $\mathrm{div}\mathbf{J}'_\psi = \mathrm{div}\mathbf{J}_\psi$ with $\mathbf{J}_\psi = \Re[\psi^*(\hat{\mathbf{p}} - e\,\mathbf{A})\psi]/m$. Gauge transforming the quantities in brackets yields $(\psi')^*\,\hat{\mathbf{p}}\,\psi' = \psi^*\,\hat{\mathbf{p}}\,\psi + |\psi|^2\,\hbar\,\mathrm{grad}\sigma$ and $(\psi')^*\,e\,\mathbf{A}'\,\psi' = |\psi|^2\,e\,\mathbf{A} + |\psi|^2\,e\,\mathrm{grad}\vartheta$, whose difference provides

$$\mathbf{J}'_\psi - \mathbf{J}_\psi = \frac{1}{m}\,|\psi|^2\,\mathrm{grad}\,(\hbar\,\sigma - e\,\vartheta). \tag{9.38}$$

In (9.38) it is $|\psi|^2 \neq 0$ and $\mathrm{grad}|\psi|^2 \neq 0$; also, $\hbar\,\sigma - e\,\vartheta$ is independent of ψ. It follows that, for $\mathrm{div}(\mathbf{J}'_\psi - \mathbf{J}_\psi) = 0$ to hold, from (A.17) it must be $\mathrm{grad}\,(\hbar\sigma - e\vartheta) = 0$, namely, $\hbar\sigma - e\vartheta$ is an arbitrary function of time only. Setting the latter to zero finally yields the expression for the exponent in (9.37), that reads[3]

$$\sigma = \frac{e}{\hbar}\,\vartheta. \tag{9.39}$$

Problems

9.1 Using the one-dimensional form of (9.26) determine the envelope function $A(x - u\,t)$ corresponding to $c_k = \sqrt{\sigma/\sqrt{\pi}}\,\exp(-\sigma^2\,k^2/2)$, with σ a length. Noting that $\int_{-\infty}^{+\infty} |c_k|^2\,\mathrm{d}k = 1$, show that A is normalized to 1 as well.

9.2 Using the envelope function $A(x - u\,t)$ obtained from Prob. 9.1 and the one-dimensional form of definition (9.23), show that $x_0(t) = u\,t$.

---

[3] The units of ϑ are $[\vartheta] = V\,s$.

Chapter 10
General Methods of Quantum Mechanics

10.1 Introduction

The preceding chapters have provided the introductory information about Quantum Mechanics. Here the general principles of the theory are illustrated, and the methods worked out for the Hamiltonian operator are extended to the operators associated to dynamical variables different from energy. The important concept of separable operator is introduced, and the property of some operators to commute with each other is related to the mutual compatibility of measurements of the corresponding dynamical variables. Then, the concepts of expectation value and uncertainty are introduced, and the Heisenberg uncertainty principle is worked out. This leads in turn to the analysis of the time derivative of the expectation values, showing that the latter fulfill relations identical to those of Classical Mechanics. The form of the minimum-uncertainty wave packet is worked out in the complements.

10.2 General Methods

The discussion carried out in Sect. 9.2 has led to a number of conclusions regarding the eigenvalues of the time-independent Schrödinger equation (7.45). They are:

- The energy of a particle subjected to a conservative force is one of the eigenvalues of the time-independent equation $\mathcal{H}w = Ew$, where $\mathcal{H}$ is derived from the corresponding Hamiltonian function by replacing p_i with $-i\hbar\,\partial/\partial x_i$. Any other energy different from an eigenvalue is forbidden.
- The wave function of a conservative case (taking by way of example the discrete-eigenvalue case) is $\psi = \sum_n c_n w_n \exp(-iE_n t/\hbar)$. The particle's localization is given by $|\psi|^2$, where it is assumed that ψ is normalized to unity. The probability that a measurement of energy finds the eigenvalue E_m is $|c_m|^2$; an energy measurement that finds the eigenvalues E_m forces c_n to become δ_{nm}.
- The time evolution of ψ is found by solving the time-dependent Schrödinger equation $i\hbar\,\partial\psi/\partial t = \mathcal{H}\psi$. The latter holds also in the non-conservative situations; although in such cases the wave function ψ is not expressible as a superposition of monochromatic waves, it can still be expanded using an orthonormal set like in (9.10) or (9.11).

© Springer Science+Business Media New York 2015
M. Rudan, *Physics of Semiconductor Devices,*
DOI 10.1007/978-1-4939-1151-6_10

An important issue is now extending the conclusions listed above to the dynamical quantities different from energy (e.g., momentum, angular momentum, and so on). The extension is achieved by analogy, namely, it is assumed that for any dynamical variable one can construct an eigenvalue equation whose solution provides the possible values of the variable itself. This line of reasoning yields the procedures listed below, that are called *general methods of Quantum Mechanics*:

1. Given a dynamical variable A, an operator $\mathcal{A}$ is associated to it. It is found, first, by expressing A in terms of canonical coordinates q_i, p_i (Sect. 1.6), then, by replacing the momentum's components p_i with $\hat{p}_i = -i\hbar \, \partial/\partial q_i$ in such a way that $\mathcal{A}$ is Hermitean.
2. It is checked whether the eigenvalue equation $\mathcal{A}v = Av$ possesses a complete, orthonormal set of eigenfunctions. If the check fails, the operator is not considered; otherwise it is accepted, and is called *observable* [78, Chap. V.9]. The eigenvalue equation is subjected to the same boundary or asymptotic conditions as $\mathcal{H}w = Ew$.
3. Let A_n or A_β be the eigenvalues of $\mathcal{A}v = Av$, with n (β) a set of discrete (continuous) indices. Such eigenvalues are the only values that a measure of the dynamical variable A can find.
4. Thanks to completeness, the wave function ψ describing the particle's localization can be written, respectively for discrete or continuous spectra,

$$\psi = \sum_n a_n(t)\, v_n(\mathbf{r}), \qquad \psi = \int_\beta a_\beta(t)\, v_\beta(\mathbf{r})\, d\beta, \qquad (10.1)$$

with $a_n = \langle v_n | \psi \rangle$, $a_\beta = \langle v_\beta | \psi \rangle$.
5. If the wave function in (10.1) is normalizable, then $\sum_n |a_n|^2 = 1$, $\int_\beta |a_\beta|^2 \, d\beta = 1$ at all times. For a discrete spectrum, $P_n = |a_n(t_A)|^2$ is the probability that a measurement of A finds the eigenvalue A_n at $t = t_A$. For a continuous spectrum, the infinitesimal probability that at $t = t_A$ the domain of A_β is found in the interval $d\beta$ around β is $dP = |a_\beta(t_A)|^2 \, d\beta$.
6. When the measurement is carried out at $t = t_A$ and an eigenvalue, say, A_m, is found, the coefficients of the first expansion in (10.1) are forced by the measurement to become $|a_n(t_A^+)|^2 = \delta_{mn}$, and the wave function at that instant[1] becomes $\psi(\mathbf{r}, t_A^+) = v_m(\mathbf{r})$. The time evolution of ψ starting from t_A^+ is prescribed by the time-dependent Schrödinger equation $i\hbar \, \partial\psi/\partial t = \mathcal{H}\psi$, with $\psi(\mathbf{r}, t_A^+)$ as the initial condition. In this respect there is a neat parallel with Classical Mechanics, where the time evolution of the canonical variables starting from the initial conditions is prescribed by the Hamilton equations (1.42).

According to the general methods listed above, the eigenvalues of $\mathcal{A}$ are the only possible outcome of a measurement of the dynamical variable A. As the eigenvalues

---

[1] Measurements are not instantaneous (refer to the discussion in Sect. 9.2). Here it is assumed that the duration of a measurement is much shorter than the time scale of the whole experiment.

represent a physical quantity, they must be real; this makes the requirement that A must be Hermitean easily understood: if an operator is Hermitean, then its eigenvalues are real (Sect. 8.4.1). The inverse is also true: if the eigenfunctions of A form a complete set and its eigenvalues are real, then A is Hermitean. In fact, for any pair of functions f, g, considering the discrete spectrum by way of example, one has

$$\langle g|Af\rangle - \langle Ag|f\rangle = \sum_n \sum_m g_n^* f_m \left[\langle v_n|Av_m\rangle - \langle Av_n|v_m\rangle\right] =$$

$$= \sum_n \sum_m g_n^* f_m \langle v_n|v_m\rangle \left(A_m - A_n^*\right) = \sum_n g_n^* f_n \left(A_n - A_n^*\right) = 0, \qquad (10.2)$$

which holds for all f, g because the eigenfunctions v_n are mutually orthogonal and the eigenvalues A_n are real.

As indicated at point 1 above, the dynamical variable A is transformed into the operator A by replacing p_i with $\hat{p}_i$. The operator obtained from such a replacement is not necessarily Hermitean: its hermiticity must be checked on a case-by-case basis. For instance, the dynamical variable $A = x\,p_x$ can be written in equivalent ways as $x\,p_x$, $p_x\,x$, and $(x\,p_x + p_x\,x)/2$. However, their quantum counterparts

$$-i\,\hbar x\,\frac{\partial}{\partial x}, \qquad -i\,\hbar\frac{\partial}{\partial x}\,x, \qquad -i\,\frac{\hbar}{2}\left(x\,\frac{\partial}{\partial x} + \frac{\partial}{\partial x}\,x\right) \qquad (10.3)$$

are different from each other, and only the third one is Hermitean (compare with Sect. 9.5).

10.3 Separable Operators

Let A be an operator acting only on the x coordinate. Similarly, let B and C two operators acting only on y and z, respectively. The eigenvalue equations for the discrete-spectrum case read

$$Au_k = A_k\,u_k, \qquad Bv_m = B_m\,v_m, \qquad Cw_n = C_n\,w_n, \qquad (10.4)$$

where $u_k(x)$, $v_m(y)$, and $w_n(z)$ are three complete and orthonormal sets of eigenfunctions. Given a function $f(x, y, z)$, thanks to the completeness of the three sets the following expansion holds:

$$f(x, y, z) = \sum_n a_n(x, y)\,w_n = \sum_n \left[\sum_m b_{mn}(x)\,v_m\right] w_n =$$

$$= \sum_n \left\{\sum_m \left[\sum_k c_{kmn}\,u_k\right] v_m\right\} w_n = \sum_{kmn} c_{kmn}\,u_k\,v_m\,w_n, \qquad (10.5)$$

showing that the set made of the products $u_k\,v_m\,w_n$ is complete. Also, for any linear combination of the above operators, with $\mathbf{a}, \mathbf{b}, \mathbf{c}$ constant vectors, it is

$$(\mathbf{a}\,A + \mathbf{b}\,B + \mathbf{c}\,C)\,u_k\,v_m\,w_n = (\mathbf{a}\,A_k + \mathbf{b}\,B_m + \mathbf{c}\,C_n)\,u_k\,v_m\,w_n, \qquad (10.6)$$

that is, $u_k v_m w_n$ is an eigenfunction corresponding to eigenvalue $\mathbf{a} A_k + \mathbf{b} B_m + \mathbf{c} C_n$. It is important to add that in (10.4) it is implied that the boundary conditions $\mathcal{A} u_k = A_k u_k$ depend on x alone, those of $\mathcal{B} v_m = B_m v_m$ on y alone, and the like for the third equation. In other terms, separability means that at least one set of coordinates exists, such that both the equation and boundary conditions are separable.

As a first example of application of (10.6), consider the classical position of a particle, $\mathbf{r} = x_1 \mathbf{i}_1 + x_2 \mathbf{i}_2 + x_3 \mathbf{i}_3$. Such a dynamical variable does not contain the components of momentum; as a consequence, the operator associated to it is $\mathbf{r}$ itself, and generates the eigenvalue equation $\mathbf{r} g(\mathbf{r}) = \mathbf{r}_0 g(\mathbf{r})$. Separating the latter and considering the eigenvalue equation for x_i, one finds $x_i v_i(x_i) = x_{i0} v_i(x_i)$, namely, $(x_i - x_{i0}) v_i(x_i) = 0$ for all $x_i \neq x_{i0}$. It follows $v_i = \delta(x_i - x_{i0})$, whence $\mathbf{r}_0 = x_{10} \mathbf{i}_1 + x_{20} \mathbf{i}_2 + x_{30} \mathbf{i}_3$, and

$$g_{\mathbf{r}_0}(\mathbf{r}) = \delta(x_1 - x_{10}) \delta(x_2 - x_{20}) \delta(x_3 - x_{30}) = \delta(\mathbf{r} - \mathbf{r}_0). \tag{10.7}$$

As a second example consider the classical momentum of a particle, $\mathbf{p} = p_1 \mathbf{i}_1 + p_2 \mathbf{i}_2 + p_3 \mathbf{i}_3$. Remembering the discussion of Sect. 8.5 one finds for the operator associated to $\mathbf{p}$,

$$\hat{\mathbf{p}} = -i\hbar \left(\mathbf{i}_1 \frac{\partial}{\partial x_1} + \mathbf{i}_2 \frac{\partial}{\partial x_2} + \mathbf{i}_3 \frac{\partial}{\partial x_3} \right) = -i\hbar\, \mathrm{grad}, \tag{10.8}$$

whose eigenvalue equation reads $-i\hbar\, \mathrm{grad} f = \mathbf{p}_0 f$. Separation yields for the ith eigenvalue equation, with $v_i = v_i(x_i)$, the first-order equation $-i\hbar\, dv_i/dx_i = p_{i0} v_i$ (compare with (8.49)), whence $v_i = (2\pi)^{-1/2} \exp(i k_i x_i)$, with $k_i = p_{i0}/\hbar$, so that $\mathbf{k} = \mathbf{p}/\hbar = k_1 \mathbf{i}_1 + k_2 \mathbf{i}_2 + k_3 \mathbf{i}_3$, and

$$f_{\mathbf{k}}(\mathbf{r}) = (2\pi)^{-3/2} \exp(i\mathbf{k} \cdot \mathbf{r}). \tag{10.9}$$

Neither (10.7) nor (10.9) are square integrable. The indices of the eigenvalues ($\mathbf{r}_0$ in (10.7) and $\mathbf{k}$ in (10.9)) are continuous in both cases. Also, from the results of Sects. C.2 and C.5 one finds that $g_{\mathbf{r}_0}(\mathbf{r}) = g(\mathbf{r}, \mathbf{r}_0)$ is the Fourier transform of $f_{\mathbf{k}}(\mathbf{r}) = f(\mathbf{r}, \mathbf{k})$.

10.4   Eigenfunctions of Commuting Operators

It has been shown in Sect. 10.2 that a measurement of the dynamical variable A at time t_A yields one of the eigenvalues of the equation $\mathcal{A} a = A a$. Considering for instance a discrete spectrum, let the eigenvalue be A_m. The initial condition $\psi(\mathbf{r}, t_A^+)$ for the time evolution of the particle's wave function after the measurement is one of the eigenfunctions of $\mathcal{A}$ corresponding to A_m. If a measurement of another dynamical variable B is carried out at a later time t_B, the wave function at $t = t_B$ is forced to become one of the eigenfunctions of $\mathcal{B} b = B b$, say, b_k. The latter can in turn be expanded in terms of the complete set derived from $\mathcal{A}$, namely, $b_k = \sum_n \langle a_n | b_k \rangle a_n$. As the coefficients of the expansion are in general different from zero, there is a finite probability that a new measurement of A at $t_C > t_B$ finds a value different from A_m. In principle this could be due to the fact that, if A is not conserved, its value has evolved, from the outcome A_m of the measurement carried out at $t = t_A$, into something different, as prescribed by the time-dependent Schrödinger equation having $\psi(\mathbf{r}, t_A^+)$ as initial condition.[2] However, the instant t_B of the second measurement can in principle be brought as close to t_A as we please, so that the two measurements can be thought of as simultaneous. As a consequence, the loss of information about the value of A must be ascribed to the second measurement, specifically, to its interference with the wave function, rather than to a natural evolution[3] of the value of A: the gain in information about the eigenvalue of B produces a loss of information about that of A; for this reason, the two measurements are said to be *incompatible*.

From the discussion above one also draws the conclusion that, if it were $b_k = a_m$, the two measurements of outcome A_m and B_k would be compatible. This is in itself insufficient for stating that the measurements of A and B are compatible in all cases; for this to happen it is necessary that the whole set of eigenfunctions of A coincides with that of B: in this case, in fact, the condition $b_k = a_m$ is fulfilled no matter what the outcome of the two measurements is.

It would be inconvenient to check the eigenfunctions to ascertain whether two observables A, B are compatible or not. In fact, this is not necessary thanks to the following property: if two operators $\mathcal{A}$ and $\mathcal{B}$ have a common, complete set of eigenfunctions, then they commute, and vice versa (as indicated in Sect. 8.6.2, two operators commute if their commutator (8.71) is the null operator). Still assuming a discrete spectrum, for any eigenfunction v_n it is $\mathcal{AB}v_n = \mathcal{A}B_n v_n = B_n \mathcal{A}v_n = B_n A_n v_n$. Similarly, $\mathcal{BA}v_n = \mathcal{B}A_n v_n = A_n \mathcal{B}v_n = A_n B_n v_n$, showing that $\mathcal{A}$ and $\mathcal{B}$ commute for all eigenfunctions. Then, using the completeness of the common set v_n to expand any function f as $f = \sum_n f_n v_n$, one finds

$$\mathcal{AB}f = \sum_n f_n \mathcal{AB}v_n = \sum_n f_n \mathcal{BA}v_n = \mathcal{BA}\sum_n f_n v_n = \mathcal{BA}f. \tag{10.10}$$

This proves that if two operators have a complete set of eigenfunctions in common, then they commute. Conversely, assume that $\mathcal{A}$ and $\mathcal{B}$ commute and let v_n be an eigenfunction of $\mathcal{A}$; then, $\mathcal{AB}v_n = \mathcal{BA}v_n$ and $\mathcal{A}v_n = A_n v_n$. Combining the latter relations yields $\mathcal{AB}v_n = \mathcal{B}A_n v_n$ which, letting $g_n = \mathcal{B}v_n$, is recast as $\mathcal{A}g_n = A_n g_n$. In conclusion, both v_n and g_n are eigenfunctions of $\mathcal{A}$ belonging to the same eigenvalue A_n.

If A_n is not degenerate, the eigenfunctions v_n and g_n must be the same function, apart from a multiplicative constant due to the homogeneity of the eigenvalue equation. Let such a constant be B_n; combining $g_n = B_n v_n$ with the definition $g_n = \mathcal{B}v_n$ yields $\mathcal{B}v_n = B_n v_n$, this showing that v_n is an eigenfunction of $\mathcal{B}$ as well. The

---

[2] By way of example one may think of A as the position x, that typically evolves in time from the original value $x_A = x(t_A)$ even if the particle is not perturbed.

[3] In any case, the evolution would be predicted exactly by the Schrödinger equation. Besides, the eigenvalue would not change if A were conserved.

property holds also when A_n is degenerate, although the proof is somewhat more involved [77, Chap. 8-5]. This proves that if two operators commute, then they have a complete set of eigenfunctions in common. Examples are given in Sect. 8.6.3.

10.5   Expectation Value and Uncertainty

The discussion carried out in Sect. 10.2 has led to the conclusion that the wave function ψ describing the particle's localization can be expanded as in (10.1), where v_n or v_β are the eigenfunctions of a Hermitean operator A that form a complete, orthonormal set. Considering a discrete spectrum first, the coefficients of the expansion are $a_n = \langle v_n | \psi \rangle$; assuming that the wave function is normalizable, it is $\sum_n |a_n|^2 = 1$.

The meaning of the coefficients is that $P_n = |a_n(t)|^2$ is the probability that a measurement of A finds the eigenvalue A_n at t. From this it follows that the statistical average of the eigenvalues is

$$\langle A \rangle (t) = \sum_n P_n A_n. \tag{10.11}$$

The average (10.11) is called *expectation value*.[4] It can be given a different form by observing that $P_n = a_n^* a_n = (\sum_m a_m^* \delta_{mn}) a_n$ and that, due to the orthonormality of the eigenfunctions of A, it is $\delta_{mn} = \langle v_m | v_n \rangle$; then,

$$\sum_n \left(\sum_m a_m^* \langle v_m | v_n \rangle \right) a_n A_n = \langle \sum_m a_m v_m | \sum_n a_n A_n v_n \rangle = \langle \sum_m a_m v_m | \sum_n a_n A v_n \rangle. \tag{10.12}$$

Combining (10.12) with (10.11) and remembering that A is Hermitean yields

$$\langle A \rangle = \langle \psi | A | \psi \rangle. \tag{10.13}$$

The same result holds for a continuous spectrum:

$$\langle A \rangle = \int_\alpha P_\alpha A_\alpha \, d\alpha = \int_\alpha |a_\alpha|^2 A_\alpha \, d\alpha = \langle \psi | A | \psi \rangle, \tag{10.14}$$

where

$$\int_\alpha |a_\alpha|^2 A_\alpha \, d\alpha = \int_\alpha \left(\int_\beta a_\beta^* \delta(\beta - \alpha) \, d\beta \right) a_\alpha A_\alpha \, d\alpha, \tag{10.15}$$

and $\int_\beta |a_\beta|^2 \, d\beta = 1$ at all times. The expectation values of Hermitean operators are real because they are the statistical averages of the eigenvalues, themselves real.

---

[4] If the wave function is normalized to a number different from unity, the definition of the expectation value is $\sum_n P_n A_n / \sum_n P_n$, and the other definitions are modified accordingly.

Using (8.57) one extends the definition of expectation value to the powers of the eigenvalues; for instance,

$$\langle A^2 \rangle = \langle \psi | \mathcal{A}^2 | \psi \rangle = \int_\Omega \psi^* \mathcal{A}\mathcal{A}\psi \, d\Omega = \langle \mathcal{A}\psi | \mathcal{A}\psi \rangle = ||\mathcal{A}\psi||^2 \geq 0, \qquad (10.16)$$

where the hermiticity of $\mathcal{A}$ is exploited. The *variance* of the eigenvalues is given by

$$(\Delta A)^2 = \langle (A - \langle A \rangle)^2 \rangle = \langle A^2 - 2\langle A \rangle A + \langle A \rangle^2 \rangle = \langle A^2 \rangle - \langle A \rangle^2, \qquad (10.17)$$

real and non negative by construction; as a consequence, $\langle A^2 \rangle \geq \langle A \rangle^2$. The general term used to indicate the positive square root of the variance, $\Delta A = \sqrt{(\Delta A)^2} \geq 0$, is *standard deviation*. When it is used with reference to the statistical properties of the eigenvalues, the standard deviation is called *uncertainty*.

Assume by way of example that the wave function at $t = t_A$ coincides with one of the eigenfunctions of $\mathcal{A}$. With reference to a discrete spectrum (first relation in (10.1)), let $\psi(t_A) = v_m$. From (10.13) and (10.17) it then follows $\langle A \rangle(t_A) = A_m$, $\Delta A(t_A) = 0$. The standard deviation of the eigenvalues is zero in this case, because the measurement of A can only find the eigenvalue A_m. As a second example consider a continuous spectrum in one dimension, and let $\psi(t_A) = \exp(i k x)/\sqrt{2\pi}$, namely, an eigenfunction of the momentum operator. In this case the wave function is not square integrable, so one must calculate the expectation value as

$$\langle A \rangle(t_A) = \lim_{x_0 \to \infty} \frac{\int_{-x_0}^{+x_0} \psi^*(t_A)\,\mathcal{A}\psi(t_A)\,dx}{\int_{-x_0}^{+x_0} \psi^*(t_A)\,\psi(t_A)\,dx}. \qquad (10.18)$$

If one lets $\mathcal{A} = \hat{p} = -i\hbar \, d/dx$, the result is $\langle p \rangle(t_A) = \hbar k$, $\Delta p(t_A) = 0$. In fact, like in the previous example, the wave function coincides with one of the eigenfunctions of the operator. If, however, one applies another operator to the same wave function, its variance does not necessarily vanish. A remarkable outcome stems from applying $\hat{x} = x$, that is, the operator associated to the dynamical variable canonically conjugate to p: one finds $\langle x \rangle(t_A) = 0$, $\Delta x(t_A) = \infty$.

In conclusion, the examples above show that the term "uncertainty" does not refer to an insufficient precision of the measurements (which in principle can be made as precise as we please), but to the range of eigenvalues that is covered by the form of $\psi(t_A)$. In the last example above all positions x are equally probable because $|\psi(t_A)|^2 = $ const, whence the standard deviation of position diverges.

10.6 Heisenberg Uncertainty Relation

Consider the wave function ψ describing the dynamics of a particle, and let $\mathcal{A}$ and $\mathcal{B}$ be Hermitean operators. A relation exists between the standard deviations of these operators, calculated with the same wave function. Defining the complex functions $f = (\mathcal{A} - \langle A \rangle)\psi$ and $g = (\mathcal{B} - \langle B \rangle)\psi$ yields

$$||f||^2 = (\Delta A)^2, \qquad ||g||^2 = (\Delta B)^2, \qquad \langle f|g \rangle - \langle g|f \rangle = i\langle C \rangle, \qquad (10.19)$$

where the first two relations derive from (10.17) while $\langle C \rangle$ in the third one is the expectation value of the commutator $C = -\mathrm{i}\,(\mathcal{A}\mathcal{B} - \mathcal{B}\mathcal{A})$. Letting $\mu = \mathrm{i}\,\nu$ in (8.16), with ν real, and using (10.19) provides

$$(\Delta A)^2 + \nu^2 \,(\Delta B)^2 - \nu \,\langle C \rangle \geq 0, \tag{10.20}$$

namely, a second-degree polynomial in the real parameter ν. In turn, the coefficients of the polynomial are real because they derive from Hermitean operators. For the polynomial to be non negative for all ν, the discriminant $\langle C \rangle^2 - 4\,(\Delta A)^2\,(\Delta B)^2$ must be non positive. The relation between the standard deviations then reads

$$\Delta A\,\Delta B \geq \frac{1}{2}\,|\langle C \rangle|. \tag{10.21}$$

The interpretation of this result follows from the discussion carried out at the end of Sect. 10.5. If $\mathcal{A}$ and $\mathcal{B}$ commute, then their commutator is the null operator, whose eigenvalue is zero. As a consequence it is $\Delta A\,\Delta B \geq 0$, namely, the minimum of the product is zero. Remembering the result of Sect. 10.4, when two operators commute they have a common, complete set of eigenfunctions. If the wave function used for calculating the variance (10.17) is an eigenfunction of $\mathcal{A}$ and $\mathcal{B}$, then both standard deviations ΔA and ΔB vanish and $\Delta A\,\Delta B = 0$, namely, the minimum can in fact be attained. If, instead, $\mathcal{A}$ and $\mathcal{B}$ do not commute, the minimum of the product $\Delta A\,\Delta B$ must be calculated on a case-by-case basis. The most interesting outcome is found when the two operators are associated to conjugate dynamical variables: $\mathcal{A} = q_i$ and $\mathcal{B} = -\mathrm{i}\,\hbar\,\partial/\partial q_i$. Remembering (8.72) one finds $C = \hbar\,\mathcal{I}$, $C = \hbar$, $\langle C \rangle = \hbar$, whence

$$\Delta A\,\Delta B \geq \frac{\hbar}{2}. \tag{10.22}$$

Inequality (10.22) is also called *Heisenberg principle* or *uncertainty principle*, because it was originally deduced by Heisenberg from heuristic arguments [48].[5] The more formal deduction leading to (10.21) was given shortly after in [61] and [116].

10.7　Time Derivative of the Expectation Value

The expectation value (10.11) of a Hermitean operator is a real function of time. In Classical Mechanics, the generalized coordinates and momenta are also functions of time, whose evolution is given by the Hamilton equations (1.42); the latter express the time derivatives of coordinates and momenta in terms of the Hamiltonian function. Then, for an arbitrary function ρ of the canonical coordinates, the total derivative with respect of time is expressed through the Poisson bracket as in (1.53). A relation

---

[5] Namely, (10.22) is a theorem rather than a principle. A similar comment applies to the Pauli principle (Sect. 15.6). The English translation of [48] is in [117].

of the same form as (1.53) is found in Quantum Mechanics by calculating the time derivative of the expectation value (10.13). It is assumed that operator $\mathcal{A}$ depends on time, but does not operate on it; as a consequence, the symbol $\partial \mathcal{A}/\partial t$ indicates the operator resulting from differentiating $\mathcal{A}$ with respect to its functional dependence on t. With these premises one finds

$$\frac{d}{dt} \int_\Omega \psi^* \mathcal{A} \psi \, d\Omega = \int_\Omega \left(\frac{\partial \psi^*}{\partial t} \mathcal{A} \psi + \psi^* \frac{\partial \mathcal{A}}{\partial t} \psi + \psi^* \mathcal{A} \frac{\partial \psi}{\partial t} \right) d\Omega. \qquad (10.23)$$

The time derivative of ψ is obtained from the time-dependent Schrödinger equation (9.9). Considering the case where $\mathcal{H}$ is real yields $\partial \psi/\partial t = -i\mathcal{H}\psi/\hbar$ and $\partial \psi^*/\partial t = i\mathcal{H}\psi^*/\hbar$, whence

$$\frac{d}{dt}\langle A \rangle = \int_\Omega \psi^* \frac{\partial \mathcal{A}}{\partial t} \psi \, d\Omega + \frac{i}{\hbar} \int_\Omega \psi^* \left(\mathcal{H}\mathcal{A} - \mathcal{A}\mathcal{H} \right) \psi \, d\Omega, \qquad (10.24)$$

which has the same structure as (1.53). Other relations similar to those of Sect. 1.8 are also deduced from (10.24). For instance, letting $\mathcal{A} = \mathcal{H}$ yields

$$\frac{d}{dt}\langle H \rangle = \left\langle \frac{\partial \mathcal{H}}{\partial t} \right\rangle, \qquad (10.25)$$

similar to (1.44). If $\langle A \rangle$ is a constant of motion, then

$$\int_\Omega \psi^* \frac{\partial \mathcal{A}}{\partial t} \psi \, d\Omega + \frac{i}{\hbar} \int_\Omega \psi^* \left(\mathcal{H}\mathcal{A} - \mathcal{A}\mathcal{H} \right) \psi \, d\Omega = 0, \qquad (10.26)$$

similar to (1.54) while, if $\mathcal{A}$ does not depend on time, (10.24) yields

$$\frac{d}{dt}\langle A \rangle = \frac{i}{\hbar} \int_\Omega \psi^* \left(\mathcal{H}\mathcal{A} - \mathcal{A}\mathcal{H} \right) \psi \, d\Omega, \qquad (10.27)$$

similar to (1.55). Finally, if $\mathcal{A}$ does not depend on time and commutes with $\mathcal{H}$, (10.24) yields $d\langle A \rangle/dt = 0$, namely, the expectation value $\langle A \rangle$ is a constant of motion.

10.8 Ehrenfest Theorem

An important application of (10.27) is found by replacing $\mathcal{A}$ with either a position operator or a momentum operator. The calculation is shown here with reference to the Hamiltonian operator $\mathcal{H} = -\hbar^2/(2m)\nabla^2 + V$, where the potential energy V is independent of time. Letting first $\mathcal{A} = x$ yields

$$(\mathcal{H}x - x\mathcal{H})\psi = \frac{\hbar^2}{2m} \left(x \frac{\partial^2 \psi}{\partial x^2} - \frac{\partial^2 x \psi}{\partial x^2} \right) = \frac{\hbar^2}{2m} \left(-2 \frac{\partial \psi}{\partial x} \right) \qquad (10.28)$$

whence, using $\hat{p}_x = -i\hbar \, \partial/\partial x$, it follows

$$\frac{d}{dt}\langle x \rangle = \frac{i}{\hbar} \int_\Omega \psi^* \frac{\hbar^2}{2m} \left(-2 \frac{\partial \psi}{\partial x} \right) d\Omega = \frac{1}{m} \langle \psi | \hat{p}_x | \psi \rangle = \frac{\langle p_x \rangle}{m}. \qquad (10.29)$$

In conclusion, the relation $d\langle x\rangle/dt = \langle p_x\rangle/m$ holds, similar to the one found in a classical case when the Hamiltonian function has the form $H = p^2/(2\,m) + V$ (compare with the second relation in (1.33)). Still with $\mathcal{H} = -\hbar^2/(2\,m)\nabla^2 + V$, consider as a second example $\mathcal{A} = \hat{p}_x = -i\,\hbar\,\partial/\partial x$, to find

$$\left(\mathcal{H}\,\hat{p}_x - \hat{p}_x\,\mathcal{H}\right)\psi = -i\,\hbar\left(V\frac{\partial\psi}{\partial x} - \frac{\partial(V\,\psi)}{\partial x}\right) = i\,\hbar\,\psi\,\frac{\partial V}{\partial x}. \tag{10.30}$$

From this, letting $F_x = -\partial V/\partial x$ be the component of the force along x, it follows

$$\frac{d}{dt}\langle p_x\rangle = \frac{i}{\hbar}\int_\Omega \psi^* i\,\hbar\,\psi\,\frac{\partial V}{\partial x}\,d\Omega = \langle F_x\rangle, \tag{10.31}$$

also in this case similar to the classical one. Combining (10.29) and (10.31) shows that the expectation values fulfill a relation similar to Newton's law,

$$m\,\frac{d^2}{dt^2}\langle x\rangle = \langle F_x\rangle. \tag{10.32}$$

This result is called *Ehrenfest theorem*. If the dependence of F_x on position is weak in the region where ψ is significant, the normalization of the wave function yields

$$\frac{d}{dt}\langle p_x\rangle \simeq F_x\int_\Omega \psi^*\,\psi\,d\Omega = F_x. \tag{10.33}$$

In this case, called *Ehrenfest approximation*, the expectation value of position fulfills Newton's law exactly. If, on the contrary, F_x depends strongly on position in the region where ψ is significant (as happens, e.g., when the potential energy has the form of a step or a barrier), then the outcome of the quantum calculation is expected to be different from the classical one (see, e.g., Sects. 11.2 and 11.3).

10.9   Complements

10.9.1   Minimum-Uncertainty Wave Function

It has been shown in Sect. 10.6 that when the two operators are associated to conjugate dynamical variables, $\mathcal{A} = q$ and $\mathcal{B} = -i\,\hbar\,d/dq$, the relation between their standard deviations is given by (10.22). It is interesting to seek a form of the wave function such that the equality $\Delta A\,\Delta B = \hbar/2$ holds. If it exists, such a form is called *minimum-uncertainty wave function*. To proceed one notes that the equality yields the value $v_m = (|\langle C\rangle|/2)/(\Delta B)^2$ corresponding to the minimum of the polynomial (10.20); moreover, such a minimum is zero. On the other hand, imposing the equality in (8.16) after letting $\mu = i\,v_m$ yields the more compact form $\|f + i\,v_m\,g\|^2 = 0$, equivalent to $f + i\,v_m\,g = 0$. Remembering the definitions given in Sect. 10.6, it is $f = (\mathcal{A} - \langle A\rangle)\,\psi$, $g = (\mathcal{B} - \langle B\rangle)\,\psi$. Now, letting $q_0 = \langle A\rangle$, $p_0 = \hbar\,k_0 = \langle B\rangle$, from the relation $f + \mathrm{i}\, v_m\, g = 0$ one obtains the first-order differential equation $v_m\, \hbar\, \mathrm{d}\psi/\mathrm{d}q = [\mathrm{i}\, v_m\, p_0 - (q - q_0)]\, \psi$, whose solution is

$$\psi(q) = \psi_0 \exp\left[\mathrm{i}\, k_0\, q - (q - q_0)^2/(2\, v_m\, \hbar)\right]. \tag{10.34}$$

The normalization condition $\langle\psi|\psi\rangle = 1$ yields $\psi_0 = (\pi\, v_m\, \hbar)^{-1/4}$. Using $\langle C\rangle = \hbar$ and combining the expression of v_m with the equality $\Delta A\, \Delta B = \hbar/2$ provides $v_m = 2\,(\Delta A)^2/\hbar$ whence, letting $\Delta q = \Delta A$,

$$\psi(q) = \frac{1}{\sqrt[4]{2\,\pi}\,\sqrt{\Delta q}} \exp\left[\mathrm{i}\, k_0\, q - \frac{(q - q_0)^2}{(2\,\Delta q)^2}\right]. \tag{10.35}$$

The minimum-uncertainty wave function turns out to be proportional to a Gaussian function centered at q_0. The factor $\exp(\mathrm{i}\, k_0\, q)$ disappears from $|\psi|^2$, whose peak value and width are determined by Δq. Note that this calculation leaves the individual values of Δq and $\Delta p = \Delta B$ unspecified.

Problems

10.1 Starting from the wave packet (9.5) describing a free particle, determine the time evolution of its position without resorting to the approximation used in Sect. 9.6.

10.2 Using the results of Prob. 10.1, determine the time evolution of the standard deviation of position.

10.3 Starting from the wave packet (9.5) describing a free particle, determine the time evolution of its momentum without resorting to the approximation used in Sect. 9.6.

10.4 Using the results of Prob. 10.3, determine the time evolution of the standard deviation of momentum.

10.5 Consider a one-dimensional wave function that at some instant of time is given by a minimum-uncertainty packet (10.35) whose polar form is

$$\alpha = \frac{1}{\sqrt[4]{2\,\pi}\,\sqrt{\sigma}} \exp\left[-\frac{(x - x_0)^2}{4\,\sigma^2}\right], \qquad \beta = k_0\, x. \tag{10.36}$$

The wave packet is normalized to 1. Using the concepts introduced in Sect. 9.7.3, find the "convective" and "thermal" parts of the expectation value of the kinetic energy.

Part III
Applications of the Schrödinger Equation

Chapter 11
Elementary Cases

11.1 Introduction

The time-independent Schrödinger equation is a linear, second-order equation with a coefficient that depends on position. An analytical solution can be found in a limited number of cases, typically, one-dimensional ones. This chapter illustrates some of these cases, starting from the step-like potential energy followed by the potential-energy barrier. In both of them, the coefficient of the Schrödinger equation is approximated with a piecewise-constant function. Despite their simplicity, the step and barrier potential profiles show that the quantum-mechanical treatment may lead to results that differ substantially from the classical ones: a finite probability of transmission may be found where the classical treatment would lead to a reflection only, or vice versa. The transmission and reflection coefficients are defined by considering a plane wave launched towards the step or barrier. It is shown that the definition of the two coefficients can be given also for a barrier of a general form, basing on the formal properties of the second-order linear equations in one dimension. Finally, the case of a finite well is tackled, showing that in the limit of an infinite depth of the well one recovers the results of the particle in a box illustrated in a preceding chapter.

11.2 Step-Like Potential Energy

Consider a one-dimensional, step-like potential energy as shown in Fig. 11.1, with $V = 0$ for $x < 0$ and $V = V_0 > 0$ for $x > 0$. From the general properties of the time-independent Schrödinger equation (Sect. 8.2.3) it follows $E \geq 0$. To proceed, it is convenient to consider the two cases $0 < E < V_0$ and $E > V_0$ separately.

© Springer Science+Business Media New York 2015
M. Rudan, *Physics of Semiconductor Devices,*
DOI 10.1007/978-1-4939-1151-6_11

Fig. 11.1 The example of the
step-like potential energy
analyzed in Sect. 11.2. Only
the case $0 \le E \le V_0$ is shown

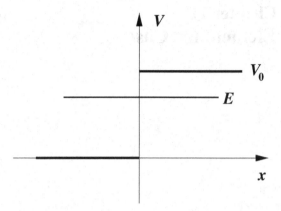

11.2.1   Case A: $0 < E < V_0$

The Schrödinger equation is split over the two partial domains to yield

$$\begin{cases} x < 0: & -w'' = k^2 w, & k = \sqrt{2mE}/\hbar \\ x > 0: & w'' = \alpha^2 w, & \alpha = \sqrt{2m(V_0 - E)}/\hbar \end{cases} \tag{11.1}$$

where the derivatives are indicated with primes. The solutions on the left and right of the origin are respectively given by

$$w = \begin{cases} w_- = a_1 \exp(ikx) + a_2 \exp(-ikx), & x < 0 \\ w_+ = a_3 \exp(-\alpha x) + a_4 \exp(\alpha x), & x > 0 \end{cases} \tag{11.2}$$

where it must be set $a_4 = 0$ to prevent w_+ from diverging. Using the continuity of w and w' in the origin yields

$$\begin{cases} w_+(0) = w_-(0), & a_1 + a_2 = a_3 \\ w'_+(0) = w'_-(0), & ik(a_2 - a_1) = \alpha a_3 \end{cases} \tag{11.3}$$

Eliminating a_3 provides the relation $\alpha(a_1 + a_2) = ik(a_2 - a_1)$ whence

$$\frac{a_2}{a_1} = \frac{ik + \alpha}{ik - \alpha} = \frac{k - i\alpha}{k + i\alpha}, \qquad \frac{a_3}{a_1} = 1 + \frac{a_2}{a_1} = \frac{2k}{k + i\alpha}, \tag{11.4}$$

that determine a_2, a_3 apart from the arbitrary constant a_1. This should be expected as w is not normalizable. From $|a_2/a_1| = |k - i\alpha|/|k + i\alpha| = 1$ one finds $a_2/a_1 = \exp(-i\varphi)$, with $\varphi = 2\arctan(\alpha/k)$. The solution of the time-independent Schrödinger equation is then recast as

$$w = \begin{cases} w_- = 2a_1 \exp(-i\varphi/2) \cos(kx + \varphi/2), & x < 0 \\ w_+ = [2k/(k + i\alpha)]a_1 \exp(-\alpha x), & x > 0 \end{cases} \tag{11.5}$$

The eigenvalues E are continuous in the range $0 \leq E < V_0$. The monochromatic wave function corresponding to w_k is $\psi_k(x,t) = w(x)\exp(-\mathrm{i}E_k t/\hbar)$, with $E_k = \hbar^2 k^2/(2m)$.

The quantity $R = |a_2/a_1|^2$ is called *reflection coefficient* of the monochromatic wave. As shown in Sect. 11.4 from a more general point of view, R is the probability that the wave is reflected by the step. In the case investigated here, where a particle with $0 < E < V_0$ is launched towards a step, it is $a_2/a_1 = \exp(-\mathrm{i}\varphi)$ whence $R = 1$.

The solution (11.3) or (11.5) shows that w becomes vanishingly small on the right of the origin as $x \to +\infty$. When a wave packet is built up using a superposition of monochromatic solutions, chosen in such a way that each energy E_k is smaller than V_0, its behavior is similar; as a consequence, the probability of finding the particle on the right of the origin becomes smaller and smaller as the distance from the origin increases. In conclusion, if a particle described by such a packet is launched from $x < 0$ towards the step, the only possible outcome is the same as in a calculation based on Classical Mechanics, namely, a reflection. A difference between the classical and quantum treatment exists though: in the former the reflection occurs at $x = 0$, whereas in the latter the reflection abscissa is not defined. This is better understood by considering a wave packet of standard deviation Δx approaching the origin from $x < 0$. Considering the approximate form of the packet described in Sect. 9.6, the incident envelope has the form $A_i = A(x - ut)$, and the localization of the incident particle is described by $|\psi_i|^2 = |A_i|^2$. Due to its finite width, the packet crosses the origin during a time $\Delta t \sim \Delta x/u$ starting, e.g., at $t = 0$. At a later instant $\Delta t + t_0$, where $t_0 \geq 0$ is the time that the wave packet takes to move away from the step to the extent that the interaction with it is practically completed, only the reflected packet exists, described by $|\psi_r|^2 = |A_r|^2$, with $A_r = A[x + u(t - t_0)]$. For $0 \leq t \leq \Delta t + t_0$ both incident and reflected packets exist. One could think that the reflection abscissa is given by the product ut_0; however, t_0 depends on the form of the packet, so the reflection abscissa is not well defined. Before the particle starts interacting with the step, only the incident packet exists and the normalization $\int_{-\infty}^{0} |\psi_i|^2 \mathrm{d}x = 1$ holds; similarly, after the interaction is completed, only the reflected packet exists and $\int_{-\infty}^{0} |\psi_r|^2 \mathrm{d}x = 1$. For $0 \leq t \leq \Delta t + t_0$ the normalization is achieved by a superposition of the incident and reflected packets.

11.2.2 Case B: $E > V_0$

Still considering the one-dimensional step of Fig. 11.1, let $E > V_0$. In this case the time-independent Schrödinger equation reads

$$
\begin{cases}
x < 0: & -w'' = k^2 w, & k = \sqrt{2mE}/\hbar \\
x > 0: & -w'' = k_1^2 w, & k_1 = \sqrt{2m(E - V_0)}/\hbar
\end{cases}
\qquad (11.6)
$$

whose solution is

$$
w = \begin{cases}
w_- = a_1 \exp(ikx) + a_2 \exp(-ikx), & x < 0 \\
w_+ = a_3 \exp(ik_1 x) + a_4 \exp(-ik_1 x), & x > 0
\end{cases} \tag{11.7}
$$

Remembering the discussion of Sect. 9.6, function w_- in (11.7) describes a superposition of two planar and monochromatic waves, belonging to the $x < 0$ region, that propagate in the forward and backward direction, respectively; a similar meaning holds for w_+ with reference to the $x > 0$ region. Now one assumes that an extra information is available, namely, that the particle was originally launched from $x < 0$ towards the origin; it follows that one must set $a_4 = 0$, because a wave that propagates in the backward direction can not exist in the region $x > 0$. By the same token one should set $a_1 = 0$ if the particle were launched from $x > 0$ towards the origin. From the continuity of w and w' in the origin it follows

$$
\begin{cases}
w_+(0) = w_-(0), & a_1 + a_2 = a_3 \\
w'_+(0) = w'_-(0), & k(a_1 - a_2) = k_1 a_3
\end{cases} \tag{11.8}
$$

Eliminating a_3 yields $k_1(a_1 + a_2) = k(a_1 - a_2)$ whence

$$
\frac{a_2}{a_1} = \frac{k - k_1}{k + k_1}, \qquad \frac{a_3}{a_1} = 1 + \frac{a_2}{a_1} = \frac{2k}{k + k_1}, \tag{11.9}
$$

that determine a_2, a_3 apart from the arbitrary constant a_1. The eigenvalues are continuous in the range $E > V_0$. The monochromatic, time-dependent wavefunction reads

$$
\psi = \begin{cases}
\psi_- = w_- \exp(-iE_- t/\hbar), & x < 0 \\
\psi_+ = w_+ \exp(-iE_+ t/\hbar), & x > 0
\end{cases} \tag{11.10}
$$

where $E_- = E(k) = \hbar^2 k^2/(2m)$ and $E_+ = E(k_1) = \hbar^2 k_1^2/(2m) + V_0$. Note that $k > k_1 > 0$, namely, the modulus of the particle's momentum in the $x > 0$ region is smaller than in the $x < 0$ region; this is similar to what happens in the classical treatment. On the other hand, from $k > k_1$ it follows $a_2 \neq 0$, showing that a monochromatic plane wave propagating in the backward direction exists in the $x < 0$ region. In other term, a finite probability of reflexion is present, which whould be impossible in the classical treatment. As before, the reflection coefficient of the monochromatic wave is defined as $R = |a_2/a_1|^2 < 1$. In turn, the *transmission coefficient* is defined as $T = 1 - R$ whence, from (11.9),

$$
R = \left| \frac{a_2}{a_1} \right|^2 = \frac{(k - k_1)^2}{(k + k_1)^2}, \qquad T = \frac{4kk_1}{(k + k_1)^2} = \frac{k_1}{k} \left| \frac{a_3}{a_1} \right|. \tag{11.11}
$$

Like in the $0 < E < V_0$ case one may build up a wave packet of standard deviation Δx. Still considering a packet approaching the origin from $x < 0$, the envelope has the form $A_i = A(x - ut)$. Before the particle starts interacting with the step, its localization is given by $|\psi_i|^2 = |A_i|^2$. The packet crosses the origin in a time $\Delta t \sim \Delta x/u$ and, using the symbols k_0, k_{10} to indicate the center of the packet in the momentum space for $x < 0$ and $x > 0$, respectively, from the first of (11.11) the reflected packet is described by

$$|\psi_r|^2 = \frac{(k_0 - k_{10})^2}{(k_0 + k_{10})^2}|A(x + ut)|^2. \tag{11.12}$$

It has the same group velocity, hence the same width, as $|\psi_i|^2$. The transmitted packet has the form

$$|\psi_t|^2 = \frac{(2k_0)^2}{(k_0 + k_{10})^2}|A(k_0 x/k_{10} - ut)|^2, \tag{11.13}$$

and its group velocity is $u_1 = dx/dt = k_{10}u/k_0 < u$. As all packets cross the origin in the same time interval Δt it follows $\Delta x/u = \Delta x_1/u_1$ whence $\Delta x_1 = (k_{10}/k_0)\Delta x < \Delta x$. This result shows that the transmitted packet is slower and narrower than the incident packet (if the incident packet were launched from $x > 0$ towards the origin, the transmitted packet would be faster and wider). From $\int_{-\infty}^{0} |\psi_i|^2 dx = 1$ it follows

$$P_r = \int_{-\infty}^{0} |\psi_r|^2 dx = \frac{(k_0 - k_{10})^2}{(k_0 + k_{10})^2}, \tag{11.14}$$

$$P_t = \frac{k_{10}}{k_0} \int_{0}^{\infty} |\psi_t|^2 d\frac{k_0 x}{k_{10}} = \frac{k_{10}}{k_0}\frac{(2k_0)^2}{(k_0 + k_{10})^2} = \frac{4k_0 k_{10}}{(k_0 + k_{10})^2}. \tag{11.15}$$

The two numbers P_r, P_t fulfill the relations $0 < P_r, P_t < 1$, $P_r + P_t = 1$, and are the reflection and transmission probabilities of the wave packet. The treatment outlined above for the case $E > V_0$ still holds when $V_0 < 0$, $E > 0$. In particular, if $V_0 < 0$ and $|V_0| \gg E$, like in Fig. 11.2, from (11.16) it turns out $P_r \simeq 1$. This result is quite different from that obtained from a calculation based on Classical Mechanics; in fact, its features are similar to those of the propagation of light across the interface between two media of different refraction index [9, Sects. 1.5,1.6]. The oddity of the result lies, instead, in the fact that term $\sqrt{2m}/\hbar$ cancels out in the expressions of P_r and P_t, so that one finds

$$P_r = \frac{(\sqrt{E_0} - \sqrt{E_0 - V_0})^2}{(\sqrt{E_0} + \sqrt{E_0 - V_0})^2}, \qquad P_t = \frac{\sqrt{E_0(E_0 - V_0)}}{(\sqrt{E_0} + \sqrt{E_0 - V_0})^2}, \tag{11.16}$$

with E_0 the total energy corresponding to k_0 and k_{10}. Thus, the classical result $P_r = 0$, $P_t = 1$ cannot be recovered by making, e.g., m to increase: the discontinuity in the potential energy makes it impossible to apply the Ehrenfest approximation (10.33) no matter what the value of the mass is. The same happens for the monochromatic wave: in fact, using (11.6) and replacing E_0 with E in (11.16) makes the latter equal to (11.11). To recover the classical result it is necessary to consider a potential energy whose asymptotic values 0 and V_0 are connected by a smooth function, and solve the corresponding Schrödinger equation. The classical case is then recovered by letting m increase (Prob. 11.1).

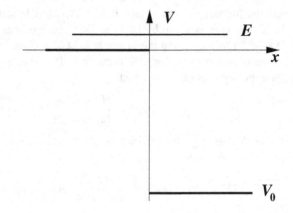

Fig. 11.2 Another example of the step-like potential energy analyzed in Sect. 11.2, with $V_0 < 0$ and $|V_0| \gg E$

11.3   Energy Barrier

Consider a one-dimensional energy barrier as shown in Fig. 11.3, with $V = V_0 > 0$ for $0 < x < s$ and $V = 0$ elsewhere. From the general properties of the time-independent Schrödinger equation (Sect. 8.2.3) it follows $E \geq 0$. To proceed, it is convenient to consider the two cases $0 < E < V_0$ and $E > V_0$ separately.

11.3.1   Case A: $0 < E < V_0$

The Schrödinger equation is split over the three domains to yield

$$
\begin{cases}
x < 0: & -w'' = k^2 w, & k = \sqrt{2m\,E}/\hbar \\
0 < x < s: & w'' = \alpha^2 w, & \alpha = \sqrt{2m(V_0 - E)}/\hbar \\
s < x: & -w'' = k^2 w, & k = \sqrt{2m\,E}/\hbar
\end{cases}
\qquad (11.17)
$$

where the derivatives are indicated with primes. The solutions of (11.17) are, respectively,

$$
w = \begin{cases}
w_- = a_1 \exp(ikx) + a_2 \exp(-ikx), & x < 0 \\
w_B = a_3 \exp(\alpha x) + a_4 \exp(-\alpha x), & 0 < x < s \\
w_+ = a_5 \exp(ikx) + a_6 \exp(-ikx), & s < x
\end{cases}
\qquad (11.18)
$$

Using the continuity of w and w' in the origin,

$$
\begin{cases}
w_-(0) = w_B(0), & a_1 + a_2 = a_3 + a_4 \\
w'_-(0) = w'_B(0), & ik(a_1 - a_2) = \alpha(a_3 - a_4)
\end{cases}
\qquad (11.19)
$$

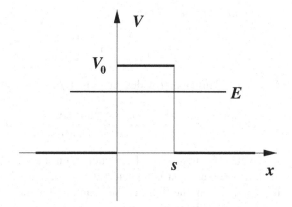

Fig. 11.3 The example of the one-dimensional energy barrier analyzed in Sect. 11.3. Only the case $0 \leq E \leq V_0$ is shown

Solving for a_1, a_2 and letting $\vartheta = 1 + i\alpha/k$ yields $2\,a_1 = \vartheta\,a_4 + \vartheta^*\,a_3$, $2\,a_2 = \vartheta\,a_3 + \vartheta^*\,a_4$. Using the same reasoning as in Sect. 11.2.2, one now assumes that an extra information is available, namely, that the particle was originally launched from $x < 0$ towards the barrier; it follows that one must set $a_6 = 0$ in (11.18), because a wave that propagates in the backward direction can not exist in the region $x > s$. By the same token one should set $a_1 = 0$ if the particle were launched from $x > s$ towards the barrier. Taking the first case ($a_6 = 0$) and using the continuity of w and w' at $x = s$ provides

$$\begin{cases} w_B(s) = w_+(s), & a_3 \exp(\alpha s) + a_4 \exp(-\alpha s) = a_5 \exp(iks) \\ w'_B(s) = w'_+(s), & \alpha[a_3 \exp(\alpha s) - a_4 \exp(-\alpha s)] = ika_5 \exp(iks) \end{cases} \tag{11.20}$$

whence $2a_3 = a_5\sigma\,\exp(iks - \alpha s)$, $2a_4 = a_5\sigma^*\exp(iks + \alpha s)$, with $\sigma = 1 + ik/\alpha$. In summary, a_1, a_2 are linear combinations of a_3, a_4; the latter, in turn, are proportional to a_5. It follows that, if it were $a_5 = 0$, then it would also be $a_3 = a_4 = 0$ and $a_1 = a_2 = 0$. However, this is impossible because w_- can not vanish identically; as a consequence it is necessarily $a_5 \neq 0$. This shows that $w_+ = a_5 \exp(ikx)$ differs from zero, namely, that a wave propagating in the forward direction exists for $x > s$.

As the relations involving the coefficients a_i are homogeneous, they determine a_2, a_3, a_4, a_5 apart from the arbitrary constant a_1. This should be expected as w is not normalizable. As shown below, the determination of the ratios between the coefficient does not impose any constraint on the total energy; as a consequence, the eigenvalues $0 \leq E < V_0$ are continuous. The ratio a_5/a_1 is found from

$$4\frac{a_1}{a_5} = 2\frac{a_3}{a_5}\vartheta^* + 2\frac{a_4}{a_5}\vartheta = \sigma\,\exp(iks - \alpha s)\vartheta^* + \sigma^*\exp(iks + \alpha s)\vartheta. \tag{11.21}$$

Letting $\mu = \vartheta\,\sigma^* = 2 + i(\alpha/k - k/\alpha)$ one finds $4a_1/a_5 \exp(-iks) = \mu \exp(\alpha s) + \mu^*\exp(-\alpha s)$, whence $a_1/a_5 \exp(-iks) = \cosh(\alpha s) + i(\alpha/k - k/\alpha)\sinh(\alpha s)/2$. Using the identity $\cosh^2\zeta - \sinh^2\zeta = 1$ finally yields for the transmission coefficient of the monochromatic wave

$$\frac{1}{T} = \left|\frac{a_1}{a_5}\right|^2 = 1 + \frac{1}{4}\left(\frac{\alpha}{k} + \frac{k}{\alpha}\right)^2 \sinh^2(\alpha s). \tag{11.22}$$

A similar calculation provides the ratio a_2/a_1; it is found

$$R = \left|\frac{a_2}{a_1}\right|^2 = \frac{(\alpha/k + k/\alpha)^2 \sinh^2(\alpha s)/4}{1 + (\alpha/k + k/\alpha)^2 \sinh^2(\alpha s)/4} = 1 - \left|\frac{a_5}{a_1}\right|^2 = 1 - T. \tag{11.23}$$

In the classical treatment, if a particle with $0 < E < V_0$ is launched from the left towards the barrier, it is reflected at $x = 0$; similarly, it is reflected at $x = s$ if it is launched from the right. In the quantum treatment it is $T > 0$: in contrast to the classical case, the particle can cross the barrier. For a given width s of the barrier and total energy E of the particle, the transmission coefficient T decreases when V_0 increases; for E and V_0 fixed, T becomes proportional to $\exp(-2\alpha s)$ when s increases to the extent that $\alpha s \gg 1$. Finally, $T \to 1$ as $s \to 0$; this was expected because the potential energy becomes equal to zero everywhere.[1]

The interaction of the particle with the barrier is better understood by considering a wave packet approaching the origin from $x < 0$. The incident envelope has the form $A_i = A(x - ut)$. Before reaching the origin the particle's localization is described by $|\psi_i|^2 = |A_i|^2$. After the interaction with the barrier is completed, both the reflected and transmitted packet exist, that move in opposite directions with the same velocity. Letting $P_r = \int_{-\infty}^0 |\psi_r|^2 dx$, $P_t = \int_s^\infty |\psi_t|^2 dx$, and observing that $\int_{-\infty}^0 |\psi_i|^2 dx = 1$, it follows that the two numbers P_r, P_t fulfill the relations $0 < P_r, P_t < 1$, $P_r + P_t = 1$, and are the reflection and transmission probabilities of the wave packet. In summary, the solution of the Schrödinger equation for the energy barrier shows that a particle with $0 < E < V_0$ has a finite probability of crossing the barrier, which would be impossible in the classical treatment. The same result holds when the form of the barrier is more complicated than the rectangular one (Sect. 11.4). The phenomenon is also called *tunnel effect*.

11.3.2 Case B: $0 < V_0 < E$

Still considering the one-dimensional barrier of Fig. 11.3, let $0 < V_0 < E$. The Schrödinger equation over the three domains reads

---

[1] If the potential energy were different on the two sides of the barrier, namely, $V = V_0 > 0$ for $0 < x < s$, $V = V_L$ for $x < 0$, and $V = V_R \neq V_L$ for $x > s$, with $V_0 > V_L, V_R$, the limit $s \to 0$ would yield the case discussed in Sect. 11.2 (compare also with Sects. 11.4 and 17.8.4).

$$\begin{cases} x < 0: & -w'' = k^2 w, & k = \sqrt{2m\,E}/\hbar \\ 0 < x < s: & -w'' = k_1^2\, w, & k_1 = \sqrt{2\,m\,(E - V_0)}/\hbar \\ s < x: & -w'' = k^2\, w, & k = \sqrt{2\,m\,E}/\hbar \end{cases} \tag{11.24}$$

where the derivatives are indicated with primes. The solutions of (11.24) are, respectively,

$$w = \begin{cases} w_- = a_1 \exp(ikx) + a_2 \exp(-ikx), & x < 0 \\ w_B = a_3 \exp(ik_1 x) + a_4 \exp(-ik_1 x), & 0 < x < s \\ w_+ = a_5 \exp(ikx) + a_6 \exp(-ikx), & s < x \end{cases} \tag{11.25}$$

As in Sect. 11.3.1 one assumes that the particle was originally launched from $x < 0$, so that $a_6 = 0$. The calculation follows the same line as in Sect. 11.3.1, and yields that w_- and w_+ are the same as in (11.18), whereas w_B is found by replacing α with ik_1 there. The determination of the ratios between the coefficient a_i does not impose any constraint on the total energy; as a consequence, the eigenvalues $E > V_0$ are continuous. Using $\cosh(i\zeta) = \cos(\zeta)$, $\sinh(i\zeta) = i\sin(\zeta)$ then yields

$$\frac{1}{T} = \left| \frac{a_1}{a_5} \right|^2 = 1 + \frac{1}{4} \left(\frac{k}{k_1} - \frac{k_1}{k} \right)^2 \sin^2(k_1 s) \tag{11.26}$$

where, from (11.24), $k_1/k = \sqrt{1 - V_0/E}$. Similarly,

$$R = \left| \frac{a_2}{a_1} \right|^2 = \frac{(k/k_1 - k_1/k)^2 \sin^2(k_1 s)/4}{1 + (k/k_1 - k/k_1)^2 \sin^2(k_1 s)/4} = 1 - \left| \frac{a_5}{a_1} \right|^2 . \tag{11.27}$$

In the classical treatment, if a particle with $0 < V_0 < E$ is launched towards the barrier, it is always transmitted. In the quantum treatment it may be $R > 0$: in contrast to the classical case, the particle can be reflected.[2] The barrier is transparent ($R = 0$) for $k_1 s = i\pi$, with i any integer. Letting $\lambda_1 = 2\pi/k_1$ yields $s = i\lambda_1/2$, which is equivalent to the optical-resonance condition in a sequence of media of refractive indices n_1, n_2, n_1 [9, Sect. 1.6].

---

[2] The reflection at the barrier for $k_1 s \neq i\pi$ explains why the experimental value of the Richardson constant A is lower than the theoretical one. Such a constant appears in the expression $J_s = A T^2 \exp[-E_W/(k_B T)]$ of the vacuum-tube characteristics [21]. This is one of the cases where the tunnel effect is evidenced in macroscopic-scale experiments. Still considering the vacuum tubes, another experimental evidence of the tunnel effect is the lack of saturation of the forward current-voltage characteristic at increasing bias.

11.4   Energy Barrier of a General Form

The interaction of a particle with an energy barrier has been discussed in Sect. 11.3 with reference to the simple case of Fig. 11.3. Here it is extended to the case of a barrier of a general form—still considering the one-dimensional, time-independent Schrödinger equation for a particle of mass m, to the purpose of calculating the transmission coefficient T. The equation reads

$$\frac{d^2w}{dx^2} + qw = 0, \qquad q(x) = \frac{2m}{\hbar^2}(E - V), \tag{11.28}$$

where the potential energy $V(x)$ is defined as follows:

$$V = V_L = \text{const.}, \quad x < 0; \qquad V = V_R = \text{const.}, \quad 0 < s < x. \tag{11.29}$$

In the interval $0 \le x \le s$ the potential energy is left unspecified, with the only provision that its form is such that (11.28) is solvable. It will also be assumed that $E > V_L, V_R$; as a consequence, the total energy is not quantized and all values of E larger than V_L and V_R are allowed. For a given E the time-dependent wave function takes the form $\psi(x,t) = w(x)\exp(-iEt/\hbar)$, where

$$w(x) = a_1 \exp(ik_L x) + a_2 \exp(-ik_L x), \qquad x < 0, \tag{11.30}$$

$$w(x) = a_5 \exp(ik_R x) + a_6 \exp(-ik_R x), \qquad s < x. \tag{11.31}$$

The real parameters $k_L, k_R > 0$ are given by $k_L = \sqrt{2m(E - V_L)}/\hbar$ and, respectively, $k_R = \sqrt{2m(E - V_R)}/\hbar$. Like in the examples of Sect. 11.3 it is assumed that the particle is launched from $-\infty$, so that the plane wave corresponding to the term multiplied by a_6 in (11.31) does not exist. As a consequence one lets $a_6 = 0$, whereas a_1, a_2, a_5 are left undetermined. In the interval $0 \le x \le s$ the general solution of (11.28) is

$$w(x) = a_3 u(x) + a_4 v(x), \qquad 0 \le x \le s, \tag{11.32}$$

where u, v are two linearly-independent solutions. The continuity equation for the wave function, (9.13), becomes in this case $dJ_\psi/dx = 0$, namely, $J_\psi = \text{const.}$ In turn, the density of the probability flux reads

$$J_\psi = \frac{i\hbar}{2m}\left(w\frac{dw^*}{dx} - w^*\frac{dw}{dx}\right). \tag{11.33}$$

Applying (11.33) to (11.30) and (11.31) yields, respectively,

$$J_\psi = \frac{\hbar k_L}{m}\left(|a_1|^2 - |a_2|^2\right), \quad x < 0; \qquad J_\psi = \frac{\hbar k_R}{m}|a_5|^2, \quad s < x. \tag{11.34}$$

As J_ψ is constant, one may equate the two expressions in (11.34) to obtain

$$\left|\frac{a_2}{a_1}\right|^2 + \frac{k_R}{k_L}\left|\frac{a_5}{a_1}\right|^2 = 1. \tag{11.35}$$

The division by $|a_1|^2$ leading to (11.35) is allowed because, by hypotesis, the particle is launched from $-\infty$, so that $a_1 \neq 0$. From (11.35) one defines the reflection and transmission coefficients

$$R = \left|\frac{a_2}{a_1}\right|^2, \qquad T = \frac{k_R}{k_L}\left|\frac{a_5}{a_1}\right|^2. \qquad (11.36)$$

Given E, V_L, V_R, and a_1, the transmission and reflection coefficients depend on the form of the potential energy in the interval $0 \leq x \leq s$. By way of example, if $V_L = 0$ and the potential energy in the interval $0 \leq x \leq s$ is equal to some constant $V_B \geq 0$, then R is expected to vary from 0 to 1 as V_B varies from 0 to $+\infty$. On the other hand, R and T can not vary independently from each other because of the relation $R + T = 1$. As a consequence, it suffices to consider only one coefficient, say, T. From the discussion above it follows that the coefficient T depends on the shape of the potential energy that exists within the interval $0 \leq x \leq s$, namely, it is a functional of V: $0 \leq T = T[V] \leq 1$. One may also note that the relation $R + T = 1$ derives only from the constancy of J_ψ due to the one-dimensional, steady-state condition. In other terms, the relation $R + T = 1$ does not depend on the form of the potential energy within the interval $0 \leq x \leq s$. It must then reflect some invariance property intrinsic to the solution of the problem. In fact, the invariance is that of the density of the probability flux J_ψ, which is intrinsic to the form of the Schrödinger equation and leads to the relation (11.35).

For the sake of generality one provisionally considers a slightly more general equation than (11.28), built by the linear operator

$$\mathcal{L} = \frac{\mathrm{d}^2}{\mathrm{d}x^2} + p(x)\frac{\mathrm{d}}{\mathrm{d}x} + q(x), \qquad (11.37)$$

where the functions p and q are real. If u is a solution of the differential equation $\mathcal{L}w = 0$ in the interval $0 \leq x \leq s$, let $P(x)$ be any function such that $p = \mathrm{d}P/\mathrm{d}x$, and define

$$v(x) = u(x) \int_a^x \frac{\exp[-P(\xi)]}{u^2(\xi)}\mathrm{d}\xi, \qquad (11.38)$$

where a, x belong to the same interval. It is found by inspection that $\mathcal{L}v = 0$ in the interval $0 \leq x \leq s$, namely, v is also a solution. Moreover, the Wronskian of u and v (Sect. A.12) reads

$$W(x) = uv' - u'v = \exp(-P). \qquad (11.39)$$

As the Wronskian never vanishes, u and v are linearly independent. This shows that, for any solution u of the differential equation $\mathcal{L}w = 0$ in a given interval, (11.38) provides another solution which is linearly independent from u. As the differential equation is linear and of the second order, the general solution is then given by a linear combination of u and v.

Being the equation $\mathcal{L}w = 0$ homogeneous, the solution u may be replaced by λu, with $\lambda \neq 0$ a constant. In this case v must be replaced by v/λ due to (11.38).

It follows that the Wronskian (11.39) is invariant under scaling of the solutions. Another consequence of the homogeneity of $\mathcal{L}w = 0$ is that the dimensions of w may be chosen arbitrarily. The same holds for the dimensions of u. Once the latter have been chosen, the dimensions of v follow from (11.38); in fact, the product uv has the dimensions of a length. From (11.32) it then follows that the products $a_3 u$ and $a_4 v$ have the same dimensions.

The linear independency allows one to choose for u and v the two fundamental solutions, namely, those having the properties [51, Sect. 5.2]

$$u(0) = 1, \qquad u'(0) = 0, \qquad v(0) = 0, \qquad v'(0) = 1, \tag{11.40}$$

so that the Wronskian W equals 1 everywhere. Then, letting $a_6 = 0$ in (11.31) and prescribing the continuity of the solution and its derivative at $x = 0$ and $x = s$ yields, from (11.32),

$$a_3 = a_1 + a_2, \qquad a_4 = i k_L (a_1 - a_2), \tag{11.41}$$

$$a_5 \exp(jk_R s) = a_3 u_s + a_4 v_s, \qquad i k_R a_5 \exp(i k_R s) = a_3 u'_s + a_4 v'_s, \tag{11.42}$$

where suffix s indicates that the functions are calculated at $x = s$. Eliminating $a_5 \exp(i k_R s)$ yields $(i k_R u_s - u'_s) a_3 = (v'_s - i k_R v_s) a_4$ whence, from (11.41),

$$\frac{a_2}{a_1} = \frac{k_L k_R v_s + u'_s + j (k_L v'_s - k_R u_s)}{k_L k_R v_s - u'_s + j (k_L v'_s + k_R u_s)} = \frac{A + jB}{C + jD}. \tag{11.43}$$

In conclusion, $T = 1 - |a_2/a_1|^2 = (C^2 - A^2 + D^2 - B^2)/(C^2 + D^2)$. Using $W = 1$ transforms the numerator into $C^2 - A^2 + D^2 - B^2 = 4k_L k_R (u_s v'_s - v_s u'_s) = 4k_L k_R$. In turn, the denominator reads $C^2 + D^2 = 2k_L k_R + (k_R u_s)^2 + (u'_s)^2 + (k_L v'_s)^2 + (k_L k_R v_s)^2$, whence

$$T = \frac{4k_L k_R}{2k_L k_R + (k_R u_s)^2 + (u'_s)^2 + (k_L v'_s)^2 + (k_L k_R v_s)^2}. \tag{11.44}$$

The expression of the transmission coefficient may be recast as $T = 1/(1 + F)$, with

$$F = \frac{1}{4k_L k_R} \left[(u'_s)^2 + (k_R u_s)^2 \right] + \frac{k_L}{4k_R} \left[(v'_s)^2 + (k_R v_s)^2 \right] - \frac{1}{2}. \tag{11.45}$$

It is easily shown that $F > 0$; in fact, this condition is equivalent to $(k_R u_s - k_L v'_s)^2 + (u'_s + k_L k_R v_s)^2 > 0$. In conclusion, (11.44) is the expression of the transmission coefficient across a barrier of any form, with no approximation [88]. To calculate T it is necessary to determine the four quantities u_s, u'_s, v_s, and v'_s. Actually only three of them suffice thanks to the condition $u_s v'_s - u'_s v_s = 1$.

Repeating the calculation of this section after letting $a_1 = 0$, $a_6 \neq 0$ provides the known result that, for a given barrier, the transmission probability for a particle of energy E is the same whether the particle is launched from the left or from the right. The property holds also in the relativistic case [79].

Fig. 11.4 The example of the one-dimensional energy well analyzed in Sect. 11.5. Only the case $V_0 < E < 0$ is shown

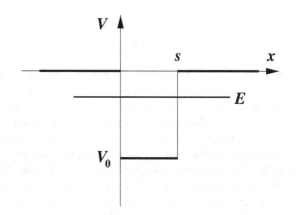

11.5 Energy Well

Taking a one-dimensional case, let $V = V_0 < 0$ for $0 < x < s$ and $V = 0$ elsewhere. From the general properties of the time-independent Schrödinger equation it follows $E > V_0$. The case $E > 0$ is treated in the same way as that of Sect. 11.3.2 and leads to similar results. The case $V_0 < E < 0$, shown in Fig. 11.4, is instead different from those investigated above: the total energy is quantized and the wave function is square integrable. The Schrödinger equation over the three domains reads

$$\begin{cases} x < 0: & w'' = \alpha^2 w, & \alpha = \sqrt{-2\,m\,E}/\hbar \\ 0 < x < s: & -w'' = k^2\,w, & k = \sqrt{2\,m\,(E - V_0)}/\hbar \\ s < x: & w'' = \alpha^2\,w, & \alpha = \sqrt{-2\,m\,E}/\hbar \end{cases} \qquad (11.46)$$

where the derivatives are indicated with primes. The solutions of (11.46) are, respectively,

$$w = \begin{cases} w_- = a_1 \exp(\alpha x) + a_5 \exp(-\alpha x), & x < 0 \\ w_W = a_2 \exp(ikx) + a_3 \exp(-ikx), & 0 < x < s \\ w_+ = a_4 \exp(-\alpha x) + a_6 \exp(\alpha x), & s < x \end{cases} \qquad (11.47)$$

where it must be set $a_5 = a_6 = 0$ to prevent w_- and w_+ from diverging. Using the continuity of w and w' in the origin,

$$\begin{cases} w_-(0) = w_W(0), & a_1 = a_2 + a_3 \\ w'_-(0) = w'_W(0), & \alpha a_1 = ik(a_2 - a_3) \end{cases} \qquad (11.48)$$

Solving for a_2, a_3 and letting $\vartheta = 1 + i\alpha/k$ yields $2a_2 = \vartheta^* a_1$, $2a_3 = \vartheta a_1$ whence $a_2/a_3 = \vartheta^*/\vartheta$. Then, using the continuity of w and w' at $x = s$ yields

$$\begin{cases} w_W(s) = w_+(s), & a_4 \exp(-\alpha s) = a_2 \exp(iks) + a_3 \exp(-iks) \\ w'_W(s) = w'_+(s), & -\alpha a_4 \exp(-\alpha s) = ik[a_2 \exp(iks) - a_3 \exp(-iks)] \end{cases}$$

Solving for a_2, a_3 yields $2a_2 = a_4 \vartheta \exp(-\alpha s - iks)$, $2a_3 = a_4 \vartheta^* \exp(-\alpha s + iks)$, whence $a_2/a_3 = (\vartheta/\vartheta^*) \exp(-2iks)$. In summary, a_2, a_3 are proportional to a_1 and to a_4. It follows that, if it were $a_1 = 0$ or $a_4 = 0$, then it would also be $a_2 = a_3 = 0$. However, this is impossible because w can not vanish identically; as a consequence it is necessarily $a_1 \neq 0$, $a_4 \neq 0$. This shows that, in contrast to the classical case, the particle penetrates the boundaries of the well. The relations found so far determine two different expressions for a_2/a_3. For them to be compatible, the equality $\vartheta^2 \exp(-iks) = (\vartheta^*)^2 \exp(iks)$ must hold, which represents the condition of a vanishing determinant of the 4×4, homogeneous algebraic system whose unknowns are a_1, a_2, a_3, and a_4. Using $\vartheta = 1 + i\alpha/k$, the equality is recast as

$$\left(1 - \frac{\alpha^2}{k^2}\right) \sin(ks) = 2\frac{\alpha}{k} \cos(ks), \qquad \frac{k^2 - \alpha^2}{2\alpha k} = \cot(ks). \qquad (11.49)$$

Finally, replacing the expressions of α and k provides the transcendental equation

$$\frac{E - V_0/2}{\sqrt{-E(E - V_0)}} = \cot\left(s\frac{\sqrt{2m}}{\hbar}\sqrt{E - V_0}\right), \qquad V_0 < E < 0, \qquad (11.50)$$

in the unknown E, whose roots fulfill the compatibility condition. As a consequence, such roots are the eigenvalues of E. Given m, s, and V_0, let $n \geq 1$ be an integer such that

$$(n-1)\pi < s\sqrt{-(2m/\hbar^2)V_0} \leq n\pi. \qquad (11.51)$$

Such an integer always exists, and indicates the number of branches of $\cot(ks)$ that belong (partially or completely) to the interval $V_0 < E < 0$. In such an interval the left hand side of (11.50) increases monotonically from $-\infty$ to $+\infty$; as a consequence, (11.50) has n roots $V_0 < E_1, E_2, \ldots, E_n < 0$. An example with five roots is shown in Fig. 11.5; the corresponding calculation is carried out in Prob. 11.2. When an eigenvalue, say E_i, is introduced into (11.46), it provides α_i, k_i, and $\vartheta_i = 1 + i\alpha_i/k_i$; in conclusion,

$$a_{i2} = \frac{1}{2}\vartheta_i^* a_{i1}, \qquad a_{i3} = \frac{1}{2}\vartheta_i a_{i1}, \qquad a_{i4} = \frac{\vartheta_i^*}{\vartheta_i} \exp(\alpha_i s + ik_i s)a_{i1}. \qquad (11.52)$$

The ith eigenfunction w_i can thus be expressed, from (11.50), in terms of a_{i1} alone. The latter, in turn, is found from the normalization condition $\int_{-\infty}^{+\infty} |w_i|^2 dx = 1$.

The case of the box treated in Sect. 8.2.2 is obtained by letting $V_0 \to -\infty$ here; at the same time one lets $E \to -\infty$ in such a way that the difference $E - V_0$ is kept finite. In this way, w_- and w_+ in (11.47) vanish identically and yield the boundary conditions for w_W used in Sect. 8.2.2.

Fig. 11.5 Graphic solution of
(11.50) using the auxiliary
variable η. The solutions
$\eta_1, \ldots, \eta_5$ are the intercepts
of the left hand side (thicker
line) with the branches of the
right hand side. The data are
given in Prob. 11.2

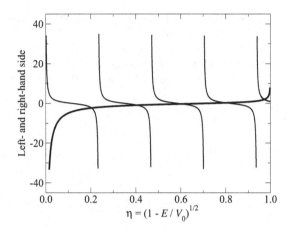

Fig. 11.6 The smooth
potential energy considered in
Prob. 11.1, with $V_0 = 2$ and
$E = 2.5$ (arbitrary units).

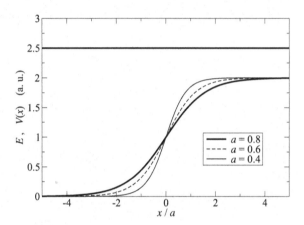

Problems

11.1 Consider a smooth potential energy described by

$$V(x) = \frac{V_0}{1 + \exp(-x/a)}, \qquad (11.53)$$

with $V_0 > 0$, $a > 0$ (Fig. 11.6). The limit of $V(x)$ for $a \to 0$ yields the
discontinuous step of Sect. 11.2. Considering a monochromatic wave with
$E > V_0$ launched from the left towards the barrier, the reflection coefficient is
found to be [41, Sect. 2.2].

$$R(a) = \frac{\sinh^2[\pi a\sqrt{2m}(\sqrt{E} - \sqrt{E - V_0})/\hbar]}{\sinh^2[\pi a\sqrt{2m}(\sqrt{E} + \sqrt{E - V_0})/\hbar]}. \qquad (11.54)$$

Discuss the limiting cases $a \to 0$ and $m \to \infty$ and compare the results with those found in Sect. 11.2.

11.2 Find the eigenvalues of the Schrödinger equation for an energy well like that of Fig. 11.5, having a width $s = 15\,\text{Å} = 1.5 \times 10^{-9}$ m and a depth[3] $-V_0 = 3\,\text{eV} \simeq 4.81 \times 10^{-19}$ J. Use $m \simeq 9.11 \times 10^{-31}$ kg, $\hbar \simeq 1.05 \times 10^{-34}$ J s.

---

[3] The *electron Volt* (eV) is a unit of energy obtained by multiplying 1 J by a number equal to the modulus of the electron charge expressed in C (Table D.1).

Chapter 12
Cases Related to the Linear Harmonic Oscillator

12.1 Introduction

The chapter is devoted to the solution of the Schrödinger equation for the linear harmonic oscillator, and to a number of important application of the results. The importance of the problem has already been outlined in the sections devoted to the classical treatment: its results can in fact be applied, with little or no modification, to mechanical situations where the positional force acting on the particle can be replaced with a first-order expansion, or to more complicate systems whose degrees of freedom can be separated into a set of Hamiltonian functions of the linear-harmonic-oscillator type. Such systems are not necessarily mechanical: for instance, the energy of the electromagnetic field *in vacuo* is amenable to such a separation, leading to the formal justification of the concept of photon. Similarly, a system of particles near a mechanical-equilibrium point can be separated in the same manner, providing the justification of the concept of phonon. An interesting property of the Fourier transform of the Schrödinger equation for the linear harmonic oscillator is shown in the complements.

12.2 Linear Harmonic Oscillator

An example of paramount importance is that of the linear harmonic oscillator, whose classical treatment is given in Sect. 3.3. The potential energy is shown in Fig. 12.1. From the general properties of the time-independent Schrödinger equation (Sect. 8.2.3) it follows $E > 0$. Also, the time-independent wave function w is expected to be square integrable. The Hamiltonian operator is found by replacing p with $\hat{p} = -i\hbar\, d/dx$ in (3.1), yielding the Schrödinger equation

$$-\frac{\hbar^2}{2m}\frac{d^2 w}{dx^2} + \frac{1}{2}m\omega^2 x^2 w = E\,w, \qquad \omega = \sqrt{c/m}, \qquad (12.1)$$

© Springer Science+Business Media New York 2015
M. Rudan, *Physics of Semiconductor Devices,*
DOI 10.1007/978-1-4939-1151-6_12

Fig. 12.1 The potential
energy of the linear harmonic
oscillator (Sect. 12.2)

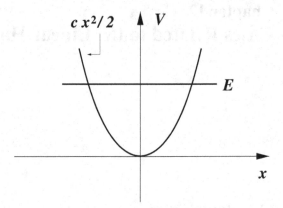

with c the elastic constant. The equation is conveniently recast in normalized form
by defining the dimensionless quantities $\varepsilon = E/(\hbar\omega)$, $\xi = (m\omega/\hbar)^{1/2} x$:

$$\mathcal{H}'w = \varepsilon\, w, \qquad \mathcal{H}' = \frac{1}{2}\left(\xi^2 - \frac{\mathrm{d}^2}{\mathrm{d}\xi^2}\right). \tag{12.2}$$

The eigenvalues and eigenfunctions of (12.2) are found by means of the factorization
method illustrated in Sect. 13.3. To begin, one defines the first-order operator

$$\hat{a} = \frac{1}{\sqrt{2}}\left(\xi + \frac{\mathrm{d}}{\mathrm{d}\xi}\right), \tag{12.3}$$

whence, for all f and g vanishing at infinity, an integration by parts yields

$$\int_{-\infty}^{+\infty} g^*\hat{a} f\,\mathrm{d}\xi = \int_{-\infty}^{+\infty} \left(\hat{a}^\dagger g\right)^* f\,\mathrm{d}\xi, \tag{12.4}$$

with $\hat{a}^\dagger = (\xi - \mathrm{d}/\mathrm{d}\xi)/\sqrt{2}$ the adjoint operator of $\hat{a}$. As $\hat{a}^\dagger \neq \hat{a}$, $\hat{a}$ is not Hermitean.
Also, $\hat{a}$ and $\hat{a}^\dagger$ do not commute; using the definitions of $\hat{a}$ and $\hat{a}^\dagger$ yields

$$\hat{a}\hat{a}^\dagger - \hat{a}^\dagger\hat{a} = \mathcal{I}, \qquad \hat{a}\hat{a}^\dagger + \hat{a}^\dagger\hat{a} = 2\mathcal{H}', \tag{12.5}$$

with $\mathcal{I}$ the identity operator. From (12.5), after defining the operator $\mathcal{N} = \hat{a}^\dagger\hat{a}$, called
number operator, one finds $2\,\mathcal{H}' = (\mathcal{I} + \hat{a}^\dagger\hat{a}) + \hat{a}^\dagger\hat{a} = 2\mathcal{N} + \mathcal{I}$, whence

$$\mathcal{H}'w = \mathcal{N}w + \mathcal{I}w/2 = \varepsilon w, \qquad \mathcal{N}w = \left(\varepsilon - \frac{1}{2}\right)w = \nu w, \tag{12.6}$$

where $\nu = \varepsilon - 1/2$. The second relation in (12.6) shows that $\mathcal{H}'$ and $\mathcal{N}$ have the
same eigenfunctions, while their eigenvalues differ by $1/2$. Using the same relation
and observing that $\hat{a}$ and $\hat{a}^\dagger$ are real, one finds

$$\int_{-\infty}^{+\infty} w^*\hat{a}^\dagger\hat{a}w\,\mathrm{d}\xi = \int_{-\infty}^{+\infty} |\hat{a}w|^2\,\mathrm{d}\xi = \nu\int_{-\infty}^{+\infty} |w|^2\mathrm{d}\xi, \tag{12.7}$$

showing that $\mathcal{N} = \hat{a}^\dagger \hat{a}$, in contrast to $\hat{a}$ and $\hat{a}^\dagger$, is Hermitean, hence its eigenvalues v are real. Moreover, they are non negative: in fact, the integral of $|w|^2$ in (12.7) is strictly positive because an eigenfunction can not vanish identically, whereas the integral of $|\hat{a}w|^2$ is strictly positive if $\hat{a}w \neq 0$, whereas it vanishes if $\hat{a}w = 0$ identically. As a consequence it is $v = 0$ if and only if $\hat{a}w = 0$, otherwise it is $v > 0$. Another relation is found by left multiplying by $\hat{a}^\dagger$ the first relation in (12.5), to find $\mathcal{N}\hat{a}^\dagger - \hat{a}^\dagger\mathcal{N} = \hat{a}^\dagger$, whence a new operator is defined: $\mathcal{N}\hat{a}^\dagger = \hat{a}^\dagger (\mathcal{N} + \mathcal{I})$. The action of the latter upon an eigenfunction w of $\mathcal{N}$ results in

$$\mathcal{N}\hat{a}^\dagger w = \hat{a}^\dagger (\mathcal{N} + \mathcal{I}) w = \hat{a}^\dagger v w + \hat{a}^\dagger w = (v + 1) \hat{a}^\dagger w. \tag{12.8}$$

In other terms, if v is an eigenvalue of $\mathcal{N}$ and w is an eigenfunction belonging to v, then $v + 1$ is also an eigenvalue of $\mathcal{N}$, with eigenfunction $\hat{a}^\dagger w$. This reasoning can be repeated indefinitely: $\mathcal{N}\hat{a}^\dagger\hat{a}^\dagger w = (v + 2)\hat{a}^\dagger\hat{a}^\dagger w, \ldots$; its conclusion is that, if a number $v \geq 0$ is known to be an eigenvalue of $\mathcal{N}$, then all numbers $v + 1$, $v + 2, \ldots$ belonging to the unlimited ladder[1] beginning with v are also eigenvalues of $\mathcal{N}$. Also, if w is an eigenfunction belonging to v, the eigenfunctions belonging to $v + 1, v + 2, \ldots$ are $\hat{a}^\dagger w, \hat{a}^\dagger\hat{a}^\dagger w, \ldots$.

One may argue that the same kind of reasoning should be applicable in the backward direction as well, to check whether $v - 1, v - 2$ are eigenvalues of $\mathcal{N}$. In fact, right multiplying by $\hat{a}$ the first relation in (12.5) yields $\hat{a}\mathcal{N} - \mathcal{N}\hat{a} = \hat{a}$, whence $\hat{a}(\mathcal{N} - \mathcal{I}) = \mathcal{N}\hat{a}$. The action of $\mathcal{N}\hat{a}$ upon an eigenfunction w of $\mathcal{N}$ results in

$$\mathcal{N}\hat{a}w = \hat{a}(\mathcal{N} - \mathcal{I})w = \hat{a}vw - \hat{a}w = (v - 1)\hat{a}w. \tag{12.9}$$

Remembering that the eigenvalues of $\mathcal{N}$ are non negative, (12.9) shows that, if v is an eigenvalue of $\mathcal{N}$ and $v \geq 1$, than $v - 1$ is also an eigenvalue of $\mathcal{N}$, to which $\hat{a}w$ belongs. By induction, $\mathcal{N}\hat{a}\hat{a}w = (v - 2)\hat{a}\hat{a}w$, and so on. However, the backward process cannot be repeated indefinitely because one of the numbers $v - 2, v - 3, \ldots$ will eventually be found to be negative: this result contradicts (12.7), that shows that the eigenvalues can not be negative. The contradiction is due to the implicit assumption that the eigenvalue v can be any real number, and is readily eliminated by specifying that $v = n = 0, 1, 2, \ldots$; consider in fact the eigenvalue $n = 1$ and let w be an eigenfunction belonging to it. Applying (12.9) shows that $w_0 = \hat{a}w$ is an eigenfunction of $\mathcal{N}$ belonging to $n = 0$, namely, $\mathcal{N}w_0 = 0$. The next step in the backward process would yield $\hat{a}w_0$ which, however, is not an eigenfunction because it vanishes identically: this is easily found by combining $\langle w_0 | \mathcal{N} w_0 \rangle = ||\hat{a}w_0||^2$ with $\mathcal{N}w_0 = 0$. In other terms, the backward process comes automatically to an end when the eigenvalue $n = 0$ is reached.

The above reasoning shows that only the numbers $n = 0, 1, 2, \ldots$ are eigenvalues of $\mathcal{N}$. It also provides an easy method to calculate the eigenfunctions, that starts from the result $\hat{a}w_0 = 0$ just found. Such a relation is a differential equation of the

---

[1] The term "ladder" is introduced in Sect. 13.3.

form

$$\frac{1}{\sqrt{2}}\left(\xi + \frac{d}{d\xi}\right)w_0 = 0, \qquad \frac{dw_0}{w_0} = -\xi d\xi, \qquad w_0 = c_0 \exp\left(-\frac{\xi^2}{2}\right). \quad (12.10)$$

The normalization constant is found from (C.27), which yields $c_0 = \pi^{-1/4}$. The eigenfunctions corresponding to $n = 1, 2, \ldots$ are found recursively with $w_1 = \hat{a}^\dagger w_0$, $w_2 = \hat{a}^\dagger w_1 = \hat{a}^\dagger \hat{a}^\dagger w_0$, ... For example,

$$w_1 = \frac{1}{\sqrt{2}}\left(\xi w_0 - \frac{dw_0}{d\xi}\right) = \frac{w_0}{\sqrt{2}}2\xi. \quad (12.11)$$

From this construction it is easily found that the eigenvalues are not degenerate, and that w_n is even (odd) if n is even (odd). Also, it can be shown that w_n has the form [78, Chap. XII.7]

$$w_n(\xi) = \left(n! 2^n \sqrt{\pi}\right)^{-1/2} \exp\left(-\xi^2/2\right) H_n(\xi), \quad (12.12)$$

where H_n is the nth *Hermite polynomial*

$$H_n(\xi) = (-1)^n \exp(\xi^2)\frac{d^n}{d\xi^n} \exp(-\xi^2). \quad (12.13)$$

The eigenfunctions of the linear harmonic oscillator form a real, orthonormal set:

$$\int_{-\infty}^{+\infty} w_n w_m \, dx = \delta_{nm}. \quad (12.14)$$

By remembering that $\nu = \varepsilon - 1/2 = n$ and $\varepsilon = E/(\hbar\omega)$, one finds for the energy E of the linear harmonic oscillator the eigenvalues

$$E_n = \left(n + \frac{1}{2}\right)\hbar\omega, \qquad n = 0, 1, 2, \ldots \quad (12.15)$$

In conclusion, the energy of the linear harmonic oscillator is the sum of the minimum energy $E_0 = \hbar\omega/2 > 0$, also called *zero point* energy, and of an integer number of elementary quanta of energy $\hbar\omega$. The paramount importance of the example of the linear harmonic oscillator has already been emphasized in the classical treatment of Sect. 3.13.1. Examples of application of the quantization of the linear harmonic oscillator are given in Sects. 12.3, 12.4, and 12.5.

The Hermite polynomials (12.13) fulfill the recursive relation [44]

$$H_{n+1} - 2\xi H_n + 2n H_{n-1} = 0, \qquad n = 1, 2, \ldots, \quad (12.16)$$

which is useful to determine some properties of the $\hat{a}$, $\hat{a}^\dagger$ operators. For instance, combining the definition of $\hat{a}^\dagger$ with (12.12) one obtains

$$\hat{a}^\dagger w_n = \frac{1/\sqrt{2}}{\left(n! 2^n \sqrt{\pi}\right)^{1/2}} \exp\left(-\xi^2/2\right)\left(2\xi H_n - \frac{dH_n}{d\xi}\right). \quad (12.17)$$

On the other hand one finds from (12.13) that $\mathrm{d}H_n/\mathrm{d}\xi = 2\xi H_n - H_{n+1}$. Replacing the derivative in (12.17) yields

$$\hat{a}^\dagger w_n = \frac{1/\sqrt{2}}{\left(n!2^n\sqrt{\pi}\right)^{1/2}} \exp\left(-\xi^2/2\right) H_{n+1} = \sqrt{n+1}\, w_{n+1}. \tag{12.18}$$

This result shows that, apart from the multiplicative constant, $\hat{a}^\dagger$ transforms the state of index n into that of index $n+1$. Due to (12.15), the transformation corresponds to an increase $E_{n+1} - E_n = \hbar\omega$ in the total energy of the oscillator. Since its action "creates" a quantum of energy, $\hat{a}^\dagger$ is called *creation operator*. Using the same procedure, combined with the recursive relation (12.16), one finds

$$\hat{a}w_n = \sqrt{n}\, w_{n-1}. \tag{12.19}$$

Due to the above, $\hat{a}$ is called *destruction operator* or *annihilation operator*. Note that, due to (12.18, 12.19), the successive application of $\hat{a}$ and $\hat{a}^\dagger$ to w_n is equivalent to $\mathcal{N}w_n = n\, w_n$.

12.3 Quantization of the Electromagnetic Field's Energy

The energy of the electromagnetic field within a finite volume V free of charge has been calculated in Sect. 5.6 in terms of the field's modes. In such a calculation the shape of V was chosen as that of a box whose sides d_1, d_2, d_3 are aligned with the coordinate axes and start from the origin (Fig. 5.1), so that $V = d_1 d_2 d_3$. The energy reads

$$W_{\mathrm{em}} = \sum_{\mathbf{k}\sigma} W_{\mathbf{k}\sigma}, \qquad W_{\mathbf{k}\sigma} = 2\varepsilon_0 V\omega^2 s_{\mathbf{k}\sigma} s_{\mathbf{k}\sigma}^*, \tag{12.20}$$

where ε_0 is the vacuum permittivity, $\mathbf{k} = \sum_{i=1}^3 2\pi n_i \mathbf{i}_i/d_i$, $n_i = 0, \pm1, \pm2, \ldots$, $\sigma = 1, 2$, and $\omega = c\,|\mathbf{k}|$, with c the speed of light and $\mathbf{i}_i$ the unit vector of the ith axis. Finally, $s_{\mathbf{k}\sigma}(t)$ is one of the two components of the complex vector defined by (5.29). The energy $W_{\mathbf{k}\sigma}$ of the degree of freedom $\mathbf{k}, \sigma$ is written in Hamiltonian form by introducing the canonical coordinates $q_{\mathbf{k}\sigma}, p_{\mathbf{k}\sigma}$ such that

$$2\sqrt{\varepsilon_0 V}\omega s_{\mathbf{k}\sigma} = \omega q_{\mathbf{k}\sigma} + i p_{\mathbf{k}\sigma}. \tag{12.21}$$

Using (12.21) transforms (12.20) into

$$W_{\mathbf{k}\sigma} = \frac{1}{2}(\omega q_{\mathbf{k}\sigma} + i p_{\mathbf{k}\sigma})(\omega q_{\mathbf{k}\sigma} - i p_{\mathbf{k}\sigma}) = \frac{1}{2}\left(p_{\mathbf{k}\sigma}^2 + \omega^2 q_{\mathbf{k}\sigma}^2\right). \tag{12.22}$$

The Hamiltonian operator is obtained by replacing $q_{\mathbf{k}\sigma}$ with $\hat{q}_{\mathbf{k}\sigma} = q_{\mathbf{k}\sigma}$ and $p_{\mathbf{k}\sigma}$ with $\hat{p}_{\mathbf{k}\sigma} = -i\hbar\, \mathrm{d}/\mathrm{d}q_{\mathbf{k}\sigma}$ in the last expression on the right of (12.22):

$$\mathcal{H}_{\mathbf{k}\sigma}^0 = -\frac{\hbar^2}{2}\frac{\mathrm{d}^2}{\mathrm{d}q_{\mathbf{k}\sigma}^2} + \frac{\omega^2}{2}q_{\mathbf{k}\sigma}^2. \tag{12.23}$$

It should be noted, however, that if the intermediate expression in (12.22), instead of that on the right, is used for the replacement of the classical coordinates with the corresponding operators, a different result is obtained. In fact, $\hat{q}_{k\sigma}$ and $\hat{p}_{k\sigma}$ do not commute so that, from (8.72), one obtains

$$\mathcal{H}_{k\sigma}^- = \mathcal{H}_{k\sigma}^0 - \frac{1}{2}\hbar\omega. \tag{12.24}$$

A third form of the Hamiltonian operator, different from the two above, is obtained by exchanging the order of factors in the intermediate expression in (12.22) prior to the replacement of the coordinates with operators:

$$\mathcal{H}_{k\sigma}^+ = \mathcal{H}_{k\sigma}^0 + \frac{1}{2}\hbar\omega. \tag{12.25}$$

To proceed one considers $\mathcal{H}_{k\sigma}^0$ first. The Schrödinger equation generated by it,

$$\mathcal{H}_{k\sigma}^0 w_{k\sigma}^0 = E_{k\sigma}^0 w_{k\sigma}^0 \tag{12.26}$$

is the Eq. (12.1) of the linear harmonic oscillator, whose eigenvalues (12.15) read $E^0(n_{k\sigma}) = (n_{k\sigma} + 1/2)\hbar\omega$, with $n_{k\sigma} = 0, 1, 2 \ldots$, and are non degenerate. The second form (12.24) of the Hamiltonian operator generates the Schrödinger equation $\mathcal{H}_{k\sigma}^- w_{k\sigma}^- = E_{k\sigma}^- w_{k\sigma}^-$, namely,

$$\mathcal{H}_{k\sigma}^0 \, w_{k\sigma}^- = \left(E_{k\sigma}^- + \frac{1}{2}\hbar\omega \right) w_{k\sigma}^-, \tag{12.27}$$

again the equation for the linear harmonic oscillator, whose operator is identical to that of (12.26). As a consequence, its eigenvalues are $E^-(n_{k\sigma}) + \hbar\omega/2 = (n_{k\sigma} + 1/2)\hbar\omega$, whence $E^-(n_{k\sigma}) = n_{k\sigma}\hbar\omega$. By the same token the eigenvalues of (12.25) are found to be $E^+(n_{k\sigma}) = (n_{k\sigma} + 1)\hbar\omega$. From (12.20), the energy of the electromagnetic field is the sum of the energy of the single modes; the three cases considered above then yield:

$$W_{em}^0 = \sum_{k\sigma} \left(n_{k\sigma} + \frac{1}{2} \right) \hbar\omega, \qquad W_{em}^- = \sum_{k\sigma} n_{k\sigma}\hbar\omega, \qquad W_{em}^+ = \sum_{k\sigma} (n_{k\sigma} + 1)\hbar\omega,$$

with $\omega = c|\mathbf{k}|$. In the expression of W_{em}^0 and W_{em}^+ the sum over $\mathbf{k}$ of the terms $\hbar\omega/2$ and, respectively, $\hbar\omega$, diverges. This is not acceptable because the total energy within V is finite. On the contrary, for the expression of W_{em}^- to converge it is sufficient that $n_{k\sigma}$ vanishes from some $|\mathbf{k}|$ on; the correct Hamiltonian is thus $\mathcal{H}_{k\sigma}^-$. Grouping for each $\mathbf{k}$ the summands corresponding to $\sigma = 1$ and $\sigma = 2$, and letting $n_k = n_{k1} + n_{k2}$, provides

$$W_{em}^- = \sum_{k} n_k \hbar\omega. \tag{12.28}$$

In conclusion, the energy of each mode of oscillation is a multiple (0 included) of the elementary quantum of energy $\hbar\,\omega(\mathbf{k})$. This result provides the formal justification of the concept of *photon*. The integer $n_{\mathbf{k}\sigma}$ is the *occupation number* of the pair $\mathbf{k},\sigma$, whereas $n_{\mathbf{k}}$ is the number of photons[2] of the mode corresponding to $\mathbf{k}$. Like in the classical treatment, the energy of the electromagnetic field is the sum of the energies of each mode of oscillation.

12.4 Quantization of the Electromagnetic Field's Momentum

The momentum of the electromagnetic field within a finite volume V free of charge has been calculated in Sect. 5.7 in terms of the field's modes. Premises and symbols here are the same as in Sect. 12.3. The momentum reads

$$\int_V \frac{\mathbf{S}}{c^2}\,\mathrm{d}V = 2\varepsilon_0 V \sum_{\mathbf{k}} \omega\,\mathbf{s}_{\mathbf{k}}\cdot\mathbf{s}_{\mathbf{k}}^*\mathbf{k} = \sum_{\mathbf{k}\sigma}\frac{1}{c}W_{\mathbf{k}\sigma}\frac{\mathbf{k}}{k}, \tag{12.29}$$

with $\mathbf{S}$ the Poynting vector and $k = |\mathbf{k}|$. For each pair $\mathbf{k},\sigma$, the same quantization procedure used for the energy in Sect. 12.3 is applicable here, and yields the operator $\mathcal{H}_{\mathbf{k}\sigma}^{-}\,\mathbf{k}/(ck)$. As the latter differs by a constant vector from the Hamiltonian operator (12.24) corresponding to the same pair $\mathbf{k},\sigma$, the eigenvalues of the ith component of momentum turn out to be

$$\frac{1}{c}n_{\mathbf{k}\sigma}\,\hbar\omega\frac{k_i}{k} = n_{\mathbf{k}\sigma}\hbar k_i, \qquad n_{\mathbf{k}\sigma} = 0,1,2,\dots \tag{12.30}$$

In conclusion, the eigenvalues of momentum corresponding to $\mathbf{k},\sigma$ are $n_{\mathbf{k}\sigma}\hbar\mathbf{k}$. Letting as in Sect. 12.3 $n_{\mathbf{k}} = n_{\mathbf{k}1} + n_{\mathbf{k}2}$, the momentum of the electromagnetic field is expressed in quantum terms as $\sum_{\mathbf{k}} n_{\mathbf{k}}\hbar\,\mathbf{k}$. This result shows that the momentum of each mode of oscillation is a multiple (0 included) of the elementary quantum of momentum $\hbar\,\mathbf{k}$, and completes the formal justification of the concept of *photon* started in Sect. 12.3. Each photon has energy and momentum, given by $\hbar\omega$ and $\hbar\mathbf{k}$ respectively. Like in the classical treatment, the momentum of the electromagnetic field is the sum of the momenta of each mode of oscillation.

---

[2] To complete the description of the photon it is necessary to work out also the quantum expression of its momentum. This is done in Sect. 12.4. The concept of photon was introduced by Einstein in 1905 [34] (the English translation of [34] is in [107]). The quantization procedure shown here is given in [30].

12.5   Quantization of a Diagonalized Hamiltonian Function

A system of particles near an equilibrium point has been investigated in Sects. 3.9 and 3.10. The analysis led to the separation of the Hamiltonian function, that reads

$$H_a - V_{a0} = \sum_{\sigma=1}^{3N} H_\sigma, \qquad H_\sigma = \frac{1}{2}\dot{b}_\sigma^2 + \frac{1}{2}\omega_\sigma^2 b_\sigma^2. \tag{12.31}$$

In (12.31), $3N$ is the number of degrees of freedom, b_σ the normal coordinate of index σ, ω_σ the angular frequency corresponding to b_σ, and V_{a0} the minimum of the system's potential energy. Apart from the constant V_{a0}, the Hamiltonian function H_a is given by a sum of terms, each associated with a single degree of freedom. In turn, each summand H_σ is identical to the Hamiltonian function of a linear harmonic oscillator with $m = 1$. As a consequence, the quantum operator corresponding to (12.31) takes the form

$$\mathcal{T}_a + V_a = \sum_{\sigma=1}^{3N} \mathcal{H}_\sigma + V_{a0}, \qquad \mathcal{H}_\sigma = -\frac{\hbar^2}{2}\frac{\partial^2}{\partial b_\sigma^2} + \frac{1}{2}\omega_\sigma^2 b_\sigma^2, \tag{12.32}$$

and generates the eigenvalue equation $(\mathcal{T}_a + V_a)v = Ev$, where

$$\left(\sum_{\sigma=1}^{3N} \mathcal{H}_\sigma\right) v = E'v, \qquad E' = E - V_{a0}. \tag{12.33}$$

Being $\mathcal{H}_a$ the sum of operators acting on individual degrees of freedom, the Schrödinger equation is separable (Sect. 10.3) and splits into $3N$ equations of the form

$$\mathcal{H}_\sigma v_{\sigma\zeta(\sigma)} = E_{\sigma\zeta(\sigma)} v_{\sigma\zeta(\sigma)}, \qquad E' = \sum_{\sigma=1}^{3N} E_{\sigma\zeta(\sigma)}, \tag{12.34}$$

where $\sigma = 1, 2, \dots, 3N$ refers to the degrees of freedom, whereas $\zeta(\sigma) = 0, 1, 2, \dots$ counts the set of eigenvalue indices corresponding to a given σ. Remembering the solution of the Schrödinger equation for a linear harmonic oscillator (Sect. 12.2), the energy of the individual degree of freedom is

$$E_{\sigma\zeta(\sigma)} = \left[\zeta(\sigma) + \frac{1}{2}\right]\hbar\omega_\sigma, \qquad \zeta(\sigma) = 0, 1, 2, \dots . \tag{12.35}$$

The total energy of the system then reads

$$E = V_{a0} + \sum_{\sigma=1}^{3N} \left[\zeta(\sigma) + \frac{1}{2}\right]\hbar\omega_\sigma. \tag{12.36}$$

As indicated in Sect. 3.10, the oscillation of the normal coordinate of index σ is called *mode* of the vibrating system. The classical expression of the energy associated to each mode has the same form as that of a mode of the electromagnetic field (compare (12.31) with (12.22)). By analogy with the electromagnetic case, a particle of energy $\hbar \omega_\sigma$, called *phonon*, is introduced in the quantum description, and the energy of the mode is ascribed to the set of phonons belonging to the mode. The integers $\zeta(\sigma)$ are the *occupation numbers* of the normal modes of oscillation.

Note that the quadratic form (12.31) of the total energy H_σ of each degree of freedom of the oscillating system was derived directly, in contrast with that of the electromagnetic field where the product of two linear forms was involved (compare with (12.22)). For this reason, the discussion about three possible forms of the Hamiltonian operator, carried out in Sect. 12.3, is not necessary here. The total energy (12.36) does not diverge because the number of degrees of freedom is finite.

12.6 Complements

12.6.1 Comments About the Linear Harmonic Oscillator

The normalized form (12.2) of the Schrödinger equation for the linear harmonic oscillator is $-d^2 w_n / d\xi^2 + \xi^2 w_n = 2\varepsilon_n w_n$. Due to the exponential decay at infinity, the Fourier transform (C.16) of the eigenfunction exists. Let

$$u_n(\eta) = \mathcal{F} w_n = \frac{1}{\sqrt{2\pi}} \int_{-\infty}^{+\infty} w_n(\xi) \exp(-i\eta\xi) \, d\xi. \tag{12.37}$$

Thanks to the property (C.22) of the Fourier transform it is $\mathcal{F} d^2 w_n / d\xi^2 = -\eta^2 u_n$, $\mathcal{F} \xi^2 w_n = -d^2 u_n / d\eta^2$. Fourier transforming (12.2) thus yields

$$\frac{1}{2} \eta^2 u_n - \frac{1}{2} \frac{d^2 u_n}{d\eta^2} = \varepsilon_n u_n, \tag{12.38}$$

namely, an equation identical to (12.2), having the same eigenvalue. As ε_n is not degenerate, it follows $u_n \propto w_n$, namely, the eigenfunctions of the harmonic oscillator are equal to their own Fourier transforms apart from a multiplicative constant at most.[3]

---

[3] Compare with (C.82), where the property is demonstrated for the Gaussian function; the latter, apart from scaling factors, coincides with the eigenfunction of the linear harmonic oscillator belonging to the eigenvalue corresponding to $n = 0$.

Chapter 13
Other Examples of the Schrödinger Equation

13.1 Introduction

A number of properties of the one-dimensional, time-independent Schrödinger equation can be worked out without specifying the form of the coefficient. To this purpose one examines the two fundamental solutions, which are real because the coefficient is such. One finds that the fundamental solutions do not have multiple zeros and do not vanish at the same point; more precisely, the zeros of the first and second fundamental solution separate each other. It is also demonstrated that the character of the fundamental solutions within an interval is oscillatory or non oscillatory depending on the sign of the equation's coefficient in such an interval. After completing this analysis, the chapter examines an important and elegant solution method, consisting in factorizing the operator. The theory is worked out for the case of localized states, corresponding to discrete eigenvalues. The procedure by which the eigenfunctions' normalization is embedded into the solution scheme is also shown. The chapter continues with the analysis of the solution of a Schrödinger equation whose coefficient is periodic; this issue finds important applications in the case of periodic structures like, e.g., crystals. Finally, the solution of the Schrödinger equation for a particle subjected to a central force is worked out; the equation is separated and the angular part is solved first, followed by the radial part whose potential energy is specified in the Coulomb case. The first complements deal with the operator associated to the angular momentum and to the solution of the angular and radial equations by means of the factorization method. The last complement generalizes the solution method for the one-dimensional Schrödinger equation in which the potential energy is replaced with a piecewise-constant function, leading to the concept of transmission matrix.

13.2 Properties of the One-Dimensional Schrödinger Equation

In the general expression (11.44) for the transmission coefficient, the fundamental solutions u, v appear in the denominator. It is then necessary to investigate the zeros of the solutions of (11.28). Due to the $u(0) = 1$ prescription, the possible zeros of u belong to the interval $0 < x \leq s$, while those of v belong to the interval $0 \leq x \leq s$.

© Springer Science+Business Media New York 2015
M. Rudan, *Physics of Semiconductor Devices*,
DOI 10.1007/978-1-4939-1151-6_13

If one or more zero exist, they can not be multiple. In fact, if u had a multiple zero at x_m it would be $u(x_m) = 0$, $u'(x_m) = 0$, hence $u = 0$ would be a solution of (11.28) compatible with such conditions. In fact, because of the uniqueness of the solution, $u = 0$ would be the only solution. Observing that u is continuous, this would contradict the condition $u(0) \neq 0$. Similarly, if v had a multiple zero it would vanish identically. Remembering that the derivative of the solution is continuous, this would contradict the condition $v'(0) = 1$ of (11.40). Another property is that u and v cannot vanish at the same point. This is apparent from the relation $W(x) = u v' - u' v = 1$ demonstrated in Sect. A.12. For the same reason, u' and v' cannot vanish at the same point.

If one of the solutions, say u, has more than one zero in $0 < x \leq s$, then the following property holds: between two consecutive zeros of u there is one and only one zero of v. Let x_L, x_R be two consecutive zeros of u, with $0 < x_L < x_R \leq s$. The property is demonstrated by showing, first, that a contradiction would arise if there were no zeros of v between x_L and x_R (that is to say, at least one zero must exist there) and, second, that if a zero of v exists between x_L and x_R, it must be unique [102]. To proceed one considers the function u/v in the interval $x_L \leq x \leq x_R$. By definition u/v vanishes at the boundaries of such an interval while, as shown above, v cannot vanish at the boundaries. If one assumes that there are no zeros of v inside the interval, then u/v exists everywhere in the interval, and is also everywhere continuous with a continuous first derivative because u and v are solutions of the second-order differential Eq. (11.28). As u/v vanishes at x_L and x_R, its derivative must vanish at least once in the open interval $x_L < x < x_R$. However, this is impossible because $d(u/v)/dx = -W/v^2 = -1/v^2 \neq 0$. This shows that v must have at least one zero between x_L and x_R. Such a zero is also unique because, if v had two zeros in $x_L < x < x_R$, then by the same reasoning u would have one zero between them, so x_L and x_R would not be consecutive. The property may be restated as *the zeros of two real linearly-independent solutions of a second-order linear differential equation separate each other*. The property does not hold for complex solutions.

So far the properties demonstrated in this section did not consider the sign of the coefficient $q(x) = 2 m (E - V)/\hbar^2$ of (11.28). The coefficient separates the interval $0 \leq x \leq s$ into subintervals where q is alternatively positive or negative. If q is continuous the extrema of the subintervals are the zeros of q, otherwise they may be discontinuity points of q. In either case the behavior of the solution u within each subinterval depends on the sign of q there. To show this, consider the function $d(u u')/dx = (u')^2 - q u^2$, where the expression at the right hand side has been found by means of (11.28). If $q \leq 0$ in the subinterval, then $d(u u')/dx$ is non negative. It follows that $u u'$ is a non-decreasing function in the subinterval, hence it has one zero at most. Remembering that u and u' can not vanish at the same point, one of the following holds: *i*) neither u nor u' vanish in the subinterval, *ii*) either u or u' vanishes once in the subinterval. For a given interval a function is called *non oscillatory* if its derivative vanishes at most once. It follows that the solution u is non oscillatory in those subintervals where $q \leq 0$. The case $V = V_0 > E > 0$ in the interval $0 < x < s$, considered in Sect. 11.3.1, is of this type.

Fig. 13.1 Form of the
potential energy that gives
rise to localized states
(Sect. 13.3). Only one state E
is shown

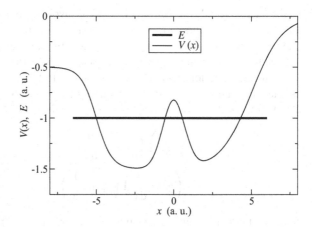

13.3   Localized States—Operator's Factorization

It may happen that the form of the potential energy V in the interval $0 \le x \le s$
is such that V has one or more negative minima (Fig. 13.1). In this case negative
eigenvalues of E may exist, giving rise to localized states. To treat this case one
must preliminarily observe that the eigenfunctions do not vanish identically outside
the interval $0 \le x \le s$, because the minima of V are finite. As a consequence, it is
convenient to replace the above interval with $x_1 \le x \le x_2$, where the two boundaries
may be brought (independently from each other) to $-\infty$ or $+\infty$, respectively. Letting

$$\lambda = \frac{2\,m}{\hbar^2}\,E, \qquad r_l(x) = -\frac{2\,m}{\hbar^2}\,V, \tag{13.1}$$

the Schrödinger Eq. (11.28) becomes

$$w'' + r_l\,w + \lambda\,w = 0. \tag{13.2}$$

The integer index $l = 0, 1, \ldots$ is attached to $r(x)$ for convenience. In fact, in typical
applications the form of the potential energy in (13.1) may be prescribed by a previous separation procedure in which the index is involved. As will be shown after the
analysis of the solution procedure, index l may eventually be disposed of. For the
time being, the solutions of (13.2) must be considered as dependent on the eigenvalue
λ and on the index l, namely, $w = w_{\lambda l}(x)$. Also, for a given pair λ, l the eigenfunctions are non degenerate due to the normalization condition. As a consequence, two
eigenfunctions belonging to the same pair differ by a multiplicative constant at most.

13.3.1   Factorization Method

A possible method for solving (13.2) is expanding w into a power series, replacing the
series into (13.2), collecting the terms of equal power, and letting their coefficients

vanish. This provides a recurrence relation involving the coefficients of the series; then, the latter is suitably truncated to obtain a square-integrable solution. Another method for solving (13.2), that greatly reduces the calculation effort and brings about a procedure of supreme elegance, is the operator's factorization. The conditions that make the factorization possible are illustrated in [52] and are briefly reported here. In essence they amount to finding a function $g_l(x)$ and a parameter L_l such that (13.2) may be recast as

$$\left(g_{l+1} + \frac{d}{dx}\right)\left(g_{l+1} - \frac{d}{dx}\right) w_{\lambda l} = (\lambda - L_{l+1}) w_{\lambda l}, \tag{13.3}$$

$$\left(g_l - \frac{d}{dx}\right)\left(g_l + \frac{d}{dx}\right) w_{\lambda l} = (\lambda - L_l) w_{\lambda l}. \tag{13.4}$$

Note that both (13.3) and (13.4) must be identical to (13.2). An additional constraint is that, for a given integer n, it must be $L_{n+1} > L_{l+1}$, $l = 0, 1, \ldots, n - 1$. To proceed, the boundary conditions $w_{\lambda l}(x_1) = w_{\lambda l}(x_2) = 0$ will be assumed. If one or both boundaries are at infinity, the condition $\int |w_{\lambda l}|^2 \, dx < \infty$ will also be assumed. Now, imposing that (13.3) is identical to (13.2) yields $g_{l+1}^2 + g'_{l+1} + L_{l+1} = -r_l$ whence, letting $l \leftarrow l - 1$,

$$g_l^2 + g'_l + L_l = -r_{l-1}. \tag{13.5}$$

Similarly, imposing that (13.4) is identical to (13.2) leads to

$$g_l^2 - g'_l + L_l = -r_l. \tag{13.6}$$

Adding (13.6) to (13.5) and subtracting (13.6) from (13.5) yields, respectively,

$$g_l^2 + L_l = -\frac{1}{2}(r_{l-1} + r_l), \qquad g'_l = -\frac{1}{2}(r_{l-1} - r_l). \tag{13.7}$$

Differentiating the first relation of (13.7) with respect to x and replacing g'_l from the second one provides

$$g_l = \frac{1}{2} \frac{r'_{l-1} + r'_l}{-2 g'_l} = \frac{1}{2} \frac{r'_{l-1} + r'_l}{r_{l-1} - r_l}. \tag{13.8}$$

Finally, replacing (13.8) into the first relation of (13.7),

$$L_l = -\frac{1}{2}(r_{l-1} + r_l) - \frac{1}{4}\left(\frac{r'_{l-1} + r'_l}{r_{l-1} - r_l}\right)^2. \tag{13.9}$$

In conclusion, the factorization is possible if L_l given by (13.9) is independent of x. In this case, g_l is given by (13.8). As r_l is real, both L_l and g_l are real as well.

13.3.2 First-Order Operators

If the factorization (13.3, 13.4) succeeds, it is useful to define the first-order, real operators

$$A_l^+ = g_l + \frac{d}{dx}, \qquad A_l^- = g_l - \frac{d}{dx}, \tag{13.10}$$

so that (13.3, 13.4) are rewritten as

$$A_{l+1}^+ A_{l+1}^- w_{\lambda l} = (\lambda - L_{l+1}) w_{\lambda l}, \qquad A_l^- A_l^+ w_{\lambda l} = (\lambda - L_l) w_{\lambda l}. \tag{13.11}$$

The two operators (13.10) are mutually adjoint. In fact, for any pair of functions f_1, f_2 fulfilling the same boundary conditions as $w_{\lambda l}$ one finds

$$\int_{x_1}^{x_2} f_1^* A_l^+ f_2 \, dx = \left[f_1^* f_2 \right]_{x_1}^{x_2} + \int_{x_1}^{x_2} (A_l^- f_1)^* f_2 \, dx, \tag{13.12}$$

where the integrated part vanishes due to the boundary conditions. From the above result one finds a property of the eigenvalue λ. In fact, multiplying (13.11) by $w_{\lambda l}^*$ and integrating, one finds

$$\int_{x_1}^{x_2} |A_{l+1}^- w_{\lambda l}|^2 \, dx = (\lambda - L_{l+1}) \int_{x_1}^{x_2} |w_{\lambda l}|^2 \, dx, \tag{13.13}$$

where (13.12) has been used after letting $f_1 = w_{\lambda l}$, $f_2 = A_{l+1}^- w_{\lambda l}$. From (13.13) it follows that, if $w_{\lambda l}$ is an eigenfunction, that is, if $w_{\lambda l}$ does not vanish identically, then the case $\lambda < L_{l+1}$ is impossible. In fact, the integral at the right hand side of (13.13) is strictly positive, while that at the left hand side is non negative. There remain two possibilities, namely:

1. $A_{l+1}^- w_{\lambda l}$ does not vanish identically, whence both integrals of (13.13) are stricly positive. It follows that $\lambda > L_{l+1}$. Also, as will be shown later, $A_{l+1}^- w_{\lambda l}$ is another eigenfunction of (13.2). The opposite is also true, namely, $\lambda > L_{l+1}$ implies that $A_{l+1}^- w_{\lambda l}$ does not vanish identically.
2. $A_{l+1}^- w_{\lambda l}$ vanishes identically, whence $\lambda = L_{l+1}$. The opposite is also true, namely, $\lambda = L_{l+1}$ implies that $A_{l+1}^- w_{\lambda l}$ vanishes identically. In this case $A_{l+1}^- w_{\lambda l}$ is not an eigenfunction of (13.2).

The discussion above allows one to identify the eigenvalue λ. In fact, there must be a value of the index l, say $l = n$, such that the equality $\lambda = L_{n+1}$ holds. It follows that the eigenvalue is identified by the index n, $\lambda = \lambda_n$.

As mentioned before the condition $L_{n+1} > L_{l+1}, l = 0, 1, \ldots, n - 1$ holds. As a consequence, the eigenfunction corresponding to the pair λ_n, l may be indicated with w_{nl} instead of $w_{\lambda l}$. In particular, the eigenfunction corresponding to $l = n$ is w_{nn}. As shown in case 2 above, such an eigenfunction corresponds to the equality $\lambda_n = L_{n+1}$ which, in turn, implies the condition $A_{n+1}^- w_{nn} = 0$. Remembering the second relation of (13.10), such a condition yields the first-order equation $(g_{n+1} - d/dx)\, w_{nn} = 0$, whose solution is real and reads

$$w_{nn} = c_{nn} \exp\left[\int_{x_1}^{x} g_{n+1}(\xi)\, d\xi\right], \quad \frac{1}{c_{nn}^2} = \int_{x_1}^{x_2} \exp\left[\int_{x_1}^{x} 2\, g_{n+1}(\xi)\, d\xi\right] dx,$$

(13.14)

with $c_{nn} = \sqrt{c_{nn}^2} > 0$.

13.3.3   The Eigenfunctions Corresponding to $l < n$

The result given in (13.14) shows that, if the factorization is achieved, the eigenfunction corresponding to $l = n$ is found by solving a first-order equation. It remains to determine the eigenfunctions corresponding to $l = 0, 1, \ldots, n - 1$. For this, left multiplying the first relation in (13.11) by $\mathcal{A}_{l+1}^{-}$, letting $l \leftarrow l + 1$ in the second relation of (13.11), and remembering that $\lambda = L_{n+1}$, one finds

$$\mathcal{A}_{l+1}^{-}\mathcal{A}_{l+1}^{+}\mathcal{A}_{l+1}^{-}w_{nl} = (L_{n+1} - L_{l+1})\,\mathcal{A}_{l+1}^{-}w_{nl},$$

(13.15)

$$\mathcal{A}_{l+1}^{-}\mathcal{A}_{l+1}^{+}w_{\lambda,l+1} = (L_{n+1} - L_{l+1})\,w_{n,l+1}.$$

(13.16)

The above show that both $w_{n,l+1}$ and $\mathcal{A}_{l+1}^{-}w_{nl}$ are eigenfunctions of the operator $\mathcal{A}_{l+1}^{-}\mathcal{A}_{l+1}^{+}$ belonging to the same eigenvalue. As the eigenfunctions are non degenerate, it must be

$$\mathcal{A}_{l+1}^{-}w_{nl} = \text{const} \times w_{n,l+1},$$

(13.17)

where the constant may be determined by imposing the normalization condition. The result shows that, if w_{n0} is known, one may calculate a sequence of eigenfunctions belonging to $\lambda_n = L_{n+1}$ (apart from the normalization constant) by successive applications of first-order operators: $\mathcal{A}_1^{-}w_{n0} = \text{const} \times w_{n1}$, $\mathcal{A}_2^{-}w_{n1} = \text{const} \times w_{n2}, \ldots$. The process stops for $l = n$ because, as shown earlier, $\mathcal{A}_{n+1}^{-}w_{nn} = 0$ is not an eigenfunction anymore, so any further application of the operator beyond $l = n$ provides a sequence of zeros. In a similar manner, left multiplying the second relation in (13.11) by $\mathcal{A}_l^{+}$ and letting $l \leftarrow l - 1$ in the first relation of (13.11), one finds

$$\mathcal{A}_l^{+}\mathcal{A}_l^{-}w_{n,l-1} = (L_{n+1} - L_l)\,w_{n,l-1},$$

(13.18)

$$\mathcal{A}_l^{+}\mathcal{A}_l^{-}\mathcal{A}_l^{+}w_{nl} = (L_{n+1} - L_l)\,\mathcal{A}_l^{+}w_{nl}.$$

(13.19)

From the above one finds that both $w_{n,l-1}$ and $\mathcal{A}_l^{+}w_{nl}$ are eigenfunctions of the operator $\mathcal{A}_l^{+}\mathcal{A}_l^{-}$ belonging to the same eigenvalue, whence

$$\mathcal{A}_l^{+}w_{nl} = \text{const} \times w_{n,l-1}.$$

(13.20)

The result shows that, if w_{nn} is known, one may calculate a sequence of eigenfunctions belonging to $\lambda_n = L_{n+1}$ (apart from the normalization constant) by successive applications of first-order operators: $\mathcal{A}_n^+ w_{nn} = \text{const} \times w_{n,n-1}$, $\mathcal{A}_{n-1}^+ w_{n,n-1} = \text{const} \times w_{n,n-2}, \ldots$. The process stops for $l = 0$ which, by hypothesis, is the minimum of l. The derivation also shows that, since w_{nn} and the operators are real, all the eigenfunctions found using the factorization method are real as well.

13.3.4 Normalization

The results of this section may be summarized as follows: (13.17) shows that the application of the first-order operator $\mathcal{A}_{l+1}^-$ to an eigenfunction of indices n, l provides an eigenfunction of indices $n, l+1$. Similarly, (13.20) shows that the application of the first-order operator $\mathcal{A}_l^+$ to an eigenfunction of indices n, l provides an eigenfunction of indices $n, l - 1$. These results may be described as a process of going up or down along a ladder characterized by an index $n \geq 0$, whose steps are numbered by a second index $l = 0, 1, \ldots, n$. It follows that by applying two suitably-chosen operators one may go up and down (or down and up) one step in the ladder and return to the same eigenfunction apart from a multiplicative constant. This is indeed true, as shown by (13.11), that also indicate that the multiplicative constant to be used at the end of the two steps starting from w_{nl} is $L_{n+1} - L_l$ when the operators' index is l. It follows that the constants in (13.17, 13.20) must be chosen as $\sqrt{L_{n+1} - L_{l+1}}$ and $\sqrt{L_{n+1} - L_l}$, respectively. This provides a method for achieving the normalization of the eigenfunctions, starting from an eigenfunction w_{nn} normalized to unity as in (13.14). For this one defines the auxiliary, mutually adjoint operators

$$\mathcal{B}_{nl}^+ = \frac{\mathcal{A}_l^+}{\sqrt{L_{n+1} - L_l}}, \qquad \mathcal{B}_{nl}^- = \frac{\mathcal{A}_l^-}{\sqrt{L_{n+1} - L_l}}, \qquad (13.21)$$

so that (13.11) become

$$\mathcal{B}_{n,l+1}^+ \mathcal{B}_{n,l+1}^- w_{nl} = w_{nl}, \qquad \mathcal{B}_{nl}^- \mathcal{B}_{nl}^+ w_{nl} = w_{nl}. \qquad (13.22)$$

Thanks to the auxiliary operators the multiplicative constant at the end of the two steps becomes unity. Remembering that the eigenfunctions and operators are real, multiplying both of (13.22) by w_{nl} and integrating yields

$$\int_{x_1}^{x_2} (\mathcal{B}_{n,l+1}^- w_{nl})^2 \, dx = \int_{x_1}^{x_2} (\mathcal{B}_{nl}^+ w_{nl})^2 = \int_{x_1}^{x_2} w_{nl}^2 \, dx, \qquad (13.23)$$

On the other hand, replacing the constant in (13.17), (13.20) with $\sqrt{L_{n+1} - L_{l+1}}$, $\sqrt{L_{n+1} - L_l}$, respectively, one derives

$$\mathcal{B}_{n,l+1}^- w_{nl} = w_{n,l+1}, \qquad \mathcal{B}_{nl}^+ w_{nl} = w_{n,l-1}. \qquad (13.24)$$

Comparing (13.24) with (13.23) shows that, if one of the eigenfunctions of the ladder is normalized to unity, all the others have the same normalization. In particular, if w_{nn} is normalized to unity as in (13.14), then the whole ladder of normalized eigenfunction is found by repeatedly applying the same procedure:

$$B_{nn}^+ w_{nn} = w_{n,n-1}, \quad B_{n,n-1}^+ w_{n,n-1} = w_{n,n-2}, \quad \ldots, \quad B_{n1}^+ w_{n1} = w_{n0}. \quad (13.25)$$

13.4   Schrödinger Equation with a Periodic Coefficient

An important case of (11.28) occurs when the coefficient q is periodic [37], with a period that will be denoted with 2ω. The independent variable (not necessarily a Cartesian one) will be denoted with z:

$$w''(z) + q(z)w(z) = 0, \qquad q(z + 2\omega) = q(z), \qquad (13.26)$$

where primes indicate derivatives. Here the variable z is considered real; the theory, however, can be extended to the case of a complex variable. Let $u(z)$, $v(z)$ be fundamental solutions (Sect. 11.4), with $u(0) = 1$, $u'(0) = 0$, $v(0) = 0$, $v'(0) = 1$. As (13.26) holds for any z, it holds in particular for $z + 2\omega$. From the periodicity of q it follows

$$w''(z + 2\omega) + q(z)w(z + 2\omega) = 0, \qquad (13.27)$$

namely, $w(z + 2\omega)$ is also a solution. Similarly, $u(z + 2\omega)$, $v(z + 2\omega)$ are solutions. As the equation has only two independent solutions it must be

$$u(z + 2\omega) = a_{11} u(z) + a_{12} v(z), \qquad v(z + 2\omega) = a_{21} u(z) + a_{22} v(z), \quad (13.28)$$

with a_{ij} suitable constants. The values of the latter are readily related to those of u, v by letting $z = 0$ and using the initial conditions: $u(2\omega) = a_{11}$, $u'(2\omega) = a_{12}$, $v(2\omega) = a_{21}$, $v'(2\omega) = a_{22}$. As the Wronskian of u, v equals unity it follows $a_{11}a_{22} - a_{12}a_{21} = 1$. One now seeks a constant s such that

$$u(z + 2\omega) = s u(z), \qquad v(z + 2\omega) = s v(z). \qquad (13.29)$$

This is equivalent to diagonalizing (13.28), namely, s must fulfill for any z the following relations:

$$(a_{11} - s) u(z) + a_{12} v(z) = 0, \qquad a_{21} u(z) + (a_{22} - s) v(z) = 0. \qquad (13.30)$$

Equating to zero the determinant of the coefficients in (13.30) yields

$$s = \frac{a_0}{2} \pm \sqrt{\frac{a_0^2}{4} - 1}, \qquad a_0 = a_{11} + a_{22}. \qquad (13.31)$$

If $a_0 = 2$ the two solutions of (13.31) are real and take the common value $s^- = s^+ = 1$. In this case, as shown by (13.29), the functions u, v are periodic with period 2ω.

Similarly, if $a_0 = -2$ the two solutions of (13.31) are real and take the common value $s^- = s^+ = -1$. In this case, as shown by (13.29), the functions u, v are periodic with period 4ω, whereas their moduli $|u|, |v|$ are periodic with period 2ω. As the moduli $|u|, |v|$ do not diverge as further and further periods are added to z, the case $a_0 = \pm 2$ is stable. If $a_0 > 2$ the two solutions of (13.31) are real and range over the intervals $0 < s^- < 1$ and, respectively, $s^+ > 1$. In particular it is $s^- \to 0$ for $a_0 \to \infty$. If $a_0 < -2$ the two solutions of (13.31) are real and range over the intervals $s^+ < -1$ and, respectively, $-1 < s^- < 0$. In particular it is $s^- \to 0$ for $a_0 \to -\infty$. From the relations (13.29) it follows that the moduli of the solutions corresponding to s^+ diverge: in fact one has $u(z + n\, 2\, \omega) = (s^+)^n\, u(z)$ and $v(z + n\, 2\, \omega) = (s^+)^n\, v(z)$, so the case $|a_0| > 2$ is unstable. When $|a_0| < 2$ the two solutions are complex, namely, $s^\pm = \exp(\pm i\,\mu)$ with $\tan^2 \mu = 4/a_0^2 - 1$. As the modulus of the solutions is unity, the case $|a_0| < 2$ is stable.

The discussion about stability may seem incomplete because the possible cases depend on the value of $a_0 = a_{11} + a_{22} = u(2\,\omega) + v'(2\,\omega)$, which depends on the fundamental solutions that are yet to be found. On the other hand, the analysis of stability must eventually reduce to considering only the properties of the coefficient of (13.26). In fact, it can be shown that if $q(z)$ is positive for all values of z and the absolute value of $2\,\omega \int_0^{2\omega} q(z)\,dz$ is not larger than 4, then $|a_0| < 2$ [74]. Multiplying both sides of the first relation in (13.29) by $\exp[-\alpha\,(z+2\,\omega)]$, with α an undetermined constant, yields

$$\tilde{u}(z + 2\,\omega) = s\,\exp(-2\,\omega\alpha)\,\tilde{u}(z), \qquad \tilde{u}(z) = \exp(-\alpha\,z)\,u(z). \qquad (13.32)$$

A similar treatment is applied to $v(z)$, to yield another auxiliary function $\tilde{v}$. Now one exploits the undetermined constant to impose that the auxiliary functions be periodic of period $2\,\omega$: for this one lets $s\,\exp(-2\,\omega\alpha) = 1$, whence

$$\alpha = \frac{\log s}{2\,\omega}. \qquad (13.33)$$

The constant α defined by (13.33) is termed *characteristic exponent* or *Floquet exponent*. Three of the cases listed in the discussion above about stability, namely, $s^\pm = \exp(\pm i\,\mu)$, $s^\pm = 1$, and $s^\pm = -1$ lead now to the single expression $\alpha^\pm = \pm i\,\mu/(2\,\omega)$, with $0 \le \mu \le \pi$, showing that α is purely imaginary. The cases $0 < s^- < 1$ and $s^+ > 1$ lead to real values of α (negative and positive, respectively) and, finally, the cases $s^+ < -1$ and $-1 < s^- < 0$ lead to complex values of α. Considering the stable cases only, one transforms (13.26) by replacing $w(z)$ with, e.g., $\tilde{u}(z)\,\exp[\pm i\,\mu\,z/(2\,\omega)]$ to find

$$\tilde{u}''(z) \pm 2\,i\,\frac{\mu}{2\,\omega}\,\tilde{u}'(z) + \left[q(z) - \frac{\mu^2}{4\,\omega^2}\right]\tilde{u}(z) = 0. \qquad (13.34)$$

The coefficients and the unknown function of (13.34) are periodic functions of period $2\,\omega$. As a consequence it suffices to solve the equation within the single period, say, $0 \le z \le 2\,\omega$. An example is given in Sect. 17.8.4; a different approach leading to the generalization to three dimensions is shown in Sect. 17.6.

13.5   Schrödinger Equation for a Central Force

In the investigation about the properties of atoms it is important to analyze the dynamics of particle subjected to a central force in the case where the motion is limited. The treatment based on the concepts of Classical Mechanics is given in Sects. 3.4 (where the general properties are illustrated), 3.7 (for the two-particle interaction), 3.8 (for a Coulomb field in the repulsive case), and 3.13.6 (for a Coulomb field in the attractive case). To proceed one considers a particle of mass[1] m_0 acted upon by a force deriving from a potential energy of the central type, $V = V(r)$, and expresses the time-independent Schrödinger equation $-\hbar^2/(2\,m_0)\,\nabla^2 w + V(r)\,w = E\,w$ in spherical coordinates r, ϑ, φ (Sect. B.1). Remembering the transformation (B.25) of the ∇^2 operator one obtains

$$\frac{1}{r}\frac{\partial^2(r\,w)}{\partial r^2} + \frac{1}{r^2}\,\hat{\Omega}w + \frac{2\,m_0}{\hbar^2}\,[E - V(r)]\,w = 0, \qquad (13.35)$$

where operator $\hat{\Omega}$ is defined as

$$\hat{\Omega} = \frac{1}{\sin^2\vartheta}\left[\sin\vartheta\,\frac{\partial}{\partial\vartheta}\left(\sin\vartheta\,\frac{\partial}{\partial\vartheta}\right) + \frac{\partial^2}{\partial\varphi^2}\right]. \qquad (13.36)$$

The r coordinate is separated by letting $w = \varrho(r)\,Y(\vartheta,\varphi)$ in (13.35) and dividing both sides by w/r^2:

$$r^2\left[\frac{1}{r\,\varrho}\frac{d^2(r\,\varrho)}{dr^2} + \frac{2\,m_0}{\hbar^2}\,(E - V)\right] = -\frac{1}{Y}\,\hat{\Omega}Y. \qquad (13.37)$$

Each side of (13.37) must equal the same dimensionless constant, say, c, whence the original Schrödinger equation separates into the pair

$$\hat{\Omega}Y = -c\,Y, \quad \left[-\frac{\hbar^2}{2\,m_0}\frac{d^2}{dr^2} + V_e(r)\right]r\,\varrho = E\,r\,\varrho, \quad V_e = V + \frac{c\,\hbar^2}{2\,m_0\,r^2}. \qquad (13.38)$$

The first equation in (13.38), called *angular equation*, does not depend on any parameter specific to the problem in hand. As a consequence, its eigenvalues c and eigenfunctions Y can be calculated once and for all. Being the equation's domain two dimensional, the eigenfunctions Y are expected to depend onto two indices, say, l, m. After the angular equation is solved, inserting each eigenvalue c into the second equation of (13.38), called *radial equation*, provides the eigenvalues and eigenfunctions of the latter. For the radial equation, the solution depends on the form

---

[1] To avoid confusion with the azimuthal quantum number m, the particle's mass is indicated with m_0 in the sections dealing with the angular momentum in the quantum case.

of $V(r)$. It is also worth noting the similarity of V_e with its classical counterpart (3.5), that reads

$$V_e = V + \frac{M^2}{2 m_0 r^2}, \qquad M^2 = \text{const.} \tag{13.39}$$

To tackle the solution of the angular equation $\hat{\Omega} Y = -c\, Y$ one associates an operator $\mathcal{L}_x$, $\mathcal{L}_y$, $\mathcal{L}_z$ to each component of the classical angular momentum $\mathbf{M} = \mathbf{r} \wedge \mathbf{p}$, and another operator $\mathcal{L}^2$ to its square modulus M^2. The procedure, illustrated in Sect. 13.6.1, shows that the three operators $\mathcal{L}_x$, $\mathcal{L}_y$, $\mathcal{L}_z$ do not commute with each other, whereas $\mathcal{L}^2$ commutes with each of them. Also, it is found that $\mathcal{L}^2$ is proportional to $\hat{\Omega}$, specifically, $\mathcal{L}^2 = -\hbar^2\, \hat{\Omega}$. In conclusion, the Schrödinger equation in the case of a central force reads

$$\mathcal{H} w = E\, w, \qquad \mathcal{H} = -\frac{\hbar^2}{2\, m\, r} \frac{\partial^2}{\partial r^2} r + \frac{\mathcal{L}^2}{2\, m_0\, r^2} + V(r). \tag{13.40}$$

As the r coordinate does not appear in $\mathcal{L}^2$, the latter commutes with $\mathcal{H}$; moreover, $\mathcal{L}^2$ does not depend on time. As a consequence, its expectation value is a constant of motion (Sect. 10.7). Similarly, $\mathcal{L}_z$ commutes with $\mathcal{L}^2$ and does not contain r or t, so it commutes with $\mathcal{H}$ as well and its expectation value is also a constant of motion. As $\mathcal{H}$, $\mathcal{L}^2$, and $\mathcal{L}_z$ commute with each other, they have a common set of eigenfunctions.

13.5.1 Angular Part of the Equation

The conservation of the expectation values of $\mathcal{L}^2$ and $\mathcal{L}_z$ is the counterpart of the classical result of the conservation of M^2 and M_z (Sect. 2.8). In contrast, the expectation values of $\mathcal{L}_x$ and $\mathcal{L}_y$ are not constants of motion. To determine the eigenfunctions w of $\mathcal{H}$, $\mathcal{L}^2$ and $\mathcal{L}_z$ it is convenient to solve the eigenvalue equation for $\mathcal{L}_z$ first:

$$\mathcal{L}_z w = L_z\, w, \qquad -\mathrm{i}\, \hbar \frac{\partial w}{\partial \varphi} = L_z\, w, \qquad w = v(r, \vartheta)\, \exp\left(\mathrm{i}\, L_z\, \varphi / \hbar\right), \tag{13.41}$$

with v yet undetermined. For an arbitrary value of L_z, the exponential part of w is a multi-valued function of φ. This is not acceptable because w should not vary when φ is changed by integer multiples of 2π. A single-valued function is achieved by letting $L_z / \hbar = m$, with m an integer. In conclusion, the eigenvalues and eigenfunctions of $\mathcal{L}_z$ are

$$L_z = m\, \hbar, \qquad w = v(r, \vartheta)\, \exp\left(\mathrm{i}\, m\, \varphi\right). \tag{13.42}$$

Combining (13.42) with (13.36) provides

$$\hat{\Omega} w = \frac{\exp\left(\mathrm{i}\, m\, \varphi\right)}{\sin^2 \vartheta} \left[\sin \vartheta \frac{\partial}{\partial \vartheta} \left(\sin \vartheta \frac{\partial}{\partial \vartheta} \right) - m^2 \right] v(r, \vartheta). \tag{13.43}$$

Table 13.1 The lowest-order spherical harmonics

Y_l^m	Form of the function
Y_0^0	$1/\sqrt{4\pi}$
Y_1^{-1}	$\sqrt{3/(8\pi)}\ \sin\vartheta\ \exp(-i\varphi)$
Y_1^0	$-\sqrt{3/(4\pi)}\ \cos\vartheta$
Y_1^1	$-\sqrt{3/(8\pi)}\ \sin\vartheta\ \exp(i\varphi)$

This result shows that in (13.40) the factor $\exp(i\,m\,\varphi)$ cancels out, so that $\hat\Omega$ actually involves the angular coordinate ϑ only. This suggests to seek the function $v(r,\vartheta)$ by separation. Remembering that w was originally separated as $w = \varrho(r)\,Y(\vartheta,\varphi)$ one finds

$$v(r,\vartheta) = \varrho(r)\,P(\vartheta), \qquad Y(\vartheta,\varphi) = P(\vartheta)\exp(i\,m\,\varphi). \tag{13.44}$$

As the separation transforms $\hat\Omega Y = -cY$ into $\hat\Omega P = -cP$, the equation to be solved for a given integer m reduces to

$$\frac{1}{\sin^2\vartheta}\left[\sin\vartheta\,\frac{d}{d\vartheta}\left(\sin\vartheta\,\frac{d}{d\vartheta}\right) - m^2\right]P = -cP. \tag{13.45}$$

From $\mathcal{L}^2 = -\hbar^2\,\hat\Omega$, it follows that the eigenvalue of $\mathcal{L}^2$ is $\lambda = \hbar^2 c$. The eigenvalues c of (13.45) are found by the factorization method described in Sect. 13.3.1; they have the form $c = l(l+1)$, with l a non-negative integer, called *orbital* (or *total*) *angular momentum quantum number*. For a given l, the allowed values of m, called *azimuthal* (or *magnetic*) *quantum number*, are the $2l+1$ integers $-l,\dots,0,\dots,l$. The details of the eigenvalue calculation are given in Sect. 13.6.2.

The factorization method provides also the eigenfunctions of $\mathcal{L}^2$ and $\mathcal{L}_z$, that are called *spherical harmonics* and, as expected, depend on the two indices l, m. The details of the calculation of the eigenfunctions Y_l^m are given in Sect. 13.6.3. The lowest-order spherical harmonics are shown in Table 13.1. As the eigenfunctions fulfill the equations

$$\mathcal{L}^2 Y_l^m = \hbar^2\,l(l+1)\,Y_l^m, \qquad \mathcal{L}_z Y_l^m = \hbar\,m\,Y_l^m, \tag{13.46}$$

the only possible results of a measurement of M^2 are $\hbar^2\,l(l+1)$, with $l = 0,1,2,\dots$ and, for a given l, the only possible results of a measurement of M_z are $\hbar\,m$, with $m = -l,\dots,-1,0,1,\dots,l$. It follows that the only possible results of a measurement of M are $\hbar\sqrt{l(l+1)}$. For any $l > 0$ it is max$(|M_z|) = \hbar l < \hbar\sqrt{l(l+1)} = M$; as a consequence, for $l > 0$ the angular momentum $\mathbf{M}$ lies on a cone centered on the z axis. The half amplitude $\alpha = \arccos[m/\sqrt{l(l+1)}]$ of such a cone is strictly positive, showing that the other two components M_x, M_y can not vanish together when $M \neq 0$. A geometrical construction describing the relation between $\mathbf{M}$ and M_z is given in Fig. 13.2. The groups of states corresponding to the orbital quantum numbers $l = 0,1,2,3$ are denoted with special symbols and names, that originate from spectroscopy [7, Chap. 15] and are listed in Table 13.2. For $l \geq 4$ the symbols continue in alphabetical order ("g", "h", ...), while no special names are used.

Fig. 13.2 Geometrical construction showing the relation between $\mathbf{M}$ and M_z. The $l = 3$ is case considered, whence one finds $m = -3, \ldots, 0, \ldots, 3$ and $\sqrt{l(l+1)} \simeq 3.46$

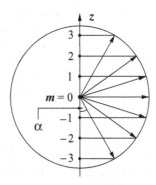

Table 13.2 Symbols and names for the states corresponding to $l = 0, 1, 2, 3$

l	m	Symbol	Name
0	0	s	Sharp
1	$-1, 0, 1$	p	Principal
2	$-2, -1, 0, 1, 2$	d	Diffuse
3	$-3, -2, -1, 0, 1, 2, 3$	f	Fundamental

13.5.2 Radial Part of the Equation in the Coulomb Case

To solve the radial part of the Schrödinger equation (second and third relation in (13.38)) one uses the eigenvalue $c = l(l+1)$ to find

$$\left[-\frac{\hbar^2}{2 m_0} \frac{d^2}{dr^2} + V_e(r) \right] r \varrho(r) = E \, r \, \varrho(r), \qquad V_e = V + \frac{\hbar^2 l(l+1)}{2 m_0 r^2}. \quad (13.47)$$

As anticipated above, the solution of (13.47) depends on the form of $V(r)$. Of particular interest is the Coulomb potential (3.31), that is considered here in the attractive case

$$V(r) = -\frac{Z q^2}{4 \pi \varepsilon_0 r}, \quad (13.48)$$

with ε_0 the vacuum permittivity, $q > 0$ the elementary electric charge, and $Z q$ the charge whence the central force originates. This form of the potential energy is typical of the case of an electron belonging to a hydrogen or hydrogen-like atom. As usual, the arbitrary constant inherent in the definition of the potential energy is such that $\lim_{r \to \infty} V = 0$. As a consequence, the electron is bound if $E < 0$ (in other terms, according to the definition given in Sect. 3.13.6, the classical motion is limited).

The eigenvalues E of (13.47, 13.48) are found by the factorization method described in Sect. 13.3.1; they have the form

$$E = E_n = -\frac{m_0}{2 \hbar^2} \left(\frac{Z q^2}{4 \pi \varepsilon_0} \right)^2 \frac{1}{n^2}, \quad (13.49)$$

where n is an integer, called *principal quantum number*, fulfilling the relation $n \geq l + 1$. The details of the eigenvalue calculation are given in Sect. 13.6.4. As $l \geq 0$, the minimum value of n is 1. For a given n the possible values of the orbital quantum number are $l = 0, 1, \ldots, n - 1$; also, as found earlier, for each l the possible values of the azimuthal quantum number are $m = -l, \ldots, 0, \ldots, l$. It follows that, for a given n the number of different pairs l, m is

$$\sum_{l=0}^{n-1} (2l + 1) = n^2, \tag{13.50}$$

namely, each eigenvalue E_n of the energy corresponds to n^2 possible combinations[2] of the eigenvalues of M and M_z. As for the radial part of the equation, the factorization method provides also the eigenfunctions of (13.47); the details are given in Sect. 13.6.5.

13.6   Complements

13.6.1   Operators Associated to Angular Momentum

Consider the classical angular momentum $\mathbf{M} = \mathbf{r} \wedge \mathbf{p}$ (Sect. 2.6), whose components in rectangular coordinates are given by (2.38), namely,

$$M_x = y\,p_z - z\,p_y, \qquad M_y = z\,p_x - x\,p_z, \qquad M_z = x\,p_y - y\,p_x. \tag{13.51}$$

The operators corresponding to (13.51) are

$$\begin{cases} \mathcal{L}_x = -i\hbar\,(y\,\partial/\partial z - z\,\partial/\partial y) \\ \mathcal{L}_y = -i\hbar\,(z\,\partial/\partial x - x\,\partial/\partial z) \\ \mathcal{L}_z = -i\hbar\,(x\,\partial/\partial y - y\,\partial/\partial x) \end{cases} \tag{13.52}$$

It is easily found that $\mathcal{L}_x$, $\mathcal{L}_y$, $\mathcal{L}_z$ are Hermitean and fulfill the relations

$$\begin{cases} \mathcal{L}_x\,\mathcal{L}_y - \mathcal{L}_y\,\mathcal{L}_x = i\hbar\,\mathcal{L}_z \\ \mathcal{L}_y\,\mathcal{L}_z - \mathcal{L}_z\,\mathcal{L}_y = i\hbar\,\mathcal{L}_x \\ \mathcal{L}_z\,\mathcal{L}_x - \mathcal{L}_x\,\mathcal{L}_z = i\hbar\,\mathcal{L}_y \end{cases} \tag{13.53}$$

namely, $\mathcal{L}_x$, $\mathcal{L}_y$, $\mathcal{L}_z$ do not commute with each other. Left multiplying the third relation in (13.53) by $\mathcal{L}_x$ and the second one by $\mathcal{L}_y$ provides, respectively,

$$\mathcal{L}_x^2\,\mathcal{L}_z = \mathcal{L}_x\,\mathcal{L}_z\,\mathcal{L}_x - i\hbar\,\mathcal{L}_x\,\mathcal{L}_y, \qquad \mathcal{L}_y^2\,\mathcal{L}_z = \mathcal{L}_y\,\mathcal{L}_z\,\mathcal{L}_y + i\hbar\,\mathcal{L}_y\,\mathcal{L}_x. \tag{13.54}$$

---

[2] The actual degree of degeneracy of E_n is $2n^2$, where factor 2 is due to spin (Sect. 15.5.1).

Similarly, right multiplying the third relation in (13.53) by $\mathcal{L}_x$ and the second one by $\mathcal{L}_y$,

$$\mathcal{L}_z \mathcal{L}_x^2 = \mathcal{L}_x \mathcal{L}_z \mathcal{L}_x + i\hbar \mathcal{L}_y \mathcal{L}_x, \qquad \mathcal{L}_z \mathcal{L}_y^2 = \mathcal{L}_y \mathcal{L}_z \mathcal{L}_y - i\hbar \mathcal{L}_x \mathcal{L}_y. \qquad (13.55)$$

The operator associated to $M^2 = M_x^2 + M_y^2 + M_z^2$ is $\mathcal{L}^2 = \mathcal{L}_x^2 + \mathcal{L}_y^2 + \mathcal{L}_z^2$ whence, using (13.54, 13.55),

$$\mathcal{L}^2 \mathcal{L}_z - \mathcal{L}_z \mathcal{L}^2 = (\mathcal{L}_x^2 + \mathcal{L}_y^2)\mathcal{L}_z - \mathcal{L}_z (\mathcal{L}_x^2 + \mathcal{L}_y^2) = 0. \qquad (13.56)$$

Similarly,

$$\mathcal{L}^2 \mathcal{L}_x - \mathcal{L}_x \mathcal{L}^2 = 0, \qquad \mathcal{L}^2 \mathcal{L}_y - \mathcal{L}_y \mathcal{L}^2 = 0. \qquad (13.57)$$

In conclusion, the components $\mathcal{L}_x$, $\mathcal{L}_y$, $\mathcal{L}_z$ do not commute with each other, while the square modulus of the angular momentum commutes with any single component of it. To check whether $\mathcal{L}^2$ or any of the components $\mathcal{L}_x$, $\mathcal{L}_y$, $\mathcal{L}_z$ commute with the Hamiltonian operator of a central force, it is necessary to express all operators in spherical coordinates. To this purpose, using $\mathcal{L}_z$ by way of example, one finds

$$-\frac{\mathcal{L}_z}{i\hbar} = x\frac{\partial}{\partial y} - y\frac{\partial}{\partial x} = r\sin\vartheta\cos\varphi\frac{\partial}{\partial y} - r\sin\vartheta\sin\varphi\frac{\partial}{\partial x}. \qquad (13.58)$$

The partial derivatives $\partial/\partial x$ and $\partial/\partial y$ in terms of the spherical coordinates are extracted from (B.4); in particular, the first one reads $\partial/\partial x = \sin\vartheta\cos\varphi\,\partial/\partial r + (1/r)\cos\vartheta\cos\varphi\,\partial/\partial\vartheta - (1/r)(\sin\varphi/\sin\vartheta)\partial/\partial\varphi$, while the expression of $\partial/\partial y$ is obtained from that of $\partial/\partial x$ by replacing $\cos\varphi$ with $\sin\varphi$ and $\sin\varphi$ with $-\cos\varphi$. When such expressions of the partial derivatives are used within (13.58), several terms cancel out to finally yield the relation

$$\mathcal{L}_z = -i\hbar\frac{\partial}{\partial\varphi} \qquad (13.59)$$

which, consistently with the classical one, $M_z = p_\varphi$ (Sect. 2.8), shows that the operator associated to the z component of the angular momentum is conjugate to the generalized coordinate φ. The quantum relation can thus be derived directly from the classical one by letting $\mathcal{L}_z = \hat{p}_\varphi = -i\hbar\,\partial/\partial\varphi$. As already noted in Sect. 2.8, the remaining components of M_x, M_y are not conjugate momenta. The expression of $\mathcal{L}_x$ in spherical coordinates reads

$$\mathcal{L}_x = i\hbar\left(\sin\varphi\frac{\partial}{\partial\vartheta} + \frac{\cos\vartheta}{\sin\vartheta}\cos\varphi\frac{\partial}{\partial\varphi}\right), \qquad (13.60)$$

while that of $\mathcal{L}_y$ is obtained from (13.60) by replacing $\cos\varphi$ with $\sin\varphi$ and $\sin\varphi$ with $-\cos\varphi$. Combining the above finding, one calculates the expression of $\mathcal{L}^2 = \mathcal{L}_x^2 + \mathcal{L}_y^2 + \mathcal{L}_z^2$, that turns out to be

$$\mathcal{L}^2 = -\frac{\hbar^2}{\sin^2\vartheta}\left[\sin\vartheta\frac{\partial}{\partial\vartheta}\left(\sin\vartheta\frac{\partial}{\partial\vartheta}\right) + \frac{\partial^2}{\partial\varphi^2}\right] = -\hbar^2\hat{\Omega}. \qquad (13.61)$$

13.6.2 Eigenvalues of the Angular Equation

The solution of the angular equation is found by the factorization method described in Sect. 13.3.1. Remembering that $\mathcal{L}_z$ and $\mathcal{L}^2$ commute, the whole eigenfunction $Y = P(\vartheta)\exp(\mathrm{i}\,m\,\varphi)$, introduced in Sect. 13.5 and common to both operators, will be used here. The following hold:

$$\mathcal{L}_z Y = L_z\, Y, \qquad \mathcal{L}^2 Y = \lambda\, Y, \tag{13.62}$$

with $L_z = m\,\hbar$, m an integer. Applying the operator $\mathcal{L}_x \pm \mathrm{i}\,\mathcal{L}_y$ to the first equation in (13.62) yields $\left(\mathcal{L}_x \pm \mathrm{i}\,\mathcal{L}_y\right)\mathcal{L}_z Y = m\,\hbar\,\left(\mathcal{L}_x \pm \mathrm{i}\,\mathcal{L}_y\right)Y$, where the upper (lower) signs hold together. Due to the commutation rules (13.53) the left hand side of the above transforms into

$$\left(\mathcal{L}_z\mathcal{L}_x - \mathrm{i}\,\hbar\,\mathcal{L}_y\right)Y \pm \mathrm{i}\left(\mathcal{L}_z\mathcal{L}_y + \mathrm{i}\,\hbar\,\mathcal{L}_x\right)Y = \mathcal{L}_z\left(\mathcal{L}_x \pm \mathrm{i}\,\mathcal{L}_y\right)Y \mp \hbar\left(\mathcal{L}_x \pm \mathrm{i}\,\mathcal{L}_y\right)Y,$$

whence the first eigenvalue equation in (13.62) becomes

$$\mathcal{L}_z\left(\mathcal{L}_x \pm \mathrm{i}\,\mathcal{L}_y\right)Y = (m \pm 1)\,\hbar\left(\mathcal{L}_x \pm \mathrm{i}\,\mathcal{L}_y\right)Y. \tag{13.63}$$

Iterating the above reasoning shows that, if Y is an eigenfunction of $\mathcal{L}_z$ belonging to the eigenvalue $m\,\hbar$, then $(\mathcal{L}_x + \mathrm{i}\,\mathcal{L}_y)Y$, $(\mathcal{L}_x + \mathrm{i}\,\mathcal{L}_y)^2 Y,\dots$ are also eigenfunctions of $\mathcal{L}_z$ which belong, respectively, to $(m+1)\,\hbar$, $(m+2)\,\hbar,\dots$, and so on. Similarly, $(\mathcal{L}_x - \mathrm{i}\,\mathcal{L}_y)Y$, $(\mathcal{L}_x - \mathrm{i}\,\mathcal{L}_y)^2 Y,\dots$ are also eigenfunctions of $\mathcal{L}_z$ belonging, respectively, to $(m-1)\,\hbar$, $(m-2)\,\hbar,\dots$, and so on. At the same time, due to the commutativity of $\mathcal{L}^2$ with $\mathcal{L}_x$ and $\mathcal{L}_y$, it is

$$\left(\mathcal{L}_x \pm \mathrm{i}\,\mathcal{L}_y\right)\mathcal{L}^2 Y = \mathcal{L}^2\left(\mathcal{L}_x \pm \mathrm{i}\,\mathcal{L}_y\right)Y = \lambda\left(\mathcal{L}_x \pm \mathrm{i}\,\mathcal{L}_y\right)Y, \tag{13.64}$$

showing that $(\mathcal{L}_x \pm \mathrm{i}\,\mathcal{L}_y)Y$ is also an eigenfunction of $\mathcal{L}^2$, belonging to the same eigenvalue as Y. By induction, $(\mathcal{L}_x \pm \mathrm{i}\,\mathcal{L}_y)^2 Y,\dots$ are also eigenfunctions of $\mathcal{L}^2$, belonging to the same eigenvalue as Y. To summarize, if $Y = P(\vartheta)\exp(\mathrm{i}\,m\,\varphi)$ is an eigenfunction common to operators $\mathcal{L}_z$ and $\mathcal{L}^2$, belonging to the eigenvalues $L_z = m\,\hbar$ and λ, respectively, then,

1. $(\mathcal{L}_x + \mathrm{i}\,\mathcal{L}_y)Y$ is another eigenfunction of $\mathcal{L}^2$ still belonging to λ, and is also an eigenfunction of $\mathcal{L}_z$ belonging to $(m+1)\,\hbar$. Similarly, $(\mathcal{L}_x + \mathrm{i}\,\mathcal{L}_y)^2 Y$ is still another eigenfunction of $\mathcal{L}^2$ belonging to λ, and is also an eigenfunction of $\mathcal{L}_z$ belonging to $(m+2)\,\hbar$, and so on.
2. $(\mathcal{L}_x - \mathrm{i}\,\mathcal{L}_y)Y$ is another eigenfunction of $\mathcal{L}^2$ still belonging to λ, and is also an eigenfunction of $\mathcal{L}_z$ belonging to $(m-1)\,\hbar$. Similarly, $(\mathcal{L}_x - \mathrm{i}\,\mathcal{L}_y)^2 Y$ is still another eigenfunction of $\mathcal{L}^2$ belonging to λ, and is also an eigenfunction of $\mathcal{L}_z$ belonging to $(m-2)\,\hbar$, and so on.

By this reasoning, starting from a given pair λ, Y it seems possible to construct as many degenerate eigenfunctions of $\mathcal{L}^2$ as we please. This, however, leads to unbounded eigenvalues of $\mathcal{L}_z$, which are not admissible as shown below. As a consequence, the procedure depicted here can be applied only a finite number of times. To demonstrate that the eigenvalues of $\mathcal{L}_z$ are bounded one starts from the relation $\mathcal{L}^2 = \mathcal{L}_x^2 + \mathcal{L}_y^2 + \mathcal{L}_z^2$ and from a given pair λ, Y. As Y is also an eigenfunction of $\mathcal{L}_z$ belonging to, say, $m\,\hbar$, an application of $\mathcal{L}^2$ to Y followed by a left scalar multiplication by Y^* yields, thanks to (13.62),

$$\lambda - m^2 \hbar^2 = \frac{\langle \mathcal{L}_x Y | \mathcal{L}_x Y \rangle + \langle \mathcal{L}_y Y | \mathcal{L}_y Y \rangle}{\langle Y | Y \rangle} \geq 0, \qquad (13.65)$$

where the hermiticity of $\mathcal{L}_x$, $\mathcal{L}_y$ has been exploited. Inequality (13.65) provides the upper bound for $|m|$. To find the acceptable values of m one defines $m^+ = \max(m)$, and lets Y^+ be an eigenfunction of $\mathcal{L}^2$ and $\mathcal{L}_z$ belonging to λ and $m^+ \hbar$, respectively. From (13.63) one obtains $\mathcal{L}_z (\mathcal{L}_x + \mathrm{i}\, \mathcal{L}_y) Y^+ = (m^+ + 1) \hbar (\mathcal{L}_x + \mathrm{i}\, \mathcal{L}_y) Y^+$ but, as the eigenvalue $(m^+ + 1)\hbar$ is not acceptable, it must be $(\mathcal{L}_x + \mathrm{i}\, \mathcal{L}_y) Y^+ = 0$. Similarly, letting $m^- = \min(m)$, and letting Y^- be an eigenfunction of $\mathcal{L}^2$ and $\mathcal{L}_z$ belonging to λ and $m^- \hbar$, respectively, it must be $(\mathcal{L}_x - \mathrm{i}\, \mathcal{L}_y) Y^- = 0$. Due to the commutation rules it is $(\mathcal{L}_x - \mathrm{i}\, \mathcal{L}_y)(\mathcal{L}_x + \mathrm{i}\, \mathcal{L}_y) = \mathcal{L}_x^2 + \mathcal{L}_y^2 - \hbar\, \mathcal{L}_z$, whence

$$\mathcal{L}^2 = (\mathcal{L}_x - \mathrm{i}\, \mathcal{L}_y)(\mathcal{L}_x + \mathrm{i}\, \mathcal{L}_y) + \hbar\, \mathcal{L}_z + \mathcal{L}_z^2. \qquad (13.66)$$

Application of (13.66) to Y^+ and Y^- yields

$$\begin{cases} \mathcal{L}^2 Y^+ = \left(\mathcal{L}_z^2 + \hbar\, \mathcal{L}_z \right) Y^+ = \hbar^2\, m^+ \left(m^+ + 1 \right) Y^+ \\ \mathcal{L}^2 Y^- = \left(\mathcal{L}_z^2 - \hbar\, \mathcal{L}_z \right) Y^- = \hbar^2\, m^- \left(m^- - 1 \right) Y^- \end{cases} \qquad (13.67)$$

By construction, Y^+ and Y^- belong to the same eigenvalue of $\mathcal{L}^2$; as a consequence it must be $m^+ \left(m^+ + 1 \right) = m^- \left(m^- - 1 \right)$. A possible integer solution of the above is $m^- = m^+ + 1$ which, however, is not acceptable because $m^+ = \max(m)$. The only acceptable solution left is $m^- = -m^+$. In conclusion, letting $l = m^+$ (so that $m^- = -l$) and using (13.67), the eigenvalues of $\mathcal{L}^2$ take the form

$$\lambda = \hbar^2\, l\, (l + 1), \qquad (13.68)$$

with l a non-negative integer. For a given l, the allowed values of m are the $2l + 1$ integers $-l, \dots, 0, \dots, l$.

13.6.3 Eigenfunctions of the Angular Equation

Due to the findings illustrated in Sect. 13.6.2, the eigenfunctions of $\hat{\Omega}$, whose form is $Y(\vartheta, \varphi) = P(\vartheta) \exp(\mathrm{i}\, m\, \varphi)$, depend on the two indices l, m and, for this reason, will be indicated with Y_l^m. In particular, the eigenfunction Y^+ introduced in Sect. 13.6.2, which belongs to $l = \max(m)$ and fulfills the equation $(\mathcal{L}_x + \mathrm{i}\, \mathcal{L}_y) Y = 0$, will be indicated with Y_l^l. Similarly, as P depends on l and may depend on m as well, it will be indicated with P_l^m. The eigenfunction Y_l^l is readily found by solving the first-order equation $(\mathcal{L}_x + \mathrm{i}\,\mathcal{L}_y)Y_l^l = 0$, where operator $\mathcal{L}_x$ is expressed in terms of φ, ϑ through (13.60), and $\mathcal{L}_y$ is obtained from (13.60) by replacing $\cos\varphi$ with $\sin\varphi$ and $\sin\varphi$ with $-\cos\varphi$. After eliminating the factor $\hbar \exp[\mathrm{i}(l+1)\varphi]$ one finds a differential equation for P_l^l, that reads

$$\frac{\mathrm{d}P_l^l}{\mathrm{d}\vartheta} - l\,\frac{\cos\vartheta}{\sin\vartheta}\,P_l^l = 0, \qquad \frac{1}{P_l^l}\frac{\mathrm{d}P_l^l}{\mathrm{d}\vartheta} = \frac{l}{\sin\vartheta}\frac{\mathrm{d}\sin\vartheta}{\mathrm{d}\vartheta}. \tag{13.69}$$

In conclusion it is found, with a an arbitrary constant,

$$P_l^l = a\,(\sin\vartheta)^l, \qquad Y_l^l = a\,\exp(\mathrm{i}\,l\,\varphi)\,(\sin\vartheta)^l. \tag{13.70}$$

Then, remembering the discussion of Sect. 13.6.2, the remaining $2l$ eigenfunctions $Y_l^{l-1}, \ldots, Y_l^0, \ldots, Y_l^{-l}$ are found by successive applications of

$$\mathcal{L}_x - \mathrm{i}\,\mathcal{L}_y = -\hbar\,\exp(-\mathrm{i}\,\varphi)\left(\frac{\partial}{\partial\vartheta} - \mathrm{i}\,\frac{\cos\vartheta}{\sin\vartheta}\frac{\partial}{\partial\varphi}\right), \tag{13.71}$$

with the help of the auxiliary relations $-\mathrm{i}\,\partial Y_l^l/\partial\varphi = l\,Y_l^l$ and

$$\left(\frac{\partial}{\partial\vartheta} + l\,\frac{\cos\vartheta}{\sin\vartheta}\right)Y_l^l = \frac{\partial[(\sin\vartheta)^l\,Y_l^l]/\partial\vartheta}{(\sin\vartheta)^l} = a\,\frac{\exp(\mathrm{i}\,l\,\varphi)}{(\sin\vartheta)^l}\frac{\mathrm{d}}{\mathrm{d}\vartheta}(\sin\vartheta)^{2l}. \tag{13.72}$$

In fact, combining $\mathcal{L}_z\left[(\mathcal{L}_x - \mathrm{i}\,\mathcal{L}_y)\,Y_l^m\right] = (m-1)\hbar\left[(\mathcal{L}_x - \mathrm{i}\,\mathcal{L}_y)\,Y_l^m\right]$ with $\mathcal{L}_z Y_l^m = m\,\hbar\,Y_l^m$ provides the recursive relation $Y_l^{m-1} = (\mathcal{L}_x - \mathrm{i}\,\mathcal{L}_y)\,Y_l^m$. In particular, letting $m = l$, $m = l-1, \ldots$ yields

$$Y_l^{l-1} = (\mathcal{L}_x - \mathrm{i}\,\mathcal{L}_y)\,Y_l^l, \qquad Y_l^{l-2} = (\mathcal{L}_x - \mathrm{i}\,\mathcal{L}_y)\,Y_l^{l-1}, \qquad \ldots \tag{13.73}$$

where Y_l^l is given by (13.70) while

$$(\mathcal{L}_x - \mathrm{i}\,\mathcal{L}_y)\,Y_l^l = Y_l^{l-1} = a\,\frac{\exp[\mathrm{i}(l-1)\varphi]}{(\sin\vartheta)^{l-1}}\frac{-\hbar}{\sin\vartheta}\frac{\mathrm{d}}{\mathrm{d}\vartheta}(\sin\vartheta)^{2l}. \tag{13.74}$$

The denominator $(\sin\vartheta)^l$ in the above has been split into two parts for the sake of convenience. The next functions are found from

$$Y_l^{l-s-1} = (\mathcal{L}_x - \mathrm{i}\,\mathcal{L}_y)\,Y_l^{l-s} = \frac{-\hbar}{\exp(\mathrm{i}\,\varphi)}\left[\frac{\partial}{\partial\vartheta} + (l-s)\frac{\cos\vartheta}{\sin\vartheta}\right]Y_l^{l-s} =$$

$$= \frac{\exp(-\mathrm{i}\,\varphi)}{(\sin\vartheta)^{l-s-1}}\frac{-\hbar}{\sin\vartheta}\frac{\partial}{\partial\vartheta}\left[(\sin\vartheta)^{l-s}\,Y_l^{l-s}\right], \tag{13.75}$$

where the product $(\sin\vartheta)^{l-s}\,Y_l^{l-s}$ is taken from the previously-calculated expression of Y_l^{l-s}. Iterating the procedure yields

$$Y_l^{l-s} = a\,\frac{\exp[\mathrm{i}(l-s)\varphi]}{(\sin\vartheta)^{l-s}}\underbrace{\frac{-\hbar}{\sin\vartheta}\frac{\mathrm{d}}{\mathrm{d}\vartheta}\cdots\frac{-\hbar}{\sin\vartheta}\frac{\mathrm{d}}{\mathrm{d}\vartheta}}_{s\ \text{times}}(\sin\vartheta)^{2l}. \tag{13.76}$$

As Y_l^{l-s} is a solution of the linear, homogeneous equations $\mathcal{L}_z Y = L_z Y$ and $\mathcal{L}^2 Y = \lambda Y$, the constant $a \, \hbar^s$ that builds up in the derivation can be dropped. Letting $m = l - s$ one finds

$$Y_l^m = c_{lm} \frac{\exp(i \, m \, \varphi)}{(\sin \vartheta)^m} \underbrace{\frac{-1}{\sin \vartheta} \frac{d}{d\vartheta} \cdots \frac{-1}{\sin \vartheta} \frac{d}{d\vartheta}}_{l-m \text{ times}} (\sin \vartheta)^{2l}, \tag{13.77}$$

where the coefficient c_{lm} has been added for normalization purposes. One may recast (13.77) in a more compact form by letting $\zeta = \cos \vartheta$, whence $-1 \le \zeta \le 1$ and $d\zeta = -\sin \vartheta \, d\vartheta$. As a consequence,

$$Y_l^m = c_{lm} \frac{\exp(i \, m \, \varphi)}{(1 - \zeta^2)^{m/2}} \frac{d^{l-m}}{d\zeta^{l-m}} (1 - \zeta^2)^l. \tag{13.78}$$

The eigenfunctions Y_l^m are square integrable and mutually orthogonal [78, App. B.10]. To examine some of their properties it is convenient to introduce some special functions; to begin with, the *associate Legendre functions* are defined in the interval $-1 \le \zeta \le 1$ by

$$P_l^m(\zeta) = \frac{(-1)^m}{2^l \, l!} (1 - \zeta^2)^{m/2} \frac{d^{l+m}}{d\zeta^{l+m}} (\zeta^2 - 1)^l, \tag{13.79}$$

with $l = 0, 1, \ldots$ and, for each l, $m = -l, \ldots, -1, 0, 1, \ldots, l$. As $(1 - \zeta^2)^{m/2}$ and $(\zeta^2 - 1)^l$ are even functions of ζ, P_l^m is even (odd) if $l + m$ is even (odd): $P_l^m(-\zeta) = (-1)^{l+m} P_l^m(\zeta)$. Furthermore, it is

$$P_l^m(\zeta) = (-1)^m \frac{(l+m)!}{(l-m)!} P_l^{-m}. \tag{13.80}$$

Replacing m with $-m$ in (13.79) shows that Y_l^m is proportional to P_l^{-m} which, in turn, is proportional to P_l^m due to (13.80). In conclusion, using $\zeta = \cos \vartheta$,

$$Y_l^m(\vartheta, \varphi) = c_{lm} \exp(i \, m \, \varphi) P_l^m(\cos \vartheta), \tag{13.81}$$

with $0 \le \vartheta \le \pi$ and $0 \le \varphi \le 2\pi$. As for the indices it is $l = 0, 1, \ldots$ and $m = -l, \ldots, -1, 0, 1, \ldots, l$. The functions Y_l^m defined by (13.81) are called *spherical harmonics*. Combining the definition of Y_l^m with the properties of P_l^m shown above yields $Y_l^{-m} = (-1)^m (Y_l^m)^*$. Note that Y_l^m and Y_l^{-m} are linearly independent, whereas P_l^m and P_l^{-m} are not. Letting

$$c_{lm} = \left[\frac{(2l+1)(l-m)!}{4\pi(l+m)!} \right]^{1/2}, \tag{13.82}$$

the set made of the spherical harmonics is orthonormal, namely,

$$\int_0^{2\pi} \int_0^{\pi} \left(Y_\lambda^\mu\right)^* Y_l^m \, d\vartheta \, d\varphi = \begin{cases} 1, & \lambda = l \text{ and } \mu = m \\ 0, & \text{otherwise} \end{cases} \tag{13.83}$$

and complete (Sect. 8.4.3), namely,

$$F(\vartheta,\varphi) = \sum_{l=0}^{\infty} \sum_{m=-l}^{l} a_{lm} Y_l^m(\vartheta,\varphi), \qquad a_{lm} = \int_0^{2\pi} \int_0^{\pi} \left(Y_l^m\right)^* F \, \mathrm{d}\vartheta \, \mathrm{d}\varphi, \quad (13.84)$$

where F is a sufficiently regular function of the angles. The inner sum of (13.84),

$$Y_l(\vartheta,\varphi) = \sum_{m=-l}^{l} a_{lm} Y_l^m(\vartheta,\varphi) \qquad (13.85)$$

is also called *general spherical harmonic of order l*, whereas the special case $m = 0$ of the associate Legendre function,

$$P_l^0(\zeta) = \frac{1}{2^l \, l!} \frac{\mathrm{d}^l}{\mathrm{d}\zeta^l} (\zeta^2 - 1)^l, \qquad (13.86)$$

is a polynomial of degree l called *Legendre polynomial*.

13.6.4   Eigenvalues of the Radial Equation—Coulomb Case

The case $E < 0$ of the radial Eq. (13.47) is considered here, corresponding to a limited motion. As a consequence, the eigenvectors are expected to depend on a discrete index, and the eigenfunctions are expected to be square integrable. Calculating the derivative and multiplying both sides of (13.47) by $-2m_0/(\hbar^2 r)$ yields

$$\frac{\mathrm{d}^2 \varrho}{\mathrm{d}r^2} + \frac{2}{r} \frac{\mathrm{d}\varrho}{\mathrm{d}r} - \frac{2m_0}{\hbar^2} V_e \varrho + \frac{2m_0}{\hbar^2} E \varrho = 0. \qquad (13.87)$$

To proceed one scales the independent variable by multiplying both sides of (13.87) by a^2, where a is a length. The term involving V_e becomes

$$-a^2 \frac{2m_0}{\hbar^2} V_e \varrho = \left[\frac{2m_0 Z q^2 a}{4\pi \varepsilon_0 \hbar^2 (r/a)} - \frac{l(l+1)}{(r/a)^2} \right] \varrho, \qquad (13.88)$$

where both fractions in brackets are dimensionless. As a may be chosen arbitrarily, it is convenient to select for it a value that makes the first fraction equal to $2/(r/a)$, namely,

$$a = \frac{4\pi \varepsilon_0 \hbar^2}{m_0 Z q^2}. \qquad (13.89)$$

As a consequence, the term involving E becomes

$$a^2 \frac{2m_0}{\hbar^2} E \varrho = \lambda \varrho, \qquad \lambda = \left(\frac{4\pi \varepsilon_0}{Z q^2} \right)^2 \frac{2\hbar^2}{m_0} E. \qquad (13.90)$$

Adopting the dimensionless variable $x = r/a$ and using the relations $a^2 \, d^2\varrho/dr^2 = d^2\varrho/dx^2$, $(2\,a^2/r)\,d\varrho/dr = (2/x)\,d\rho/dx$ yields the radial equation in scaled form,

$$\frac{d^2\varrho}{dx^2} + \frac{2}{x}\frac{d\varrho}{dx} + \left[\frac{2}{x} - \frac{l(l+1)}{x^2}\right]\varrho + \lambda\varrho = 0. \tag{13.91}$$

The range of the independent variable is $0 \leq x < \infty$, so the equation has a double pole in the origin: it is necessary to select solutions that vanish in the origin in such a way as to absorb the pole (more on this in Sect. 13.6.5). The replacement $\varrho = \sigma/x$ gives (13.91) the simpler form

$$\frac{d^2\sigma}{dx^2} + \left[\frac{2}{x} - \frac{l(l+1)}{x^2}\right]\sigma + \lambda\sigma = 0, \tag{13.92}$$

which is identical to (13.2). The factorization of (13.92) is then accomplished following the scheme shown in Sect. 13.3, and is based upon the function $g_l = l/x - 1/l$; in this case operators (13.10) and parameter (13.9) read, respectively,

$$\mathcal{A}_l^+ = \frac{l}{x} - \frac{1}{l} + \frac{d}{dx}, \qquad \mathcal{A}_l^- = \frac{l}{x} - \frac{1}{l} - \frac{d}{dx}, \qquad L_l = -1/l^2. \tag{13.93}$$

The latter depends only on l and fulfills the relation $L_{n+1} > L_{l+1}, l = 0, 1, \ldots, n-1$. As a consequence, remembering the second relation in (13.90), the eigenvalues[3] $\lambda = \lambda_n = L_{n+1}$ of (13.92) and those of (13.87) are, respectively,

$$\lambda_n = -\frac{1}{(n+1)^2}, \qquad E_n = -\left(\frac{Z\,q^2}{4\,\pi\,\varepsilon_0}\right)^2 \frac{m_0/(2\,\hbar^2)}{(n+1)^2}, \qquad n = 0, 1, \ldots. \tag{13.94}$$

13.6.5 Eigenfunctions of the Radial Equation—Coulomb Case

The eigenfunction corresponding to $l = n$ is found by applying (13.14). As the eigenfunction is indicated here with σ_{nn}, one must solve $\mathcal{A}_{n+1}^-\sigma_{nn} = 0$, namely, using (13.93), $[(n+1)/x - 1/(n+1) - d/dx]\,\sigma_{nn} = 0$, whose solution is

$$\sigma_{nn} = c_{nn}\, x^{n+1}\,\exp\left(-\frac{x}{n+1}\right), \qquad n = 0, 1, \ldots, \tag{13.95}$$

which vanishes both in the origin and at infinity, this fulfilling the requirements stated in Sect. 13.6.4. The eigenfunction (13.95) is also square integrable with, from (C.88, C.90),

$$\frac{1}{c_{nn}^2} = \int_0^\infty x^{2n+2}\,\exp\left(-\frac{2x}{n+1}\right)\,dx = (2n+2)!\left(\frac{n+1}{2}\right)^{2n+3}. \tag{13.96}$$

---

[3] In (13.49) a more convenient notation is used, obtained from (13.94) through the replacements $n + 1 \leftarrow n' \leftarrow n$, with $n' = 1, 2, \ldots$.

Combining the first definition in (13.21) with the third relation in (13.93), the auxiliary operator $\mathcal{B}_{nl}^+$ reads

$$\mathcal{B}_{nl}^+ = \frac{l(n+1)}{\sqrt{(n+1-l)(n+1+l)}} \, \mathcal{A}_l^+. \tag{13.97}$$

Then, from the second of (13.24), the normalized eigenfunctions corresponding to $l < n$ are found recursively from

$$\sigma_{n,n-1} = \mathcal{B}_{nn}^+ \sigma_{nn}, \qquad \sigma_{n,n-2} = \mathcal{B}_{n,n-1}^+ \sigma_{n,n-1}, \qquad \ldots \tag{13.98}$$

The last eigenfunction found by the recursive procedure is $\sigma_{n0} = \mathcal{B}_{n1}^+ \sigma_{n1}$ as expected. In fact, a further iteration would not yield an eigenfunction because $\mathcal{B}_{n0}^+ = 0$.

The eigenfunction of (13.87) corresponding to the lowest total energy $E_{\min} = E(n = 0)$ is found by combining (13.95, 13.96) with $\sigma = \varrho/x$ and $x = r/a$, this yielding $\varrho(r) = (1/2)\exp(-r/a)$. There is only one spherical harmonic compatible with this energy eigenvalue, specifically, $Y_0^0 = 1/\sqrt{4\pi}$ (Table 13.1). Thus, the product $w(E_{\min}) = (c/2)\exp(-r/a)/\sqrt{4\pi}$, with c a normalization constant, is the eigenfunction of the Schrödinger Eq. (13.35) corresponding to the lowest total energy. The normalization constant is necessary because ϱ is obtained from scaling another function σ, originally normalized to unity. Taking the Jacobian determinant $J = r^2 \sin\vartheta$ from (B.3) one finds

$$\frac{1}{c^2} = \frac{1}{16\pi} \int_0^\infty \int_0^\pi \int_0^{2\pi} \exp\left(-\frac{2r}{a}\right) r^2 \sin\vartheta \, dr \, d\vartheta \, d\varphi = \frac{a^3}{16}, \tag{13.99}$$

whence $w(E_{\min}) = \exp(-r/a)/\sqrt{\pi a^3}$.

13.6.6 Transmission Matrix

The one-dimensional, time-independent Schrödinger Eq. (11.28) is solvable analytically in a limited number of cases, some of which have been illustrated in the sections above. When the analytical solution is not known one must resort to approximate methods; an example is given here, with reference to a finite domain $0 \le x \le s$. The latter is tessellated by selecting N points $x_1 < x_2 < \ldots < x_N$, internal to the domain, called *nodes*. The boundaries of the domain are indicated with $0 = x_0 < x_1$ and $s = x_{N+1} > x_N$. The segment bounded by x_i and x_{i+1} is indicated with h_{i+1} and is called *element*. The same symbol indicates the length of the element, $h_{i+1} = x_{i+1} - x_i$. Finally, a subdomain Ω_i, called *cell*, is associated to each node. For the internal nodes $x_1, \ldots, x_N$ the cell is bounded by $x_i - h_i/2$ and $x_i + h_{i+1}/2$. The same symbol is used to indicate also the cell length, $\Omega_i = (h_i + h_{i+1})/2$ (Fig. 13.3). The left boundary x_0 is associated to the cell Ω_0 of length $h_1/2$ placed on the right of x_0, while the right boundary x_{N+1} is associated to the cell Ω_{N+1} of length $h_{N+1}/2$ placed on the left of x_{N+1}.

Fig. 13.3 Illustration of the concepts of node, element, and cell (Sect. 13.6.6)

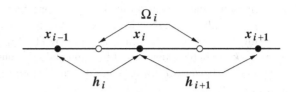

The approximation methods that are applicable to a given tessellation are numerous. The method depicted in this section replaces the coefficient $q(x)$ of (11.28) over each element h_i with an approximating function $q_i(x)$ such that the solution $w_i(x)$ to (11.28) over h_i can be found analytically. The approximating functions q_i may differ from an element to another, this yielding different analytical solutions. Then, the continuity of the analytical solutions and their derivatives is imposed at each node; finally, the same continuity is imposed at the boundaries 0 and s, where the form of the wave function is supposed to be known.

To proceed, consider an internal node $i = 1, 2, \ldots, N$ and the two elements h_i, h_{i+1} adjacent to it. The solutions w_i, w_{i+1} over the two elements is expressed in terms of the fundamental solutions u, v (compare with Sect. 11.4):

$$w_i(x) = a_i^u u_i(x) + a_i^v v_i(x), \qquad w_{i+1}(x) = a_{i+1}^u u_{i+1}(x) + a_{i+1}^v v_{i+1}(x), \quad (13.100)$$

with $a_i^u, a_i^v, a_{i+1}^u, a_{i+1}^v$ undetermined constants. The fundamental solutions fulfill the boundary conditions

$$u_{i+1}(x_i) = 1, \qquad u'_{i+1}(x_i) = 0, \qquad v_{i+1}(x_i) = 0, \qquad v'_{i+1}(x_i) = 1, \quad (13.101)$$

$i = 1, 2, \ldots, N$, where primes indicate derivatives. Imposing the continuity of w, w' at x_i yields

$$a_{i+1}^u = a_i^u u_i(x_i) + a_i^v v_i(x_i), \qquad a_{i+1}^v = a_i^u u'_i(x_i) + a_i^v v'_i(x_i). \quad (13.102)$$

Letting

$$\mathbf{a}_i = \begin{bmatrix} a_i^u \\ a_i^v \end{bmatrix}, \qquad \mathbf{N}_i = \begin{bmatrix} u_i(x_i) & v_i(x_i) \\ u'_i(x_i) & v'_i(x_i) \end{bmatrix}, \quad (13.103)$$

the relations (13.102) take the form

$$\mathbf{a}_{i+1} = \mathbf{N}_i \, \mathbf{a}_i. \quad (13.104)$$

Matrix $\mathbf{N}_i$ is known by construction, and provides the link between the unknown vectors $\mathbf{a}_i$ and $\mathbf{a}_{i+1}$. Vector $\mathbf{a}_i$ belongs to element h_i only, whereas matrix $\mathbf{N}_i$ belongs to element h_i (due to u_i, v_i) and also to node x_i (because u_i, v_i are calculated at x_i). Iterating (13.104) yields

$$\mathbf{a}_{N+1} = \mathbf{N}_I \mathbf{a}_1, \qquad \mathbf{N}_I = \mathbf{N}_N \mathbf{N}_{N-1} \ldots \mathbf{N}_2 \mathbf{N}_1. \quad (13.105)$$

Remembering the discussion in Sect. A.12 one finds $\det \mathbf{N}_i = W = 1$, whence $\det \mathbf{N}_I = \det \mathbf{N}_N \ldots \det \mathbf{N}_1 = 1$. Now it is necessary to link the solution over h_1 with that over $x < 0$, which is given by (11.30). Although the two functions $\exp(\pm i k_L x)$ in the latter are not fundamental solutions, it is convenient to keep the form (11.30) because a_2/a_1 provides the information about the reflection coefficient directly. Letting

$$\mathbf{a}_L = \begin{bmatrix} a_1 \\ a_2 \end{bmatrix}, \qquad \mathbf{N}_L = \begin{bmatrix} 1 & 1 \\ i k_L & -i k_L \end{bmatrix}, \qquad (13.106)$$

the continuity of w and w' at $x = 0$ yields $\mathbf{a}_1 = \mathbf{N}_L \mathbf{a}_L$. Similarly, it is necessary to link the solution over h_{N+1} with that over $x > s$, which is given by (11.31). Again, the two functions $\exp(\pm i k_R x)$ in the latter are not fundamental solutions, however, they are kept here because a_5/a_1 provides the information about the transmission coefficient directly. Letting

$$\mathbf{a}_R = \begin{bmatrix} a_5 \\ a_6 \end{bmatrix}, \qquad \mathbf{N}_R = \begin{bmatrix} \exp(i k_R s) & \exp(-i k_R s) \\ i k_R \exp(i k_R s) & -i k_R \exp(-i k_R s) \end{bmatrix}, \qquad (13.107)$$

the continuity of w and w' at $x = s$ yields $\mathbf{N}_R \mathbf{a}_R = \mathbf{N}_{N+1} \mathbf{a}_{N+1}$, with $\det \mathbf{N}_{N+1} = 1$, $\det \mathbf{N}_L = -2 i k_L$, $\det \mathbf{N}_R = -2 i k_R$. Combining the relations found so far,

$$\mathbf{a}_R = \mathbf{N} \mathbf{a}_L, \qquad \mathbf{N} = \mathbf{N}_R^{-1} \mathbf{N}_{N+1} \mathbf{N}_I \mathbf{N}_L, \qquad (13.108)$$

where

$$\mathbf{N}_R^{-1} = \begin{bmatrix} \exp(-i k_R s)/2 & \exp(-i k_R s)/(2 i k_R) \\ \exp(i k_R s)/2 & i \exp(i k_R s)/(2 k_R) \end{bmatrix}, \qquad \det \mathbf{N}_R^{-1} = -\frac{1}{2 i k_R}. \qquad (13.109)$$

whence $\det \mathbf{N} = \det \mathbf{N}_R^{-1} \det \mathbf{N}_{N+1} \det \mathbf{N}_I \det \mathbf{N}_L = k_L/k_R$. Matrix $\mathbf{N}$ (also called *transmission matrix* in [15]) provides the link between $\mathbf{a}_L$ and $\mathbf{a}_R$. Splitting the first relation of (13.108) into its components gives

$$a_5 = N_{11} a_1 + N_{12} a_2, \qquad a_6 = N_{21} a_1 + N_{22} a_2. \qquad (13.110)$$

If the particle is launched, e.g., from $-\infty$ one lets $a_6 = 0$, whence

$$\frac{a_2}{a_1} = -\frac{N_{21}}{N_{22}}, \qquad \frac{a_5}{a_1} = \frac{\det \mathbf{N}}{N_{22}} = \frac{k_L/k_R}{N_{22}}. \qquad (13.111)$$

Combining (13.111) with (11.35) yields[4] $|N_{21}|^2 + k_L/k_R = |N_{22}|^2$.

---

[4] Within a numerical solution of the Schrödinger equation, the relation $|N_{21}|^2 + k_L/k_R = |N_{22}|^2$ may be exploited as a check for the quality of the approximation.

Fig. 13.4 Example of a
potential energy $V(x)$
replaced with a
piecewise-constant function
V_i (Sect. 13.6.6)

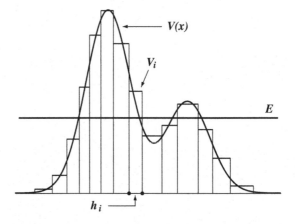

The derivation of the transmission matrix has been carried out here without specifying the form of the fundamental solutions u_i, v_i over the corresponding element
h_i. In the practical cases, to easily find an analytical solution over each element one
approximates the coefficient $q(x)$ of (11.28) with a constant, $q_i = \text{const}$ in h_i; this is
equivalent to replacing the potential energy $V(x)$ with a piecewise-constant function
V_i (Fig. 13.4). Depending on the sign of $q_i = 2m(E - V_i)/\hbar^2$ the possible cases for
u_i, v_i are:

$$\begin{cases} q_i = -\alpha_i^2 < 0 & u_i = \cosh[\alpha_i(x - x_{i-1})] & v_i = \sinh[\alpha_i(x - x_{i-1})]/\alpha_i \\ q_i = k_i^2 > 0 & u_i = \cos[k_i(x - x_{i-1})] & v_i = \sin[k_i(x - x_{i-1})]/k_i \\ q_i = 0 & u_i = 1 & v_i = x - x_{i-1} \end{cases}$$

$$(13.112)$$

with α_i, k_i real. As the potential energy is replaced with a piecewise-constant function,
the accuracy of the approximation is not very high.

Problems

13.1 Letting $Z = 1$ in (13.49) one finds the expression of the energy levels of
the hydrogen atom in a bound state, consistently with that obtained from the
Bohr hypothesis (Sect. 7.4.4). Use the same equation to calculate the minimum
energy that must be given to the electron to extract it from the hydrogen atom
(*ionization energy*).

13.2 With reference to the hydrogen atom, calculate the expectation value of the
radius r corresponding to the eigenfunction $w(E_{\min}) = \exp(-r/a)/\sqrt{\pi a^3}$
found in Sect. 13.6.5.

Chapter 14
Time-Dependent Perturbation Theory

14.1 Introduction

In many physical problems it is necessary to consider the collision of a particle with another particle or system of particles. The treatment based on Classical Mechanics is given in Sects. 3.5, 3.6 with reference to the motion's asymptotic conditions, without considering the form of the interaction, while Sect. 3.8 shows a detailed treatment of the Coulomb interaction. Here the approach based on Quantum Mechanics is shown, dealing with the following problem: a particle in a conservative motion enters at $t = 0$ an interaction with another particle or system of particles; such an interaction has a finite duration t_P, at the end of which the particle is in a conservative motion again. The perturbation produced by the interaction, which is described by a suitable Hamiltonian operator, may change the total energy of the particle; the analysis carried out here, called *time-dependent perturbation theory*, allows one to calculate such an energy change. The other particle or system, with which the particle under consideration interacts, is left unspecified. However, it is implied that the larger system, made of the particle under consideration and the entity with which it interacts, form an isolated system, so that the total energy is conserved: if the particle's energy increases due to the interaction, then such an energy is absorbed from the other entity, or vice versa. As in Classical Mechanics, other dynamical properties of an isolated system are conserved; an example of momentum conservation is given in Sect. 14.8.3.

The discussion is carried out first for the case where the eigenvalues prior and after the perturbation are discrete and non degenerate. Starting from the general solution of the perturbed problem, a first-order approximation is applied, which holds for small perturbations, and the probability per unit time of the transition from a given initial state to another state is found. The analysis is repeated for the degenerate case (still for discrete eigenvalues) and, finally, for the situation where both the initial and final state belong to a continuous set. The last section shows the calculation of the perturbation matrix for a screened Coulomb perturbation. The complements deal with the important problems of a perturbation constant in time and a harmonic perturbation; a short discussion about the Fermi golden rule and the transitions from discrete to continuous levels follows.

© Springer Science+Business Media New York 2015
M. Rudan, *Physics of Semiconductor Devices*,
DOI 10.1007/978-1-4939-1151-6_14

14.2   Discrete Eigenvalues

Let $\mathcal{H}$ be the Hamiltonian operator that describes the dynamics of the particle when the perturbation is absent. Such an operator is assumed to be conservative, namely, $\mathcal{H} = -\hbar^2 \nabla^2/(2\,m) + V(\mathbf{r})$. When the perturbation is present, namely, for $0 \leq t \leq t_P$, the Hamiltonian operator is referred to as $\mathcal{H}'$. The two operators, respectively called *unperturbed Hamiltonian* and *perturbed Hamiltonian*, are Hermitean, so their difference $\delta \mathcal{H} = \mathcal{H}' - \mathcal{H}$, called *perturbation Hamiltonian*, is Hermitean as well. Also, it is assumed that $\mathcal{H}$ and $\mathcal{H}'$ are real, and that $\delta \mathcal{H}$ does not act on time; however, it is $\delta \mathcal{H} = \delta \mathcal{H}(t)$ because the perturbation is present only when $0 \leq t \leq t_P$.

For $t < 0$ and $t > t_P$ the wave function is unperturbed; remembering the concepts introduced in Sect. 9.2 and considering the case of discrete eigenvalues, it reads

$$\psi = \sum_n c_n\, w_n(\mathbf{r})\, \exp\left(-\,\mathrm{i}\, E_n\, t/\hbar\right), \tag{14.1}$$

with w_n the solutions of $\mathcal{H} w_n = E_n\, w_n$. As usual, n stands for a single index or a set of indices. For the sake of simplicity, here it is assumed provisionally that the energy eigenvalues are not degenerate, so that a one-to-one correspondence exists between E_n and w_n; this assumption will be removed later (Sect. 14.5). The wave function is assumed to be square integrable and normalized to unity, whence $|c_n|^2$ is the probability that the particle under consideration is found in the nth state, and $\langle w_s | w_n \rangle = \delta_{sn}$. As noted above, expansion (14.1) holds for both $t < 0$ and $t > t_P$. However, prior to the perturbaton and as a consequence of a measurement, the additional information about the energy state is available; assuming that the outcome of the measurement was the rth state, it follows (Sect. 9.2) that for $t < 0$ it is $c_n = 0$ when $n \neq r$, and[1]

$$|c_r|^2 = 1, \qquad c_r = 1, \qquad \psi = w_r(\mathbf{r})\, \exp\left(-\,\mathrm{i}\, E_r\, t/\hbar\right). \tag{14.2}$$

When $t \rightarrow 0$ the last relation in (14.2) yields $\psi(\mathbf{r}, t = 0) = w_r$ which, by continuity, provides the initial condition for the time-dependent Schrödinger equation to be solved in the interval $0 \leq t \leq t_P$; such an equation reads

$$(\mathcal{H} + \delta \mathcal{H})\, \psi = \mathrm{i}\, \hbar\, \frac{\partial \psi}{\partial t}. \tag{14.3}$$

Thanks to the completeness of the w_ns one expands the wave function as $\psi = \sum_n b_n(t)\, w_n(\mathbf{r})$, where functions b_n are unknown (compare with (9.10)). However, it is convenient to transfer the role of unknowns to a new set of functions $a_n(t) = b_n(t)\, \exp\left(\mathrm{i}\, E_n\, t/\hbar\right)$, so that the expansion reads

$$\psi = \sum_n a_n(t)\, w_n(\mathbf{r})\, \exp\left(-\,\mathrm{i}\, E_n\, t/\hbar\right). \tag{14.4}$$

---

[1] The phase of c_r is irrelevant and is set to zero.

By this token, the initial condition yields $a_n(0) = \delta_{nr}$. Replacing (14.4) in (14.3),

$$\sum_n a_n \exp(-i E_n t/\hbar) (\mathcal{H} w_n + \delta\mathcal{H} w_n) =$$

$$= i\hbar \sum_n w_n \exp(-i E_n t/\hbar) (da_n/dt - i E_n a_n/\hbar), \qquad (14.5)$$

where the first and last terms cancel out due to $\mathcal{H} w_n = E_n w_n$. The next step is a scalar multiplication of the remaining terms by one of the eigenfunctions of the unperturbed problem, say, w_s. The sum $i\hbar \sum_n w_n \exp(-i E_n t/\hbar) da_n/dt$ at the right hand side of (14.5) transforms, due to $\langle w_s | w_n \rangle = \delta_{sn}$, into the single term $i\hbar \exp(-i E_s t/\hbar) da_s/dt$. In conclusion, the time-dependent Schrödinger equation (14.3) becomes a set of infinite, coupled linear equations in the unknowns a_s:

$$\frac{da_s}{dt} = \frac{1}{i\hbar} \sum_n a_n h_{ns} \exp(-i\omega_{ns} t), \qquad a_s(0) = \delta_{sr}, \qquad (14.6)$$

with

$$h_{ns}(t) = \int_\Omega w_s^* \delta\mathcal{H} w_n \, d^3 r, \qquad \omega_{ns} = (E_n - E_s)/\hbar. \qquad (14.7)$$

The coefficients of (14.6) embed the eigenvalues and eigenfunctions of the unperturbed problem. Due to its form, the set of elements $h_{ns}(t)$ is called *perturbation matrix*; remembering that $\delta\mathcal{H}$ is real, the definition (14.7) of the elements shows that $h_{sn} = h_{ns}^*$.

14.3 First-Order Perturbation

The differential equations (14.6) are readily transformed into a set of integral equations by integrating from $t = 0$ to $t \leq t_P$ and using the initial condition:

$$a_s = \delta_{sr} + \frac{1}{i\hbar} \int_0^t \sum_n a_n(t') h_{ns}(t') \exp(-i\omega_{ns} t') dt'. \qquad (14.8)$$

As mentioned above, the solution of the Schrödinger equation for $t > t_P$ is (14.1), with $|c_n|^2$ the probability that a measurement carried out at $t = t_P$ yields the eigenvalue E_n. The coefficients c_n are found by observing that, after the solution of (14.6) or (14.8) is calculated, the time-continuity of ψ and the uniqueness of the expansion yield $c_n = a_n(t_P)$. It follows that the probability that at $t = t_P$ an energy measurement finds the eigenvalue E_s is $|a_s(t_P)|^2$. On the other hand, the energy state prior to the perturbation was assumed to be E_r, and the functions $a_n(t)$ inherit this assumption through the initial condition $a_n(0) = \delta_{nr}$; as a consequence, the quantity

$|a_s(t_P)|^2 = |b_s(t_P)|^2$ can be thought of as the probability that the perturbation brings the particle from the initial state E_r to the final state E_s: for this reason, $P_{rs} = |a_s(t_P)|^2$ is called *transition probability from state r to state s*. Thanks to the normalization condition it is $\sum_s P_{rs} = \int_\Omega |\psi(t_P)|^2 \, d^3r = 1$; the term of equal indices, P_{rr}, is the probability that the perturbation leaves the particle's state unchanged.

The two forms (14.6) and (14.8) are equivalent to each other; however, the second one is better suited for an iterative-solution procedure, that reads

$$a_s^{(k+1)} = \delta_{sr} + \frac{1}{i\hbar} \int_0^t \sum_n a_n^{(k)} h_{ns} \exp\left(-i\omega_{ns} t'\right) dt', \qquad (14.9)$$

where $a_n^{(k)}(t)$ is the kth iterate. The iterations are brought to an end when $||a_s^{(k+1)} - a_s^{(k)}|| < \varepsilon$, where the bars indicate a suitable norm and ε is a small positive constant. To start the procedure it is necessary to choose the iterate of order zero, $a_n^{(0)}(t)$, which is typically done by letting $a_n^{(0)}(t) = a_n^{(0)}(0) = \delta_{nr}$; in other terms, the initial iterate of a_n is selected as a constant equal to the initial condition of a_n. Replacing this value into the integral of (14.9) yields the first-order iterate

$$a_r^{(1)} = 1 + \frac{1}{i\hbar} \int_0^t h_{rr} \, dt', \qquad a_s^{(1)} = \frac{1}{i\hbar} \int_0^t h_{rs} \exp\left(-i\omega_{rs} t'\right) dt', \qquad (14.10)$$

$s \neq r$. If the perturbation is small enough, the first-order iterate is already close to the solution, so that $a_r \simeq a_r^{(1)}$, $a_s \simeq a_s^{(1)}$. This case happens when the norm of the integrals in (14.10) is much smaller than unity; it follows $P_{rr} \simeq 1$, $P_{rs} \ll 1$. The approximate solution thus found is called *first-order solution*, or *first-order perturbation*. Note that, as h_{rr} is real, the iterate $a_r^{(1)}$ is a complex number whose real part equals unity; as a consequence it is $|a_r^{(1)}|^2 > 1$. This non-physical result is due to the approximation.

14.4   Comments

Considering the case where the initial and final states are different, and observing that the entries of the perturbation matrix vanish for $t < 0$ and $t > t_P$, one can calculate $a_s(t_P)$ by replacing the integration limits 0 and t_P with $-\infty$ and $+\infty$, respectively. This shows that $a_s^{(1)}$ is proportional to the Fourier transform (C.16) of h_{rs} evaluated at $\omega_{rs} = (E_r - E_s)/\hbar$,

$$a_s^{(1)} = \frac{1}{i\hbar} \int_{-\infty}^{+\infty} h_{rs} \exp\left(-i\omega_{rs} t'\right) dt' = \frac{\sqrt{2\pi}}{i\hbar} \left. \mathcal{F} h_{rs} \right|_{\omega=\omega_{rs}}. \qquad (14.11)$$

In conclusion, the first-order solution of the time-dependent Schrödinger equation (14.3) yields the following probability of a transition from state r to state s:

$$P_{rs} = \frac{2\pi}{\hbar^2} \left| \mathcal{F} h_{rs} \right|^2. \qquad (14.12)$$

The units of h_{ns} are those of an energy. It follows that the units of $\mathcal{F}h_{rs}$ are those of an action, and P_{rs} is dimensionless, as expected. Some important consequences derive from (14.12):

1. It may happen that for a given perturbation Hamiltonian $\delta\mathcal{H}$ the eigenfunctions w_r, w_s ($s \neq r$) are such that $h_{rs} = 0$. In this case $\delta\mathcal{H}$ is not able to induce the transition from state r to state s: the transition is *forbidden*. Basing on this observation one can determine the pairs of indices for which the transitions are permitted, thus providing the so-called *transition rules* or *selection rules*. For this analysis it is sufficient to consider the symmetry of the integrand in the definition (14.7) of h_{rs}, without the need of calculating the integral itself.

2. By exchanging r and s one finds $h_{sr} = h_{rs}^*$, while ω_{rs} becomes $-\omega_{rs}$. From (14.11) it follows that $\mathcal{F}h_{sr} = (\mathcal{F}h_{rs})^*$, whence $P_{sr} = P_{rs}$: for a given perturbation Hamiltonian $\delta\mathcal{H}$ the probability of the $r \rightarrow s$ and $s \rightarrow r$ transitions are the same.

The transition from an energy state to a different one entails a change in the total energy of the particle under consideration. Such a change is due to the interaction with another particle or system of particles whence $\delta\mathcal{H}$ originates. Examples are given in Sects. 14.8.1 and 14.8.2.

The replacement of the integration limits 0 and t_P with $-\infty$ and $+\infty$, carried out above, has the advantage of making the presence of the Fourier transform clearer; however, remembering that the duration t_P of the perturbation is finite, one observes that the probability P_{rs} is a function of t_P proper. From this, the *probability per unit time* of the transition is defined as

$$\dot{P}_{rs} = \frac{\mathrm{d}P_{rs}}{\mathrm{d}t_P}. \tag{14.13}$$

14.5 Degenerate Energy Levels

In Sect. 14.2 non-degenerate energy levels have been assumed for the sake of simplicity. The case of degenerate levels is considered here, still assuming that the indices are discrete. By way of example, let each energy value E_n correspond to a set $w_{n1}, \ldots, w_{n\gamma}, \ldots$ of linearly-independent, mutually-orthogonal eigenfunctions. An example of this is given by the eigenvalues (13.49) of the Schrödinger equation for a central force of the Coulomb type, whose degree of degeneracy in the spinless case is given by (13.50). Expression (14.1) of the unperturbed wave function, that holds for $t < 0$ and $t > t_P$, becomes in this case

$$\psi = \sum_{n\gamma} c_{n\gamma}\, w_{n\gamma}(\mathbf{r})\, \exp(-\mathrm{i}\, E_n t/\hbar), \tag{14.14}$$

with $w_{n\gamma}$ the solutions of $\mathcal{H}w_{n\gamma} = E_n w_{n\gamma}$. As before, the wave function is assumed to be square integrable and normalized to unity, whence $|c_{n\gamma}|^2$ is the

probability that the particle under consideration is found in the state labeled by
n, γ, and $\langle w_{s\beta}|w_{n\gamma}\rangle = \delta_{sn}\delta_{\beta\gamma}$. Prior to the perturbaton and as a consequence
of measurements, the additional information about the energy state is available,
along with that of the observable associated to index γ, whose measurement is
compatible with that of energy (compare with Sect. 10.4); assuming that the outcome of the measurements was the state labeled r, α, it follows that for $t < 0$
it is $c_{r\alpha} = 1$, $\psi = w_{r\alpha}(\mathbf{r})\exp(-iE_r t/\hbar)$, while all other coefficients vanish.
As a consequence, the initial condition for the time-dependent Schrödinger equa-
tion to be solved in the interval $0 \le t \le t_P$ is $\psi(\mathbf{r}, t = 0) = w_{r\alpha}$. Following
the same reasoning as in Sect. 14.2 shows that in such an interval the expansion
$\psi = \sum_{n\gamma} a_{n\gamma}(t) w_{n\gamma}(\mathbf{r})\exp(-iE_n t/\hbar)$ holds, and the time-dependent Schrödinger
equation transforms into the set of infinite, coupled linear equations

$$\frac{da_{s\beta}}{dt} = \frac{1}{i\hbar}\sum_{n\gamma} a_{n\gamma}\, h_{ns}^{\gamma\beta}\exp(-i\omega_{ns}t), \qquad a_{s\beta}(0) = \delta_{sr}\delta_{\beta\alpha}, \tag{14.15}$$

with

$$h_{ns}^{\gamma\beta}(t) = \int_\Omega w_{s\beta}^* \,\delta\mathcal{H} w_{n\gamma}\, d^3 r, \qquad \omega_{ns} = (E_n - E_s)/\hbar. \tag{14.16}$$

The first-order perturbative solution of (14.15) is obtained following the same path as
in Sect. 14.3. Within this approximation, and considering a final state s, β different
from the initial one, the probability of a $r, \alpha \to s, \beta$ transition induced by the
perturbation is

$$P_{rs}^{\alpha\beta} = \frac{1}{\hbar^2}\left|\int_0^{t_P} h_{rs}^{\alpha\beta}(t)\exp(-i\omega_{ns}t)\,dt\right|^2. \tag{14.17}$$

Thanks to the normalization condition it is $\sum_{s\beta} P_{rs}^{\alpha\beta} = \int_\Omega |\psi(t_P)|^2\, d^3 r = 1$, which
can be expressed as

$$\sum_s P_{rs}^\alpha = 1, \qquad P_{rs}^\alpha = \sum_\beta P_{rs}^{\alpha\beta}. \tag{14.18}$$

This shows that the inner sum P_{rs}^α is the probability that the perturbation induces a
transition from the initial state r, α to any final state whose energy is $E_s \ne E_r$.

14.6   Continuous Energy Levels

When the spectrum is continuous, the wave packet describing a particle in a three-
dimensional space and in a conservative case is given by (9.5), namely

$$\psi(\mathbf{r}, t) = \iiint_{-\infty}^{+\infty} c_\mathbf{k}\, w_\mathbf{k}(\mathbf{r})\exp(-iE_\mathbf{k} t/\hbar)\, d^3 k, \tag{14.19}$$

with E_k, w_k the eigenvalues and eigenfunctions of $\mathcal{H}w_k = E_k w_k$, and $\mathbf{k}$ a three-dimensional vector whose components are continuous. If the wave function is square integrable and normalized to unity, (9.7) holds:

$$\int_\Omega |\psi|^2 \, d^3r = \iiint_{-\infty}^{+\infty} |c_k|^2 \, d^3k = 1. \tag{14.20}$$

Remembering the discussion of Sects. 9.2, 9.6, and 10.2, the product $|\psi(\mathbf{r},t)|^2 \, d^3r$ is the infinitesimal probability that at time t the particle is localized within d^3r around $\mathbf{r}$, and the product $|c_k|^2 \, d^3k$ is the infinitesimal probability that the outcome of an energy measurement belongs to the range of $E(\mathbf{k})$ values whose domain is d^3k.

To proceed one assumes that the unperturbed Hamiltonian operator is that of a free particle, $\mathcal{H} = -(\hbar^2/2\,m)\,\nabla^2$; it follows that the wave function and energy corresponding to a wave vector $\mathbf{k}$ read (Sect. 9.6)

$$w_k(\mathbf{r}) = \frac{1}{(2\pi)^{3/2}} \exp(i\,\mathbf{k} \cdot \mathbf{r}), \qquad E_k = \frac{\hbar^2}{2m}\,(k_1^2 + k_2^2 + k_3^2) = \hbar\,\omega_k. \tag{14.21}$$

Thanks to the completeness of the eigenfunctions (14.21), during the perturbation the wave function is given by

$$\psi(\mathbf{r},t) = \iiint_{-\infty}^{+\infty} a_k(t)\,w_k(\mathbf{r})\,\exp(-i\,E_k\,t/\hbar)\,d^3k. \tag{14.22}$$

Due to (14.20), the units of $|c_k|^2$ and, consequently, of $|a_k|^2$, are those of a volume. The same reasoning as in Sect. 14.2 yields in this case

$$\int \frac{da_k}{dt}\,w_k(\mathbf{r})\,\exp(-i\,E_k\,t/\hbar)\,d^3k = \int \frac{a_k}{i\hbar}\,\delta\mathcal{H}w_k(\mathbf{r})\,\exp(-i\,E_k\,t/\hbar)\,d^3k \tag{14.23}$$

(for the sake of simplicity, the symbol of triple integral over $\mathbf{k}$ or $\mathbf{r}$ is replaced with $\int$ in (14.23) and in the relations below). Considering a state $\mathbf{g}$, a scalar multiplication of (14.23) by the corresponding eigenfunction w_g is carried out; performing the integration over $\mathbf{r}$ first, yields at the right hand side the entry of the perturbation matrix of labels $\mathbf{k}$ and $\mathbf{g}$:

$$h_{kg}(t) = \frac{1}{(2\pi)^3} \int \exp(-i\,\mathbf{g} \cdot \mathbf{r})\,\delta\mathcal{H} \exp(i\,\mathbf{k} \cdot \mathbf{r})\,d^3r, \tag{14.24}$$

where the units of $h_{kg}(t)$ are those of an energy times a volume. At the left hand side of (14.23), still performing the integration over $\mathbf{r}$ first, and using (C.56), provides

$$\int \frac{da_k}{dt}\,\delta(\mathbf{k} - \mathbf{g})\,\exp(-i\,E_k\,t/\hbar)\,d^3k = \frac{da_g}{dt}\,\exp\left(-i\,E_g\,t/\hbar\right) \tag{14.25}$$

which, combined with (14.24) and (14.23), yields

$$\frac{da_g}{dt} = \frac{1}{i\hbar} \int a_k\,h_{kg}\,\exp\left[-i(E_k - E_g)\,t/\hbar\right]\,d^3k, \tag{14.26}$$

the analogue of (14.6) and (14.15). However, a difference with respect to the discrete case exists, because a_n in (14.6) and $a_{n\gamma}$ in (14.15) are dimensionless quantities, whereas a_k in (14.26) is not. As a consequence, when the first-order perturbation method is used, and a_k within the integral of (14.26) is replaced with the initial condition, its expression contains one or more parameters whose values enter the final result. Given these premises, choose for the initial condition, e.g., a Gaussian function centered on some vector $\mathbf{b} \neq \mathbf{g}$,

$$a_k(0) = \pi^{-3/4} \lambda^{3/2} \exp(-\lambda^2 |\mathbf{k} - \mathbf{b}|^2/2), \tag{14.27}$$

with $\lambda > 0$ a length. Inserting (14.27) into (14.22) yields the initial condition for the wave function (compare with (C.82)),

$$\psi(\mathbf{r}, 0) = \pi^{-3/4} \lambda^{-3/2} \exp[-r^2/(2\lambda^2) + i\mathbf{b} \cdot \mathbf{r}]. \tag{14.28}$$

Both (14.27) and (14.28) are square integrable and normalized to unity for any positive λ; when the latter becomes large, $\psi(\mathbf{r}, 0)$ becomes more and more similar to a plane wave, while the peak of $a_k(0)$ around $\mathbf{b}$ becomes narrower and higher. Assuming that the $\mathbf{k}$-dependence of h_{kg} is weaker than that of $a_k(0)$, one replaces $\mathbf{k}$ with $\mathbf{b}$ in h_{kg} and E_k, so that in (14.26) only the integral of $a_k(0)$ is left, which yields $(2\sqrt{\pi}/\lambda)^{3/2}$. Completing the calculation as in Sects. 14.4 and 14.5, and remembering that $\mathbf{g} \neq \mathbf{b}$, provides

$$a_g(t_P) \simeq \frac{(2\sqrt{\pi}/\lambda)^{3/2}}{i\hbar} \int_0^{t_P} h_{bg}(t) \exp\left(-i\omega_{bg} t\right) dt, \qquad \omega_{bg} = \frac{E_b - E_g}{\hbar}. \tag{14.29}$$

The product $dP_{bg} = |a_g(t_P)|^2 d^3g$ is the infinitesimal probability that at the end of the perturbation the outcome of an energy measurement belongs to the range of $E(\mathbf{g})$ values whose domain is d^3g.

Typical applications of (14.26) are encountered in the cases where the perturbation matrix is independent of time, $h_{bg} = h_{bg}^{(0)} = \text{const} \neq 0$. A calculation similar to that of Sect. 14.8.1 yields in this case

$$dP_{bg} = \frac{8\pi^{3/2} |h_{bg}^{(0)}|^2}{\lambda^3 \hbar^2} f(\omega_{bg}) d^3g, \qquad f(\omega_{bg}) = \left[\frac{\sin(\omega_{bg} t_P/2)}{\omega_{bg}/2}\right]^2. \tag{14.30}$$

In place of the domain d^3g one can consider the corresponding range of energy dE_g; for this, one profits by the concept of density of states introduced in Sect. B.5. Here the calculation is simple because the $E = E(\mathbf{g})$ relation is given by (14.21) so that, by the same calculation leading to (B.34), one obtains $d^3g = (1/2) \sin\vartheta \, d\vartheta \, d\varphi (2m/\hbar^2)^{3/2} \sqrt{E_g} \, dE_g$. Considering that the initial state $\mathbf{b}$ and the duration t_P are prescribed, the factor $f(\omega_{bg})$ in (14.30) depends only on E_g, while $h_{bg}^{(0)}$ may depend on the angles ϑ, φ (compare with Prob. 14.1). Integrating (14.30) over the angles and letting

$$H_b^{(0)}(E_g) = \int_0^\pi \int_0^{2\pi} |h_{bg}^{(0)}|^2 \sin\vartheta \, d\vartheta \, d\varphi, \tag{14.31}$$

yields the infinitesimal probability that a perturbation constant in time induces a transition from the initial condition (14.28) to a final state whose energy belongs to the range dE_g:

$$dP_b = \int_0^\pi \int_0^{2\pi} dP_{bg} \sin\vartheta \, d\vartheta \, d\varphi = \left(\frac{2\pi m}{\hbar^2}\right)^{3/2} \frac{4 f(\omega_{bg}) H_b^{(0)}}{\lambda^3 \hbar^2} \sqrt{E_g} \, dE_g.$$

(14.32)

14.7 Screened Coulomb Perturbation

An important case of perturbation is that of a charged particle deflected by another charged particle fixed in the origin. The perturbation Hamiltonian is independent of time and, *in vacuo*, takes the form (3.31) of the Coulomb potential energy.[2] Though, the perturbation matrix (14.24) calculated using (3.31) diverges. Such an outcome is explained by observing that, as the vanishing behavior of the Coulomb potential energy away from the origin is weak, the particle is actually subjected to it at large distances from the origin; as a consequence, using the free particle's eigenfunctions $\exp(i\mathbf{k} \cdot \mathbf{r})/(2\pi)^{3/2}$ as solutions of the unperturbed Schrödinger equation is too strong an approximation. A more appropriate approach adopts for the perturbation Hamiltonian the *screened Coulomb potential energy*

$$\delta\mathcal{H} = \frac{A}{4\pi r} \exp(-q_c r), \qquad r > 0, \qquad A = \frac{\kappa Z e^2}{\varepsilon_0}, \tag{14.33}$$

with $e > 0$ the elementary electric charge, Z a positive integer, ε_0 the vacuum permittivity, $q_c > 0$ the *inverse screening length* and, finally, $\kappa = 1(-1)$ in the repulsive (attractive) case. The asymptotic vanishing of (14.33) is much stronger than that of the pure Coulomb case, and the resulting matrix elements are finite, as shown below. Although the choice of a screened potential energy is not realistic *in vacuo*, an expression like (14.33) is more appropriate than (3.31) when a solid material is considered (Sect. 20.5).

To calculate (14.24) one lets $\mathbf{q} = \mathbf{k} - \mathbf{g}$ and chooses a Cartesian reference such that $\mathbf{q}$ is aligned with the z axis: turning to spherical coordinates (B.1) transforms $d^3 r$ into $r^2 \sin\vartheta \, d\vartheta \, d\varphi \, dr$ and $\mathbf{q} \cdot \mathbf{r}$ into $q r \cos\vartheta$. Letting $\mu = \cos\vartheta$, and observing that the integration over φ yields a 2π factor, one gets

$$h_{kg}^{(0)} = \frac{A/2}{(2\pi)^3} \int_0^\infty \left[\int_{-1}^{+1} \exp(-i q r \mu) d\mu\right] \exp(-q_c r) r \, dr = \frac{A/(2\pi)^3}{q_c^2 + q^2}.$$

(14.34)

---

[2] This case is the quantum analogue of that treated in classical terms in Sect. 3.8.

Fig. 14.1 Form of $f(\omega_{rs})/t_P$, with f given by the second expression in (14.36), for different values of t_P (in arbitrary units)

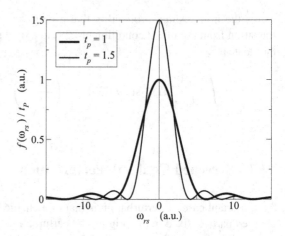

14.8   Complements

14.8.1   Perturbation Constant in Time

The simplest example of the time-dependent perturbation theory of Sect. 14.1 occurs when the matrix elements h_{rs} are constant in time during the perturbation. In this case one lets $h_{rs} = h_{rs}^{(0)} = \text{const} \neq 0$ for $0 \leq t \leq t_P$, and $h_{rs} = 0$ elsewhere. From (14.10) it follows

$$h_{rs}^{(0)} \int_0^{t_P} \exp(-\,\mathrm{i}\,\omega_{rs}\,t)\,\mathrm{d}t = h_{rs}^{(0)} \exp(-\,\mathrm{i}\,\omega_{rs}\,t_P/2)\,\frac{\sin(\omega_{rs}\,t_P/2)}{\omega_{rs}/2}, \qquad (14.35)$$

$$P_{rs} = \frac{|h_{rs}^{(0)}|^2}{\hbar^2}\,f(\omega_{rs}), \qquad f(\omega_{rs}) = \left[\frac{\sin(\omega_{rs}\,t_P/2)}{\omega_{rs}/2}\right]^2. \qquad (14.36)$$

The form of $f(\omega_{rs})/t_P$ is shown in Fig. 14.1 in arbitrary units. The zeros of f nearest to $\omega_{rs} = 0$ are $\omega^+ = 2\pi/t_P$ and $\omega^- = -2\pi/t_P$, which provides the width $\omega^+ - \omega^- = 4\pi/t_P$ of the peak; in turn, the height of the peak is $f(\omega_{rs} = 0) = t_P^2$, this indicating that the area of the peak is proportional to t_P. In fact, from (C.15) one obtains

$$\int_{-\infty}^{+\infty} f(\omega_{rs})\,\mathrm{d}\omega_{rs} = 2\pi\,t_P. \qquad (14.37)$$

The form of $f(\omega_{rs})/t_P$ suggests that, if t_P is sufficiently large, such a ratio may be approximated with a Dirac delta (Sect. C.4), namely, $f(\omega_{rs}) \approx 2\pi\,t_P\,\delta(\omega_{rs})$, where the coefficient $2\pi\,t_P$ is chosen for consistency with (14.37). To this purpose one also notes that, due to the smallness of $\hbar$, the modulus of $\omega_{rs} = (E_r - E_s)/\hbar$ is very large (whence $f(\omega_{rs})$ is very small) unless $E_s = E_r$. Using the approximate form within the probability's definition (14.36) yields

$$P_{rs} \approx 2\pi \frac{|h_{rs}^{(0)}|^2}{\hbar^2} t_P \delta(\omega_{rs}) = 2\pi \frac{|h_{rs}^{(0)}|^2}{\hbar} t_P \delta(E_r - E_s). \tag{14.38}$$

As expected, P_{rs} is invariant when r and s are interchanged (compare with (C.55) and comments therein). Differentiating (14.38) with respect to t_P yields the probability[3] per unit time of the transition from state r to state s:

$$\dot{P}_{rs} \approx 2\pi \frac{|h_{rs}^{(0)}|^2}{\hbar^2} \delta(\omega_{rs}) = 2\pi \frac{|h_{rs}^{(0)}|^2}{\hbar} \delta(E_r - E_s). \tag{14.39}$$

This shows that the particle's energy is approximately conserved when the perturbation lasts for a long time. The result is intuitive, because in the limit $t_P \to \infty$ a constant perturbation is equivalent to a shift in the potential energy, which makes the Hamiltonian operator conservative at all times. On the other hand, the conservation of energy does not imply the conservation of other dynamic quantities like, e.g., momentum (compare with the analysis of the two-particle collision carried out in classical terms in Sect. 3.6 and in quantum terms in Sect. 14.6).

14.8.2 Harmonic Perturbation

Another important example is that of the harmonic perturbation at an angular frequency $\omega_0 > 0$: in this case the matrix elements read $h_{rs} = h_{rs}^{(0)} \cos(\omega_0 t)$ for $0 \le t \le t_P$, $h_{rs}^{(0)} = \text{const} \ne 0$, and $h_{rs} = 0$ elsewhere. From (14.10) it follows

$$\int_0^{t_P} h_{rs} \exp(-i\omega_{rs} t)\, dt = \frac{h_{rs}^{(0)}}{2} \left[\exp\left(\frac{S t_P}{2i}\right) \sigma(S) + \exp\left(\frac{D t_P}{2i}\right) \sigma(D) \right], \tag{14.40}$$

with $S = \omega_{rs} + \omega_0$, $D = \omega_{rs} - \omega_0$, $\sigma(\eta) = \sin(\eta t_P/2)/(\eta/2)$. Comparing the definition of σ with (14.36) shows that $f = \sigma^2$, whence

$$P_{rs} = \frac{|h_{rs}^{(0)}|^2}{4\hbar^2} F(\omega_{rs}), \qquad F = f(S) + f(D) + 2\sigma(S)\sigma(D) \cos(\omega_0 t_P). \tag{14.41}$$

The form of $F(\omega_{rs})/t_P$ is shown in Fig. 14.2 in arbitrary units. The largest peaks correspond to $\omega_{rs} = \omega_0$ and $\omega_{rs} = -\omega_0$; if $t_P \gg 1/\omega_0$ holds, the two peaks are practically separate whence, using as in Sect. 14.8.1 the approximation $f \simeq 2\pi t_P \delta$, one finds

$$P_{rs} \simeq 2\pi \frac{|h_{rs}^{(0)}|^2}{4\hbar^2} [\delta(\omega_{rs} + \omega_0) + \delta(\omega_{rs} - \omega_0)] t_P. \tag{14.42}$$

---

[3] The expressions in terms of energy in (14.38, 14.39) are obtained from $\delta(\omega)\, d\omega = \delta(E)\, dE = \delta(E)\, d\hbar\omega$. Compare with the comments about the dimension of Dirac's δ made in Sect. C.5.

Fig. 14.2 Form of $F(\omega_{rs})/t_P$, with F given by the second expression in (14.41), with $t_P = 1$, $\omega_0 = 5$ (in arbitrary units)

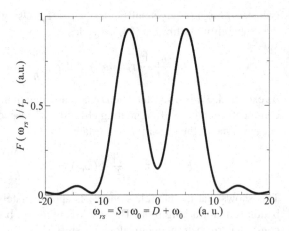

Remembering that $\omega_{rs} = (E_r - E_s)/\hbar$, the transition probability per unit time due to the harmonic perturbation is finally found to be

$$\dot{P}_{rs} = 2\pi \frac{|h_{rs}^{(0)}|^2}{4\hbar} \left[\delta(E_r - E_s + \hbar\omega_0) + \delta(E_r - E_s - \hbar\omega_0)\right]. \qquad (14.43)$$

When r and s are interchanged, the two summands between brackets replace each other. As $\omega_0 \neq 0$, the arguments of δ in (14.42) or (14.43) cannot vanish simultaneously. If the first vanishes it is $E_s = E_r + \hbar\omega_0 > E_r$, namely, the final energy E_s is larger than the initial one: the particle acquires the quantum of energy $\hbar\omega_0$ from the perturbing entity (*absorption*). If the second argument vanishes it is $E_s = E_r - \hbar\omega_0 < E_r$: the particle releases the energy quantum to the perturbing entity (*emission*). For example, the energy can be absorbed from, or emitted towards, an electromagnetic field; in this case the particle interacts with the mode at angular frequency ω_0 by absorbing or emitting a photon (Sect. 12.28). The total energy of the particle and field is conserved in both cases.[4] The same description applies to the interaction with a vibrational field; in this case the particle absorbs or emits a phonon (Sect. 12.5).

----

[4] The spatial dependence of the field is embedded in $h_{rs}^{(0)}$.

14.8.3 Fermi's Golden Rule

Expression (14.32) gives the infinitesimal probability that a perturbation constant in time induces a transition from the initial state **b** to a final state whose energy belongs to the range dE_g; it is an example of a more general expression denoted with *Fermi's Golden Rule*. Remembering from Sect. 14.8.1 that, for a sufficiently large value of t_P, it is $f(\omega_{bg}) \approx 2\pi\, t_P\, \delta(\omega_{bg})$, one finds from (14.32)

$$dP_b \approx \left(\frac{2\pi m}{\hbar^2}\right)^{3/2} \frac{8\pi\, t_P\, \delta(E_b - E_g)\, H_b^{(0)}}{\lambda^3\, \hbar} \sqrt{E_g}\, dE_g. \qquad (14.44)$$

Dividing (14.44) by t_P provides the infinitesimal probability per unit time. Factor $\delta(E_b - E_g)$ entails the conservation of energy; as the unperturbed Hamiltonian operator, upon which the derivation of (14.44) is based, is that of a free particle, the relation $E_g = E_b$ combined with the second relation in (14.21) implies $g^2 = b^2$, namely, the modulus of momentum is also conserved. The result is the same as in the case of the classical treatment of a particle's collision with another particle having a much larger mass (Sect. 3.6).

14.8.4 Transitions from Discrete to Continuous Levels

The transition probability is calculated following a reasoning similar to that of Sect. 14.5 also in the case where the initial state is labeled by a discrete index and the final state belongs to a continuous set. A physical situation where such a transition may occur is that of a particle initially trapped within a well: with reference to Sect. 11.5, the energy levels are discrete if $E < 0$, whereas they are continuous for $E > 0$, namely, the spectrum is mixed (Sect. 8.4). A particle whose initial state belongs to the discrete set may absorb from the perturbation an amount of energy sufficient for reaching the continuous set of states, thus leaving the well. As the final energy belongs to a continuous set, the outcome of the calculation is the expression of an infinitesimal probability like in Sect. 14.6.

Problems

14.1 Using (14.31) and (14.34), find $H_b^{(0)}$ for the screened Coulomb perturbation. Assume for simplicity that the condition $g = b$ holds (Sect. 14.8.3).

Part IV
Systems of Interacting Particles— Quantum Statistics

Chapter 15
Many-Particle Systems

15.1 Introduction

The chapter illustrates the properties of many-particle systems. The quantum-mechanical description of the latter is obtained by solving the time-dependent Schrödinger equation. After commenting the simplifications that occur when the Hamiltonian operator is separable, the important issue of the symmetry or antisymmetry of the wave function is introduced, to the purpose of illustrating the peculiar properties possessed by the systems of identical particles. Then, the concept of spin and the exclusion principle are introduced. After a general discussion, the above concepts are applied to the important case of a conservative system, and further properties related to the separability of the Hamiltonian operator are worked out. The remaining part of the chapter is devoted to the derivation of the equilibrium statistics in the quantum case (Fermi–Dirac and Bose–Einstein statistics). The connection between the microscopic statistical concepts and the macroscopic thermodynamic properties is illustrated in the complements, where two important examples of calculation of the density of states are also given.

15.2 Wave Function of a Many-Particle System

The quantum-mechanical concepts outlined in Parts II and III dealt with wave functions ψ describing a single particle. In such a case, if ψ is normalized to unity, the product $|\psi(\mathbf{r}, t)|^2 \, \mathrm{d}^3 r$ is the infinitesimal probability that at time t the particle's position belongs to the elementary volume $\mathrm{d}^3 r = \mathrm{d}x \, \mathrm{d}y \, \mathrm{d}z$ centered on $\mathbf{r}$; specifically, the x coordinate belongs to $\mathrm{d}x$, and so on. It is now necessary to extend the treatment to the case of many-particle systems. This is readily accomplished by considering, first, a system made of two particles: the wave function ψ describing such a system depends on two sets of coordinates, $\mathbf{r}_1$, $\mathbf{r}_2$, and time. The first set labels one of the

© Springer Science+Business Media New York 2015
M. Rudan, *Physics of Semiconductor Devices,*
DOI 10.1007/978-1-4939-1151-6_15

particles, the second set labels the other particle. Assume that ψ is normalized to unity,

$$\int |\psi(\mathbf{r}_1, \mathbf{r}_2, t)|^2 \, \mathrm{d}^3 r_1 \, \mathrm{d}^3 r_2 = 1, \tag{15.1}$$

where $\int$ is a short-hand notation for a six-fold integral over $\mathrm{d}x_1 \ldots \mathrm{d}z_2$. Then, the product $|\psi(\mathbf{r}_1, \mathbf{r}_2, t)|^2 \, \mathrm{d}^3 r_1 \, \mathrm{d}^3 r_2$ is the infinitesimal probability that, at time t, set $\mathbf{r}_1$ belongs to $\mathrm{d}^3 r_1$ and set $\mathbf{r}_2$ belongs to $\mathrm{d}^3 r_2$. The wave function in (15.1) is the solution of the time-dependent Schrödinger equation

$$\mathrm{i}\hbar \frac{\partial \psi}{\partial t} = \mathcal{H}\psi, \qquad \psi(\mathbf{r}_1, \mathbf{r}_2, 0) = \psi_0(\mathbf{r}_1, \mathbf{r}_2), \tag{15.2}$$

where the initial condition ψ_0 is prescribed. In turn, the Hamiltonian operator in (15.2) is derived, following the procedure illustrated in Sect. 10.2, from the Hamiltonian function that describes the two-particle system in the classical case. Considering by way of example a case where the forces acting on the two particles derive from a potential energy $V = V(\mathbf{r}_1, \mathbf{r}_2, t)$, the Hamiltonian function and the Hamiltonian operator read, respectively,

$$H = \frac{p_1^2}{2\,m_1} + \frac{p_2^2}{2\,m_2} + V, \qquad \mathcal{H} = -\frac{\hbar^2}{2\,m_1}\nabla_1^2 - \frac{\hbar^2}{2\,m_2}\nabla_2^2 + V, \tag{15.3}$$

where m_1, m_2 are the particles' masses, while $p_1^2 = p_{x1}^2 + p_{y1}^2 + p_{z1}^2$, $\nabla_1^2 = \partial^2/\partial x_1^2 + \partial^2/\partial y_1^2 + \partial^2/\partial z_1^2$, and the same for label 2.

It may happen that the Hamiltonian operator is separable with respect to the two sets $\mathbf{r}_1$, $\mathbf{r}_2$, namely, $\mathcal{H} = \mathcal{H}_1 + \mathcal{H}_2$ such that $\mathcal{H}_1$ does not contain any component of $\mathbf{r}_2$ and $\mathcal{H}_2$ does not contain any component[1] of $\mathbf{r}_1$. Also, let $\psi_1 = \psi_1(\mathbf{r}_1, t)$, $\psi_2 = \psi_2(\mathbf{r}_2, t)$ be solutions, respectively, of

$$\mathrm{i}\hbar \frac{\partial \psi_1}{\partial t} = \mathcal{H}_1\psi_1, \qquad \mathrm{i}\hbar \frac{\partial \psi_2}{\partial t} = \mathcal{H}_2\psi_2, \tag{15.4}$$

with the initial conditions $\psi_{10} = \psi_1(\mathbf{r}_1, 0)$, $\psi_{20} = \psi_2(\mathbf{r}_2, 0)$. Letting $\psi = \psi_1\psi_2$ and using (15.4) yields

$$\mathrm{i}\hbar \frac{\partial \psi}{\partial t} - \mathcal{H}\psi = \psi_2\left(\mathrm{i}\hbar \frac{\partial \psi_1}{\partial t} - \mathcal{H}_1\psi_1\right) + \psi_1\left(\mathrm{i}\hbar \frac{\partial \psi_2}{\partial t} - \mathcal{H}_2\psi_2\right) = 0, \tag{15.5}$$

showing that $\psi_1\psi_2$ solves the Schrödinger equation for the particles' system, with $\psi_{10}\psi_{20}$ as initial condition. The concepts introduced in this section are readily extended to the case of larger systems. Letting $N > 2$ be the number of particles, and still assuming that the system's wave function is normalized to unity,

$$\int |\psi(\mathbf{r}_1, \mathbf{r}_2, \ldots, \mathbf{r}_N, t)|^2 \, \mathrm{d}^3 r_1 \, \mathrm{d}^3 r_2 \ldots \mathrm{d}^3 r_N = 1, \tag{15.6}$$

---

[1] By way of example, (15.3) is separable if $V = V_1(\mathbf{r}_1, t) + V_2(\mathbf{r}_2, t)$.

the product $|\psi(\mathbf{r}_1, \mathbf{r}_2, \ldots, \mathbf{r}_N, t)|^2 \, d^3r_1 \, d^3r_2 \ldots d^3r_N$ is the infinitesimal probability that, at time t, set $\mathbf{r}_i$ belongs to d^3r_i, with $i = 1, 2, \ldots, N$. If the Hamiltonian operator is separable, the solution of the time-dependent Schrödinger equation of the system has the form $\psi = \psi_1(\mathbf{r}_1, t) \ldots \psi_N(\mathbf{r}_N, t)$. From (15.6) one also notes that the units of ψ depend on the number of particles involved; specifically, in (15.6) it is $[\psi] = \mathrm{cm}^{-3N/2}$ (compare with the discussion of Sect. 9.7.1).

15.3 Symmetry of Functions and Operators

The Hamiltonian operator and the wave function describing a many-particle system contain sets of coordinates like $\mathbf{r}_1$, $\mathbf{r}_2$ .... It is important to introduce a number of properties related to the exchange of two such sets within the operator or the wave function. The problem is tackled first in a rather abstract way; the applications to specific cases of interest are shown in Sect. 15.6.

Consider a function $f = f(q_1, q_2, \ldots, q_n)$, where q_k represents a group of coordinates.[2] Let $\mathcal{S}_{ij}$ be an operator such that [78, Chap. XIV.3]

$$\mathcal{S}_{ij} f(q_1, \ldots, q_i, \ldots, q_j, \ldots, q_n) = f(q_1, \ldots, q_j, \ldots, q_i, \ldots, q_n), \qquad (15.7)$$

namely, $\mathcal{S}_{ij}$ exchanges the names of the ith and jth group, leaving the rest unchanged. From the definition it follows $\mathcal{S}_{ij}^2 = \mathcal{S}_{ij} \mathcal{S}_{ij} = \mathcal{I}$. Now, let λ be an eigenvalue of $\mathcal{S}_{ij}$, and w an eigenfunction corresponding to it: $\mathcal{S}_{ij} w = \lambda w$. The following relations hold together:

$$\mathcal{S}_{ij}^2 w = w, \qquad \mathcal{S}_{ij}^2 w = \lambda^2 w, \qquad (15.8)$$

the first due to the general property shown before, the second to the definition of λ and w. As a consequence, $\lambda = \pm 1$, namely, $\mathcal{S}_{ij}$ has two eigenvalues. As their modulus equals unity, $\mathcal{S}_{ij}$ is unitary (Sect. 8.6.2), namely, $\mathcal{S}_{ij}^{-1} = \mathcal{S}_{ij}^{\dagger}$.

The properties of the operator's eigenfunctions are found by letting $w^s = \mathcal{S}_{ij} w$, so that w^s is the function that results from exchanging the names of the ith and jth group of coordinates. Depending on the eigenvalue, two cases are possible: the first one is $\lambda = +1$, whence $\mathcal{S}_{ij} w = +1 \times w$ and $\mathcal{S}_{ij} w = w^s$ hold together, so that $w^s = w$; the second case is $\lambda = -1$, whence $\mathcal{S}_{ij} w = -1 \times w$ and $\mathcal{S}_{ij} w = w^s$ hold together, so that $w^s = -w$. A function such that $w^s = w$ is called *symmetric* with respect to indices ij, while a function such that $w^s = -w$ is called *antisymmetric* with respect to indices ij. In conclusion,

- all symmetric functions are eigenfunctions of $\mathcal{S}_{ij}$ belonging to $\lambda = +1$;
- all antisymmetric functions are eigenfunctions of $\mathcal{S}_{ij}$ belonging to $\lambda = -1$.

---

[2] A "group" of coordinates may also consist of a single coordinate.

The set of eigenfunctions of $\mathcal{S}_{ij}$ is complete; in fact, for any function f it is

$$f = \frac{1}{2}\left(f + \mathcal{S}_{ij} f\right) + \frac{1}{2}\left(f - \mathcal{S}_{ij} f\right), \tag{15.9}$$

where the first term at the right hand side is symmetric and the second one is antisymmetric, so that both terms at the right hand side are eigenfunctions of $\mathcal{S}_{ij}$. This shows that any function is expressible as a linear combination of eigenfunctions of $\mathcal{S}_{ij}$.

Only a specific pair ij of coordinate group has been considered so far. On the other hand it may happen that a function is symmetric (antisymmetric) with respect to all pairs of indices; in this case it is called *symmetric* (*antisymmetric*) with no further specification.

The definitions above extend to operators. For instance, an operator $\mathcal{A}$ is symmetric with respect to ij if $\mathcal{A}^s = \mathcal{S}_{ij}\mathcal{A} = \mathcal{A}$; it is symmetric without further specification if $\mathcal{A}^s = \mathcal{A}$ for any pair ij. Given a function f and an operator $\mathcal{A}$, and letting $f^s = \mathcal{S}_{ij} f$, it is for all f,

$$(\mathcal{A}f)^s = \mathcal{S}_{ij}\mathcal{A}f = \mathcal{S}_{ij}\mathcal{A}\mathcal{S}_{ij}^{-1}\mathcal{S}_{ij} f = \mathcal{A}^s f^s. \tag{15.10}$$

If $\mathcal{A}$ is symmetric, replacing $\mathcal{A}^s = \mathcal{A}$ in (15.10) shows that $\mathcal{S}_{ij}\mathcal{A}f = \mathcal{A}\mathcal{S}_{ij} f$ namely, $\mathcal{S}_{ij}$ commutes with all symmetric operators.

The operator whose symmetry properties are of interest is typically the Hamiltonian one. Considering for instance a system of N particles interacting with each other through Coulomb interactions *in vacuo*, one has

$$\mathcal{H} = -\sum_{k=1}^{N} \frac{\hbar^2}{2\,m_k}\,\nabla_k^2 + \frac{1}{2}\sum_{k=1}^{N}\sum_{\substack{s=1 \\ s \neq k}}^{N} \frac{e_k\,e_s}{4\,\pi\,\varepsilon_0\,|\mathbf{r}_k - \mathbf{r}_s|} \tag{15.11}$$

with ε_0 the vacuum permittivity, $\nabla_k^2 = \partial^2/\partial x_k^2 + \partial^2/\partial y_k^2 + \partial^2/\partial z_k^2$, and m_k, e_k the mass and charge of the kth particle, respectively. In general, this operator has no particular symmetry property; however, it is symmetric with respect to the groups of coordinates x_k, y_k, z_k when the particles are identical to each other $\left(m_1 = m_2 = \ldots = m_N,\ e_1 = e_2 = \ldots = e_N\right)$.

15.4 Conservation of Symmetry in Time

Consider a wave function ψ expanded into a complete set of orthonormal functions w_k. Using as in Sect. 15.3 the symbols q_i for the groups of coordinates, one has

$$\psi(q_1,\ldots,q_n,t) = \sum_k a_k(t)\,w_k(q_1,\ldots,q_n). \tag{15.12}$$

The wave function is assumed to be normalized to unity, so that from Parseval theorem (8.41) it follows

$$\langle\psi|\psi\rangle = \sum_k |a_k|^2 = 1 \tag{15.13}$$

at all times. Now assume that ψ is the solution of a Schrödinger equation deriving from a symmetric Hamiltonian operator, and that ψ itself is symmetric at some instant t' with respect to the pair ij. As the functions w_k of (15.12) are linearly independent, the symmetry of ψ entails that of w_k for all k. As a consequence, for the pair ij, w_k is an eigenfunction of the operator S_{ij} corresponding to $\lambda = 1$. Combining (15.13) with the definition (10.13) of the expectation value of the eigenvalues yields $\langle \lambda \rangle = 1$ at $t = t'$. In turn, due to symmetry, $\mathcal{H}$ commutes with S_{ij}; this yields, for the time derivative (10.27) of the average value of the eigenvalues of S_{ij},

$$\frac{d}{dt}\langle \lambda \rangle = -i\hbar \int \psi^* \left(\mathcal{H}S_{ij} - S_{ij}\mathcal{H} \right) \psi \, dq_1 \ldots dq_n = 0, \tag{15.14}$$

namely, $\langle \lambda \rangle$ is conserved in time. The above calculation can be summarized as follows:

- Given a symmetric Hamiltonian $\mathcal{H}$, select a pair of indices ij. Due to commutativity, a complete set of eigenfuctions w_k of $\mathcal{H}$ exists, that belongs also to operator S_{ij}.
- The eigenfunctions w_k of $\mathcal{H}$ can thus be separated into two sets, made of symmetric and antisymmetric functions, respectively.
- Let $\psi(q_1, \ldots, t) = \sum_k a_k(t)w_k(q_1, \ldots)$ be the wave function of the system described by $\mathcal{H}$, and let ψ be symmetric at some instant t'. It follows that the non vanishing coefficients $a_k(t')$ in the expansion of $\psi(q_1, \ldots, q_n, t')$ are only those multiplying the symmetric eigenfunctions.

As $\langle \lambda \rangle = \int \psi^* S_{ij} \psi \, dq_1 \ldots q_n = 1$ at all times, the expansion of ψ is made in terms of the symmetric w_ks at all times. Hence, ψ is always symmetric with respect to the groups of coordinates of indices ij. The above reasoning can be repeated for all pairs of indices for which ψ is symmetric. Note that, in order to repeat the reasoning for different pairs of indices, say, ij and jk, one needs not assume that the corresponding operators S_{ij}, S_{jk} commute with each other (in fact, they typically do not commute). The analysis holds equally for the case where ψ, at time t', is antisymmetric with respect to two indices. In such a case, it remains antisymmetric at all times with respect to them.

15.5 Identical Particles

It is interesting to ascertain whether the mathematical properties related to symmetry or antisymmetry, briefly discussed in Sects. 15.3 and 15.4, correspond to some physical property. This is indeed so, and is especially important when a system of identical particles is considered.

Take for instance a system of two identical particles interacting with each other.[3] In classical terms, the two identical objects that form the system can always be

---

[3] The reasoning outlined here does not apply to systems where the particles are different: they can be distinguished in $|\psi|^2$, e.g., by the mass or electric charge.

Fig. 15.1 Schematic
description of a system made
of two identical particles

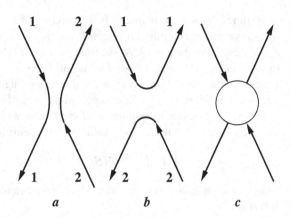

a b c

made distinguishable from each other, without disturbing their motion. The typical
example is that of two identical billiard balls, that are made distinguishable by a
different coloring; although the latter has no influence on the balls' dynamics, it
allows one to distinguish them from each other irrespective of the number of collisions
they undergo. As a consequence, the conjugate variables describing the motion of
each particle (e.g., position and momentum) are exactly known at each instant of
time. By way of example consider the collision of two identical, charged particles
schematically illustrated in Fig. 15.1: it is assumed that the particles are initially far
apart, with equal moduli of the initial velocities; the initial velocity of the particle
labeled 1 is described by the upper-left arrow visible in cases a and b of the figure,
while the initial velocity of the particle labeled 2 is described by the lower-right arrow.
As they come closer, the particles repel each other due to the Coulomb interaction,
and their classical motion is described as in Sects. 3.5–3.8. If the initial velocities
were exactly aligned (this case is not shown in the figure), the two particles would
bounce back along the same direction; if, however, the initial velocity of particle
1 were slightly misaligned to the left, the collision would yield case a, whereas it
would yield case b for a right misalignment. Even if the misalignment is made as
small as we please, either case a or b occurs, and the two possible outcomes are
distinguishable from each other.

In the quantum-mechanical description, instead, it is not possible to track each
particle separately, because the dynamical information about the system derives from
the wave function; when the particles come closer, the norm of the wave function
in (15.1) is significantly different from zero in a finite region of space, which is
schematically indicated by the circle of case c in Fig. 15.1. Due to the Heisenberg
principle (10.22), it is impossible to determine the position and momentum of each
particle at the same time with arbitrary precision. It follows that, for identical particles, the collision is described as an event where two particles enter the circle and
two particles eventually leave it, without the possibility of distinguishing between
cases a and b: the two cases must in fact be counted as one, and the wave function
describing the system must be consistent with it. This requires $|\psi|^2$ be symmetric

with respect to the groups $\mathbf{r}_1, \mathbf{r}_2$:

$$|\psi^s|^2 = |\mathcal{S}_{12}\psi|^2 = |\psi|^2, \tag{15.15}$$

which implies $\psi^s = \exp(i\alpha)\,\psi$ with α a real constant. On the other hand, remembering the first equation in (15.8),

$$\mathcal{S}_{12}^2\psi = \exp(2\,i\alpha)\,\psi, \qquad \mathcal{S}_{12}^2\psi = \psi, \tag{15.16}$$

whence $\exp(j\alpha) = \pm 1$. In conclusion, when the two particles of the system are indistinguishable from each other, the system's wave function is either symmetric or antisymmetric. One may argue that, when the system is made of more than two identical particles, its wave function could be symmetric with respect to some pairs of indices and antisymmetrical with respect to other pairs. However this is not possible, as the simple case of three identical particles shows [69, Sect. 26]. Assume that ψ is symmetric with respect to $\mathbf{r}_1, \mathbf{r}_2$ and antisymmetric with respect to $\mathbf{r}_1, \mathbf{r}_3$; it follows

$$\psi\left(\mathbf{r}_1, \mathbf{r}_2, \mathbf{r}_3, t\right) = \psi\left(\mathbf{r}_2, \mathbf{r}_1, \mathbf{r}_3, t\right) = -\psi\left(\mathbf{r}_2, \mathbf{r}_3, \mathbf{r}_1, t\right) = -\psi\left(\mathbf{r}_1, \mathbf{r}_3, \mathbf{r}_2, t\right) = \tag{15.17}$$

$$= \psi\left(\mathbf{r}_3, \mathbf{r}_1, \mathbf{r}_2, t\right) = \psi\left(\mathbf{r}_3, \mathbf{r}_2, \mathbf{r}_1, t\right) = -\psi\left(\mathbf{r}_1, \mathbf{r}_2, \mathbf{r}_3, t\right), \tag{15.18}$$

that is, a contradiction. In other terms, ψ can either be symmetric with respect to all the identical particles of the system, or antisymmetric with respect to all of them. This result has a far-reaching consequence, namely, the class of all particles is made of two subclasses: the first one collects the types of particles whose systems are described by symmetric wave functions; they are called *Bose particles* or *bosons*. The second subclass collects the types of particles whose systems are described by antisymmetric wave functions; they are called *Fermi particles* or *fermions*. This applies to all known particles, including composite ones (e.g., atoms) and particles describing collective motions (e.g., phonons, photons).[4]

15.5.1 Spin

The properties discussed so far in this section bring about an interesting question: consider a single particle, e.g., an electron. Although electrons are fermions,[5] here the concept of symmetry or antisymmetry does not apply because the wave function of a

---

[4] The names "bosons", "fermions" of the two subclasses have this origin: when a system of identical particles is in thermodynamic equilibrium, the particles' energy follows a statistical distribution whose expression is named after Bose and Einstein (Sect. 15.8.2) and, respectively, Fermi and Dirac (Sect. 15.8.1).

[5] It must be noted, however, that in condensed-matter physics two electrons or other fermions may bind together at low temperatures to form a so-called Cooper pair, which turns out to have an integer spin, namely, it is a composite boson [20].

single electron contains only one group of coordinates. Then add a second electron, so that a system of two identical particles is formed: how do these electrons "know" that, when paired, the wave function of their system must be antisymmetric? It is reasonable to assume that each particle in itself must possess a property that makes it to behave like a fermion or a boson within a system of particles identical to it. Such a property, called *spin*, does in fact exist; as its existence can be proven only within the frame of the relativistic quantum theory [31], [32], [99, Sect. 15], which is beyond the scope if this book, only a brief illustration of spin's properties of interest will be given.

In contrast with the other dynamic quantities considered so far, there is no classical counterpart of spin. Therefore, the latter can not be derived from the expression of a dynamic variable by replacing conjugate coordinates with suitable operators. It can be shown that the eigenvalues of spin are derived in a manner similar to that of angular momentum: this leads, like in Sect. 13.5.1, to determining the square modulus of spin, σ^2, and its component along one of the coordinate axes, say, σ_z. Their values are given by expressions similar to (13.46), specifically,

$$S^2 = \hbar^2 s (s+1), \qquad S_z = \hbar s_z. \tag{15.19}$$

The important difference with (13.46) is that s, instead of being a non-negative integer, is a non-negative half integer: $s = 0, \frac{1}{2}, 1, \frac{3}{2}, 2, \ldots$; in turn, s_z can take the $2s + 1$ values $-s, -s + 1, \ldots, s - 1, s$.

The introduction of spin must be accounted for in the expression of the wave function: the latter, in the case of a single particle, must be indicated with $\psi(\mathbf{r}, s_z, t)$, and its normalization to unity, if existing, is expressed by

$$\sum_{s_z} \int_\Omega |\psi(\mathbf{r}, s_z, t)|^2 \, d^3 r = 1. \tag{15.20}$$

If (15.20) holds, the product $|\psi(\mathbf{r}, s_z, t)|^2 \, d^3 r$ is the probability that at time t the particle is in the elementary volume $d^3 r$ centered on $\mathbf{r}$, and the component of its spin along the z axis is $S_z = \hbar s_z$.

The connection between spin and boson-like or fermion-like behavior is the following: the quantum number s is integer for bosons, half integer for fermions [81]. It is then meaningful to use the terms "boson" or "fermion" for an individual particle. All known fermions have $s = 1/2$, whence $2s + 1 = 2$. It follows that for fermions the z-component of spin has two possible values, $\hbar/2$ (*spin up*) and $-\hbar/2$ (*spin down*). As anticipated above, electrons are fermions. Photons are bosons with $s = 1$.

The similarity between the expressions of the quantum numbers for spin and those of the angular momentum (Eqs. (13.46) cited above) is the origin of the qualitative visualization of spin in classical terms: spin is described as an intrinsic angular momentum of the particle, as if the particle were a sphere spinning on its axis.

15.6 Pauli Exclusion Principle

Consider a system of identical particles, so that its Hamiltonian operator $\mathcal{H}$ is symmetric with respect to each pair of particle labels ij. Its wave function, in turn, is either symmetric or antisymmetric depending on the nature of the particles forming the system. It may happen that, when solving the Schrödinger equation of the system, a solution is found, say φ, that does not possess the necessary symmetry properties. One can then exploit a relation like (15.9) to construct from φ another solution which is either symmetric or antisymmetric. For this, one must remember that φ depends on the groups of coordinates $(\mathbf{r}_1, s_{z1})$, $(\mathbf{r}_2, s_{z2})$, ... which, for the sake of conciseness, will be indicated with the symbol $\mathbf{q}_i = (\mathbf{r}_i, s_{zi})$. Remembering from Sect. 15.3 that the Hamiltonian operator, due to its symmetries, commutes with any operator $\mathcal{S}_{ij}$, one finds

$$\mathcal{S}_{ij} \left(\mathcal{H} - i\hbar \frac{\partial}{\partial t} \right) \varphi = \left(\mathcal{H} - i\hbar \frac{\partial}{\partial t} \right) \varphi^s, \tag{15.21}$$

where $\mathcal{S}_{ij}$ exchanges $\mathbf{q}_i$ with $\mathbf{q}_j$. The parenthesis on the left hand side of (15.21) is zero because φ solves the Schrödinger equation; it follows that φ^s is also a solution. Due to the linearity of the Schrödinger equation, the two functions

$$\varphi + \varphi^s, \qquad \varphi - \varphi^s, \tag{15.22}$$

are solutions of the Schrödinger equation, which are also symmetric and, respectively, antisymmetric with respect to the pair ij. The procedure is easily generalized to obtain a wave function that is symmetric or antisymmetric with respect to all pairs of indices. To this purpose, one considers the $N!$ permutations of the system's N particles; let ν be an index representing the order of each permutation with respect to the fundamental one ($1 \leq \nu \leq N!$), and $\mathcal{S}_\nu$ the operator that achieves the νth permutation of the particles' coordinates within φ. The functions

$$\psi = a \sum_\nu \mathcal{S}_\nu \varphi, \qquad \psi = b \sum_\nu (-1)^\nu \mathcal{S}_\nu \varphi \tag{15.23}$$

are solutions of the Schrödinger equation, which are also symmetric and, respectively, antisymmetric with respect to all particles' permutations. Symbols a and b denote two constants, that can be used to normalize ψ if φ is normalizable. The above constructions can be worked out at any instant, as the symmetry or antisymmetry of the wave function is conserved in time (Sect. 15.4). The second relation in (15.23) lends itself to an interesting derivation. Considering for simplicity the case $N = 2$ one finds

$$\psi = b \left[\varphi(\mathbf{r}_1, s_{z1}, \mathbf{r}_2, s_{z2}, t) - \varphi(\mathbf{r}_2, s_{z2}, \mathbf{r}_1, s_{z2}, t) \right], \tag{15.24}$$

If it were $\mathbf{r}_2 = \mathbf{r}_1$, $s_{z2} = s_{z1}$, the wave function (15.24) would vanish, which is not acceptable. The same unphysical results is found by letting $\mathbf{r}_j = \mathbf{r}_i$, $s_{zj} = s_{zi}$ in the second expression in (15.23). The conclusion is that in a system of identical fermions, two (or more) particles with the same spin can not occupy the same position; this finding derives solely from the antisymmetry of the wave function for a system of identical fermions, and is called *Pauli principle* or *exclusion principle*.[6] As shown in Sect. 15.7 it can be restated in different forms depending on the system under consideration. No similar restriction applies to system of identical bosons, as the form of the first relation in (15.23) shows.

15.7 Conservative Systems of Particles

An important example of a system of N interacting particles occurs when the forces are conservative. To begin, the general case of non-identical particles is considered. The Hamiltonian function and the corresponding Hamiltonian operator read, respectively:

$$H = \sum_{i=1}^{N} \frac{p_i^2}{2m_i} + V, \qquad \mathcal{H} = -\sum_{i=1}^{N} \frac{\hbar^2}{2m_i} \nabla_i^2 + V, \qquad (15.25)$$

where the symbols are the same as in (15.3). Here the potential energy depends only on the spatial coordinates, $V = V(\mathbf{r}_1, \ldots, \mathbf{r}_N)$. If the system is in a state of definite and constant energy E_S, its wave function reads

$$\varphi = W \exp(-i\, E_S t/\hbar), \qquad W = W(\mathbf{q}_1, \ldots \mathbf{q}_N), \qquad (15.26)$$

where E_S is an eigenvalue of $\mathcal{H}W = E\, W$. Extending to this case the definition of Sect. 15.2, the system is separable if $V = \sum_i V_i(\mathbf{r}_i)$, which gives the Hamiltonian operator the form

$$\mathcal{H} = \sum_{i=1}^{N} \mathcal{H}_i, \qquad \mathcal{H}_i = -\frac{\hbar^2}{2m_i} \nabla_i^2 + V_i(\mathbf{r}_i). \qquad (15.27)$$

Assuming that the eigenvalues are discrete, the ith Hamiltonian yields the single-particle equations

$$\mathcal{H}_i w_{n(i)} = E_{n(i)} w_{n(i)}, \qquad (15.28)$$

where index $n(i)$ denotes the nth eigenvalue of the ith particle. From the general properties of operators (Sect. 10.3) it follows that each eigenfunction of the whole system is the product of eigenfunctions like $w_{n(i)}$,

$$W = w_{n(1)}(\mathbf{q}_1)\, w_{n(2)}(\mathbf{q}_2)\, \ldots\, w_{n(N)}(\mathbf{q}_N), \qquad (15.29)$$

---

[6] Like the Heisenberg principle illustrated in Sect. 10.6, that of Pauli was originally deduced from heuristic arguments. The analysis of this section shows in fact that it is a theorem rather than a principle.

and the eigenvalue of $\mathcal{H}$ is the sum of eigenvalues like $E_{n(i)}$:

$$E_S = E_{n(1)} + E_{n(2)} + \ldots + E_{n(N)}. \tag{15.30}$$

If the particles are identical, $m_1 = m_2 = \ldots = m_N$, then the single-particle Hamiltonian operators $\mathcal{H}_i$ become identical to each other; as a consequence, each eigenvalue Eq. (15.28) produces the same set of eigenvalues and eigenfunctions. It follows that all $N!$ permutations of the indices $1, 2, \ldots, N$ in (15.30) leave the total energy unchanged. On the other hand, as for any pair of groups $\mathbf{q}_r, \mathbf{q}_s$ it is

$$w_{n(r)}(\mathbf{q}_r)\, w_{n(s)}(\mathbf{q}_s) \neq w_{n(s)}(\mathbf{q}_r)\, w_{n(r)}(\mathbf{q}_s), \tag{15.31}$$

the total eigenfuction is changed by a permutation of the coordinate indices. Thus, to E_S there correspond $N!$ eigenfunctions w, namely, the eigenvalues of $\mathcal{H}W = E\,W$ for a system of identical particles are $N!$-fold degenerate.

As noted in Sect. 15.6, the solution (15.26) of the system's Schrödinger equation is not necessarily symmetric or antisymmetric. A solution with the correct symmetry property is found from (15.23) and has the form

$$\psi = a \, \exp(-\mathrm{i}\, E_S\, t/\hbar) \sum_\nu S_\nu w_{n(1)}(\mathbf{q}_1) \ldots w_{n(N)}(\mathbf{q}_N) \tag{15.32}$$

in the symmetric case, and

$$\psi = b \, \exp(-\mathrm{i}\, E_S\, t/\hbar) \sum_\nu (-1)^\nu S_\nu w_{n(1)}(\mathbf{q}_1) \ldots w_{n(N)}(\mathbf{q}_N) \tag{15.33}$$

in the antisymmetric one. It is worth specifying that S_ν acts on the coordinate groups $\mathbf{q}_1, \ldots, \mathbf{q}_N$, not on the indices $n(1), \ldots, n(N)$.

When the wave function has the form (15.32) or (15.33), and the eigenfunctions $w_{n(1)}, \ldots, w_{n(N)}$ are normalized to unity, the constants a and b are readily found. Considering the symmetric case with $N = 2$, the wave function reads

$$\psi = a \, \exp(-\mathrm{i}\, E_S\, t/\hbar) \left[w_{n(1)}(\mathbf{q}_1)\, w_{n(2)}(\mathbf{q}_2) + w_{n(2)}(\mathbf{q}_1)\, w_{n(1)}(\mathbf{q}_2) \right], \tag{15.34}$$

where it is assumed that the single-particle eigenfunctions are normalized to unity:

$$\sum_{s_{z1}} \int |w_{n(i)}(\mathbf{r}_1, s_{z1})|^2 \, \mathrm{d}^3 r_1 = 1, \qquad \sum_{s_{z2}} \int |w_{n(i)}(\mathbf{r}_2, s_{z2})|^2 \, \mathrm{d}^3 r_2 = 1. \tag{15.35}$$

In (15.35) it is $i = 1, 2$, and the indication of the domain of $\mathbf{r}_1, \mathbf{r}_2$ is omitted. As $n(2) \neq n(1)$, the pairs of eigenfunctions with such indices are mutually orthogonal (Sect. 8.4.1), whence

$$\sum_{s_{z1}} \sum_{s_{z2}} \iint |\psi(\mathbf{r}_1, s_{z1}, \mathbf{r}_2, s_{z2})|^2 \, \mathrm{d}^3 r_1 \, \mathrm{d}^3 r_2 = 2\,|a|^2. \tag{15.36}$$

Imposing the normalization of ψ to unity, and observing that the phase factor in a is irrelevant, yields $a = 1/\sqrt{2}$. The treatment of (15.33) is identical and yields the same result for b. By the same token, one finds $a = b = 1/\sqrt{N!}$ in the N-particle case. Still with reference to the antisymmetric wave function (15.33), one notes that its spatial part can be recast as a determinant,

$$\sum_\nu (-1)^\nu S_\nu w_{n(1)}(\mathbf{q}_1) \ldots w_{n(N)}(\mathbf{q}_N) = \begin{bmatrix} w_{n(1)}(\mathbf{q}_1) & \cdots & w_{n(N)}(\mathbf{q}_1) \\ \vdots & \ddots & \vdots \\ w_{n(1)}(\mathbf{q}_N) & \cdots & w_{n(N)}(\mathbf{q}_N) \end{bmatrix},$$

(15.37)

that is called *Slater determinant*. A transposition of two particles involves the exchange of the corresponding coordinate sets, but not of the eigenfunction indices; this is equivalent to exchanging two rows of the determinant, whence the change of sign. Also, if two or more particles belonged to the same state (including spin), two or more columns of the Slater determinant would be equal to each other and the wave function (15.33) would vanish. This is another form of the proof of Pauli's exclusion principle.

15.8 Equilibrium Statistics in the Quantum Case

This section illustrates the quantum-mechanical treatment of a system of particles in a condition of macroscopic equilibrium. The approach is the same as that outlined in Sect. 6.3 for the classical case; however, the constraints to which the particles are subjected are different. Here the term "particle" is used in a broader meaning, incorporating, e.g., also the case of photons (Sect. 12.3) and phonons (Sect. 16.6). As in Sect. 6.3 one considers a conservative system of identical particles, having a total energy E_S, enclosed in a stationary container of volume Ω. The conservation of the total energy introduces a first constraint, identical to (6.10):

$$F_E(N_1, N_2, \ldots) = 0, \qquad F_E = -E_S + \sum_i N_i E_i.$$

(15.38)

The constraint identical to (6.9),

$$F_N(N_1, N_2, \ldots) = 0, \qquad F_N = -N + \sum_i N_i,$$

(15.39)

describing the conservation of the total number of particles, may, instead, be fulfilled or not depending on the type of particles. For instance, a system of photons does not fulfill it: in fact, a photon may be absorbed by the container's wall and, say, two photons may be emitted by it, such that the energy of the emitted photons equals that of the absorbed one. In this way, constraint (15.38) applies, whereas constraint (15.39) does not. Another difference from the classical treatment is that, as remarked in Sect. 15.5, identical quantum particles belonging to a system are not distinguishable from each other; as a consequence, the method of counting their

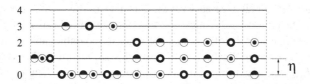

Fig. 15.2 Placement of three identical particles into equally-spaced energy states. The particles' total energy equals three energy units η. Different graphic symbols are used for the particles to make the classical treatment clearer

placement into the cells of the phase space (illustrated in Sect. 6.3 for the classical case), is different here. A further distinction must be made between the cases of systems made of fermions, to which the exclusion principle applies, and systems made of bosons, to which it does not apply. To appreciate the strong differences that are introduced by the constraints due to indistinguishability and exclusion, the example below is of help.

Consider a system made of three identical particles, whose total energy E_S is, in some units η, equal to 3η. For simplicity, the particles are considered spinless and, instead of the phase space, a space made of energy cells[7] is used, where the energies of the particles are assumed to be quantized starting from a ground level. As a consequence, the energies allowed to the particles are $0, \eta, 2\eta, 3\eta, \dots$; the energy levels are shown in Fig. 15.2, where the particles are drawn with different graphics to make them provisionally distinguishable. If the system is considered as classical, it is easily found that there are ten possible ways of placing the three particles into the available states in such a way as to fulfill the energy constraint $E_S = 3\eta$. One way is to place all three particles in the η state. Another choice is to place one particle in the 3η state and the remaining two in the ground state; this provides three different ways, as there are three distinct possibilities to choose the particle to be placed in the 3η state. The last choice is to place one particle in the ground state, another particle in the η state, and the remaining one in the 2η state; this yields 3! ways as shown in the figure. If, instead, the particles are bosons, the three former combinations with energies $(0, 0, 3\eta)$ reduce to a single one because of the particles' indistinguishability; by the same token, the six former combinations with energies $(0, \eta, 2\eta)$ reduce to a single one as well. This gives a total of three ways for bosons, in contrast with the ten ways of the classical particles. Finally, if the particles are fermions, the combinations (η, η, η) and $(0, 0, 3\eta)$ must be excluded because the exclusion principle forbids one or more fermions to occupy the same state; as a consequence, the only way left for fermions is $(0, \eta, 2\eta)$, that fulfills both indistinguishability and exclusion.

Coming back to the general case, consider a system in thermodynamic equilibrium at temperature T, with E_S the system's total energy. As in the classical case outlined in Sect. 6.3, the system is considered dilute, namely such that the mutual interaction

---

[7] The use of energy intervals does not entail a loss of generality, as the subsequent treatment of the general case will show.

among the particles, albeit necessary for the existence of the system, is weak enough to assume that the energy of the interaction among the particles is negligible within the Hamiltonian operator. It follows that the latter is separable, and the expressions found in Sect. 15.7 are applicable. As the particles are identical, the single-particle eigenvalues $E_{n(i)}$ and eigenfunctions $w_{n(i)}(\mathbf{q}_i)$ obtained by solving (15.28) are the same for all particles. Also, the indices[8] $n(i)$ are those of an energy, so that the procedure of placing the particles into the available states can be carried out directly in the energy space. To account for spin, states corresponding to the same energy and different spin are to be considered as distinct. The minimum eigenvalue of (15.28) is fixed by the form of the potential energy within the Hamiltonian operator. Given these premises, the energy axis is divided into equal intervals of length ΔE; as in the classical case, the partitioning has the advantage that the set of intervals is countable. The intervals are numbered starting from the one containing the minimum eigenvalue mentioned above, and their size ΔE is such that each of them contains a number of eigenvalues of (15.28). Let g_r be the number of eigenvalues within the rth interval, $r = 1, 2 \ldots$, and N_r the number of particles whose eigenvalues belong to the same interval; if the size ΔE is taken small, one can approximate the energy of the N_r particles with the product $N_r E_r$, where E_r is the energy at the interval's center. Following the same procedure as in Sect. 6.4, one then constructs the function

$$F\left(N_1, N_2, \ldots, \alpha, \beta\right) = \log W + \alpha F_N + \beta F_E, \qquad (15.40)$$

where α, β are the Lagrange multipliers, respectively related to the total number of particles and total energy of the system, and the form of W depends on the type of particles, as shown below. If the constraint on the total number of particles is not applicable, one lets $\alpha = 0$. Using the numbers $N_1, N_2, \ldots$ as continuous variables, taking the derivative of F with respect to N_r and equating it to zero yields

$$\frac{\partial}{\partial N_r} \log W = \alpha + \beta E_r. \qquad (15.41)$$

On the other hand it is $W = W_1 W_2 \ldots, W_r \ldots$, where W_r is the number of ways in which N_r particles can be placed into the g_r states of the rth interval, subjected to the constraints of the type of particles under consideration. As in the left hand side of (15.41) only the rth summand depends on N_r, the relation to be worked out is eventually

$$\frac{\partial}{\partial N_r} \log W_r = \alpha + \beta E_r. \qquad (15.42)$$

The expression of W_r depends on the type of particles; it is given in Sect. 15.8.1 for the case of fermions and in Sect. 15.8.2 for that of bosons.

---

[8] Here the eigenvalues of the Hamiltonian operator are discrete because the system is enclosed in a container, hence the wave function is normalizable. As usual, the notation $n(i)$ stands for a group of indices.

15.8.1   Fermi–Dirac Statistics

For a system of fermions it is $N_r \leq g_r$ due to the exclusion principle. To calculate the number of ways of placing N_r particles into g_r states one provisionally assumes that the particles are distinguishable. There are g_r possibilities to place the first particle; after the latter has been placed, there remain $g_r - 1$ states due to the exclusion principle, hence the different ways of placing the first two (distinct) particles are $g_r (g_r - 1)$. The process continues until all N_r particles are used up, this leading to $g_r (g_r - 1) \ldots (g_r - N_r + 1)$ ways of placing them. On the other hand the particles are not distinct; as a consequence, after the placement is completed, any of the $N_r!$ permutations of the particles corresponds to the same placement. In conclusion, the product above must be divided by $N_r!$, this leading to

$$W_r = \frac{g_r (g_r - 1) \ldots (g_r - N_r + 1)}{N_r!} = \frac{g_r!}{N_r! (g_r - N_r)!} = \binom{g_r}{N_r}. \tag{15.43}$$

Using the same procedure as in Sect. 6.4 yields

$$\frac{d \log W_r}{dN_r} = \frac{d \log [(g_r - N_r)!]}{d(-N_r)} - \frac{d \log (N_r!)}{dN_r} \simeq \log (g_r - N_r) - \log (N_r) \tag{15.44}$$

which, combined with (15.42), provides $N_r = g_r / [\exp (\alpha + \beta E_r) + 1]$. For convenience, the total number of particles is indicated here with N_S instead of N; the constraints then read

$$\sum_{r=1}^{U} \frac{g_r}{\exp (\alpha + \beta E_r) + 1} = N_S, \qquad \sum_{r=1}^{U} \frac{g_r E_r}{\exp (\alpha + \beta E_r) + 1} = E_S, \tag{15.45}$$

by which the two Lagrange multipliers α and β are determined. The sums are carried out up to a maximum energy E_U; in fact, being the total energy of the system prescribed, there exists a maximum energy that the single particle can not exceed. As outlined in Sect. 15.9.1, a comparison with the findings of Thermodynamics shows that, in full analogy with the classical case treated in Sect. 6.4, it is

$$\beta = \frac{1}{k_B T}, \tag{15.46}$$

where $k_B \simeq 1.38 \times 10^{-23}$ J K^{-1} is Boltzmann's constant and T the system temperature. As a consequence it is $\alpha = \alpha(T)$. Defining the *Fermi energy* or *Fermi level* $E_F(T) = -k_B T \alpha(T)$, the expression of N_r becomes

$$N_r = \frac{g_r}{\exp [(E_r - E_F) / (k_B T)] + 1}. \tag{15.47}$$

In general the number of energy levels is large and their separation small so that, instead of using the number g_r of states at energy E_r, one disposes of the index and

Fig. 15.3 The Fermi–Dirac statistics as a function of energy for different values of the system's temperature. For simplicity the temperature dependence of the Fermi level E_F is not considered

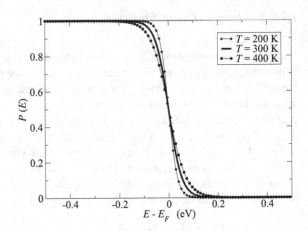

considers the number $g(E)\,dE$ of states in the infinitesimal interval dE around E. Thus, $g(E)$ indicates the number of states per unit energy, and is called *density of states in energy* (compare with Sect. B.5). The constraints now read

$$\int_{E_1}^{E_U} \frac{g(E)\,dE}{\exp(\alpha + \beta E) + 1} = N_S, \qquad \int_{E_1}^{E_U} \frac{E\,g(E)\,dE}{\exp(\alpha + \beta E) + 1} = E_S, \qquad (15.48)$$

with $\beta = 1/(k_B T)$. Consistently, N_r is replaced with $N(E)\,dE$, where $N(E)$ is the number of particles per unit energy.[9] As a consequence, the relation $N(E) = g(E)\,P(E)$ is fulfilled, where function

$$P(E) = \frac{1}{\exp[(E - E_F)/(k_B T)] + 1} \qquad (15.49)$$

is called *Fermi–Dirac statistics*. As $0 < P(E) < 1$, the Fermi–Dirac statistics bears also the meaning of occupation probability of a state at energy E. Its high-energy tail $(E - E_F \gg k_B T)$ identifies with the Maxwell–Boltzmann distribution law (6.14). The dependence of (15.49) on E is shown in Fig. 15.3 at different temperatures. The states whose energy is substantially lower than the Fermi level $(E - E_F \ll -k_B T)$ are mostly filled with particles, those whose energy is substantially higher than the Fermi level are mostly empty. The energy states in the vicinity of the Fermi level have a probability around $1/2$ of being filled. For the sake of simplicity the dependence of E_F on T is not considered in the figure; in fact, it is influenced by the form of $g(E)$ and must therefore be determined on a case-by-case basis. In the limit $T \to 0$ the function becomes discontinuous, specifically it is $P = 1$ for $E < E_F$ and $P = 0$ for $E > E_F$.

---

[9] The units are $[g], [N] = \mathrm{J}^{-1}$.

15.8.2 Bose–Einstein Statistics

For a system of Bosons the exclusion principle does not hold, hence it may be $N_r \geq g_r$. Also in this case, to calculate the number of ways of placing N_r particles, not subject to the exclusion principle, into g_r states, one provisionally assumes that the particles are distinguishable. This yields a number of ways equal to $(g_r + N_r - 1)(g_r + N_r - 2) \ldots g_r$. Then, to account for indistinguishability one divides the result by $N_r!$, to find

$$W_r = \frac{(g_r + N_r - 1) \ldots g_r}{N_r!} = \frac{(g_r + N_r - 1)!}{N_r!\,(g_r - 1)!} = \binom{g_r + N_r - 1}{N_r}. \tag{15.50}$$

Using the same procedure as in Sect. 6.4 yields

$$\frac{\mathrm{d} \log W_r}{\mathrm{d} N_r} = \frac{\mathrm{d} \log \left[(g_r + N_r - 1)!\right]}{\mathrm{d} N_r} - \frac{\mathrm{d} \log \left(N_r!\right)}{\mathrm{d} N_r} \simeq \log \left(g_r + N_r - 1\right) - \log N_r \tag{15.51}$$

which, combined with (15.42) after neglecting the unity in $g_r + N_r - 1$, yields $N_r = g_r/\left[\exp(\alpha + \beta E_r) - 1\right]$. As in the latter it is $N_r, g_r > 0$, it follows $\alpha + \beta E_r > 0$. Still indicating the number of particles with N_S instead of N, the constraints read

$$\sum_{r=1}^{U} \frac{g_r}{\exp\left(\alpha + \beta\,E_r\right) - 1} = N_S, \qquad \sum_{r=1}^{U} \frac{g_r\,E_r}{\exp\left(\alpha + \beta\,E_r\right) - 1} = E_S, \tag{15.52}$$

by which the two Lagrange multipliers α and β are determined. The explanation of the upper summation limit U in (15.52) is similar to that given in Sect. 15.8.1. As outlined in Sect. 15.9.5, a comparison with the findings of Thermodynamics shows that, in full analogy with the cases of Sects. 6.4 and 15.8.1, β is given by (15.46), whence it is $\alpha = \alpha(T)$. Defining $E_B(T) = -k_B\,T\,\alpha(T)$, the expression of N_r becomes

$$N_r = \frac{g_r}{\exp\left[(E_r - E_B)/(k_B\,T)\right] - 1}. \tag{15.53}$$

The inequality $\alpha + \beta\,E_r > 0$ found above implies $E_r > E_B$. If the constraint on the number of particles does not hold, one lets $\alpha = 0$ whence $E_B = 0$, $E_r > 0$. In general the number of energy levels is large and their separation small so that, instead of using the number g_r of states at energy E_r, one disposes of the index and considers the number $g(E)\,\mathrm{d}E$ of states in the infinitesimal interval $\mathrm{d}E$ around E. The constraints now read

$$\int_{E_1}^{E_U} \frac{g(E)\,\mathrm{d}E}{\exp\left(\alpha + \beta\,E\right) - 1} = N_S, \qquad \int_{E_1}^{E_U} \frac{E\,g(E)\,\mathrm{d}E}{\exp\left(\alpha + \beta\,E\right) - 1} = E_S, \tag{15.54}$$

with $\beta = 1/(k_B T)$. Consistently, N_r is replaced with $N(E)\,dE$, where $N(E)$ is the number of particles per unit energy. As a consequence, the relation $N(E) = g(E)\,P(E)$ is fulfilled, where function

$$P(E) = \frac{1}{\exp\left[(E - E_B)\,/\,(k_B\,T)\right] - 1} \tag{15.55}$$

is called *Bose–Einstein statistics*. As $P(E)$ may be larger than unity, the Bose–Einstein statistics is not a probability; rather, it represents the occupation number of a state at energy E. Its high-energy tail $\left(E - E_B \gg k_B\,T\right)$ identifies with the Maxwell–Boltzmann distribution law (6.14).

15.9 Complements

15.9.1 Connection with Thermodynamic Functions

The calculation of the equilibrium distribution carried out in Sects. 6.4, 15.8.1, and 15.8.2 respectively for classical particles, fermions, and bosons, entails the maximization of the function $\log W$, subjected to suitable constraints. On the other hand, from the second principle of Thermodynamics one derives that the equilibrium state of a system corresponds to the condition that the system's entropy S has a maximum. For this reason one expects that a functional dependence $W(S)$ exists; to identify its form one notes that, if W_1 and W_2 indicate the value of W of two independent systems, the value for the composite system is $W_1\,W_2$ due to the definition of W. On the other hand, entropy is additive, so that the functional dependence sought must be such that $S(W_1\,W_2) = S(W_1) + S(W_2)$, namely, of the logarithmic type. In fact it is

$$S = k_B\,\log W, \tag{15.56}$$

with $k_B = 1.38 \times 10^{-23}$ J K^{-1} the Boltzmann constant. The choice of the constant makes (15.56) consistent with the definition $dS = dQ/T$ of entropy in Thermodynamics, with dQ the heat absorbed during an infinitesimal transformation.[10]

Now, consider a system subjected to both constraints (15.38) and (15.39), and assume that the container where the system is placed undergoes an infinitesimal volume change[11] $d\Omega$ which, in turn, makes all variables N_r and g_r to change; due to (15.41) it is $\partial \log W/\partial N_r = \alpha + \beta\,E_r$ so that [27]

$$\frac{1}{k_B}\,dS = d\log W = \sum_{r=1}^{U} \left[(\alpha + \beta\,E_r)\,dN_r + \frac{\partial \log W}{\partial g_r}\,dg_r\right]. \tag{15.57}$$

---

[10] Compare with the non-equilibrium definition of entropy introduced in Sect. 6.6.3 and the note therein.

[11] The geometrical configuration is kept similar to the original one during the change in volume.

Using the constraints and the relation $dS = dQ/T$ transforms (15.57) into

$$dQ = k_B T \alpha \, dN + k_B T \beta \, dE_S + k_B T \sum_{r=1}^{U} \left(\frac{\partial \log W}{\partial g_r} \frac{dg_r}{d\Omega} \right) d\Omega. \qquad (15.58)$$

Assuming that during the change in volume there is no exchange of matter with the environment, one lets $dN = 0$ in (15.58); the first principle of Thermodynamics shows that for this type of transformation it is $dQ = dE_S + P \, d\Omega$, with P the pressure at the boundary of Ω. A comparison with (15.58) then yields

$$k_B T \beta = 1, \qquad k_B T \sum_{r=1}^{U} \left(\frac{\partial \log W}{\partial g_r} \frac{dg_r}{d\Omega} \right) = P. \qquad (15.59)$$

The first of (15.59) coincides with (6.26), and provides the expected relation between one of the Lagrange multipliers of (15.45) and a state function of Thermodynamics.

15.9.2 Density of States for a Particle in a Three-Dimensional Box

The density of states $g(E)$ that has been introduced in (15.48) and (15.54) is the number of states per unit energy. Its form depends on the system under consideration. Two examples of the derivation of $g(E)$ are given here (with reference to the problem of an electron in a box) and in Sect. 15.9.4 (with reference to photons).

The problem of the particle confined within a one-dimensional box has been tackled in Sect. 8.2.2. The set of eigenvalues is given by the first relation in (8.6), and the corresponding eigenfunctions are (8.8). For a three-dimensional box whose sides have lengths d_1, d_2, d_3, due to the properties of separable operators (Sect. 10.3), the eigenfunctions are products of one-dimensional eigenfunctions,

$$w_{n_1 n_2 n_3}(\mathbf{r}) = \sqrt{8/V} \, \sin\left(k_{n_1} x_1\right) \sin\left(k_{n_2} x_2\right) \sin\left(k_{n_3} x_3\right), \qquad (15.60)$$

where $V = d_1 d_2 d_3$ is the volume of the box. In turn, the eigenvalues are

$$E_{n_1 n_2 n_3} = E_{n_1} + E_{n_2} + E_{n_3}, \qquad E_{n_i} = \frac{\hbar^2 k_{n_i}^2}{2m}, \qquad k_{n_i} = n_i \frac{\pi}{d_i}, \qquad (15.61)$$

namely,

$$E = \frac{\hbar^2 k^2}{2m}, \qquad \mathbf{k} = n_1 \frac{\pi}{d_1} \mathbf{i}_1 + n_2 \frac{\pi}{d_2} \mathbf{i}_2 + n_3 \frac{\pi}{d_3} \mathbf{i}_3, \qquad n_i = 1, 2, \ldots \qquad (15.62)$$

with $\mathbf{i}_1, \ldots$ the unit vectors of the Cartesian axes and $k^2 = \mathbf{k} \cdot \mathbf{k}$. In contrast with the one-dimensional case, the eigenvalues (15.61) are degenerate, because different triads n_1, n_2, n_3 correspond to the same energy.

The density of states could be calculated by the procedure depicted in Sect. (B.5), with the provision that in this case the variables k_{n_i} belong to a discrete set whereas those in Sect. B.5 are continuous. On the other hand, the relations involved here are simple enough to be tackled by a direct calculation. One observes, first, that the distance between two consecutive projections of $\mathbf{k}$ along the ith axis is π/d_i; as a consequence, one may partition the $k_1 k_2 k_3$ space into equal volumes[12] π^3/V, so that each $\mathbf{k}$ vector is associated to one and only one volume: this shows that the density of $\mathbf{k}$ vectors in the $k_1\, k_2\, k_3$ space is $Q_k = V/\pi^3$. Given the electron's energy E, from the geometrical point of view the first relation in (15.62) describes a sphere of radius $(2\, m\, E/\hbar^2)^{1/2}$ in the $k_1\, k_2\, k_3$ space; thus, the total number N_k of π^3/V volumes contained within the sphere is obtained by multiplying the sphere's volume by the density Q_k. Clearly the volumes that are near the boundary produce a ragged surface; the latter is identified with that of the sphere by assuming that the distribution of $\mathbf{k}$ vectors belonging to the spherical surface is very dense. With this provision one finds

$$N_k = Q_k \frac{4}{3}\pi k^3 = \frac{V}{\pi^3}\frac{4}{3}\pi \left(\frac{2\, m\, E}{\hbar^2}\right)^{3/2} = \frac{V}{\pi^2}\frac{8\sqrt{2}\, m^{3/2}}{3\,\hbar^3}E^{3/2}. \qquad (15.63)$$

As indices n_1, n_2, n_3 take only positive values, the $\mathbf{k}$ vectors belong to $1/8$ of the sphere only; it follows that their number is $N_k/8$. Each $\mathbf{k}$ vector is associated to a triad of quantum numbers n_1, n_2, n_3; however, to completely define the state of the electron a fourth quantum number is necessary, related to spin (Sect. 15.5.1). Remembering that electrons have two possible spin states, one finds that the number of electron states within the sphere is twice the number of $\mathbf{k}$ vectors, namely, $N_k/4$. Finally, the number of states per unit energy is found by differentiating the latter with respect to energy,[13]

$$\mathrm{d}\frac{N_k}{4} = \frac{\sqrt{2}\,V m^{3/2}}{\pi^2 \hbar^3}E^{1/2}\,\mathrm{d}E = g(E)\,\mathrm{d}E, \qquad g(E) = \frac{\sqrt{2}\,V\,m^{3/2}}{\pi^2 \hbar^3}E^{1/2}. \quad (15.64)$$

Apart from the constants involved, this results is consistent with (B.34). Along with the density of state in energy it is useful to define a combined density of states in energy and coordinate space, that is indicated with γ. In the case of the electron within a box one finds, from (15.64),

$$\gamma(E) = \frac{g(E)}{V} = \frac{\sqrt{2}\, m^{3/2}}{\pi^2 \hbar^3}E^{1/2}. \qquad (15.65)$$

---

[12] Here the term "volume" is used in a broader meaning; in fact, the units of π^3/V are m^{-3}.

[13] As noted above the $\mathbf{k}$ vectors, hence the values of energy corresponding to them, are distributed very densely. This makes it possible to treat E as a continuous variable.

15.9.3 Density of States for a Two- or One-Dimensional Box

As observed in Sect. B.5, the functional dependence of g on E is influenced by the number of spatial dimensions: considering by way of example the case of an electron within a two-dimensional box, one must associate to each vector $\mathbf{k}$ an area of the $k_1 k_2$ space equal to $\pi^2/(d_1 d_2) = \pi^2/A$, where $A = d_1 d_2$ is the area of the box. The density of the $\mathbf{k}$ vectors in the $k_1 k_2$ space is $Q_k = A/\pi^2$, whence the total number of π^2/A areas in a circle of radius $(2 m E/\hbar^2)^{1/2}$ is

$$N_k = Q_k \pi k^2 = \frac{A}{\pi^2} \pi \frac{2 m E}{\hbar^2}. \tag{15.66}$$

As indices n_1, n_2 take only positive values, the $\mathbf{k}$ vectors belong to $1/4$ of the circle only; it follows that their number is $N_k/4$. Accounting for spin one finds that the number of states within the circle is twice the number of $\mathbf{k}$ vectors, namely, $N_k/2$. Finally, the number of states per unit energy is found by differentiating the latter with respect to energy,

$$g(E) = \frac{\mathrm{d} N_k/2}{\mathrm{d} E} = \frac{A m}{\pi \hbar^2}. \tag{15.67}$$

When the energy dependence on the k coordinates is quadratic, the density of states of a two-dimensional case is constant (compare with (B.33)).

Finally, considering the case of an electron within a one-dimensional box, one must associate to each vector $\mathbf{k}$ a segment π/d_1 of the k_1 space, to find $Q_k = d_1/\pi$ for the density of the $\mathbf{k}$ vectors. The total number of π/d_1 segments in a domain of length $(2 m E/\hbar^2)^{1/2}$ is

$$N_k = \frac{d_1}{\pi} \left(\frac{2 m E}{\hbar^2} \right)^{1/2}. \tag{15.68}$$

Accounting for spin one finds that the number of states within the domain is twice the number of $\mathbf{k}$ vectors, namely, $2 N_k$. Finally, the number of states per unit energy is found by differentiating the latter with respect to energy,

$$g(E) = \frac{\mathrm{d} 2 N_k}{\mathrm{d} E} = \frac{d_1 (2 m)^{1/2}}{\pi \hbar E^{1/2}}, \tag{15.69}$$

to be compared with (B.32). Expression (15.67) is useful, e.g., for treating the problem of a two-dimensional charge layer in the channel of a semiconductor device (Sect. 17.6.7); in turn, expression (15.69) is used with reference to nanowires (Sect. 17.6.7);

15.9.4 Density of States for Photons

Consider the case of the electromagnetic field within a box whose sides d_1, d_2, d_3 are aligned with the coordinate axes and start from the origin. It is assumed that no charge is present within the box, so that the calculation of the modes is the same as that illustrated in Sect. 5.5. The wave vectors have the form

$$\mathbf{k} = n_1 \frac{2\pi}{d_1} \mathbf{i}_1 + n_2 \frac{2\pi}{d_2} \mathbf{i}_2 + n_3 \frac{2\pi}{d_3} \mathbf{i}_3, \qquad n_i = 0, \pm 1, \pm 2, \ldots, \qquad (15.70)$$

the angular frequency of the mode corresponding to $\mathbf{k}$ is $\omega = c\,k$, with k the modulus of $\mathbf{k}$, and the energy of each photon of that mode is $E = \hbar\,\omega = \hbar\,c\,k$, with c the speed of light. The calculation of the density of states associated to this case follows the same line as that used in Sect. 15.9.2 for a particle in a three-dimensional box; the differences are that the distance between two consecutive projections of $\mathbf{k}$ along the ith axis is $2\pi/d_i$ instead of π/d_i, the indices n_i are not limited to the positive values, and the $E(\mathbf{k})$ relation is linear instead of being quadratic.

With these premises, and with reference to Fig. 15.4, the volume associated to each $\mathbf{k}$ is $(2\,\pi)^3/(d_1\,d_2\,d_3) = (2\,\pi)^3/V$, where V is the volume of the box, so that the density of the $\mathbf{k}$ vectors in the k space is $Q_k = V/(2\,\pi)^3$; in turn, the total number of $\mathbf{k}$ vectors within a sphere of radius k like that shown in the figure is[14]

$$N_k = Q_k \frac{4}{3}\pi\,k^3 = \frac{V}{(2\,\pi)^3}\frac{4}{3}\pi\frac{E^3}{\hbar^3\,c^3} = \frac{V}{2\,\pi^2}\frac{E^3}{3\,\hbar^3\,c^3}. \qquad (15.71)$$

Due to spin, the total number of states is $2N_k$ so that, differentiating with respect to energy,

$$\mathrm{d}2\,N_k = \frac{V}{\pi^2}\frac{E^2}{\hbar^3\,c^3}\,\mathrm{d}E = g(E)\,\mathrm{d}E, \qquad g(E) = \frac{V}{\pi^2\,\hbar^3\,c^3}\,E^2. \qquad (15.72)$$

The result expressed by (15.72) can be recast in equivalent forms by using another variable proportional to energy; for instance, letting $G(\omega)$ denote the density of states with respect to angular frequency and $\tilde{g}(\nu)$ denote the density of states with respect to frequency, from the relations

$$\tilde{g}(\nu)\,\mathrm{d}\nu = G(\omega)\,\mathrm{d}\omega = g(E)\,\mathrm{d}E, \qquad E = \hbar\,\omega = \hbar\,2\,\pi\,\nu \qquad (15.73)$$

one obtains

$$G(\omega) = \frac{V}{\pi^2\,c^3}\,\omega^2, \qquad \tilde{g}(\nu) = 8\,\pi\frac{V}{c^3}\,\nu^2. \qquad (15.74)$$

---

[14] The calculation shown here is equivalent to counting the number of elements of volume $(2\,\pi)^3/(d_1\,d_2\,d_3)$ that belong to the spherical shell drawn in Fig. 15.4. The result is then multiplied by 2 to account for spin.

Fig. 15.4 Constant-energy sphere of the **k** space illustrating the procedure for determining the density of states

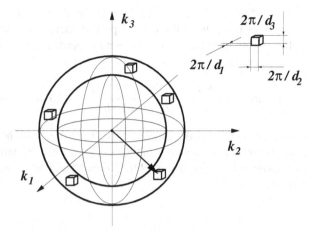

15.9.5   Derivation of Planck's Law

As illustrated in Sect. 7.4.1, Planck's law provides the black-body's spectral energy density u at equilibrium. From its definition it follows that the product $u\,dv$ is the electromagnetic energy per unit volume in the elementary interval dv; using the quantum concepts, the electromagnetic energy in dv may, in turn, be written as the product of the number dN_{ph} of photons belonging to dv by the energy $h\,v$ of each photon. Considering that photons are bosons, and that the equilibrium case is considered, the number dN_{ph} is given in turn by the product $\tilde{g}(v)\,dv\,P(E = h\,v)$, with P the Bose–Einstein statistics (15.55). In summary, using the second relation in (15.74), one finds

$$u\,dv = \frac{1}{V}\,h\,v\,dN_{\mathrm{ph}} = \frac{h\,v}{V}\,\frac{\tilde{g}(v)\,dv}{\exp(\beta\,h\,v) - 1} = \frac{8\,\pi\,h\,v^3/c^3}{\exp(\beta\,h\,v) - 1}\,dv. \qquad (15.75)$$

The expression of the Bose–Einstein statistic used in (15.75) accounts for the fact that the number of photons is not conserved, so that the only constraint to be considered is that on the total electromagnetic energy within volume V. It follows that the Bose–Einstein statistics has only one Lagrange multiplier, whose value is provisionally left undetermined. From (15.75) one derives Planck's law

$$u(v, \beta) = \frac{8\,\pi\,h\,v^3/c^3}{\exp(\beta\,h\,v) - 1}, \qquad (15.76)$$

to be compared with (7.19). Like in Sect. 15.9.1, the undetermined parameter β is obtained by comparing the result of the microscopic derivation carried out above with those of Thermodynamics. Letting $W_{\mathrm{em}}^{\mathrm{eq}}$ be the electromagnetic energy within V at equilibrium, the following relations hold:

$$\frac{W_{\mathrm{em}}^{\mathrm{eq}}}{V} = \int_0^\infty u\,dv, \qquad \frac{W_{\mathrm{em}}^{\mathrm{eq}}}{V} = \frac{4}{c}\,\sigma\,T^4, \qquad (15.77)$$

with $\sigma \simeq 5.67 \, 10^{-12}$ W cm^{-2}K^{-4} the *Stefan–Boltzmann constant*. In fact, the first one derives from the definition of spectral energy density, while the second one (called *Stefan–Boltzmann law*) is found experimentally, On the other hand, using (C.127) one finds

$$\int_0^\infty u \, d\nu = \frac{8 \pi h}{c^3 (\beta h)^4} \int_0^\infty \frac{(\beta h \nu)^3}{\exp(\beta h \nu) - 1} \, d(\beta h \nu) = \frac{8 \pi}{c^3 h^3 \beta^4} \frac{\pi^4}{15}. \quad (15.78)$$

Combining (15.78) with (15.77) yields $1/(\beta T)^4 = 15 \, c^2 \, h^3 \, \sigma/(2 \pi^5)$; replacing the constants at the right hand side of the latter shows that the units and numerical value of it are those of k_B^4; it follows that $1/(\beta T)^4 = k_B^4$ and the expected result $\beta = 1/(k_B T)$ is found.

Problems

15.1 Estimate the extension of the energy region where the main variation of the Fermi statistics (15.49) occurs.

Chapter 16
Separation of Many-Particle Systems

16.1 Introduction

The chapter illustrates a number of steps that are necessary to reduce the many-particle problem to a tractable form. The analysis starts from a system of interacting electrons and nuclei; such a system is not made of identical particles and its Hamiltonian operator is not necessarily separable. Besides that, the number of particles that are present in the typical volume of, e.g., a solid-state device is so huge that the solution of the Schrödinger equation of such a system in the original form is a hopeless task. The first step consists in the application of the adiabatic approximation, by which the system made of the electrons is separated from that of the nuclei. The way in which such a separation is accomplished has the inconvenience that the nuclei are kept fixed in the equilibrium positions; this approximation is too strong, because it prevents the exchange of energy and momentum between the two systems from occurring: in fact, it is used provisionally and is removed at a later stage. The next step deals with the electron system which, despite the separation from the nuclei, is still too complicate to be dealt with directly; using the Ritz method, the Schrödinger equation for the electron system is separated into single-particle equations, in which each electron is subjected to the average field of the others. This step yields the Hartree equations and greatly simplifies the problem; in fact, the equations, besides being separated, are also identical to each other, so that the set of eigenvalues and eigenfunction obtained from one of them is applicable to all electrons. The Hartree equations do not comply with the exclusion principle, which must necessarily be fulfilled because the system under consideration is made of identical particles; a further modification, yielding the Hartree–Fock equations, provides the wave function with the expected antisymmetry property. Finally, the system of nuclei is taken again into consideration to the purpose of eliminating the simplification that the nuclei are fixed in the equilibrium positions: considering the fact that the nuclei are strongly bound together, so that their displacement from the equilibrium position is small, the nuclei are treated as a system of linear harmonic oscillators. In this way, the interaction between an electron and the nuclei is described (in a later chapter) using the quantum-mechanical, first-order perturbation theory applied to the two-particle collision of an electron with a phonon.

© Springer Science+Business Media New York 2015
M. Rudan, *Physics of Semiconductor Devices*,
DOI 10.1007/978-1-4939-1151-6_16

16.2   System of Interacting Electrons and Nuclei

The analysis of a conservative system of identical particles, started in Sect. 15.7, is based on the single-particle Hamiltonian operator (15.28). It must be noted, however, that the typical systems to be dealt with are not made of identical particles but, rather, of a mixture of subsystems; the particles of each subsystem are identical to each other, but different from those of the other subsystems. Moreover, the Hamiltonian operator of each subsystem is not necessarily separable. It follows that the existence of a single-particle Hamiltonian operator like (15.28) is by no means obvious. It is then necessary to tackle the problem from a more general point of view, starting from the consideration of a mixture of systems, and determining the approximations that may simplify the problem to a level that is practically affordable.

To this purpose, consider a conservative system made of K electrons and N nuclei, interacting with each other. The particles are bound together, so that the system's wave function can be assumed to be normalized to unity. The coordinates associated to the electrons are indicated with small letters and grouped into a single vector $\mathbf{r} = (\mathbf{r}_1, \dots, \mathbf{r}_K)$; those of the nuclei are indicated with capital letters: $\mathbf{R} = (\mathbf{R}_1, \dots, \mathbf{R}_N)$. The interaction among the particles is assumed to be of the Coulomb type, so that the contributions to the potential energy due to the electron–electron and nucleus–nucleus interactions are, respectively,

$$U_e(\mathbf{r}) = \frac{1}{2} \sum_{i,j=1}^{K} \frac{q^2}{4\pi\varepsilon_0 |\mathbf{r}_i - \mathbf{r}_j|}, \qquad U_a(\mathbf{R}) = \frac{1}{2} \sum_{i,j=1}^{N} \frac{Z_i Z_j q^2}{4\pi\varepsilon_0 |\mathbf{R}_i - \mathbf{R}_j|}, \qquad (16.1)$$

with $j \neq i$, where $q > 0$ is the electron charge, ε_0 the vacuum permittivity, and $Z_j q$ the charge of the jth nucleus; factor $1/2$ is introduced to avoid a double summation. In addition one must consider the electron–nucleus interaction, whose contribution to the potential energy is

$$U_{ea}(\mathbf{r}, \mathbf{R}) = \sum_{i=1}^{K} \sum_{j=1}^{N} \frac{-Z_j q^2}{4\pi\varepsilon_0 |\mathbf{r}_i - \mathbf{R}_j|}. \qquad (16.2)$$

Remembering (15.3), letting $\mathbf{r}_i = (x_{i1}, x_{i2}, x_{i3})$, $\mathbf{R}_i = (X_{j1}, X_{j2}, X_{j3})$, and defining

$$\nabla_i^2 = \frac{\partial^2}{\partial x_{i1}^2} + \frac{\partial^2}{\partial x_{i2}^2} + \frac{\partial^2}{\partial x_{i3}^2}, \qquad \nabla_j^2 = \frac{\partial^2}{\partial X_{j1}^2} + \frac{\partial^2}{\partial X_{j2}^2} + \frac{\partial^2}{\partial X_{j3}^2}, \qquad (16.3)$$

the kinetic parts of the system's Hamiltonian operator are

$$\mathcal{T}_e = -\sum_{i=1}^{K} \frac{\hbar^2}{2m} \nabla_i^2, \qquad \mathcal{T}_a = -\sum_{j=1}^{N} \frac{\hbar^2}{2m_j} \nabla_j^2. \qquad (16.4)$$

The time-independent Schrödinger equation of the system then reads

$$\mathcal{H} w = E w, \qquad \mathcal{H} = \mathcal{T}_e + \mathcal{T}_a + U_e + U_a + U_{ea} + U_{\text{ext}}, \qquad (16.5)$$

where the eigenfunctions depend on all variables, $w = w(\mathbf{r}, \mathbf{R})$. In (16.5), $U_{\text{ext}} = U_{\text{ext}}(\mathbf{r}, \mathbf{R})$ is the potential energy due to an external, conservative field acting on the system. If it were $U_{ea} + U_{\text{ext}} = 0$, the Hamiltonian would be separable with respect to the two subsystems, $\mathcal{H} = (\mathcal{T}_e + U_e) + (\mathcal{T}_a + U_a)$. The time-independent equations resulting from the separation would be

$$(\mathcal{T}_e + U_e)u = E_e\,u, \qquad (\mathcal{T}_a + U_a)v = E_a v, \qquad (16.6)$$

with $u = u(\mathbf{r})$, $v = v(\mathbf{R})$; due to the general properties of separable operators (compare with Sect. 10.3), the eigenvalues and eigenfunctions of (16.5) would have the form $w = u\,v$, $E = E_e + E_a$. As in general it is $U_{ea} + U_{\text{ext}} \neq 0$, the Schrödinger Eq. (16.5) is not actually separable, and its solution may often present a formidable challenge: considering by way of example that the concentration of atoms in solid matter is of the order of 5×10^{22} cm^{-3}, and that to a nucleus there correspond Z electrons, with Z the atomic number, the number of scalar coordinates necessary to describe a cubic centimeter of solid matter is of the order of $15\,(Z + 1) \times 10^{22}$. It is then necessary to make a number of approximations; they are introduced step by step and, as illustrated in Sects. 16.3, 16.4, and 16.5, they are capable to bring the problem back to the solution of single-particle equations.

16.3 Adiabatic Approximation

The problem of solving the Schrödinger Eq. (16.5) for a system of interacting electrons and nuclei is simplified by observing that the mass of a nucleus is much larger than that of the electron. As a first consequence, one expects that the interaction with an individual electron influences little the motion of a nucleus; rather, the latter is expected to be influenced by the average interaction with many electrons. A second consequence is that, if the particles' dynamics is provisionally considered in classical terms, the motion of the nuclei is much slower than that of the electrons; as a consequence, the classical positions $\mathbf{R}$ of the nuclei can be considered as slowly-varying parameters when the electron dynamics is investigated. For this reason, the procedure shown below, based on the latter observation, is called *adiabatic approximation* or also *Born-Oppenheimer approximation* [62, Sect. 8]. In quantum terms, the approximation leads to the splitting of the original Eq. (16.5) into a set of two coupled equations. To this purpose, let

$$\mathcal{H}_e = \mathcal{T}_e + U_e + U_{ea} + U_{\text{ext}}, \qquad (16.7)$$

where the dependence on $\mathbf{R}$ is algebraic only. Considering such a dependence, the Hamiltonian operator $\mathcal{H}_e$ provides an eigenvalue equation whose eigenvalues and eigenfunctions depend on $\mathbf{R}$; they read

$$\mathcal{H}_e u = E_e u, \qquad E_e = E_e(\mathbf{R}), \qquad u = u(\mathbf{r}, \mathbf{R}). \qquad (16.8)$$

Also, from the Schrödinger Eq. (16.8) one finds that, for any function v that depends only on $\mathbf{R}$, the following holds: $\mathcal{H}_e u\,v = E_e\,u\,v$. As v is undetermined, one may seek a form of v such that $w = u\,v$ is also an eigenfunction of the full Hamiltonian operator $\mathcal{H}$ of (16.5). From the definition of $\mathcal{H}_e$ it follows $\mathcal{H} = \mathcal{T}_a + \mathcal{U}_a + \mathcal{H}_e$, whence the original Schrödinger Eq. (16.5) is recast as

$$(\mathcal{T}_a + \mathcal{U}_a + \mathcal{H}_e)uv = Euv. \tag{16.9}$$

The second relation in (16.4) shows that each Laplacian operator in $\mathcal{T}_a$ acts on both u and v, specifically, $\nabla_j^2\, u\,v = u\,\nabla_j^2\, v + v\,\nabla_j^2\, u + 2\,\nabla_j\, u \cdot \nabla_j\, v$. Multiplying the latter by $-\hbar^2/(2m_j)$ and adding over j transforms (16.9) into

$$u(\mathcal{T}_a + \mathcal{U}_a + E_e)v + v\mathcal{T}_a u + \mathcal{G}_a u\,v = Euv, \tag{16.10}$$

where the short-hand notation $\mathcal{G}_a u v = -\sum_{j=1}^{N} (\hbar^2/m_j)\,\nabla_j u \cdot \nabla_j v$ has been used. To proceed one notes that the coefficients in the Schrödinger Eq. (16.8) are real, so that u can be taken real as well. Remembering that the particles described by u are bound, one finds that u is normalizable so that, for any $\mathbf{R}$,

$$\int u^2(\mathbf{r}, \mathbf{R})dr = 1, \qquad dr = d^3r_1 \ldots d^3r_K, \tag{16.11}$$

with the integral extended to the whole K-dimensional space $\mathbf{r}$. Remembering that ∇_j acts on $\mathbf{R}_j$, from (16.11) one derives

$$\int u\,\nabla_j u\,dr = \frac{1}{2}\int \nabla_j\, u^2 dr = \frac{1}{2}\nabla_j \int u^2 dr = 0, \tag{16.12}$$

whence

$$\int u\,\mathcal{G}_a uv\,dr = -\sum_{j=1}^{N} \frac{\hbar^2}{m_j}\nabla_j v \cdot \int u\nabla_j u\,dr = 0. \tag{16.13}$$

Left multiplying (16.10) by u, integrating over $\mathbf{r}$, and using (16.11, 16.13), yields

$$(\mathcal{T}_a + \mathcal{U}_a + E_e + \mathcal{U}_u)v = Ev, \qquad \mathcal{U}_u(\mathbf{R}) = \int u\mathcal{T}_a u\,dr. \tag{16.14}$$

In this way, the original Schrödinger Eq. (16.5) is eventually split into two coupled equations, (16.8) and (16.14); the former is written in extended form as

$$(\mathcal{T}_e + \mathcal{U}_e + \mathcal{U}_{ea} + \mathcal{U}_{\text{ext}})u = E_e u. \tag{16.15}$$

The first equation, (16.14), is the equation for the nuclei, in fact its kinetic operator $\mathcal{T}_a$ acts on $\mathbf{R}$ only, and its potential-energy term $\mathcal{U}_a + E_e + \mathcal{U}_u$ depends on $\mathbf{R}$ only; this equation is coupled with the second one, (16.15), because E_e is an eigenvalue of (16.15) and $\mathcal{U}_u$ is given by the integral in (16.14) that involves the eigenfunctions of (16.15). In turn, (16.15) is the equation for the electrons, in fact its kinetic operator $\mathcal{T}_e$ acts on $\mathbf{r}$ only; in turn, the part $\mathcal{U}_{ea} + \mathcal{U}_{\text{ext}}$ of its potential-energy term couples (16.15) with (16.14) due to the dependence on $\mathbf{R}$.

In principle, one may solve (16.15) after fixing $\mathbf{R}$ and, from the solution, calculate E_e and U_u to be used in (16.14). From v one then determines the expectation value of $\mathbf{R}$, that updates the starting point for the solution of (16.15). Considering the case of solid matter, and still reasoning in classical terms, a zero-order solution for the iterative procedure depicted above is found by observing that the nuclei, being massive and tightly bound together, are expected to depart little from their equilibrium positions $\mathbf{R}_0$. One then fixes $\mathbf{R} = \mathbf{R}_0$ in (16.15) to find, for the electrons' Schrödinger equation, the approximate form

$$\left(\mathcal{T}_e + V_e\right)u = E_e u, \qquad V_e(\mathbf{r}) = U_e(\mathbf{r}) + U_{ea}\left(\mathbf{r}, \mathbf{R}_0\right) + U_{\text{ext}}\left(\mathbf{r}, \mathbf{R}_0\right). \qquad (16.16)$$

This separates completely the equation for the electrons, that may be thought of as forming a separate system of total energy E_e. Admittedly, keeping the positions of the nuclei fixed is rather crude an approximation; in this way, in fact, the nuclei can not exchange momentum and energy with the electrons any more. On the other hand it can be shown that the solution of (16.16) provides an acceptable approximation for the energy of the electrons. Another advantage of keeping the nuclei in the equilibrium positions occurs in the investigation of materials where the nuclei are arranged in periodic structures; in fact, one exploits in this case the properties of differential equations having periodic coefficients (compare with Sects. 13.4 and 17.5.1). In conclusion, the present analysis will continue by provisionally considering only the system of electrons as described by (16.16); obviously the problem of the exchange of momentum and energy with the system of nuclei can not be dispensed with, and will be resumed at a later stage (Sect. 16.6).

16.4   Hartree Equations

Equation (16.16), describing the separate system of K electrons after fixing the nuclei to the equilibrium positions $\mathbf{R}_0$, lends itself to the application of the Ritz method outlined in Sect. 16.7.1. For the sake of generality, one starts by assuming that the K particles are distinguishable; for this reason, the symbols $\mathcal{T}_e$, V_e, u, and E_e of (16.16) are not used. Also, the application of the Ritz method will be carried out in the case where the external forces are absent, $U_{\text{ext}} = 0$; such forces will be introduced again into the problem in Sect. 19.2.1, where the single-particle dynamics is described. Given these premises, the operator $\mathcal{A}$, its minimum eigenvalue A_1, and the corresponding eigenfunction v_1 used in Sect. 16.7.1 for describing the Ritz method are indicated here with $\mathcal{A} = \mathcal{H}$, $A_1 = E_1$, $v_1 = w_1$, where

$$\mathcal{H} = \sum_{i=1}^{K} \mathcal{T}_i + \sum_{i=1}^{K}\sum_{j<i} V_{ij}, \qquad \mathcal{T}_i = -\frac{\hbar^2 \, \nabla_i^2}{2\, m_i}, \qquad V_{ij} = \frac{\sigma_{ij}\, q^2 \, Z_i \, Z_j}{4\,\pi\varepsilon_0 \, |\mathbf{r}_i - \mathbf{r}_j|},$$

$$(16.17)$$

with $\sigma_{ij} = \pm 1$. The above describe a system of different particles interacting through Coulomb potentials. Introducing the auxiliary function f as in (16.37), one minimizes the functional $\langle f_1 \ldots f_K | \mathcal{H} | f_1 \ldots f_K \rangle$ subjected to the K constraints $\langle f_i | f_i \rangle = 1$. Using the method of Lagrange multipliers, this is equivalent to finding the absolute minimum of

$$F_{\mathcal{H}}[f_1, \ldots, f_K] = \langle f_1 \ldots f_K | \mathcal{H} | f_1 \ldots f_K \rangle - \sum_{i=1}^{K} \varepsilon_i \langle f_i | f_i \rangle, \qquad (16.18)$$

with ε_i the multipliers. The terms related to $\mathcal{T}_i$ are separable, while those related to V_{ij} are separable in pairs. From the orthonormalization condition it follows

$$F_{\mathcal{H}} = \sum_{i=1}^{K} \left(\langle f_i | \mathcal{T}_i | f_i \rangle - \varepsilon_i \langle f_i | f_i \rangle + \sum_{j=1}^{i-1} \langle f_i f_j | V_{ij} | f_i f_j \rangle \right). \qquad (16.19)$$

The minimum of $F_{\mathcal{H}}$ is found by letting $\delta F_{\mathcal{H}} = F_{\mathcal{H}}[f_1 + \delta f_1, \ldots, f_K + \delta f_K] - F_{\mathcal{H}}[f_1, \ldots, f_K] = 0$. Neglecting the second-order terms and observing that $\mathcal{T}_i$ is Hermitean yields

$$\delta \langle f_i | \mathcal{T}_i | f_i \rangle = \langle \delta f_i | \mathcal{T}_i | f_i \rangle + \langle f_i | \mathcal{T}_i | \delta f_i \rangle = 2 \,\Re \langle \delta f_i | \mathcal{T}_i | f_i \rangle. \qquad (16.20)$$

By the same token one finds $\delta \varepsilon_i \langle f_i | f_i \rangle = \varepsilon_i \, 2 \,\Re \langle \delta f_i | f_i \rangle$ and $\delta \langle f_i f_j | V_{ij} | f_i f_j \rangle = 2 \,\Re \langle f_i \, \delta f_j | V_{ij} | f_i f_j \rangle + 2 \,\Re \langle f_j \, \delta f_i | V_{ij} | f_i f_j \rangle$. The symmetry of the Coulomb terms yields $V_{ji} = V_{ij}$, whence

$$\sum_{j=1}^{i-1} \left(\langle f_i \, \delta f_j | V_{ij} | f_i f_j \rangle + \langle f_j \, \delta f_i | V_{ij} | f_i f_j \rangle \right) = \sum_{j=1}^{K} \langle f_j \, \delta f_i | V_{ij} | f_i f_j \rangle, \qquad (16.21)$$

with $j \neq i$ at the right hand side. The minimization condition of (16.19) then reads

$$2 \,\Re \sum_{i=1}^{K} \left(\langle \delta f_i | \mathcal{T}_i | f_i \rangle - \varepsilon_i \, \langle \delta f_i | f_i \rangle + \sum_{j=1}^{K} \langle f_j \, \delta f_i | V_{ij} | f_i f_j \rangle \right) = 0, \qquad (16.22)$$

with $j \neq i$ in the inner sum. As the variations δf_i are independent of each other, the term within parentheses in (16.22) must vanish for each i, this yielding a system of K coupled equations. The inner sum has a double integral in it, and is recast as

$$\sum_{j=1}^{K} \langle f_j \, \delta f_i | V_{ij} | f_i f_j \rangle = \langle \delta f_i | U_i | f_i \rangle, \qquad U_i(\mathbf{r}_i) = \sum_{j=1}^{K} \langle f_j | V_{ij} | f_j \rangle, \qquad (16.23)$$

still with $j \neq i$. The ith equation of the system then reads

$$\langle \delta f_i | \mathcal{T}_i | f_i \rangle - \varepsilon_i \, \langle \delta f_i | f_i \rangle + \langle \delta f_i | U_i | f_i \rangle = \langle \delta f_i | \, (\mathcal{T}_i + U_i - \varepsilon_i) \, f_i \rangle = 0. \quad (16.24)$$

As the above holds for any δf_i, the terms on the right hand side of the scalar products of (16.24) must cancel each other. In conclusion, the minimization condition provides a set of K single-particle equations coupled through the terms U_i:

$$(\mathcal{T}_i + U_i) \, f_i = \varepsilon_i \, f_i, \qquad i = 1, \ldots, K. \qquad (16.25)$$

The above are called *Hartree equations*, and constitute a set of Schrödinger equations whose potential energy is given by the second relation in (16.23):

$$U_i\left(\mathbf{r}_i\right) = \sum_{j=1}^{K} \int \frac{\sigma_{ij}\, q^2\, Z_i\, Z_j}{4\,\pi\,\varepsilon_0\,|\mathbf{r}_i - \mathbf{r}_j|}\, |f_j|^2\, dr_j, \qquad j \neq i. \qquad (16.26)$$

The potential energy of the ith particle is the sum of two-particle potential energies averaged with the localization probabilities of the particles different from the ith. The eigenvalue ε_i is then the energy of the ith particle in the field of the other $K-1$ particles (for this reason, the sum $\varepsilon_1 + \ldots + \varepsilon_N$ is not the total energy of the system).

The solution of (16.25) is found by iteration; one starts with $i = 1$ and provides an initial guess for $f_2, \ldots, f_K$, so that the initial guess for U_1 is calculated from (16.26) and used in (16.25). The solution of the latter yields the first iteration for f_1 and, remembering the Ritz method outlined in Sect. 16.7.1, the parameters embedded in f_1 are exploited at this stage to lower the eigenvalue; then, one proceeds with $i = 2$, using the first iteration for f_1 and the initial guess for $f_3, \ldots, f_K$, and so on. It must be noted that the initial guess is used within an integral, whose effect is that of averaging the difference between the initial guess and the actual solution; therefore, it may happen that the accuracy of the first iteration is sufficient, to the extent that the iterative process may be brought to an end. In this case, the K Eqs. (16.25) become independent of each other and the solution effort is greatly reduced.

16.5   Hartree–Fock Equations

It is now necessary to investigate the problem originally introduced in Sect. 16.3, namely, that of Eq. (16.16), which describes the separate system of electrons after fixing the nuclei to the equilibrium positions $\mathbf{R}_0$. The Hartree Eqs. (16.25) can not be applied as they stand, because they have been deduced for a system made of distinguishable particles. To treat a system of electrons, instead, it is necessary to account for the particles' spin, that was not considered in Sects. 16.3 and 16.4, and ensure the antisymmetry of the wave function (Sect. 15.7). The procedure is similar to that depicted in Sect. 16.4, the difference being that the auxiliary function f is not expressed as a product like in (16.37), but is given by a Slater determinant (15.37), whose entries depend on the position and spin coordinates of the corresponding particle [62, Sect. 8], [99, Sect. 16.3].

The calculations are rather involved and are not reported here. It is important to mention that, like in Sect. 16.4, the derivation is carried out by assuming that the external forces are absent ($U_{\text{ext}} = 0$). The procedure yields a set of K equations, coupled with each other, that generalize the Hartree Eqs. (16.25) and are called *Hartree–Fock equations*. If the accuracy of the first iteration is sufficient, the original Hamiltonian operator of (16.16), describing the system of electrons as a whole, is separated into K single-particle, identical operators:

$$\mathcal{T}_e + V_e = \sum_{i=1}^{K} \left[-\frac{\hbar^2}{2\,m}\nabla_i^2 + V_{ei}(\mathbf{r}_i) \right]. \qquad (16.27)$$

In this way the form (15.28) of the Schrödinger equation for the electrons, sought at the beginning of Sect. 16.2, is recovered.

16.6 Schrödinger Equation for the Nuclei

In the process of separating the Schrödinger equation for the electrons, (16.15), from that of the nuclei, (16.14), that is carried out in Sect. 16.3, the positions of the nuclei are fixed to the equilibrium values $\mathbf{R} = \mathbf{R}_0$. This is done to the purpose of calculating the coefficients E_e and U_u (compare with (16.6) and (16.14), respectively); such coefficients depend on $\mathbf{R}$ and can be obtained only by solving the equation for the electrons. After this step is accomplished, one turns to the equation for the nuclei, in which E_e and U_u are fixed to constants from the previous iteration; the potential energy of (16.14) then becomes

$$V_a(\vec{R}) = U_a(\vec{R}) + E_e(\vec{R}_0) + U_u(\vec{R}_0). \tag{16.28}$$

If the positions of the nuclei are kept fixed, that is, the iterative procedure outlined in Sect. 16.3 is brought to an end without solving (16.14), the exchange of momentum and energy between the system of electrons and that of nuclei can not take place. It is then necessary to proceed to the solution of (16.14); such a solution is obtained by means of an approximation shown below, and the iterative procedure for the solution of (16.14) and (16.15) is stopped right after. In this way, one keeps for the energy of the electrons the eigenvalues obtained with the nuclei fixed at $\mathbf{R}_0$, which is a convenient choice due to the advantages illustrated in Sect. 16.3.

As for the nuclei, it has already been observed in Sect. 16.3 with respect to the case of solid matter, that the nuclei, being massive and tightly bound together, are expected to depart little from their equilibrium positions $\mathbf{R}_0$. To improve with respect to the zero-order approximation $\mathbf{R} = \mathbf{R}_0$, and provisionally reasoning in classical terms, one assumes that the instantaneous displacement $\mathbf{R} - \mathbf{R}_0$ with respect to the equilibrium point is small, so that V_a in (16.28) can be approximated with a second-order Taylor expansion around $\mathbf{R}_0$. The classical form of the problem is thus brought to the case already solved in Sects. 3.9 and 3.10: indicating with $T_a + V_a$ the classical equivalent of the operator $\mathcal{T}_a + \mathcal{V}_a$, the vibrational state of the nuclei is described in terms of the normal coordinates b_σ, whose conjugate moments are $\dot{b}_\sigma$, and the total energy of the nuclei reads (compare with (3.50)):

$$T_a + V_a = \sum_{\sigma=1}^{3N} H_\sigma + V_{a0}, \qquad H_\sigma = \frac{1}{2} \dot{b}_\sigma^2 + \frac{1}{2} \omega_\sigma^2 b_\sigma^2, \tag{16.29}$$

with $\omega_\sigma > 0$ the angular frequency of the mode. As a consequence, the quantum operator takes the form

$$\mathcal{T}_a + \mathcal{V}_a = \sum_{\sigma=1}^{3N} \mathcal{H}_\sigma + V_{a0}, \qquad \mathcal{H}_\sigma = -\frac{\hbar^2}{2} \frac{\partial^2}{\partial b_\sigma^2} + \frac{1}{2} \omega_\sigma^2 b_\sigma^2 \tag{16.30}$$

which, introduced into (16.14), yields

$$\sum_{\sigma=1}^{3N} \mathcal{H}_\sigma v = E'v, \qquad E' = E - V_{a0}. \tag{16.31}$$

This procedure provides the separation of the degrees of freedom of the nuclei, and completes the separation process whose usefulness was anticipated in Sect. 16.2. The solution of (16.31) is illustrated in Sect. 12.5, and shows that the mode's energy is ascribed to the set of phonons belonging to the mode itself. As the phonons are bosons, the equilibrium distribution of their occupation numbers is given by the Bose–Einstein statistics. The description of the interaction between an electron and the nuclei is obtained from the quantum-mechanical, first-order perturbation theory applied to the two-particle collision of an electron with a phonon.

16.7 Complements

16.7.1 Ritz Method

Let $\mathcal{A}$ be a Hermitean operator with a discrete spectrum and a complete, orthonormal set of eigenfunctions. For the sake of simplicity, the eigenvalues are assumed non degenerate:

$$\mathcal{A}v_n = A_n v_n, \qquad \langle v_m | v_n \rangle = \delta_{mn}. \tag{16.32}$$

Consider the expansion of a function f in terms of the eigenfunctions of $\mathcal{A}$, $f = \sum_{n=1}^{\infty} c_n v_n$, with $c_n = \langle v_n | f \rangle$. From Parseval theorem (8.41) one finds

$$\langle f | \mathcal{A} | f \rangle = \sum_{n=1}^{\infty} A_n |c_n|^2. \tag{16.33}$$

As the eigenvalues are real, one orders them in a non-decreasing sequence: $A_1 \leq A_2 \leq A_3 \leq \ldots$, whence

$$\langle f | \mathcal{A} | f \rangle \geq A_1 \sum_{n=1}^{\infty} |c_n|^2 = A_1 \langle f | f \rangle. \tag{16.34}$$

More generally, if f is orthogonal to the first $s - 1$ eigenfunctions, then $c_1 = c_2 = \ldots = c_{s-1} = 0$, and

$$f = \sum_{n=s}^{\infty} c_n v_n, \qquad \langle f | \mathcal{A} | f \rangle \geq A_s \langle f | f \rangle, \tag{16.35}$$

where the equality holds if and only if $f = \text{const} \times v_s$. In conclusion, the functional

$$G_{\mathcal{A}}[f] = \frac{\langle f | \mathcal{A} | f \rangle}{\langle f | f \rangle} \tag{16.36}$$

has a minimum for $f = v_s$, whose value is A_s. The findings above are the basis of the *Ritz method*, that provides approximations for the eigenvalues and eigenfunctions of $\mathcal{A}$. For instance, the minimum eigenvalue A_1 and the corresponding eigenfunction $v_1(\mathbf{r})$ are approximated by letting $v_1 \simeq f(\mathbf{r}, \alpha_1, \alpha_2, \dots)$, where the form of f is prescribed and $\alpha_1, \alpha_2, \dots$ are parameters. The latter are then used to minimize $G_{\mathcal{A}}[f]$. The constraint $\langle f | f \rangle = 1$ is imposed along the calculation, yielding $A_1 \simeq \min_\alpha G_{\mathcal{A}}[f] = \langle f | \mathcal{A} | f \rangle$. When A_1, v_1 have been found, the next pair A_2, v_2 is determined by using an approximating function orthonormal to v_1, and so on. For a system made of N particles, the eigenfunctions depend on $3N$ coordinates: $v_n = v_n(\mathbf{r}_1, \dots, \mathbf{r}_N)$. It is convenient to approximate them by separating the variables:

$$f = f_1(\mathbf{r}_1) \dots f_N(\mathbf{r}_N), \tag{16.37}$$

where each f_i may also depend on parameters. The normalization constraint then yields

$$\langle f_i | f_i \rangle = 1, \qquad i = 1, \dots, N. \tag{16.38}$$

Part V
Applications to Semiconducting Crystals

Chapter 17
Periodic Structures

17.1 Introduction

The chapter outlines a number of concepts that are useful in the description of periodic structures. The first sections describe the geometrical entities (characteristic vectors, direct and reciprocal lattices, translation vectors, cells, and Brillouin zones) used for describing a lattice. The analysis focuses on the lattices of cubic type, because silicon and other semiconductor materials used in the fabrication of integrated circuits have this type of structure. The next sections introduce the mathematical apparatus necessary for solving the Schrödinger equation within a periodic structure, specifically, the translation operators, Bloch theorem, and periodic boundary conditions. Basing on such tools, the Schrödinger equation is solved, leading to the dispersion relation of the electrons, whose properties are worked out and discussed. The analogy between a wave packet in a periodic potential and in free space is also outlined. Then, the parabolic-band approximation is introduced, leading to the concept of effective mass and to the explicit calculation of the density of states. Examples of the structure of the conduction and valence bands of silicon, germanium, and gallium arsenide are provided. Considering the importance of two-dimensional and one-dimensional structures in modern technological applications, a detailed derivation of the subbands and the corresponding density of states is also given. Then, the same mathematical concepts used for solving the Schrödinger equation (Bloch theorem and periodic boundary conditions) are applied to the calculation of the lattice's vibrational spectrum in the harmonic approximation. The properties of the eigenvalues and eigenfunctions of the problem are worked out, leading to the expression of the vibrational modes. The complements provide a number of details about the description of crystal planes and directions, and about the connection between the symmetries of the Hamiltonian operator and the properties of its eigenvalues. A number of examples of application of the concepts outlined in the chapter are given in the last part of the complements; specifically, the Kronig–Penney model, showing a one-dimensional calculation of the electrons' dispersion relation, and the derivation of the dispersion relation of the one-dimensional monatomic and diatomic chains. The complements are concluded by a discussion about some analogies between the energy of the electromagnetic field and that of the vibrating lattice, and between the dispersion relation of the electrons and that of the lattice.

© Springer Science+Business Media New York 2015
M. Rudan, *Physics of Semiconductor Devices,*
DOI 10.1007/978-1-4939-1151-6_17

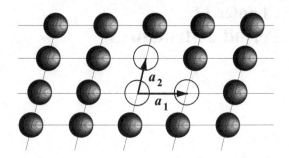

17.2 Bravais Lattice

The concepts illustrated in the previous chapters will now be applied to study the properties of a specific class of materials, the *crystals*. In fact, in the majority of cases the solid-state devices are manufactured using a crystal as basic material.[1] A crystal is made of a periodic arrangement of an atom, or a group of atoms, called *basis*. As a periodic structure is unlimited in all spatial directions, the general properties of crystals are derived using the provisional hypothesis that the material extends to infinity. The more realistic case of a finite crystal is considered at a later stage.

To describe the properties of a crystal it is convenient to superimpose to it a geometric structure, called *Bravais lattice*, made of an infinite array of discrete points generated by translation operations of the form [10]

$$\mathbf{l} = m_1 \mathbf{a}_1 + m_2 \mathbf{a}_2 + m_3 \mathbf{a}_3, \tag{17.1}$$

with m_1, m_2, m_3 any integers. In (17.1), $\mathbf{l}$ is called *translation vector* while $\mathbf{a}_1$, $\mathbf{a}_2$, $\mathbf{a}_3$ are the *characteristic vectors*; the discrete points generated by (17.1) are called *nodes*. The set of vectors $\mathbf{l}$ is closed under vector addition. Although the characteristic vectors are not necessarily of equal length, nor are they orthogonal to each other, they form a complete set; it follows that any vector $\mathbf{r}$ of the three-dimensional space is expressible as

$$\mathbf{r} = \mu_1 \mathbf{a}_1 + \mu_2 \mathbf{a}_2 + \mu_3 \mathbf{a}_3, \tag{17.2}$$

with μ_1, μ_2, μ_3 real numbers. In a zero-dimensional or one-dimensional space only one Bravais lattice is possible. In a two-dimensional space there are five Bravais lattices, respectively called *oblique, rectangular, centered rectangular (rhombic), hexagonal*, and *square* [63]. An example of oblique lattice is shown in Fig. 17.1.

---

[1] Some important exceptions exist. *Thin-Film Transistors* (TFT), commonly used in flat-screen or liquid-crystal displays, are obtained by depositing a semiconductor layer (typically, silicon) over a non-conducting substrate; due to the deposition process, the structure of the semiconductor layer is amorphous or polycrystalline. *Phase-Change Memories* (PCM) exploit the property of specific materials like chalcogenides (for example, $Ge_2Sb_2Te_5$), that switch from the crystalline to the amorphous state, and vice versa, in a controlled way when subjected to a suitable electric pulse.

Fig. 17.2 Examples of cells
in a two-dimensional Bravais
lattice of the oblique type

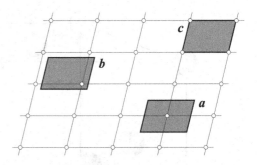

Another important concept is that of *cell*. Still considering a two-dimensional
lattice of the oblique type, a cell is a two-dimensional surface that is associated to
each node and has the following properties: (*i*) the association is one-to-one, (*ii*) the
cells are equal to each other, and (*iii*) the union of the cells covers the lattice exactly.
The cell so defined is not unique; by way of examples, all the shaded surfaces shown
in Fig. 17.2 fulfill the properties listed above. It may seem that the cell of case (*c*)
is not correct, because it touches four nodes; however, each node is shared by four
cells, so that the one-to-one correspondence is maintained. In fact, the type of cell
shown in case (*c*) is most useful to extend the definitions given so far to the more
realistic three-dimensional case. One notes in passing that the common value of the
area of the cells depicted in Fig. 17.2 is $A = |\mathbf{a}_1 \wedge \mathbf{a}_2|$, with $\mathbf{a}_1$, $\mathbf{a}_2$ the characteristic
vectors indicated in Fig. 17.1.

In a three-dimensional space the simplest cells are three-dimensional volumes,
that fulfill the same properties as in the two-dimensional case and have the atoms
placed at the corners. Such an arrangement is called *primitive*. It can be shown that
seven primitive arrangements are possible; each of them may further by enriched in
five possible ways, by adding (*a*) one atom at the center of the cell, or (*b*) one atom
at the center of each face of the cell, or (*c*) one atom at the center of each pair of
cell faces (this can be done in three different ways). The addition of extra atoms to
the primitive arrangement is called *centering*. The total number of three-dimensional
arrangements, including the primitive ones, is thus $7 \times 6 = 42$; however, it is found
that not all of them are distinct, so that the actual number of distinct three-dimensional
cells reduces to 14 [63].

A three-dimensional lattice whose characteristic vectors are mutually orthogonal
and of equal length is called *cubic*. Besides the primitive one, other two arrangements
of cubic type are possible: the first one has one additional atom at the cell's center
and is called *body-centered cubic* (BCC), the second one has one additional atom at
the center of each cell's face and is called *face-centered cubic* (FCC). A portion of a
lattice of the FCC type is shown in Fig. 17.3. Examples of chemical species whose
crystalline phase has a cubic cell are carbon (C) in the diamond phase, silicon (Si),
and germanium (Ge); in fact, this type of crystallization state is also collectively
indicated with *diamond structure*. Elements like C, Si, and Ge belong to the fourth
column of the periodic table of elements; Si and Ge are semiconductors, while C in

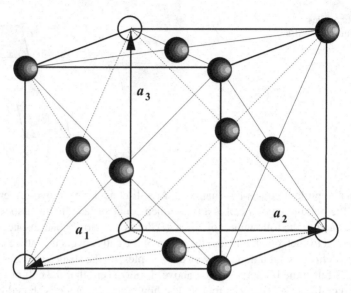

Fig. 17.3 Schematic description of a three-dimensional Bravais lattice of the FCC type. Four atoms have been removed to better show the characteristic vectors. The latter are orthogonal to each other, and of equal length

Table 17.1 Crystal constants of silicon and germanium

Material	Lattice constant (nm)	Interatomic distance (nm)
Si	0.543	0.233
Ge	0.566	0.244

the diamond phase is an insulator.[2] The FCC cell is also exhibited by some compound materials, an example of which is gallium arsenide (GaAs), a semiconductor of the so-called *III-V type*.[3] Some properties of GaAs, along with those of other III-V semiconductors, are listed in Tables 17.1 and 17.2. The type of crystallization state of GaAs and the other materials of the table is called *zincblende structure*.[4]

---

[2] The meaning of "insulator" or "semiconductor", as well as that of "conductor", is specified in Sect. 18.2.

[3] The term "III-V" derives from the fact the Ga and As belong, respectively, to the third and fifth column of the periodic table of elements.

[4] *Zincblende* is another name for *Sphalerite*, an economically important mineral whence zinc is extracted. It consists of zinc sulphide in crystalline form with some contents of iron, (Zn,Fe)S.

Table 17.2 Crystal constants of some III-V semiconductors

Material	Lattice constant (nm)	Interatomic distance (nm)
GaAs	0.563	0.244
GaP	0.545	0.236
GaSb	0.609	0.264
InAs	0.605	0.262
InP	0.586	0.254
InSb	0.647	0.280
AlSb	0.613	0.265

17.3 Reciprocal Lattice

As indicated in Sect. 17.2, the type of cell having the nodes at the corners is the most useful one for introducing a number of definitions. For instance, observing that the characteristic vectors coincide with the cell's edges (this property is evident for the FCC cell as shown in Fig. 17.3, however, it applies also to the other types of cells), one obtains for the cell's volume

$$\tau_l = \mathbf{a}_1 \cdot \mathbf{a}_2 \wedge \mathbf{a}_3 = \mathbf{a}_2 \cdot \mathbf{a}_3 \wedge \mathbf{a}_1 = \mathbf{a}_3 \cdot \mathbf{a}_1 \wedge \mathbf{a}_2, \tag{17.3}$$

where the orientation of the characteristic vectors is chosen in such a way as to make τ_l positive.

The major advantage of dealing with a periodic structure is the possibility of using the Fourier series. In fact, the functions describing the physical properties of the structure are expected to be periodic, so that one can take advantage of the Fourier expansion. The latter, in turn, entails a transformation from the coordinate space to another space reciprocal to it. Considering, for instance, the one-dimensional case of the Fourier transform given in (C.17), there the k space is the reciprocal of the x space; similarly, in an n-dimensional space like that considered in (C.20), the $\mathbf{k}$ space is the reciprocal of the $\mathbf{x}$ space: both vectors $\mathbf{k}, \mathbf{x}$ are linear combinations of the same set of mutually-orthogonal, unit vectors $\mathbf{i}_1, \ldots, \mathbf{i}_n$, so that the scalar product $\mathbf{k} \cdot \mathbf{x}$ yields $k_1 x_1 + \cdots + k_n x_n$. In the case considered here, instead, one must account for the fact that the reference $\mathbf{a}_1, \mathbf{a}_2, \mathbf{a}_3$ of the space under consideration is made of vectors that, in general, are neither mutually orthogonal nor of equal length. For this reason it is necessary to introduce a *reciprocal lattice* by defining its characteristic vectors as

$$\mathbf{b}_1 = \frac{\mathbf{a}_2 \wedge \mathbf{a}_3}{\tau_l}, \qquad \mathbf{b}_2 = \frac{\mathbf{a}_3 \wedge \mathbf{a}_1}{\tau_l}, \qquad \mathbf{b}_3 = \frac{\mathbf{a}_1 \wedge \mathbf{a}_2}{\tau_l}. \tag{17.4}$$

From the definition (17.3) of τ_l and the mixed-product property (A.32) one finds that the characteristic vectors fulfill the orthogonality and normalization relation

$$\mathbf{a}_i \cdot \mathbf{b}_j = \delta_{ij}, \tag{17.5}$$

with δ_{ij} the Kronecker symbol (A.18). The translation vectors and the general vectors of the reciprocal lattice are linear combinations of $\mathbf{b}_1$, $\mathbf{b}_2$, $\mathbf{b}_3$ whose coefficients are, respectively, integer numbers or real numbers. To distinguish the lattice based on $\mathbf{b}_1$, $\mathbf{b}_2$, $\mathbf{b}_3$ from the one based on $\mathbf{a}_1$, $\mathbf{a}_2$, $\mathbf{a}_3$, the latter is also called *direct lattice*. The common value of the cells' volume in the reciprocal lattice is

$$\tau_G = \mathbf{b}_1 \cdot \mathbf{b}_2 \wedge \mathbf{b}_3 = \mathbf{b}_2 \cdot \mathbf{b}_3 \wedge \mathbf{b}_1 = \mathbf{b}_3 \cdot \mathbf{b}_1 \wedge \mathbf{b}_2. \tag{17.6}$$

Observing that $\mathbf{a}_i$ is a length, from (17.4) it follows that $\mathbf{b}_j$ is the inverse of a length; as a consequence, the units of τ_G are m^{-3}, and the product $\gamma = \tau_l \, \tau_G$ is a dimensionless constant. The value of γ is found by combining the first relation in (17.4), which yields $\tau_l \, \mathbf{b}_1 = \mathbf{a}_2 \wedge \mathbf{a}_3$, with (17.6), so that

$$\gamma = \tau_l \, \tau_G = \tau_l \, \mathbf{b}_1 \cdot \mathbf{b}_2 \wedge \mathbf{b}_3 = \mathbf{a}_2 \wedge \mathbf{a}_3 \cdot (\mathbf{b}_2 \wedge \mathbf{b}_3) = \mathbf{a}_2 \cdot \mathbf{a}_3 \wedge (\mathbf{b}_2 \wedge \mathbf{b}_3), \tag{17.7}$$

where the last equality is due to the invariance of the mixed product upon interchange of the "wedge" and "dot" symbols (Sect. A.7). Then, using (A.33) to resolve the double vector product,

$$\tau_l \, \tau_G = \mathbf{a}_2 \cdot (\mathbf{a}_3 \cdot \mathbf{b}_3 \, \mathbf{b}_2 - \mathbf{a}_3 \cdot \mathbf{b}_2 \, \mathbf{b}_3) = \mathbf{a}_2 \cdot (\delta_{33} \, \mathbf{b}_2 - \delta_{32} \, \mathbf{b}_3) = \mathbf{a}_2 \cdot \mathbf{b}_2 = 1. \tag{17.8}$$

Also, after defining $\mathbf{e}_1 = \mathbf{b}_2 \wedge \mathbf{b}_3 / \tau_G$ one finds, by a similar calculation,

$$\mathbf{e}_1 = \frac{\mathbf{b}_2 \wedge \mathbf{b}_3}{\tau_G} = \frac{\mathbf{b}_2 \wedge (\mathbf{a}_1 \wedge \mathbf{a}_2)}{\tau_l \, \tau_G} = \mathbf{b}_2 \cdot \mathbf{a}_2 \, \mathbf{a}_1 - \mathbf{b}_2 \cdot \mathbf{a}_1 \, \mathbf{a}_2 = \mathbf{a}_1. \tag{17.9}$$

From the properties that provide the definition of cell (Sect. 17.2) it follows that, for a given lattice, the volume of the cell does not depend on its form. It follows that $\tau_l \, \tau_G = 1$ no matter what the cell's form is and, from (17.9), that the direct lattice is the reciprocal of the reciprocal lattice.

Given a direct lattice of characteristic vectors $\mathbf{a}_1$, $\mathbf{a}_2$, $\mathbf{a}_3$, it is convenient to introduce, besides the reciprocal lattice of characteristic vectors $\mathbf{b}_1$, $\mathbf{b}_2$, $\mathbf{b}_3$ defined by (17.4), another lattice called *scaled reciprocal lattice*, whose characteristic vectors are $2\pi \, \mathbf{b}_1, 2\pi \, \mathbf{b}_2, 2\pi \mathbf{b}_3$. A translation vector of the scaled reciprocal lattice has the form

$$\mathbf{g} = n_1 \, 2\pi \, \mathbf{b}_1 + n_2 \, 2\pi \, \mathbf{b}_2 + n_3 \, 2\pi \, \mathbf{b}_3, \tag{17.10}$$

with n_1, n_2, n_3 any integers, whereas a general vector of the scaled reciprocal lattice has the form

$$\mathbf{k} = \nu_1 \, 2\pi \, \mathbf{b}_1 + \nu_2 \, 2\pi \, \mathbf{b}_2 + \nu_3 \, 2\pi \, \mathbf{b}_3, \tag{17.11}$$

with ν_1, ν_2, ν_3 any real numbers. From the definitions (17.1, 17.10) of $\mathbf{l}$ and $\mathbf{g}$ one finds

$$\mathbf{l} \cdot \mathbf{g} = \sum_{is=1}^{3} m_i \, \mathbf{a}_i \cdot n_s \, 2\pi \, \mathbf{b}_s = 2\pi \sum_{is=1}^{3} m_i \, n_s \, \delta_{is} = 2\pi \sum_{i=1}^{3} m_i \, n_i, \tag{17.12}$$

namely, $l \cdot g = 2 \pi M$ with M an integer. It follows that

$$\exp\left[i\,\mathbf{g} \cdot (\mathbf{r} + \mathbf{l})\right] = \exp\left(i\,\mathbf{g} \cdot \mathbf{r}\right) \exp\left(i\,2\,\pi\,M\right) = \exp\left(i\,\mathbf{g} \cdot \mathbf{r}\right), \qquad (17.13)$$

that is, $\exp(i\,\mathbf{g} \cdot \mathbf{r})$ is periodic in the $\mathbf{r}$ space. This shows the usefulness of the scaled reciprocal lattice for treating problems related to periodic structures. In fact, given a periodic function in the $\mathbf{r}$ space, $F(\mathbf{r}+\mathbf{l}) = F(\mathbf{r})$, the Fourier expansion is generalized to the non-orthogonal case as

$$F(\mathbf{r}) = \sum_{\mathbf{g}} F_{\mathbf{g}} \exp\left(i\,\mathbf{g} \cdot \mathbf{r}\right), \qquad F_{\mathbf{g}} = \frac{1}{\tau_l} \int_{\tau_l} F(\mathbf{r}) \exp\left(-i\,\mathbf{g} \cdot \mathbf{r}\right) \, \mathrm{d}^3 r, \qquad (17.14)$$

with $\sum_{\mathbf{g}} = \sum_{n_1} \sum_{n_2} \sum_{n_3}$. The property holds also in reverse, namely,

$$\exp\left[i\,\mathbf{l} \cdot (\mathbf{k} + \mathbf{g})\right] = \exp\left(i\,\mathbf{l} \cdot \mathbf{k}\right), \qquad (17.15)$$

so that, given a periodic function in the $\mathbf{k}$ space, $\Phi(\mathbf{k} + \mathbf{g}) = \Phi(\mathbf{k})$, the following expansion holds:

$$\Phi(\mathbf{k}) = \sum_{\mathbf{l}} \Phi_{\mathbf{l}} \exp\left(i\,\mathbf{l} \cdot \mathbf{k}\right), \qquad \Phi_{\mathbf{l}} = \frac{1}{\tau_g} \int_{\tau_g} \Phi(\mathbf{k}) \exp\left(-i\,\mathbf{l} \cdot \mathbf{k}\right) \, \mathrm{d}^3 k, \qquad (17.16)$$

with $\sum_{\mathbf{l}} = \sum_{m_1} \sum_{m_2} \sum_{m_3}$. From (17.8) one also finds that in the scaled reciprocal lattice the volume of the cell is given by

$$\tau_g = (2\,\pi)^3 \, \tau_G = \frac{(2\,\pi)^3}{\tau_l}. \qquad (17.17)$$

The origin of the reference of the direct or reciprocal space has not been identified so far. After selecting the origin, consider the cell of the $\mathbf{r}$ space whose sides emanate from it; these sides are made to coincide with the characteristic vectors (compare, e.g., with Fig. 17.3), so that, to any point $\mathbf{r} = \mu_1\,\mathbf{a}_1 + \mu_2\,\mathbf{a}_2 + \mu_3\,\mathbf{a}_3$ that belongs to the interior or the boundary of the cell, the following restriction apply: $0 \leq \mu_i \leq 1$. Similarly, if one considers the cell of the $\mathbf{k}$ space whose sides emanate from the origin, for any point $\mathbf{k} = \nu_1\,2\,\pi\,\mathbf{b}_1 + \nu_2\,2\,\pi\,\mathbf{b}_2 + \nu_3\,2\,\pi\,\mathbf{b}_3$ that belongs to the interior or the boundary of the cell it is $0 \leq \nu_i \leq 1$.

17.4 Wigner–Seitz Cell—Brillouin Zone

It has been mentioned in Sect. 17.2 that the cell properties do not identify the cell uniquely. Cells of a form different from that shown, e.g., in the two-dimensional example of Fig. 17.2 can be constructed. Among them, of particular interest is the *Wigner–Seitz* cell, whose construction is shown in Fig. 17.4 still considering a two-dimensional lattice of the oblique type. First, the node to be associated to the cell is connected to its nearest neighbors as shown in the figure (the connecting segments are

Fig. 17.4 A Wigner–Seitz
cell in a two-dimensional,
oblique lattice

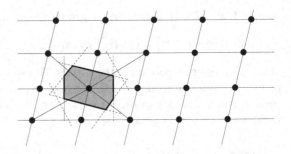

the continuous lines); then, the axis of each segment is drawn (dashed lines). As the axis is the locus of points having the same distance from the extrema of the connecting segment, one can select a contour, made of portions of the axes (thick line), such that the following property holds: any point inside the contour is closer to the node associated to the cell than to any other lattice node. From the construction described above it also follows that the shaded surface so obtained fulfills the requisites listed in Sect. 17.2, so it is a cell proper; such a cell, named after Wigner and Seitz, fulfills the additional property of the closeness of the internal points. Its construction in the three-dimensional case is similar, the only difference being that the axis is replaced with the plane normal to the connecting segment at the midpoint.

The concept of Wigner–Seitz cell, that has been introduced above using the direct lattice by way of example, is applicable also to the reciprocal and scaled-reciprocal lattices. In the scaled reciprocal lattice, the Wigner-Seitz cell associated to the origin is called *first Brillouin zone*. The set of Wigner-Seitz cells adjacent to the first Brilluoin zone is called *second Brillouin zone*, and so on. The first Brilluoin zone of the FCC lattice is shown in Fig. 17.5; its boundary is made of six square faces and eight hexagonal faces. The center of the zone is called Γ *point*; the k_1 axis belongs to the [100] crystal direction[5] and intercepts the boundary of the Brilluoin zone at two opposite positions called X *points*, that coincide with the center of square faces; the k_2, k_3 axes belong to the [010] and [001] directions, respectively, and intercept the boundary at X points as well. In turn, the {111} directions intercept the boundary at positions called L *points*, that coincide with the center of the hexagonal faces. There is a total of eight L points, because the set {111} is made of the [111], [$\bar{1}$11], [$\bar{1}\bar{1}$1], [1$\bar{1}$1] directions along with those of complementary signs.

17.5 Translation Operators

The typical procedure by which the physical properties of periodic structures, like crystals, are investigated, entails the solution of eigenvalue equations generated by quantum-mechanical operators. In this respect it is useful to introduce a class of

---

[5] The symbols indicating the crystal directions are illustrated in Sect. 17.8.1.

Fig. 17.5 The first Brillouin
zone of the FCC lattice

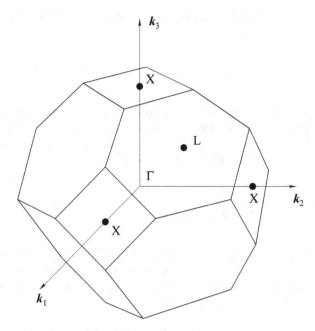

operators, called *translation operators*, that are associated to the direct-lattice vectors
l defined by (17.1). To ease the notation, the translation operator associated to the
ith lattice vector $\mathbf{l}_i$ is indicated with $\mathcal{T}_i = \mathcal{T}(\mathbf{l}_i)$. A translation operator is defined by
the property

$$\mathcal{T}_i f(\mathbf{r}) = f(\mathbf{r} + \mathbf{l}_i) \tag{17.18}$$

for all functions f defined over the direct lattice. It is easily found that translation
operators are linear and non-Hermitean. Also, they commute with each other; in fact,

$$\mathcal{T}_i \mathcal{T}_s f(\mathbf{r}) = f(\mathbf{r} + \mathbf{l}_i + \mathbf{l}_s) = \mathcal{T}_s \mathcal{T}_i f(\mathbf{r}) \tag{17.19}$$

for all functions f and all indices i, s. Remembering the property derived in
Sect. 10.4, it follows that all translation operators have a common set of eigenfunctions v. Combining this result with definition (17.18) provides the form of the
eigenvalues α of the translation operators. For this, consider the operators associated
to three arbitrary vectors $\mathbf{l}_i$, $\mathbf{l}_s$, and $\mathbf{l}_u$, and generate the eigenvalue equations

$$\mathcal{T}_i v = \alpha(\mathbf{l}_i) v, \qquad \mathcal{T}_s v = \alpha(\mathbf{l}_s) v, \qquad \mathcal{T}_u v = \alpha(\mathbf{l}_u) v. \tag{17.20}$$

If one now lets $\mathbf{l}_u = \mathbf{l}_i + \mathbf{l}_s$, it follows $\mathcal{T}_i \mathcal{T}_s v(\mathbf{r}) = v(\mathbf{r} + \mathbf{l}_i + \mathbf{l}_s) = \mathcal{T}_u v(\mathbf{r})$, that is,
$\mathcal{T}_i \mathcal{T}_s = \mathcal{T}_u$. On the other hand, the first two eigenvalue equations in (17.20) provide
$\mathcal{T}_i \mathcal{T}_s v = \mathcal{T}_i \alpha(\mathbf{l}_s) v = \alpha(\mathbf{l}_i) \alpha(\mathbf{l}_s) v$ which, combined with the third one, yields

$$\alpha(\mathbf{l}_i) \alpha(\mathbf{l}_s) = \alpha(\mathbf{l}_u) = \alpha(\mathbf{l}_i + \mathbf{l}_s). \tag{17.21}$$

The result shows that the functional dependence of α on the translation vector must be of the exponential type:

$$\alpha(\mathbf{l}) = \exp{(\mathbf{c} \cdot \mathbf{l})}, \qquad \mathbf{c} = \sum_{s=1}^{3} (\Re\chi_s + i\,\Im\chi_s)\,2\pi\,\mathbf{b}_s, \qquad (17.22)$$

where $\mathbf{c}$ is a complex vector whose units are the inverse of a length. For this reason $\mathbf{c}$ is given the general form shown in the second relation of (17.22), with $\mathbf{b}_s$ a characteristic vector of the reciprocal lattice, $s = 1, 2, 3$, and χ_s a dimensionless complex number that is provisionally left unspecified.

17.5.1   Bloch Theorem

The expression (17.22) of the eigenvalues of the translation operators makes it possible to specify some property of the eigenfunctions; in fact, combining the definition of translation operator, $\mathcal{T}v(\mathbf{r}) = v(\mathbf{r} + \mathbf{l})$, with the eigenvalue equation $\mathcal{T}v(\mathbf{r}) = \exp{(\mathbf{c} \cdot \mathbf{l})}\,v(\mathbf{r})$ yields

$$v_{\mathbf{c}}(\mathbf{r} + \mathbf{l}) = \exp{(\mathbf{c} \cdot \mathbf{l})}\,v_{\mathbf{c}}(\mathbf{r}), \qquad (17.23)$$

called *Bloch's theorem (first form)*. The importance of this result can be appreciated by observing that, if v is known within a lattice cell, and $\mathbf{c}$ is given, then the eigenfunction can be reconstructed everywhere else. The index in (17.23) reminds one that the eigenfunction depends on the choice of $\mathbf{c}$. The theorem can be recast differently by defining an auxiliary function $u_{\mathbf{c}}(\mathbf{r}) = v_{\mathbf{c}}(\mathbf{r})\exp{(-\mathbf{c} \cdot \mathbf{r})}$, so that

$$v_{\mathbf{c}}(\mathbf{r} + \mathbf{l}) = \exp{(\mathbf{c} \cdot \mathbf{l})}\,v_{\mathbf{c}}(\mathbf{r}) = \exp{(\mathbf{c} \cdot \mathbf{l})}\,u_{\mathbf{c}}(\mathbf{r})\exp{(\mathbf{c} \cdot \mathbf{r})}. \qquad (17.24)$$

In turn, from the definition of $u_{\mathbf{c}}$ one draws $v_{\mathbf{c}}(\mathbf{r} + \mathbf{l}) = u_{\mathbf{c}}(\mathbf{r} + \mathbf{l})\exp{[\mathbf{c} \cdot (\mathbf{r} + \mathbf{l})]}$ which, combined with (17.24), yields the *Bloch theorem (second form)*:

$$v_{\mathbf{c}}(\mathbf{r}) = u_{\mathbf{c}}(\mathbf{r})\exp{(\mathbf{c} \cdot \mathbf{r})}, \qquad u_{\mathbf{c}}(\mathbf{r} + \mathbf{l}) = u_{\mathbf{c}}(\mathbf{r}). \qquad (17.25)$$

The second form of Bloch's theorem shows that the auxiliary function $u_{\mathbf{c}}$ is periodic in the direct lattice, so that the eigenfunctions of the translation operators are the product of an exponential function times a function having the lattice periodicity. One notes the similarity of this result with that expressed by (13.32); in fact, the Bloch theorem is a form of the Floquet theorem (Sect. 13.4).

The eigenfunctions of the translation operators play an important role in the description of the physical properties of periodic structures. For this reason, vectors $\mathbf{c}$ of the general form are not acceptable because their real part would make $v_{\mathbf{c}}(\mathbf{r}) = u_{\mathbf{c}}(\mathbf{r})\exp{(\mathbf{c} \cdot \mathbf{r})}$ to diverge as $\mathbf{r}$ departs more and more from the origin.[6]

---

[6] This aspect is further elaborated in Sect. 17.5.3.

It is then necessary to impose the restriction $\mathbf{c} = i\,\mathbf{k}$, with $\mathbf{k}$ real. This is achieved by letting $\Re\chi_s = 0$ and $\Im\chi_s = v_s$ in the second relation of (17.22), so that the eigenvalues of the translation operators become

$$\alpha(\mathbf{l}) = \exp(i\,\mathbf{k}\cdot\mathbf{l}), \qquad \mathbf{k} = \sum_{s=1}^{3} v_s\, 2\pi\,\mathbf{b}_s. \tag{17.26}$$

Remembering (17.15), such eigenvalues are periodic in the scaled reciprocal lattice. In turn, the first and second form of the Bloch theorem become, respectively,

$$v_{\mathbf{k}}(\mathbf{r}+\mathbf{l}) = \exp(i\,\mathbf{k}\cdot\mathbf{l})\,v_{\mathbf{k}}(\mathbf{r}), \tag{17.27}$$

$$v_{\mathbf{k}}(\mathbf{r}) = u_{\mathbf{k}}(\mathbf{r})\,\exp(i\,\mathbf{k}\cdot\mathbf{r}), \qquad u_{\mathbf{k}}(\mathbf{r}+\mathbf{l}) = u_{\mathbf{k}}(\mathbf{r}). \tag{17.28}$$

Eigenfunctions of the form (17.27, 17.28) are also called *Bloch functions*. They fulfill the eigenvalue equation $\mathcal{T}v_{\mathbf{k}}(\mathbf{r}) = \exp(i\,\mathbf{k}\cdot\mathbf{l})\,v_{\mathbf{k}}(\mathbf{r})$ so that, observing that $\mathcal{T}$ is real and taking the conjugate of the eigenvalue equation yields

$$\mathcal{T}v_{\mathbf{k}}^*(\mathbf{r}) = \exp(-i\,\mathbf{k}\cdot\mathbf{l})\,v_{\mathbf{k}}^*(\mathbf{r}). \tag{17.29}$$

If, instead, one replaces $\mathbf{k}$ with $-\mathbf{k}$ in the original equation, the following is found:

$$\mathcal{T}v_{-\mathbf{k}}(\mathbf{r}) = \exp(-i\,\mathbf{k}\cdot\mathbf{l})\,v_{-\mathbf{k}}(\mathbf{r}). \tag{17.30}$$

Comparing (17.30) with (17.29) shows that $v_{\mathbf{k}}^*$ and $v_{-\mathbf{k}}$ belong to the same eigenvalue. Moreover, comparing the second expression in (17.26) with (17.11) shows that $\mathbf{k}$ is a vector of the scaled reciprocal lattice.

A further reasoning demonstrates that the variability of $\mathbf{k}$ in (17.27, 17.28) can be limited to a single cell of the scaled reciprocal lattice; for instance, to the first Brillouin zone or, alternatively, to the cell of the $\mathbf{k}$ space whose sides emanate from the origin, so that the coefficients of (17.11) fulfill the relation $0 \le v_i \le 1$ as shown in Sect. 17.3. The property derives from the periodicity of the eigenvalues in the scaled reciprocal lattice, $\exp[i\,(\mathbf{k}+\mathbf{g})\cdot\mathbf{l}] = \exp(i\,\mathbf{k}\cdot\mathbf{l})$, due to which the values of $\exp(i\,\mathbf{k}\cdot\mathbf{l})$, with $\mathbf{k}$ ranging over a single cell, provide the whole set of the operator's eigenvalues. As $\exp[i\,(\mathbf{k}+\mathbf{g})\cdot\mathbf{l}]$ is the same eigenvalue as $\exp(i\,\mathbf{k}\cdot\mathbf{l})$, the Bloch function $v_{\mathbf{k}+\mathbf{g}}(\mathbf{r})$ is the same as $v_{\mathbf{k}}(\mathbf{r})$. Note that the reasoning does not prevent the eigenvalue from being degenerate: if this is the case, one finds

$$v_{\mathbf{k}+\mathbf{g}}^{(1)} = v_{\mathbf{k}}^{(1)}, \qquad v_{\mathbf{k}+\mathbf{g}}^{(2)} = v_{\mathbf{k}}^{(2)}, \qquad \ldots \tag{17.31}$$

17.5.2   Periodic Operators

An operator $\mathcal{A}$ is *periodic* in the direct lattice if $\mathcal{A}(\mathbf{r}+\mathbf{l}) = \mathcal{A}(\mathbf{r})$ for all vectors $\mathbf{r}$ and all translation vectors $\mathbf{l}$ of the direct lattice. Periodic operators commute with translation operators: this is shown by letting $v' = \mathcal{A}v$, so that

$$\mathcal{T}(\mathbf{l})\mathcal{A}(\mathbf{r})v(\mathbf{r}) = \mathcal{T}(\mathbf{l})v'(\mathbf{r}) = v'(\mathbf{r}+\mathbf{l}) =$$

Fig. 17.6 A finite block of
material obtained by
sectioning a crystal by means
of three pairs of parallel
crystal planes

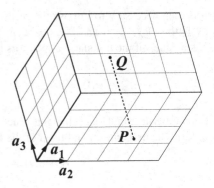

$$= \mathcal{A}(\mathbf{r}+\mathbf{l})v(\mathbf{r}+\mathbf{l}) = \mathcal{A}(\mathbf{r})v(\mathbf{r}+\mathbf{l}) = \mathcal{A}(\mathbf{r})\mathcal{T}(\mathbf{l})v(\mathbf{r}). \qquad (17.32)$$

From the commutativity property $\mathcal{T}\mathcal{A} = \mathcal{A}\mathcal{T}$ it follows that $\mathcal{T}$ and $\mathcal{A}$ have a common set of eigenfunctions, so that the eigenfunctions of a periodic operator are Bloch functions; letting $A_\mathbf{k}$ be the eigenvalue, one has

$$\mathcal{A}v_\mathbf{k} = A_\mathbf{k}\, v_\mathbf{k}, \qquad (17.33)$$

with $\mathbf{k}$ belonging to a single cell of the scaled reciprocal lattice, and $v_{\mathbf{k}+\mathbf{g}} = v_\mathbf{k}$. Since an eigenfunction belongs to one eigenvalue only, it follows

$$A_{\mathbf{k}+\mathbf{g}} = A_\mathbf{k}, \qquad (17.34)$$

namely, if an operator is periodic in the direct lattice, its eigenvalues are periodic in the scaled reciprocal lattice.

17.5.3   Periodic Boundary Conditions

In the derivation of the Bloch theorem, carried out in Sect. 17.5.1, it has been observed that complex vectors $\mathbf{c}$ in the expression $v_\mathbf{c}(\mathbf{r}) = u_\mathbf{c}(\mathbf{r})\exp{(\mathbf{c}\cdot\mathbf{r})}$ of the eigenfunctions are not acceptable because their real part would make the function to diverge. In fact, as noted in Sect. 17.5.2, such eigenfunctions belong also to the operators that commute with the translation operators and may describe physical properties of the crystal, so that diverging solutions must be discarded. On the other hand, such a divergence is due to the assumption that the crystal is unlimited; in the more realistic case of a finite block of material, diverging solutions would not appear. Unfortunately, a finite block of material is not periodic, hence the useful concepts and properties worked out so far in this chapter are not applicable to it.

To further investigate on the subject one notes that a finite block may be thought of as derived from the original crystal by sectioning the latter using three pairs of parallel crystal planes, as shown in Fig. 17.6. One of the vertices of the block coincides with the origin of the reference of the direct lattice, and the block's sides are aligned with the characteristic vectors. Also, the type of cell chosen here is the one whose sides coincide with the characteristic vectors themselves. The relation between the total number N_c of the block's cells, the block's volume Ω, and that of the cell is easily found to be

$$N_c = N_1 N_2 N_3 = \frac{\Omega}{\tau_l}, \tag{17.35}$$

with N_s the number of cell sides that match the side of the block in the sth direction. In a finite volume of material, the number of cells that belong to the interior is typically much larger than the number of cells that are adjacent to the boundaries; when solving a single-electron Schrödinger equation within such a structure, it is found that in the interior of it the eigenfunctions practically coincide with Bloch functions, whereas the effect of the real part of vector $\mathbf{c}$ in $v_\mathbf{c}(\mathbf{r}) = u_\mathbf{c}(\mathbf{r}) \exp(\mathbf{c} \cdot \mathbf{r})$ becomes relevant only when the position under investigation is close to a boundary. In fact, the real part of $\mathbf{c}$ is such that the eigenfunctions become vanishingly small far away from the volume considered [15, Sects. F-XI, O-III].

The considerations above show that for practical purposes one may keep the analysis based on the original periodicity of the unlimited structure by replacing the vanishing boundary conditions with a different type of conditions, able to formally restore periodicity in a finite structure. This is accomplished by imposing the identity of the Bloch functions corresponding to two boundary points facing each other along the direction of a characteristic vector. This prescription, called *periodic boundary condition* or *Born–Von Karman boundary condition*, is illustrated with the aid of Fig. 17.6. Consider for instance point $\mathbf{r} = \mu_1 \mathbf{a}_1 + \mu_2 \mathbf{a}_2$ (labeled P in the figure), that belongs to the boundary plane defined by $\mathbf{a}_1$, $\mathbf{a}_2$. Point Q facing P on the opposite boundary plane is such that $Q - P = \mathbf{l} = N_3 \mathbf{a}_3$ whence, applying the first form (17.23) of Bloch's theorem, one obtains $v_c(\mathbf{r} + N_3 \mathbf{a}_3) = \exp(N_3 \mathbf{c} \cdot \mathbf{a}_3) v_c(\mathbf{r})$. Imposing $v_c(\mathbf{r} + N_3 \mathbf{a}_3) = v_c(\mathbf{r})$ yields $N_3 \mathbf{c} \cdot \mathbf{a}_3 = i n_3 2 \pi$, with n_3 any integer, so that, using expression (17.22) for $\mathbf{c}$,

$$N_3 \mathbf{c} \cdot \mathbf{a}_3 = 2 \pi N_3 (\Re\chi_3 + i \Im\chi_3) = i n_3 2 \pi. \tag{17.36}$$

In conclusion, $\Re\chi_3 = 0$ and $\Im\chi_3 = 2 \pi n_3/N_3$. The same reasoning is repeated along the other directions, to finally yield

$$\mathbf{c} = i \mathbf{k}, \qquad \mathbf{k} = \sum_{s=1}^{3} \frac{n_s}{N_s} 2 \pi \mathbf{b}_s. \tag{17.37}$$

In summary, the application of the periodic boundary conditions gives $\mathbf{c}$ the same imaginary form $\mathbf{c} = i \mathbf{k}$ that was found in an unlimited structure, the difference being that in a finite structure the components of $\mathbf{k}$ are discrete instead of being continuous: given the size of the structure, which is prescribed by N_1, N_2, N_3, each $\mathbf{k}$ vector of the scaled reciprocal lattice is associated to a triad of integers n_1, n_2, n_3.

Note that the reasoning carried out at the end of Sect. 17.5.1 about the variability of $\mathbf{k}$ still holds; as a consequence, $\mathbf{k}$ can be restricted to a single cell of the scaled reciprocal lattice, so that its coefficients $v_s = n_s/N_s$ fulfill the relation $0 \leq v_s \leq 1$ as shown in Sect. 17.3. In fact, as $v_s = 0$ and $v_s = 1$ are redundant, the above relation must more appropriately be recast as $0 \leq n_s/N_s < 1$, corresponding to $n_s = 0, 1, \ldots, N_s - 1$ or, alternatively, $0 < n_s/N_s \leq 1$, corresponding to $n_s = 1, 2, \ldots, N_s$. In both cases, n_s can take N_s distinct values, so that the total number of distinct $\mathbf{k}$ vectors in a cell of the scaled reciprocal lattice is $N_1 N_2 N_3 = N_c$. From (17.35) one finds that such a number equals the number of the structure's cells in the direct lattice. Also, as the $\mathbf{k}$ vectors are equally spaced in each direction, their density in the reciprocal scaled lattice is uniform; it is given by the ratio N_c/τ_g, with τ_g the cell's volume. Remembering that the latter is invariant (Sect. 17.3), one may think of $\mathbf{k}$ as restricted to the first Brillouin zone. Combining (17.8) with (17.17) and (17.35) yields for the density

$$\frac{N_c}{\tau_g} = \frac{\Omega/\tau_l}{\tau_g} = \frac{\Omega}{(2\pi)^3}. \tag{17.38}$$

One can also define a combined density of the $\mathbf{k}$ vectors in the $\mathbf{r}, \mathbf{k}$ space, which is obtained by dividing (17.38) by the volume Ω. This yields the dimensionless combined density

$$\frac{1}{\Omega} \frac{N_c}{\tau_g} = \frac{1}{(2\pi)^3}. \tag{17.39}$$

17.6 Schrödinger Equation in a Periodic Lattice

The concepts introduced in the previous sections of this chapter are applied here to the solution of the Schrödinger equation in a periodic lattice. It is assumed provisionally that the lattice is unlimited; as a consequence, the components of the $\mathbf{k}$ vector are continuous. The Schrödinger equation to be solved derives from the single-electron operator (16.27) obtained from the separation procedure outlined in Sects. 16.2–16.5. This means the nuclei are kept fixed and the force acting on the electron derives from a potential energy[7] having the periodicity of the direct lattice: $V(\mathbf{r}+\mathbf{l}) = V(\mathbf{r})$, with $\mathbf{l}$ given by (17.1). As mentioned in Sects. 16.4 and 16.5, the external forces are absent ($U_{ext} = 0$). The equation then reads

$$\mathcal{H}w = E w, \qquad \mathcal{H} = -\frac{\hbar^2}{2m} \nabla^2 + V. \tag{17.40}$$

Replacing $\mathbf{r}$ with $\mathbf{r}+\mathbf{l}$ is equivalent to add a constant to each component of $\mathbf{r}$, say, $x_i \leftarrow x_i + l_i$, hence the partial derivatives in (17.40) are unaffected. As a

---

[7] The indices of (16.27) are dropped for simplicity.

consequence, the Hamiltonian operator as a whole has the lattice periodicity, so that its eigenfunctions are Bloch functions. Remembering (17.28), they read

$$w_{\mathbf{k}}(\mathbf{r}) = u_{\mathbf{k}}(\mathbf{r}) \exp(i\,\mathbf{k}\cdot\mathbf{r}), \qquad u_{\mathbf{k}}(\mathbf{r}+\mathbf{l}) = u_{\mathbf{k}}(\mathbf{r}), \tag{17.41}$$

with $\mathbf{k}$ belonging to the first Brillouin zone (Sect. 17.5.3). Letting $k^2 = |\mathbf{k}|^2$, (17.41) yields $\nabla^2 w_{\mathbf{k}} = \exp(i\,\mathbf{k}\cdot\mathbf{r})(-k^2+\nabla^2+2\,i\,\mathbf{k}\cdot\mathrm{grad})\,u_{\mathbf{k}}$ whence, if $E_{\mathbf{k}}$ is the eigenvalue corresponding to $w_{\mathbf{k}}$, the Schrödinger equation (17.40) becomes

$$V\,u_{\mathbf{k}} = \left[E_{\mathbf{k}} + \frac{\hbar^2}{2\,m}\left(-k^2 + \nabla^2 + 2\,i\,\mathbf{k}\cdot\mathrm{grad}\right) \right] u_{\mathbf{k}}. \tag{17.42}$$

As both V and $u_{\mathbf{k}}$ have the periodicity of the lattice, they can be expanded in terms of the translation vectors of the scaled reciprocal lattice $\mathbf{g} = n_1\,2\,\pi\,\mathbf{b}_1 + n_2\,2\,\pi\,\mathbf{b}_2 + n_3\,2\,\pi\,\mathbf{b}_3$:

$$V(\mathbf{r}) = \sum_{\mathbf{g}} V_{\mathbf{g}} \exp(i\,\mathbf{g}\cdot\mathbf{r}), \qquad u_{\mathbf{k}}(\mathbf{r}) = \sum_{\mathbf{g}} s_{\mathbf{kg}} \exp(i\,\mathbf{g}\cdot\mathbf{r}), \tag{17.43}$$

where $\sum_{\mathbf{g}} = \sum_{n_1}\sum_{n_2}\sum_{n_3}$ and

$$V_{\mathbf{g}} = \frac{1}{\tau_l}\int_{\tau_l} V(\mathbf{r})\exp(-i\,\mathbf{g}\cdot\mathbf{r})\,\mathrm{d}^3r, \qquad s_{\mathbf{kg}} = \frac{1}{\tau_l}\int_{\tau_l} u_{\mathbf{k}}(\mathbf{r})\exp(-i\,\mathbf{g}\cdot\mathbf{r})\,\mathrm{d}^3r. \tag{17.44}$$

Letting $g^2 = |\mathbf{g}|^2$, from the expansion of $u_{\mathbf{k}}$ it follows $(\nabla^2 + 2\,i\,\mathbf{k}\cdot\mathrm{grad})u_{\mathbf{k}} = -\sum_{\mathbf{g}}(g^2 + 2\,\mathbf{k}\cdot\mathbf{g})s_{\mathbf{kg}}\exp(i\,\mathbf{g}\cdot\mathbf{r})$ whence, using $g^2 + 2\,\mathbf{k}\cdot\mathbf{g} + k^2 = |\mathbf{g}+\mathbf{k}|^2$,

$$V\,u_{\mathbf{k}} = \sum_{\mathbf{g}} \left[E_{\mathbf{k}} - \frac{\hbar^2}{2\,m}|\mathbf{g}+\mathbf{k}|^2 \right] s_{\mathbf{kg}}\exp(i\,\mathbf{g}\cdot\mathbf{r}). \tag{17.45}$$

In turn, the left hand side of (17.45) reads

$$V\,u_{\mathbf{k}} = \sum_{\mathbf{g}'}\sum_{\mathbf{g}''} V_{\mathbf{g}'}\,s_{\mathbf{kg}''}\exp[i\,(\mathbf{g}'+\mathbf{g}'')\cdot\mathbf{r}] = \sum_{\mathbf{g}'}\sum_{\mathbf{g}-\mathbf{g}'} V_{\mathbf{g}'}\,s_{\mathbf{k},\mathbf{g}-\mathbf{g}'}\exp(i\,\mathbf{g}\cdot\mathbf{r}), \tag{17.46}$$

with $\mathbf{g} = \mathbf{g}'+\mathbf{g}''$. Note that the last expression on the right of (17.46) is left unchanged if $\sum_{\mathbf{g}-\mathbf{g}'}$ is replaced with $\sum_{\mathbf{g}}$. In fact, as for each vector $\mathbf{g}'$ the indices n_i of $\mathbf{g}$ span from $-\infty$ to $+\infty$, all ∞^6 combinations of indices of $\mathbf{g}$ and $\mathbf{g}'$ are present in either form of the expansion; using $\sum_{\mathbf{g}}$ instead of $\sum_{\mathbf{g}-\mathbf{g}'}$ merely changes the order of summands. Combining (17.45) with (17.46) then yields

$$\sum_{\mathbf{g}}\exp(i\,\mathbf{g}\cdot\mathbf{r})\left\{ \sum_{\mathbf{g}'} V_{\mathbf{g}'}\,s_{\mathbf{k},\mathbf{g}-\mathbf{g}'} - \left[E_{\mathbf{k}} - \frac{\hbar^2}{2\,m}|\mathbf{g}+\mathbf{k}|^2 \right] s_{\mathbf{kg}} \right\} = 0. \tag{17.47}$$

As the factors $\exp(i\,\mathbf{g}\cdot\mathbf{r})$ are linearly independent from each other for all $\mathbf{r}$, to fulfill (17.47) it is necessary that the term in braces vanishes. To proceed it is useful to associate[8] a single index b to the triad (n_1, n_2, n_3) defining $\mathbf{g}$, and another single index b' to the triad (n_1', n_2', n_3') defining $\mathbf{g}'$. Remembering that at the beginning of this section the assumption of an unlimited lattice has been made, $\mathbf{k}$ must be considered a continuous variable, so that s and E become functions of $\mathbf{k}$ proper. In conclusion, (17.47) transforms into

$$\sum_{b'} s_{b-b'}(\mathbf{k})\, V_{b'} = [E(\mathbf{k}) - T_b(\mathbf{k})]\, s_b(\mathbf{k}), \qquad b = 0, \pm1, \pm2, \ldots, \qquad (17.48)$$

with $T_b(\mathbf{k})$ the result of the association $b \leftrightarrow (n_1, n_2, n_3)$ in $\hbar^2\,|\mathbf{g}+\mathbf{k}|^2/(2\,m)$. For each $\mathbf{k}$, (17.48) is a linear, homogeneous algebraic system in the infinite unknowns s_b and coefficients $V_{b'}$, $E(\mathbf{k}) - T_b(\mathbf{k})$, with $E(\mathbf{k})$ yet undetermined.

The solution of (17.48) provides an infinite set of eigenvalues $E_1(\mathbf{k})$, $E_2(\mathbf{k})$, $\ldots, E_i(\mathbf{k}), \ldots$ associated to the given $\mathbf{k}$. As the latter ranges over the first Brillouin zone, the functions $E_i(\mathbf{k})$ are thought of as branches[9] of a many-valued function. For each branch-index i, the function $E_i(\mathbf{k})$ is called *dispersion relation*, and the set of values spanned by $E_i(\mathbf{k})$ as $\mathbf{k}$ runs over the Brillouin zone is called *energy band* of index i. Being an eigenvalue of a periodic operator, $E_i(\mathbf{k})$ is periodic within the reciprocal, scaled lattice (compare with (17.34)); also, it can be shown that $E_i(\mathbf{k})$ is even with respect to $\mathbf{k}$ (Sect. 17.8.3):

$$E_i(\mathbf{k}+\mathbf{g}) = E_i(\mathbf{k}), \qquad E_i(-\mathbf{k}) = E_i(\mathbf{k}). \qquad (17.49)$$

When a finite structure is considered, supplemented with the periodic boundary condition discussed in Sect. 17.5.3, vector $\mathbf{k}$ is discrete. On the other hand, for the derivation of (17.47) it is irrelevant whether $\mathbf{k}$ is continuous or discrete; hence, the analysis carried out in this section still holds for a discrete $\mathbf{k}$, provided the additional relations derived in Sect. 17.5.3, that describe the form of $\mathbf{k}$ and the corresponding densities, are accounted for. It must be remarked that the number N_s of cells along each direction in the direct lattice is typically very large.[10] As a consequence, a change by one unity of n_s in (17.37) is much smaller than the corresponding denominator N_s, so that for all practical purposes $E_i(\mathbf{k})$ is treated as a function of continuous variables when the derivatives with respect to the components of $\mathbf{k}$ enter the calculations.

---

[8] The association $b \leftrightarrow (n_1, n_2, n_3)$ can be accomplished in a one-to-one fashion by, first, distributing the triads into groups having a common value of $d = |n_1| + |n_2| + |n_3|$, then ordering the groups in ascending order of d: for example, $d = 0$ corresponds to $(0,0,0)$, $d = 1$ to $[(0,0,1),(0,1,0),(1,0,0),(0,0,-1),(0,-1,0),(-1,0,0)]$, and so on. As each group is made by construction of a finite number of triads, the latter are numbered within each group using a finite set of values of b; in order to have b ranging from $-\infty$ to $+\infty$, one associates a positive (negative) value of b to the triads in which the number of negative indices is even (odd).

[9] Typically, a graphic representation of $E_i(\mathbf{k})$ is achieved by choosing a crystal direction and drawing the one-dimensional restriction of E_i along such a direction. Examples are given in Sect. 17.6.5.

[10] For instance, in a cube of material with an atomic density of 6.4×10^{27} m^{-3}, the number of atoms per unit length in each direction is 4000 μm^{-1}.

The analysis carried out in this section clarifies the role of $\mathbf{k}$. In fact, for a given band index i, $\mathbf{k}$ labels the energy eigenvalue; for this reason, remembering the discussion about spin carried out in Sect. 15.5, $\mathbf{k}$ and the quantum number associated to spin determine the state of the particle. For fermions, the quantum number associated to spin has two possible values, so that two states with opposite spins are associated to each $\mathbf{k}$ vector. When the periodic boundary conditions are considered, the density of $\mathbf{k}$ vectors in the $\mathbf{k}$ space is given by (17.38), and the combined density of $\mathbf{k}$ vectors in the $\mathbf{r}, \mathbf{k}$ space is given by (17.39). As a consequence, the *density of states* in the $\mathbf{k}$ space and in the $\mathbf{r}, \mathbf{k}$ space are given, respectively, by

$$Q_k = 2\,\frac{N_c}{\tau_g} = \frac{\Omega}{4\,\pi^3}, \qquad Q = 2\,\frac{1}{\Omega}\,\frac{N_c}{\tau_g} = \frac{1}{4\,\pi^3}. \tag{17.50}$$

17.6.1 Wave Packet in a Periodic Potential

From the solution of the Schrödinger equation worked out from (17.48) one reconstructs the periodic part of the Bloch function using the second relation in (17.43). Such a function inherits the band index i, so that the Bloch functions[11] read $w_{i\,\mathbf{k}} = \zeta_{i\,\mathbf{k}}\,\exp{(\mathrm{i}\,\mathbf{k}\cdot\mathbf{r})}$; they form a complete set so that, letting $\omega_{i\,\mathbf{k}} = \omega_i(\mathbf{k}) = E_i(\mathbf{k})/\hbar$, the expansion of the wave function $\psi(\mathbf{r}, t)$ in terms of the eigenfunctions of the periodic Hamiltonian operator (17.40) reads

$$\psi(\mathbf{r}, t) = \sum_{i\,\mathbf{k}} c_{i\,\mathbf{k}}\,w_{i\,\mathbf{k}}(\mathbf{r})\,\exp{(-\,\mathrm{i}\,\omega_{i\,\mathbf{k}}\,t)} = \sum_{i\,\mathbf{k}} c_{i\,\mathbf{k}}\,\zeta_{i\,\mathbf{k}}(\mathbf{r})\,\exp{[\mathrm{i}\,(\mathbf{k}\cdot\mathbf{r} - \omega_{i\,\mathbf{k}}\,t)]},$$

$$\tag{17.51}$$

with $c_{i\,\mathbf{k}} = \langle w_{i\,\mathbf{k}}|\psi\rangle_{t=0}$ a set of constants. The expansion (17.51) bears a strong similarity with that of the wave packet describing a free particle (compare with (9.1)), the only difference between the two expansions being the periodic factor $\zeta_{i\,\mathbf{k}}(\mathbf{r})$. The similarity suggests that an approximate expression of the wave packet is achieved by following the same reasoning as in Sect. 9.6, namely, by expanding $\omega_i(\mathbf{k})$ around the average value $\mathbf{k}_0$ of the wave vector and retaining the first-order term of the expansion:[12]

$$\omega_i(\mathbf{k}) \simeq \omega_i(\mathbf{k}_0) + \mathbf{u}_i(\mathbf{k}_0)\cdot(\mathbf{k} - \mathbf{k}_0), \qquad \mathbf{u}_i(\mathbf{k}_0) = \big(\mathrm{grad}_{\mathbf{k}}\omega_i\big)_0, \tag{17.52}$$

with $\mathbf{u}_i$ the group velocity of the ith band. The approximation holds as long as $|R_i|\,t \ll 2\,\pi$, where R_i is the rest of the expansion. Letting $\omega_{i\,0} = \omega_i(\mathbf{k}_0)$, $\zeta_{i\,0} = \zeta_i(\mathbf{k}_0)$, and $\Phi_{i\,0} = \mathbf{k}_0\cdot\mathbf{r} - \omega_{i\,0}\,t$, the approximate expression of ψ reads

$$\psi(\mathbf{r}, t) \simeq \sum_{i\,\mathbf{k}} c_{i\,\mathbf{k}}\,\zeta_{i\,0}\,\exp{(\mathrm{i}\,\Phi_{i\,0})}\,\exp{[\mathrm{i}\,(\mathbf{r} - \mathbf{u}_i\,t)\cdot(\mathbf{k} - \mathbf{k}_0)]}. \tag{17.53}$$

---

[11] Here the periodic part of $w_{i\,\mathbf{k}}$ is indicated with $\zeta_{i\,\mathbf{k}}$ to avoid confusion with the group velocity.

[12] In the case of a free particle (Sect. 9.6) the approximation neglects only the second order because $\omega(\mathbf{k})$ has a quadratic dependence on the components of $\mathbf{k}$. Here, instead, the expansion has in general all terms due to the more complicate form of $\omega_i(\mathbf{k})$, so the neglected rest R_i contains infinite terms.

Fig. 17.7 A one-dimensional example of the periodic factor ζ_{n0} of (17.56)

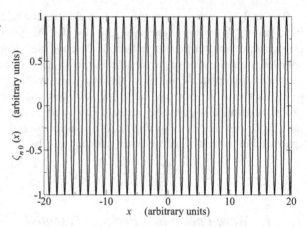

The envelope function is now defined as in (9.26), the difference being that a sum is used here instead of an integral:

$$A\left(\mathbf{r} - \mathbf{u}_i\,t\,;\mathbf{k}_0\right) = \sum_{\mathbf{k}} c_{i\mathbf{k}} \exp\left[\mathrm{i}\,\left(\mathbf{r} - \mathbf{u}_i\,t\right)\cdot\left(\mathbf{k} - \mathbf{k}_0\right)\right], \tag{17.54}$$

so that

$$\psi(\mathbf{r}, t) \simeq \sum_i \zeta_{i0}\,\exp\left(\mathrm{i}\,\Phi_{i0}\right) A\left(\mathbf{r} - \mathbf{u}_i\,t\,;\mathbf{k}_0\right). \tag{17.55}$$

As a further approximation one considers the fact that the number of $\mathbf{k}$ vectors of the first Brillouin zone is in general very large, because it equals the number N_c of direct-lattice cells. It follows that, although the set of eigenfunctions belonging to a single branch is not complete, such a set is still able to provide an acceptable description of the wave packet. In this case one fixes the branch index, say, $i = n$, so that $\psi(\mathbf{r}, t) \simeq \zeta_{n0}\,\exp\left(\mathrm{i}\,\Phi_{n0}\right) A(\mathbf{r} - \mathbf{u}_n\,t\,;\mathbf{k}_0)$. It follows

$$|\psi(\mathbf{r}, t)|^2 \simeq |\zeta_{n0}(\mathbf{r})|^2\,|A\left(\mathbf{r} - \mathbf{u}_n\,t\,;\mathbf{k}_0\right)|^2. \tag{17.56}$$

In (17.56), the periodic factor $|\zeta_{n0}(\mathbf{r})|^2$ is a rapidly-oscillating term whose period is of the order of the lattice constant; such a term does not provide any information about the particle's localization. This information, in fact, is carried by the envelope function, like in the case of a free particle outlined in Sect. 9.6. A one-dimensional example about how (17.56) is built up is given in Figs. 17.7, 17.8, 17.9, and 17.10.

17.6.2 Parabolic-Band Approximation

The dispersion relation $E_n(\mathbf{k})$ obtained from the solution of the Schrödinger equation (17.40) in a periodic lattice fulfills the periodicity condition given by the

Fig. 17.8 A one-dimensional example of the envelope function $A(\mathbf{r} - \mathbf{u}_n t ; \mathbf{k}_0)$ of (17.56)

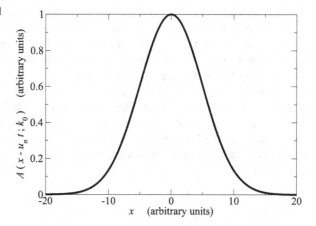

Fig. 17.9 Product of the two functions shown in Figs. 17.7 and 17.8

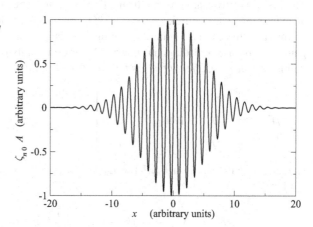

Fig. 17.10 The function of Fig. 17.9 squared

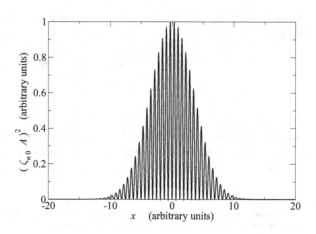

first expression in (17.49). As a consequence, $E_n(\mathbf{k})$ has necessarily a number of extremum points within the first Brillouin zone or at the boundary of it.[13]

In view of further developments of the theory it is useful to investigate the form of $E_n(\mathbf{k})$ in the vicinity of such extremum points. To this purpose, the absolute minima are considered first; for a given branch index n assume that the number of such minima is M_C, and let $\mathbf{k}_a$ be the value of $\mathbf{k}$ at the ath minimum, $a = 1, \ldots, M_C$, with $E_C = E_n(\mathbf{k}_a)$. At $\mathbf{k} = \mathbf{k}_a$ the Hessian matrix of $E_n(\mathbf{k})$ is symmetric and positive definite, hence it can be diagonalized with positive real eigenvalues. In other terms, the reference in the $\mathbf{k}$ space can be chosen in such a way as to make the Hessian matrix of $E_n(\mathbf{k})$ diagonal; using such a reference, the second-order expansion of $E_n(\mathbf{k})$ around $\mathbf{k}_a$ reads

$$E_n(\mathbf{k}) \simeq E_C + \frac{1}{2} \sum_{i=1}^{3} \left(\frac{\partial^2 E_n}{\partial k_i^2} \right)_a (k_i - k_{ia})^2 \geq E_a, \quad a = 1, \ldots, M_C. \quad (17.57)$$

The first derivatives are missing from (17.57) because the expansion is carried out at an extremum. The coefficients $(\partial^2 E_n / \partial k_i^2)_a$ are in general different from each other, so that the sum in (17.57) may be thought of as a positive-definite quadratic form generated by a 3×3 diagonal matrix. Noting the units of the matrix entries one defines the *inverse, effective-mass tensor* of the ath minimum as

$$(\hat{m}_a)^{-1} = \begin{bmatrix} 1/m_{1a} & 0 & 0 \\ 0 & 1/m_{2a} & 0 \\ 0 & 0 & 1/m_{3a} \end{bmatrix}, \qquad \frac{1}{m_{ia}} = \frac{1}{\hbar^2} \left(\frac{\partial^2 E_n}{\partial k_i^2} \right)_a > 0 \quad (17.58)$$

so that, using the notation $E_{ne}(\mathbf{k}) = E_n(\mathbf{k}) - E_C \geq 0$, $\delta k_i = k_i - k_{ia}$, (17.57) takes the form

$$E_{ne} = \sum_{i=1}^{3} \frac{\hbar^2}{2 m_{ia}} (k_i - k_{ia})^2 = \frac{1}{2} \hbar \, \delta \mathbf{k} \cdot (\hat{m}_a)^{-1} \hbar \, \delta \mathbf{k}. \quad (17.59)$$

Being the inverse, effective-mass tensor diagonal, the *effective-mass tensor* $\hat{m}_a$ is given by

$$\hat{m}_a = \begin{bmatrix} m_{1a} & 0 & 0 \\ 0 & m_{2a} & 0 \\ 0 & 0 & m_{3a} \end{bmatrix}. \quad (17.60)$$

The approximation shown above, that consists in replacing the dispersion relation with its second-order expansion near an extremum, is called *parabolic-band approximation*. The group velocity to be associated with a $\mathbf{k}$ vector in the vicinity of a

---

[13] As mentioned in Sect. 17.6, $E_n(\mathbf{k})$ is considered as a function of a continuous vector variable $\mathbf{k}$ even when the periodic boundary conditions are assumed.

minimum is found by applying to (17.59) the second relation in (17.52):

$$\mathbf{u}_n(\mathbf{k}) = \frac{1}{\hbar} \operatorname{grad}_\mathbf{k} E_n(\mathbf{k}) = \frac{1}{\hbar} \operatorname{grad}_\mathbf{k} E_{ne}(\mathbf{k}) = (\hat{m}_a)^{-1} \hbar \, \delta\mathbf{k}. \qquad (17.61)$$

The calculation in the vicinity of an absolute maximum is similar.[14] Assume that the number of maxima in the nth branch of the dispersion relation is M_V, and let $\mathbf{k}_a$ be the value of $\mathbf{k}$ at the ath maximum, $a = 1, \ldots, M_V$, with $E_V = E_n(\mathbf{k}_a)$. The second-order expansion of $E_n(\mathbf{k})$ around $\mathbf{k}_a$ reads

$$E_n(\mathbf{k}) \simeq E_V + \frac{1}{2} \sum_{i=1}^{3} \left(\frac{\partial^2 E_n}{\partial k_i^2} \right)_a (k_i - k_{ia})^2 \le E_a, \quad a = 1, \ldots, M_V, \qquad (17.62)$$

where the Hessian matrix is negative definite. For this reason, the inverse, effective-mass tensor at the ath maximum is defined as

$$(\hat{m}_a)^{-1} = \begin{bmatrix} 1/m_{1a} & 0 & 0 \\ 0 & 1/m_{2a} & 0 \\ 0 & 0 & 1/m_{3a} \end{bmatrix}, \qquad \frac{1}{m_{ia}} = -\frac{1}{\hbar^2} \left(\frac{\partial^2 E_n}{\partial k_i^2} \right)_a > 0$$

$$(17.63)$$

so that, using the notation $E_{nh}(\mathbf{k}) = E_V - E_n(\mathbf{k}) \ge 0$, (17.57) takes the form

$$E_{nh} = \sum_{i=1}^{3} \frac{\hbar^2}{2 \, m_{ia}} (k_i - k_{ia})^2 = \frac{1}{2} \hbar \, \delta\mathbf{k} \cdot (\hat{m}_a)^{-1} \hbar \, \delta\mathbf{k}. \qquad (17.64)$$

The group velocity to be associated with a $\mathbf{k}$ vector in the vicinity of a maximum reads

$$\mathbf{u}_n(\mathbf{k}) = \frac{1}{\hbar} \operatorname{grad}_\mathbf{k} E_n(\mathbf{k}) = -\frac{1}{\hbar} \operatorname{grad}_\mathbf{k} E_{nh}(\mathbf{k}) = -(\hat{m}_a)^{-1} \hbar \, \delta\mathbf{k}. \qquad (17.65)$$

It is important to note that the expressions of the parabolic-band approximation given in this section have been worked out in a specific reference of the $\mathbf{k}$ space, namely, the reference where the Hessian matrix is diagonal. In so doing, the reference of the direct space $\mathbf{r}$ has been fixed as well, because the two references are reciprocal to each other (Sect. 17.3). In other terms, when diagonal expressions like (17.59) or (17.64) are used in a dynamical calculation, the reference in the $\mathbf{r}$ space can not be chosen arbitrarily.

---

[14] The parabolic-band approximation is not necessarily limited to absolute minima or absolute maxima; here it is worked out with reference to such cases because they are the most interesting ones. However, it applies as well to relative minima and relative maxima. The different values of the inverse, effective-mass tensor's entries between an absolute and a relative minimum of a branch in GaAs give rise to interesting physical effects (Sect. 17.6.6).

17.6.3 Density of States in the Parabolic-Band Approximation

Calculations related to many-particle systems often involve the density of states in energy (e.g., Sects. 15.8.1, 15.8.2). The calculation of this quantity is relatively simple for the dispersion relation $E_n(\mathbf{k})$ in the parabolic-band approximation, because the dispersion relation is quadratic in the components of $\mathbf{k}$ and, in turn, the density of the $\mathbf{k}$ vectors is constant. In fact, it is found from (B.34) that in the three-dimensional case a quadratic expression $A = u^2 + v^2 + w^2$ yields a density of states equal to $A^{1/2}$. It is then sufficient to reduce (17.59) and (17.64) to the quadratic expression above. Taking (17.59) by way of example, one disposes with the multiplicative factors and the shift in the origin by applying the *Herring–Vogt transformation*

$$\eta_i = \frac{\hbar}{\sqrt{2 m_{ia}}} (k_i - k_{ia}), \qquad (17.66)$$

to find

$$E_{ne} = \sum_{i=1}^{3} \eta_i^2 = \eta^2, \qquad dk_i = \frac{\sqrt{2 m_{ia}}}{\hbar} d\eta_i, \qquad (17.67)$$

with $\eta > 0$, and

$$d^3 k = dk_1 \, dk_2 \, dk_3 = 2 \frac{\sqrt{2}}{\hbar^3} m_{ea}^{3/2} d^3 \eta, \qquad m_{ea} = (m_{1a} m_{2a} m_{3a})^{1/3}. \qquad (17.68)$$

Turning to spherical coordinates $\eta_1 = \eta \sin \vartheta \cos \varphi$, $\eta_2 = \eta \sin \vartheta \sin \varphi$, $\eta_3 = \eta \cos \vartheta$ yields $d^3 \eta = \eta^2 d\eta \sin \vartheta \, d\vartheta \, d\varphi$ (Sect. B.1), where the product $\eta^2 \, d\eta$ is found by combining the relations $2 \eta \, d\eta = dE_{ne}$ and $d\eta = dE_{ne}/(2 \sqrt{E_{ne}})$:

$$\eta^2 \, d\eta = \frac{1}{2} \sqrt{E_{ne}} \, dE_{ne}. \qquad (17.69)$$

The number of states belonging to the elementary volume $d^3 k$ is $dN = Q_k \, d^3 k$, with Q_k the density of states in the $\mathbf{k}$ space given by the first expression in (17.50). If the elementary volume is centered on a $\mathbf{k}$ vector close to the ath minimum of $E_n(\mathbf{k})$, so that the parabolic-band approximation holds, one has

$$dN_a = Q_k \, d^3 k = \frac{\Omega}{4 \pi^3} 2 \frac{\sqrt{2}}{\hbar^3} m_{ea}^{3/2} \frac{1}{2} \sqrt{E_{ne}} \, dE_{ne} \sin \vartheta \, d\vartheta \, d\varphi. \qquad (17.70)$$

The integral over the angles yields 4π, whence

$$\int_{\varphi=0}^{2\pi} \int_{\vartheta=0}^{\pi} dN_a = \Omega \frac{\sqrt{2}}{\pi^2 \hbar^3} m_{ea}^{3/2} \sqrt{E_{ne}} \, dE_{ea} = g_a(E_{ne}) dE_{ne} \qquad (17.71)$$

where, by construction, $g_a(E_{ne})$ is the density of states in energy around the ath minimum. Adding g_a over the M_C absolute minima yields the total density of states in energy,

$$g(E_{ne}) = \sum_{a=1}^{M_C} g_a = \Omega \frac{\sqrt{2}}{\pi^2 \hbar^3} M_C m_e^{3/2} \sqrt{E_{ne}}, \qquad m_e = \left(\frac{1}{M_C} \sum_{a=1}^{M_C} m_{ea}^{3/2} \right)^{2/3},$$

(17.72)

with m_e the *average effective mass* of the absolute minima. The *combined density of states* in the energy and **r** spaces then reads

$$\gamma(E_{ne}) = \frac{g(E_{ne})}{\Omega} = \frac{\sqrt{2}}{\pi^2 \hbar^3} M_C m_e^{3/2} \sqrt{E_{ne}}.$$

(17.73)

Note that, apart from the different symbol used to indicate the volume in the **r** space, and the replacement of m with $M_C m_e^{3/2}$, the relations (17.72) and (17.73) are identical, respectively, to (15.64) and (15.65), expressing the density of states and combined density of states in a box.

The calculation of the density of states in energy in the vicinity of the M_V absolute maxima is identical to the above, and yields

$$g(E_{nh}) = \sum_{a=1}^{M_V} g_a = \Omega \frac{\sqrt{2}}{\pi^2 \hbar^3} M_V m_h^{3/2} \sqrt{E_{nh}}, \qquad m_h = \left(\frac{1}{M_V} \sum_{a=1}^{M_V} m_{ha}^{3/2} \right)^{2/3},$$

(17.74)

where m_h is the average effective mass of the absolute maxima. In turn it is $m_{ha} = (m_{1a} m_{2a} m_{3a})^{1/3}$, with m_{ia} given by the second relation in (17.63).

17.6.4 Crystals of Si, Ge, and GaAs

Among semiconductors, silicon (Si), germanium (Ge), and gallium arsenide (GaAs) are very important for the electronic industry. This section is devoted to illustrating some properties of their crystal and energy-band structures. The crystals of silicon and germanium are of the face-centered, cubic type; the reciprocal lattices have the body-centered, cubic structure. The *lattice constants*, that is, the physical sizes of the unit cell, are the same in the [100], [010], and [001] directions (Sect. 17.8.1). Their values at $T = 300$ K are given in Table 17.1 [80]. The crystals of the materials under consideration are formed by elementary blocks like that shown in Fig. 17.11. Each atom has four electrons in the external shell, so that it can form four chemical bonds with other identical atoms; the latter place themselves symmetrically in space, to build up the tetrahedral structure shown in the figure. In this structure, which is of the body-centered cubic type with a side equal to one half the lattice constant a, the chemical bonds of the central atom are saturated, whereas the atoms placed at

Fig. 17.11 Tetrahedral
organization of the
elementary, body-centered
cubic block of silicon or
germanium. The side of the
cube is one half the lattice
constant a

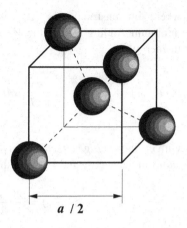

$a/2$

the vertices still have three bonds unsaturated; as a consequence, they may behave
as centers of new tetrahedral structures identical to the original one.

An example of this is given in Fig. 17.12: the top half of the figure shows two
replicas of the elementary block of Fig. 17.11 sharing an atom belonging to an
upper corner, while the bottom half of the figure shows again two replicas, this time
sharing an atom belonging to a lower corner. The atoms drawn in white do not belong
to any of the elementary blocks considered in the figure, and serve the purpose of
demonstrating how the rest of the crystal is connected to them. Note that the structure
in the bottom half of Fig. 17.12 is identical to that of the top half, the difference being
simply that one structure is rotated by 90° with respect to the other on a vertical axis.
The construction is now completed by bringing the two halves together, as shown
in Fig. 17.13; this provides the diamond structure mentioned in Sect. 17.2. Such a
structure is of the face-centered, cubic type, with an additional atom at the center of
each tetrahedral block.

The minimum distance d among the atoms (*interatomic distance*) is the distance
from the atom in the center of the tetrahedral elementary block to any of the atoms
at its vertices; its relation with the lattice constant is easily found to be

$$d = \frac{\sqrt{3}}{4}\, a. \tag{17.75}$$

The description is similar for gallium arsenide [80], and for a number of semiconductors of the III-V type, whose crystal constants are listed in Table 17.2.

17.6.5 Band Structure of Si, Ge, and GaAs

Coming now to the description of the band structure, it is important to focus on the
bands that are able to contribute to the electric conduction of the material. In fact,
considering the aim of manufacturing electronic devices out of these materials, the

Fig. 17.12 Diamond structure. The *top* and *bottom* halves are shown separately

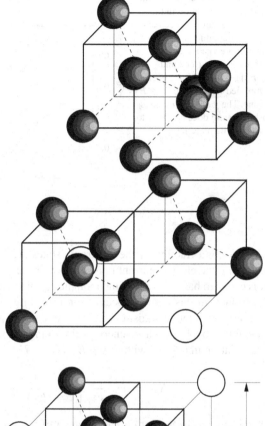

Fig. 17.13 Diamond structure obtained by joining together the *top* and *bottom* halves shown separately in Fig. 17.12

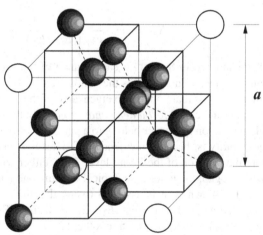

bands that do not contribute to the electric current are not relevant. It is intuitive that a band with no electrons, that is, whose states have a zero probability of being occupied, is not able to provide any conduction; it is less intuitive (in fact, this is demonstrated in Sect. 19.3) that a band whose states are fully occupied does not provide any conduction either. It follows that the only bands of interest are those

Fig. 17.14 Calculation of the particles' population in the conduction and valence bands of a semiconductor. To make them more visible, the products $g(E) P(E)$ and $g(E) [1 - P(E)]$ have been amplified with respect to $g(E)$ alone. The gap's extension is arbitrary and does not refer to any specific material

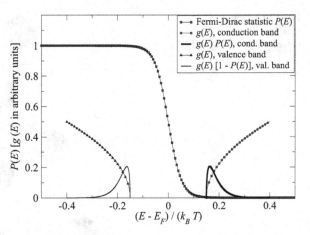

where only a fraction of the electronic states are occupied. Although a discussion involving the electric current must necessarily refer to a non-equilibrium condition, it is easier to base the reasoning upon the equilibrium condition at some temperature T; in fact, in this case the occupation probability of the electronic states is given by the Fermi–Dirac statistics (15.49). As a consequence, the number of electrons belonging to a band whose energy values range, say, from E_a to E_b, is given by the first relation in (15.48) with $\alpha + \beta E = (E - E_F)/(k_B T)$, namely,

$$N_{ab} = \int_{E_a}^{E_b} \frac{g(E)}{\exp\left[(E - E_F)/(k_B T)\right] + 1} \, \mathrm{d}E. \qquad (17.76)$$

As mentioned in Sect. 17.6 each branch of the dispersion relation $E_i(\mathbf{k})$ spans an energy band. In many cases the bands are disjoint from each other, namely, energy intervals exist that contain no eigenvalue of the Schrödinger equation (17.40). Such intervals are called *forbidden bands* or *gaps*. In the equilibrium condition the energy of an electron can never belong to a gap, no matter what the value of the occupation probability is, because the density of states is zero there. Also, at a given temperature the position of the Fermi level E_F is either within a band (edges included), or within a gap; the latter case, typical of semiconductors, is illustrated with the aid of Fig. 17.14, where it is assumed (using the units of $(E - E_F)/(k_B T)$) that a gap exists between the energies E_V, E_C such that $(E_V - E_F)/(k_B T) = -0.15$ and $(E_C - E_F)/(k_B T) = +0.15$. In other terms, E_V is the upper energy edge of a band, and E_C the lower energy edge of the next band. These assumptions also imply that the Fermi level coincides with the gap's midpoint. As will become apparent below, the two bands that are separated by the Fermi level are especially important; for this reason they are given specific names: the band whose absolute maximum is E_V is called *valence band*, that whose absolute minimum is E_C is called *conduction band*.

As shown in the figure, the case is considered (typical of Si, Ge, and GaAs) where the gap's width contains the main variation of the Fermi–Dirac statistics;[15] as a consequence, the occupation probability becomes vanishingly small as the difference $E - E_C$ becomes larger, so that only the energy states near the absolute minimum E_C have a non-vanishing probability of being occupied. Thank to this reasoning, to the purpose of calculating (17.76) one can replace the density of states $g(E)$ with the simplified expression (17.72) deduced from the parabolic-band approximation; such an expression, $g(E) \propto \sqrt{E - E_C}$, is shown in Fig. 17.14 in arbitrary units, along with the $g(E) P(E)$ product (thick line), that represents the integrand of (17.76) with reference to the conduction band. To make it more visible, the $g(E) P(E)$ product is drawn in a scale amplified by 10^3 with respect to that of $g(E)$ alone. The number of electrons belonging to the conduction band is proportional to the area subtended by the $g(E) P(E)$ curve.

Coming now to the valence band, the probability $1 - P(E)$ that a state at energy E is empty becomes vanishingly small as the difference $E_V - E$ becomes larger, so that only the energy states near the absolute maximum E_V have a non-vanishing probability of being empty. Empty states are also called *holes*. The number of holes is given by an integral similar to (17.76), where $P(E)$ is replaced with $1 - P(E)$. This calculation is made easier by observing that, due to the form of the Fermi–Dirac statistics, it is

$$1 - \frac{1}{\exp\left[(E - E_F)/(k_B T)\right] + 1} = \frac{1}{\exp\left[(E_F - E)/(k_B T)\right] + 1}. \tag{17.77}$$

Also in this case one can use for the density of states the parabolic-band approximation; such an expression, $g(E) \propto \sqrt{E_V - E}$, is shown in Fig. 17.14 in arbitrary units, along with the $g(E)[1 - P(E)]$ product (thin line). As before, the product is drawn in a scale amplified by 10^3 with respect to that of $g(E)$ alone. The number of holes belonging to the valence band is proportional to the area subtended by the $g(E)[1 - P(E)]$ curve.

Thanks to the spatial uniformity of the crystal, the concentration of the electrons in the conduction band is obtained by dividing their number by the crystal volume Ω or, equivalently, by replacing the density of states in energy g with the combined density of states in energy and volume γ given by (17.73). A similar reasoning holds for holes. The explicit expressions of the concentrations are given in Sect. 18.3. Here it is important to remark that the perfect symmetry of the curves $g(E) P(E)$ and $g(E)[1 - P(E)]$ in Fig. 17.14 is due to the simplifying assumptions that E_F coincides with the gap's midpoint and that $M_V m_h^{3/2} = M_C m_e^{3/2}$ (compare with (17.73) and (17.74)). Neither hypothesis is actually true, so that in real cases the two curves are not symmetric; however, as shown in Sect. 18.3, the areas subtended by them are nevertheless equal to each other.

---

[15] The extension of the energy region where the main variation of the Fermi statistics occurs is estimated in Prob. 15.1.

Fig. 17.15 Schematic view of the two branches of the valence band of Si, Ge, or GaAs in the [100] direction

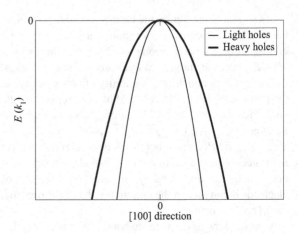

[100] direction

17.6.5.1   Valence Band

The valence band of Si, Ge, and GaAs is made of two branches of $E(\mathbf{k})$, having the same absolute maximum E_V at $\mathbf{k} = 0$ (so that $M_V = 2$), but different curvatures. They are shown in Fig. 17.15, where the horizontal axis coincides with the [100] direction in the $\mathbf{k}$ space, corresponding to the scalar variable k_1. As a consequence, the origin of the horizontal axis coincides with the Γ point (Sect. 17.4); the axis intersects the boundary of the first Brillouin zone at the X points (not shown in the figure). The origin of the vertical axis coincides with E_V. The two branches are not spherically symmetric; in fact, letting $E_V = 0$, the dependence of each of them on the spherical coordinates k, ϑ, φ has the form [54, Sect. 8.7]

$$-\frac{\alpha}{2} k^2 \left[1 \pm j(\vartheta, \varphi)\right], \qquad \alpha > 0, \tag{17.78}$$

called *warped*. In the parabolic-band approximation the angular part j is neglected with respect to unity, and the two branches become spherically symmetric around $\mathbf{k} = 0$; still with $E_V = 0$, the dependence on k_1 of each branch has the form $E = -\alpha k_1^2/2$, where the constant α is smaller in the upper branch (indicated by the thick line in Fig. 17.15), and larger in the lower one. As a consequence, the corresponding component of the effective-mass tensor (17.63), that reads in this case $m_1 = \hbar^2/\alpha$, is larger in the upper branch and smaller in the lower one. For this reason, the holes associated to the energy states of the upper branch are called *heavy holes*, those associated to the lower branch are called *light holes*.

The analysis is identical in the other two directions [010] and [001] so that, for each branch of the valence band, the diagonal entries of the effective-mass tensor are equal to each other. Such tensors then read $m_{hh}\,\mathcal{I}$, $m_{hl}\,\mathcal{I}$, with $\mathcal{I}$ the identity tensor; the first index of the scalar effective mass stands for "hole", while the second one stands for "heavy" or "light". The second-order expansions around E_V take

Table 17.3 Normalized effective masses of the valence band of Si, Ge, and GasAs

Material	$m_{hh}(T_a)/m_0$	$m_{hl}(T_a)/m_0$
Si	0.5	0.16
Ge	0.3	0.04
GaAs	0.5	0.12

respectively the form[16]

$$E_V - E_h(\mathbf{k}) = \frac{\hbar^2}{2\,m_{hh}} \sum_{i=1}^{3} k_i^2, \qquad E_V - E_l(\mathbf{k}) = \frac{\hbar^2}{2\,m_{hl}} \sum_{i=1}^{3} k_i^2. \qquad (17.79)$$

Due to (17.79), the constant-energy surfaces $E_V - E_h(\mathbf{k}) = $ const and $E_V - E_l(\mathbf{k}) = $ const are spheres, whose radius squared is $2\,m_{hh}\,(E_V - E_h)/\hbar^2$ and $2\,m_{hl}\,(E_V - E_l)/\hbar^2$, respectively. The values of m_{hh} and m_{hl} at room temperature T_a are listed in Table 17.3 [103, Sect. 2-3]; they are normalized to the rest mass of the free electron, $m_0 \simeq 9.11 \times 10^{-31}$ kg. The effective masses depend in general on temperature because a change in the latter modifies the lattice constants: as a consequence, the characteristic vectors of the reciprocal lattice change as well, this deforming the dispersion relation $E(\mathbf{k})$; on the other hand, the variation of the effective masses with temperature is weak, so it is typically neglected.[17]

17.6.5.2 Conduction Band

The conduction band of Si, Ge, and GaAs has only one branch. However, the absolute minima (also called *valleys*) are placed differently. In GaAs there is only one absolute minimum at $\mathbf{k} = 0$, with spherical symmetry. In the parabolic-band approximation, the constant-energy surface is given by

$$E(\mathbf{k}) - E_C = \frac{\hbar^2}{2\,m_e} \sum_{i=1}^{3} k_i^2, \qquad (17.80)$$

namely, a sphere whose radius squared is $2\,m_e\,(E - E_C)/\hbar^2$. The band exhibits also secondary minima at $E_C + \Delta E$, with $\Delta E \simeq 0.36$ eV (Fig. 17.16).

The conduction band of Si has six absolute minima ($M_C = 6$), grouped into three pairs. The latter belong to the [100], [010], and [001] directions, respectively, and are symmetrically placed with respect to the Γ point $\mathbf{k} = 0$. Their coordinates are

$$[100] : (\pm k_m, 0, 0), \qquad [010] : (0, \pm k_m, 0), \qquad [001] : (0, 0, \pm k_m), \qquad (17.81)$$

where $k_m \simeq 0.85\,k_B > 0$, with k_B the distance between the Γ and X points (Fig. 17.17).

---

[16] From now on the band index n introduced in (17.56) is omitted from the notation.

[17] In contrast, the temperature dependence of the energy gap, due to the deformation of the dispersion relation, can not be neglected because of its strong effect on the carrier concentration (Sect. 18.3).

Fig. 17.16 Schematic view of
the conduction band of GaAs
in the [100] direction

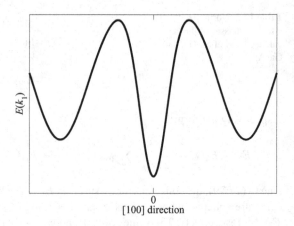

$E(k_1)$

0
[100] direction

Fig. 17.17 Schematic view of
the conduction band of Si in
the [100] direction

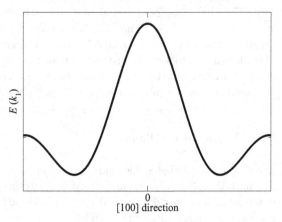

$E(k_1)$

0
[100] direction

In the parabolic-band approximation, the surfaces at constant energy of the conduction band of Si are ellipsoids of revolution about the [100], [010], or [001] axes. Their expressions are

$$[100]: \quad E_{e1} = E(\mathbf{k}) - E_C = \frac{\hbar^2}{2}\left[\frac{(k_1 - k_m)^2}{m_l} + \frac{k_2^2}{m_t} + \frac{k_3^2}{m_t}\right], \quad (17.82)$$

$$[010]: \quad E_{e2} = E(\mathbf{k}) - E_C = \frac{\hbar^2}{2}\left[\frac{k_1^2}{m_t} + \frac{(k_2 - k_m)^2}{m_l} + \frac{k_3^2}{m_t}\right], \quad (17.83)$$

$$[001]: \quad E_{e3} = E(\mathbf{k}) - E_C = \frac{\hbar^2}{2}\left[\frac{k_1^2}{m_t} + \frac{k_2^2}{m_t} + \frac{(k_3 - k_m)^2}{m_l}\right]. \quad (17.84)$$

Similarly, E_{e4}, E_{e5}, E_{e6} are derived from E_{e1}, E_{e2}, E_{e3}, respectively, by letting $k_m \leftarrow -k_m$. The effective masses m_l and m_t are called *longitudinal* and *transverse* mass, respectively.

Fig. 17.18 Schematic view of the conduction band of Ge in the [100] direction.

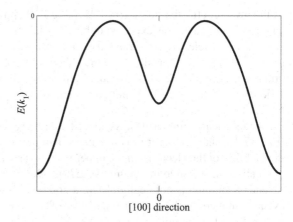

$E(k_1)$

0

[100] direction

Table 17.4 Normalized effective masses of the conduction band of Si, Ge, and GasAs

Material	$m_l(T_a)/m_0$	$m_t(T_a)/m_0$
Si	0.97	0.19
Ge	1.6	0.082
GaAs[a]	0.068	0.068

[a]The effective masses of GaAs are equal to each other due to the band's isotropy

The conduction band of Ge has eight absolute minima, grouped into four pairs. The pairs belong to the four $\{111\}$ directions and are placed at the boundary of the first Brillouin zone (Fig. 17.18); thus, only four absolute minima must be accounted for ($M_C = 4$). In the parabolic-band approximation, the surfaces at constant energy of the conduction band of Ge are ellipsoids of revolution about the corresponding axis; like in silicon, for each ellipsoid the longitudinal mass corresponds to the direction of the axis itself, while the transverse masses correspond to the directions normal to it. The values of m_l and m_t at room temperature T_a, normalized to the rest mass of the free electron, are listed in Table 17.4 [103, Sect. 2-3].

17.6.6 Further Comments About the Band Structure

As better detailed in Sects. 19.5.2 and 19.5.3, among the coefficients of the equations describing the transport phenomena in a semiconductor are the electron and hole *mobilities*, that enter the relation between current density and electric field in a uniform material. For the conduction band of the semiconductors considered here, and in the parabolic-band approximation, the electron mobility μ_n turns out to be proportional to $1/m_n = (2/m_t + 1/m_l)/3$, that is, a weighed average of the entries of the inverse, effective-mass tensor.[18] Table 17.4 shows that GaAs has the largest value of $1/m_n$; thus, it is expected to have the largest mobility, which is indeed the case. As far as holes are concerned, the effective masses of heavy holes of Si, Ge, and GaAs are similar to each other; also the effective masses of light holes have the same order of magnitude. Besides, considering that the valence band has two branches of $E(\mathbf{k})$, the effective masses do not combine in the simple way as for the conduction band.

The secondary minima of GaAs, placed at an energy $E_C + \Delta E$ with $\Delta E \simeq 0.36\,\mathrm{eV}$ (Fig. 17.16), have a larger effective mass than the absolute minimum; due to this, the mobility of the electrons in the upper valleys is smaller than that of the electrons populating the absolute minimum. As ΔE is relatively small, the population of the secondary minima is not negligible; in a non-equilibrium condition, the scattering events tend increase the electron population of the upper valleys at the expense of that of the absolute minimum, with a ratio between the upper and lower population that depends on the applied electric field. This gives rise to a negative differential resistivity in the current-to-voltage curve of the material, i.e., an operating region exists where the current density decreases as the electric field increases. The phenomenon is called *Ridley–Watkins–Hilsum mechanism* [103, Sect. 14-3].

In semiconductors, the absorption of energy from an electromagnetic field may induce the transition of an electron from a state belonging to the valence band to a state belonging to the conduction band. Such a transition increases by one the number of electrons in the conduction band and, at the same time, increases by one the number of holes in the valence band; for this reason it is called *generation of an electron-hole pair*. The opposite phenomenon may also occur, namely, a release of electromagnetic energy due the transition of an electron from a state belonging to the conduction band to a state belonging to the valence band. Such a transition decreases by one the number of electrons in the conduction band and, at the same time, decreases by one the number of holes in the valence band (*recombination of an electron-hole pair*). It is worth pointing out that generation and recombination events may also occur with an energy absorption from, or release to, an external agent different from the electromagnetic field (e.g., the agent could be a vibrational mode of the lattice); for this reason, the phenomena considered here are better specified as *generations-recombinations of the radiative type*. In GaAs, the minimum of the conduction band and the maxima of the two branches of the valence band correspond to the same value of $\mathbf{k}$; semiconductors fulfilling this condition are called *direct-gap semiconductors*. Instead, Si and Ge are *indirect-gap semiconductors*, because the maxima of the valence band correspond to $\mathbf{k} = 0$, whereas the minima of the conduction band correspond to $\mathbf{k} \neq 0$. Direct- and indirect-gap semiconductors behave differently as far as generations-recombinations of the radiative type are concerned; in fact, the probability of such events is much higher in direct-gap semiconductors. This explains

---

[18] If the magnitudes of m_t and m_l are significantly different, the smaller effective mass dictates the magnitude of m_n.

why some classes of solid-state optical devices like, e.g., lasers, are manufactured using direct-gap semiconductors.[19]

17.6.7 Subbands

The calculations of the density of states carried out so far have been based on the assumption that all components of the $\mathbf{k}$ vector can be treated as continuous variables. In particular, the adoption of the parabolic-band approximation in the case of a periodic lattice (Sect. 17.6) leads to expressions for the density of states $g(E)$ and combined density of states $\gamma(E)$ that are formally identical to those obtained for a particle in a three-dimensional box (Sect. 15.9.2). However, in some situations it happens that not all components of $\mathbf{k}$ may be treated as continuous. To describe this case it is convenient to use the example of the box first; that of the periodic lattice is worked out later, in the frame of the parabolic-band approximation.

To proceed, consider like in Sect. 15.9.2 a three-dimensional box whose sides have lengths d_1, d_2, d_3, so that the eigenvalues of the Schrödinger equation are $E_{n_1 n_2 n_3} = \hbar^2 k^2/(2\,m)$, where k^2 is the square of

$$\mathbf{k} = n_1 \frac{\pi}{d_1} \mathbf{i}_1 + n_2 \frac{\pi}{d_2} \mathbf{i}_2 + n_3 \frac{\pi}{d_3} \mathbf{i}_3, \qquad n_i = 1, 2, \ldots \qquad (17.85)$$

The distance between two consecutive projections of $\mathbf{k}$ along the ith side is $\Delta k_i = \pi/d_i$, and the volume associated to each $\mathbf{k}$ is $\Delta k_1 \Delta k_2 \Delta k_3 = \pi^3/V$, with $V = d_1 d_2 d_3$ the volume of the box in the $\mathbf{r}$ space. The density of the $\mathbf{k}$ vectors in the $\mathbf{k}$ space is $Q_k = V/\pi^3$.

17.6.7.1 Two-Dimensional Layer

Now, in contrast to what was implicitly assumed in Sect. 15.9.2, let one side of the box be much different from the others, for instance, $d_2 \sim d_1, d_3 \ll d_1, d_2$. It follows that $\Delta k_3 \gg \Delta k_1, \Delta k_2$. If the magnitudes involved are such that k_1, k_2 may still be considered continuous variables, while k_3 can not, one must calculate the density of states by treating k_1, k_2 differently from k_3. Considering $k_1 = n_1 \pi/d_1, k_2 = n_2 \pi/d_2$ as continuous, fix E and s in the relations

$$\frac{2m}{\hbar^2} E = k_1^2 + k_2^2 + n_3^2 \frac{\pi^2}{d_3^2}, \qquad n_3 = s < \frac{d_3}{\pi} \frac{\sqrt{2mE}}{\hbar}. \qquad (17.86)$$

---

[19] The reasoning seems to contradict the fact the large-area, solid-state optical sensors used in cameras and video cameras, based on the CCD or CMOS architecture, are made of silicon. In fact, the complex structure of these several-megapixel sensors and related signal-management circuitry can be realized only with the much more advanced technology of silicon. The relative ease of fabricating complex structures largely compensates for the poorer optical properties of the material.

For each integer $s = 1, 2, \ldots$ the two relations (17.86) determine in the k_1, k_2 plane a circumference of radius $c_s = \sqrt{c_s^2}$, with

$$c_s^2 = k_1^2 + k_2^2, \qquad c_s^2 = \frac{2mE}{\hbar^2} - s^2 \frac{\pi^2}{d_3^2}. \tag{17.87}$$

It is $\min_s c_s > 0$ because $s_{max} < d_3 \sqrt{2mE}/(\pi \hbar)$, and $\max_s c_s = c_1 > 0$. For a fixed s the states are distributed over the circumference of radius c_s: such a set of states is also called *subband*.

The density of states in energy of the subband thus defined is calculated following the same reasoning as in Sect. 15.9.3: in fact, one observes that the density of **k** vectors in the two-dimensional space k_1, k_2 is $d_1 d_2/\pi^2$, namely, the inverse of the area $\pi^2/(d_1 d_2)$ associated to each **k** belonging to the given circumference. Then, the total number of **k** vectors in a circle of radius c_s is

$$N_{ks} = \frac{d_1 d_2}{\pi^2} \pi c_s^2 = \frac{d_1 d_2}{\pi} \left(\frac{2mE}{\hbar^2} - s^2 \frac{\pi^2}{d_3^2} \right). \tag{17.88}$$

Remembering that indices n_1, n_2 are positive, it is necessary to consider only the first quadrant; as a consequence, N_{ks} must be divided by 4. Further, it is necessary to multiply it by 2 to account for electron spin. In conclusion, the density of states of the two-dimensional subbands is

$$g_{2D}(E) = \frac{d(2 N_{ks}/4)}{dE} = \frac{d_1 d_2 m}{\pi \hbar^2} = \text{const}, \tag{17.89}$$

to be compared with (15.67). Note that (17.89) is independent of index s. This result is useful, e.g., for treating the problem of a two-dimensional charge layer in the channel of a semiconductor device.

17.6.7.2 Wire

Now, assume that $d_2 \sim d_3$, and $d_2, d_3 \ll d_1$. It follows that $\Delta k_2, \Delta k_3 \gg \Delta k_1$. If the magnitudes involved are such that k_1 may still be considered a continuous variable, while k_2, k_3 can not, one must calculate the density of states by treating k_1 differently from k_2, k_3. Considering $k_1 = n_1 \pi/d_1$ as continuous, fix E, r, s in the relations

$$\frac{2m}{\hbar^2} E = k_1^2 + n_2^2 \frac{\pi^2}{d_2^2} + n_3^2 \frac{\pi^2}{d_3^2}, \qquad n_2 = r, \quad n_3 = s, \tag{17.90}$$

with $r^2/d_2^2 + s^2/d_3^2 \le 2mE/(\pi^2 \hbar^2)$. For each pair of integers $r, s = 1, 2, \ldots$, (17.90) determine in the **k** space two points given by the relation

$$\kappa_{rs}^2 = \frac{2mE}{\hbar^2} - \frac{r^2 \pi^2}{d_2^2} - \frac{s^2 \pi^2}{d_3^2}, \qquad \kappa_{rs} = \sqrt{\kappa_{rs}^2}. \tag{17.91}$$

It is $\min_{rs} \kappa_{rs} > 0$ and $\max_{rs} \kappa_{rs} = \kappa_{11}$. For a fixed pair r, s the states are placed at the ends of the segment $[- \kappa_{rs}, +\kappa_{rs}]$ parallel to k_1. The density of states in energy of such a segment is calculated following the same reasoning as in Sect. 15.9.3: in fact, one observes that the density of $\mathbf{k}$ vectors in the one-dimensional space k_1 is d_1/π, namely, the inverse of the length π/d_1 associated to each $\mathbf{k}$ belonging to the segment. Then, the total number of $\mathbf{k}$ vectors in the segment of length $2\kappa_{rs}$ is

$$N_{krs} = \frac{d_1}{\pi} 2\kappa_{rs} = \frac{2d_1}{\pi} \left(\frac{2mE}{\hbar^2} - r^2 \frac{\pi^2}{d_2^2} - s^2 \frac{\pi^2}{d_3^2} \right)^{1/2} . \tag{17.92}$$

Remembering that index n_1 is non negative, it is necessary to consider only the positive half of the segment. As a consequence, N_{krs} must be divided by 2. Further, it is necessary to multiply it by 2 to account for electron spin. The density of states of the one-dimensional case then reads

$$g_{1D}(E) = \frac{d(2 N_{krs}/2)}{dE} = \frac{2d_1 m}{\pi \hbar^2 \kappa_{rs}} , \tag{17.93}$$

to be compared with (15.69). Note that, in contrast with the two-dimensional case (17.89), here the result depends on both indices r, s. A device with $d_2, d_3 \ll d_1$ is also called *wire*. When the device size is such that the transport of a particle in it must be studied by means of Quantum Mechanics, it is also called *quantum wire*. The $E(\kappa_{rs})$ relation may be recast as

$$\frac{\hbar^2}{2m} \kappa_{rs}^2 = E - E_{rs}, \qquad E_{rs} = \frac{\pi^2 \hbar^2}{2m} \left(\frac{r^2}{d_2^2} + \frac{s^2}{d_3^2} \right) . \tag{17.94}$$

As E_{rs} is an increasing function of the indices, its minimum is attained for $r = s = 1$ and represents the ground state in the variables k_2, k_3. It is interesting to note that, if the total energy E is prescribed, e.g., by injecting the particle from an external source, such that $E_{11} < E < \min(E_{12}, E_{21})$, then the particle's wave function has the form

$$\psi = \sqrt{\frac{8}{V}} \sin(\kappa_{11} x_1) \sin(\pi x_2/d_2) \sin(\pi x_3/d_3) \exp(- i E t/\hbar) \tag{17.95}$$

(compare with (15.60)). Remembering the expression (15.64) of the density of states in a box where $d_1 \sim d_2 \sim d_3$, the results obtained so far are summarized as:

$$g_{3D}(E) = \frac{V \sqrt{2 m^3 E}}{\pi^2 \hbar^3} , \qquad g_{2D}(E) = \frac{d_1 d_2 m}{\pi \hbar^2} , \qquad g_{1D}(E) = \frac{d_1 \sqrt{2m}}{\pi \hbar \sqrt{E - E_{rs}}} . \tag{17.96}$$

17.6.8 Subbands in a Periodic Lattice

The calculations leading to (17.96) consider the case of a box within which the potential energy is zero (as a consequence, the total energy E is purely kinetic), and prescribe a vanishing wave function at the boundaries. If a periodic lattice is present, with the provisions indicated in Sect. 17.5.3 one can apply the periodic boundary conditions. In this case, the spacing between the components of **k** in each direction doubles ($n_i \pi/d_i \leftarrow 2n_i \pi/d_i$), but the number of components doubles as well ($n_i = 1, 2, \ldots \leftarrow n_i = 0, \pm 1, \pm 2, \ldots$), so the density of states remains the same. In a semiconductor the calculation leading to the density of states is made more complicated by the presence of the lattice. However, the analysis may be brought to a simple generalization of that carried out in a box by means of the following simplifications:

- It is assumed that a band structure exists even if the size of the device is small in one or two spatial directions. In fact, it can be shown that the presence of a number of atomic planes of the order of ten is sufficient to form a band structure.
- The analysis is limited to the case of parabolic bands.

The case of the conduction band of silicon is considered by way of example, with the k_1, k_2, k_3 axes placed along the [100], [010], [001] directions. The parabolic-band approximation yields for the kinetic energies E_{e1}, E_{e2}, $E_{e3} \geq 0$ the expressions given in (17.82, 17.83, 17.84); the other three kinetic energies E_{e4}, E_{e5}, E_{e6} are derived from E_{e1}, E_{e2}, E_{e3}, respectively, by letting $k_m \leftarrow -k_m$. Apart from the constant E_C, the energies $E_{e1}, E_{e2}, \ldots$ are simplified forms of the eigenvalues of the Schrödinger equation (17.40). Conversely, to the purpose of determining the corresponding eigenfunctions, one may view $E_{e1}, E_{e2}, \ldots$ as the exact eigenvalues of simplified forms of the original Hamiltonian operator (17.40), that hold near the band's minima; such simplified forms are expected to be of the purely kinetic type. They are found by replacing k_i with $-i\,d/dx_i$ in (17.82, 17.83, 17.84), this yielding

$$[100] : \quad \mathcal{H}_{e1} = \frac{\hbar^2}{2} \left[+\frac{1}{m_l} \left(i\frac{\partial}{\partial x_1} + k_m \right)^2 - \frac{1}{m_t} \frac{\partial^2}{\partial x_2^2} - \frac{1}{m_t} \frac{\partial^2}{\partial x_3^2} \right], \quad (17.97)$$

$$[010] : \quad \mathcal{H}_{e2} = \frac{\hbar^2}{2} \left[-\frac{1}{m_t} \frac{\partial^2}{\partial x_1^2} + \frac{1}{m_l} \left(i\frac{\partial}{\partial x_2} + k_m \right)^2 - \frac{1}{m_t} \frac{\partial^2}{\partial x_3^2} \right], \quad (17.98)$$

$$[001] : \quad \mathcal{H}_{e3} = \frac{\hbar^2}{2} \left[-\frac{1}{m_t} \frac{\partial^2}{\partial x_1^2} - \frac{1}{m_t} \frac{\partial^2}{\partial x_2^2} + \frac{1}{m_l} \left(i\frac{\partial}{\partial x_3} + k_m \right)^2 \right], \quad (17.99)$$

with m_l and m_t the longitudinal and transverse masses. Considering $\mathcal{H}_{e1}$ first, the solution of the time-independent Schrödinger equation generated by it,

$$\mathcal{H}_{e1} w_1(\mathbf{r}, n_1, n_2, n_3) = E(n_1, n_2, n_3)\, w_1(\mathbf{r}, n_1, n_2, n_3), \quad (17.100)$$

is found by separation, specifically, by letting $E = E_\alpha(n_1) + E_\beta(n_2) + E_\gamma(n_3)$, $w_1 = \exp(i\,k_m\,x_1)\,\alpha(x_1, n_1)\,\beta(x_2, n_2)\,\gamma(x_3, n_3)$. One finds for the **k** vector the expression

$$\mathbf{k} = n_1 \frac{\pi}{d_1} \mathbf{i}_1 + n_2 \frac{\pi}{d_2} \mathbf{i}_2 + n_3 \frac{\pi}{d_3} \mathbf{i}_3, \quad (17.101)$$

with $n_i = 1, 2, \ldots$, while the eigenfunctions and eigenvalues read

$$w_1 = \sqrt{\frac{8}{V}} \, \exp\left(i\, k_m \, x_1\right) \sin\left(\frac{n_1 \, \pi}{d_1} x_1\right) \sin\left(\frac{n_2 \, \pi}{d_2} x_2\right) \sin\left(\frac{n_3 \, \pi}{d_3} x_3\right), \quad (17.102)$$

$$E = \frac{\hbar^2}{2\, m_l} n_1^2 \frac{\pi^2}{d_1^2} + \frac{\hbar^2}{2\, m_t} n_2^2 \frac{\pi^2}{d_2^2} + \frac{\hbar^2}{2\, m_t} n_3^2 \frac{\pi^2}{d_3^2}. \quad (17.103)$$

The eigenvalues and eigenfunctions of $\mathcal{H}_{e2}$, $\mathcal{H}_{e3}$ are found by a cyclic permutation of the indices, $1 \leftarrow 2 \leftarrow 3 \leftarrow 1$, while those of $\mathcal{H}_{e4}$, $\mathcal{H}_{e5}$, $\mathcal{H}_{e6}$ are derived from those of $\mathcal{H}_{e1}$, $\mathcal{H}_{e2}$, $\mathcal{H}_{e3}$, respectively, by letting $k_m \leftarrow -k_m$.

One notes that the $\mathbf{k}$ vectors and the eigenfunctions are not influenced by the effective masses, whereas the eigenvalues are. As a consequence, the density of states is affected as well. In the case where $d_1 \sim d_2 \sim d_3$ the density of states associated to the minimum of index 1 is found by the same procedure as that leading to the first relation in (17.96); the result is

$$g_{3D}^{(1)}(E) = \frac{d_1\, d_2\, d_3\, \sqrt{2\, m_l\, m_t^2}}{\pi^2\, \hbar^3} \sqrt{E}. \quad (17.104)$$

Such a density of states is not affected by interchanging the effective masses; thus, the total density of states is found by adding over the densities of states of the M_C minima of the conduction band:

$$g_{3D}(E) = M_C \frac{d_1\, d_2\, d_3\, \sqrt{2\, m_l\, m_t^2}}{\pi^2\, \hbar^3} \sqrt{E}. \quad (17.105)$$

As in the case of a box, the distance between two consecutive projections of $\mathbf{k}$ along the ith side is $\Delta k_i = \pi/d_i$, and the volume associated to each $\mathbf{k}$ is $\Delta k_1\, \Delta k_2\, \Delta k_3 = \pi^3/V$, with $V = d_1 d_2 d_3$. The density of the $\mathbf{k}$ vectors in the $\mathbf{k}$ space is $Q_k = V/\pi^3$.

Consider now the case of a two-dimensional layer, namely, $d_2 \sim d_1$, while $d_3 \ll d_1, d_2$. Let $k_1 = n_1\pi/d_1$, $k_2 = n_2\pi/d_2$, and fix $n_3 = 1$ whence, for the minima of indices 1 and 4,

$$E = \frac{\hbar^2}{2\, m_l} k_1^2 + \frac{\hbar^2}{2\, m_t} k_2^2 + \frac{\hbar^2}{2\, m_t} \frac{\pi^2}{d_3^2}, \qquad E \geq \frac{\hbar^2}{2\, m_t} \frac{\pi^2}{d_3^2}. \quad (17.106)$$

A calculation similar to that carried out in a box provides, for the minima of indices 1 and 4, an expression similar to that of the second relation in (17.96):

$$g_{2D}^{(1)} = g_{2D}^{(4)} = \frac{d_1\, d_2 \sqrt{m_l\, m_t}}{\pi\, \hbar^2}. \quad (17.107)$$

For the other pairs of minima one finds

$$g_{2D}^{(2)} = g_{2D}^{(5)} = \frac{d_1\, d_2\, \sqrt{m_l\, m_t}}{\pi\, \hbar^2}, \qquad g_{2D}^{(3)} = g_{2D}^{(6)} = \frac{d_1\, d_2\, m_t}{\pi\, \hbar^2}. \quad (17.108)$$

Fig. 17.19 Normalized, two-dimensional density of states (17.109) for the $1, 2, 4, 5$ valleys of silicon, as a function of E/E_t, in the parabolic-band approximation

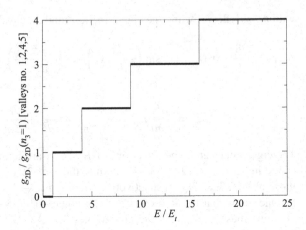

In conclusion, for a two-dimensional layer with $d_3 \ll d_1, d_2$ and $n_3 = 1$, within the parabolic-band approximation, the density of states for the minima of indices 1, 2, 4, and 5 is the same constant for all energies $E \geq \hbar^2 \pi^2 / (2 m_t d_3^2)$. The total density of states for these minima is

$$g_{2D}^{(1,2,4,5)} = 4 \frac{d_1 d_2 \sqrt{m_l m_t}}{\pi \hbar^2}. \qquad E \geq E_t = \frac{\hbar^2 \pi^2}{2 m_t d_3^2}, \qquad n_3 = 1, \qquad (17.109)$$

while $g_{2D,n_3=1}^{(1,2,4,5)} = 0$ for $E < E_t$. Similarly, still with $n_3 = 1$, the density of states for the minima of indices 3 and 6 is another constant for all energies $E \geq \hbar^2 \pi^2 / (2 m_l d_3^2)$. The total density of states for these minima is

$$g_{2D}^{(3,6)} = 2 \frac{d_1 d_2 m_t}{\pi \hbar^2}, \qquad E \geq E_l = \frac{\hbar^2 \pi^2}{2 m_l d_3^2}, \qquad n_3 = 1, \qquad (17.110)$$

while $g_{2D,n_3=1}^{(3,6)} = 0$ for $E < E_l$. Now, let $n_3 = 2$; it is easily found that the value of $g_{2D,n_3=2}^{(1,2,4,5)}$ is the same as above, however, it holds for $E \geq 4 E_t$. It adds up to the value found for $n_3 = 1$, giving rise to a stair-like form of $g_{2D}^{(1,2,4,5)}$ as a function of energy. The same is obtained for $g_{2D,n_3=2}^{(3,6)}$ when $E \geq 4 E_l$, and so on. An example of such a density of states is sketched in Fig. 17.19, where the ratio $g_{2D}^{(1,2,4,5)} / g_{2D,n_3=1}^{(1,2,4,5)}$ is shown as a function of E/E_t. The total density of states is found by adding up the two stair-like functions. From Table 17.4 one finds that in silicon at room temperature it is $m_l \simeq 0.97 \, m_0$, $m_t \simeq 0.19 \, m_0$, whence $E_t \simeq 5.1 \, E_l$ and $g_{2D,n_3=1}^{(1,2,4,5)} \simeq 4.47 \, g_{2D,n_3=1}^{(3,6)}$.

As shown by Fig. 17.19, the derivative of the density of states with respect to energy diverges at some points. Such divergences are called *Van Hove singularities* [2, Chap. 8].

Finally, consider the case of a wire, namely, $d_2 \sim d_3$, while $d_2, d_3 \ll d_1$. Let $k_1 = n_1 \pi / d_1$ and fix $n_2 = n_3 = 1$ whence, for the minima of indices 1 and 4,

$$E = \frac{\hbar^2}{2\,m_l} k_1^2 + \frac{\hbar^2}{2\,m_t} \frac{\pi^2}{d_2^2} + \frac{\hbar^2}{2\,m_t} \frac{\pi^2}{d_3^2}, \qquad E \geq \frac{\pi^2 \hbar^2}{2\,m_t} \left(\frac{1}{d_2^2} + \frac{1}{d_3^2} \right) = E_{11}^{(1,4)}.$$

(17.111)

A calculation similar to that carried out in a box provides, for the minima of indices 1 and 4, an expression similar to that of the third relation in (17.96):

$$g_{1D}^{(1)} = g_{1D}^{(4)} = \frac{2\,d_1\,m_l}{\pi\,\hbar^2\,\kappa_{11}^{(1,4)}} = \frac{d_1\sqrt{2\,m_l}}{\pi\,\hbar\sqrt{E - E_{11}^{(1,4)}}},$$

(17.112)

while $g_{1D}^{(1,4)} = 0$ if $E < E_{11}^{(1,4)}$. For the other minima one finds

$$g_{1D}^{(2,5)} = \frac{d_1\sqrt{2\,m_t}}{\pi\,\hbar\sqrt{E - E_{11}^{(2,5)}}}, \qquad E \geq E_{11}^{(2,5)} = \frac{\pi^2\hbar^2}{2}\left(\frac{1}{m_l\,d_2^2} + \frac{1}{m_t\,d_3^2}\right),$$

(17.113)

with $g_{1D}^{(2,5)} = 0$ if $E < E_{11}^{(2,5)}$, and

$$g_{1D}^{(3,6)} = \frac{d_1\sqrt{2\,m_t}}{\pi\,\hbar\sqrt{E - E_{11}^{(3,6)}}}, \qquad E \geq E_{11}^{(3,6)} = \frac{\pi^2\hbar^2}{2}\left(\frac{1}{m_t\,d_2^2} + \frac{1}{m_l\,d_3^2}\right),$$

(17.114)

with $g_{1D}^{(3,6)} = 0$ if $E < E_{11}^{(3,6)}$. In conclusion, for a wire with $d_2, d_3 \ll d_1$ and $n_2 = n_3 = 1$, within the parabolic-band approximation, the density of states of each pair of minima is the sum of expressions of the form (17.112, 17.113, 17.114); the latter are complicate because all possible pairs of indices r, s combine with the two lengths d_2, d_3, that in general are not commensurable with each other. A somewhat easier description is obtained by considering (17.112) alone and letting $d_2 = d_3$ in it; this yields $E_{21} = E_{12} = 2.5\,E_{11}$, $E_{22} = 4\,E_{11}$, $E_{31} = E_{13} = 5\,E_{11}$, and so on, and

$$g_{1D}^{(1,4)} = \frac{g_{1D}^{(1,4)}(E = 2\,E_{11})}{\sqrt{E/E_{11} - 1}}, \qquad E_{11} < E \leq E_{21}.$$

(17.115)

In the next interval $E_{21} < E \leq E_{22}$, the density of states is the sum of (17.115) and $2/\sqrt{E/E_{11} - 2.5}$, where factor 2 accounts for the $(r = 2, s = 1)$, $(r = 1, s = 2)$ degeneracy; in the interval $E_{22} < E \leq E_{31}$ one adds the further summand $1/\sqrt{E/E_{11} - 4}$, and so on. The normalized density of states $g_{1D}(E)/g_{1D}(2\,E_{11})$ is shown in Fig. 17.20 as a function of E/E_{11}.

Fig. 17.20 Normalized, one-dimensional density of states for the 1, 4 valleys of silicon, as a function of E/E_{11}, in the parabolic-band approximation and with $d_2 = d_3$

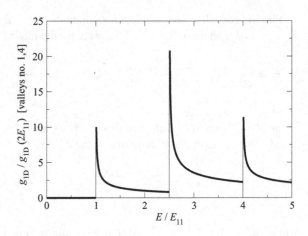

Also in this case the Van Hove singularities are present; in addition, the density of states itself diverges at such points. However, such divergences are integrable; consider for instance an integral of the form

$$\int_{E_0}^{\infty} \frac{c}{\sqrt{E - E_0}} P(E)\, dE, \tag{17.116}$$

with c a constant and $0 < P < 1$ a distribution function. Splitting the integration domain into two intervals $E_0 \leq E \leq E'$ and $E' \leq E < \infty$, with $E' > E_0$, one finds for the first integral, that contains the singularity,

$$\int_{E_0}^{E'} \frac{c}{\sqrt{E - E_0}} P(E)\, dE \leq \int_{E_0}^{E'} \frac{c}{\sqrt{E - E_0}}\, dE < \infty. \tag{17.117}$$

17.7   Calculation of Vibrational Spectra

The discussion carried out in Sect. 16.6 has led to the conclusion that in the case of solid matter the nuclei, being massive and tightly bound together, are expected to depart little from their equilibrium positions $\mathbf{R}_0$. The classical description of the nuclear motion is thus brought to the case already solved in Sects. 3.9 and 3.10: the vibrational state of the nuclei is described in terms of the normal coordinates b_σ, whose conjugate moments are $\dot{b}_\sigma$, and the total energy of the nuclei reads (compare with (3.50))

$$T_a + V_a = \sum_{\sigma=1}^{3N} H_\sigma + V_{a0}, \qquad H_\sigma = \frac{1}{2} \dot{b}_\sigma^2 + \frac{1}{2} \omega_\sigma^2 b_\sigma^2, \tag{17.118}$$

where each H_σ corresponds to one degree of freedom and $\omega_\sigma > 0$ is the angular frequency of the corresponding mode. The system is completely separable in the

Fig. 17.21 Definition of the labels used to identify the degrees of freedom in a periodic lattice

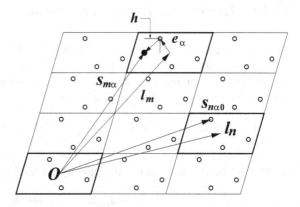

normal coordinates, and each normal coordinate evolves in time as a linear harmonic oscillator. The calculation is based on Classical Mechanics; it is carried out in this chapter because it exploits the periodicity properties of the material and, in this respect, presents several analogies with the solution of the Schrödinger equation in a periodic lattice. To determine the vibrational frequencies ω_σ it is necessary to solve the eigenvalue equation (3.43), namely,

$$\mathbf{C}\,\mathbf{g}_\sigma = \omega_\sigma^2\,\mathbf{M}\,\mathbf{g}_\sigma, \qquad \sigma = 1,\ldots,3\,N, \qquad (17.119)$$

with $\mathbf{g}_\sigma$ the eigenvectors. The entries of $\mathbf{C}$, $\mathbf{M}$ are given by

$$c_{kn} = [\mathbf{C}]_{kn} = \left(\frac{\partial^2 V_a}{\partial h_k\,\partial h_n}\right)_0, \qquad [\mathbf{M}]_{kn} = \mu_n\,\delta_{kn}, \qquad (17.120)$$

where V_a is the potential energy, h_k the displacement of the kth degree of freedom with respect to the equilibrium position, μ_k the mass associated to the kth degree of freedom, and δ_{kn} the Kronecker delta.

The calculation is in principle the same for any system of particles; however, if the system has special properties, they reflect into the form of the eigenvalues and eigenvectors. A particularly important case is that of a periodic structure, such as a crystal. Considering this case, let the crystal be made of N_c elementary cells, with a basis made of N_b nuclei (the definition of basis is in Sect. 17.2). It follows that the total number of nuclei is $N = N_b\,N_c$, and the total number of degrees of freedom is $3\,N$. With respect to a given origin O (Fig. 17.21), the mth cell of the lattice is identified by the corresponding translation vector of the direct lattice, $\mathbf{l}_m$; the latter determines a local origin within the mth cell. In turn, the equilibrium position of the αth nucleus of the mth cell with respect to the local origin is identified by a vector $\mathbf{e}_\alpha$ of the direct lattice.

17.7.1   Labeling the Degrees of Freedom— Dynamic Matrix

To proceed, it is convenient to label the degrees of freedom in such a way as to distinguish the indices of the cells from those of the basis and of the coordinate axes. To this purpose, one observes that the component along the uth coordinate axis of the equilibrium position of the jth nucleus is

$$X_{ju0} = s_{q0}, \qquad q = u + 3(j-1), \qquad j = \alpha + N_b(m-1), \qquad (17.121)$$

with $u = 1, 2, 3$; $\alpha = 1, \ldots, N_b$; $m = 1, \ldots, N_c$. The same applies to the displacements, which are more conveniently expressed in terms of three indices:

$$h_q \longleftarrow h_{m\alpha u}, \qquad h_r \longleftarrow h_{n\beta w}. \qquad (17.122)$$

The entries of $\mathbf{C}$ are identified in the same manner:

$$c_{qr} = \left(\frac{\partial^2 V_a}{\partial h_q \partial h_r}\right)_0 \qquad \longleftarrow \qquad c_{m\alpha u}^{n\beta w} = \left(\frac{\partial^2 V_a}{\partial h_{m\alpha u}\, \partial h_{n\beta w}}\right)_0, \qquad (17.123)$$

with

$$m, n = 1, \ldots, N_c, \qquad \alpha, \beta = 1, \ldots, N_b, \qquad u, w = 1, 2, 3. \qquad (17.124)$$

The order of derivation is irrelevant, so that $c_{m\alpha u}^{n\beta w} = c_{n\beta w}^{m\alpha u}$. As the number of nuclei is finite, the crystal is not actually periodic; as indicated above, periodicity is recovered by imposing periodic boundary conditions to the quantities of interest (Sect. 17.5.3).[20] With this provision, the entries of $\mathbf{C}$ are invariant with respect to the lattice translations. The latter are related only to the cell indices m, n and are obtained by the replacements $\mathbf{l}_m \leftarrow \mathbf{l}_m + \mathbf{l}_\nu$, $\mathbf{l}_n \leftarrow \mathbf{l}_n + \mathbf{l}_\nu$, with ν any integer. In particular, taking $\mathbf{l}_\nu = -\mathbf{l}_n$ yields

$$c_{m\alpha u}^{n\beta w} = c_{\alpha u}^{\beta w}(\mathbf{l}_m, \mathbf{l}_n) = c_{\alpha u}^{\beta w}(\mathbf{l}_m - \mathbf{l}_n, 0) = c_{\alpha u}^{\beta w}(\mathbf{l}_m - \mathbf{l}_n). \qquad (17.125)$$

The above shows that the entries of $\mathbf{C}$ depend on the relative positions of the cells. Due to the invariance of $\mathbf{C}$ with respect to the lattice translations one sees that, given α, u and β, w, there are only N_c distinct entries of $\mathbf{C}$ out of N_c^2, namely, the distinct entries are those such that $m - n = 0$, $m - n = 1$, $\ldots$, $m - n = N_c - 1$. In fact, all remaining $N_c^2 - N_c$ entries are derived from the first N_c ones by suitable translations of the indices. In turn, using the new indices (17.124) the entries of $\mathbf{M}$ read

$$\mu_r \delta_{qr} \qquad \longleftarrow \qquad \mu_{n\beta w} \delta_{m\alpha u}^{n\beta w} = \mu_\beta \delta_{m\alpha u}^{n\beta w}, \qquad (17.126)$$

---

[20] As mentioned in Sect. 17.5.3, the periodic boundary conditions are actually an approximation; however, the interatomic interactions typically give rise to short-range forces, hence the above reasoning holds for all the cells that are not too close to the boundaries.

where the last equality is due to the fact that the mass of a given nucleus of the cell does not depend on the cell position within the crystal nor on the coordinate axis. In the new indices the eigenvalue equation (17.119) becomes

$$\sum_{n\beta w} c_{m\alpha u}^{n\beta w} g_{n\beta w} = \omega^2 \mu_\alpha g_{m\alpha u}, \tag{17.127}$$

where the indices' ranges are given in (17.124). The indices of the eigenvalue and eigenvector have been omitted for simplicity. Defining

$$d_{m\alpha u}^{n\beta w} = \frac{c_{m\alpha u}^{n\beta w}}{\sqrt{\mu_\alpha \mu_\beta}}, \qquad z_{m\alpha u} = \sqrt{\mu_\alpha} g_{m\alpha u}, \qquad z_{n\beta w} = \sqrt{\mu_\beta} g_{n\beta w}, \tag{17.128}$$

transforms (17.127) into

$$\sum_{n\beta w} d_{m\alpha u}^{n\beta w} z_{n\beta w} = \omega^2 z_{m\alpha u}. \tag{17.129}$$

The latter form of the eigenvalue equation is more convenient because it eliminates the coefficient μ_α from the right hand side. Matrix $\mathbf{D}$ of entries $d_{m\alpha u}^{n\beta w}$ is called *dynamic matrix* and, due to the properties of $\mathbf{C}$, is symmetric ($d_{m\alpha u}^{n\beta w} = d_{n\beta w}^{m\alpha u}$) and translationally invariant:

$$d_{m\alpha u}^{n\beta w} = d_{\alpha u}^{\beta w}(\mathbf{l}_m, \mathbf{l}_n) = d_{\alpha u}^{\beta w}(\mathbf{l}_m - \mathbf{l}_n, 0) = d_{\alpha u}^{\beta w}(\mathbf{l}_m - \mathbf{l}_n). \tag{17.130}$$

17.7.2 Application of the Bloch Theorem

As a consequence of the translational invariance of $\mathbf{D}$, Bloch's theorem (17.23) applies,[21] namely, for any eigenvector of indices $k\gamma e$, and letting $\mathbf{l}_0 = 0$, the following holds:

$$\mathbf{z}_{\gamma e}(\mathbf{l}_k) = \exp(\mathbf{c} \cdot \mathbf{l}_k) \mathbf{z}_{\gamma e}(0). \tag{17.131}$$

In (17.131), $\mathbf{c}$ is any complex vector of the reciprocal lattice, and $k = 0, \ldots, N_c - 1$; $\gamma = 1, \ldots, N_b$; $e = 1, 2, 3$. The complex form of the eigenvectors is adopted for convenience; at the end of the calculation, a set of real eigenvectors is recovered from suitable combinations of the complex ones. Using the periodic boundary conditions, the expression of $\mathbf{c}$ is found to be

$$\mathbf{c} = \mathbf{i}\mathbf{q}, \qquad \mathbf{q} = \sum_{s=1}^{3} \frac{v_s}{N_s} 2\pi \mathbf{b}_s, \tag{17.132}$$

---

[21] The Bloch theorem was derived in Sect. 17.5.1 with reference to the eigenfunctions of a translation operator in the continuous case; the theorem equally holds for a translation operator in the discrete case, like the dynamic matrix considered here.

with N_1, N_2, N_3 the number of cells along the directions of the characteristic vectors of the direct lattice, $\mathbf{b}_1, \mathbf{b}_2, \mathbf{b}_3$ the characteristic vectors of the reciprocal lattice, and v_1, v_2, v_3 integers, with $v_s = 0, 1, \ldots, N_s - 1$. The total number of distinct $\mathbf{q}$ vectors is thus $N_1 N_2 N_3 = N_c$. Comparing (17.132) with (17.37) shows that the structure of the $\mathbf{q}$ vector is the same as that of the $\mathbf{k}$ vector found in the solution of the Schrödinger equation (Sect. 17.5.3). Inserting (17.130) into (17.129) yields, for the line of indices $m\alpha u$ of the eigenvalue equation,

$$\sum_{n\beta w} A_{m\alpha u}^{n\beta w} z_{\beta w}(0) = \omega^2 z_{\alpha u}(0), \tag{17.133}$$

with

$$A_{m\alpha u}^{n\beta w} = \frac{1}{\sqrt{\mu_\alpha \mu_\beta}} c_{\alpha u}^{\beta w}(\mathbf{l}_m - \mathbf{l}_n) \exp\left[i\,\mathbf{q}\cdot(\mathbf{l}_n - \mathbf{l}_m)\right]. \tag{17.134}$$

As the eigenvalues ω^2 are real, the matrix made of the entries $A_{m\alpha u}^{n\beta w}$ must be Hermitean; in fact, this is easily found by observing that $\mathbf{D}$ is real and symmetric:

$$A_{n\beta w}^{m\alpha u} = d_{\beta w}^{\alpha u}(\mathbf{l}_n - \mathbf{l}_m) \exp\left[i\,\mathbf{q}\cdot(\mathbf{l}_m - \mathbf{l}_n)\right] = \left(A_{m\alpha u}^{n\beta w}\right)^*. \tag{17.135}$$

Another property stems from the expression at the left hand side of (17.133),

$$\sum_{n\beta w} A_{m\alpha u}^{n\beta w} z_{\beta w}(0) = \sum_{\beta w} \left(\sum_n A_{m\alpha u}^{n\beta w}\right) z_{\beta w}(0), \tag{17.136}$$

where $A_{m\alpha u}^{n\beta w}$ is translationally invariant because it depends on the cell indices only through the difference $\mathbf{l}_m - \mathbf{l}_n$. It follows that $\sum_n A_{m\alpha u}^{n\beta w}$ does not depend on m. This is easily verified by carrying out the sum first with, say, $m = 1$, then with $m = 2$, and observing that the terms of the second sum are the same as in the first one, displaced by one position. In summary, letting $\mathbf{A}$ be the $3 N_b \times 3 N_b$, Hermitean matrix of entries

$$A_{\alpha u}^{\beta w}(\mathbf{q}) = \sum_{n=1}^{N_c} d_{\alpha u}^{\beta w}(\mathbf{l}_m - \mathbf{l}_n) \exp\left[i\,\mathbf{q}\cdot(\mathbf{l}_n - \mathbf{l}_m)\right], \tag{17.137}$$

(17.133) becomes

$$\sum_{\beta w} A_{\alpha u}^{\beta w}(\mathbf{q})\, z_{\beta w}(0) = \omega^2 z_{\alpha u}(0). \tag{17.138}$$

For a given $\mathbf{q}$, (17.138) is an eigenvalue equation of order $3 N_b$, whose eigenvalues are found by solving the algebraic equation

$$\det(\mathbf{A} - \omega^2 \mathbf{I}) = 0, \tag{17.139}$$

with $\mathbf{I}$ the identity matrix. As the entries of $\mathbf{A}$ depend on $\mathbf{q}$, the calculation of the $3 N_b$ eigenvalues of (17.138) must be repeated for each distinct value of $\mathbf{q}$, namely, N_c times. The total number of eigenvalues thus found is $3 N_b \times N_c = 3 N$, as should be. This result shows that, while the translational invariance eliminates the dependence on $\mathbf{l}_m$, it introduces that on $\mathbf{q}$. As the number of different determinations of the two vectors $\mathbf{l}_m$ and $\mathbf{q}$ is the same, namely, N_c, the total number of eigenvalues is not affected. Letting the N_c determinations of $\mathbf{q}$ be numbered as $\mathbf{q}_1, \mathbf{q}_2 \ldots \mathbf{q}_k \ldots$, the algebraic system (17.138) is recast as

$$\sum_{\beta w} A_{\alpha u}^{\beta w}(\mathbf{q}_k) z_{\beta w}(0, \mathbf{q}_k) = \omega^2(\mathbf{q}_k) z_{\alpha u}(0, \mathbf{q}_k), \qquad k = 1, 2, \ldots N_c \qquad (17.140)$$

which, for each $\mathbf{q}_k$, yields $3 N_b$ eigenvalues ω^2 and $3 N_b$ eigenvectors of length $3 N_b$; as a consequence, the set of column vectors made of the eigenvectors associated to $\mathbf{q}_k$ forms a $3 N_b \times 3 N_b$ matrix, indicated here with $\mathbf{Z}_{1k}$. By letting $\mathbf{q}_k$ span over all its N_c determinations, the total number of eigenvalues turns out to be $3 N$, namely,

$$\omega_{\gamma e}^2(\mathbf{q}_1), \; \ldots, \; \omega_{\gamma e}^2(\mathbf{q}_{N_c}), \qquad \gamma = 1, \ldots, N_b, \qquad e = 1, 2, 3. \qquad (17.141)$$

Similarly, the total number of eigenvectors (of order $3 N_b$) turns out to be $3 N$,

$$\mathbf{z}_{\gamma e}(0, \mathbf{q}_1), \; \ldots, \; \mathbf{z}_{\gamma e}(0, \mathbf{q}_{N_c}), \qquad \gamma = 1, \ldots, N_b, \qquad e = 1, 2, 3. \qquad (17.142)$$

They provide the set of N_c square matrices of order $3 N_b$, indicated with $\mathbf{Z}_{11}, \mathbf{Z}_{12}, \ldots,$ $\mathbf{Z}_{1N_c}$. Finally, each $\mathbf{z}_{\gamma e}(0, \mathbf{q}_k)$ provides an eigenvector of order $3 N$ whose entries are

$$z_{\gamma e}^{\alpha u}(\mathbf{l}_m, \mathbf{q}_k) = \exp(\mathrm{i} \, \mathbf{q}_k \cdot \mathbf{l}_m) \, z_{\gamma e}^{\alpha u}(0, \mathbf{q}_k), \qquad (17.143)$$

where, as usual, $\alpha, \gamma = 1, \ldots, N_b$; $u, e = 1, 2, 3$ and, in turn, $m = 0, \ldots, N_c - 1$; $k = 1, \ldots, N_c$. The first index of matrices $\mathbf{Z}_{11}, \mathbf{Z}_{12}, \ldots$ corresponds to $m = 0$. Similarly, index $m = 1$ provides a new set of matrices $\mathbf{Z}_{21}, \mathbf{Z}_{22}, \ldots$, and so on. The whole set of N_c^2 matrices $\mathbf{Z}_{mk}$ is equivalent to the $3 N \times 3 N$ matrix $\mathbf{Z}$ of the eigenvectors of the dynamic matrix, according to the following scheme:

$$\mathbf{Z} = \begin{bmatrix} \mathbf{Z}_{11} & \mathbf{Z}_{12} & \cdots & \mathbf{Z}_{1N_c} \\ \mathbf{Z}_{21} & \mathbf{Z}_{22} & \cdots & \mathbf{Z}_{2N_c} \\ \vdots & \vdots & \ddots & \vdots \\ \mathbf{Z}_{N_c 1} & \mathbf{Z}_{N_c 2} & \cdots & \mathbf{Z}_{N_c N_c} \end{bmatrix}. \qquad (17.144)$$

17.7.3 Properties of the Eigenvalues and Eigenvectors

Remembering that $\mathbf{A}$ (defined in (17.137)) is Hermitean, and ω^2 is real, one finds $(\mathbf{A} - \omega^2 \mathbf{I})^* = \mathbf{A}^* - \omega^2 \mathbf{I} = \mathbf{A}^T - \omega^2 \mathbf{I} = (\mathbf{A} - \omega^2 \mathbf{I})^T$, whence

$$\det[(\mathbf{A} - \omega^2 \mathbf{I})^*] = \det[(\mathbf{A} - \omega^2 \mathbf{I})^T] = \det(\mathbf{A} - \omega^2 \mathbf{I}). \qquad (17.145)$$

This shows that the eigenvalue equation $\mathbf{A}(\mathbf{q}_k)\,\mathbf{z}(0,\mathbf{q}_k) = \omega^2\,(\mathbf{q}_k)\,\mathbf{z}(0,\mathbf{q}_k)$, and its conjugate, $\mathbf{A}^*(\mathbf{q}_k)\,\mathbf{z}^*(0,\mathbf{q}_k) = \omega^2\,(\mathbf{q}_k)\,\mathbf{z}^*(0,\mathbf{q}_k)$ have the same eigenvalues. Moreover, as the entries (17.137) of $\mathbf{A}$ are polynomials in $\exp[\mathrm{i}\,\mathbf{q}_k \cdot (\mathbf{l}_n - \mathbf{l}_m)]$ with real coefficients, the following hold:

$$A_{\alpha u}^{\beta w}(-\mathbf{q}_k) = \left[A_{\alpha u}^{\beta w}(\mathbf{q}_k)\right]^*, \qquad \mathbf{A}(-\mathbf{q}_k) = \mathbf{A}^*(\mathbf{q}_k). \tag{17.146}$$

The above properties give rise to other important consequences for the eigenvalues and eigenvectors. In fact, from the property $\mathbf{A}(-\mathbf{q}_k) = \mathbf{A}^*(\mathbf{q}_k)$ and the hermiticity of $\mathbf{A}$ one finds

$$\det\left[\mathbf{A}(-\mathbf{q}_k) - \omega^2\,\mathbf{I}\right] = \det\{[\mathbf{A}(\mathbf{q}_k) - \omega^2\,\mathbf{I}]^T\} = \det\left[\mathbf{A}(\mathbf{q}_k) - \omega^2\,\mathbf{I}\right], \tag{17.147}$$

showing that the eigenvalues calculated from $\mathbf{A}(-\mathbf{q}_k)$ are the same as those calculated from $\mathbf{A}(\mathbf{q}_k)$. It follows that ω is an even function of $\mathbf{q}_k$:

$$\omega(-\mathbf{q}_k) = \omega(\mathbf{q}_k), \qquad \mathbf{A}(-\mathbf{q}_k)\,\mathbf{z}(0,-\mathbf{q}_k) = \omega^2(\mathbf{q}_k)\,\mathbf{z}(0,-\mathbf{q}_k). \tag{17.148}$$

Taking the conjugate of the second equation in (17.148) and using again the relation $\mathbf{A}(-\mathbf{q}_k) = \mathbf{A}^*(\mathbf{q}_k)$ yields $\mathbf{A}(\mathbf{q}_k)\,\mathbf{z}^*(0,-\mathbf{q}_k) = \omega^2(\mathbf{q}_k)\,\mathbf{z}^*(0,-\mathbf{q}_k)$. Comparing the above with the original eigenvalue equation $\mathbf{A}(\mathbf{q}_k)\,\mathbf{z}(0,\mathbf{q}_k) = \omega^2\,(\mathbf{q}_k)\,\mathbf{z}(0,\mathbf{q}_k)$ provides a relation between the eigenvectors:

$$\mathbf{z}(0,-\mathbf{q}) = \mathbf{z}^*(0,\mathbf{q}). \tag{17.149}$$

From Bloch's theorem (17.131) it follows $z_{\gamma e}^{\alpha u}(\mathbf{l}_m,\mathbf{q}_k) = \exp(\mathrm{i}\,\mathbf{q}_k \cdot \mathbf{l}_m)\,z_{\gamma e}^{\alpha u}(0,\mathbf{q}_k)$ which, combined with (17.149), allows one to recover a set of real eigenvectors of the dynamic matrix:

$$z_{\gamma e}^{\alpha u}(\mathbf{l}_m,\mathbf{q}_k) + z_{\gamma e}^{\alpha u}(\mathbf{l}_m,-\mathbf{q}_k) = z_{\gamma e}^{\alpha u}(0,\mathbf{q}_k)\,\exp(\mathrm{i}\,\mathbf{q}_k \cdot \mathbf{l}_m) + z_{\gamma e}^{*\ \alpha u}(0,\mathbf{q}_k)\,\exp(-\mathrm{i}\,\mathbf{q}_k \cdot \mathbf{l}_m)$$

where, as usual, the indices $k\gamma e$ count the eigenvectors and the indices $m\alpha u$ count the entries. Using the results of Sect. 3.10, the displacements of the particles from the equilibrium position are given by $\mathbf{h} = \mathbf{G}\mathbf{b}$, where $\mathbf{G}$ is the matrix of the eigenvalues of (17.119) and the entries of $\mathbf{b}$ have the form (3.49), namely,

$$b_{k\gamma e}(t) = \frac{1}{2}\left\{\tilde{b}_{k\gamma e0}\,\exp[-\mathrm{i}\,\omega_{\gamma e}(\mathbf{q}_k)\,t] + \tilde{b}_{k\gamma e0}^*\,\exp[\mathrm{i}\,\omega_{\gamma e}(\mathbf{q}_k)\,t]\right\}, \tag{17.150}$$

with $\tilde{b}_{k\gamma e0}$ depending on the initial conditions $b_{k\gamma e0}(0)$, $\dot{b}_{k\gamma e0}(0)$. In turn, the entries of matrix $\mathbf{g}$ are $g_{k\gamma e}^{m\alpha u}$, where the lower indices refer to the columns and count the eigenvectors, the upper ones refer to the rows and count the entries of each eigenvector. Due to (17.128), such entries equal the corresponding terms of the real eigenvector of the dynamic matrix, divided by $\sqrt{\mu_\alpha}$. In conclusion, from $\mathbf{h} = \mathbf{G}\mathbf{b}$, the displacements are given by

$$h_{m\alpha u} = \sum_{k\gamma e} g_{k\gamma e}^{m\alpha u}\,b_{k\gamma e} = \sum_{k\gamma e} \frac{1}{\sqrt{\mu_\alpha}}\,z_{k\gamma e}^{m\alpha u}\,b_{k\gamma e}. \tag{17.151}$$

Using (17.150) yields

$$h_{m\alpha u} = \frac{1}{\sqrt{\mu_\alpha}} \, \Re \sum_{k\gamma e} z_{\gamma e}^{\alpha u}(0, \mathbf{q}_k) \left[\tilde{b}_{k\gamma e 0} \, \exp(\mathrm{i} \, \Phi_{k\gamma e}^m) + \tilde{b}_{k\gamma e 0}^* \, \exp(\mathrm{i} \, \Psi_{k\gamma e}^m) \right],$$

(17.152)

where the phases are defined by

$$\Phi_{k\gamma e}^m = \mathbf{q}_k \cdot \mathbf{l}_m - \omega_{\gamma e}(\mathbf{q}_k) t, \qquad \Psi_{k\gamma e}^m = \mathbf{q}_k \cdot \mathbf{l}_m + \omega_{\gamma e}(\mathbf{q}_k) t. \qquad (17.153)$$

The above result shows that, in the harmonic approximation, the displacements have the form of a superposition of plane and monochromatic waves, whose wave vector and angular frequency are $\mathbf{q}_k$, $\omega_{\gamma e}(\mathbf{q}_k)$. The wave corresponding to a given $\mathbf{q}_k$ is called *vibrational mode*. Typically, the number of $\mathbf{q}_k$ vectors is very large; in such cases, the same reasoning made in Sect. 17.6 with reference to the $\mathbf{k}$ vectors holds, and $\mathbf{q}$ is considered a continuous variable ranging over the first Brillouin zone.

Function $\omega_{\gamma e}(\mathbf{q})$ is also called *dispersion relation*, and is viewed as a multi-valued function of $\mathbf{q}$ having $3N_b$ branches. For each branch, letting $q = |\mathbf{q}|$, the *wavelength* is defined by $\lambda = 2\pi/q$ and the *phase velocity* by $u_f = \omega/q = \lambda \nu$, with $\nu = \omega/(2\pi)$ the frequency.

The *group velocity* is defined by $\mathbf{u} = \mathrm{grad}_{\mathbf{q}}\omega$. As shown in Sect. 3.10, the total energy of the system is the sum of the mode energies, and in the classical description is expressed in terms of the initial conditions as

$$T_a + V_a = V_{a0} + \sum_{\sigma=1}^{3N} E_\sigma, \qquad E_\sigma = \frac{1}{2} \dot{b}_\sigma^2(0) + \frac{1}{2} \omega_\sigma^2 b_\sigma^2(0). \qquad (17.154)$$

As remarked in Sect. 12.5, the classical expression of the energy associated to each mode has the same form as that of a mode of the electromagnetic field. In turn, the energy quantization shows that each mode energy is made of terms of the form $\hbar\omega_{\gamma e}(\mathbf{q})$, this leading to the concept of phonon (Eqs. (12.35, 12.36)).

17.8 Complements

17.8.1 Crystal Planes and Directions in Cubic Crystals

From the general definition (17.1) of the translation vector, that provides the positions of all nodes of the crystal, it follows that a *crystal plane* is defined by the set of all triads of integers m_1, m_2, m_3 such that, for a given vector $\mathbf{g}_0$ of the scaled reciprocal lattice, the quantity $(m_1 \mathbf{a}_1 + m_2 \mathbf{a}_2 + m_3 \mathbf{a}_3) \cdot \mathbf{g}_0/(2\pi)$ equals a fixed integer. Such a plane is normal to $\mathbf{g}_0$. In turn, given two crystal planes defined as above using, respectively, two non-parallel vectors $\mathbf{g}_1$ and $\mathbf{g}_2$, a *crystal direction* is defined by the set of all triads of integers m_1, m_2, m_3 that belong to the two crystal planes so

Fig. 17.22 Example of node labeling in the cubic lattice

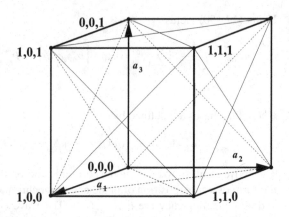

prescribed. In cubic crystals, the typical method by which the crystal planes are identified is outlined below [103, Sect. 2.2].

Let the plane be indicated with Π. After labeling the nodes by the respective triads of integers m_1, m_2, m_3, as shown in Fig. 17.22, one starts by finding the intercepts of Π with the directions of the characteristic vectors. Letting such intercepts be (m_1^*, m_2^*, m_3^*), the triad $(r\, m_1^*, r\, m_2^*, r\, m_3^*)$ with $r \neq 0$ an integer, spans a set of planes parallel to Π. If M is the largest divisor of m_1^*, m_2^*, m_3^*, then the new triad $m_i' = m_i^*/M$ identifies the plane Π' parallel to Π and closest to the origin. Then, the inverse of the triad's elements are taken: $1/m_1'$, $1/m_2'$, $1/m_3'$. This avoids the occurrence of infinities; in fact, if Π were parallel to one of the characteristic vectors, say, $\mathbf{a}_i$, then m_i^* and m_i' would become infinite. One the other hand, using the inverse indices may bring to fractional numbers, a circumstance that must be avoided as well; so, as the last step, the new elements $1/m_i'$ are multiplied by the least multiple N of the m_i' that are not infinite:

$$(m_1'', m_2'', m_3'') = \left(\frac{N}{m_1'}, \frac{N}{m_2'}, \frac{N}{m_3'} \right). \tag{17.155}$$

The elements m_i'' thus found are the *Miller indices* of Π. They are enclosed in parentheses as in (17.155). By way of example, if $m_1^* = \infty$, $m_2^* = 2$, $m_3^* = 4$, then $M = 2$ so that $m_1' = \infty$, $m_2' = 1$, $m_3' = 2$. Calculating the inverse indices yields $1/m_1' = 0$, $1/m_2' = 1$, $1/m_3' = 1/2$; the least multiple is $N = 2$, so that the Miller indices are found to be $(0, 2, 1)$.

The indices that turn out to be negative are marked with a bar; for instance, in $(h\bar{k}l)$ the second index is negative. Some planes have the same symmetry; in cubic crystals this happens, for instance, to planes (100), (010), (001), ($\bar{1}$00), (0$\bar{1}$0), and (00$\bar{1}$). A set of planes with the same symmetry is indicated with braces, e.g., {100}. Examples of the (111), (001), and (010) planes are given in Fig. 17.23. As remarked in Sect. 24.4 about the silicon-oxidation process, the (111) plane has the highest concentration of atoms, followed by the {100} planes.

Symbols using three integers are also used to identify the crystal directions. To distinguish them from the symbols introduced so far, such triads of integers are

Fig. 17.23 Schematic representation of the (111) plane (*top left*) and of the (001) and (010) planes (*bottom right*) in a cubic crystal

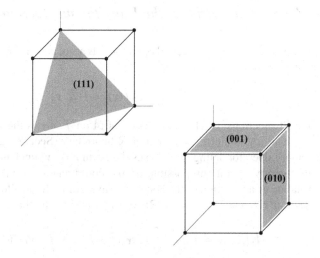

enclosed in brackets. Consider, for instance, the line connecting nodes P and Q, oriented from P to Q. Letting m_{1P}, m_{2P}, m_{3P} be the coordinates of P, and the like for those of Q, one forms the new triad $m_1' = m_{1Q} - m_{1P}$, $m_2' = m_{2Q} - m_{2P}$, and $m_3' = m_{2Q} - m_{2P}$. Then, the indices of the crystal direction are obtained as

$$[m_1'', m_2'', m_3''] = \left[\frac{m_1'}{M}, \frac{m_2'}{M}, \frac{m_3'}{M}\right], \qquad (17.156)$$

with M the largest divisor of m_1', m_2', m_3'. Also in this case, negative indices are marked with a bar. By way of examples, the characteristic vectors $\mathbf{a}_1$, $\mathbf{a}_2$, and $\mathbf{a}_3$ in Fig. 17.22 are aligned, respectively, with the [100], [010], and [001] directions.

17.8.2 Examples of Translation Operators

A one-dimensional example of translation operator is easily found by considering the Taylor expansion of a function f around some position x:

$$f(x + l) = \sum_{n=0}^{\infty} \frac{l^n}{n!} \left(\frac{\mathrm{d}^n f}{\mathrm{d}x^n}\right)_{l=0} = \sum_{n=0}^{\infty} \frac{l^n}{n!} \frac{\mathrm{d}^n}{\mathrm{d}x^n} f(x) = \exp\left(l\frac{\mathrm{d}}{\mathrm{d}x}\right) f(x), \quad (17.157)$$

where the expression on the right stems from a formal application of the Taylor expansion of the exponential function, in which a numerical factor within the exponent is replaced with the operator $\mathrm{d}/\mathrm{d}x$. Extending the above reasoning to three dimensions yields

$$\mathcal{T}(\mathbf{l}) = \exp\left(\mathbf{l} \cdot \mathrm{grad}\right). \qquad (17.158)$$

17.8.3   Symmetries of the Hamiltonian Operator

Given an operator $\mathcal{R}$, a second operator $\mathcal{A}$ is associated to $\mathcal{R}$ in the following manner [58, Sect. 1.5]:

$$\mathcal{A}f(\mathbf{r}) = f(\mathcal{R}^\dagger \mathbf{r}) \tag{17.159}$$

for all functions f. Thus, the action of $\mathcal{A}$ on f at $\mathbf{r}$ is the same as calculating the original function at $\mathbf{r}' = \mathcal{R}^\dagger \mathbf{r}$. Let $\mathcal{R}$ be unitary (Sect. 8.6.2), whence $\mathbf{r} = \mathcal{R}\mathbf{r}'$. A unitary operator acting on $\mathbf{r}$ leaves the norm $r = |\mathbf{r}|$ unchanged; as a consequence, the unitary operations possible on the coordinates are only those that perform a rotation or a reflexion of the coordinate axes, or both. It follows that the unit volume $\mathrm{d}\tau = \mathrm{d}^3 r$ is also invariant: $\mathrm{d}^3 \mathcal{R}r = \mathrm{d}^3 r$, this showing that $\mathcal{A}$ is unitary as well:

$$\int_\tau |\mathcal{A}f(\mathbf{r})|^2 \, \mathrm{d}^3 r = \int_{\tau'} |f(\mathbf{r}')|^2 \, \mathrm{d}^3 \mathcal{R}r' = \int_{\tau'} |f(\mathbf{r}')|^2 \, \mathrm{d}^3 r', \tag{17.160}$$

where τ' is the transformed domain. This reasoning does not apply to the translation operators $\mathcal{T}$. In fact, the operation $\mathcal{T}\mathbf{r} = \mathbf{r} + \mathbf{l}$ does not leave the norm of $\mathbf{r}$ unchanged. This shows in passing that $\mathcal{T}$ is not unitary. Other consequences of the above definitions and of the proof that $\mathcal{A}$ is unitary are

$$\mathcal{A}f(\mathcal{R}\mathbf{r}') = f(\mathbf{r}'), \qquad \mathcal{A}^\dagger f(\mathbf{r}') = f(\mathcal{R}\mathbf{r}'). \tag{17.161}$$

Also, $\mathcal{A}$ commutes with the operators that are invariant under the transformation $\mathbf{r} \leftarrow \mathcal{R}^\dagger \mathbf{r}$. In fact, if $\mathcal{B}$ is such an operator,

$$\mathcal{A}\mathcal{B}(\mathbf{r})f(\mathbf{r}) = \mathcal{B}(\mathcal{R}^\dagger \mathbf{r})f(\mathcal{R}^\dagger \mathbf{r}) = \mathcal{B}(\mathbf{r})\mathcal{A}f(\mathbf{r}) \tag{17.162}$$

for all functions f. As $\mathcal{R}^\dagger$ is the inverse of $\mathcal{R}$, then $\mathcal{B}$ is also invariant under the transformation $\mathbf{r} \leftarrow \mathcal{R}\mathbf{r}$. As a consequence, $\mathcal{B}$ commutes also with $\mathcal{A}^\dagger$.

Let $\mathcal{B}v_n = b_n v_n$ be the eigenvalue equation for $\mathcal{B}$ (a discrete spectrum is assumed for the sake of simplicity). If b_n is s-fold degenerate, and $v_n^{(1)}, v_n^{(2)}, \ldots, v_n^{(s)}$ are s linearly-independent eigenfunctions corresponding to b_n, then

$$\mathcal{B}\sum_{i=1}^s c_i v_n^{(i)} = \sum_{i=1}^s c_i \mathcal{B}v_n^{(i)} = \sum_{i=1}^s c_i b_n v_n^{(i)} = b_n \sum_{i=1}^s c_i v_n^{(i)}, \tag{17.163}$$

namely, any non-vanishing linear combination of the form $\varphi_n = \sum_{i=1}^s c_i v_n^{(i)}$ is also an eigenfunction of $\mathcal{B}$ belonging to b_n. Let M be the space of all linear combinations of the form of φ_n; from (17.163) it follows that all members of M are eigenfunctions of $\mathcal{B}$ belonging to b_n. Conversely, all eigenfunctions of $\mathcal{B}$ belonging to b_n are members of M: letting $q_n \neq 0$ be one such eigenfunction, if q_n were not a member of M it would be $q_n - \sum_{i=1}^s c_i v_n^{(i)} \neq 0$ for all choices of the coefficients c_i. But this would imply that $q_n, v_n^{(1)}, v_n^{(2)}, \ldots, v_n^{(s)}$ are $s + 1$ linearly-independent eigenfunctions of b_n, thus contradicting the hypothesis that the latter's degeneracy is of order s. Finally, if $\mathcal{A}$ commutes with $\mathcal{B}$ it is

$$\mathcal{B}\mathcal{A}\varphi_n = \mathcal{A}\mathcal{B}\varphi_n = \mathcal{A}b_n\varphi_n = b_n\mathcal{A}\varphi_n, \tag{17.164}$$

namely, $\mathcal{A}\varphi_n$ belongs to M.

In crystals, the unitary coordinate transformations $\mathbf{r}' = \mathcal{R}\mathbf{r}$ that leave the Hamiltonian operator $\mathcal{H}$ invariant are of particular interest. In fact, such coordinate transformations provide a method to study the degenerate eigenvalues of $\mathcal{H}$.

Let $\mathcal{B} = \mathcal{H}$, and let $\mathcal{H}$ be invariant under a coordinate transformation $\mathcal{R}\mathbf{r}$. If, in addition, $\mathcal{H}$ is translationally invariant and the periodic boundary conditions apply (Sect. 17.5.3), then the eigenfunctions w of $\mathcal{H}$ are Bloch functions, namely, they fulfill the Bloch theorem

$$w_i(\mathbf{r} + \mathbf{l}, \mathbf{k}) = \exp(i\,\mathbf{k} \cdot \mathbf{l})\,w_i(\mathbf{r}, \mathbf{k}), \tag{17.165}$$

with $\mathbf{l}$ a translation vector and i the band index. Let $\mathcal{A}$ be the operator associated to $\mathcal{R}$. Then, from $\mathcal{H}\mathcal{A}^\dagger = \mathcal{A}^\dagger\mathcal{H}$,

$$\mathcal{H}\mathcal{A}^\dagger w_i(\mathbf{r}, \mathbf{k}) = E_i(\mathbf{k})\,\mathcal{A}^\dagger w_i(\mathbf{r}, \mathbf{k}), \tag{17.166}$$

with $E_i(\mathbf{k})$ the eigenvalue. One infers from (17.166) that, if $w_i(\mathbf{r}, \mathbf{k})$ and $\mathcal{A}^\dagger w_i(\mathbf{r}, \mathbf{k})$ are linearly independent, then the eigenvalue is degenerate. Such a degeneracy does not depend on the detailed form of the Hamiltonian operator, but only on its symmetry properties. For this reason, the degeneracy is called *essential*. If further degeneracies exist, that depend on the detailed form of $\mathcal{H}$, they are called *accidental*.

Let $M(\mathbf{k})$ be the space made of the linearly-independent eigenfunctions of $E(\mathbf{k})$, and of any non-vanishing linear combination of them, and define

$$v_i(\mathbf{r}, \mathbf{k}') = \mathcal{A}^\dagger w_i(\mathbf{r}, \mathbf{k}) = w_i(\mathcal{R}\mathbf{r}, \mathbf{k}), \tag{17.167}$$

where symbol $\mathbf{k}'$ accounts for a possible influence on $\mathbf{k}$ of the coordinate transformation $\mathcal{R}\mathbf{r}$. Being an eigenfunction of $\mathcal{H}$, $v_i(\mathbf{r}, \mathbf{k}')$ is a Bloch function,

$$v_i(\mathbf{r} + \mathbf{l}, \mathbf{k}') = \exp(i\,\mathbf{k}' \cdot \mathbf{l})\,v_i(\mathbf{r}, \mathbf{k}'), \tag{17.168}$$

where $v_i(\mathbf{r} + \mathbf{l}, \mathbf{k}') = w_i(\mathcal{R}\mathbf{r} + \mathcal{R}\mathbf{l}, \mathbf{k})$. On the other hand, Bloch's theorem applied to $w_i(\mathcal{R}\mathbf{r} + \mathcal{R}\mathbf{l}, \mathbf{k})$ yields

$$w_i(\mathcal{R}\mathbf{r} + \mathcal{R}\mathbf{l}, \mathbf{k}) = \exp(j\,\mathbf{k} \cdot \mathcal{R}\mathbf{l})\,w_i(\mathcal{R}\mathbf{r}, \mathbf{k}), \tag{17.169}$$

where the equality $\mathbf{k} \cdot \mathcal{R}\mathbf{l} = \mathcal{R}^\dagger\mathbf{k} \cdot \mathbf{l}$ holds due to the definition of adjoint operator. Comparison with the expression of the Bloch theorem applied to $v_i(\mathbf{r}+\mathbf{l}, \mathbf{k}')$ provides $\mathbf{k}' = \mathcal{R}^\dagger\mathbf{k}$, whence

$$w_i(\mathcal{R}\mathbf{r}, \mathbf{k}) = v_i(\mathbf{r}, \mathcal{R}^\dagger\mathbf{k}). \tag{17.170}$$

In conclusion, if $w_i(\mathbf{r}, \mathbf{k})$ is a Bloch function belonging to $M(\mathbf{k})$ and $\mathcal{R}\mathbf{r}$ a coordinate transformation that leaves the Hamiltonian operator invariant, then the eigenfunction obtained by such a transformation also belongs to $M(\mathbf{k})$ and is labeled by $\mathcal{R}^\dagger \mathbf{k}$. The following also holds true,

$$\mathcal{H}v_i(\mathbf{r}, \mathcal{R}^\dagger \mathbf{k}) = E_i(\mathcal{R}^\dagger \mathbf{k})\, v_i(\mathbf{r}, \mathcal{R}^\dagger \mathbf{k}) \tag{17.171}$$

which, compared with $\mathcal{H}w_i(\mathbf{r}, \mathbf{k}) = E_i(\mathbf{k})\, w_i(\mathbf{r}, \mathbf{k})$, shows that

$$E_i(\mathcal{R}^\dagger \mathbf{k}) = E_i(\mathbf{k}). \tag{17.172}$$

The theory of this section is applied by way of example to the Hamiltonian operator of a system of K electrons and N nuclei, interacting through electrostatic forces, that was introduced in Sect. 16.2. The potential energy is (compare with (16.5))

$$U_e(\mathbf{r}) + U_a(\mathbf{r}) + U_{ea}(\mathbf{r}, \mathbf{R}) + U_{\text{ext}}(\mathbf{r}, \mathbf{R}), \tag{17.173}$$

with

$$U_e(\mathbf{r}) = \sum_{i,j=1}^{K} \frac{q^2}{4\pi\,\varepsilon_0\,|\mathbf{r}_i - \mathbf{r}_j|}, \qquad j \neq i. \tag{17.174}$$

Similar expressions hold for U_a and U_{ea} (the second relation in (16.1) and (16.2), respectively). If $U_{\text{ext}} = 0$, the potential energy is invariant upon the reflexion transformation $\mathcal{R}\mathbf{r} = -\mathbf{r}$, $\mathcal{R}\mathbf{R} = -\mathbf{R}$. Clearly, the kinetic part of the Hamiltonian operator is also invariant. In the adiabatic approximation (Sect. 16.3), the coordinates of the nuclei are fixed to the equilibrium positions $\mathbf{R}_0$, which preserves the reflexion invariance. Finally, the reflexion invariance is still preserved in the Hartree and Hartree–Fock approximations (Sects. 16.4 and 16.5, respectively), which also provide single-electron Hamiltonian operators that are translationally invariant. Due to lattice periodicity, the eigenfunctions of the Hamiltonian operator are Bloch functions. Denoting now with $\mathbf{r}$ the coordinates associated to a single electron, the transformation $\mathcal{R}\mathbf{r} = -\mathbf{r}$ corresponds to $\mathcal{R}^\dagger \mathbf{k} = -\mathbf{k}$ whence, from (17.172),

$$E_i(-\mathbf{k}) = E_i(\mathbf{k}). \tag{17.175}$$

This type of degeneracy is accidental because it depends on the detailed form of the Hamiltonian operator. If the crystal has also a reflection symmetry, then the reflexion invariance of the single-electron Hamiltonian operators occurs irrespective of the form of the interactions. In this case, the degeneracy is essential.

17.8.4   Kronig–Penney Model

The general method for solving the Schrödinger equation in a periodic lattice, shown in Sect. 17.6, is applied here to a one-dimensional case, where the potential energy is described as the series of equal barriers shown in Fig. 17.24. The approach is

Fig. 17.24 Potential energy
in the Kronig–Penney model

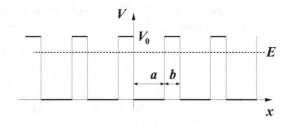

called *Kronig–Penney model*; it is amenable to an analytical solution and, despite its
simplicity, is able to capture the main properties of the dispersion relation $E(\mathbf{k})$.

As shown in the figure, the potential energy is prescribed as $V = 0$ for $n\,(a+b) <
x < n\,(a+b) + a$, and $V = V_0 > 0$ for $n\,(a+b) - b < x < n\,(a+b)$,
with $n = 0, \pm 1, \pm 2 \ldots$ There is only one characteristic vector in the direct lattice,
$\mathbf{a}_1 = (a+b)\,\mathbf{i}_1$; the corresponding characteristic vector of the reciprocal lattice is

$$\mathbf{b}_1 = \frac{\mathbf{i}_1}{a+b}. \tag{17.176}$$

As a consequence, the first Brillouin zone extends from $-\pi/(a+b)$ to $+\pi/(a+b)$
in the $\mathbf{i}_1$ direction. From the general properties of the time-independent Schrödinger
equation (Sect. 8.2.3) it follows $E \geq 0$. As shown in Fig. 17.24, the case $0 < E < V_0$
is considered. A non-localized wave function w is expected even in the $E < V_0$ case
due to the tunnel effect. From the Bloch theorem, the wave function has the form

$$w_k = u_k \exp(\mathrm{i}\,k\,x), \qquad u_k(x + a + b) = u_k(x), \tag{17.177}$$

where k belongs to the first Brillouin zone. In the intervals where $V = 0$ the
Schrödinger equation reads

$$-w'' = \alpha^2\,w, \qquad \alpha = \sqrt{2\,m\,E}/\hbar > 0. \tag{17.178}$$

Replacing (17.177) into (17.178) yields

$$u_k'' + 2\,\mathrm{i}\,k\,u_k' - (k^2 - \alpha^2)\,u_k = 0, \tag{17.179}$$

whose associate algebraic equation has the roots

$$s = -\mathrm{i}\,k \pm \sqrt{-k^2 + (k^2 - \alpha^2)} = -\mathrm{i}\,k \pm \mathrm{i}\,\alpha. \tag{17.180}$$

The solution of (17.178) then reads

$$u_k^+ = c_1 \exp[\mathrm{i}\,(\alpha - k)\,x] + c_2 \exp[-\mathrm{i}\,(\alpha + k)\,x], \tag{17.181}$$

with c_1, c_2 undetermined coefficients. The procedure is similar in the intervals where $V = V_0$, and yields

$$w'' = \beta^2 \, w, \quad \beta = \sqrt{2\,m\,(V_0 - E)}/\hbar, \quad u_k'' + 2\,i\,k\,u_k' - (k^2 + \beta^2)\,u_k = 0,$$
$$\tag{17.182}$$

$$s = -i\,k \pm \sqrt{-k^2 + (k^2 + \beta^2)} = -i\,k \pm \beta, \tag{17.183}$$

whence

$$u_k^- = c_3 \, \exp\left[(\beta - i\,k)\,x\right] + c_4 \, \exp\left[-(\beta + i\,k)\,x\right], \tag{17.184}$$

with c_3, c_4 undetermined coefficients. The regional solutions u_k^+, u_k^- must fulfill the continuity conditions imposed by the general properties of the Schrödinger equation; in addition, they must fulfill the periodicity condition prescribed by the Bloch theorem (second relation in (17.177)). To proceed, one focuses on the period $-b \le x \le a$, so that the continuity conditions at $x = 0$ for the function, $u_k^+(0) = u_k^-(0)$, and first derivative, $(u_k^+)'(0) = (u_k^-)'(0)$, provide

$$c_1 + c_2 = c_3 + c_4, \qquad i\,\alpha\,(c_1 - c_2) = \beta\,(c_3 - c_4). \tag{17.185}$$

Combining (17.185),

$$c_1 = \sigma c_3 + \sigma^* c_4, \quad c_2 = \sigma^* c_3 + \sigma c_4, \quad 2\,\sigma = 1 - i\,\beta/\alpha. \tag{17.186}$$

In turn, from the periodicity of u, namely, $u_k^+(a) = u_k^-(-b)$, and of u', namely, $(u_k^+)'(a) = (u_k^-)'(-b)$, one finds

$$c_1 A + \frac{c_2}{A} = K L \left(\frac{c_3}{B} + c_4 \, B\right), \quad c_1 A - \frac{c_2}{A} = -K L \left(\frac{c_3}{B} - c_4 \, B\right) i \frac{\beta}{\alpha},$$
$$\tag{17.187}$$

with

$$A = \exp(i\,\alpha\,a), \quad B = \exp(\beta\,b), \quad K = \exp(i\,k\,a), \quad L = \exp(i\,k\,b). \tag{17.188}$$

Combining (17.187),

$$c_1 = \frac{KL}{A} \left(\frac{\sigma}{B} c_3 + \sigma^* B \, c_4\right), \quad c_2 = A\,K\,L \left(\frac{\sigma^*}{B} c_3 + \sigma\,B\,c_4\right). \tag{17.189}$$

Eliminating c_1, c_2 between (17.186) and (17.189), finally provides an algebraic system in the two unknowns c_3, c_4:

$$\sigma \left(1 - \frac{KL}{AB}\right) c_3 + \sigma^* \left(1 - \frac{BKL}{A}\right) c_4 = 0, \tag{17.190}$$

$$\sigma^* \left(1 - \frac{AKL}{B}\right) c_3 + \sigma \, (1 - ABKL) \, c_4 = 0. \tag{17.191}$$

Fig. 17.25 Graphic solution of (17.194), with $\vartheta = 10$. The two *vertical lines* mark the values of $\alpha\,a$ delimiting the lowest band

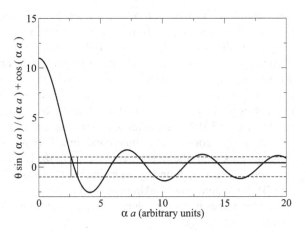

As expected, the system is homogeneous, so a solution is possible only if the determinant vanishes. This in turn determines a relation between $\alpha(E)$, $\beta(E)$, and k, that eventually provides the dispersion relation $E(k)$. The determinant vanishes if

$$(\sigma^{*2} - \sigma^2)\left(KL + \frac{1}{KL}\right) = \sigma^{*2}\left(\frac{A}{B} + \frac{B}{A}\right) - \sigma^2\left(AB + \frac{1}{AB}\right). \quad (17.192)$$

Introducing the expressions (17.186, 17.188) of σ, A, B, K, L transforms (17.192) into

$$\frac{\beta^2 - \alpha^2}{2\alpha\beta}\sin(\alpha\,a)\sinh(\beta\,b) + \cos(\alpha\,a)\cosh(\beta\,b) = \cos[k\,(a+b)], \quad (17.193)$$

which has the form $F(E) = G(k)$. From this, the relation $E = E(k)$ can be determined. Note that $G(-k) = G(k)$ and $G[k+2\pi/(a+b)] = G(k)$. As a consequence, the function $E(k)$ is even and has the periodicity of the reciprocal scaled lattice, as should be.

To the purpose of completing the analysis one may simplify (17.193) by considering a limiting case, namely, $V_0 \gg E$ so that, from (17.178, 17.182), the limit $\beta^2 \gg \alpha^2$ would result. This, however, would eliminate the tunnel effect and reduce the problem to that of a series of boxes. To avoid this outcome, the proper limiting case is $b \to 0$ and $V_0 \to \infty$, in such a way as to leave the area $b\,V_0$ of each barrier unchanged.[22] In other terms, one lets $b = \text{const}/V_0$, so that $\beta^2 b \to \text{const} \neq 0$ while $\beta\,b \to 0$. It follows $\sinh(\beta\,b) \to \beta\,b$, $\cosh(\beta\,b) \to 1$ so that, letting $\vartheta = \lim(a\,b\,\beta^2/2)$, the $F(E) = G(k)$ relation (17.193) simplifies to

$$\vartheta\,\frac{\sin(\alpha\,a)}{\alpha\,a} + \cos(\alpha\,a) = \cos(k\,a), \qquad \vartheta > 0, \qquad \alpha = \frac{\sqrt{2mE}}{\hbar}. \quad (17.194)$$

---

[22] The same type of limit is applicable to the single-barrier case, whose transmission coefficient is given in (11.22).

The function $E = E(k)$ can be determined by inverting (17.194); alternatively, it may be obtained in graphic form as shown in Fig. 17.25, where ϑ has been fixed to 10: given k, the right hand side of (17.194) is fixed at some value $-1 \le \cos(k\,a) \le 1$. The energy E is then found by seeking $\alpha\,a$ such that the two sides become equal. The horizontal, dashed lines in the figure correspond to $\cos(ka) = 1$ and $\cos(k\,a) = -1$; they limit the interval where (17.194) has real solutions. The horizontal, continuous line corresponds to $\cos(k\,a) = 0.4$, while the oscillating curve represents the left hand side of (17.194). The latter intercepts the $\cos(k\,a) = 0.4$ line at infinite points $\alpha_1\,a,\ \alpha_2\,a, \ldots$; from each α_i thus found, one determines the energy corresponding to the given k from the relation $\alpha_i = \sqrt{2m\,E_i}/\hbar$. Each branch of the multi-valued function $E(k)$ is then found by repeating the procedure for all values of k within the first Brillouin zone, this making $\cos(k\,a)$ to range from -1 to 1. In the figure, the two vertical lines mark the values of $\alpha\,a$ delimiting the lowest band. The following are also worth noting:

- Letting λ indicate the left hand side of (17.194), there are no real solutions for $\lambda > 1$ or $\lambda < -1$; the intervals with no real solutions are the forbidden bands. In fact, the k solutions in the forbidden bands are complex: it is $k\,a = \pm i \log(\lambda + \sqrt{\lambda^2 - 1})$ when $\lambda > 1$, and $k\,a = \pi \pm i \log(|\lambda| + \sqrt{\lambda^2 - 1})$ when $\lambda < -1$.
- At large energies the (17.194) relation tends to $\cos(\alpha a) = \cos(ka)$, namely, to the free-particle one: $k = \alpha = \sqrt{2mE}/\hbar$.
- Like in the general case, for a finite structure where the periodic boundary conditions are applied, the above calculation still holds, with k a discrete variable.

17.8.5 Linear, Monatomic Chain

The calculation of vibrational spectra has been carried out in general form in Sect. 17.7. Simple examples of application, with reference to a one-dimensional lattice, are given in this section and in the next one. Like the Kronig–Penney model used in Sect. 17.8.4 for determining the dispersion relation of electrons, the one-dimensional models of the lattice vibrations are amenable to analytical solutions; the latter, as shown below, are able to provide the explicit expression of the dispersion relation.

To begin, consider a one-dimensional monatomic lattice made of N_c cells (*linear, monatomic chain*). Let the lattice be aligned with the x axis, and the corresponding characteristic vector be $\mathbf{a} = a\,\mathbf{i}$, $a > 0$, with $\mathbf{i}$ the unit vector of the x axis. Finally, let the positions of the $N_c + 1$ nodes be $0, a, 2\,a, \ldots, n\,a, \ldots$. The translation vector associated to the nth node is $\mathbf{l}_n = n\,a\,\mathbf{i}$. Finally, it is assumed that the motion of each atom is constrained to the x axis, and the periodic boundary conditions are applied.

Due to the periodic boundary conditions the nodes of indices $n = 0$ and $n = N_c$ are actually the same node. As a one-dimensional case is considered, with $N_b = 1$, the total number of atoms is $N = N_c$. The number of the lattice's degrees of freedom is N_c, and the correspondence with the indices used in the general theory (compare with (17.124)) is

$$m, n = 1, \ldots, N_c, \qquad \alpha, \beta = 1, \qquad u, w = 1. \qquad (17.195)$$

As only one atom per cell is present, one may assume that the equilibrium position of each nucleus coincides with that of a node. In the harmonic approximation the force acting on the rth nucleus is a linear function of the displacements:

$$F_r = -\sum_{k=1}^{N_c} c_{rk} \, h_k, \qquad (17.196)$$

where all coefficients c_{rk} in general differ from 0. In real crystals, however, the interaction between nuclei becomes rapidly negligible as the distance increases. As a consequence, the dynamics of a nucleus may be tackled in a simplified manner by considering only the interaction with the neighboring nuclei to be effective. This is equivalent to letting $c_{rk} = 0$ when $|r - k| > 1$, whence

$$F_r = -c_r^{r-1} \, h_{r-1} - c_r^r \, h_r - c_r^{r+1} \, h_{r+1} = F_r(h_{r-1}, h_r, h_{r+1}). \qquad (17.197)$$

In the coefficients of (17.197), the lower index refers to the node being acted upon by the force at the left hand side, the upper index refers to the node whose displacement contributes to such a force. When the nuclei of indices $r - 1, r, r + 1$ are in the equilibrium positions it is $F_r(0, 0, 0) = 0$ for all r. On the other hand, it is also $F_r(\delta, \delta, \delta) = 0$, with $\delta \neq 0$ an arbitrary displacement. In fact, when all displacements are equal, the interatomic distance remains the same as in the equilibrium condition. From $F_r(\delta, \delta, \delta) = 0$ it follows that the coefficients are connected by the relation

$$c_r^{r-1} + c_r^r + c_r^{r+1} = 0. \qquad (17.198)$$

Moreover, on account of the fact that all atoms are identical and all equilibrium distances are also identical, it is $F_r(-\delta, 0, \delta) = 0$, $\delta \neq 0$, whence

$$c_r^{r-1} = c_r^{r+1}. \qquad (17.199)$$

From (17.198, 17.199) it follows

$$c_r^r = -c_r^{r-1} - c_r^{r+1} = -2 \, c_r^{r-1} = -2 \, c_r^{r+1}. \qquad (17.200)$$

Finally, the relation $F_r(0, \delta, 0) = -c_r^r \, \delta$, on account of the fact that $(0, 0, 0)$ is an equilibrium condition, shows that $c_r^r > 0$. As shown in Sect. 17.7 for the general case, due to the translational invariance the elastic coefficients do not depend on the cell index, but on the difference between cell indices (compare with (17.125)); in conclusion, letting

$$\chi = -c_r^{r-1} = -c_r^{r+1} > 0, \qquad c_r^r = 2\chi, \qquad (17.201)$$

and letting μ be the common mass of the nuclei, the dynamics of the rth nucleus is described by the equation

$$\mu \ddot{h}_r = -\chi \ (2\,h_r - h_{r+1} - h_{r-1}). \tag{17.202}$$

The general theory shows that the displacement has the form

$$h_r = h_0 \exp\left(i\,q\,r\,a - i\,\omega\,t\right), \tag{17.203}$$

(compare with (17.152)), where h_0 is a complex constant, $q\,r\,a = \mathbf{q} \cdot \mathbf{l}_r = q\,\mathbf{i} \cdot r\,a\,\mathbf{i}$, and $\omega = \omega(q)$. Replacing (17.203) in (17.202) and dividing by h_r yields

$$\mu\,\omega^2 = \chi \left[2 - \exp\left(i\,q\,a\right) - \exp\left(-i\,qa\right)\right] = 4\,\chi\,\sin^2\left(qa/2\right). \tag{17.204}$$

Defining $\tilde{\omega} = \sqrt{\chi/\mu}$ and remembering that ω is non negative, one finds the dispersion relation

$$\omega(q) = 2\tilde{\omega}\,|\sin\left(qa/2\right)|. \tag{17.205}$$

From the periodic boundary condition $h(r = N_c) = h(r = 0)$ one finds, with ν an integer,

$$\exp\left(i\,q\,N_c\,a\right) = 1, \qquad q\,N_c\,a = 2\,\pi\,\nu, \qquad q = \frac{\nu}{N_c}\,\frac{2\,\pi}{a}. \tag{17.206}$$

Replacing this form of q within (17.203) and (17.205) shows that using $\nu + N_c$ instead of ν leaves h_r and ω unchanged. As a consequence, it is sufficient to consider only N_c consecutive values of ν, say, $\nu = 0, 1, \ldots, N_c - 1$, which in turn limit the possible values of q to an interval of length $2\pi/a$. This was expected, because the values of the indices in (17.195) are such that the number of eigenvalues of the problem is N_c. Thus, the dispersion relation has only one branch, given by (17.205). One also notes that $2\pi/a$ is the size of the first Brillouin zone in the one-dimensional case. Typically, the interval of q is made to coincide with the first Brillouin zone, namely, $-\pi/a \leq q < +\pi/a$. Also, as mentioned in Sect. 17.7, in most cases q is treated as a continuous variable. The phase and group velocities are

$$u_f = \frac{\omega}{q} = \pm a\,\tilde{\omega}\,\frac{\sin\left(q\,a/2\right)}{qa/2}, \qquad u = \frac{d\omega}{dq} = \pm a\,\tilde{\omega}\,\cos\left(qa/2\right), \tag{17.207}$$

respectively, where the positive (negative) sign holds when q is positive (negative). At the boundary of the Brillouin zone it is $qa/2 = \pi/2$, whence $\omega = 2\,\tilde{\omega}$, $u_f = \pm a\tilde{\omega}/\pi$, $u = 0$. Near the center of the Brillouin zone it is $\omega \simeq a\tilde{\omega}\,|q|$, $u_f \simeq u \simeq \pm a\tilde{\omega}$. At the center it is $\omega = 0$.

The dispersion relation (17.205) normalized to $\tilde{\omega} = \sqrt{\chi/\mu}$ is shown in Fig. 17.26 as a function of $qa/2$. The range of the first Brillouin zone is $-\pi/2 \leq qa/2 \leq +\pi/2$. Remembering that the wavelength corresponding to q is $\lambda = 2\pi/q$, the interval near the origin, where the phase and group velocities are equal to each other and independent of q, corresponds to the largest values of the vibrations' wavelength. As some of these wavelengths fall in the audible range, the branch is called *acoustic branch*.

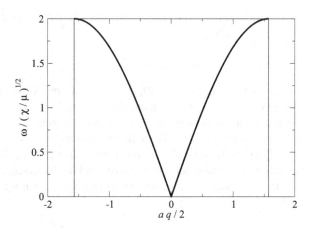

Fig. 17.26 Normalized dispersion relation of a linear, monatomic chain. The *vertical lines*, placed at $a\,q/2 = \pm\pi/2$, are the limits of the first Brillouin zone

17.8.6   Linear, Diatomic Chain

As a second example, consider a one-dimensional lattice made of N_c cells, with a two-atom basis. Let the lattice be aligned with the x axis, and the corresponding characteristic vector be $\mathbf{a} = a\,\mathbf{i}$, $a > 0$, with $\mathbf{i}$ the unit vector of the x axis. Finally, let the positions of the $N_c + 1$ nodes be $0, a, 2\,a, \ldots, n\,a, \ldots$. The translation vector associated to the nth node is $\mathbf{l}_n = n\,a\,\mathbf{i}$. Finally, it is assumed that the motion of each atom is constrained to the x axis, and the periodic boundary conditions are applied.

Due to the periodic boundary conditions the nodes of indices $n = 0$ and $n = N_c$ are actually the same node. As a one-dimensional case is considered, with $N_b = 2$, the total number of atoms is $N = 2\,N_c$. The number of degrees of freedom of the lattice is $2\,N_c$, and the correspondence with the indices (17.124) used in the general theory of Sect. 17.7 is

$$m, n = 1, \ldots, N_c, \qquad \alpha, \beta = 1, 2, \qquad u, w = 1. \qquad (17.208)$$

As two atoms per cell are present, one may assume that the equilibrium position of one type of nucleus coincides with that of a node. Such nuclei will be given the index $\alpha, \beta = 1$, while the other nuclei will be given the index $\alpha, \beta = 2$. In the harmonic approximation the force acting on a nucleus is a linear function of the displacements:

$$F_r = -\sum_{k=1}^{2\,N_c} c_{rk}\,h_k. \qquad (17.209)$$

In real crystals, the interaction between nuclei becomes rapidly negligible as the distance increases. Following the same reasoning as in Sect. 17.8.5, the dynamics of a nucleus is tackled in a simplified manner by considering only the interaction with the neighboring nuclei to be effective. This is equivalent to letting $c_{rk} = 0$ when $|r - k| > 1$. For a nucleus of type 1 the neighboring nuclei are of type 2. It is assumed that the node numbering is such, that the neighbors of interest belong to the cells of indices $r - 1$ and r. It follows

$$F_{r,1} = -c_{r,1}^{r-1,2} \, h_{r-1,2} - c_{r,1}^{r,1} \, h_{r,1} - c_{r,1}^{r,2} \, h_{r,2} = F_{r,1}(h_{r-1,2}, h_{r,1}, h_{r,2}). \quad (17.210)$$

In the coefficients of (17.210), the left-lower index refers to the node being acted upon by the force at the left hand side, the left-upper index refers to the node whose displacement contributes to such a force, the right-lower and right-upper indices refer to the nucleus type. When the nuclei involved are in the equilibrium positions it is $F_{r,1}(0, 0, 0) = 0$. On the other hand, it is also $F_{r,1}(\delta, \delta, \delta) = 0$, with $\delta \neq 0$ an arbitrary displacement. In fact, when all displacements are equal the interatomic distance remains the same as in the equilibrium condition. From $F_{r,1}(\delta, \delta, \delta) = 0$ it follows

$$c_{r,1}^{r-1,2} + c_{r,1}^{r,1} + c_{r,1}^{r,2} = 0. \quad (17.211)$$

As on the other hand there is no special symmetry in the interaction of the nucleus of indices $r, 1$ with the neighboring ones, it is in general (in contrast to the case of a monatomic linear chain) $c_{r,1}^{r-1,2} \neq c_{r,1}^{r,2}$. Finally, the relation $F_{r,1}(0, \delta, 0) = -c_{r,1}^{r,1} \delta$, on account of the fact that $(0, 0, 0)$ is an equilibrium condition, shows that $c_{r,1}^{r,1} > 0$.

The calculation is then repeated for a nucleus of type 2, whose neighboring nuclei are of type 1. Due to the node numbering chosen here, the neighbors of interest belong to the cells of indices r and $r + 1$. As a consequence,

$$F_{r,2} = -c_{r,2}^{r,1} \, h_{r,1} - c_{r,2}^{r,2} \, h_{r,2} - c_{r,2}^{r+1,1} \, h_{r+1,1} = F_{r,2}(h_{r,1}, h_{r,2}, h_{r+1,1}). \quad (17.212)$$

By the same reasoning leading to (17.211) one finds

$$c_{r,2}^{r,1} + c_{r,2}^{r,2} + c_{r,2}^{r+1,1} = 0, \quad (17.213)$$

where, like in the case of (17.211), it is in general $c_{r,2}^{r,1} \neq c_{r,2}^{r+1,1}$. Finally, the relation $F_{r,2}(0, \delta, 0) = -c_{r,2}^{r,2} \delta$, on account of the fact that $(0, 0, 0)$ is an equilibrium condition, shows that $c_{r,2}^{r,2} > 0$. Due to the lattice periodicity the coefficients do not depend on the cell index, whence one lets

$$-\chi_1 = c_{r-1,1}^{r-1,2} = c_{r,1}^{r,2} = c_{r+1,1}^{r+1,2} = \ldots \quad -\chi_2 = c_{r,1}^{r-1,2} = c_{r+1,1}^{r,2} = c_{r+2,1}^{r+1,2} = \ldots \quad (17.214)$$

Remembering the invariance relation $c_{m\alpha u}^{n\beta w} = c_{n\beta w}^{m\alpha u}$ of the general theory one also finds

$$c_{r,2}^{r,1} = c_{r,1}^{r,2} = -\chi_1, \qquad c_{r,2}^{r+1,1} = c_{r+1,1}^{r,2} = c_{r,1}^{r-1,2} = -\chi_2, \quad (17.215)$$

whence

$$c_{r,1}^{r,1} = c_{r,2}^{r,2} = \chi_1 + \chi_2 > 0. \quad (17.216)$$

Finally, from the relations $F_{r,1}(0,0,\delta) = \chi_1 \delta$, $F_{r,1}(\delta,0,0) = \chi_2 \delta$, it follows that $\chi_1 > 0$, $\chi_2 > 0$, on account of the fact that $(0,0,0)$ is an equilibrium condition. Letting μ_1, μ_2 be the masses of the two types of nuclei, the dynamics of the rth nuclei is described by the equations

$$\mu_1 \ddot{h}_{r,1} = -\chi_1 \left(h_{r,1} - h_{r,2} \right) - \chi_2 \left(h_{r,1} - h_{r-1,2} \right), \tag{17.217}$$

$$\mu_2 \ddot{h}_{r,2} = -\chi_1 \left(h_{r,2} - h_{r,1} \right) - \chi_2 \left(h_{r,2} - h_{r+1,1} \right), \tag{17.218}$$

where the displacements have the form

$$h_{r,1(2)} = h_{0,1(2)} \exp\left(i\, q r a - i\, \omega t \right), \tag{17.219}$$

thanks to the general theory. In (17.219), $h_{0,1}$, $h_{0,2}$ are complex constants, $q\,r\,a = \mathbf{q} \cdot \mathbf{l}_r = q\, \mathbf{i} \cdot r\, a\, \mathbf{i}$, and $\omega = \omega(q)$. Replacing (17.219) in (17.217, 17.218) and dividing by $h_{r,1}$, $h_{r,2}$, respectively, yields

$$\mu_1 \omega^2 h_{0,1} = \chi_1 \left(h_{0,1} - h_{0,2} \right) + \chi_2 \left[h_{0,1} - h_{0,2} \exp(-i\, qa) \right], \tag{17.220}$$

$$\mu_2 \omega^2 h_{0,2} = \chi_1 \left(h_{0,2} - h_{0,1} \right) + \chi_2 \left[h_{0,2} - h_{0,1} \exp(+i\, qa) \right]. \tag{17.221}$$

Defining A_{11}, A_{12}, A_{21}, A_{22} such that

$$\mu_1 A_{11} = \chi_1 + \chi_2, \qquad -\mu_1 A_{12} = \chi_1 + \chi_2 \exp(-i\, qa), \tag{17.222}$$

$$\mu_2 A_{22} = \chi_1 + \chi_2, \qquad -\mu_2 A_{21} = \chi_1 + \chi_2 \exp(+i\, qa), \tag{17.223}$$

the homogeneous algebraic system (17.220, 17.221) transforms into

$$\left(A_{11} - \omega^2 \right) h_{0,1} + A_{12} h_{0,2} = 0, \qquad A_{21} h_{0,1} + \left(A_{22} - \omega^2 \right) h_{0,2} = 0. \tag{17.224}$$

The trace $T = A_{11} + A_{22}$ and determinant $D = A_{11} A_{22} - A_{12} A_{21}$ of the matrix formed by A_{11}, A_{12}, A_{21}, A_{22} read

$$T = \frac{\mu_1 + \mu_2}{\mu_1 \mu_2} \left(\chi_1 + \chi_2 \right), \qquad D = 2 \frac{\chi_1 \chi_2}{\mu_1 \mu_2} \left[1 - \cos(qa) \right]. \tag{17.225}$$

The eigenvalues ω^2 are found by solving the algebraic equation $(\omega^2)^2 - T\, \omega^2 + D = 0$, whose discriminant is

$$\Delta(q) = T^2 - 4D = \left[\frac{(\chi_1 + \chi_2)(\mu_1 + \mu_2)}{\mu_1 \mu_2} \right]^2 + 8 \frac{\chi_1 \chi_2}{\mu_1 \mu_2} \left[\cos(qa) - 1 \right]. \tag{17.226}$$

Remembering that $\chi_1, \chi_2 > 0$, the minimum Δ_m of (17.226) occurs for $q = \pm \pi/2$. Letting $K_\chi = (\chi_1 - \chi_2)^2/(4\, \chi_1 \chi_2) \geq 0$ and $K_\mu = (\mu_1 - \mu_2)^2/(4\, \mu_1 \mu_2) \geq 0$, one finds the relation

$$\Delta(q) \geq \Delta_m = 16 \frac{\chi_1 \chi_2}{\mu_1 \mu_2} \left[\left(1 + K_\chi \right) \left(1 + K_\mu \right) - 1 \right] \geq 0, \tag{17.227}$$

Fig. 17.27 Normalized
dispersion relation of a linear,
diatomic chain with
$\mu_1 = \mu_2 = \mu$ and $\chi_1 = 3\,\chi_2$.
The *vertical lines*, placed at
$a\,q/2 = \pm\pi/2$, are the limits
of the first Brillouin zone

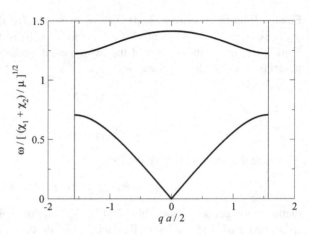

showing that the discriminant is non negative. It follows that the eigenvalues ω^2 are
real, as should be. The solution of the algebraic equation provides two branches of
the dispersion relation, to be found by taking the square root of

$$\omega^2 = \frac{T}{2} \pm \frac{1}{2}\sqrt{\Delta(q)}. \tag{17.228}$$

Observing that $\Delta(0) = T^2$, one finds that selecting the minus sign in (17.228)
provides the branch that contains $\omega = 0$. As in the case of the monatomic chain, this
branch is called *acoustic branch*. In the other branch it is always $\omega > 0$; in ionic
crystals like, e.g., sodium chloride, the frequencies typical of this branch are excited
by infrared radiation. For this reason, the branch is called *optical branch*.

The acoustic and optical branch of a linear, diatomic chain are shown in Fig. 17.27,
where $\mu_1 = \mu_2 = \mu$ is assumed for simplicity. Letting $\tilde{\omega} = \sqrt{(\chi_1 + \chi_2)/\mu}$, one
finds for the acoustic branch

$$\frac{\omega_{\text{ac}}^2}{\tilde{\omega}^2} = 1 - \left[1 - 4\,\frac{\chi_1\,\chi_2}{(\chi_1 + \chi_2)^2}\sin^2\left(\frac{qa}{2}\right)\right]^{1/2}. \tag{17.229}$$

At the center of the first Brillouin zone it is $\omega_{\text{ac}} = 0$, while the maximum of ω_{ac} is
reached at the boundary $qa/2 = \pm\pi/2$ of the zone. For the optical branch one finds

$$\frac{\omega_{\text{op}}^2}{\tilde{\omega}^2} = 1 + \left[1 - 4\,\frac{\chi_1\,\chi_2}{(\chi_1 + \chi_2)^2}\sin^2\left(\frac{qa}{2}\right)\right]^{1/2}. \tag{17.230}$$

At the center of the first Brillouin zone ω_{op} reaches its maximum. The minimum of
ω_{op} is reached at the boundary $qa/2 = \pm\pi/2$ of the zone. The discretization of q
due to the periodic boundary conditions, and the definitions of the phase and group
velocity, are the same as those already given for the monatomic lattice.

From (17.229, 17.230), the distance $G = \omega_{\text{op}}(q = \pm\pi/a) - \omega_{\text{ac}}(q = \pm\pi/a)$
between the minimum of the optical branch and the maximum of the acoustic branch

fulfills the relation

$$\frac{G}{\tilde{\omega}} = \left(1 + \frac{|\chi_1 - \chi_2|}{\chi_1 + \chi_2}\right)^{1/2} - \left(1 - \frac{|\chi_1 - \chi_2|}{\chi_1 + \chi_2}\right)^{1/2}. \tag{17.231}$$

It is $G > 0$ if $\chi_2 \neq \chi_1$, whereas $G = 0$ if $\chi_2 = \chi_1$. The latter case is called *degenerate*.

17.8.7 Analogies

It is interesting to note the analogy between the expression of the energy of the electromagnetic field *in vacuo*, described as a superposition of modes (Eqs. (5.38, 5.40)), and that of a system of vibrating nuclei (Eqs. (3.48) and (17.118)). In essence, the two expressions derive from the fact the in both cases the energy is a positive-definite, symmetric quadratic form. In the case of the electromagnetic field the form is exact because of the linearity of the Maxwell equations; for the vibrating nuclei the form is approximate because of the neglect of the anharmonic terms (Sect. 3.13.1).

Other analogies exist between the dispersion relation $E(\mathbf{k})$ of the electrons subjected to a periodic potential energy, worked out in Sect. 17.6, and the dispersion relation $\omega(\mathbf{q})$, worked out in Sect. (17.7). Both relations are even and periodic in the reciprocal, scaled lattice; both have a branch structure, the difference being that the number of branches of $\omega(\mathbf{q})$ is finite because the number of degrees of freedom of the vibrating lattice is finite, whereas that of $E(\mathbf{k})$ is infinite.

Chapter 18
Electrons and Holes in Semiconductors at Equilibrium

18.1 Introduction

Purpose of this chapter is providing the equilibrium expressions of the electron and hole concentrations in a semiconductor. For comparison, the cases of insulators and conductors are discussed qualitatively, and the concepts of conduction band, valence band, and generation of an electron-hole pair are introduced. The important issue of the temperature dependence of the concentrations is also discussed. Then, the general expressions of the concentrations in an intrinsic semiconductor are worked out, followed by an estimate of the Fermi level's position. Next, the equilibrium expressions are worked out again, this time in the case where substitutional impurities of the donor or acceptor type are present within the semiconductor. The mechanism by which donor-type dopants provide electrons to the conduction band, and acceptor-type dopants provide holes to the valence band, is explained. An important outcome of the analysis is that the introduction of suitable dopants makes the concentration of majority carriers practically independent of temperature, at least in a range of temperatures of practical interest for the functioning of integrated circuits. The simplifications due to the complete-ionization and non-degeneracy conditions are illustrated, along with the compensation effect. Finally, the theory is extended to the case of a non-uniform doping distribution, where the concentrations must be calculated self-consistently with the electric potential by solving the Poisson equation. The last section illustrates the band-gap narrowing phenomenon. In the complements, after a brief description of the relative importance of germanium, silicon, and gallium arsenide in the semiconductor industry, a qualitative analysis of the impurity levels is carried out by an extension of the Kronig–Penney model, and the calculation of the position of the impurity levels with respect to the band edges is carried out.

© Springer Science+Business Media New York 2015
M. Rudan, *Physics of Semiconductor Devices,*
DOI 10.1007/978-1-4939-1151-6_18

Fig. 18.1 Description of the particles' population in the conduction and valence bands of an insulator. To make them more visible, the products $g(E) P(E)$ and $g(E)[1 - P(E)]$ have been amplified, with respect to $g(E)$ alone, by a factor 2×10^{31} (compare with Figs. 17.14 and 18.2). The gap's extension is arbitrary and does not refer to any specific material

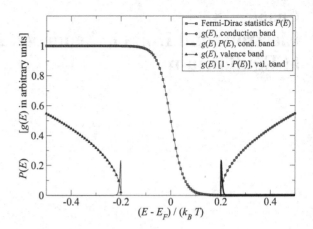

18.2   Equilibrium Concentration of Electrons and Holes

The expressions worked out in this chapter are obtained by combining the information about the band structure of the crystal under investigation, given in Chap. 17, with that of the equilibrium distribution of fermions (Chap. 15). A qualitative description is given first, starting from the simplified case where the structures of the conduction and valence band are symmetric (Fig. 17.14).

Considering again a case where the Fermi level E_F coincides with the gap's midpoint and $M_V m_h^{3/2} = M_C m_e^{3/2}$, let the only difference with respect to the material of Fig. 17.14 be that the energy gap is larger (Fig. 18.1). Despite the fact that the situations illustrated in the two figures may appear similar to each other, it must be realized that a 2×10^{31} amplification factor, with respect to the scale of $g(E)$ alone, is necessary to make the products $g(E) P(E)$, $g(E)[1 - P(E)]$ visible in Fig. 18.1, in contrast to the 10^3 factor used in Fig. 17.14. In practice, for the material of Fig. 18.1 the states in the conduction band are empty and those of the valence band are full.[1] Remembering that a band whose states are empty and, similarly, a band whose states are fully occupied, do not provide any conduction, it turns out that the electrical conductivity of a material like that of Fig. 18.1 is expected to vanish: the material is an *insulator*.

A different case is found when the Fermi level is inside a band, like in the crystal illustrated in Fig. 18.2. The name *conduction band* is given in this case to the band where the Fermi level belongs; the band beneath is called *valence band* also in this case. Due to the position of the Fermi level, the parabolic-band approximation is

---

[1] The curves of Figs. 17.14, 18.1 are drawn in arbitrary units. To better appreciate the difference in a practical situation, one may use the equilibrium concentration of electrons in silicon at $T = 300$ K, which is about 10^{16} m^{-3}. Thus, if Fig. 17.14 is thought of as representing silicon, the ratio of the amplification factors used in Figs. 17.14 and 18.1 produces, in the latter, a concentration of about one electron in 1000 km^3.

Fig. 18.2 Description of the electron population in the conduction band of a conductor. The product $g(E) P(E)$ is drawn in the same scale as $g(E)$ alone (compare with Figs. 17.14 and 18.1). The gap's extension is arbitrary and does not refer to any specific material

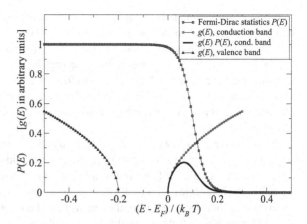

grossly mistaken, and is used for a qualitative discussion only. In the figure, the $g(E) P(E)$ product is drawn without using any amplification factor, this showing that the electron concentration in the conduction band is much larger than for a semiconductor; the latter band, in turn, is the only one that contributes to conduction, because the concentration of holes in the valence band is negligible. Due to the much larger concentration of charges one expects that the conductivity of the crystal under investigation be large; in fact, this is found to be the case, and the crystal is a *conductor*.

In summary, the combination of a few factors: position of the Fermi level with respect to the bands, gap's width, and structure of the density of states, dictates the presence of partially-filled bands and the concentration of charges in them. Other situations may occur besides those depicted in Figs. 17.14, 18.1, and 18.2, e.g., the Fermi level may be positioned within the gap, but closer to one of the band edges than to the other. They are illustrated in Sects. 18.4.1 and 18.4.2.

Another important issue is the dependence of the electron and hole populations on temperature. Considering the case of a semiconductor first, the discussion is carried out with reference to Figs. 17.14 and 15.3; the latter illustrates the temperature dependence of the Fermi-Dirac statistics, using the simplifying hypothesis that the position of the Fermi level does not change with temperature.[2] Following the reasoning carried out in Sect. 17.6.5, the temperature dependence of the effective masses is neglected as well. When temperature increases, the occupation probability of the conduction-band states increases, this making the concentration of electrons in the conduction band to increase; at the same time, the occupation probability of the valence-band states decreases, this making the concentration of holes in the valence band to increase as well. This outcome is easily understood if one thinks that the increase in temperature is achieved by transferring energy from an external reservoir

---

[2] The temperature dependence of the Fermi level is influenced by the shape of the density of states [54]. Such a dependence can be neglected in the qualitative discussion carried out here.

to the semiconductor; part of this energy is absorbed by the nuclei and produces a change in the equilibrium distribution of phonons (Sect. 16.6), while the other part is absorbed by the electrons and produces a redistribution of the latter within the available energy states. The absorption of energy by electrons that, prior to the temperature increase, belonged to valence-band states, makes some of them to transit to the conduction band. As each electron that makes a transition leaves a hole in the valence band, this is an example of *generation of an electron–hole pair* (compare with Sect. 17.6.6).

As shown in Sect. 19.5.5, the conductivity of a semiconductor is proportional to $\mu_n\, n + \mu_p\, p$, where n is the concentration of conduction-band electrons, p the concentration of valence-band holes, and μ_n, μ_p the electron and hole mobilities, respectively. Mobilities account for the scattering events undergone by electrons and holes during their motion within the material, and are found to decrease when temperature increases. It follows that the decrease in mobility competes with the increase of the concentrations in determining the temperature dependence of conductivity in a semiconductor. In practice, the increase in the concentrations is much stronger due to the exponential form of the Fermi-Dirac statistics, so that the conductivity of a semiconductor strongly increases with temperature.[3]

The qualitative analysis of conductivity is the same for a conductor where, as holes are absent, the conductivity is proportional to $\mu_n\, n$. However, the outcome is different; in fact, while μ_n, like that of a semiconductor, decreases when temperature increases, the electron concentration n is unaffected by temperature. In fact, from Fig. 18.2 one finds that the deformation of the Fermi-Dirac statistics due to a temperature variation produces a rearrangement of the electron distribution within the conduction band itself, and no electron-hole generation; as a consequence, the energies of some electrons of the conduction band change, while the electron number (hence the concentration n) does not. As mobility depends weakly on temperature, the conductivity of a conductor turns out to slightly decrease as temperature increases.

The calculation of the equilibrium electron concentration in the conduction band of a semiconductor is based on (17.76), where the density of states in energy g is replaced with the combined density of states in energy and volume γ given by (17.73). The concentration of electrons and the Fermi level are indicated here with n_i, E_{Fi} instead of n, E_F to remark the fact that the semiconductor is free from impurities (for this reason, the semiconductor is called *intrinsic*). From the discussion of Sect. 17.6.5, the parabolic-band approximation is acceptable, so that the combined density of states is given by (17.73).[4] Remembering that the lower edge of the conduction band is indicated with E_C, and letting E_{CU} be the upper edge of the same band, one finds

$$n_i = \int_{E_C}^{E_{CU}} \gamma(E)\, P(E)\, \mathrm{d}E \simeq \int_{E_C}^{+\infty} \frac{\sqrt{2}\, M_C\, m_e^{3/2}\, \sqrt{E - E_C}/(\pi^2\, \hbar^3)}{\exp\left[(E - E_{Fi})/(k_B\, T)\right] + 1}\, \mathrm{d}E, \quad (18.1)$$

---

[3] This is a negative aspect because the electrical properties of the material are strongly influenced by the ambient temperature. The drawback is absent in doped semiconductors, at least in the range of temperatures that are typical of the operating conditions of semiconductor devices (Sect. 18.4.1).

[4] As before, the band index is dropped from (17.73).

where the upper integration limit has been replaced with $+\infty$ on account of the fact that the integrand vanishes exponentially at high energies. Using the auxiliary variables

$$\zeta_e = E_C - E_{Fi}, \qquad x = \frac{E - E_C}{k_B T} \geq 0, \qquad \xi_e = -\frac{\zeta_e}{k_B T}, \tag{18.2}$$

with x, ξ_e dimensionless, transforms (18.1) into

$$n_i = \frac{\sqrt{2}}{\pi^2 \hbar^3} M_C m_e^{3/2} (k_B T)^{3/2} \int_0^{+\infty} \frac{\sqrt{x}}{\exp(x - \xi_e) + 1} \, dx. \tag{18.3}$$

From the definition (C.109) of the Fermi integral of order $1/2$ it then follows

$$n_i = N_C \, \Phi_{1/2}(\xi_e), \qquad N_C = 2 M_C \left(\frac{m_e}{2 \pi \hbar^2} k_B T\right)^{3/2}, \tag{18.4}$$

with N_C the *effective density of states* of the conduction band. Observing that $\Phi_{1/2}$ is dimensionless, the units of N_C are m^{-3}.

The concentration of holes in the valence band is determined in a similar manner, namely, starting from an integral of the form (17.76), where $P(E)$ is replaced with $1 - P(E)$ and the density of states in energy g is replaced with the combined density of states in energy and volume γ. The concentration of holes is indicated here with p_i and, as for the electrons, the parabolic-band approximation is adopted, so that the combined density of states is obtained from (17.74). Finally, $1 - P(E)$ is expressed through (17.77). Remembering that the upper edge of the valence band is indicated with E_V, and letting E_{VL} be the lower edge of the same band, one finds

$$p_i = \int_{E_{VL}}^{E_V} \gamma(E) [1 - P(E)] \, dE \simeq \int_{-\infty}^{E_V} \frac{\sqrt{2} \, M_V \, m_h^{3/2} \, \sqrt{E_V - E} / (\pi^2 \hbar^3)}{\exp[(E_{Fi} - E)/(k_B T)] + 1} \, dE, \tag{18.5}$$

where the lower integration limit has been replaced with $-\infty$ on account of the fact that the integrand vanishes exponentially at low energies. Using the auxiliary variables

$$\zeta_h = E_{Fi} - E_V, \qquad x = \frac{E_V - E}{k_B T} \geq 0, \qquad \xi_h = -\frac{\zeta_h}{k_B T}, \tag{18.6}$$

transforms (18.5) into

$$p_i = \frac{\sqrt{2}}{\pi^2 \hbar^3} M_V m_h^{3/2} (k_B T)^{3/2} \int_0^{+\infty} \frac{\sqrt{x}}{\exp(x - \xi_h) + 1} \, dx, \tag{18.7}$$

whence, introducing the *effective density of states* of the valence band, N_V,

$$p_i = N_V \, \Phi_{1/2}(\xi_h), \qquad N_V = 2 M_V \left(\frac{m_h}{2 \pi \hbar^2} k_B T\right)^{3/2}. \tag{18.8}$$

Table 18.1 Gap and average effective masses of silicon, germanium, and gallium arsenide[a]

Material	E_{G0} (eV)	α (eV/K)	β (K)	$E_G(T_a)$	m_e/m_0	m_h/m_0
Si	1.160	7.02×10^{-4}	1108	1.12	0.33	0.56
Ge	0.741	4.56×10^{-4}	210	0.66	0.22	0.31
GaAs	1.522	5.80×10^{-4}	300	1.43	0.68	0.50

[a]Symbol T_a indicates the room temperature

18.3   Intrinsic Concentration

Equations (18.2, 18.4) and (18.6, 18.8) express the equilibrium concentrations of electron and holes in a semiconductor in terms of temperature and of the distance of the Fermi level from the edge E_C of the conduction band or, respectively, from the edge E_V of the valence band. Obiouvsly the two distances are not independent from each other; from (18.2, 18.6) one finds in fact

$$-(\xi_e + \xi_h) = \frac{\zeta_e + \zeta_h}{k_B T} = \frac{E_C - E_{Fi} + E_{Fi} - E_V}{k_B T} = \frac{E_G}{k_B T}, \tag{18.9}$$

where $E_G = E_C - E_V$ is the extension of the semiconductor's gap, known from the calculation of the band structure, or also from electrical or optical measurements. As the band structure is influenced by temperature, the gap depends on temperature as well (Sect. 17.6.5); such a dependence is important because it strongly influences the concentration of electrons and holes. The results of gap's calculations or measurements, that show that E_G decreases when temperature increases, are usually rendered in compact form by means of interpolating expressions, an example of which is

$$E_G(T) \approx E_{G0} - \alpha \frac{T^2}{T + \beta}, \tag{18.10}$$

where E_{G0} is the gap's width extrapolated at $T = 0$ and $\alpha > 0$, $\beta > 0$ are material's parameters. Table 18.1 reports the parameters related to the gap's width for Si, Ge, and GaAs, along with the values of the average effective masses normalized to the rest mass of the free electron [103, Chap. 2–3]. The plot of $E_G(T)$ is shown in Fig. 18.3 for the three semiconductors of Table 18.1.

Note that expressions (18.4, 18.8) can be used only if the position of the Fermi level is known; in fact, the latter (which is unknown as yet) enters the definitions (18.2, 18.6) of parameters ξ_e, ξ_h. To proceed one remembers that in the $T \to 0$ limit the Fermi-Dirac statistics becomes discontinuous, specifically it is $P = 1$ for $E < E_{Fi}$ and $P = 0$ for $E > E_{Fi}$ (Sect. 15.8.1); on the other hand, the experimental evidence shows that in the $T \to 0$ limit the conductivity of a semiconductor vanishes: this corresponds to a situation where all states of the conduction band are empty while those of the valence bands are filled. In conclusion, when $T \to 0$ the Fermi level is still positioned in the gap and it is $n_i = p_i = 0$. If, starting from this situation, the temperature is brought again to some finite value $T > 0$, such that some of the

Fig. 18.3 Plot of the gap
as a function of temperature
for Ge, Si, and GaAs. The
vertical line marks $T = 300$ K

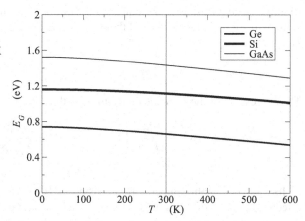

valence-band electrons transit to the conduction band, the total number of holes thus
formed equals that of the transited electrons. Due to the spatial uniformity of the
material, the concentrations are equal to each other as well; in conclusion it is[5]

$$n_i = p_i, \qquad M_C \, m_e^{3/2} \, \Phi_{1/2}(\xi_e) = M_V \, m_h^{3/2} \, \Phi_{1/2}(\xi_h), \qquad (18.11)$$

the second of which has been obtained by deleting the common factors from (18.4)
and (18.8). Relations (18.9) and (18.11) are a set of two equations in the unknowns
ξ_e, ξ_h, whose solution allows one to determine the position of the Fermi level through
(18.2) or (18.6).

The second relation in (18.11) can also be exploited for carrying out an estimate
of ξ_e, ξ_h, basing on the values of the masses given in Table 18.1. To this purpose
one observes[6] that the ratio $M_C \, m_e^{3/2}/(M_V \, m_h^{3/2})$ is about 1.4, 1.2, 0.8 for Si, Ge,
and GaAs, respectively, so that a crude estimate is obtained by letting $\Phi_{1/2}(\xi_e) \simeq$
$\Phi_{1/2}(\xi_h)$. As the Fermi integral is a monotonic function of the argument (Sect. C.13),
it follows $\xi_e \simeq \xi_h$. Replacing from (18.2, 18.6) yields $E_{Fi} \simeq (E_C + E_V)/2$, and
$\zeta_e \simeq \zeta_h \simeq E_G/2 > 0$.

The usefulness of this estimate actually lies in that it simplifies the Fermi integrals.
Taking for instance the case of room temperature, it is $k_B \, T_a \simeq 26$ meV; using the
values of $E_G(T_a)$ from Table 18.1 shows that $E_G \gg k_B T_a$, whence $-\xi_e \gg 1$ and
$-\xi_h \gg 1$, so that the approximate expression (C.105) applies: $\Phi_{1/2}(\xi) \simeq \exp(\xi)$. In
conclusion, (18.4, 18.8) simplify to $n_i \simeq N_C \exp(\xi_e)$, $p_i \simeq N_V \exp(\xi_h)$, namely,

$$n_i \simeq N_C \exp\left(-\frac{E_C - E_{Fi}}{k_B \, T}\right), \qquad p_i \simeq N_V \exp\left(-\frac{E_{Fi} - E_V}{k_B \, T}\right). \qquad (18.12)$$

---

[5] Note that the reasoning leading to the first relation in (18.11) is not limited to the case of the
parabolic-band approximation, but holds for a general form of the densities of states.

[6] Remembering the discussion carried out in Sect. 17.6.5 it is $M_C(\text{Si}) = 6$, $M_C(\text{Ge}) = 4$,
$M_C(\text{GaAs}) = 1$, $M_V = 2$.

As the two concentrations are equal to each other it is customary to use the same symbol n_i for both; the product of the two expressions (18.12) combined with (18.9) yields

$$n_i \, p_i = n_i^2 \simeq N_C \, N_V \, \exp\left(\xi_e + \xi_h\right) = N_C \, N_V \, \exp\left[- E_G/(k_B \, T)\right]. \qquad (18.13)$$

The expression of the *intrinsic concentration* thus reads

$$n_i \simeq \sqrt{N_C \, N_V} \, \exp\left(-\frac{E_G}{2 \, k_B \, T}\right), \qquad N_C \, N_V = \frac{M_C \, M_V}{2 \, \pi^3 \, \hbar^6} \, (m_e \, m_h)^{3/2} \, (k_B \, T)^3 \, .$$
$$(18.14)$$

Its values at room temperature are listed in Table 18.2 along with those of the effective densities of states, for Si, Ge, and GaAs. As for the temperature dependence of n_i one notes that, besides appearing in the exponent's denominator in the first relation of (18.14), the lattice temperature also influences the numerator E_G and the $N_C \, N_V$ factor. Among these dependencies, that of the exponent's denominator is by far the strongest; it follows that a first-hand description of $n_i(T)$ can be given by considering only the latter. This yields the Arrhenius plots shown in Fig. 18.4, where the relations $n_i(T)$ reduce to straight lines. It is important to note that, despite the similarity of the effective densities of states for the three semiconductors considered here (Table 18.2), the intrinsic concentrations differ by orders of magnitude. This is due to the exponential dependence of n_i on E_G, which amplifies the differences in E_G (visible in Table 18.1) of the three semiconductors.

The estimate of the position of the Fermi level in an intrinsic semiconductor carried out above has led to the conclusion that the Fermi integrals used for calculating the intrinsic concentrations can be replaced with exponentials.[7] Such a conclusion can now be exploited for a more precise calculation of the Fermi level's position, where the coefficients of (18.11) are kept. Using the exponentials in (18.11) one finds

$$M_C \, m_e^{3/2} \, \exp\left(\xi_e\right) = M_V \, m_h^{3/2} \, \exp\left(\xi_h\right). \qquad (18.15)$$

Taking the logarithm of both sides and using (18.9) yields $\xi_h - \xi_e = \xi_h + \xi_e - 2\,\xi_e = \log\left[M_C \, m_e^{3/2}/(M_V \, m_h^{3/2})\right]$, whence

$$E_C - E_{Fi} = \frac{E_G}{2} + \frac{k_B \, T}{2} \, \log\left(\frac{M_C \, m_e^{3/2}}{M_V \, m_h^{3/2}}\right). \qquad (18.16)$$

The second term at the right hand side of (18.16) is the correction with respect to the estimate carried out earlier. In the $T \to 0$ limit, the Fermi level in the intrinsic semiconductors under consideration coincides with the gap's midpoint. When the temperature increases, if $M_C \, m_e^{3/2} > M_V \, m_h^{3/2}$ the distance $E_C - E_{Fi}$ becomes

---

[7] This is not necessarily true for an *extrinsic* semiconductor, where suitable impurity atoms are introduced into the semiconductor lattice (Sects. 18.4.1, 18.4.2).

Table 18.2 Intrinsic concentrations of silicon, germanium, and gallium arsenide[a]

Material	$N_C(T_a)$ (cm^{-3})	$N_V(T_a)$ (cm^{-3})	$n_i(T_a)$ (cm^{-3})
Si	2.82×10^{19}	1.05×10^{19}	7.61×10^{9}
Ge	1.04×10^{19}	0.43×10^{19}	1.39×10^{12}
GaAs	4.45×10^{19}	0.99×10^{19}	2.40×10^{6}

[a]In this field it is customary to express the concentrations in cm^{-3} instead of m^{-3}

Fig. 18.4 Arrhenius plot of the intrinsic concentration in Ge, Si, and GaAs. The *vertical line* marks $T = 300$ K

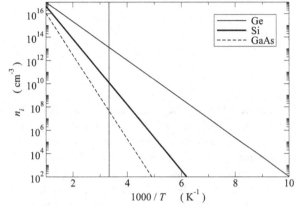

larger, that is, the Fermi level moves towards the valence band; the opposite happens if $M_C\, m_e^{3/2} < M_V\, m_h^{3/2}$. For all practical purposes, considering that the argument of the logarithm in (18.16) is close to unity and the coefficient $k_B\, T$ is always small with respect to E_G, the position of the Fermi level can be thought of as coinciding with the gap's midpoint.

18.4   Uniform Distribution of Impurities

As described in Chap. 23, the fabrication of integrated circuits (IC) requires the introduction into the semiconductor material of atoms (called *impurities* or *dopants*) belonging to specifically-selected chemical species. Dopants are divided into two classes, termed *n-type* and *p-type*. With reference to silicon (Si), the typical *n*-type dopants are phosphorus (P), arsenic (As), and antimony (Sb), while the typical *p*-type dopants are boron (B), aluminum (Al), gallium (Ga), and Indium (In). For a qualitative introduction to the effects of dopants it is instructive to start with the case of an intrinsic semiconductor; the analysis is based on the simplified picture in which the original arrangement of the atoms in space, like that shown in Fig. 17.11, is deformed in such a way as to become two-dimensional. This representation, shown in Fig. 18.5 with reference to silicon, is convenient for the description carried out below. Each silicon atom has four electrons in the external shell, so that it can form

Fig. 18.5 Two-dimensional representation of the intrinsic silicon lattice. The *upper-left* part of the figure shows the $T \to 0$ limit

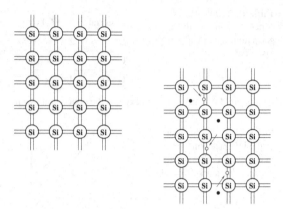

four covalent bonds[8] with other identical atoms; each pair of lines connecting two atoms in Fig. 18.5 stands for a pair of shared electrons. In the $T \to 0$ limit, the electrons are permanently bound to the atoms because the energy necessary to ionize is not available; this situation is drawn in the upper-left part of the figure. If, instead, the temperature is brought to some finite value $T > 0$ by transferring energy from an external reservoir to the semiconductor, part of this energy is absorbed by some of the electrons; the latter break the bond and become free to move within the material.[9] This situation is depicted in the lower-right part of Fig. 18.5, where the free electrons are represented with black dots.

When an electron becomes free and departs from an atom, the unbalanced positive charge left behind in the nucleus deforms the shape of the potential energy in the vicinity of the nucleus itself. The deformation is such that the potential-energy barrier that separates the nucleus from the neighboring ones becomes lower and thinner; this, in turn, enhances the probability of tunneling across the barrier by an electron belonging to a shared pair. Such a tunneling event restores the original pair, but leaves an unbalanced positive charge behind; instead of considering it as the motion of an electron from a complete to an incomplete pair, the tunneling event is more conveniently described as the opposite motion of an empty state, that is, a hole. This is indicated in Fig. 18.5 by the combinations of arrows and white dots.

The above description, based on a spatial picture of the material, is able to provide a qualitative explanation of the existence of electrons and holes in a semiconductor,[10] and completes the description given in Sect. 18.3, that focuses on an energy picture. It also constitutes the basis for analyzing the case of a doped semiconductor, as shown in the following sections.

---

[8] In a *covalent* bond atoms share their outermost electrons.

[9] Remembering that an equilibrium situation is considered here, the contributions to the electric current of these electrons cancel each other.

[10] A similar description could as well apply to a conductor. However, in such as case the barrier deformation is small and tunneling does not occur.

Fig. 18.6 Two-dimensional representation of the *n*-doped silicon lattice. The *upper-left* part of the figure shows the $T \to 0$ limit

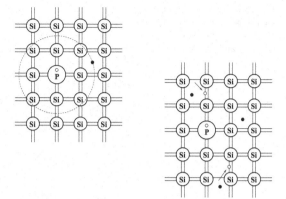

18.4.1   Donor-Type Impurities

As shown in Chap. 23, when a dopant atom is introduced into the semiconductor lattice, in order to properly act as a dopant it must replace an atom of the semiconductor, namely, it must occupy a lattice position (*substitutional impurity*). The concentration of the dopant atoms that are introduced into a semiconductor is smaller by orders of magnitude than the concentration of the semiconductor atoms themselves. As a consequence, the average distance between dopant atoms within the lattice is much larger than that between the semiconductor atoms. Due to this, when the band structure is calculated, the modification in the potential energy introduced by the dopant atoms can be considered as a perturbation with respect to the periodic potential energy of the lattice; the resulting band structure is therefore the superposition of that of the intrinsic semiconductor and of a set of additional states, whose characteristics will be described later. For the moment being, the spatial description is considered, still using the two-dimensional picture, where it is assumed that phosphorus is used as dopant material (Fig. 18.6).

As the dopant concentration is small with respect to the semiconductor's, each phosphorus atom is surrounded by silicon atoms. Phosphorus has five electrons in the external shell: thus, it forms four covalent bonds with silicon, while the remaining electron does not participate in any bond. In the $T \to 0$ limit, the electrons are permanently bound to the atoms as in the intrinsic case; this situation is drawn in the upper-left part of the figure, where the electron that does not form any bond is represented, in a particle-like picture, as orbiting around the phosphorus atom. The orbit's radius is relatively large because the binding force is weak.[11] The white dot inside the phosphorus atom indicates the positive nuclear charge that balances the orbiting electron. If, instead, the temperature is brought to some finite value $T > 0$ by transferring energy from an external reservoir to the semiconductor, a fraction of

---

[11] In a more precise, quantum-mechanical description the electron is described as a stationary wave function extending over several lattice cells.

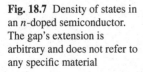

Fig. 18.7 Density of states in an n-doped semiconductor. The gap's extension is arbitrary and does not refer to any specific material

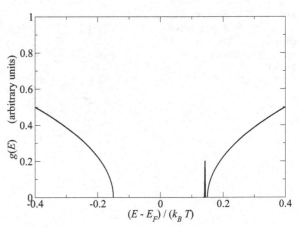

the orbiting electrons break the bond and become free to move within the material. At the same time, electrons belonging to shared pairs may also break their bonds, like in intrinsic silicon. This situation is depicted in the lower-right part of Fig. 18.6, where the free electrons are represented with black dots.

The deformation in the shape of the potential energy in the vicinity of a phosphorus nucleus is different from that near a silicon nucleus. In the latter case, the phenomenon is the same as in the intrinsic material: a series of tunneling events takes place, leading to the motion of holes from one site to a neighboring one. This is still represented by the combinations of arrows and white dots in Fig. 18.6. In the case of phosphorus, instead, the barrier deformation is small and tunneling does not occur: thus, the phosphorus atoms provide free electrons to the lattice, but no holes.

It is intuitive that the insertion of a prescribed amount of n-type dopants into the semiconductor lattice provides a method for controlling the number of free electrons and, through them, the material's conductivity. Moreover, if the ionization of the dopant atoms is not influenced by temperature, at least in a range of temperatures of practical interest, and the number of electrons made available by the dopant is dominant with respect to that provided by the intrinsic semiconductor, the drawback mentioned in Sect. 18.2 is eliminated: conductivity becomes temperature independent and its value is controlled by the fabrication process. This analysis is better specified below, starting from the consideration of the density of states in an n-type semiconductor, shown in Fig. 18.7.

18.4.1.1   Concentration of Ionized Impurities (Donor Type)

Atoms like that of phosphorus contribute electrons to the lattice; for this reason they are also called *donors*. The concentration of donors is indicated with N_D and, in this section, is assumed to be uniform in space. When donor atoms are present, the equilibrium concentrations of electrons and holes are different from those of the intrinsic case; they are indicated with n and p, respectively, and are termed *extrinsic concentrations*. Their derivation is identical to that of the intrisic case and yields

$$n = N_C \, \Phi_{1/2}(\xi_e), \qquad p = N_V \, \Phi_{1/2}(\xi_h), \tag{18.17}$$

with N_C, N_V given by (18.4, 18.8), respectively, and

$$\xi_e = -\frac{\zeta_e}{k_B T} = -\frac{E_C - E_F}{k_B T}, \qquad \xi_h = -\frac{\zeta_h}{k_B T} = -\frac{E_F - E_V}{k_B T}. \tag{18.18}$$

The above expression of ξ_e, ξ_h are similar to (18.2, 18.6), the only difference being that the intrinsic Fermi level E_{Fi} is replaced here with the *extrinsic Fermi level* E_F.

In addition to the electrons of the conduction band and holes of the valence band, a third population of particles must be accounted for, namely, that associated to the dopants. Let $N_D^+ \le N_D$ be the spatially-constant concentration of the ionized donors.[12] The difference $n_D = N_D - N_D^+$ is the concentration of the donor atoms that have not released the orbiting electron to the lattice. Equivalently, n_D may be thought of as the concentration of orbiting electrons that have not been released; such a concentration is given by

$$n_D = \int_{\Delta E_D} \gamma_D(E) \, P_D(E) \, \mathrm{d}E, \tag{18.19}$$

where $\gamma_D(E)$ is the combined density of states produced by the donor atoms, ΔE_D the energy range where $\gamma_D(E) \ne 0$, and $P_D(E)$ the occupation probability of such states. The form of γ_D depends on the concentration N_D; for low or moderately-high values of the donor concentration, γ_D has the form shown in Fig. 18.7. Such a density of states can be approximated by a Dirac delta centered at an energy E_D, called *donor level*, positioned within the gap at a small distance from the edge E_C of the conduction band:

$$\gamma_D(E) \simeq N_D \, \delta(E - E_D). \tag{18.20}$$

The coefficient in (18.20) is such that (18.19) yields $n_D = N_D$ when $P_D = 1$. The distance of the donor level from the minimum of the conduction band is calculated in Sect. 18.7.3; for the typical n-type dopants used with silicon it is $E_C - E_D \simeq 0.033\,\mathrm{eV}$. Another important feature of the donor levels (still in the case of low or moderately-high values of the impurity concentration) is that they are localized in space, as schematically illustrated in Fig. 18.8. For this reason, the Fermi-Dirac statistics describing the equilibrium distribution of the electrons within the donor states is slightly different from that used for the band electrons, and reads

---

[12] A phosphorous atom that has not released the orbiting electron is electrically neutral; when the orbiting electron breaks the bond and becomes free to move within the lattice, the atom becomes positively ionized. The symbol N_D^+ reminds one of that. A second ionization does not occur because the energy necessary for it is too high.

Fig. 18.8 Schematic
representation of the donor
states

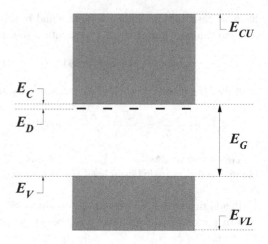

$$P_D(E) = \frac{1}{(1/d_D)\,\exp\left[(E - E_F)/(k_B\,T)\right] + 1}, \qquad (18.21)$$

with d_D the *donors' degeneracy coefficient* [23, Sect. 3.4]. In silicon, germanium,
and gallium arsenide it is $d_D = 2$ [103, Sect. 2.4]. Combining (18.19) with (18.20)
and (18.21) yields $n_D = N_D\,P_D(E_D)$, namely,

$$N_D^+ = N_D\,[1 - P_D(E_D)] = \frac{N_D}{d_D\,\exp\left[(E_F - E_D)/(k_B\,T)\right] + 1}. \qquad (18.22)$$

Like in the intrinsic case discussed in Sect. 18.3, expressions (18.17, 18.22) can
be used only if the position of the Fermi level is known. To proceed one considers
again the $T \to 0$ limit, where the Fermi-Dirac statistics becomes discontinuous and
the experimental evidence shows that the conductivity of a doped semiconductor
vanishes: this corresponds to a situation where all states of the conduction band
are empty, those of the valence bands are filled, and the donor states are filled as
well (in other terms, no dopant atoms are ionized). In conclusion, when $T \to 0$
the Fermi level of an *n*-doped semiconductor is positioned between E_D and E_C, so
that $n = p = N_D^+ = 0$. If, starting from this situation, the temperature is brought
again to some finite value $T > 0$, such that some of the valence-band electrons and
some of the electrons belonging to the dopant atoms transit to the conduction band,
the total number of holes and ionized donors thus formed is equal to that of the
transited electrons. Due to the spatial uniformity of the material, the same relation
holds among the concentrations, so that

$$n = p + N_D^+, \qquad (18.23)$$

whose limit for $N_D \to 0$ yields (18.11) as should be. Now, inserting (18.17, 18.22)
into (18.23) after letting $\xi_D = (E_D - E_C)/(k_B\,T) < 0$, provides

$$N_C\,\Phi_{1/2}(\xi_e) = N_V\,\Phi_{1/2}(\xi_h) + \frac{N_D}{d_D\,\exp\,(\xi_e - \xi_D) + 1}. \qquad (18.24)$$

The latter, along with the relation $\xi_e + \xi_h = -E_G/(k_B T)$, form a system in the two unknowns ξ_e, ξ_h, whose solution determines the position of E_F with respect to the band edges E_C and E_V at a given temperature $T > 0$ and donor concentration N_D. It is easily found that, for a fixed temperature, the argument $\xi_e = (E_F - E_C)/(k_B T)$ increases when N_D increases. In fact, using the short-hand notation $f(\xi_e) = d_D \exp(\xi_e - \xi_D) + 1 > 1$ one finds from (18.24)

$$\frac{dN_D}{d\xi_e} = N_D \frac{f-1}{f} + f \left[N_C \frac{d\Phi_{1/2}(\xi_e)}{d\xi_e} - N_V \frac{d\Phi_{1/2}(\xi_h)}{d\xi_h} \frac{d\xi_h}{d\xi_e} \right], \qquad (18.25)$$

where the right hand side is positive because $d\xi_h/d\xi_e = -1$ and $\Phi_{1/2}$ is a monotonically-increasing function of the argument (Sect. C.13). In the intrinsic case $N_D = 0$ it is $E_F = E_{Fi}$ and $\xi_e < 0$; as N_D increases, the Fermi level moves towards E_C, this making ξ_e less and less negative. At extremely high concentrations of the donor atoms, the Fermi level may reach E_C and even enter the conduction band, this making ξ_e positive. As n, p, and N_D^+ are non negative, it is intuitive that an increasing concentration of donor atoms makes n larger than p; for this reason, in an n-doped semiconductor electrons are called *majority carriers* while holes are called *minority carriers*.[13]

The behavior of the Fermi level with respect to variations in temperature, at a fixed dopant concentration, can not be discussed as easily as that with respect to the dopant variations, because (18.24) is not expressible in the form $T = T(E_F)$ or its inverse. The analytical approach is made easier when some approximations are introduced into (18.24), as shown below. From a qualitative standpoint, one may observe that at very high temperatures the concentration of electron–hole pairs generated by the semiconductor prevails over the concentration of electrons provided by the dopant atoms, so that the position of the Fermi level must be close to the intrinsic one; in contrast, when $T \to 0$ the Fermi level is positioned between E_D and E_C as remarked above. In conclusion, for an n-doped semiconductor one expects that $dE_F/dT < 0$.

In view of further elaborations it is convenient to associate to the Fermi level an electric potential φ_F, called *Fermi potential*, defined by

$$q\,\varphi_F = E_{Fi} - E_F, \qquad (18.26)$$

with $q > 0$ the elementary charge. In an n-type semiconductor it is $E_F > E_{Fi}$; thus, the Fermi potential is negative.

18.4.1.2   Complete Ionization and Non-Degenerate Condition (Donor Type)

When the Fermi level's position and temperature are such that

$$\xi_e - \xi_D < -1, \qquad E_F < E_D - k_B T, \qquad (18.27)$$

---

[13] The electrons of the conduction band and the holes of the valence band are collectively indicated as *carriers*.

the exponential term at the right hand side of (18.22) is negligible with respect to unity, so that $N_D^+ \simeq N_D$. This condition is indicated with *complete ionization*. Note that, since $\xi_D < 0$, the complete-ionization condition in an n-doped semiconductor also implies $\xi_e < -1$; as a consequence, the approximation $\Phi_{1/2}(\xi_e) \simeq \exp(\xi_e)$ for the Fermi integral holds (Sect. C.13). Moreover, it is $\xi_h < \xi_e < -1$ because the Fermi level belongs to the upper half of the gap; thus, the same approximation applies to $\Phi_{1/2}(\xi_h)$ as well. When both equilibrium concentrations n and p are expressible through the approximation $\Phi_{1/2}(\xi) \simeq \exp(\xi)$, the semiconductor is called *non degenerate* and the balance relation (18.24) reduces to

$$N_C \exp\left(-\frac{E_C - E_F}{k_B T}\right) \simeq N_V \exp\left(-\frac{E_F - E_V}{k_B T}\right) + N_D. \qquad (18.28)$$

From the exponential expressions of n and p it follows, using (18.13),

$$n\,p \simeq N_C N_V \exp\left(-\frac{E_C - E_V}{k_B T}\right) = N_C N_V \exp\left(-\frac{E_G}{k_B T}\right) = n_i^2. \qquad (18.29)$$

Letting $N_D^+ = N_D$ in (18.23) and multiplying both sides of it by n provides, thanks to (18.29), an easy algebraic derivation of the electron and hole concentrations in the non-degenerate and spatially-uniform case:

$$n^2 - N_D\,n - n_i^2 = 0, \qquad n = \frac{N_D}{2} + \sqrt{\frac{N_D^2}{4} + n_i^2}, \qquad p = \frac{n_i^2}{n}. \qquad (18.30)$$

The range of validity of (18.30) is quite vast; in silicon at room temperature the approximation (18.28) holds up to $N_D \simeq 10^{17}$ cm^{-3}; also, even for the lowest doping concentrations (about 10^{13} cm^{-3}) it is $N_D \gg n_i$, so that (18.30) can further be approximated as $n \simeq N_D$, $p \simeq n_i^2/N_D$. Thus, the concentration of majority carriers turns out to be independent of temperature, while that of minority carriers still depends on temperature through n_i. Assuming by way of example $N_D = 10^{15}$ cm^{-3} and taking a temperature such that $n_i = 10^{10}$ cm^{-3} (compare with Table 18.2), one finds $n \simeq 10^{15}$ cm^{-3}, $p \simeq 10^5$ cm^{-3}. This result is very important because it demonstrates that the concentration of minority carriers is negligible; as anticipated above, the inclusion of a suitable concentration of dopants makes the material's conductivity independent of temperature. The Arrhenius plot of $n(T)$ for an n-type semiconductor is shown in Fig. 18.9 in arbitrary units. The left part of the curve, called *intrinsic range*, corresponds to the situation where the intrinsic concentration prevails due to the high temperature; the plateau, called *saturation range*, corresponds to the situation where $n \simeq N_D$; finally, the right part of the curve, called *freeze-out range*, corresponds to the case where $n \simeq N_D^+ < N_D$, with N_D^+ decreasing as temperature decreases. From the practical standpoint, the important outcome is that the saturation region covers the range of temperatures within which the integrated circuits operate.[14]

---

[14] For instance, for silicon with $N_D = 10^{15}$ cm^{-3} the saturation region ranges from $T_{\min} \simeq 125$ K to $T_{\max} \simeq 370$ K [103, Sect. 2.4].

Fig. 18.9 Arrhenius plot of $n(T)$ for an n-type semiconductor, in arbitrary units

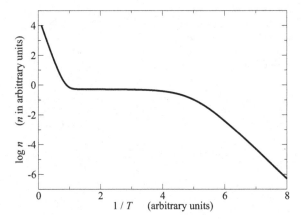

Coupling $n = N_D$ with the expression of n given by the left hand side of (18.28), and considering the values of N_C given in Table 18.2, yields

$$E_F = E_C - k_B T \log\left(\frac{N_C}{N_D}\right) < E_C, \qquad (18.31)$$

whence, as anticipated,

$$\frac{dE_F}{dN_D} = \frac{k_B T}{N_D} > 0, \qquad \frac{dE_F}{dT} = k_B \log\left(\frac{N_D}{N_C}\right) = -\frac{E_C - E_F}{T} < 0. \qquad (18.32)$$

From definition (18.26) of the Fermi potential it follows $E_C - E_F = E_C - E_{Fi} + q \, \varphi_F$; replacing the latter in $n = N_C \exp\left[-(E_C - E_F)/(k_B T)\right]$ yields an alternative expression of the equilibrium concentration and of the Fermi potential itself,

$$n = n_i \exp\left(-\frac{q \, \varphi_F}{k_B T}\right), \qquad \varphi_F = -\frac{k_B T}{q} \log\left(\frac{N_D}{n_i}\right) < 0. \qquad (18.33)$$

18.4.2 Acceptor-Type Impurities

The analysis carried out in Sect. 18.4.1 is repeated here in a shorter form with reference to a substitutional impurity of the acceptor type like, e.g., boron (Fig. 18.10).

As the dopant concentration is small with respect to the semiconductor's, each boron atom is surrounded by silicon atoms. Boron has three electrons in the external shell: while these electrons form covalent bonds with silicon, the remaining unsaturated bond deforms the shape of the potential energy in the vicinity of the boron atom. This attracts an electron from a shared pair of a neighboring silicon. In other terms, to form four covalent bonds with silicon, boron generates an electron–hole pair as shown in the figure. In the $T \to 0$ limit, the holes are permanently bound to

Fig. 18.10 Two-dimensional representation of the *p*-doped silicon lattice. The *upper-left* part of the figure shows the $T \to 0$ limit

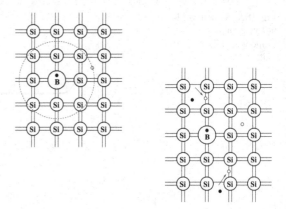

Fig. 18.11 Density of states in a *p*-doped semiconductor. The gap's extension is arbitrary and does not refer to any specific material

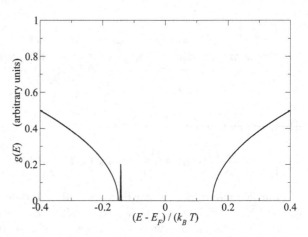

the atoms as in the intrinsic case; this situation is drawn in the upper-left part of the figure, where the hole is represented, in a particle-like picture, as orbiting around the boron atom. The orbit's radius is relatively large because the binding force is weak. The black dot inside the boron atom indicates the negative charge that balances the orbiting hole. If, instead, the temperature is brought to some finite value $T > 0$ by transferring energy from an external reservoir to the semiconductor, a fraction of the orbiting holes break the bond and become free to move within the material. At the same time, electrons belonging to shared pairs may also break their bonds, like in intrinsic silicon. This situation is depicted in the lower-right part of Fig. 18.10, where the free holes are represented with white dots. The negative charge within the boron atom remains trapped within the atom itself: thus, the boron atoms provide free holes to the lattice, but no electrons.

The analysis is better specified below, starting from the consideration of the density of states in a *p*-type semiconductor, shown in Fig. 18.11.

18.4.2.1 Concentration of Ionized Impurities (Acceptor Type)

Atoms like that of boron trap electrons from the lattice; for this reason they are also called *acceptors*. The concentration of acceptors is indicated with N_A and, in this section, is assumed to be uniform in space. When acceptor atoms are present, the equilibrium concentrations of electrons and holes are different from those of the intrinsic semiconductor; their derivation is identical to that of the intrisic or n-type case and yields again (18.17), with N_C, N_V given by (18.4, 18.8), respectively, and ξ_e, ξ_h given by (18.18).

In addition to the electron of the conduction band and holes of the valence band, a third population of particles must be accounted for, namely, that associated to the dopants. Let $N_A^- \leq N_A$ be the spatially-constant concentration of the acceptors that have released the orbiting hole to the lattice.[15] Equivalently, N_A^- may be thought of as the concentration of electrons that have been captured by the acceptor atoms;

$$N_A^- = \int_{\Delta E_A} \gamma_A(E) \, P_A(E) \, dE, \tag{18.34}$$

where $\gamma_A(E)$ is the combined density of states produced by the acceptor atoms, ΔE_A the energy range where $\gamma_A(E) \neq 0$, and $P_A(E)$ the occupation probability of such states. The form of γ_A depends on the concentration N_A; for low or moderately-high values of the acceptor concentration, γ_A has the form shown in Fig. 18.11. Such a density of states can be approximated by a Dirac delta centered at an energy E_A, called *acceptor level*, positioned within the gap at a small distance from the edge E_V of the conduction band:

$$\gamma_A(E) \simeq N_A \, \delta(E - E_A). \tag{18.35}$$

The coefficient in (18.35) is such that (18.34) yields $N_A^- = N_A$ when $P_A = 1$. The distance of the acceptor level from the maximum of the valence band is calculated in Sect. 18.7.3; for the typical p-type dopants used with silicon it is $E_A - E_V \simeq 0.05 \, \mathrm{eV}$ for the heavy holes and $E_A - E_V \simeq 0.016 \, \mathrm{eV}$ for the light holes. Another important feature of the acceptor levels (still in the case of low or moderately-high values of the impurity concentration) is that they are localized in space, as schematically illustrated in Fig. 18.12. For this reason, the Fermi-Dirac statistics describing the equilibrium distribution of the electrons within the acceptor states is slightly different from that used for the band electrons, and reads

$$P_A(E) = \frac{1}{(1/d_A) \exp\left[(E - E_F)/(k_B T)\right] + 1}, \tag{18.36}$$

---

[15] A boron atom that has not released the orbiting hole is electrically neutral; when the orbiting hole breaks the bond and becomes free to move within the lattice, the atom becomes negatively ionized, because the release of a hole is actually the capture of a valence-band electron by the boron atom. The symbol N_A^- reminds one of that. The release of a second hole does not occur.

Fig. 18.12 Schematic
representation of the acceptor
states

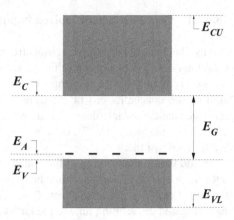

with d_A the *acceptors' degeneracy coefficient*. In silicon, germanium, and gallium arsenide it is $d_A = 4$ [103, Sect. 2.4]. Combining (18.34) with (18.35) and (18.36) yields

$$N_A^- = \frac{N_A}{(1/d_A)\,\exp\left[(E_D - E_F)/(k_B\,T)\right] + 1}.\qquad(18.37)$$

The position of the Fermi level is calculated by the same token as for the n-type dopant and is based on the balance relation

$$n + N_A^- = p,\qquad(18.38)$$

whose limit for $N_A \to 0$ yields (18.11) as should be. Now, inserting (18.17, 18.37) into (18.38) after letting $\xi_A = (E_V - E_A)/(k_B\,T) < 0$, provides

$$N_C\,\Phi_{1/2}(\xi_e) + \frac{N_A}{(1/d_A)\,\exp\,(\xi_h - \xi_A) + 1} = N_V\,\Phi_{1/2}(\xi_h).\qquad(18.39)$$

The latter, along with the relation $\xi_e + \xi_h = -E_G/(k_B\,T)$, form a system in the two unknowns ξ_e, ξ_h, whose solution determines the position of E_F with respect to the band edges E_C and E_V at a given temperature $T > 0$ and acceptor concentration N_A. It is easily found that, for a fixed temperature, the argument $\xi_h = (E_V - E_F)/(k_B\,T)$ increases when N_A increases. The demonstration is identical to that leading to (18.25). In the intrinsic case $N_A = 0$ it is $E_F = E_{Fi}$ and $\xi_h < 0$; as N_A increases, the Fermi level moves towards E_V, this making ξ_h less and less negative. At extremely high concentrations of the acceptor atoms, the Fermi level may reach E_V and even enter the valence band, this making ξ_h positive. As n, p, and N_A^- are non negative, it is intuitive that an increasing concentration of acceptor atoms makes p larger than n; for this reason, in a p-doped semiconductor holes are called *majority carriers* while electrons are called *minority carriers*.

To discuss from a qualitative standpoint the dependence of the Fermi level with respect to variations in temperature one may observe that, like in an n-doped material, at very high temperatures the concentration of electron–hole pairs generated by the semiconductor prevails over the concentration of holes provided by the dopant atoms, so that the position of the Fermi level must be close to the intrinsic one; in contrast, when $T \to 0$ the Fermi level is positioned between E_V and E_A. In conclusion, for a p-doped semiconductor one expects that $dE_F/dT > 0$. Also, in a p-type semiconductor it is $E_F < E_{Fi}$; thus, the Fermi potential (18.26) is positive.

18.4.2.2   Complete Ionization and Non-Degenerate Condition (Acceptor Type)

When the Fermi level's position and temperature are such that

$$\xi_h - \xi_A < -1, \qquad E_A < E_F - k_B T, \tag{18.40}$$

the exponential term at the right hand side of (18.37) is negligible with respect to unity, so that $N_A^- \simeq N_A$. This condition is indicated with *complete ionization*. Note that, since $\xi_A < 0$, the complete-ionization condition in a p-doped semiconductor also implies $\xi_h < -1$; as a consequence, the approximation $\Phi_{1/2}(\xi_h) \simeq \exp(\xi_h)$ for the Fermi integral holds (Sect. C.13). Moreover, it is $\xi_e < \xi_h < -1$ because the Fermi level belongs to the lower half of the gap; thus, the same approximation applies to $\Phi_{1/2}(\xi_e)$ as well. In conclusion, the semiconductor is non degenerate and the balance relation (18.39) reduces to

$$N_C \exp\left(-\frac{E_C - E_F}{k_B T}\right) + N_A \simeq N_V \exp\left(-\frac{E_F - E_V}{k_B T}\right). \tag{18.41}$$

From the exponential expressions of n and p it follows, like in the n-type case, that the product of the equilibrium concentrations fulfills (18.29). The algebraic derivation of the electron and hole concentrations in the non-degenerate and spatially-uniform case is also similar:

$$p^2 - N_A\,p - n_i^2 = 0, \qquad p = \frac{N_A}{2} + \sqrt{\frac{N_A^2}{4} + n_i^2}, \qquad n = \frac{n_i^2}{p}, \tag{18.42}$$

with the further approximation $p \simeq N_A$, $n \simeq n_i^2/N_A$ holding within the low and moderately-high range of dopant concentrations, and around room temperature.

Coupling $p = N_A$ with the expression of p given by the right hand side of (18.41), and considering the values of N_V given in Table 18.2, yields

$$E_F = E_V + k_B T \log\left(\frac{N_V}{N_A}\right) > E_V, \tag{18.43}$$

whence, as anticipated,

$$\frac{dE_F}{dN_A} = -\frac{k_B T}{N_A} < 0, \qquad \frac{dE_F}{dT} = k_B \log\left(\frac{N_V}{N_A}\right) = \frac{E_F - E_V}{T} > 0. \tag{18.44}$$

Fig. 18.13 Schematic representation of a semiconductor with both donor and acceptor states

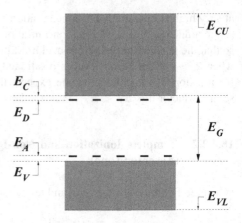

From definition (18.26) of the Fermi potential it follows $E_F - E_V = E_{Fi} - E_V - q\,\varphi_F$; replacing the latter in $p = N_V \exp[-(E_F - E_V)/(k_B T)]$ yields an alternative expression of the equilibrium concentration and of the Fermi potential itself,

$$p = n_i \, \exp\left(\frac{q\,\varphi_F}{k_B T}\right), \qquad \varphi_F = -\frac{k_B T}{q} \, \log\left(\frac{N_A}{n_i}\right) > 0. \qquad (18.45)$$

18.4.3 Compensation Effect

The architecture of semiconductor devices is such that donor and acceptor dopants are present in the same region (Fig. 18.13). Letting N_D, N_A be the corresponding concentrations, still assumed uniform in space, the equilibrium concentrations of electrons and holes are expressed by (18.17) as in the other cases discussed above, with N_C, N_V and ξ_e, ξ_h given by (18.4, 18.8) and (18.18), respectively. In turn, the concentrations of ionized donors and acceptors are given by (18.22) and (18.37), respectively. The balance equation reads

$$n + N_A^- = p + N_D^+, \qquad (18.46)$$

namely, $N_C \, \Phi_{1/2}(\xi_e) + N_A \, P_A(E_A) = N_V \, \Phi_{1/2}(\xi_h) + N_D \, [1 - P_D(E_D)]$. One observes that if $N_D^+ = N_A^-$, the balance equation (18.46) coincides with that of an intrinsic semiconductor (compare with (18.11)). As a consequence, the position of the Fermi level is given by (18.16), and the electrons released by the donor atoms are trapped by the acceptor ones. In this case the semiconductor is *fully compensated*. If, instead, it is $N_D^+ \neq N_A^-$, the semiconductor is *partially compensated*, and one must distinguish between two cases of the balance relation (18.46), depending on the sign of the *net ionized impurity concentration* $N = N_D^+ - N_A^-$. In the first case,

$$N > 0, \qquad n = p + N, \qquad (18.47)$$

the balance relation is identical to that of an n-doped semiconductor (compare with (18.23)), with an effective donor dopant equal to N. In the second case,

$$N < 0, \qquad n + |N| = p, \tag{18.48}$$

the balance relation is identical to that of a p-doped semiconductor (compare with (18.38)), with an effective acceptor dopant equal to $|N|$. In the non-degenerate case, (18.29) still holds. If complete ionization also occurs, then $N = N_D - N_A$; when the donor-type dopant prevails, electrons are the majority carriers, and the same calculation as that leading to (18.30) yields

$$n = \frac{N}{2} + \sqrt{\frac{N^2}{4} + n_i^2}. \tag{18.49}$$

If $N \gg n_i$, then

$$n \simeq N, \qquad p \simeq \frac{n_i^2}{N}, \qquad \varphi_F = -\frac{k_B T}{q} \log\left(\frac{N}{n_i}\right) < 0. \tag{18.50}$$

If, on the contrary, the acceptor-type dopant prevails, holes are the majority carriers, and the same calculation as that leading to (18.42) yields

$$p = -\frac{N}{2} + \sqrt{\frac{N^2}{4} + n_i^2}. \tag{18.51}$$

If $-N \gg n_i$, then

$$p \simeq -N = |N|, \qquad n \simeq \frac{n_i^2}{|N|}, \qquad \varphi_F = \frac{k_B T}{q} \log\left(\frac{|N|}{n_i}\right) < 0. \tag{18.52}$$

18.5 Non-Uniform Distribution of Dopants

This section deals with the more realistic case where the donor concentration N_D, the acceptor concentration N_A, or both, depend on position. A qualitative reasoning shows that the balance equations (18.23), (18.38), or (18.46) do not hold any more. To show this, consider the simple case where only the n-type dopant is present, $N_D = N_D(\mathbf{r})$, $N_A = 0$, and select two nearby positions, say, $\mathbf{r}_1$ and $\mathbf{r}_2$, in the limiting case $T \to 0$; then, let the temperature increase such that a fraction of the donor atoms ionizes. If $N_D(\mathbf{r}_1) \neq N_D(\mathbf{r}_2)$, the numbers of electrons transiting to the conduction band at $\mathbf{r}_1$ and $\mathbf{r}_2$ is different, so that the concentration in one position, say, $\mathbf{r}_1$ is larger than that in the other; thus, some electrons diffuse[16] from the former position to the latter. On the other hand, as the position of the positive charges

---

[16] Diffusive transport is introduced in Sect. 23.3.

within the ionized donors is fixed, the spatial rearrangement of the conduction-band electrons unbalances the negative charges with respect to the positive ones, so that the local charge density differs from zero. The same reasoning applies when both donor and acceptor dopants are present, so that in a general non-uniform case it is

$$\varrho(\mathbf{r}) = q \left[p(\mathbf{r}) - n(\mathbf{r}) + N_D^+(\mathbf{r}) - N_A^-(\mathbf{r}) \right] \neq 0. \tag{18.53}$$

A non-vanishing charge density produces in turn an electric field $\mathbf{E} = \mathbf{E}(\mathbf{r})$ whose action balances the diffusion. In conclusion, the equilibrium condition is kept by an exact balance of the two transport mechanisms. Considering that the equilibrium condition is time-independent, the Maxwell equations in the semiconductor reduce to $\varepsilon_{sc} \operatorname{div}\mathbf{E} = \varrho$, with $\mathbf{E} = -\operatorname{grad}\varphi$ (Sect. 4.4), φ and ε_{sc} being the electric potential and semiconductor permittivity, respectively.[17] In other terms, the electric field due to the non-uniformity of ϱ is found by solving the Poisson equation.

The effect onto the total energy of the electrons due to the presence of an electric field also influences the statistical distribution of the electrons in the energy states. It will be demonstrated in Sect. 19.2.2 that the statistical distribution to be used here is a modified form of the Fermi-Dirac statistics where E is replaced with $E - q\,\varphi(\mathbf{r})$:

$$P(E, \mathbf{r}) = \frac{1}{\exp\left[(E - q\,\varphi(\mathbf{r}) - E_F)\,/\,(k_B\,T)\right] + 1}; \tag{18.54}$$

similarly, (18.21) and (18.36) become

$$P_{D(A)}(E, \mathbf{r}) = \frac{1}{(1/d_{D(A)})\,\exp\left[(E - q\,\varphi(\mathbf{r}) - E_F)/(k_B\,T)\right] + 1}. \tag{18.55}$$

Note that the calculations leading to the concentrations are carried out in the same manner as in Sects. 18.4.1, 18.4.2, because they involve integrals over energy only. As a consequence, (18.17) generalize to

$$n(\mathbf{r}) = N_C\,\Phi_{1/2}\left(\xi_e(\mathbf{r})\right), \qquad \xi_e(\mathbf{r}) = -\frac{\zeta_e(\mathbf{r})}{k_B\,T} = -\frac{E_C - E_F - q\,\varphi(\mathbf{r})}{k_B\,T}, \tag{18.56}$$

and

$$p(\mathbf{r}) = N_V\,\Phi_{1/2}\left(\xi_h(\mathbf{r})\right), \qquad \xi_h(\mathbf{r}) = -\frac{\zeta_h(\mathbf{r})}{k_B\,T} = -\frac{E_F + q\,\varphi(\mathbf{r}) - E_V}{k_B\,T}. \tag{18.57}$$

Similarly, (18.22, 18.34) become

$$N_D^+(\mathbf{r}) = N_D(\mathbf{r})\,[1 - P_D(E_D, \mathbf{r})], \qquad N_A^-(\mathbf{r}) = N_A(\mathbf{r})\,P_A(E_A, \mathbf{r}). \tag{18.58}$$

Note that the summands at the right hand side of (18.53) depend on position through the electric potential φ and also through the explicit dependence of N_D and N_A; as a

---

[17] Note that the material's permittivity must be used here instead of vacuum's. This is coherent with the use of charge density and, in a non-equilibrium situation, of current density, which entail averages over volumes of space.

consequence, inserting the expression of the charge density into the right hand side of the Poisson equation yields

$$-\varepsilon_{sc} \, \nabla^2 \varphi = \varrho(\varphi, \mathbf{r}), \tag{18.59}$$

namely, a second-order, partial differential equation in the unknown φ, with position-dependent coefficients. The equation must be supplemented with suitable boundary conditions.[18] Equation (18.59) is the generalization of the balance equation (18.46) to the non-uniform case: in the case of (18.46), the problem is algebraic and yields E_F, whereas in the case of (18.59) it is differential and yields $E_F + q \, \varphi$. After solving (18.59) one reconstructs n, p, N_D^+, and N_A^- at each point through (18.56), (18.57), and (18.58). In the non-degenerate case, (18.56) becomes

$$n = n^{(0)} \, \exp\left(\frac{q \, \varphi}{k_B \, T}\right), \quad n^{(0)} = N_C \, \exp\left(\frac{E_F - E_C}{k_B \, T}\right) = n_i \, \exp\left(\frac{-q\varphi_F}{k_B \, T}\right), \tag{18.60}$$

with $n^{(0)}$ the value of the electron concentration in the position(s) where $\varphi = 0$. The last expression of $n^{(0)}$ is obtained by combining definition (18.26) with first relation in (18.12). Similarly,

$$p = p^{(0)} \, \exp\left(\frac{-q \, \varphi}{k_B \, T}\right), \quad p^{(0)} = N_V \, \exp\left(\frac{E_V - E_F}{k_B \, T}\right) = n_i \, \exp\left(\frac{q\varphi_F}{k_B \, T}\right). \tag{18.61}$$

From (18.60, 18.61) one finally obtains

$$n = n_i \, \exp\left[\frac{q \, (\varphi - \varphi_F)}{k_B \, T}\right], \quad p = n_i \, \exp\left[\frac{q \, (\varphi_F - \varphi)}{k_B \, T}\right]. \tag{18.62}$$

One observes that (18.29) still holds, namely, $n \, p = n^{(0)} \, p^{(0)} = n_i^2$: in the non-degenerate case, the equilibrium product does not depend on position. If complete ionization also occurs, then $N_D^+(\mathbf{r}) = N_D(\mathbf{r})$, $N_A^-(\mathbf{r}) = N_A(\mathbf{r})$: the ionized-dopant concentrations do not depend on the electric potential.

18.6 Band-Gap Narrowing

When the dopant concentration is large, the density of states associated to the dopant atoms can no longer be described as in Figs. 18.7, 18.11. Rather, considering by way of example an n-type dopant, the form of the density of states is similar to that shown in Fig. 18.14, namely, it overlaps the lower portion of the conduction band forming

---

[18] Such boundary conditions must be coeherent with the choice of the zero point of the total energy. An example is given in Sect. 21.2.

Fig. 18.14 Density of states in an *n*-doped semiconductor, where the high concentration of the dopant produces the band-gap narrowing. The gap's extension is arbitrary and does not refer to any specific material

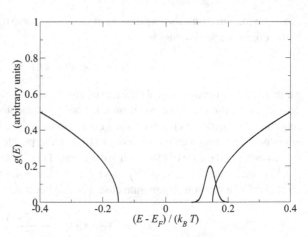

the so-called *impurity band*. In addition, the dopant atoms are close to each other, to the extent that the probability of tunneling of an electron, from a neutral to a nearby, ionized donor atom, is not negligible (Sect. 18.7.2). In a non-equilibrium condition, the tunneling electrons give rise to a current density; the phenomenon is referred to as *impurity-band conduction*. From the practical standpoint, the union of the conduction and impurity bands is viewed as a broader conduction band whose lower edge is shifted with respect to the undoped, or moderately-doped case; the effect is also called *band-gap narrowing*. The analysis is similar when a large concentration of acceptor atoms is present. In conclusion, band-gap narrowing is in general produced by the lowering of the conduction-band edge, $\Delta E_C(\mathbf{r}) > 0$, combined with the lifting of the valence-band edge, $\Delta E_V(\mathbf{r}) > 0$. Both quantities are position dependent because in general the dopant concentrations are such.

Indicating with E_{Ci}, E_{Vi} the lower edge of the conduction band and, respectively, the upper edge of the valence band in the undoped or moderately-doped case, and observing that the variations due to heavy doping are positive, one has for the actual positions of the band edges:

$$E_C(\mathbf{r}) = E_{Ci} - \Delta E_C(\mathbf{r}), \qquad E_V(\mathbf{r}) = E_{Vi} + \Delta E_V(\mathbf{r}), \tag{18.63}$$

whence

$$E_G(\mathbf{r}) = E_{Gi} - \Delta E_G(\mathbf{r}), \quad E_{Gi} = E_{Ci} - E_{Vi}, \quad \Delta E_G = \Delta E_C + \Delta E_V > 0. \tag{18.64}$$

To calculate the carrier concentrations when band-gap narrowing is present, one must replace E_C with $E_{Ci} - \Delta E_C(\mathbf{r})$ in the second relation of (18.56), and E_V with $E_{Vi} + \Delta E_V(\mathbf{r})$ in the second relation of (18.57), to find

$$\xi_e(\mathbf{r}) = -\frac{E_{Ci} - E_F - q\,\varphi(\mathbf{r})}{k_B T} + \frac{\Delta E_C(\mathbf{r})}{k_B T}, \tag{18.65}$$

Fig. 18.15 Band-gap narrowing as a function of the total doping concentration, in normalized form, using the experimental expression 18.67

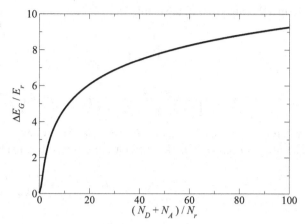

$$\xi_h(\mathbf{r}) = -\frac{E_F - E_{Vi} + q\,\varphi(\mathbf{r})}{k_B\,T} + \frac{\Delta E_V(\mathbf{r})}{k_B\,T}. \tag{18.66}$$

Band-gap narrowing makes ξ_e, ξ_h to increase with respect to the moderately-doped case; as a consequence, the equilibrium carrier concentrations n and p are larger as well. As mentioned in Sect. 18.3, band gap is measured by either electrical or optical methods. The results of gap's measurements, that show that E_G decreases when the dopant concentration exceeds some limiting value N_r, are usually rendered in compact form by means of interpolating expressions, an example of which is [100, 101]

$$\Delta E_G = E_r \left(F + \sqrt{F^2 + 0.5} \right), \qquad F = \log\left(\frac{N_D + N_A}{N_r}\right), \tag{18.67}$$

with $E_r = 9$ meV, $N_r = 10^{17}$ cm^{-3}. The function described by (18.67) is shown in normalized form in Fig. 18.15. Expressions like (18.67) describe the cumulative effect of the total doping concentration, without distinguishing between the donor or acceptor contribution to band-gap narrowing. For this reason, when band-gap narrowing is accounted for in numerical calculations, ΔE_G is equally distributed between the two bands, namely, $\Delta E_C = \Delta E_V = \Delta E_G/2$.

It is interesting to note that the onset of band-gap narrowing corresponds to a total dopant concentration of about 10^{17} cm^{-3}, where the non-degeneracy condition still holds. As a consequence, a range of dopant concentrations exists where the exponential approximation can be used for the equilibrium carrier concentrations, whereas the band-gap narrowing effect must be accounted for.[19] The non-degeneracy condition reads in this case

$$E_{Ci} - E_F - q\,\varphi - \Delta E_C > k_B\,T, \qquad E_F - E_{Vi} + q\,\varphi - \Delta E_V > k_B\,T. \tag{18.68}$$

---

[19] Such a range may be quite large if one considers the compensation effect (Sect. 18.4.3).

Remembering (18.62), the equilibrium concentrations become

$$n = n_e \exp\left[\frac{q\ (\varphi - \varphi_F)}{k_B\ T}\right], \qquad n_e = n_i \exp\left(\frac{\Delta E_C}{k_B\ T}\right) > n_i, \qquad (18.69)$$

$$p = p_e \exp\left[\frac{q\ (\varphi_F - \varphi)}{k_B\ T}\right], \qquad p_e = n_i \exp\left(\frac{\Delta E_V}{k_B\ T}\right) > n_i, \qquad (18.70)$$

where $p_e = n_e$ on account of $\Delta E_V = \Delta E_C$. The common value n_e is called *effective intrinsic concentration*. The equilibrium product then reads

$$n\,p = n_e^2, \qquad n_e^2 = n_i^2 \exp\left(\frac{\Delta E_G}{k_B\ T}\right). \qquad (18.71)$$

18.7 Complements

18.7.1 Si, Ge, GaAs in the Manufacturing of Integrated Circuits

As noted in Sect. 18.3, silicon, germanium, and gallium arsenide have similar effective densities of states, but different gap extensions (Table 18.1); in this respect, silicon is considered as a reference material, so that germanium is indicated as a *narrow-gap material* while gallium arsenide is indicated as a *wide-gap material*. The differences in the gap extension produce huge differences in the intrinsic concentration n_i (Table 18.2); the latter, in turn, has a strong influence on the functioning of the integrated circuits. In fact, the saturation current of a *p-n* junction is proportional to n_i^2 (Sect. 21.3.1); as a consequence, this parameter determines the current of the junction when a reverse bias is applied to it. When many junctions are present, as is typically the case in integrated circuits, the inverse currents may build up and give rise to a substantial parasitic current. From this standpoint, gallium arsenide is preferable with respect to silicon, which in turn is preferable with respect to germanium. Gallium arsenide is also preferable because of the smaller effective mass of the electrons, which makes the electron mobility larger (Sect. 17.6.6). On the other hand, silicon is much less expensive; in fact it is the second most abundant element in Earth's crust (the first one is oxygen); gallium, germanium, and arsenic are much rarer.

The historical development of the semiconductor-device manufacture has followed, instead, a different path. Until the mid sixties of the last century, germanium was preferred; the reason for this was that a technological process, able to purify the material to the level required by the electronic industry, was available for germanium first. As soon as the purification method became available for silicon as well, the latter replaced germanium in the fabrication of semiconductor devices and, soon after, of integrated circuits. The silicon technology developed with a steady pace, giving rise to decades of exponential miniaturization and development in the integrated-circuit manufacture. The miniaturization of gallium-arsenide-based circuits did not proceed with the same pace because the technology of compound materials is more complicate.

In 1980, practically 100 % of the worldwide market share of integrated circuit was silicon based, almost equally distributed between the bipolar and MOSFET technologies [13]. In the following years the MOSFET technology became dominant, reaching a market share of 88 % in the year 2000; of the remaining 12 %, the bipolar, silicon-based technology kept an 8 % share, while the remaining 4 % was taken by integrated circuits using III-V compounds.

In 1989, germanium was introduced again in the silicon integrated-circuit technology to form silicon-germanium alloys ($Si_{1-x}Ge_x$). The alloy makes a flexible band-gap tuning possible; it is used for manufacturing heterojunction bipolar transistors, yielding higher forward gain and lower reverse gain than traditional bipolar transistors. Another application of the alloy is in the Silicon-Germanium-On-Insulator (SGOI) technology. The difference in the lattice constants of germanium and silicon induces a strain in the material under the gate, that makes electron mobility to increase.

18.7.2 Qualitative Analysis of the Impurity Levels

A qualitative analysis of the impurity levels may be carried out basing on a modified version of the Kronig–Penney model discussed in Sect. 17.8.4. To this purpose, consider the case of donor atoms placed at equal distances in a one-dimensional lattice, like that of Fig. 18.16. The deeper wells are those introduced by the dopants, while the finer structure above the x axis is due to the semiconductor nuclei. Note that the relative distances in the figure are not realistic and are used only for illustrative purposes; assuming in fact that the structure represented a cross section of a three-dimensional semiconductor, where a uniform donor concentration N_D is present, the distance between two neighboring impurity atoms is found to be $(1/N_D)^{1/3}$. If the semiconductor's concentration is N_{sc}, the ratio $(N_{sc}/N_D)^{1/3}$ indicates how many semiconductor atoms are present in the interval between two neighboring impurities. Considering silicon by way of example ($N_{sc} = 5 \times 10^{22}$ cm^{-3}), and taking $N_D = 5 \times 10^{16}$ cm^{-3} yields $(N_{sc}/N_D)^{1/3} = 100$.

With this provision, let c be the width of the barrier separating two dopant-induced wells, and consider a negative value E' of the electron's energy. If c is large, the probability that the electron tunnels from a well to an adjacent, empty one is negligibly small; in this case, the electron is confined within the well where the localization probability is the largest, and the energy states $E' < 0$ are similar to those of a single well (the energy states of the finite rectangular well are worked out in Sect. 11.5). The lowest state of the well is the ground state E_D shown in Fig. 18.16. If, instead, a positive energy state E is considered, the wave function is extended over the whole lattice like in the Kronig–Penney model. In summary, the addition of dopant atoms with a low or moderately high concentration, such that their mutual distance c is large, provides a set of energy states that adds up to the band structure of the intrinsic

Fig. 18.16 Potential energy in the Kronig–Penney model modified to account for impurity atoms

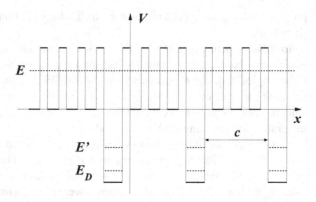

semiconductor. The states introduced by the dopants are localized in space at the positions of the dopant atoms, and are distributed in energy as discrete levels whose mutual distances depend on the form of the well.

In turn, the states with positive energy have a structure similar to that of the intrinsic semiconductor, because in this respect the dopant atoms have little effect on the lattice; in Fig. 18.16, the lower edge E_C of the conduction band coincide with $E = 0$. As said above, electrons belonging to the dopant atoms can not move as long as they are confined within wells, because tunneling is precluded; on the other hand, they may be promoted to conduction-band states by absorbing energy in a collision, and become band electrons, that is, mobile. Conversely, a band electron may loose energy due to a collision, and be trapped in an empty well.

When, due to an increasing dopant concentration, the width c of the barrier becomes smaller, the transmission coefficient increases and the electrons belonging to the wells have a non-negligible probability of moving from a well to another. When the structure of the wells becomes finer, the description of their energy states becomes more similar to that applicable to the intrinsic semiconductor: for an infinite structure one obtains a continuum of states, that fill up the well and connect to those of the conduction band. This explains the band-gap narrowing phenomenon introduced in Sect. 18.6.

18.7.3   Position of the Impurity Levels

To determine the position of the impurity levels for the case of a low or moderately high impurity concentration of the donor type, one considers the dispersion relation $E(\mathbf{k})$ of the conduction band of the intrinsic semiconductor, in the parabolic-band approximation (17.57, 17.58):

$$E(\mathbf{k}) \simeq E_C + \sum_{i=1}^{3} \frac{\hbar^2}{2\,m_{ia}} (k_i - k_{ia})^2, \qquad \frac{1}{m_{ia}} = \frac{1}{\hbar^2} \left(\frac{\partial^2 E}{\partial k_i^2} \right)_a > 0. \qquad (18.72)$$

Now, assume that a donor-type impurity is added in the origin, and that the impurity is ionized. As a consequence, it gives rise to a hydrogenic-like potential energy[20] of the form $V = -q^2/(4\pi\,\varepsilon_{sc}\,r)$. The effect of V may be considered as a local perturbation over the band structure, so that when the ionized impurity is present the total energy of an electron becomes $H = E(\mathbf{k}) + V(r)$, with $E(\mathbf{k})$ the same as in the unperturbed case. Shifting the origin of $\mathbf{k}$ to $\mathbf{k}_a$, and replacing k_i with $-i\,\partial/\partial x_i$, one obtains the Hamiltonian operator[21]

$$\mathcal{H} \simeq E_C - \sum_{i=1}^{3} \frac{\hbar^2}{2\,m_{ia}} \frac{\partial^2}{\partial x_i^2} - \frac{q^2}{4\pi\,\varepsilon_{sc}\,r}. \tag{18.73}$$

To estimate the eigenvalues of $\mathcal{H}$ one may replace each m_{ia} with the average $m_{ea} = (m_{1a}\,m_{2a}\,m_{3a})^{1/3}$ (Sect. 17.6.3), to obtain

$$-\frac{\hbar^2}{2\,m_{ea}} \nabla^2 v_n - \frac{q^2}{4\pi\,\varepsilon_{sc}\,r} v_n = (E_n' - E_C)\,v_n, \tag{18.74}$$

with v_n the eigenfunction. Apart from the coefficients, (18.74) is identical to the Schrödinger equation for the Coulomb case treated in Sect. 13.5.2, whose eigenvalues are given by (13.49). It follows

$$E_n' = E_C - \frac{m_{ea}}{2\,\hbar^2} \left(\frac{q^2}{4\pi\,\varepsilon_{sc}} \right)^2 \frac{1}{n^2}, \qquad n = 1, 2, \ldots \tag{18.75}$$

Thus, the donor impurity provides infinite levels from the minimum E_1' to the maximum E_C. Considering the case of silicon by way of example, one has $m_{1a} = m_l = 0.97\,m_0$, $m_{2a} = m_{3a} = m_t = 0.19\,m_0$ (Table 17.4), with $m_0 \simeq 9.11 \times 10^{-31}$ kg the rest mass of the electron, whence $m_{ea} = 0.33\,m_0$. Letting $E_D = E_1'$ and using $\varepsilon_{sc} \simeq 11.7\,\varepsilon_0$, with $\varepsilon_0 = 8.854 \times 10^{-14}$ F cm^{-1} the vacuum permittivity, one finds the ionization energy of the donor impurity in silicon:

$$E_C - E_D = \frac{m_{ea}}{2\,\hbar^2} \left(\frac{q^2}{4\pi\,\varepsilon_{sc}} \right)^2 \simeq 32.8 \quad \text{meV}. \tag{18.76}$$

The analysis is similar for an ionized, acceptor-type impurity. The hydrogenic-like potential energy becomes $V = q^2/(4\pi\,\varepsilon_{sc}\,r)$, and the dispersion relation around a maximum $\mathbf{k}_a$ of the valence band reads

$$E(\mathbf{k}) \simeq E_V - \sum_{i=1}^{3} \frac{\hbar^2}{2 m_{ia}} (k_i - k_{ia})^2, \qquad \frac{1}{m_{ia}} = -\frac{1}{\hbar^2} \left(\frac{\partial^2 E}{\partial k_i^2} \right)_a > 0. \tag{18.77}$$

---

[20] Compare with Sect. 17.6.6. Like in Sect. 18.5, the semiconductor permittivity is used instead of that of vacuum, because the wave function of an electron subject to the force due to V extends over several lattice cells.

[21] More comments about the procedure of obtaining an operator from a simplified form of the eigenvalues of a more general operator are made in Sect. 19.2.

When the impurity is present the Hamiltonian operator becomes

$$\mathcal{H} \simeq E_V + \sum_{i=1}^{3} \frac{\hbar^2}{2 m_{ia}} \frac{\partial^2}{\partial x_i^2} + \frac{q^2}{4 \pi \, \varepsilon_{sc} \, r}. \tag{18.78}$$

Again, to estimate the eigenvalues one replaces each m_{ia} with the average m_{ha}. One finds

$$-\frac{\hbar^2}{2 m_{ha}} \nabla^2 v_n - \frac{q^2}{4 \pi \varepsilon_{sc} \, r} v_n = (E_V - E_n'') v_n, \tag{18.79}$$

whence

$$E_n'' = E_V + \frac{m_{ha}}{2 \hbar^2} \left(\frac{q^2}{4 \pi \, \varepsilon_{sc}} \right)^2 \frac{1}{n^2}, \tag{18.80}$$

$n = 1, 2, \ldots$. The acceptor impurity provides infinite levels from the minimum E_V to the maximum E_1''. Letting $E_A = E_1''$ one finds the ionization energy $E_A - E_V$ of the donor impurity. Taking again silicon by way of example, and using the values $m_{hh} = 0.5 \, m_0$, $m_{hl} = 0.16 \, m_0$ from Table 17.3, one finds $E_A - E_V = 49.7$ meV for the heavy holes and $E_A - E_V = 15.9$ meV for the light holes.

The ionization energies of phosphorus and boron *in vacuo* are about 10.5 eV and 8.3 eV, respectively, that is, much higher than those calculated here. The strong difference is ascribed to the presence of the silicon crystal: a comparison between (13.49) and (18.75) or (18.80) shows in fact that the coefficients of $1/n^2$ in the crystal case are much smaller than that *in vacuo*, due to the presence of the effective mass in the numerator and of the square of the material permittivity in the denominator. The small distance between the ground state of the impurity atoms and the edge of the band explains the ease with which the dopants ionize at room temperature.

Part VI
Transport Phenomena in Semiconductors

Chapter 19
Mathematical Model of Semiconductor Devices

19.1 Introduction

The chapter describes the reasoning that leads from the single-particle Schrödinger equation for an electron in a crystal to the mathematical model of semiconductor devices. The latter is a set of equations describing the evolution in space and time of a number of average quantities of interest: with reference to the electrons of the conduction band or holes of the valence band, such quantities are the concentration, average velocity, current density, average kinetic energy, and so on. The model of semiconductor devices has different levels of complexity depending on the trade-off between the information that one needs to acquire about the physical behavior of the device under investigation and the computational cost of the system of differential equations to be solved. In fact, the possible models are hierarchically ordered from the drift-diffusion model, which is the simplest one, to the hydrodynamic model, and so on. In essence, these models are different approaches to the problem of solving, in a more or less simplified form, the Boltzmann Transport Equation. Those described in this chapter are the most widely adopted in the commercial simulation programs used by semiconductor Companies. Other important methods, that are not addressed in this book, are the Monte Carlo method and the spherical-harmonics expansion.

The steps leading to the mathematical model of semiconductor devices start with a form of the single-particle Schrödinger equation based on the equivalent Hamiltonian operator, where it is assumed that the external potential energy is a small perturbation superimposed to the periodic potential energy of the nuclei; this leads to a description of the collisionless electron's dynamics in terms of canonically-conjugate variables, that are the expectation values of the wave packet's position and momentum. The dynamics of the Hamiltonian type makes it possible to introduce the statistical description of a many-electron system, leading to the semiclassical Boltzmann Transport Equation. After working out the collision operator, the perturbative approximation is considered; the simplified form of the transport equation thus found is tackled by means of the moments method, whence the hydrodynamic and drift-diffusion versions of the model are derived. A detailed analysis of the derivation of the electron and hole mobility in the parabolic-band approximation is provided. Then, the semiconductor model is coupled with Maxwell's equation, and

© Springer Science+Business Media New York 2015
M. Rudan, *Physics of Semiconductor Devices,*
DOI 10.1007/978-1-4939-1151-6_19

the applicability of the quasi-static approximation is discussed. The typical boundary conditions used in the analysis of semiconductor devices are shown, and an example of analytical solution of the one-dimensional Poisson equation is given.

The complements discuss the analogy between the equivalent Hamiltonian operator and the corresponding Hamiltonian function introduced in an earlier chapter, provide a detailed description of the closure conditions of the models, and illustrate the Matthiessen's rule for the relaxation times. Finally, a short summary of the approximations leading to the derivation of the semiconductor model is given.

19.2 Equivalent Hamiltonian Operator

The separation procedure outlined in Sects. 16.2 through 16.5 has led to the single-electron Schrödinger Equation (17.40), namely, $[-\hbar^2/(2m)\nabla^2 + V]w = Ew$, where the nuclei are kept fixed and the force acting onto the electron derives from a potential energy having the periodicity of the direct lattice: $V(\mathbf{r}+\mathbf{l}) = V(\mathbf{r})$, with $\mathbf{l}$ given by (17.1); as mentioned in Sects. 16.4, 16.5, the external forces are absent ($U_{ext} = 0$).

Thanks to the periodicity of the Hamiltonian operator of (17.40) it is possible to recast the time-independent Schrödinger equation into a different form as shown below. The procedure is based on the analogy with the Schrödinger equation for a free particle, $-\nabla^2 w = k^2 w$ with $k^2 = 2mE/\hbar^2$. One notes in fact that the left hand side is obtained by replacing $\mathbf{k}$ with $-i$ grad in the right hand side; this is just another form of the transformation of momentum into the momentum operator (Sect. 8.5), which in the present case yields the whole Hamiltonian operator because for a free particle the energy is purely kinetic. This type of transformation can be pursued also for the Schrödinger equation with a periodic potential energy by observing that the eigenvalues $E_i(\mathbf{k})$, for each branch i, are periodic within the reciprocal, scaled lattice, $E_i(\mathbf{k}+\mathbf{g}) = E_i(\mathbf{k})$ (Sect. 17.6), hence they can be expanded in terms of the direct-lattice translational vectors $\mathbf{l}$ (Sect. 17.3):

$$E_i(\mathbf{k}) = \sum_{\mathbf{l}} E_{i\mathbf{l}} \exp(i\,\mathbf{l}\cdot\mathbf{k}) \qquad E_{i\mathbf{l}} = E_i(\mathbf{l}) = \frac{1}{\tau_g}\int_{\tau_g} E_i(\mathbf{k})\exp(-i\,\mathbf{l}\cdot\mathbf{k})\,d^3k.$$

$$(19.1)$$

The eigenvalue $E_i(\mathbf{k})$ is now transformed into an operator by letting $\mathbf{k} \leftarrow -i$ grad; remembering (17.158), this yields

$$E_i(\mathbf{k}) \leftarrow E_i(-i\,\text{grad}) = \sum_{\mathbf{l}} E_{i\mathbf{l}} \exp(\mathbf{l}\cdot\text{grad}) = \sum_{\mathbf{l}} E_{i\mathbf{l}}\,\mathcal{T}_{\mathbf{l}}. \qquad (19.2)$$

The form of operator (19.2) is purely kinetic, as the space coordinates do not appear in it. The shape of the potential energy V whence $E_i(\mathbf{k})$ originates is embedded in the coefficients $E_{i\mathbf{l}}$ of expansion (19.1). To complete the procedure one must show that operator (19.2) has the same eigenvalues and eigenfunctions as the original operator of $[-\hbar^2/(2\,m)\,\nabla^2 + V]\,w = E\,w$. Applying $E_i(-\mathrm{i}\,\mathrm{grad})$ to a Bloch function $w_{i\mathbf{k}}$ yields, using the first relation in (19.1) and the periodicity[1] of $\zeta_{i\mathbf{k}}$ (Sect. 17.6),

$$\sum_{\mathbf{l}} E_{i\mathbf{l}}\,\mathcal{T}_{\mathbf{l}}\,\zeta_{i\mathbf{k}}(\mathbf{r})\,\exp{(\mathrm{i}\,\mathbf{k}\cdot\mathbf{r})} = \zeta_{i\mathbf{k}}(\mathbf{r})\sum_{\mathbf{l}} E_{i\mathbf{l}}\,\exp{[\mathrm{i}\,\mathbf{k}\cdot(\mathbf{r}+\mathbf{l})]} =$$

$$= \zeta_{i\mathbf{k}}(\mathbf{r})\,\exp{(\mathrm{i}\,\mathbf{k}\cdot\mathbf{r})}\sum_{\mathbf{l}} E_{i\mathbf{l}}\,\exp{(\mathrm{i}\,\mathbf{k}\cdot\mathbf{l})} = w_{i\mathbf{k}}(\mathbf{r})\,E_i(\mathbf{k}). \tag{19.3}$$

The result

$$E_i(-\mathrm{i}\,\mathrm{grad})\,w_{i\mathbf{k}} = E_i(\mathbf{k})\,w_{i\mathbf{k}} \tag{19.4}$$

shows that, for each branch i, the purely kinetic operator $E_i(-\mathrm{i}\,\mathrm{grad})$ has the same eigenvalues and eigenfunctions as the Hamiltonian whence $E_i(\mathbf{k})$ originates. For this reason, operator $E_i(-\mathrm{i}\,\mathrm{grad})$ is called *equivalent Hamiltonian*. In summary, when the potential energy is periodic it is possible to directly reconstruct an equivalent operator by letting $\mathbf{k} \leftarrow -\mathrm{i}\,\mathrm{grad}$, in the same way as for a free particle. In this respect, the latter case may be viewed as a limiting condition of the former one, obtained by extending the period of the potential energy to infinity.[2] Another similarity between the free-particle case and that of a periodic potential energy is that the Hamiltonian operator is purely kinetic (albeit, in the latter case, at the cost of a more complicate form of the kinetic term).

19.2.1 Electron Dynamics

As illustrated in Sect. 17.6.1, the general solution of the time-dependent Schrödinger equation

$$\left[-\frac{\hbar^2}{2\,m}\nabla^2 + V(\mathbf{r})\right]\psi = \mathrm{i}\,\hbar\,\frac{\partial\psi}{\partial t}, \tag{19.5}$$

for a particle subjected to a periodic potential energy V, is (17.51), or its approximation using the wave packet of branch $i = n$,

$$\psi(\mathbf{r}, t) \simeq \zeta_{n0}\,\exp{(\mathrm{i}\,\Phi_{n0})}\sum_{\mathbf{k}} c_{n\mathbf{k}}\,\exp{[\mathrm{i}\,(\mathbf{r}-\mathbf{u}_n\,t)\cdot(\mathbf{k}-\mathbf{k}_0)]}, \tag{19.6}$$

with $\Phi_{n0} = \mathbf{k}_0\cdot\mathbf{r} - \omega_{n0}\,t$.

---

[1] Like in Sect. 17.6.1, the periodic part of the Bloch function is indicated with $\zeta_{i\mathbf{k}}$ to avoid confusion with the group velocity.

[2] This method of reconstructing the operator from the eigenvalues was anticipated in Sect. 17.6.8.

Now, consider the case where an external,[3] non periodic potential energy $U(\mathbf{r})$ is added to the periodic one; the Hamiltonian operator becomes $-\hbar^2/(2\,m)\,\nabla^2 + V + U$, yielding the eigenvalue equation

$$\left[-\frac{\hbar^2}{2\,m}\,\nabla^2 + V(\mathbf{r}) + U(\mathbf{r})\right] w'_{\mathbf{q}} = E'_{\mathbf{q}}\,w'_{\mathbf{q}}, \tag{19.7}$$

with $\mathbf{q}$ the label of the new eigenvalues. As the set $w'_{\mathbf{q}}$ is complete, a possible expansion of ψ is

$$\psi = \sum_{\mathbf{q}} c'_{\mathbf{q}}\,w'_{\mathbf{q}}\,\exp\left(-\mathrm{i}\,E'_{\mathbf{q}}\,t/\hbar\right). \tag{19.8}$$

However, expansion (19.8) is inconvenient because the Hamiltonian operator in (19.7) is not periodic; as a consequence, the properties of the eigenvalues and eigenfunctions typical of the periodic case are lost. A more suitable expansion[4] is found by using the eigenfunctions of the Hamiltonian operator corresponding to $U = 0$, namely, the Bloch functions; in this case the coefficients of the expansion depend on time:

$$\psi = \sum_{i\mathbf{k}} a_{i\mathbf{k}}(t)\,w_{i\mathbf{k}} = \sum_{i\mathbf{k}} c_{i\mathbf{k}}(t)\,w_{i\mathbf{k}}\,\exp\left(-\mathrm{i}\,\omega_{i\mathbf{k}}\,t\right), \tag{19.9}$$

where $c_{i\mathbf{k}} = a_{i\mathbf{k}}\exp(\mathrm{i}\,\omega_{i\mathbf{k}}\,t) \to$ const as $U \to 0$. This form is more convenient because it holds also in the case where the external potential energy depends on time, $U = U(\mathbf{r}, t)$. The approximate expression (19.6) of the wave function becomes

$$\psi(\mathbf{r}, t) \simeq \zeta_{n0}\,\exp\left(\mathrm{i}\,\Phi_{n0}\right) A, \qquad A = \sum_{\mathbf{k}} c_{n\mathbf{k}}(t)\,\exp\left[\mathrm{i}\,(\mathbf{r} - \mathbf{u}_n\,t)\cdot(\mathbf{k} - \mathbf{k}_0)\right], \tag{19.10}$$

with $|\psi|^2 = |\zeta_{n0}|^2\,|A|^2$ and $\int_\Omega |\psi|^2\,d^3r = 1$. As ζ_{n0} is a rapidly-varying function of $\mathbf{r}$, the physical information about the dynamics of the wave packet is given by $|A|^2$.

So far, the only approximation in (19.10) is the use of a single branch n of the dispersion relation. On the other hand it must be observed that in the sum $V + U$ the first term has the periodicity of the lattice, namely, it varies rapidly in space, whereas the external potential energy is typically a slowly-varying function; in fact, it is due to the application of external generators and/or to the presence of a non-uniform distribution of charge within the material.[5] Also, the field associated to U is weak,

---

[3] For the sake of simplicity, suffix "ext" is dropped from the symbol of the external energy.

[4] The approach is the same as that used for treating the time-dependent perturbation theory (compare with 14.4).

[5] The field produced by non uniformities in the local charge density, which is present also in an equilibrium condition if the dopant distribution is not spatially constant (compare with Sect. 18.5), is classified as "external" because it can be treated as a perturbation. Instead, rapid variations of the physical properties of the material, like those that typically occur at interfaces, can not be treated using the perturbative method and require the solution of the Schrödinger equation without approximations.

so that it does not influence the form of V. This leads to the idea of treating U as a perturbation superimposed to the periodic Hamiltonian operator $-\hbar^2/(2\,m)\,\nabla^2 + V$. Using this approximation, the Hamiltonian operator of the perturbed problem is rewritten as

$$-\frac{\hbar^2}{2\,m}\,\nabla^2 + V(\mathbf{r}) + U(\mathbf{r}, t) \simeq E_n(-\,\mathrm{i}\,\mathrm{grad}) + U(\mathbf{r}, t), \qquad (19.11)$$

where index n reminds one that the eigenfunctions of only the nth branch are used in the expansion. The approximation inherent in (19.11) consists in using the properties of the unperturbed problem in the perturbed case; in fact, the functional dependence of $E_n(-\,\mathrm{i}\,\mathrm{grad})$ on $-\mathrm{i}\,\mathrm{grad}$ derives from the unperturbed eigenvalues $E_n(\mathbf{k})$. Remembering that $E_n(-\,\mathrm{i}\,\mathrm{grad})$ is purely kinetic, (19.11) is similar to the Hamiltonian operator of a particle subjected only to the external potential U. The approximate form of the time-dependent Schrödinger equation then reads

$$\left[E_n(-\,\mathrm{i}\,\mathrm{grad}) + U\right]\psi = \mathrm{i}\,\hbar\,\frac{\partial\psi}{\partial t}. \qquad (19.12)$$

19.2.2 Expectation Values—Crystal Momentum

The solution of (19.12) consists in determining the coefficients $c_{n\mathbf{k}}(t)$ of (19.10); this can be tackled by the method illustrated in Sect. 14.2, namely, by reducing (19.12) to a system of coupled differential equations in the unknowns $c_{n\mathbf{k}}$. More interesting it is to use (19.12) for calculating the expectation values of position and momentum; remembering that $\psi \simeq \exp(\mathrm{i}\,\Phi_{n0})\,\zeta_{n0}\,A$ is normalized to unity, one readily finds $\langle\mathbf{r}\rangle = \langle\psi|\mathbf{r}|\psi\rangle = \mathbf{r}_0$, where $\mathbf{r}_0$ denotes the center of the wave packet in the position space. As for momentum, it is $\langle\mathbf{p}\rangle = \langle\psi| - \mathrm{i}\,\hbar\mathrm{grad}|\psi\rangle$, namely, using $\Phi_{n0} = \mathbf{k}_0 \cdot \mathbf{r} - \omega_{n0}\,t$,

$$\langle\mathbf{p}\rangle = -\mathrm{i}\,\hbar\int_\Omega \psi^*\left[\mathrm{i}\,\mathbf{k}_0\,\psi + \exp(\mathrm{i}\,\Phi_{n0})\,\mathrm{grad}(\zeta_{n0}\,A)\right]\mathrm{d}^3r =$$

$$= \hbar\,\mathbf{k}_0\int_\Omega |\psi|^2\,\mathrm{d}^3r - \mathrm{i}\,\hbar\int_\Omega (\zeta_{n0}A)^*\,\mathrm{grad}(\zeta_{n0}\,A)\,\mathrm{d}^3r. \qquad (19.13)$$

The first term at the right hand side of (19.13) yields $\hbar\,\mathbf{k}_0$ due to normalization. Letting $\zeta_{n0}\,A = a + \mathrm{i}\,b$, one finds in the second term $(\zeta_{n0}\,A)^*\,\mathrm{grad}(\zeta_{n0}\,A) = (1/2)\,\mathrm{grad}(a^2 + b^2) + \mathrm{i}\,(a\,\mathrm{grad}b - b\,\mathrm{grad}a)$. The contribution of $\mathrm{grad}(a^2 + b^2)$ to the integral is zero due to normalization; since A is slowly varying and normalizable, while ζ_{n0} oscillates rapidly, it follows

$$\langle\mathbf{p}\rangle = \hbar\,\mathbf{k}_0 + \hbar\int_\Omega (a\,\mathrm{grad}b - b\,\mathrm{grad}a)\,\mathrm{d}^3r \simeq \hbar\,\mathbf{k}_0, \qquad (19.14)$$

with $\mathbf{k}_0$ the center of the wave packet in the $\mathbf{k}$ space. The product $\hbar\,\mathbf{k}_0$ is called *crystal momentum*. As for the time derivatives of $\langle\mathbf{r}\rangle$ and $\langle\mathbf{p}\rangle$ one finds, from (17.52),

$$\dot{\mathbf{r}}_0 = \mathbf{u}_n = \frac{1}{\hbar}\,\left(\mathrm{grad}_{\mathbf{k}}\,E_n\right)_{\mathbf{k}_0} = \frac{i}{\hbar}\,\sum_{\mathbf{l}}\mathbf{l}\,E_{i\mathbf{l}}\,\exp\left(i\,\mathbf{l}\cdot\mathbf{k}_0\right), \qquad (19.15)$$

where the last expression derives from (19.1). For the time derivative of momentum one preliminarily observes that $E_n(-i\,\mathrm{grad})$ commutes with the gradient operator; in fact,

$$E_n(-i\,\mathrm{grad})\,\mathrm{grad}\,\psi = \sum_{\mathbf{l}} E_{n\mathbf{l}}\,\mathcal{T}_{\mathbf{l}}\,\mathrm{grad}\,\psi(\mathbf{r},t) = \sum_{\mathbf{l}} E_{n\mathbf{l}}\,\mathrm{grad}\,\psi(\mathbf{r}+\mathbf{l},t), \quad (19.16)$$

$$\mathrm{grad}\,E_n(-i\,\mathrm{grad})\,\psi = \mathrm{grad}\,\sum_{\mathbf{l}} E_{n\mathbf{l}}\,\psi(\mathbf{r}+\mathbf{l},t) = \sum_{\mathbf{l}} E_{n\mathbf{l}}\,\mathrm{grad}\,\psi(\mathbf{r}+\mathbf{l},t).$$
$$(19.17)$$

Then, using definition (10.24) of the time derivative of an expectation value,[6] and remembering that the operator associated to $\mathbf{p}$ is $-i\,\hbar\,\mathrm{grad}$, one finds

$$\hbar\,\dot{\mathbf{k}}_0 = \frac{d\langle\mathbf{p}\rangle}{dt} = \langle\psi|\left[E_n(-i\,\mathrm{grad})+U\right]\mathrm{grad} - \mathrm{grad}\left[E_n(-i\,\mathrm{grad})+U\right]|\psi\rangle.$$
$$(19.18)$$

Moreover it is $U\,\mathrm{grad}\,\psi - \mathrm{grad}(U\,\psi) = -\psi\,\mathrm{grad}\,U$, so that (19.18) eventually reduces to

$$\hbar\dot{\mathbf{k}}_0 = \frac{d\langle\mathbf{p}\rangle}{dt} = -\int_{\Omega}|\psi|^2\,\mathrm{grad}\,U\,d^3r. \qquad (19.19)$$

As U is slowly-varying in space, Ehrenfest approximation (10.33) applies, whence

$$\hbar\,\dot{\mathbf{k}}_0 = \frac{d\langle\mathbf{p}\rangle}{dt} \simeq -\,(\mathrm{grad}\,U)_{r_0}\,. \qquad (19.20)$$

Introducing the function $H_n(\mathbf{r}_0,\mathbf{k}_0,t) = E_n(\mathbf{k}_0) + U(\mathbf{r}_0,t)$, one finds that (19.15) and (19.20) are equivalent, respectively, to

$$\dot{x}_{i0} = \frac{\partial H_n}{\partial(\hbar\,k_{i0})}, \qquad \hbar\,\dot{k}_{i0} = -\frac{\partial H_n}{\partial x_{i0}}, \qquad i = 1,2,3. \qquad (19.21)$$

Relations (19.21) are of paramount importance in solid-state theory. They show in fact that within a periodic lattice the dynamics of the expectation values of a wave packet, subjected to an external potential energy that varies slowly in space, is described by Hamilton equations (compare with (1.42)), where $\mathbf{r}_0 = \langle\mathbf{r}\rangle$ and $\hbar\,\mathbf{k}_0 = \langle\mathbf{p}\rangle$ play the role of position and momentum, respectively. It follows that H_n is a Hamiltonian

---

[6] Definition (10.24) could be used also for deriving (19.15).

function proper. Another important observation is that the time variations of the wave packet's momentum are due to the external force only; as a consequence, if $U = \text{const}$ one has $\hbar \dot{\mathbf{k}}_0 = 0$, namely, the crystal momentum is a constant of motion.

A further insight into the structure of H_n is obtained by calculating the work exerted onto the wave packet by the external force $-\text{grad}_{\mathbf{r}_0} U = \hbar \dot{\mathbf{k}}_0$ during an elementary time dt:

$$dW = \hbar \dot{\mathbf{k}}_0 \cdot d\mathbf{r}_0 = \hbar \dot{\mathbf{k}}_0 \cdot \mathbf{u}_n \, dt = \hbar \mathbf{u}_n \cdot d\mathbf{k}_0 = \left(\text{grad}_{\mathbf{k}} E_n\right)_{\mathbf{k}_0} \cdot d\mathbf{k}_0 = dE_n.$$

(19.22)

The work equals the variation of E_n; it follows that E_n, apart from an additive constant, is the kinetic energy of the wave packet. In turn, U is the potential energy which, as mentioned above, derives from the external force only. If the force acting on the electron is due to an electric field, then $U = -q \, \varphi$; this justifies the modified form (18.54) of the Fermi–Dirac statistics to be used when an electric field is present.[7] In the more general case where a magnetic field is also acting on the electron, $\hbar \, \delta \mathbf{k}$ is given by the Lorentz force

$$\hbar \, \delta \dot{\mathbf{k}} = \mathbf{F} = -q \left(\mathbf{E} + \mathbf{u}_n \wedge \mathbf{B}\right),$$

(19.23)

and the Hamiltonian operator in (19.12) must be modified accordingly (compare with (9.19)).

It is important to remark again that the description of the wave packet's dynamics given in this section holds when the force is a weak perturbation with respect to the unperturbed situation. As a consequence, the description does not apply when the electron undergoes a collision; in fact, the force acting during a collision is strong and can not be treated as a perturbation.

19.2.3 Dynamics in the Parabolic-Band Approximation

When the wave packet is centered onto a wave vector $\mathbf{k}_0$ near the ath minimum of the conduction band, the diagonal expansion of $E_n(\mathbf{k})$ yields (17.57). Dropping the branch index n and letting $\mathbf{k} = \mathbf{k}_0$, $\delta k_i = k_{i0} - k_{ia}$ yields

$$E_e = E(\mathbf{k}_0) - E_C \simeq \frac{1}{2} \sum_{i=1}^{3} \frac{\hbar^2}{m_{ia}} (k_{i0} - k_{ia})^2 = \frac{1}{2} \hbar \, \delta \mathbf{k} \cdot \left(\hat{m}_a\right)^{-1} \hbar \, \delta \mathbf{k} \geq 0,$$

(19.24)

with $\left(\hat{m}_a\right)^{-1}$ given by (17.58). Expression (19.24) bears a strong similarity with the kinetic energy of the classical case. The same comment applies to the expression of group velocity given by (17.61), namely,

$$\mathbf{u} = \left(\hat{m}_a\right)^{-1} \hbar \, \delta \mathbf{k}.$$

(19.25)

---

[7] More comments about the analogy with the perturbation theory in the classical case are made in Sect. 19.6.1.

Replacing (19.25) into (19.24) yields $E_e = (1/2)\hat{m}_a \mathbf{u} \cdot \mathbf{u}$. When the expectation value $\hbar \mathbf{k}_0$ of momentum coincides with $\hbar \mathbf{k}_a$, corresponding to an absolute minimum E_C of the conduction band, it is $E_e = 0$. Such a value is also the minimum of the positive-definite quadratic form at the right hand side of (19.24). This shows that E_e is the kinetic energy of the electron, and allows one to identify E_C as the additive constant mentioned above.

In general, the relation between force and acceleration within a crystal is anisotropic. For the sake of simplicity consider the case of the parabolic-band approximation; the time derivative of (19.25) then yields

$$\dot{\mathbf{u}} = (\hat{m}_a)^{-1} \hbar \, \delta\dot{\mathbf{k}} = (\hat{m}_a)^{-1} \mathbf{F}. \tag{19.26}$$

If the entries of the mass tensor are different from each other, the acceleration is not parallel to the force; the physical reason for this is easily understood if one thinks that the forces due to the crystal structure are embedded in the mass tensor through the second derivatives of $E(\mathbf{k})$. The mass tensor becomes a scalar only if the branch E is isotropic: $\hat{m}_a = m_a \mathcal{I}$, with $\mathcal{I}$ the identity tensor. More comments about this issue are made in Sect. 19.6.2.

The analysis for the valence band is similar. Again, the branch index n is dropped and symbols $\mathbf{k} = \mathbf{k}_0$, $\delta k_i = k_{i0} - k_{ia}$ are used,[8] to find (17.64), namely,

$$E_h = E_V - E(\mathbf{k}_0) \simeq \frac{1}{2} \sum_{i=1}^{3} \frac{\hbar^2}{m_{ia}} (k_{i0} - k_{ia})^2 = \frac{1}{2} \hbar \, \delta\mathbf{k} \cdot (\hat{m}_a)^{-1} \hbar \, \delta\mathbf{k} \geq 0,$$

$$\tag{19.27}$$

with $(\hat{m}_a)^{-1}$ given by (17.63), $m_{ia} > 0$. For the group velocity one finds

$$\mathbf{u} = \frac{1}{\hbar} \left(\text{grad}_\mathbf{k} E \right)_{\mathbf{k}_0} = - (\hat{m}_a)^{-1} \hbar \, \delta\mathbf{k}. \tag{19.28}$$

The work exerted onto the wave packet by the external force $-\text{grad}_{\mathbf{r}_0} U = \hbar \dot{\mathbf{k}}_0$ during an elementary time dt is

$$dW = \hbar \dot{\mathbf{k}}_0 \cdot \mathbf{u} \, dt = d \left[-\sum_{i=1}^{3} \frac{\hbar^2}{2 m_{ia}} \delta k_i^2 \right] = dE, \tag{19.29}$$

which, again, shows that E is the kinetic energy of the electron apart from an additive constant. The negative signs in (19.27) and (19.28) make the discussion of the valence-band case somewhat awkward; however, the difficulty is readily eliminated if one refers to holes instead of electrons. For example, consider the case of an electron whose expectation value of momentum, initially equal to $\hbar \mathbf{k}_a$, is brought by the action of an external field to some other value $\hbar \mathbf{k}_0'$ in the vicinity of $\hbar \mathbf{k}_a$. For

---

[8] For Si, Ge, and GaAs it is $k_{ia} = 0$ (Sect. 17.6.5).

this transition to occur it is implied that the initial state $\mathbf{k}_a$ is occupied[9] and the final state $\mathbf{k}_0'$ is empty. As a consequence of (19.27), E changes from E_V to $E(\mathbf{k}_0') < E_V$, namely, it decreases during the time interval Δt during which the energy variation occurs; hence, the external field has exerted in Δt a negative work onto the electron, in fact, energy has been absorbed from the electron by the field. If a hole is considered instead, the initial and final states of the transition exchange roles; however, from the standpoint of the energy balance nothing changes, namely, the field still absorbs energy from the particle. It follows that the hole's energy must decrease due to the transition: this is possible only if the energy axis associated to the hole is reversed with respect to that of the electron, so that, apart from an additive constant, the hole's kinetic energy is $-E$. From this point on, the reasoning becomes identical to that out-lined above for the electron of the valence band: using (19.27), when the expectation value $\hbar\,\mathbf{k}_0$ of momentum coincides with $\hbar\,\mathbf{k}_a$, corresponding to an absolute maxi-mum E_V of the valence band, it is $E_h = 0$. Such a value is also the minimum of the positive-definite quadratic form at the right hand side of (19.27). This shows that E_h is the kinetic energy of the hole, and allows one to identify E_V as the additive constant.

19.3   Dynamics in the Phase Space

The theory outlined in Sects. 19.2, 19.2.1, and 19.2.2 has led to the conclusion that the dynamics of the expectation values of a wave packet describing an electron's motion, subjected to an external potential energy that varies slowly in space, is described by the Hamilton equations (19.21) where $\langle \mathbf{r} \rangle$ and $\langle \mathbf{p} \rangle$ play the role of position and momentum.

For a system made of a large number of electrons, the description of the dynamics of the individual wave packets is impossible from the practical standpoint. In this case one resorts to the same device as that used in Sect. 6.2 for a system of classical particles, namely, the distribution function. Being the formal apparatus identical to that of Sect. 6.2, only the relevant differences will be remarked. The μ-type phase space is defined here by the variables

$$
\mathbf{s} = \begin{bmatrix} x_1 \\ x_2 \\ x_3 \\ k_1 \\ k_2 \\ k_3 \end{bmatrix}, \qquad
\mathbf{e} = \begin{bmatrix} \partial H/\partial k_1 \\ \partial H/\partial k_2 \\ \partial H/\partial k_3 \\ -\partial H/\partial x_1 \\ -\partial H/\partial x_2 \\ -\partial H/\partial x_3 \end{bmatrix}, \qquad (19.30)
$$

---

[9] For the sake of simplicity, spin is not considered here.

(compare with (1.57)) so that the distribution function[10] reads $f = f(\mathbf{r}, \mathbf{k}, t)$. Note that the units of f are different from those of the classical distribution function. For the latter, in fact, it is $[f_\mu] = (\text{J s})^{-3}$, so that the product $f_\mu \, \mathrm{d}^3 r \, \mathrm{d}^3 p$ is dimensionless (compare with (6.1)); in the present case, instead, both $f \, \mathrm{d}^3 r \, \mathrm{d}^3 k$ and $\mathrm{d}^3 r \, \mathrm{d}^3 k$ are dimensionless, hence the distribution function itself is dimensionless.

The system considered for the investigation is that of the electrons belonging to the conduction band. Remembering the first relation in (6.3), the concentration and average velocity of such electrons are given by

$$n(\mathbf{r}, t) = \iiint_{-\infty}^{+\infty} f(\mathbf{r}, \mathbf{k}, t) \, \mathrm{d}^3 k, \qquad \mathbf{v}(\mathbf{r}, t) = \frac{1}{n} \iiint_{-\infty}^{+\infty} \mathbf{u}(\mathbf{k}) \, f(\mathbf{r}, \mathbf{k}, t) \, \mathrm{d}^3 k,$$

$$(19.31)$$

with $\mathbf{u}$ the electron's group velocity. In the equilibrium condition it is $f^{\text{eq}} = Q \, P$, where the Fermi–Dirac statistics P depends on $\mathbf{k}$ only through $E(\mathbf{k})$, namely, it is even with respect to $\mathbf{k}$. In turn, $\mathbf{u} = (1/\hbar) \operatorname{grad}_{\mathbf{k}} E$ is odd, so that the whole integrand in the second definition of (19.31) is odd. As the integration domain is symmetric with respect to $\mathbf{k} = 0$, it is $\mathbf{v}^{\text{eq}} = 0$ as should be. In a non-equilibrium condition it is $f = Q \, \Phi$, with $\Phi(\mathbf{r}, \mathbf{k})$ the occupation probability of a state. If the band is completely filled, then $\Phi = 1$, and the electron flux $n \, \mathbf{v}$ becomes proportional to the integral of $\mathbf{u}$; as the latter is odd, the flux vanishes: this explains why a completely filled band does not contribute to the material's conduction, as anticipated in Sect. 17.6.5.

The Boltzmann collisionless equation in the $\mathbf{r}, \mathbf{k}$ space is derived in the same manner as for (6.28); it reads

$$\frac{\partial f}{\partial t} + \dot{\mathbf{r}} \cdot \operatorname{grad}_{\mathbf{r}} f + \dot{\mathbf{k}} \cdot \operatorname{grad}_{\mathbf{k}} f = 0. \qquad (19.32)$$

The effects of collisions may be grouped into two classes: the collisions of the first class induce transitions that change the number of electrons of the band. Such transitions are the generations and recombinations introduced in Sect. 17.6.6, where the initial state of the electron belongs to the conduction band and the final one belongs to the valence band, or vice versa.[11] The transitions of this class are collectively called *inter-band transitions*.

The collisions of the second class are those where the initial and final state belong to the same band, and are called *intra-band transitions*; they do not change the number of electrons of the band. The distinction between the two classes is useful because the inter-band transitions exhibit characteristic times that are much larger than those of the intra-band transitions. In turn, the intra-band transitions are further divided into two subclasses: the *intra-valley transitions*, where the initial and final

---

[10] Suffix μ is dropped to distinguish this distribution function from that of the classical case.

[11] In addition to this one must also consider the trapping-detrapping phenomena involving localized states. So far, only the localized states due to dopants have been considered (Sect. 18.4); other types of localized states are introduced in Chap. 20.

states are in the vicinity of the same extremum of the band, and the *inter-valley transitions*, where the initial and final state are in the vicinity of different extrema.[12]

Within each class, the transitions are further grouped depending on the entity with which the collision occurs; typical examples of collisions are those with phonons, impurities, defects, and photons. Like in the classical case, collisions are not accounted for in the derivation of (19.32), where the effect of only the slowly-varying external potential energy is present; the further time change of f due to collisions is more conveniently kept separate from that of the external potential energy. Also, it is assumed that the system under consideration is dilute, so that each wave packet spends a relatively large fraction of time without suffering any collision; in other terms, the time during which an electron is subjected to the external field is much longer than that involved in a collision. For this reason it is preferable to write the Boltzmann equation, when the collisions are accounted for, as

$$\frac{\partial f}{\partial t} + \mathbf{u} \cdot \operatorname{grad}_r f - \frac{q}{\hbar}\,(\mathbf{E} + \mathbf{u} \wedge \mathbf{B}) \cdot \operatorname{grad}_k f = C \qquad (19.33)$$

(compare with (6.29) and (6.31)). To derive (19.33), the expression (19.23) of the Lorentz force acting on the electron is used, after dropping index n from the group velocity. Term C embeds the forces acting during the collisions; such forces are short ranged and much more intense than those due to the external field; as a consequence, the Ehrenfest approximation (10.33) does not apply, so that a full quantum-mechanical approach is necessary to treat the collision term.

In conclusion, the relations involving the expectation values at the left hand side of (19.33) are formally identical to those of the classical case, whereas the right hand side is calculated by quantum methods. The form (19.33) of the Boltzmann Transport Equation (BTE) is also called *semiclassical*.

19.3.1   Collision Term

The left hand side of (19.33) equals df/dt, whence the right hand side is the rate of change of f due to collisions. As $f(\mathbf{r}, \mathbf{k}, t)\,d^3r\ d^3k$ is the number of electrons of the conduction band that at time t belong to the elementary volume $d^3r\ d^3k$ centered on $(\mathbf{r}, \mathbf{k})$, the rate of change can be expressed as

$$C = C_{\text{in}} - C_{\text{out}}, \qquad (19.34)$$

where $C_{\text{in}}\,d^3r\ d^3k$ is the number of electrons entering $d^3r\ d^3k$ per unit time, due to collisions, and $C_{\text{out}}\,d^3r\ d^3k$ is the number of electrons leaving $d^3r\ d^3k$ per unit time, due to collisions. To illustrate the reasoning it is convenient to refer to Fig. 19.1,

---

[12] Here the extrema are the minima of the conduction band. The region near an extremum of a band is also called *valley*.

Fig. 19.1 Example of the
time evolution of a
phase-space domain in a
one-dimensional case. The
situation with no external
force is considered

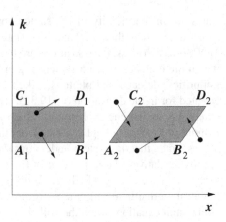

where a one-dimensional case is illustrated using the x, k coordinates. Instead of an elementary volume $dx\, dk$, a finite, rectangular cell is considered, whose position at time t_1 is fixed by the vertices A_1, B_1, C_1, and D_1. For simplicity it is assumed that no external force is present ($U = \mathrm{const}$), so that the crystal momentum is conserved. In particular, the vertices' momenta at $t = t_1$ are $\hbar k_m = \hbar k(A_1) = \hbar k(B_1)$ and $\hbar k_M = \hbar k(C_1) = \hbar k(D_1)$. The corresponding positions are $x(A_1) = x(C_1) = 0$, $x_0 = x(B_1) = x(D_1)$. Letting m^* indicate the effective mass, the position of the vertices at a subsequent time $t_2 = t_1 + \Delta t$ is

$$x(A_2) = \frac{\hbar k_m}{m^*}\, \Delta t, \ x(B_2) = x_0 + x(A_2), \ x(C_2) = \frac{\hbar k_M}{m^*}\, \Delta t, \ x(D_2) = x_0 + x(C_2),$$

this giving rise to the parallelogram also shown in Fig. 19.1. If no collisions occur, the electrons inside the parallelogram at $t = t_2$ are the same as those that were inside the rectangle at $t = t_1$; in contrast, when collisions occur, some electrons leave the rectangle without reaching the parallelogram (hence, $C_{\mathrm{out}} \neq 0$), while the parallelogram is reached by other electrons that originally did not belong to the rectangle ($C_{\mathrm{in}} \neq 0$). This is schematically indicated by the arrows in Fig. 19.1. In general it is $C_{\mathrm{out}} \neq C_{\mathrm{in}}$, so that $df/dt \neq 0$. The description is the same also in the case when the external force is present, the difference being that the trajectories in the phase space are not rectilinear and the deformation of the domain is more complicate.

To give the analysis a more formal aspect it is necessary to determine how the population of the elementary domain $d^6 s = d^3 r\, d^3 k$ evolves in the elementary time interval dt. To begin, one introduces the *scattering probability per unit time and unit phase volume*, S, from an initial state to a final state of the phase space.[13] The initial

---

[13] The units of S are $[S] = \mathrm{s}^{-1}$.

(final) state is indicated by the first (second) pair of arguments of S, namely,

$$S\left(\mathbf{r}, \mathbf{k} \to \mathbf{r}', \mathbf{k}'\right) \mathrm{d}^3 r' \, \mathrm{d}^3 k' \tag{19.35}$$

is the probability per unit time that an electron scatters from $(\mathbf{r}, \mathbf{k})$ to the elementary volume $\mathrm{d}^3 r' \, \mathrm{d}^3 k'$ centered at $(\mathbf{r}', \mathbf{k}')$. Then, let $\mathrm{d}N_{\mathrm{in}} = C_{\mathrm{in}} \, \mathrm{d}^6 s$, $\mathrm{d}N_{\mathrm{out}} = C_{\mathrm{out}} \, \mathrm{d}^6 s$, and $\mathbf{s} = (\mathbf{r}, \mathbf{k})$, $\mathbf{s}' = (\mathbf{r}', \mathbf{k}')$. The number $\mathrm{d}N_{\mathrm{in}}$ is determined by observing that the electrons contributing to it are those that initially belong to elementary phase-space volumes, say, $\mathrm{d}^6 s'$, different from $\mathrm{d}^6 s$. The population of $\mathrm{d}^6 s'$ at time t is $f(\mathbf{s}', t) \, \mathrm{d}^6 s'$; if the latter is multiplied by the scattering probability per unit time from $\mathbf{s}'$ to $\mathrm{d}^6 s$, given by $S\left(\mathbf{s}' \to \mathbf{s}\right) \mathrm{d}^6 s$, the unconditional number of transitions from $\mathrm{d}^6 s'$ to $\mathrm{d}^6 s$ is obtained. The actual number of such transitions is then found by remembering that electrons are fermions, so that transitions towards $\mathrm{d}^6 s$ are possible only if the final states are empty; in other terms, the unconditional number of $\mathbf{s}' \to \mathbf{s}$ transitions must be multiplied by $1 - \Phi(\mathbf{s}, t)$, where $\Phi(\mathbf{s}, t)$ is the probability that the final state is full. Finally, the contributions of all elementary volumes $\mathrm{d}^6 s'$ must be added up, to find[14]

$$\mathrm{d}N_{\mathrm{in}} = \int_{\mathbf{s}'} \left[f(\mathbf{s}', t) \, \mathrm{d}^6 s' \right] \left[S\left(\mathbf{s}' \to \mathbf{s}\right) \mathrm{d}^6 s \right] \left[1 - \Phi(\mathbf{s}, t) \right]. \tag{19.36}$$

The derivation of $\mathrm{d}N_{\mathrm{out}}$ is similar; one obtains

$$\mathrm{d}N_{\mathrm{out}} = \int_{\mathbf{s}'} \left[f(\mathbf{s}, t) \, \mathrm{d}^6 s \right] \left[S\left(\mathbf{s} \to \mathbf{s}'\right) \mathrm{d}^6 s' \right] \left[1 - \Phi(\mathbf{s}', t) \right]. \tag{19.37}$$

The collision term $C = C_{\mathrm{in}} - C_{\mathrm{out}}$ is now determined by subtracting (19.37) from (19.36) and dividing the result by $\mathrm{d}^6 s$. This shows that C is the sum of two terms; the first one is linear with respect to f and reads

$$\int_{\mathbf{s}'} \left[f(\mathbf{s}', t) \, S\left(\mathbf{s}' \to \mathbf{s}\right) - f(\mathbf{s}, t) \, S\left(\mathbf{s} \to \mathbf{s}'\right) \right] \mathrm{d}^6 s'. \tag{19.38}$$

As for the second term, one must preliminarily observe that $f = Q \Phi$, with $Q = 1/(4\pi^3)$ the density of states in the phase space, (17.50); then, the second term of C turns out to be quadratic with respect to Φ or f:

$$Q \int_{\mathbf{s}'} \Phi(\mathbf{s}, t) \, \Phi(\mathbf{s}', t) \left[S\left(\mathbf{s} \to \mathbf{s}'\right) - S\left(\mathbf{s}' \to \mathbf{s}\right) \right] \mathrm{d}^6 s'. \tag{19.39}$$

---

[14] For the sake of conciseness, in Sects. 19.3.1, 19.3.2, and 19.3.3 the six-fold integrals over $\mathrm{d}^3 r' \, \mathrm{d}^3 k'$ and the three-fold integrals over $\mathrm{d}^3 k'$ are indicated with $\int_{\mathbf{s}'}$ and $\int_{\mathbf{k}'}$, respectively.

Fig. 19.2 Qualitative picture
of a collision between an
electron and a
negatively-ionized impurity.
The latter is schematically
represented by the black
circle, whereas the gray area
indicates the screening region.
The initial and final state of
the electron are indicated with
$(\mathbf{r}, \mathbf{k})$ and $(\mathbf{r}', \mathbf{k}')$, respectively

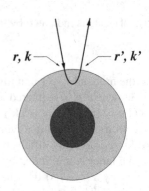

r, k ─   ─ r', k'

19.3.2   Point-Like Collisions

The two summands (19.38), (19.39) in the expression of C are substantially simplified thanks to the property that the collisional forces, albeit very strong, are short
ranged; as a consequence, whereas the momentum of the colliding electron may
undergo a large change due to the collision, the electron's position changes little.
The issue is illustrated with the aid of Fig. 19.2, that schematically describes an
electron collision with a negatively-ionized dopant. The latter is represented by the
black circle, whereas the gray region around it indicates the positive charge attracted
by the negative ion; such a positive charge acts like an *electric screen* that tends to
neutralize the ion. As a consequence of the screen, the decay of the electrostatic potential acting between the ion and the incoming electron, when the relative distance
increases, is much stronger than in the pure Coulomb case.[15] In practice, one can
assume that the electron-ion repulsion is non negligible only when the electron is
inside the screen. This makes the dynamics of the interaction rather different from
that of the pure Coulomb case, treated in Sect. 3.8. As shown in the figure, the final
momentum $\mathbf{k}'$ may differ largely from the initial one, $\mathbf{k}$; in contrast, considering the
atomic scale of the phenomenon, the final position $\mathbf{r}'$ may be thought of as coinciding
with the initial one, $\mathbf{r}$. To vest this observation with mathematical form, considering
that the scattering probability in (19.38) and (19.39) undergoes an integration over
$\mathbf{r}'$, one lets[16]

$$S\left(\mathbf{s} \rightarrow \mathbf{s}'\right) = S_0\left(\mathbf{r}, \mathbf{k} \rightarrow \mathbf{k}'\right)\delta(\mathbf{r}' - \mathbf{r}). \qquad (19.40)$$

Another important consequence of the above discussion is that, although the
duration of the interaction is very short, the force acting on the electron due the
interaction is much stronger than the external forces, to the extent that the effects of

---

[15] The derivation and treatment of the screened Coulomb interaction are carried out in Sects. 20.6.4
and 14.7, respectively.

[16] Note that the units of S_0 are different from those of S: in fact, $[S_0] = \mathrm{cm}^3/\mathrm{s}$. Examples of
calculations of phonon scattering and ionized-impurity scattering are given in Sects. 20.5.1 and
20.5.2, respectively.

the latter can be neglected during the interaction itself. It follows that S and S_0 do not depend on the external forces; this greatly simplifies the analysis of S_0. Inserting (19.40) into (19.38) yields, for the first part of C,

$$\int_{k'} \left[f(\mathbf{r}, \mathbf{k}', t) \, S_0 \left(\mathbf{r}, \mathbf{k}' \to \mathbf{k} \right) - f(\mathbf{r}, \mathbf{k}, t) \, S_0 \left(\mathbf{r}, \mathbf{k} \to \mathbf{k}' \right) \right] \mathrm{d}^3 k' ; \qquad (19.41)$$

in turn, the second part (19.39) becomes

$$Q \int_{k'} \Phi(\mathbf{r}, \mathbf{k}, t) \, \Phi(\mathbf{r}, \mathbf{k}', t) \left[S_0 \left(\mathbf{r}, \mathbf{k} \to \mathbf{k}' \right) - S_0 \left(\mathbf{r}, \mathbf{k}' \to \mathbf{k} \right) \right] \mathrm{d}^3 k'. \qquad (19.42)$$

The term S_0 is typically calculated using the first-order perturbation theory (Sect. 14.3), which shows that the transition probability is invariant upon reversal of the initial and final states. It follows that the quantity in brackets in (19.42) vanishes, so that (19.41) is in fact the only contribution to C. The latter is recast in a more compact form by defining the *relaxation time* $\tau(\mathbf{r}, \mathbf{k})$ and the *collision operator* $\tilde{f}(\mathbf{r}, \mathbf{k})$ such that

$$\frac{1}{\tau} = \int_{k'} S_0 \left(\mathbf{r}, \mathbf{k} \to \mathbf{k}' \right) \mathrm{d}^3 k', \qquad \tilde{f} = \frac{\int_{k'} f(\mathbf{r}, \mathbf{k}', t) \, S_0 \left(\mathbf{r}, \mathbf{k}' \to \mathbf{k} \right) \mathrm{d}^3 k'}{\int_{k'} S_0 \left(\mathbf{r}, \mathbf{k} \to \mathbf{k}' \right) \mathrm{d}^3 k'},$$

$$(19.43)$$

to find $C = C_{\text{in}} - C_{\text{out}} = (\tilde{f} - f)/\tau$. The BTE (19.33) thus becomes

$$\frac{\partial f}{\partial t} + \mathbf{u} \cdot \mathrm{grad}_r f - \frac{q}{\hbar} \left(\mathbf{E} + \mathbf{u} \wedge \mathbf{B} \right) \cdot \mathrm{grad}_k f = -\frac{f - \tilde{f}}{\tau}. \qquad (19.44)$$

In the derivation of (19.44) no distinction is made between the inter-band and intra-band transitions. For a dilute system one can assume that the transitions of the two types are uncorrelated, so that the corresponding probabilities are additive: $S_0 = S_{0b} + S_{0v}$, where index b (v) stands for "inter-band" ("intra-band"). As a consequence,

$$\frac{1}{\tau} = \frac{1}{\tau_b} + \frac{1}{\tau_v}, \qquad \frac{1}{\tau_b} = \int_{k'} S_{0b} \, \mathrm{d}^3 k', \qquad \frac{1}{\tau_v} = \int_{k'} S_{0v} \, \mathrm{d}^3 k'. \qquad (19.45)$$

For the semiconductors of interest, the relaxation times defined in (19.45) differ by several orders of magnitude (e.g., in electronic-grade silicon[17] is $\tau_b > 10^{-6}$ s, $\tau_v < 10^{-12}$ s). This makes the intra-band transitions dominant ($\tau = \tau_b \tau_v / (\tau_b + \tau_v) \simeq \tau_v$); one exception exists though, where the effects of the intra-band transitions cancel each other exactly, so that the inter-band transitions only are left. Such an exception is discussed in Sect. 19.4.

---

[17] The semiconductor's purification degree necessary for manufacturing integrated circuit is called *electronic grade*; it indicates that the ratio between the concentration of impurities (different from dopants) and that of the semiconductor atoms is smaller than 10^{-9}. Lower-quality materials, with a ratio smaller than 10^{-6}, are used in the fabrication of solar cells; in this case the purification degree is called *solar grade*.

19.3.3   Perturbative Form of the BTE

The Boltzmann Transport equation (19.44) is an integral-differential equation in the phase space and time, in the unknown f. The kernel of the integral part is S_0, while the equation coefficients are $\mathbf{E}(\mathbf{r}, t)$, $\mathbf{B}(\mathbf{r}, t)$, $\tau(\mathbf{r}, \mathbf{k})$, and $\mathbf{u}(\mathbf{k}) = (1/\hbar) \operatorname{grad}_\mathbf{k} H$. In equilibrium f becomes $f^{\mathrm{eq}} = Q\,P$, with $P(\mathbf{r}, \mathbf{k})$ the Fermi–Dirac statistics; as shown in Sect. 19.2.2, P depends on position if the semiconductor is not uniform. Moreover it is $d f^{\mathrm{eq}}/dt = 0$; hence, to make the collision term to vanish at equilibrium, it must be $\tilde{f}^{\mathrm{eq}} = f^{\mathrm{eq}}$ (detailed-balance principle, Sect. 6.5).

In general, the solution of the BTE is quite a difficult task. The issue of effective solution methods for this equation will be introduced later; however, a solution procedure is outlined here which, although seldom used in practical cases, has the advantage of providing a simplified form of (19.44), upon which a number of models for semiconductor-device analysis are based. The procedure consists in setting up the iterative scheme

$$\frac{d f^{(m+1)}}{dt} = -\frac{f^{(m+1)} - \tilde{f}^{(m)}}{\tau}, \qquad \tilde{f}^{(m)} = \frac{\int_{\mathbf{k}'} f^{(m)}\, S_0\, d^3 k'}{\int_{\mathbf{k}'} S_0\, d^3 k'}. \tag{19.46}$$

with m the iteration index. In this way the $\tilde{f}^{(m)}$ term at the right hand side of the first equation in (19.46) is known from the previous iteration, so that the integral-differential equation is transformed into a number of differential-only equations. If convergence occurs, the iterations are brought to an end when a suitable norm $\| f^{(m+1)} - f^{(m)} \|$ is smaller than a prescribed value.

To start the procedure it is reasonable to choose, for the approximation of order zero, the equilibrium distribution: $f^{(0)} = f^{\mathrm{eq}}$; from the detailed-balance principle it follows $\tilde{f}^{(0)} = f^{\mathrm{eq}}$. The first step of the iteration procedure then yields $f^{(1)}$, called *first-perturbation solution*. In many cases of practical interest, the material or device under investigation is sufficiently close to the equilibrium condition to make $f^{(1)}$ an acceptable solution; one then stops the iterations at the first-perturbation solution and takes $f \simeq f^{(1)}$. This is equivalent to solving the *perturbative form of the BTE*

$$\frac{\partial f}{\partial t} + \mathbf{u} \cdot \operatorname{grad}_\mathbf{r} f - \frac{q}{\hbar}\, (\mathbf{E} + \mathbf{u} \wedge \mathbf{B}) \cdot \operatorname{grad}_\mathbf{k} f = -\frac{f - f^{\mathrm{eq}}}{\tau}. \tag{19.47}$$

It is interesting to comment on the form of (19.47). The first term at the left hand side differs from zero only if the distribution function depends explicitly on time; hence, it vanishes in a non-equilibrium condition if the latter is of the steady-state type. The second term vanishes if the distribution function is independent of the spatial coordinates, hence it describes a contribution to electron transport that originates from a spatial non uniformity. For this reason this term is called *diffusive term* (compare with Sect. 23.3). The third term vanishes if the external force is missing, hence it originates from the action of the external forces on the electrons and, for this reason, is called *drift term*. At the right hand side of (19.47), the magnitude of the relaxation time influences the amount by which the distribution function departs from equilibrium; to better show this, one recasts (19.47) as

$$f = f^{eq} - \tau \mathcal{L}f, \qquad \mathcal{L} = \frac{\partial}{\partial t} + \mathbf{u} \cdot \text{grad}_{\mathbf{r}} - \frac{q}{\hbar}(\mathbf{E} + \mathbf{u} \wedge \mathbf{B}) \cdot \text{grad}_{\mathbf{k}}, \qquad (19.48)$$

with $\mathcal{L}$ the *Liouvillian operator*. If $\tau \to 0$, then $f \to f^{eq}$; this shows that the perturbative solution is in fact acceptable if the relaxation time is sufficiently small.[18]

A final comment refers to the spatially-uniform case, where $f = f(\mathbf{k}, t)$, $\tau = \tau(\mathbf{k})$; then, (19.47) simplifies to

$$\frac{\partial f}{\partial t} - \frac{q}{\hbar}(\mathbf{E} + \mathbf{u} \wedge \mathbf{B}) \cdot \text{grad}_{\mathbf{k}} f = -\frac{f - f^{eq}}{\tau}. \qquad (19.49)$$

If the fields are set to zero at some instant of time, say, $t = 0$, (19.49) reduces to a differential equation with respect to time only, with $\mathbf{k}$ a parameter; thus, for each $\mathbf{k}$ the solution approaches the equilibrium distribution[19] according to the law

$$f = f^{eq} + (f_{t=0} - f^{eq}) \exp(-t/\tau). \qquad (19.50)$$

In passing, this result explains why τ is called "relaxation time".

19.4 Moments Expansion of the BTE

The *moments expansion of the BTE* is a transformation method that has successfully been applied to the analysis of semiconductor devices. It reduces the original equation to a set of partial-differential equations in the $\mathbf{r}, t$ space; the number of such equations can be adapted to the type of information to be acquired about the device under investigation. More specifically, applying the moments method to the BTE and truncating the series of moments at a suitable order, one extracts from the BTE a hierarchically-ordered set of models, ranging from the simplest to the more complicate ones ([90–92], and references therein).

The BTE is an equation in the $\mathbf{r}, \mathbf{k}, t$ space. The basic idea of the moments method is that, for the description of carrier-transport phenomena in semiconductor devices, it is often sufficient to rely on equations defined over the $\mathbf{r}, t$ space alone; in fact, for practical applications the information about the distribution of the crystal momentum is less important. The equations in the $\mathbf{r}, t$ space are extracted from the BTE by multiplying the latter by suitable functions $\alpha(\mathbf{k})$ and integrating the result over the $\mathbf{k}$ space. The integration saturates the $\mathbf{k}$ coordinates and, as shown in Sect. C.6, provides an equation in the $\mathbf{r}, t$ space.[20] Remembering that the electron dynamics

---

[18] On the other hand, in a collisionless case it is $S_0 \to 0$ whence, from (19.43), it follows $\tau \to \infty$. In this situation there is no limit to the departure of f from f^{eq}.

[19] Compare with the discussion carried out in Sect. 6.6.3.

[20] As indicated in Sect. C.6, term "moment" is specifically used when α is a polynomial in $\mathbf{k}$. As the dependence of α on $\mathbf{k}$ is not specified yet, it is implied that the form of α is such that the integrals in (19.51) converge.

using the equivalent Hamiltonian operator is described by expanding the electron's wave function in terms of the Bloch functions (Sect. 19.2.1), the integration over the **k** space is in fact limited to the first Brillouin zone. On the other hand, the typical behavior of the distribution function at the boundary Γ of the first Brilluoin zone is such that $\alpha(\mathbf{k}) f(\mathbf{r}, \mathbf{k}, t) \to 0$ when $\mathbf{k} \to \mathbf{k}_\Gamma$. This amount to assuming that there are no electrons at the boundary.[21] From a practical standpoint, the hypothesis has the same effect as that of replacing the first Brillouin zone with an infinite domain and assuming that the distribution function in a non-equilibrium condition vanishes at infinity in the same exponential-like fashion as it does at equilibrium, where it becomes proportional to the Fermi–Dirac statistics [86, 89]. With these premises, using the general form (19.33) of the equation, the moment of the BTE with respect to α reads

$$\iiint_{-\infty}^{+\infty} \alpha \left[\frac{\partial f}{\partial t} + \mathbf{u} \cdot \mathrm{grad}_r f - \frac{q}{\hbar} (\mathbf{E} + \mathbf{u} \wedge \mathbf{B}) \cdot \mathrm{grad}_k f \right] d^3k = \iiint_{-\infty}^{+\infty} \alpha\, C\, d^3k. \tag{19.51}$$

As the BTE is a continuity equation of the distribution function in the phase space, (19.51) is expected to be a continuity equation in the **r** space; in fact, as shown below, it is the continuity equation of the product $n\,\overline{\alpha}$, where n is the concentration of the electrons in the conduction band, given by the first relation in (19.31), and

$$\overline{\alpha}(\mathbf{r}, t) = \frac{1}{n} \iiint_{-\infty}^{+\infty} \alpha(\mathbf{k})\, f(\mathbf{r}, \mathbf{k}, t)\, d^3k \tag{19.52}$$

is the average of α over the **k** space. The continuity equation is derived below by working out separately the different terms appearing in (19.51).

Time Derivative
The derivation of this term is readily accomplished by observing that $\alpha\, \partial f/\partial t = \partial(\alpha\, f)/\partial t$, whence

$$\iiint_{-\infty}^{+\infty} \alpha \frac{\partial f}{\partial t} d^3k = \frac{\partial}{\partial t} \iiint_{-\infty}^{+\infty} \alpha\, f\, d^3k = \frac{\partial}{\partial t} (n\,\overline{\alpha}). \tag{19.53}$$

This shows that (19.51) is the continuity equation of $n\,\overline{\alpha}$, as anticipated.

Diffusion Term
To calculate this terms one starts with the relation $\alpha\, \mathbf{u} \cdot \mathrm{grad}_r f = \mathrm{div}_r(\alpha\, \mathbf{u}\, f)$, that derives from the second identity in (A.16) and from the fact that $\alpha\, \mathbf{u}$ does not depend on **r**. Thus,

$$\iiint_{-\infty}^{+\infty} \alpha\, \mathbf{u} \cdot \mathrm{grad}_r f\, d^3k = \mathrm{div}_r \iiint_{-\infty}^{+\infty} \alpha\, \mathbf{u}\, f\, d^3k = \mathrm{div}_r (n\,\overline{\alpha\,\mathbf{u}}). \tag{19.54}$$

---

[21] In the case of the conduction band of germanium, the minima are at the boundary (Sect. 17.6.5), which makes the hypothesis inconsistent as it stands; to perform the integration one must shift the origin of the **k** space and exploit the periodicity of the band structure. The hypothesis that the distribution function vanishes at the boundary of the first Brillouin zone is made also in the application of the moments method to the holes of the valence band.

Drift Term

The term containing the electric field is treated starting from the identity

$$\alpha\, \mathbf{E} \cdot \mathrm{grad}_k f = \mathbf{E} \cdot \mathrm{grad}_k(\alpha\, f) - f\, \mathbf{E} \cdot \mathrm{grad}_k \alpha. \qquad (19.55)$$

The integral of the first term at the right hand side of (19.55) vanishes due to identity (A.26) and to the asymptotic behavior of f described earlier. As a consequence,

$$-\iiint_{-\infty}^{+\infty} \alpha\, \mathbf{E} \cdot \mathrm{grad}_k f\, d^3k = \mathbf{E} \cdot \iiint_{-\infty}^{+\infty} f\, \mathrm{grad}_k \alpha\, d^3k = \mathbf{E} \cdot n\, \overline{\mathrm{grad}_k \alpha}. \qquad (19.56)$$

The term containing the magnetic induction is treated more easily by rewriting the mixed product as $\mathrm{grad}_k f \cdot \alpha\, \mathbf{u} \wedge \mathbf{B} = \mathrm{grad}_k f \wedge \alpha\, \mathbf{u} \cdot \mathbf{B}$ (compare with (A.31)) and using the first identity in (A.35), to find

$$\mathrm{grad}_k f \wedge \alpha\, \mathbf{u} \cdot \mathbf{B} = \mathrm{rot}_k(f\, \alpha\, \mathbf{u}) \cdot \mathbf{B} - f\, \mathrm{rot}_k(\alpha\, \mathbf{u}) \cdot \mathbf{B}. \qquad (19.57)$$

The integral of $\mathrm{rot}_k(f\, \alpha\, \mathbf{u}) \cdot \mathbf{B}$ over k vanishes due to identity (A.38) and to the asymptotic behavior of f; this yields

$$-\iiint_{-\infty}^{+\infty} \alpha\, \mathbf{u} \wedge \mathbf{B} \cdot \mathrm{grad}_k f\, d^3k = \mathbf{B} \cdot \iiint_{-\infty}^{+\infty} f\, \mathrm{rot}_k(\alpha\, \mathbf{u})\, d^3k. \qquad (19.58)$$

In turn, identity (A.35) transforms the integrand at the right hand side of (19.58) as $f\, \mathrm{rot}_k(\alpha\, \mathbf{u}) = f\, \alpha\, \mathrm{rot}_k \mathbf{u} + f\, \mathrm{grad}_k \alpha \wedge \mathbf{u}$ where, thanks to the definition (17.52) of the group velocity, it is $\mathrm{rot}_k \mathbf{u} = (1/\hbar)\, \mathrm{rot}_k \mathrm{grad}_k H = 0$. Thus, (19.58) becomes

$$-\iiint_{-\infty}^{+\infty} \alpha\, \mathbf{u} \wedge \mathbf{B} \cdot \mathrm{grad}_k f\, d^3k = \mathbf{B} \cdot \iiint_{-\infty}^{+\infty} f\, \mathrm{grad}_k \alpha \wedge \mathbf{u}\, d^3k =$$

$$= \iiint_{-\infty}^{+\infty} f\, \mathrm{grad}_k \alpha \cdot \mathbf{u} \wedge \mathbf{B}\, d^3k = n\, \overline{\mathrm{grad}_k \alpha \cdot \mathbf{u} \wedge \mathbf{B}}. \qquad (19.59)$$

In (19.59), the term containing the magnetic induction does not contribute to the moment if α or f depends on k through energy alone, $\alpha = \alpha(H)$ or $f = f(H)$. In fact, in the first case it is $\mathrm{grad}_k \alpha = (d\alpha/dH)\, \mathrm{grad}_k H$, whence

$$\mathrm{grad}_k \alpha \cdot \mathbf{u} \wedge \mathbf{B} = \frac{d\alpha}{dH}\, \hbar\, \mathbf{u} \cdot \mathbf{u} \wedge \mathbf{B} = 0. \qquad (19.60)$$

The same calculation holds when $f = f(H)$, starting from the integrand at the left hand side of (19.59).

Collision Term

Here it is convenient to distinguish between the inter-band and intra-band transitions, introduced in Sects. 19.3, 19.3.2. Thus, the collision term is written $C = C_b + C_v$, where as above suffix b (v) stands for "inter-band" ("intra-band"). This yields

$$\iiint_{-\infty}^{+\infty} \alpha\, C\, d^3k = W_b + W_v, \qquad W_{b(v)}[\alpha] = \iiint_{-\infty}^{+\infty} \alpha\, C_{b(v)}\, d^3k, \qquad (19.61)$$

where the functional symbol reminds one that $W_{b(v)}$ is determined by the form of α.

19.4.1   Moment Equations

Adding up (19.53), (19.54), (19.56), (19.59), and (19.61) after multiplying the drift terms by $q/\hbar$ provides the explicit form of (19.51), that reads

$$\frac{\partial}{\partial t}(n\,\overline{\alpha}) + \mathrm{div}_{\mathbf{r}}(n\,\overline{\alpha\,\mathbf{u}}) + \frac{q}{\hbar}\,n\,\overline{\mathrm{grad}_{\mathbf{k}}\alpha \cdot (\mathbf{E} + \mathbf{u} \wedge \mathbf{B})} = W_b[\alpha] + W_v[\alpha]. \quad (19.62)$$

A simple reasoning shows that the equations of the form (19.62) that are obtained from different choices of α are coupled with each other. To show this one takes for simplicity the one-dimensional case and lets $\alpha = c\,k^m$, with m a positive integer and c a constant. Also, the parabolic-band approximation is assumed to hold, so that $\mathbf{u}$ is a linear function of k. It follows that the time derivative in (19.62) contains the moment of order m of f, while the diffusion term (due to the product $\alpha\,\mathbf{u}$) contains the moment of order $m + 1$; in turn, the summand proportional to $\mathbf{E}$ in the drift term contains the moment of order $m - 1$ due to the derivative of α, while the summand proportional to $\mathbf{B}$ contains the moment of order m. These considerations are sufficient to show that the equation whose unknown is the moment of order m is coupled with those whose unknowns are the moment of order $m - 1$ and $m + 1$.

In the typical applications of the moments expansion a finite set of equations is considered, starting from the lowest-order moment $m = 0$ up to some order m_0. The system of equations thus obtained is indeterminate because, due to the coupling mentioned above, the number of equations is m_0 whereas the number of unknown moments appearing in them is $m_0 + 1$. To make the system determinate it is then necessary to add an extra condition, that is found by prescribing an approximate form of the $(m_0 + 1)$th moment.[22] Such a prescription reduces the number of unknown moments to m_0 and makes the system of differential equations determinate; for this reason it is called *closure condition*. The typical choice for the closure condition is to approximate the $(m_0 + 1)$th moment using the equilibrium distribution.

19.4.1.1   Moment of Order Zero

The *moment of order zero* is obtained by letting $\alpha = 1$ in (19.62), whose left hand side becomes $\partial n/\partial t + \mathrm{div}_{\mathbf{r}}(n\,\overline{\mathbf{u}})$; the electron concentration n is the moment of order zero and $\overline{\mathbf{u}} = \mathbf{v}$ is the average velocity, as defined in (19.31). If the zero-order moment of the collision term does not introduce further unknowns, the equation's unknowns are two: n and $\mathbf{v}$. The form of the left hand side shows that integrating the zero-order moment of the BTE over an arbitrary volume Ω of the $\mathbf{r}$ space provides the balance equation for the number of electrons of the conduction band (compare

---

[22] The choice of the highest-order moment as the function to be approximated is reasonable in view of the analysis of the moments method carried out in Sect. C.6. In fact, as the moments are the coefficients of a converging Taylor series, they become smaller and smaller as the order increases; thus, the error due to approximating the highest-order coefficient is expected to be the smallest.

with Sect. 23.2):

$$\frac{d}{dt} \int_\Omega n \, d\Omega + \int_\Sigma n \, \mathbf{v} \cdot \mathbf{s} \, d\Sigma = \int_\Omega W_b[1] \, d\Omega + \int_\Omega W_v[1] \, d\Omega, \qquad (19.63)$$

where Σ is the boundary of Ω and $\mathbf{s}$ the unit vector normal to Σ, oriented in the outward direction. The second integral at the right hand side of (19.63) does not contribute to the electrons' balance; in fact, it describes transitions that do not influence the number of electrons of the band because both the initial and final state belong to it. On the other hand, due to the arbitrariness of Ω, the integral vanishes only if $W_v[1] = 0$. This result shows that the zero-order moment of the intra-band transitions vanishes,[23] so that the only transitions of importance for the zero-order moment, despite being dominated by much larger relaxation times, are the inter-band ones. This is the exception anticipated in Sect. 19.3.2. The zero-order moment for the electrons of the conduction band then reads

$$\frac{\partial n}{\partial t} + \text{div}_\mathbf{r} (n \, \mathbf{v}) = W_b[1]. \qquad (19.64)$$

The form of the inter-band term $W_b[1]$, which is not relevant for the analysis in hand, is worked out in Chap. 20.

19.4.1.2 General Form of the Higher-Order Moments

As shown above, the contribution of the intra-band transitions vanishes for $\alpha = 1$. In contrast, it becomes dominant for other choices of α; in fact, in such cases the intra-band transitions do not cancel out any more and their scattering rates turn out to be much higher than those of the inter-band transitions. This allows one to adopt an approximation for $W_b[\alpha]$, namely,

$$W_b[\alpha] = \iiint_{-\infty}^{+\infty} \alpha \, C_b \, d^3k \simeq \overline{\alpha} \iiint_{-\infty}^{+\infty} C_b \, d^3k. \qquad (19.65)$$

In other terms it is assumed that, since the contribution of $W_b[\alpha]$ to the collision term is small when $\alpha \neq 1$, the error introduced by (19.65) is negligible. Expanding the time derivative in (19.62), and using (19.64), (19.65), yields

$$n \frac{\partial \overline{\alpha}}{\partial t} + \text{div}_\mathbf{r} (n \, \overline{\alpha \, \mathbf{u}}) - \overline{\alpha} \, \text{div}_\mathbf{r} (n \, \mathbf{v}) + \frac{q}{\hbar} n \, \overline{\text{grad}_\mathbf{k} \alpha \cdot (\mathbf{E} + \mathbf{u} \wedge \mathbf{B})} = W_v[\alpha], \qquad (19.66)$$

where only the intra-band transitions appear. Due to its simpler form, (19.66) will be used in the following to derive the balance equations with $\alpha \neq 1$.

---

[23] A similar reasoning is used to explain (20.16).

19.4.1.3   Moments of Order One, Two, and Three

The *moment of order one* of the BTE is found by letting $\alpha = u_i$ with $i = 1, 2, 3$ in (19.66); this yields the continuity equation for the ith component of the average velocity of the electrons, $\bar{u}_i = v_i$:

$$n \frac{\partial v_i}{\partial t} + \text{div}_{\mathbf{r}} \left(n \, \overline{u_i \, \mathbf{u}} \right) - v_i \, \text{div}_{\mathbf{r}} \left(n \, \mathbf{v} \right) + \frac{q}{\hbar} n \, \overline{\text{grad}_{\mathbf{k}} u_i} \cdot \left(\mathbf{E} + \mathbf{u} \wedge \mathbf{B} \right) = W_v[u_i].$$

$$(19.67)$$

To proceed it is necessary to introduce the definition of average kinetic energy and average flux of the electrons' kinetic energy,[24]

$$w(\mathbf{r}, t) = \frac{1}{n} \iiint_{-\infty}^{+\infty} E_e(\mathbf{k}) \, f(\mathbf{r}, \mathbf{k}, t) \, \mathrm{d}^3 k, \qquad (19.68)$$

$$\mathbf{b}(\mathbf{r}, t) = \frac{1}{n} \iiint_{-\infty}^{+\infty} E_e(\mathbf{k}) \, \mathbf{u}(\mathbf{k}) \, f(\mathbf{r}, \mathbf{k}, t) \, \mathrm{d}^3 k, \qquad (19.69)$$

with $E_e = E(\mathbf{k}) - E_C$. Then, the *moment of order two* of the BTE is found by letting $\alpha = E_e$ in (19.66); this yields the continuity equation for the average kinetic energy of the electrons, (19.68). In the derivation, the term containing the magnetic induction vanishes due to (19.60); using the definition (17.52) of the group velocity, the equation reads

$$n \frac{\partial w}{\partial t} + \text{div}_{\mathbf{r}} \left(n \, \mathbf{b} \right) - w \, \text{div}_{\mathbf{r}} \left(n \, \mathbf{v} \right) + q \, n \, \mathbf{v} \cdot \mathbf{E} = W_v[E_e]. \qquad (19.70)$$

The *moment of order three* of the BTE is found by letting $\alpha = E_e \, u_i$ with $i = 1, 2, 3$ in (19.66); this yields the continuity equation for the average flux of the electrons' kinetic energy, (19.69); the equation reads

$$n \frac{\partial b_i}{\partial t} + \text{div}_{\mathbf{r}} \left(n \, \overline{E_e \, u_i \, \mathbf{u}} \right) - b_i \, \text{div}_{\mathbf{r}} \left(n \, \mathbf{v} \right) + \frac{q}{\hbar} n \, \overline{\text{grad}_{\mathbf{k}} (E_e \, u_i)} \cdot \left(\mathbf{E} + \mathbf{u} \wedge \mathbf{B} \right) =$$

$$W_v[E_e \, u_i]. \qquad (19.71)$$

The choices $\alpha = 1$, $\alpha = u_i$, $\alpha = E_e$, and $\alpha = E_e \, u_i$ are such that each moment equation provides the balance relation of a dynamic quantity of interest: number of electrons, average velocity, average kinetic energy, average flux of the kinetic energy; the even-order moments yield a scalar equation, whereas the odd-order moments yield a vector equation.

---

[24] In the equilibrium condition the product $E_e \, f^{\text{eq}}$ is even with respect to $\mathbf{k}$. In turn, $\mathbf{u} = (1/\hbar) \, \text{grad}_{\mathbf{k}} E$ is odd, so that $\mathbf{b}^{\text{eq}} = 0$. Compare with the similar comment made about the average velocity in (19.31).

19.4.2 Hierarchical Models

The order-one moment (19.67) contains the new unknown $\overline{u_i\,\mathbf{u}}$ besides n and $\mathbf{v}$ already present in (19.64); the order-two moment (19.70) contains again n and $\mathbf{v}$, and the new unknowns w, $\mathbf{b}$. The order-three moment contains n, $\mathbf{v}$, $\mathbf{b}$, and the new unknown $\overline{E_e\,u_i\,\mathbf{u}}$. The drift terms and the collision terms, depending on their form, may, or may not introduce extra unknowns; even if they don't, the number of unknowns listed above exceeds that of the equations. It is worth anticipating that the finite set of balance equations indicated in Sect. 19.4.1 is obtained by taking the equations in pairs: specifically, the first pair is made of the balance equations of order zero and one, (19.64) and (19.66), that are collectively termed *drift-diffusion model*; in this case, the three unknowns n, $\mathbf{v}$, and $\overline{u_i\,\mathbf{u}}$ are reduced to two by the closure condition (Sect. 19.4.1), that consists in replacing the highest-order moment $\overline{u_i\,\mathbf{u}}$ with its equilibrium expression. A more elaborate model is obtained by taking the first two pairs, namely, the balance equations of order zero through three, (19.64), (19.66), (19.70), and (19.71), that are collectively termed *hydrodynamic model*; the six unknowns n, $\mathbf{v}$, and $\overline{u_i\,\mathbf{u}}$, w, $\mathbf{b}$, $\overline{E_e\,u_i\,\mathbf{u}}$ are reduced to five by prescribing the closure condition, then to four by determining a relation between the second-order moments $\overline{u_i\,\mathbf{u}}$ and w.

By this procedure one constructs a set of hierarchically-ordered models of increasing complexity. The type of model adopted in practical applications depends on the trade-off between the information that one needs to acquire about the physical behavior of the device under investigation and the computational cost of the system of differential equations to be solved. To date, the moments method has been investigated up to order 6 [45], and has been extended to order 21 using a scheme based on Legendre polynomial expansion [59]; the standard implementations in the commercial simulation programs used by semiconductor Companies adopt the hydrodynamic model.[25]

The balance equations derived so far are still rather cumbersome in view of the application to the analysis of semiconductor devices. A number of simplifications are illustrated below, which eventually lead to the standard form of the hydrodynamic model [38, 40, 89, 93]. To begin, one considers the time derivatives at the left hand side of the balance equations. Such derivatives differ from zero only if the distribution function depends explicitly on time, which typically happens when time-dependent boundary conditions are imposed to the device under investigation. In the practical cases, the maximum frequency of the electric signals applied to a device or an integrated circuit is lower by many orders of magnitude than the inverse relaxation times associated to the intra-band transitions; this makes it possible to neglect the time derivatives of v_i, w, and b_i. A quasi-static approximation[26] is thus assumed

---

[25] Comprehensive reviews of the solution methods for the BTE are in [55, 56] as far as the Monte Carlo method is concerned, and in [49] for deterministic methods.

[26] A similar reasoning is used to treat the time derivative of the vector potential when the semiconductor equations are coupled with the Maxwell equations (Sect. 19.5.4).

in the continuity Eqs. (19.67), (19.70), and (19.71). The argument leading to this approximation does not apply to the case of (19.64) because only the inter-band transitions take place there, whose relaxation times are much longer than those of the intra-band transitions and, in many cases, also than the inverse maximum frequency of the external signal. As a consequence, the term $\partial n/\partial t$ in (19.64) must be retained when the boundary conditions depend on time.

As a second approximation, one adopts the parabolic-band approximation; this implies that in a non-equilibrium condition the electrons of the conduction band still occupy energy states in the vicinity of the absolute minima. Such a condition is in general fulfilled as shown below.[27] Letting a indicate one of the absolute minima of the conduction band, and using (19.24) after dropping suffix "0", yields

$$u_i = \frac{\hbar (k_i - k_{ia})}{m_{ia}}, \quad \mathrm{grad}_k u_i = \frac{\hbar}{m_{ia}} \mathbf{i}_i, \quad \mathrm{grad}_k(E_e \, u_i) = \hbar \left(\frac{E_e}{m_{ia}} \mathbf{i}_i + u_i \, \mathbf{u} \right),$$
(19.72)

with $\mathbf{i}_i$ the unit vector of the ith axis. From now on, the equations derived from the parabolic-band approximation refer to the ath valley. In principle, the electron concentration, average velocity, and the other averages should be indicated with n_a, $\mathbf{v}_a$, and so on; this is not done here to avoid complicacies in the notation. The suffix will be introduced in Sect. 19.5.2, where the contributions of the valleys are added up. This comment does not apply to the moment of order zero, (19.64), because its derivation does not entail any simplifying hypothesis.

With these premises, one manipulates the terms $\overline{u_i \, \mathbf{u}}$, $E_e \, \overline{u_i \, \mathbf{u}}$ by introducing the auxiliary quantity $\mathbf{c} = \mathbf{u} - \mathbf{v}$, called *random velocity*. Clearly it is $\overline{c}_i = \overline{u}_i - v_i = 0$, so that $\overline{u_i \, \mathbf{u}} = v_i \, \mathbf{v} + \overline{u_i \, \mathbf{u}}$; it follows

$$\mathrm{div}_r(n \, \overline{u_i \, \mathbf{u}}) = \mathrm{div}_r(n \, \overline{c_i \, \mathbf{c}}) + v_i \, \mathrm{div}_r(n \, \mathbf{v}) + n \, \mathbf{v} \cdot \mathrm{grad}_r v_i.$$
(19.73)

The last term at the right hand side of (19.73) is called *convective term*. In the typical operating conditions of the semiconductor devices this term can be neglected (refer to [86] and the comments below). Replacing into (19.67) the simplified form of (19.73) along with the second relation of (19.72) yields

$$\mathrm{div}_r (n \, m_{ia} \, \overline{c_i \, \mathbf{c}}) + q \, n \, (\mathbf{E} + \mathbf{v} \wedge \mathbf{B})_i = m_{ia} \, W_v[u_i].$$
(19.74)

The latter equation contains the unknowns n and $\mathbf{v}$ already present in (19.64), and the new unknown $\overline{c_i \, \mathbf{c}}$. The drift term does not introduce extra unknowns. Note that $\overline{c_i \, \mathbf{c}}$ is actually made of three vectors, so that it may be thought of as a symmetric 3×3 tensor with components $\overline{c_i \, c_j}$, $i, j = 1, 2, 3$. To give it a more compact form, after observing that $m_{ia} \, \overline{c_i \, \mathbf{c}}$ has the units of an energy, one defines the *electron-temperature tensor* of components T_{ij} such that

$$k_B \, T_{ij} = m_{ia} \, \overline{c_i \, c_j}, \quad n \, k_B \, T_{ij} = \iiint_{-\infty}^{+\infty} m_{ia} \, c_i \, c_j \, f \, \mathrm{d}^3 k,$$
(19.75)

---

[27] The adoption of the parabolic-band approximation may be avoided at the cost of redefining the carrier temperature and introducing more relaxation times [108].

with k_B the Boltzmann constant. Letting $\mathbf{T}_i$ be the vector of entries T_{i1}, T_{i2}, T_{i3}, so that $k_B \, \mathbf{T}_i = m_{ia} \, \overline{c_i \, \mathbf{c}}$, one finds for the ith component of the moment of order one

$$\text{div}_{\mathbf{r}} \, (n \, k_B \, \mathbf{T}_i) + q \, n \, (\mathbf{E} + \mathbf{v} \wedge \mathbf{B})_i = m_{ia} \, W_v[u_i]. \tag{19.76}$$

In the equilibrium condition the electron-temperature tensor reduces to the product of a scalar coefficient times the identity tensor; in the limit of the Boltzmann distribution, the scalar coefficient identifies with the lattice temperature (Sect. 19.6.4), this providing an estimate of the modulus $|\mathbf{c}|$ of the random velocity. The modulus $|\mathbf{v}|$ of the average velocity in a non-equilibrium condition can be estimated as well, basing upon the current density and carrier concentration of the devices' operating conditions. It is found that in typical situations it is $|\mathbf{v}| \ll |\mathbf{c}|$, so that the average motion of the carriers in a non-equilibrium condition can be thought of as that of a slowly-drifting fluid. This justifies the neglect of the convective term in (19.73), and also allows one to neglect $v_i \, \mathbf{v}$ with respect to $\overline{c_i \, \mathbf{c}}$ when these terms appear in the same expression.

The simplifications used in (19.67) apply in the same manner to the moment of order three, (19.71); in fact it is $\overline{E_e \, u_i \, \mathbf{u}} = \overline{E_e \, u_i} \, \mathbf{c} + \overline{E_e \, u_i} \, \mathbf{v}$, whence

$$\text{div}_{\mathbf{r}}(n \, \overline{E_e \, u_i \, \mathbf{u}}) = \text{div}_{\mathbf{r}}(n \, \overline{E_e \, u_i} \, \mathbf{c}) + b_i \, \text{div}_{\mathbf{r}}(n \, \mathbf{v}) + n \, \mathbf{v} \cdot \text{grad}_{\mathbf{r}} b_i. \tag{19.77}$$

Using the third relation of (19.72) in the drift term of (19.71) transforms the latter into $(q/m_{ia}) \, n \, [(w \, \mathbf{i}_i + m_{ia} \, v_i \, \mathbf{v} + m_{ia} \, \overline{c_i \, \mathbf{c}}) \cdot \mathbf{E} + \mathbf{b} \wedge \mathbf{B} \cdot \mathbf{i}_i + m_{ia} \, \overline{u_i \, \mathbf{u}} \cdot \mathbf{u} \wedge \mathbf{B}]$, where the mixed product vanishes due to the repeated factor, and $m_{ia} \, v_i \, \mathbf{v}$ is negligible as shown above. Replacing (19.77) into (19.71) after neglecting the time derivative and the convective term, yields

$$\text{div}_{\mathbf{r}} \left(n \, m_{ia} \, \overline{E_e \, u_i} \, \mathbf{c} \right) + q \, n \, [(w \, \mathbf{i}_i + k_B \, \mathbf{T}_i) \cdot \mathbf{E} + \mathbf{b} \wedge \mathbf{B} \cdot \mathbf{i}_i] = m_{ia} \, W_v[E_e \, u_i]. \tag{19.78}$$

The relation between the second-order moments necessary to reduce the number of unknowns is now determined starting from the expression of E_e in the parabolic-band approximation. Using the first relation in (19.72) yields $E_e = (1/2) \sum_{i=1}^{3} m_{ia} \, u_i^2$ whence, from (19.68),

$$w = \frac{1}{2} \sum_{i=1}^{3} m_{ia} \, v_i^2 + \frac{3}{2} \, k_B \, T_e, \qquad T_e = \frac{T_{11} + T_{22} + T_{33}}{3}, \tag{19.79}$$

with T_e the *electron temperature*. The two summands of w in (19.79) are also called *convective part* and *thermal part* of the average kinetic energy, respectively. The same reasoning that has led to the neglect of $m_{ia} \, v_i \, \mathbf{v}$ with respect to $m_{ia} \, c_i \, \mathbf{c}$ also shows that the thermal part is dominant with respect to the convective part, so that $w \simeq (3/2) \, k_B \, T_e$. Moreover, it can be assumed that the electron-temperature tensor retains the same structure of the equilibrium case, so that

$$\begin{bmatrix} T_{11} & T_{12} & T_{13} \\ T_{21} & T_{22} & T_{23} \\ T_{31} & T_{32} & T_{33} \end{bmatrix} \simeq \begin{bmatrix} T_{11} & 0 & 0 \\ 0 & T_{22} & 0 \\ 0 & 0 & T_{33} \end{bmatrix} \simeq T_e(\mathbf{r}, t) \, \mathcal{I}, \tag{19.80}$$

with $\mathcal{I}$ the identity tensor. As a consequence, $(w \, \mathbf{i}_i + k_B \, \mathbf{T}_i) \cdot \mathbf{E} = (5/2) \, k_B \, T_e \, \mathbf{i}_i \cdot \mathbf{E}$ and $\mathrm{div}_\mathbf{r}(n \, k_B \, \mathbf{T}_i) = \partial(n \, k_B \, T_e)/\partial x_i$. The latter is the ith component of $\mathrm{grad}_\mathbf{r} \, (n \, k_B \, T_e)$. In summary, the balance equations for the moments of order one, two, and three read

$$\frac{\partial \, (n \, k_B \, T_e)}{\partial x_i} + q \, n \, \; (\mathbf{E} + \mathbf{v} \wedge \mathbf{B})_i = m_{ia} \, W_v[u_i], \tag{19.81}$$

$$\mathrm{div}_\mathbf{r} \, (n \, \mathbf{b}) - (3/2) \, k_B \, T_e \, \mathrm{div}_\mathbf{r} \, (n \, \mathbf{v}) + q \, n \, \mathbf{v} \cdot \mathbf{E} = W_v[E_e], \tag{19.82}$$

$$\mathrm{div}_\mathbf{r} \, \left(n \, m_{ia} \, \overline{E_e \, u_i} \, \mathbf{c} \right) + q \, n \, \; [(5/2) \, k_B \, T_e \, \mathbf{E} + \mathbf{b} \wedge \mathbf{B}]_i = m_{ia} \, W_v[E_e \, u_i]. \tag{19.83}$$

19.4.2.1   Macroscopic Relaxation Times of the Higher-Order Moments

As remarked in Sect. (19.4.1), the collision terms of the moments of order higher than zero account for the intra-band transitions only. These terms are worked out here using the perturbative form of the BTE (Sect. 19.3.3); this approach is coherent with the other approximations from which the balance Eqs. (19.81–19.83) derive. The collision term of (19.82) then becomes

$$W_v[E_e] = - \iiint_{-\infty}^{+\infty} E_e \, \frac{f - f^{eq}}{\tau} \, \mathrm{d}^3 k, \tag{19.84}$$

with $\tau \simeq \tau_v$ (Sect. 19.3.2). The equilibrium part is worked out by defining the *energy-relaxation time* τ_w such that

$$\iiint_{-\infty}^{+\infty} E_e \, \frac{f^{eq}}{\tau_v} \, \mathrm{d}^3 k = \frac{1}{\tau_w} \iiint_{-\infty}^{+\infty} E_e \, f^{eq} \, \mathrm{d}^3 k = \frac{n^{eq} \, w^{eq}}{\tau_w} \simeq \frac{3}{2} \, \frac{n^{eq} \, k_B T_e^{eq}}{\tau_w}, \tag{19.85}$$

where the definitions (19.68, 19.79) of the electrons' average kinetic energy and temperature are used. The left hand side of (19.85) does not vanish because the integrand is positive definite. The non-equilibrium part of $W_v[E_e]$ is approximated as

$$\iiint_{-\infty}^{+\infty} E_e \, \frac{f}{\tau_v} \, \mathrm{d}^3 k \simeq \frac{1}{\tau_w} \iiint_{-\infty}^{+\infty} E_e \, f \, \mathrm{d}^3 k = \frac{n \, w}{\tau_w} \simeq \frac{3}{2} \, \frac{n \, k_B T_e}{\tau_w}, \tag{19.86}$$

based on the observation that, due to the smallness of the intra-band relaxation time τ_v, the distribution function departs little from the equilibrium one (Sect. 19.3.3).

The derivation of the analogues of τ_w for the collision terms of (19.81) and (19.83) is somewhat more complicate. In fact, in most semiconductors, among which Si, Ge, and GaAs, the relaxation time τ_v is even with respect to $\mathbf{k}$ [57], which makes the integrals of $u_i \, f^{eq}/\tau_v$ and $E_e \, u_i \, f^{eq}/\tau_v$ to vanish because the integrand is odd. To overcome the difficulty one expands $f - f^{eq}$ into a Taylor series with respect to a parameter and truncates the series in such a way as to retain the first summand which is odd with respect to $\mathbf{k}$. For instance, letting λ be the parameter[28] and assuming that the first-order term of the expansion is odd, one lets $f - f^{\text{eq}} \simeq (\mathrm{d}f/\mathrm{d}\lambda)^{\text{eq}} \lambda$ whence

$$\iiint_{-\infty}^{+\infty} u_i \, \frac{(\mathrm{d}f/\mathrm{d}\lambda)^{\text{eq}}}{\tau_v} \, \mathrm{d}^3 k = \frac{1}{\tau_{pi}} \iiint_{-\infty}^{+\infty} u_i \, (\mathrm{d}f/\mathrm{d}\lambda)^{\text{eq}} \, \mathrm{d}^3 k, \tag{19.87}$$

$$\iiint_{-\infty}^{+\infty} E_e \, u_i \, \frac{(\mathrm{d}f/\mathrm{d}\lambda)^{\text{eq}}}{\tau_v} \, \mathrm{d}^3 k = \frac{1}{\tau_{bi}} \iiint_{-\infty}^{+\infty} E_e \, u_i \, (\mathrm{d}f/\mathrm{d}\lambda)^{\text{eq}} \, \mathrm{d}^3 k, \tag{19.88}$$

with τ_{pi} and τ_{bi} the *momentum-relaxation time* and *relaxation time of the energy flux*, respectively.[29] Due to their definitions, τ_{pi} and τ_{bi} are diagonal tensors. However, as their degree of anisotropy is small, they are approximated by scalar quantities, $\tau_{pi} \simeq \tau_p$ and $\tau_{bi} \simeq \tau_b$. Investigations about the relaxation times have been carried out by different techniques, specifically, the spherical-harmonics expansion method to determine the dependence on the average energy [90, 93], and the Monte Carlo method to study the anisotropy properties [11, 12, 43].

Using (19.87, 19.88) along with the definitions (19.31, 19.69) of the average velocity and average flux of kinetic energy finally yields

$$\iiint_{-\infty}^{+\infty} \frac{u_i}{\tau_v} \left(\frac{\mathrm{d}f}{\mathrm{d}\lambda} \lambda \right)^{\text{eq}} \mathrm{d}^3 k = \frac{n \, v_i}{\tau_p}, \qquad \iiint_{-\infty}^{+\infty} \frac{E_e \, u_i}{\tau_v} \left(\frac{\mathrm{d}f}{\mathrm{d}\lambda} \lambda \right)^{\text{eq}} \mathrm{d}^3 k = \frac{n \, b_i}{\tau_b}. \tag{19.89}$$

19.5 Hydrodynamic and Drift-Diffusion Models

In Sect. 19.4 the moments method has been applied to derive a set of balance equations; the general form of the latter has successively been modified by introducing a number of simplifications: among them is the parabolic-band approximation, due to which, as indicated in Sect. 19.4.2, a set of equations restricted to the ath valley of the conduction band is obtained. In order to recover the equations for the whole band, it is necessary to add up the single-valley contributions. The procedure is the same for the hydrodynamic and drift-diffusion models; it will be worked out explicitly only for the simpler case of the drift-diffusion model.

---

[28] Typically the parameter used in this procedure is the electric field [57]. An expansion truncated to the first order is coherent with the first-order perturbation approach.

[29] The term "momentum" for τ_{pi} derives from the observation that the continuity equation for the ith component of the average velocity v_i of the electrons, (19.67), may also be thought of as the continuity equation for the ith component of the average momentum, $m_{ia} \, v_i$. In turn, τ_{bi} is also called *heat-relaxation time*.

19.5.1 HD Model

As anticipated in Sect. 19.4.2, the hydrodynamic (HD) model is obtained by taking the
balance equations of order zero through three, (19.64), (19.81), (19.82), and (19.83),
and imposing the closure condition onto the fourth-order moment. For simplicity,
the latter is considered in the non-degenerate case whence, from (19.159), it is
$m_{ia} \left(\overline{E_e u_i c} \right)^{eq} \simeq \mathbf{i}_i (5/2) (k_B T)^2$. Letting $W = W_b[1]$ and using (19.85), (19.86),
(19.89) yields

$$\frac{\partial n}{\partial t} + \mathrm{div}_\mathbf{r} (n \, \mathbf{v}) = W, \qquad \frac{\partial (n \, k_B \, T_e)}{\partial x_i} + q \, n \; (\mathbf{E} + \mathbf{v} \wedge \mathbf{B})_i = -\frac{m_{ia}}{\tau_p} n \, v_i,$$

$$(19.90)$$

$$\mathrm{div}_\mathbf{r} (n \, \mathbf{b}) - \frac{3}{2} k_B \, T_e \, \mathrm{div}_\mathbf{r} (n \, \mathbf{v}) + q \, n \; \mathbf{v} \cdot \mathbf{E} = -\frac{3}{2} \frac{k_B}{\tau_w} \left[n \, T_e - (n \, T_e)^{eq} \right], \quad (19.91)$$

$$\frac{5}{2} (k_B \, T)^2 \frac{\partial n}{\partial x_i} + q \, n \; \left(\frac{5}{2} k_B \, T_e \, \mathbf{E} + \mathbf{b} \wedge \mathbf{B} \right)_i = -\frac{m_{ia}}{\tau_b} n \, b_i. \qquad (19.92)$$

which constitute a system of first-order, partial-differential equations in the unknowns
n, $\mathbf{v}$, T_e, and $\mathbf{b}$. In general, the model's equations are to be solved over a volume that
encloses the device under investigation; the boundary conditions that typically apply
are discussed in Sect. 19.5.6. Two of the equations are scalar (namely, (19.91) and
the first one in (19.90)), while the other two are vector equations. The system is non
linear because the unknowns are multiplied by each other.[30] Note, however, that the
second equation in (19.90) is linear with respect to the components of $\mathbf{v}$; the latter
can be extracted and replaced into the two scalar equations. The same procedure
is applicable to (19.92), which is linear with respect to the components of $\mathbf{b}$. After
the replacements are completed, the system reduces to two scalar equations of the
second order. An example is given in Sect. 19.5.5, with reference to the simpler case
of the drift-diffusion model. Due to the components m_{ia} of the effective-mass tensor,
the vector equations are anisotropic; however, when the contributions of the different
valleys are combined together, the anisotropy cancels out (the explicit calculation is
provided for the drift-diffusion model below).

The qualitative analysis of the model carried out above implies that the electric
field and magnetic induction are known, so that they are embedded in the model's
coefficients. In fact, this is not true, because the fields are influenced by the distribution of electric charge and current density that are some of the model's unknowns.
For this reason, as shown below, the hydrodynamic equations, and the drift-diffusion
ones as well, must be coupled with the Maxwell equations.

---

[30] Also, the generation-recombination term W embeds non-linear dependencies on some of the
unknowns, Chap. 20.

19.5.2   DD Model

The drift-diffusion (DD) model is obtained by taking the balance equations of order zero and one, (19.90), and imposing the closure condition onto the second-order moment. For simplicity, the latter is considered in the non-degenerate case, whence $T_e^{eq} = T$ (Sect. 19.6.4); the model thus reads

$$\frac{\partial n}{\partial t} + \text{div}_r (n \, \mathbf{v}) = W, \qquad k_B \, T \, \frac{\partial n_a}{\partial x_i} + q \, n_a \, (\mathbf{E} + \mathbf{v}_a \wedge \mathbf{B})_i = -\frac{m_{ia}}{\tau_p} \, n_a \, v_{ia}. \tag{19.93}$$

As indicated in Sect. 19.4.2, the first equation in (19.93) refers to the whole conduction band because its derivation did not entail any simplifying hypothesis; in contrast, the second equation refers to the ath minimum of the band due to the parabolic-band approximation. This explains the index attached to n and to the average velocity; the momentum-relaxation time, instead, does not depend on the valley [90]. As noted above, the dependence on the components v_{ia} is linear, which makes it possible to express them in terms of the other functions. In fact, it is more convenient to extract, instead of $\mathbf{v}_a$, the electron-current density of the ath minimum; remembering (4.21) and (4.22), the latter is given by $\mathbf{J}_a = -q \, n_a \, \mathbf{v}_a$. Then, the second equation in (19.93) is recast as

$$J_{ia} = J_{ia}' - \frac{q \, \tau_p}{m_{ia}} \, (\mathbf{J}_a \wedge \mathbf{B})_i \, , \qquad J_{ia}' = k_B \, T \, \frac{q \, \tau_p}{m_{ia}} \, \frac{\partial n_a}{\partial x_i} + \frac{q \, \tau_p}{m_{ia}} \, q \, n_a \, (\mathbf{E})_i, \tag{19.94}$$

with $J_{ia}' = J_{ia}(\mathbf{B} = 0)$. Letting $\mu_{ia} = q \, \tau_p / m_{ia}$, the matrix form of (19.94) reads

$$\begin{bmatrix} J_{a1} \\ J_{a2} \\ J_{a3} \end{bmatrix} = \begin{bmatrix} J_{a1}' \\ J_{a2}' \\ J_{a3}' \end{bmatrix} - \begin{bmatrix} \mu_{a1} & 0 & 0 \\ 0 & \mu_{a2} & 0 \\ 0 & 0 & \mu_{a3} \end{bmatrix} \begin{bmatrix} J_{a2} \, B_3 - J_{a3} \, B_2 \\ J_{a3} \, B_1 - J_{a1} \, B_3 \\ J_{a1} \, B_2 - J_{a2} \, B_1 \end{bmatrix}, \tag{19.95}$$

equivalent to

$$\begin{bmatrix} 1 & \mu_{a1} \, B_3 & -\mu_{a1} \, B_2 \\ -\mu_{a2} \, B_3 & 1 & \mu_{a2} \, B_1 \\ \mu_{a3} \, B_2 & -\mu_{a3} \, B_1 & 1 \end{bmatrix} \begin{bmatrix} J_{a1} \\ J_{a2} \\ J_{a3} \end{bmatrix} = \begin{bmatrix} J_{a1}' \\ J_{a2}' \\ J_{a3}' \end{bmatrix}. \tag{19.96}$$

The diagonal tensor $\hat{\mu}_a$ of entries μ_{ia} is called *mobility tensor* of the ath valley. Note that the product of a mobility by a magnetic induction is dimensionless. Letting $\mathbf{M}_a = \mu_{a1} \, \mu_{a2} \, \mu_{a3} \, (\hat{\mu}_a)^{-1} \, \mathbf{B}$, the components of the current density are found by solving the algebraic system (19.96), where the determinant of the matrix is

$$D_M = 1 + \mu_{a1} \, \mu_{a2} \, \mu_{a3} \left(\frac{B_1^2}{\mu_{1a}} + \frac{B_2^2}{\mu_{2a}} + \frac{B_3^2}{\mu_{3a}} \right) = 1 + \mathbf{B} \cdot \mathbf{M}_a. \tag{19.97}$$

The components of $\mathbf{J}_a$ are finally found to be

$$D_M \, J_{ai} = J'_{ai} + \mu_{ai} \left(\mathbf{B} \wedge \mathbf{J}'_a\right)_i + \left(\mathbf{M}_a \cdot \mathbf{J}'_a\right) B_i. \tag{19.98}$$

In typical situations the modulus of the magnetic induction is small; hence terms that are quadratic in the components of $\mathbf{B}$ may be neglected. This yields the approximate form

$$J_{ai} \simeq J'_{ai} + \mu_{ai} \left(\mathbf{B} \wedge \mathbf{J}'_a\right)_i . \tag{19.99}$$

The electron current density of the whole conduction band is thus found as

$$\mathbf{J} = \sum_{a=1}^{M_C} \mathbf{J}_a = \sum_{a=1}^{M_C} \left[k_B \, T \, \hat{\mu}_a \, \mathrm{grad} n_a + \hat{\mu}_a \, q \, n_a \, \mathbf{E} + \hat{\mu}_a \left(\mathbf{B} \wedge \mathbf{J}'_a\right)\right]. \tag{19.100}$$

In the perturbative approach followed here, it can be assumed that the total electron concentration n equally distributes[31] over the valleys, $n_a = n/M_C$. The first two summands at the right hand side of (19.100) then yield

$$\mathbf{J}' = \sum_{a=1}^{M_C} \mathbf{J}'_a = q \, \hat{\mu}_n \, n \, \mathbf{E} + q \, \hat{D}_n \, \mathrm{grad} n, \tag{19.101}$$

where the diagonal tensors $\hat{\mu}_n$, $\hat{D}_n$ are defined as

$$\hat{\mu}_n = \frac{1}{M_C} \sum_{a=1}^{M_C} \hat{\mu}_a, \qquad \hat{D}_n = \frac{k_B \, T}{q} \, \hat{\mu}_n. \tag{19.102}$$

They are called *electron-mobility* tensor and *electron-diffusivity* tensor, respectively. The second relation in (19.102), that states that diffusivity and mobility are proportional through $k_B \, T/q$, is called *Einstein relation*.[32] The form of $\hat{\mu}_n$ is specified on a case-by-case basis, depending on the semiconductor under consideration. Taking silicon by way of example ($M_C = 6$), the mass tensor is obtained from (17.82–17.84); thus, the mobility tensor $\hat{\mu}_a$ has one of the following forms:

$$\begin{bmatrix} \mu_l & 0 & 0 \\ 0 & \mu_t & 0 \\ 0 & 0 & \mu_t \end{bmatrix}, \quad \begin{bmatrix} \mu_t & 0 & 0 \\ 0 & \mu_l & 0 \\ 0 & 0 & \mu_t \end{bmatrix}, \quad \begin{bmatrix} \mu_t & 0 & 0 \\ 0 & \mu_t & 0 \\ 0 & 0 & \mu_l \end{bmatrix}, \tag{19.103}$$

---

[31] From this assumption and from (19.100) it also follows $\mathbf{J} = -q \sum_{a=1}^{M_C} n_a \mathbf{v}_a = -q \, (n/6) \sum_{a=1}^{M_C} \mathbf{v}_a$, whence $\mathbf{J} = -q \, n \, \mathbf{v}$ with $\mathbf{v} = (1/6) \sum_{a=1}^{M_C} \mathbf{v}_a$.

[32] The relation derives from Einstein's investigation on the Brownian motion [35] and has therefore a broader application. In a semiconductor it holds within the approximations of parabolic bands and non-degenerate conditions.

with $\mu_l = q\,\tau_p/m_l$, $\mu_t = q\,\tau_p/m_t$. The first form in (19.103) applies to the two minima belonging to axis k_1, and so on; thus, the electron-mobility tensor (19.102) is found to be

$$\hat{\mu}_n = \frac{1}{6}\left(2\begin{bmatrix} \mu_l & 0 & 0 \\ 0 & \mu_t & 0 \\ 0 & 0 & \mu_t \end{bmatrix} + 2\begin{bmatrix} \mu_t & 0 & 0 \\ 0 & \mu_l & 0 \\ 0 & 0 & \mu_t \end{bmatrix} + 2\begin{bmatrix} \mu_t & 0 & 0 \\ 0 & \mu_t & 0 \\ 0 & 0 & \mu_l \end{bmatrix}\right) = \mu_n\,\mathcal{I},$$

(19.104)

with $\mathcal{I}$ the identity tensor and $\mu_n = (\mu_l + 2\,\mu_t)/3$ the *electron mobility*. The second definition in (19.102) then yields $\hat{D}_n = D_n\,\mathcal{I}$, with $D_n = (k_B\,T/q)\mu_n$ the *electron diffusivity* or *electron-diffusion coefficient*. From these results and (19.101) one derives

$$\mathbf{J}' = q\,\mu_n\,n\,\mathbf{E} + q\,D_n\,\mathrm{grad}n, \qquad \mathbf{J}'_a = \frac{1}{\mu_n\,M_C}\,\hat{\mu}_a\,\mathbf{J}'. \tag{19.105}$$

From this, after a somewhat lengthy calculation, the last term at the right hand side of (19.100) is found to be

$$\sum_{a=1}^{M_C}\hat{\mu}_a\,(\mathbf{B}\wedge\mathbf{J}'_a) = a_n\,\mu_n\,\mathbf{B}\wedge\mathbf{J}', \qquad a_n = \frac{\mu_t\,(\mu_t + 2\,\mu_l)}{3\,\mu_n^2}. \tag{19.106}$$

As anticipated in the qualitative discussion about the HD model, despite the fact that each vector equation is anisotropic, when the contributions of the different valleys are combined together the anisotropy cancels out. From the definition of the electron mobility $\mu_n = (\mu_l + 2\,\mu_t)/3$ one may also extract a scalar effective mass $m_n = q\,\tau_p/\mu_n$, that fulfills $1/m_n = (2/m_t + 1/m_l)/3$. Using the room-temperature values taken from Table 17.4 yields, for silicon, $m_n/m_0 \simeq 0.26$. By the same token one finds $a_n \simeq 2.61$.

In the next sections, the current density of the electrons in the conduction band will be used in equations involving also the current density of the holes in the valence band; for this reason it is necessary to use different symbols. Specifically, $\mathbf{J}_n$ for the former and $\mathbf{J}_p$ for the latter; with this provision, the above calculation yields

$$\mathbf{J}_n = q\,\mu_n\,n\,\mathbf{E} + q\,D_n\,\mathrm{grad}n + q\,a_n\,\mu_n\,\mathbf{B}\wedge(\mu_n\,n\,\mathbf{E} + D_n\,\mathrm{grad}n), \tag{19.107}$$

which is called *drift-diffusion transport equation*. Thus the DD model for the electrons of the conduction band is given by (19.107) along with the first equation in (19.93); the latter is rewritten here as

$$\frac{\partial n}{\partial t} - \frac{1}{q}\,\mathrm{div}(\mathbf{J}_n) = W_n, \tag{19.108}$$

where a specific symbol for the generation-recombination term has been introduced as well.

19.5.3 DD Model for the Valence Band

The transport models illustrated so far are applicable to the valence band as well; here, the DD model will be worked out. Remembering the discussion of Sect. 19.2.3 about the dynamics in the parabolic-band approximation, the model is described in terms of the concentration and current density of holes. The two quantities are defined by adapting the corresponding expression for electrons, (19.31), as shown below. Letting $f = Q\,\Phi$, with $Q = 1/(4\pi^3)$ the density of states in the phase space and Φ the occupation probability, the hole concentration is

$$p(\mathbf{r},t) = \iiint_{-\infty}^{+\infty} Q\,(1 - \Phi)\,\mathrm{d}^3k. \tag{19.109}$$

In turn, the hole current density is defined starting from the definition of the electron current density of the valence band. The latter is similar to (19.31), the difference being that the integration in (19.31) is restricted to the branch of $E(\mathbf{k})$ belonging to the conduction band, whereas the integration in (19.110) below is restricted to one of the two branches of the valence band:

$$\mathbf{J}_a = -q \iiint_{-\infty}^{+\infty} \mathbf{u}(\mathbf{k})\,Q\,\Phi(\mathbf{r},\mathbf{k},t)\,\mathrm{d}^3k. \tag{19.110}$$

Letting $\Phi = 1 - (1 - \Phi)$ transforms (19.110) into

$$\mathbf{J}_a = q \iiint_{-\infty}^{+\infty} \mathbf{u}\,Q(1 - \Phi)\,\mathrm{d}^3k - q \iiint_{-\infty}^{+\infty} \mathbf{u}\,Q\,\mathrm{d}^3k, \tag{19.111}$$

where the second integral vanishes because $\mathbf{u}$ is odd with respect to $\mathbf{k}$. As a consequence, the current density of the branch under consideration may also be thought of as given by the motion of the empty states (holes), having the group velocity $\mathbf{u}(\mathbf{k})$ and the positive charge q. Moreover, one defines the average velocity of holes using $Q\,(1 - \Phi)$ as weighing function, to find

$$\mathbf{J}_a = q\,p_a\,\frac{\int_{\mathbf{k}} \mathbf{u}\,Q\,(1 - \Phi)\,\mathrm{d}^3k}{\int_{\mathbf{k}} Q\,(1 - \Phi)\,\mathrm{d}^3k} = q\,p_a\,\mathbf{v}_a \tag{19.112}$$

where the definition (19.109) of the hole concentration has been specified for the branch under consideration, and the short-hand notation $\int_{\mathbf{k}}$ has been used.

Given the above definitions, the derivation of the drift-diffusion model for holes follows the same pattern as for the electrons. Remembering the description of the band structure given in Sect. 17.6.5, for the valence band index a ranges over h and l; moreover, due to the isotropy of each branch deriving from the parabolic-band approximation (compare with 17.78), the effective mass is scalar. Then, the equivalent of (19.94) read, in vector form,

$$\mathbf{J}_a = \mathbf{J}_a' + \frac{q\,\tau_{pa}}{m_a}\,\mathbf{J}_a \wedge \mathbf{B}\,, \qquad \mathbf{J}_a' = -k_B\,T\,\frac{q\,\tau_{pa}}{m_a}\,\mathrm{grad}\,p_a + \frac{q\,\tau_{pa}}{m_a}\,q\,p_a\,\mathbf{E}\,, \tag{19.113}$$

where index a is attached also to the momentum-relaxation time because the two branches are different. Still due to such a difference, the holes do not distribute equally over the branches; the contribution of the drift and diffusion components then read, respectively,

$$\frac{q \, \tau_{ph}}{m_{hh}} q \, p_h \, \mathbf{E} + \frac{q \, \tau_{pl}}{m_{hl}} q \, p_l \, \mathbf{E} = q \, (\mu_{ph} + \mu_{pl}) \, p \, \mathbf{E}, \tag{19.114}$$

$$-k_B \, T \frac{q \, \tau_{ph}}{m_{hh}} \operatorname{grad} p_h - k_B \, T \frac{q \, \tau_{pl}}{m_{hl}} \operatorname{grad} p_l = -k_B \, T \, (\mu_{ph} + \mu_{pl}) \operatorname{grad} p, \tag{19.115}$$

where

$$\mu_{ph} = \frac{q \, \tau_{ph}}{m_{hh}} \frac{p_h}{p}, \qquad \mu_{pl} = \frac{q \, \tau_{pl}}{m_{hl}} \frac{p_l}{p}. \tag{19.116}$$

An approximate expression of μ_{ph}, μ_{ph} is obtained by replacing the concentrations with the corresponding equilibrium values $p_h = N_{Vh} \, \Phi_{1/2}(\xi_h)$, $p_l = N_{Vl} \, \Phi_{1/2}(\xi_h)$, with

$$N_{Vh} = 2 \, M_V \left(\frac{m_{hh}}{2 \, \pi \, \hbar^2} k_B \, T \right)^{3/2}, \qquad N_{Vl} = 2 \, M_V \left(\frac{m_{hl}}{2 \, \pi \, \hbar^2} k_B \, T \right)^{3/2} \tag{19.117}$$

(compare with (18.8)), whence, using $p = p_h + p_l$,

$$\mu_{ph} \simeq \frac{q \, \tau_{ph} \, m_{hh}^{1/2}}{m_{hh}^{3/2} + m_{hl}^{3/2}}, \qquad \mu_{pl} \simeq \frac{q \, \tau_{pl} \, m_{hl}^{1/2}}{m_{hh}^{3/2} + m_{hl}^{3/2}}. \tag{19.118}$$

Then, $\mathbf{J}_a$ is extracted from the first relation in (19.113), whose matrix form is

$$\begin{bmatrix} 1 & \mu_a \, B_3 & -\mu_a \, B_2 \\ -\mu_a \, B_3 & 1 & \mu_a \, B_1 \\ \mu_a \, B_2 & -\mu_a \, B_1 & 1 \end{bmatrix} \begin{bmatrix} J_{a1} \\ J_{a2} \\ J_{a3} \end{bmatrix} = \begin{bmatrix} J'_{a1} \\ J^{\ast}_{a2} \\ J'_{a3} \end{bmatrix}, \tag{19.119}$$

$\mu_a = q \, \tau_{pa} / m_a$. The determinant of the matrix in (19.119) is $D_M = 1 + \mu_a^2 \, B^2$. Still considering the case where $\mathbf{B}$ is weak, one finds

$$\mathbf{J}_a \simeq \mathbf{J}'_a - \mu_a \, \mathbf{B} \wedge \mathbf{J}'_a. \tag{19.120}$$

In turn, the contribution of the last term at the right hand side of the above yields

$$-a_p \, \mathbf{B} \wedge \left(q \, \mu_p \, p \, \mathbf{E} - q \, D_p \operatorname{grad} p \right), \qquad \mu_p = \mu_{ph} + \mu_{pl}, \tag{19.121}$$

with μ_p the *hole mobility*. In turn, the *hole diffusivity* (or *hole-diffusion coefficient*) and the dimensionless parameter a_p are given by

$$D_p = \frac{k_B \, T}{q} \mu_p, \qquad a_p = \frac{1}{\mu_p^2} \left(\frac{q \, \tau_{ph}}{m_{hh}} \mu_{ph} + \frac{q \, \tau_{pl}}{m_{hl}} \mu_{pl} \right). \tag{19.122}$$

Putting (19.114), (19.115), and (19.121) together finally provides the *drift-diffusion transport equation for the holes*,

$$\mathbf{J}_p = q\,\mu_p\,p\,\mathbf{E} - q\,D_p\,\mathrm{grad}\,p - q\,a_p\,\mu_p\,\mathbf{B}\wedge\left(\mu_p\,p\,\mathbf{E} - D_p\,\mathrm{grad}\,p\right). \quad (19.123)$$

Thus, the DD model for the holes of the valence band is given by (19.123) along with the balance equation for the holes' number, that reads

$$\frac{\partial p}{\partial t} + \frac{1}{q}\,\mathrm{div}(\mathbf{J}_p) = W_p. \quad (19.124)$$

19.5.4   Coupling with Maxwell's Equations

As anticipated in Sect. 19.5.1, as the electromagnetic field is influenced by the distribution of charge and current density, it is necessary to couple the equations describing the charge transport (in the form, e.g., of the hydrodynamic or drift-diffusion model) with the Maxwell equations. For this, one inserts the total charge density ρ and current density $\mathbf{J}$ into the right hand sides of (4.19); considering that there are different groups of charges and currents, one uses (4.22), where the charge density is given by (18.53), namely,[33]

$$\rho = q\,(p - n + N), \qquad N = N_D^+ - N_A^-. \quad (19.125)$$

In turn, the current density reads

$$\mathbf{J} = \mathbf{J}_p + \mathbf{J}_n = \rho_p\,\mathbf{v}_p + \rho_n\,\mathbf{v}_n = q\,p\,\mathbf{v}_p - q\,n\,\mathbf{v}_n, \quad (19.126)$$

with $\mathbf{J}_n$ and $\mathbf{J}_p$ given by (19.107) and (19.123), respectively. As noted in Sect. 18.5, the material's permittivity must be used here instead of vacuum's; as a consequence, the relation between electric displacement and field reads $\mathbf{D} = \varepsilon_{\mathrm{sc}}\,\mathbf{E}$.

One notes that the $\mathbf{E}$ and $\mathbf{B}$ fields are the sum of two contributions: the first one derives from the internal charge and current-density distribution as mentioned above, while the second one derives from external sources, e.g., voltage or current generators connected to the device or integrated circuits, or electric and magnetic fields present in the environment. In general, the internal contribution to $\mathbf{B}$ is negligible and is not considered in semiconductor devices or integrated circuit; it follows that $\mathbf{B}$ is to be accounted for in (19.107) and (19.123) only when it derives from external sources[34] and, due to this, it must be thought of as a prescribed function of $\mathbf{r}$ and t.

---

[33] Equation (18.53) is the definition of charge density in a semiconductor; as a consequence it holds in general, not only in the equilibrium condition considered in Sect. 18.5. In fact, it can readily be extended to account for charges trapped in energy states different from those of the dopants (Sect. 20.2.2).

[34] A typical example is found when a semiconductor device or circuit is used as a magnetic-field sensor or in specific measurement setups, like in the Hall-voltage measurement (Sect. 25.4).

With these premises, the analysis will continue here after letting $\mathbf{B} = 0$. Despite this simplification, the continuity and transport Eqs. (19.108), (19.107) and (19.123), (19.124) must be coupled with the whole set of Maxwell equations: in fact, the expression of the electric field in terms of the potentials is given by the second relation in (4.26), namely, $\mathbf{E} = -\mathrm{grad}\varphi - \partial\mathbf{A}/\partial t$; as a consequence, in a dynamic condition both the scalar and vector potential must be determined. In a steady-state or equilibrium condition, instead, the expression of the electric field reduces to $\mathbf{E} = -\mathrm{grad}\varphi$; in this case it is sufficient to couple the semiconductor model with the first equation in (4.19) only.

The presence of the vector potential $\mathbf{A}$ makes the model more complicate; thus, it is useful to ascertain whether, in the typical operating conditions, the time derivative $\partial\mathbf{A}/\partial t$ in the expression of $\mathbf{E}$ should be kept or not. As noted above, the derivative differs from zero only if the boundary conditions (e.g., the applied voltages) vary with time. To associate a characteristic time to a boundary condition one takes the period associated to the maximum frequency of the boundary condition's spectrum, $\tau_{\min} = 1/\nu_{\max}$; then, one compares $\tau_{\min}$ with the time Δt necessary for the electromagnetic perturbation produced by the boundary condition to propagate to a position internal to the semiconductor. If d is the distance between a point on the boundary and an internal point, the propagation time can be estimated to be $\Delta t = d/u_f$, with u_f the radiation's phase velocity corresponding to $\nu_{\max}$. If it happens that $\Delta t \ll \tau_{\min}$, the propagation is practically instantaneous, namely, the electromagnetic field at the internal point is consistent with the boundary condition existing at the same instant of time; as a consequence, the boundary condition is thought of as stationary, and the $\partial\mathbf{A}/\partial t$ derivative is neglected. This is called *quasi-static approximation*; the condition of its applicability is summarized as[35]

$$\Delta t = \frac{d}{u_f} \ll \tau_{\min} = \frac{1}{\nu_{\max}}, \qquad \nu_{\max} \ll \frac{u_f}{d}. \tag{19.127}$$

To estimate the condition one must fix the value of d; as a conservative choice one takes the channel length of the MOSFET transistors of the old generations, $d \approx 10^{-7}$ m. Using $u_f \approx 10^8$ m s^{-1} yields $\nu_{\max} \ll 10^{15}$ Hz, which is amply fulfilled in the present state-of-the-art integrated circuits. Note that the condition is even better verified in the last-generation devices, whose channel length is shorter than 10^{-7} m.

The choice of the channel length in the above estimate is dictated by the fact that the channel is the active region of the device. Choosing, instead, d as the (much larger) thickness of the silicon wafer would not make sense, because the phenomena taking place in the wafer's bulk are relatively unimportant. Other distances within an integrated circuit are larger by orders of magnitude than the value of d considered in the estimate; for instance, the diameter of the integrated circuit itself is of the order of 10^{-2} m, hence the quasi-static approximation is not applicable to the case of two devices placed, e.g., at opposite corners of a chip and connected by a line. In fact,

---

[35] A similar reasoning is used to treat the time derivative of v_i, w, and b_i in the derivation of the BTE's moments of order larger than zero (Sect. 19.4.2).

the propagation of signals along the lines connecting different devices on a chip is modeled using the whole set of Maxwell equations.[36]

19.5.5 Semiconductor-Device Model

Thanks to the quasi-static approximation, the equations describing the semiconductor, in the drift-diffusion case and with $\mathbf{B} = 0$, read

$$\mathrm{div}\mathbf{D} = q\,(p - n + N), \qquad \mathbf{D} = -\varepsilon_{\mathrm{sc}}\,\mathrm{grad}\varphi, \tag{19.128}$$

$$\frac{\partial n}{\partial t} - \frac{1}{q}\,\mathrm{div}\mathbf{J}_n = W_n, \qquad \mathbf{J}_n = q\,\mu_n\,n\,\mathbf{E} + q\,D_n\,\mathrm{grad}n, \tag{19.129}$$

$$\frac{\partial p}{\partial t} + \frac{1}{q}\,\mathrm{div}\mathbf{J}_p = W_p, \qquad \mathbf{J}_p = q\,\mu_p\,p\,\mathbf{E} - q\,D_p\,\mathrm{grad}p. \tag{19.130}$$

As outlined in Chap. 24, insulating layers play an essential role in the fabrication of integrated circuits; it is then necessary to extend the model to incorporate also the description of such layers. This is easily accomplished by observing that mobile charges are absent in an insulator, so that the balance equations for the number of particles and the transport equations reduce to identities, $0 = 0$. The model for the insulators then reduces to Poisson's equation only. The right hand side of the latter does not necessarily vanish because, as indicated in Sect. 24.1, contaminants may enter the insulator during the fabrication steps; some of these contaminants may ionize and act as fixed charges so that, letting N_{ox} be their density, the model for the insulator reads[37]

$$\mathrm{div}\mathbf{D} = q\,N_{\mathrm{ox}}, \qquad \mathbf{D} = -\varepsilon_{\mathrm{ox}}\,\mathrm{grad}\varphi, \qquad n = p = 0, \qquad \mathbf{J}_n = \mathbf{J}_p = 0. \tag{19.131}$$

The set of Eqs. (19.128–19.131) is commonly called *semiconductor-device model*. It is made of partial-differential equations of the first order, in the unknowns φ, n, p and $\mathbf{D}$, $\mathbf{J}_n$, $\mathbf{J}_p$. The coefficients are μ_n, μ_p, $D_n = (k_B\,T/q)\,\mu_n$, $D_p = (k_B\,T/q)\,\mu_p$. In turn, N, W_n, W_p are either known functions, or are expressed in terms of the unknowns themselves. Some of the equations are non linear because the unknowns are multiplied by each other. Each equation on the left in (19.128–19.131) contains the divergence of a vector; in turn, the expression of the vector is given by the corresponding equation on the right, in terms of the scalar unknowns (in fact it is $\mathbf{E} = -\mathrm{grad}\varphi$). It follows that, by introducing the expressions of $\mathbf{D}$, $\mathbf{J}_n$, and $\mathbf{J}_p$

---

[36] The progressive device scaling from one generation to the next is in general associated to an increase in the size of the chips. Due to this, the constraints on the circuit's speed are rather imposed by the lines connecting the devices than by the devices themselves.

[37] The insulator's permittivity is indicated with $\varepsilon_{\mathrm{ox}}$ because, in the examples shown later, silicon dioxide (SiO_2) is used as the reference insulator.

into the divergence operator, each pair of first-order equation is transformed into a single, second-order equation. This observation is useful in view of the application of numerical methods to the solution of the semiconductor-device model.

Remembering the derivation of the transport model, the terms W_n, W_p in (19.129), (19.130) are due to the generation-recombination phenomena. Specifically, W_n is the difference between the number of electrons entering the conduction band, and of those leaving it, per unit volume and time; in turn, W_p is the difference between the number of holes entering the valence band, and of those leaving it, per unit volume and time.[38] For this reason, they are also called *net generation rates*.

As mentioned in Sect. 19.3, the transitions of a given class are further grouped depending on the entity with which the particle's collision occurs. As far as the net generation rates are concerned, it is customary to separate the contribution of the phonon collisions from those of the other types (e.g., electron–electron collisions, electron–photon collisions, and so on); in fact, unless the device is kept at a very low temperature, the phonon collisions are the most important ones. Thus, the net generation rates are recast as

$$W_n = G_n - U_n, \qquad W_p = G_p - U_p, \qquad (19.132)$$

where U_n, U_p describe the transitions due to phonon collisions, while G_n, G_p describe those of the other types. The minus signs in (19.132) come from the fact that U_n is defined as the difference between the number of electrons leaving the conduction band, and of those entering it, because of phonon collisions, per unit volume and time; similarly, U_p is defined as the difference between the number of holes leaving the valence band, and of those entering it, because of phonon collisions, per unit volume and time. The terms used for U_n, U_p are *net thermal recombination rates*, those for G_n, G_p are *net non-thermal generation rates*.

Another comment about the semiconductor-device model concerns the drift terms in (19.129), (19.130). The latter can be recast as $\mathbf{J}_n^{dr} = \sigma_n \mathbf{E}$ and $\mathbf{J}_p^{dr} = \sigma_p \mathbf{E}$, where

$$\sigma_n = q \, \mu_n \, n, \qquad \sigma_p = q \, \mu_p \, p \qquad (19.133)$$

are the *electron conductivity* and *hole conductivity*, respectively. From $\mathbf{J} = \mathbf{J}_n + \mathbf{J}_p$, in a uniform material one obtains $\mathbf{J} = \sigma \mathbf{E}$, that is, *Ohm's law*, with

$$\sigma = \sigma_n + \sigma_p = q \, \left(\mu_n \, n + \mu_p \, p \right). \qquad (19.134)$$

19.5.6 Boundary Conditions

In practical applications, the equations of the semiconductor-device model, (19.128) through (19.131), are solved over a closed domain whose boundary[39] is indicated

---

[38] The units are $[W_n, W_p] = \text{m}^{-3}\,\text{s}^{-1}$.

[39] A two- or three-dimensional case is considered. In the one-dimensional case the boundary reduces to the two points enclosing the segments over which the equations are to be solved.

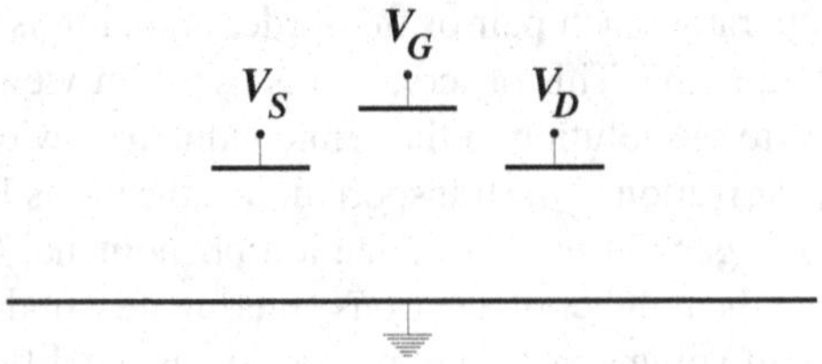

$$V_G$$
$$V_S \quad \perp \quad V_D$$

Fig. 19.3 MOS structure used to discuss the boundary conditions for the mathematical model of semiconductor devices. Only the conducting boundaries are shown. Note that the vertical scale of the drawing is not realistic

here with Γ. The boundary is partitioned into portions, some of which, indicated with $\Gamma_{i1}, \Gamma_{i2}, \ldots$, are *insulating boundaries*, namely, they can not be crossed by electrons or holes; the remaining portions, $\Gamma_{c1}, \Gamma_{c2}, \ldots$, can be crossed by the carriers and are termed *conducting boundaries*.

Considering that the domain over which the equations are solved is in general a part of a much larger domain enclosing an integrated circuit, some flexibility exists as for the choice of Γ. Thanks to this, it is possible to select $\Gamma_{i1}, \Gamma_{i2}, \ldots$ such that the normal component of the vector unknowns vanishes there; in other terms, letting **s** be the unit vector normal to an insulating boundary at some point **r**, it is

$$\mathbf{E} \cdot \mathbf{s} = 0, \qquad \frac{\partial \varphi}{\partial s} = 0, \qquad \mathbf{r} \in \Gamma_{i1}, \Gamma_{i2}, \ldots . \qquad (19.135)$$

where $\partial/\partial s$ indicates the derivative normal to the boundary at **r**. An example of how this can be accomplished is given in Figs. 19.3 and 19.4, representing the schematic cross-section of a MOSFET. In Fig. 19.3 only the conducting boundaries are shown, with V_S, V_G, and V_D indicating the voltage applied to the source, gate, and drain contact, respectively. The bulk contact is grounded, as shown by the line below; such a contact is the ground reference for the whole chip and, for this reason, extends laterally beyond the region occupied by the MOSFET under consideration. The insulating boundaries must now be selected in order to form a closed line that completes the boundary Γ; in principle, such a completion may be accomplished in different ways. Consider, however, the completion shown in the upper part of Fig. 19.4: as in general it is $V_S \neq V_G$, it is likely that one of the field lines in the region between the source and gate contacts coincides with the segment ab. As a consequence, choosing ab as the insulating boundary in that region guarantees that the component of **E** normal to such a boundary is zero, thus achieving the condition sought. The same reasoning applies to line cd. By this procedure one succeeds in prescribing the boundary conditions for the Poisson equation along the insulating boundaries. If, instead, the insulating boundaries were placed differently, like, e.g., in the lower part of Fig. 19.4, the component of **E** normal to the boundary would be different from zero; besides that, it would be impossible to determine it *a priori*, and Poisson's equation would become ill-posed.

Once the insulating boundaries are completed, the same condition as (19.135) is prescribed onto the current densities, namely, $\mathbf{J}_n \cdot \mathbf{s} = \mathbf{J}_p \cdot \mathbf{s} = 0$ whence, using the second equation in (19.129),

Fig. 19.4 The same structure as in Fig. 19.3, to which the insulating boundaries have been added (dash-dotted lines). The upper part of the figure shows the correct placement of the insulating boundaries, the lower part shows a wrong placement

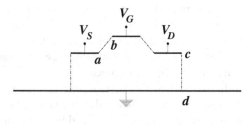

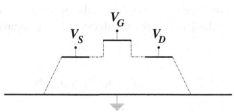

$$-q\,\mu_n\,n\frac{\partial\varphi}{\partial s}+q\,D_n\,\frac{\partial n}{\partial s}=0, \qquad \mathbf{r}\in\Gamma_{i1},\Gamma_{i2},\ldots. \qquad (19.136)$$

Combining the above with (19.135) yields $\partial n/\partial s = 0$; repeating the calculation for the holes finally shows that at the insulating boundaries the boundary condition is the same for all scalar unknowns φ, n, p:

$$\frac{\partial\varphi}{\partial s}=0, \qquad \frac{\partial n}{\partial s}=0, \qquad \frac{\partial p}{\partial s}=0, \qquad \mathbf{r}\in\Gamma_{i1},\Gamma_{i2},\ldots, \qquad (19.137)$$

i.e., a boundary condition of the *homogeneous Neumann type*.

The conducting boundaries $\Gamma_{c1}, \Gamma_{c2}, \ldots$ are typically covered with metal layers or heavily-doped polycrystalline layers that provide the electric contacts to the device. Unless the operating condition of the device departs strongly from equilibrium, a contact is able to supply the amount of charge necessary to keep the equilibrium and charge-neutrality conditions in the semiconductor layer adjacent to it. Thus at each point $\mathbf{r}$ of this layer one takes $\rho = 0$ or, more specifically,

$$\rho_c = q\,(p_c - n_c + N_c) = 0, \qquad \mathbf{r}\in\Gamma_{c1},\Gamma_{c2},\ldots, \qquad (19.138)$$

where index c indicates the conducting boundary. In most cases the metal or poly-crystalline layer is connected to a voltage generator, so that the electric potential is prescribed, or to another part of the integrated circuit, so that the electric potential can be determined from a separate calculation. In these cases, the electric potential of the contact is known. From it, one derives the electric potential φ_c of the conducting boundary adjacent to the contact; in fact, when the departure from the equilibrium condition is not too strong, the difference between the electric potential of the con-ducting boundary and that of the contact does not depend on the current density that crosses the boundary, and equals the contact's work function.[40] The latter is experi-mentally known, this yielding φ_c. In conclusion, at a conducting boundary where the

---

[40] Examples of application of this concept are given in Sects. 21.2.2 and 22.2.

voltage is prescribed, the boundary condition is the same for all scalar unknowns: $\varphi = \varphi_c, n = n_c, p = p_c, \mathbf{r} \in \Gamma_{c1}, \Gamma_{c2}, \ldots$, i.e., a boundary condition of the *Dirichlet type*. Note that the quantities in parenthesis in (19.138) depend on φ_c (compare with (18.56) and (18.57)); as a consequence, they can be calculated only after the electric potential has been determined.

In some instances a current generator is connected to a contact; as a consequence, the voltage is not prescribed at the corresponding conducting boundary. However, such a voltage can be determined by observing that the flux of the current density across the boundary equals the generator's current. This provides the extra relation that keeps the well-posedness of the mathematical problem.[41]

19.5.7 Quasi-Fermi Potentials

The drift-diffusion transport equations, given by the second relation in (19.129), (19.130), can be recast in a monomial form by defining two auxiliary functions

$$\varphi_n(\mathbf{r}, t) = \varphi - \frac{k_B T}{q} \log\left(\frac{n}{n_i}\right), \qquad \varphi_p(\mathbf{r}, t) = \varphi + \frac{k_B T}{q} \log\left(\frac{p}{n_i}\right), \quad (19.139)$$

whose inversion yields

$$n = n_i \exp\left[\frac{q\,(\varphi - \varphi_n)}{k_B T}\right], \qquad p = n_i \exp\left[\frac{q\,(\varphi_p - \varphi)}{k_B T}\right]. \quad (19.140)$$

In the equilibrium limit, (19.140) must coincide with (18.62), namely, $\varphi_n \to \varphi_F$, $\varphi_p \to \varphi_F$. It follows that the auxiliary functions (19.139) are a formal generalization of the concept of Fermi potential; they have the advantage of keeping the exponential form of the expressions of n and p in the non-equilibrium case. For this reason, φ_n and φ_p are called *quasi-Fermi potentials* for electrons and holes, respectively.[42] From (19.140) one finds $(k_B T/q)\,\mathrm{grad}\,n = n\,\mathrm{grad}(\varphi - \varphi_n)$ and $(k_B T/q)\,\mathrm{grad}\,p = p\,\mathrm{grad}(\varphi_p - \varphi)$ which, replaced into the second relation of (19.129), (19.130), respectively, yield

$$\mathbf{J}_n = -q\,\mu_n\,n\,\mathrm{grad}\varphi + q\,\frac{k_B T_L}{q}\,\mu_n\,\mathrm{grad}n = -q\,\mu_n\,n\,\mathrm{grad}\varphi_n, \quad (19.141)$$

$$\mathbf{J}_p = -q\,\mu_p\,p\,\mathrm{grad}\varphi - q\,\frac{k_B T}{q}\,\mu_p\,\mathrm{grad}p = -q\,\mu_p\,p\,\mathrm{grad}\varphi_p. \quad (19.142)$$

---

[41] This outcome becomes immediately clear by applying a numerical-discretization method to the problem. In fact, the component of the current density normal to the contact depends on the electric potential of the contact itself; thus, the extra relation provided by the flux-conservation equation embeds the extra unknown φ_c.

[42] By some authors, φ_n and φ_p are called *Imref potentials*, where "Imref" is "Fermi" read from right to left [103].

One notes that the monomial forms (19.141), (19.142) thus achieved are similar to drift-diffusion equations where the drift term only is present. This result allows one to interpret $-\mathrm{grad}\varphi_n$ and $-\mathrm{grad}\varphi_p$ as effective fields acting on the electrons (or, respectively, holes) and incorporating both drift and diffusion effects. The monomial form is useful for describing unipolar devices, where one of the current densities $\mathbf{J}_n$, $\mathbf{J}_p$ dominates over the other and is essentially solenoidal (Sect. 22.6).

The definition of the quasi-Fermi potential given above is applicable only to drift-diffusion equations of the form (19.129), (19.130), that are valid within the approximations of parabolic bands and non-degenerate conditions. However, the concept of quasi-Fermi potential can be generalized by disposing, e.g., of the non-degeneracy hypothesis. For this, one starts from the equilibrium expressions of n and p, given by (18.56) and (18.57), respectively, and replaces E_F in the definitions of ξ_e, ξ_h with $E_F = E_{Fi} - q\,\varphi_F$ (compare with (18.26)); then, the Fermi potential φ_F is replaced with φ_n in the definition of ξ_e, and with φ_p in that of ξ_h. The non-equilibrium concentrations then read

$$n = N_C\,\Phi_{1/2}\,(\xi_e)\,, \qquad \xi_e = -\frac{E_C - E_{Fi}}{k_B\,T} + \frac{q\,(\varphi - \varphi_n)}{k_B\,T}, \tag{19.143}$$

and

$$p = N_V\,\Phi_{1/2}\,(\xi_h)\,, \qquad \xi_h = -\frac{E_{Fi} - E_V}{k_B\,T} + \frac{q\,(\varphi_p - \varphi)}{k_B\,T}. \tag{19.144}$$

19.5.8 Poisson Equation in a Semiconductor

In the equilibrium condition the concentrations of electrons (19.143) and holes (19.144) depend on the electric potential only. In turn, the ionized donor and acceptor concentrations depend on the electric potential and, possibly, on position if the dopant distributions are position dependent. In summary, the general form of the equilibrium charge concentration is $\rho = \rho(\varphi, \mathbf{r})$. The semiconductor-device model reduces to the Poisson equation alone because at equilibrium it is $\partial/\partial t = 0$, $\mathbf{J}_n = \mathbf{J}_p = 0$; the equation reads $-\varepsilon_{\mathrm{sc}}\,\nabla^2\varphi = q\,(p - n + N) = \varrho(\varphi, \mathbf{r})$, which is a semi-linear partial differential equation (PDE).[43] If the explicit dependence on $\mathbf{r}$ is absent, $\rho = \rho(\varphi)$, and the problem is one dimensional, say, in the x direction, Poisson's equation can be solved analytically over a domain I where $\mathrm{d}\varphi/\mathrm{d}x \neq 0$. In

---

[43] A PDE of order s in the unknown φ is called *quasi-linear* if it is linear in the order-s derivatives of φ and its coefficients depend on the independent variables and the derivatives of φ of order $m < s$. A quasi-linear PDE where the coefficients of the order-s derivatives are functions of the independent variables alone, is called *semi-linear*. A PDE which is linear in the unknown function and all its derivatives, with coefficients depending on the independent variables alone, is called *linear*. PDE's not belonging to the classes above are *fully non-linear*.

fact, multiplying by $d\varphi/dx$ both sides of $-\varepsilon_{sc}\,d^2\varphi/dx^2 = \rho(\varphi)$, one obtains

$$-\varepsilon_{sc}\frac{d^2\varphi}{dx^2}\frac{d\varphi}{dx} = \rho(\varphi)\frac{d\varphi}{dx}, \qquad \frac{d}{dx}\left(\frac{d\varphi}{dx}\right)^2 = \frac{d}{dx}S(\varphi), \qquad (19.145)$$

where $dS/d\varphi = -2\,\rho(\varphi)/\varepsilon_{sc}$. Integrating (19.145) from x_0 to x, where both points belong to I, and letting $\varphi_0 = \varphi(x_0)$, yields

$$\left(\frac{d\varphi}{dx}\right)^2 = G^2(\varphi), \qquad G(\varphi) = \left[\left(\frac{d\varphi}{dx}\right)_0^2 + S(\varphi) - S(\varphi_0)\right]^{1/2}. \qquad (19.146)$$

Separating the variables in (19.146) then provides

$$dx = \pm\frac{d\varphi}{G(\varphi)}, \qquad x = x_0 \pm \int_{\varphi_0}^{\varphi}\frac{d\varphi}{G(\varphi)}, \qquad (19.147)$$

namely, the inverse relation $x = x(\varphi)$. The choice of the sign is made on case-by-case basis (an example is given in Sect. 21.2). The approach can be extended to the non-equilibrium case if it happens that the electric potential depends on one independent variable, say, x, while the Poisson equation contains terms that depend also on independent variables different from x. In fact, the other variables can be considered as parameters during the integration with respect to x. An example of this case is given in Sect. 22.7.1.

19.6 Complements

19.6.1 Comments on the Equivalent Hamiltonian Operator

It has been observed in Sect. 19.2.2 that, in the description of the wave-packet dynamics based on the equivalent Hamiltonian operator and crystal momentum, that the time variations of the latter are due to the external force only; as a consequence, if $U = \text{const}$ one has $\hbar\,\dot{\mathbf{k}}_0 = 0$, namely, the crystal momentum is a constant of motion. The periodic part of the potential energy is incorporated with the kinetic part, to form an equivalent operator. Thus, the expression of the Hamiltonian operator is eventually made of two terms, the equivalent-kinetic part, where the space coordinates do not explicitly appear, and the potential part due to the external force only; from this standpoint it is similar to that of a particle *in vacuo*.

   The same results were found in the analysis of the motion of a classical particle subjected to a periodic potential onto which a weak perturbation is superimposed (Sect. 3.11). The classical investigation makes it clear that the crystal momentum is in fact different from the actual momentum of the particle; also, the subsequent elaboration carried out in Sect. 3.12 shows that the concept of effective mass is not distinctive of Quantum Mechanics.

19.6.2 *Special Cases of Anisotropy*

It is interesting to note that for an anisotropic branch the acceleration $\dot{\mathbf{u}}$ may still be parallel to $\mathbf{F}$; this happens when the force is parallel to one of the coordinate axes. Taking by way of example a force parallel to the first coordinate axis, $\mathbf{F} = F\,\mathbf{i}_1$, it follows in fact

$$\dot{u}_1 = F/m_{1a}, \qquad \dot{u}_2 = 0, \qquad \dot{u}_3 = 0. \qquad (19.148)$$

This seems to imply that a suitable choice of the reference is sufficient to make the acceleration parallel to the force. However, this is not so: as noted in Sect. 17.6.2, for (19.24) to hold, the reference in the $\mathbf{k}$ space must be chosen in such a way as to make the Hessian matrix of $E_n(\mathbf{k})$ diagonal; this, in turn, fixes the reference in the $\mathbf{r}$ space, because the two references are reciprocal to each other. As a consequence, if one rotates the $\mathbf{r}$ reference to align one of its axes with the force, the $\mathbf{k}$ reference rotates as well; as the mass tensor in the new reference is not necessarily diagonal, the anisotropy present in the old reference is maintained.

19.6.3 *α-Moment at Equilibrium*

The derivation of the moment with respect to α of the BTE has been shown in Sect. 19.4.1, leading to (19.62). In the equilibrium condition the distribution function f^{eq} is independent of t and depends on $\mathbf{k}$ through the Hamiltonian function H only; since the latter is even with respect to $\mathbf{k}$, the distribution function is even as well. In turn, as the transitions balance each other, the right hand side of (19.62) vanishes. The term with the magnetic induction vanishes as well (compare with (19.60)); in conclusion, in the equilibrium condition (19.62) reduces to

$$\mathrm{div}_{\mathbf{r}}\,(n\,\overline{\alpha\,\mathbf{u}})^{\text{eq}} + \frac{q}{\hbar}\,\left(n\,\overline{\mathrm{grad}_{\mathbf{k}}\alpha \cdot \mathbf{E}}\right)^{\text{eq}} = 0. \qquad (19.149)$$

It is easily found that (19.149) yields the identity $0 = 0$ if α is even with respect to $\mathbf{k}$. This, on the contrary, is not true when α is odd; in this case the equilibrium condition consists in the balance between the diffusion term, due to the spatial non-uniformity of $(n\,\overline{\alpha\,\mathbf{u}})^{\text{eq}}$, and the second term, proportional to the carrier concentration and linearly dependent on the electric field.

19.6.4 *Closure Conditions*

The closure conditions for the drift-diffusion and hydrodynamic model are derived in this section. The former consists in calculating (19.75) using the equilibrium distribution $f^{\text{eq}} = Q\,P$, with $Q = 1/(4\,\pi^3)$ the density of states in the $\mathbf{r}, \mathbf{k}$ space and P the Fermi–Dirac statistics. In the equilibrium case it is $\mathbf{v} = 0$, whence $\mathbf{c} = \mathbf{u}$, and

$$n^{\text{eq}} k_B T_{ij}^{\text{eq}} = \frac{m_{ia}}{4 \pi^3} \iiint_{-\infty}^{+\infty} \frac{u_i \, u_j}{\exp\left[(E_e + \zeta_e)/(k_B T)\right] + 1} \, d^3k, \tag{19.150}$$

with $E_e = E - E_C$ and $\zeta_e = E_C - q \varphi - E_F$ (compare with (18.54)). In the above, E_e is even with respect to all components of $\mathbf{k}$. For $j \neq i$ the integrand is odd with respect to k_i because $u_i = (1/\hbar) \, \partial E_e / \partial k_i$, and with respect to k_j because $u_j = (1/\hbar) \, \partial E_e / \partial k_j$; as a consequence it is $T_{ij} = 0$ for $j \neq i$, while

$$n^{\text{eq}} k_B T_{ii}^{\text{eq}} = \frac{m_{ia}}{4 \pi^3} \iiint_{-\infty}^{+\infty} \frac{u_i^2}{\exp\left[(E_e + \zeta_e)/(k_B T)\right] + 1} \, d^3k, \tag{19.151}$$

namely, in the equilibrium condition the electron-temperature tensor is diagonal. For this result to hold, the parabolic-band approximation is not necessary.

In the parabolic-band approximation, (19.24) and (19.72) hold, and the temperature tensor in equilibrium is evaluated at the ath minimum of the conduction band by letting

$$\eta_i = \frac{\hbar \, \delta k_i}{\sqrt{2 \, m_{ia}}}, \qquad d^3k = 2 \frac{\sqrt{2}}{\hbar^3} m_{ea}^{3/2} d^3\eta, \qquad m_{ia} u_i^2 = 2 \eta_i^2, \tag{19.152}$$

where the first relation is the Herring-Vogt transformation (17.66) and m_{ea} is defined in (17.68). Adding up over the minima and using (17.72) yields

$$n^{\text{eq}} k_B T_{ii}^{\text{eq}} = \frac{\sqrt{2}}{\pi^3 \hbar^3} \iiint_{-\infty}^{+\infty} \frac{M_C \, m_e^{3/2} \, \eta_i^2}{\exp\left[(\eta^2 + \zeta_e)/(k_B T)\right] + 1} \, d^3\eta. \tag{19.153}$$

As the value of the integral in (19.153) does not depend on index i, it follows that the diagonal entries T_{ii}^{eq} are equal to each other; as a consequence, the common value of the three integrals can be replaced with $T_e^{\text{eq}} = (T_{11}^{\text{eq}} + T_{22}^{\text{eq}} + T_{33}^{\text{eq}})/3$, namely,

$$n^{\text{eq}} k_B T_{ii}^{\text{eq}} = \frac{\sqrt{2}}{3 \pi^3 \hbar^3} \iiint_{-\infty}^{+\infty} \frac{M_C \, m_e^{3/2} \, \eta^2}{\exp\left[(\eta^2 + \zeta_e)/(k_B T_L)\right] + 1} \, d^3\eta. \tag{19.154}$$

Turning to spherical coordinates (B.1) yields $d^3\eta = \eta^2 \, d\eta \, \sin\theta \, d\theta \, d\phi$, with $\eta^2 = \eta_1^2 + \eta_2^2 + \eta_3^2 = E_e$ and $d\eta = 1/(2\sqrt{E_e}) \, dE_e$, $\eta^2 \, d\eta = \sqrt{E_e} \, dE_e/2$. The integral over the angles equals 4π whence, using (C.104) with $\Gamma(1 + 3/2) = (3/2) \sqrt{\pi}/2$,

$$n^{\text{eq}} k_B T_{ii}^{\text{eq}} = k_B T \, N_C \, \Phi_{3/2}(\xi_e), \tag{19.155}$$

with $N_C = 2 M_C \left[m_e k_B T / (2\pi \hbar^2) \right]^{3/2}$ the effective density of states (18.4), and $\xi_e = -\zeta_e/(k_B T)$. On the other hand, from (18.17) it is $n^{\text{eq}} = N_C \, \Phi_{1/2}(\xi_e)$, whence

$$T_e^{\text{eq}}(\mathbf{r}) = T \, \frac{\Phi_{3/2}(\xi_e)}{\Phi_{1/2}(\xi_e)}. \tag{19.156}$$

The dependence on position is due to $\xi_e = (q\,\varphi - E_C + E_F)/(k_B\,T)$. However, in the non-degenerate case the approximation (C.105) holds, $\Phi_\alpha(\xi_e) = \exp(\xi_e)$, and the electron-temperature tensor at equilibrium reduces to $T_e^{\mathrm{eq}} = T$.

Coming now to the hydrodynamic model, and remembering (19.83), the closure condition is found by calculating the equilibrium value of $n\,m_{ia}\,E_e\,u_i\,\mathbf{c}$, that reads

$$n^{\mathrm{eq}}\,m_{ia}\,\overline{\left(E_e\,u_i\,\mathbf{c}\right)}^{\mathrm{eq}} = \frac{m_{ia}}{4\,\pi^3} \iiint_{-\infty}^{+\infty} E_e\,u_i\,\mathbf{c}\,P\,\mathrm{d}^3k, \tag{19.157}$$

with P the Fermi–Dirac statistics. The procedure is similar to that of the drift-diffusion model. At equilibrium one can replace $\mathbf{c}$ with $\mathbf{u}$; then, out of the three components of $\mathbf{u}$, only that of index i contributes to the integral, because those of index $j \neq i$ make the integrand odd with respect to both k_i and k_j. In other terms, the integrand in (19.157) is replaced with $E_e\,u_i^2\,P\,\mathbf{i}_i$, which differs from the integrand of (19.151) because of factor $E_e\mathbf{i}_i$; this shows that the tensor defined by (19.157) is diagonal as well.[44] When the transformation (19.152) is used, the right hand side of (19.157) becomes similar to (19.153), the only difference being the additional factor $\eta^2\,i$ that derives from $E_e\mathbf{i}_i$. The next step, replacing the common value of the three integrals with one third of their sum, yields an expression similar to (19.154), the only difference being that η^2 is replaced with $\eta^4\,\mathbf{i}$. Transforming into spherical coordinates and inserting (C.104) with $\Gamma(1 + 5/2) = (5/2)\,\Gamma(1 + 3/2) = (15/4)\,\sqrt{\pi}/2$ finally yields

$$n^{\mathrm{eq}}\,m_{ia}\,\overline{\left(E_e\,u_i\,\mathbf{c}\right)}^{\mathrm{eq}} = \mathbf{i}_i\,\frac{5}{2}\,(k_B\,T)^2\,N_C\,\Phi_{5/2}(\xi_e). \tag{19.158}$$

It follows

$$m_{ia}\,\overline{\left(E_e\,u_i\,\mathbf{c}\right)}^{\mathrm{eq}} = \mathbf{i}_i\,\frac{5}{2}\,(k_B\,T)^2\,\frac{\Phi_{5/2}(\xi_e)}{\Phi_{1/2}(\xi_e)} \simeq \mathbf{i}_i\,\frac{5}{2}\,(k_B\,T)^2, \tag{19.159}$$

where the approximation holds in the non-degenerate case.

19.6.5 Matthiessen's Rule

The effects of the inter-band and intra-band transitions have been separated in Sect. 19.3.2 under the assumption that the two types are uncorrelated. The separation may further be pursued within each class, depending on the entity with which the collision occurs. Here the collisions leading to the intra-band transitions only are considered. With reference to (19.45), and assuming that the intra-band transitions are uncorrelated, one lets $S_{0v} = S_{0v}^{(1)} + S_{0v}^{(2)} + \ldots$, whence

$$\frac{1}{\tau_v} = \frac{1}{\tau_{v1}} + \frac{1}{\tau_{v2}} + \ldots, \qquad \frac{1}{\tau_{vj}} = \iiint_{-\infty}^{+\infty} S_{0v}^{(j)}(\mathbf{r}, \mathbf{k} \to \mathbf{k}')\,\mathrm{d}^3k'. \tag{19.160}$$

---

[44] As above, for this result to hold the parabolic-band approximation is not necessary.

This way of combining the relaxation times, also called *Matthiessen's rule*, still holds in the definitions of the macroscopic relaxation times τ_p, τ_w and τ_b (Sect. 19.4.2), and in the definitions (19.104), (19.118) of electron and hole mobilities through τ_p.

19.6.6   Order of Magnitude of Mobility and Conductivity

As shown in Sect. 19.5.2, the electron and hole mobilities are expressed by relations of the form $\mu = q\,\tau_p/m^*$, where τ_p is the momentum-relaxation time and m^* an effective mass. To the purpose of estimating the order of magnitude it is not necessary to distinguish between electrons and holes. Considering the $T = 300$ K case and taking $\tau_p \approx 0.25 \times 10^{-12}$ s, $m^* \approx 0.4 \times 10^{-30}$ kg yields[45]

$$\mu = q\,\frac{\tau_p}{m^\star} \approx 1.60 \times 10^{-19}\ \mathrm{C} \times \frac{0.25 \times 10^{-12}\ \mathrm{s}}{0.4 \times 10^{-30}\ \mathrm{kg}} = 10^3\ \frac{\mathrm{cm}^2}{\mathrm{V\,s}}. \qquad (19.161)$$

The diffusion coefficient at $T = 300$ K is estimated from the Einstein relation (19.102), $D = (k_B\,T/q)\,\mu$, where

$$\frac{k_B\,T}{q} \approx \frac{1.38 \times 10^{-23}\ \mathrm{(J/K)} \times 300\ \mathrm{K}}{1.60 \times 10^{-19}\ \mathrm{C}} = 26 \times 10^{-3}\quad \mathrm{V}. \qquad (19.162)$$

One finds

$$D \approx 26 \times 10^{-3}\ \mathrm{V} \times 10^3\ \frac{\mathrm{cm}^2}{\mathrm{V\,s}} = 26\ \frac{\mathrm{cm}^2}{\mathrm{s}}. \qquad (19.163)$$

To estimate the conductivity one takes by way of example the expression for electrons from (19.133), $\sigma_n = q\,\mu_n\,n$, where, due to the estimates above, it is $q\,\mu_n \approx 1.60 \times 10^{-19}\ \mathrm{C} \times 10^3\ \mathrm{cm}^2/\mathrm{(V\,s)} = 1.60 \times 10^{-16}\ \mathrm{cm}^2/\Omega$. For silicon at $T = 300$ K it is $n = n_i \approx 10^{10}$ cm^{-3} (Table 18.2);[46] in comparison, when silicon is doped with a uniform donor concentration equal to, say, $N_D' = 10^{16}$ cm^{-3}, it is $n \simeq N_D' = 10^{16}$ cm^{-3} (compare with (18.30)). In conclusion,

$$\sigma_n[\text{intrinsic}] \approx 10^{-6}\ (\Omega\,\mathrm{cm})^{-1}, \qquad \sigma_n[N_D'] \approx 1\ (\Omega\,\mathrm{cm})^{-1}. \qquad (19.164)$$

As a further comparison, the estimates for an insulator and, respectively, a conductor, are made with $n \simeq 10^4$ cm^{-3} and $n \simeq 10^{22}$ cm^{-3}, to find

$$\sigma_n[\text{insulator}] \approx 10^{-12}\ (\Omega\,\mathrm{cm})^{-1}, \qquad \sigma_n[\text{conductor}] \approx 10^6\ (\Omega\,\mathrm{cm})^{-1}. \quad (19.165)$$

---

[45] Mobility is traditionally expressed in cm^2/(V s) instead of m^2/(V s).

[46] The equilibrium concentrations are used in the estimates.

19.6.7 A Resumé of the Transport Model's Derivation

The number of steps that lead to the hydrodynamic or drift-diffusion model for semiconductors is quite large; thus, a brief summary is of use. The starting point is the single-particle Schrödinger equation for an electron in a crystal. To reach this stage of the theory a considerable amount of work has already been spent, necessary to reduce the many-particle problem to a tractable form (Chap. 16). When the external forces are absent, the single-particle equation is recast in a form based on the equivalent Hamiltonian operator, that exploits the periodicity of the lattice (this implies in turn that the nuclei are kept fixed in the equilibrium positions). Finally, the external forces are added, assuming that the external potential energy is a small perturbation; thanks to this hypothesis, it is possible to describe the collisionless motion of a single electron by means of a Hamiltonian function whose canonical variables are the expectation values of the wave packet's position and momentum (Sects. 19.2.1 and 19.2.2).

Basing on the Hamiltonian function thus found, the analysis shifts from the description of the single-particle to the statistical treatment of a system made of a large number of electrons, following the same pattern as in the classical case; this leads to the semiclassical BTE (Sect. 19.3), for which the general form of the collision term is worked out (Sect. 19.3.1). The latter is simplified, first, by considering point-like collisions, then by taking the perturbative form of the collision operator (Sects. 19.3.2 and 19.3.3).

The perturbative form of the BTE is treated with the moments method, that provides a hierarchical set of models, e.g., the drift-diffusion and the hydrodynamic model. Important approximations at this stage are, for all moments of order larger than zero, the neglect of the inter-band transitions, of the time derivatives, and of the convective terms. The models reach the final form thanks to the hypothesis of parabolic bands and the approximation of the relaxation-time tensors with scalar quantities.

Problems

19.1 In the expressions (19.115), (19.118) defining the hole mobility μ_p, assume that $\tau_{ph} \simeq \tau_{pl}$. Letting τ_p be the common value, determine the value of the normalized effective mass $\overline{m}_h/m_0$ to be used in $\mu_p = q\,\tau_p/\overline{m}_h$ for silicon at room temperature. Also, determine the value of parameter a_p in (19.122) in the same conditions.

Chapter 20
Generation-Recombination and Mobility

20.1 Introduction

The chapter illustrates the main contributions to the transitions of the inter-band type, that give rise to the generation-recombination terms in the continuity equations for electrons and holes, and to those of the intra-band type, that give rise to the electron and hole mobilities in the current-density equations. The inter-band transitions that are considered are the net thermal recombinations (of the direct and trap-assisted type), Auger recombinations, impact-ionization generations, and net-optical recombinations. The model for each type of event is first given as a closed-form function of the semiconductor-device model's unknowns, like carrier concentrations, electric field, or current densities. Such functions contain a number of coefficients, whose derivation is successively worked out in the complements by means of a microscopic analysis. Some discussion is devoted to the optical-generation and recombination events, to show how the concepts of semiconductor laser, solar cell, and optical sensor may be derived as particular cases of non-equilibrium interactions between the material and an electromagnetic field. The intra-band transitions are treated in a similar manner: two examples, the collisions with acoustic phonons and ionized impurities, are worked out in some detail; the illustration then follows of how the contributions from different scattering mechanisms are combined together in the macroscopic mobility models. The material is supplemented with a brief discussion about advanced modeling methods.

20.2 Net Thermal Recombinations

As anticipated in Sect. 19.5.5, it is customary to separate the net generation rates W_n, W_p into two contributions, namely, those deriving from the phonon collisions and those of the other types (e.g., electron-electron collisions, electron-photon collisions, and so on). The separate contributions are defined in (19.132); this section deals with the net thermal recombination rates U_n, U_p.

© Springer Science+Business Media New York 2015
M. Rudan, *Physics of Semiconductor Devices*,
DOI 10.1007/978-1-4939-1151-6_20

Fig. 20.1 A graphic example of direct thermal recombination (*a*) and generation (*b*). The edges of the conduction and valence bands are indicated with the same symbols used in Sect. 18.2. The same drawing applies also to the description of the direct optical recombinations and generations (Sect. 20.4)

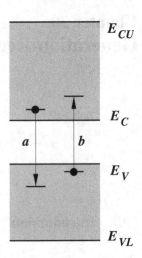

In the calculations carried out below, the non-equilibrium carrier concentrations are derived by integrating over the bands' energy. This is consistent with the general definitions (19.31) and (19.109). In fact, considering the non-equilibrium electron concentration n as defined in (19.31), one introduces the variable transformation illustrated in Sect. B.5 and replaces the quantities appearing in it as follows:

$$(u, v, w) \leftarrow (k_1, k_2, k_3), \qquad \sigma \leftarrow (\mathbf{r}, t), \qquad \eta \leftarrow E, \qquad (20.1)$$

$$S \leftarrow n, \qquad s \leftarrow f = Q \, \Phi, \qquad b \leftarrow g, \qquad \bar{s} \leftarrow P, \qquad (20.2)$$

where Q, $g(E)$ are the densities of states in the phase space $\mathbf{r}, \mathbf{k}$ and, respectively, in energy, while $\Phi(\mathbf{r}, \mathbf{k}, t)$, $P(\mathbf{r}, E, t)$ are the non-equilibrium occupation probabilities in the phase space and, respectively, in energy; the integration in energy is carried out over the range corresponding to the conduction band's branch. The hole concentration is treated in the same manner. In conclusion,

$$n(\mathbf{r}, t) = \iiint_{-\infty}^{+\infty} Q \, \Phi \, \mathrm{d}^3 k = \int_{E_C}^{E_{CU}} g \, P \, \mathrm{d}E, \qquad (20.3)$$

$$p(\mathbf{r}, t) = \iiint_{-\infty}^{+\infty} Q \, (1 - \Phi) \, \mathrm{d}^3 k = \int_{E_{VL}}^{E_V} g \, (1 - P) \, \mathrm{d}E. \qquad (20.4)$$

20.2.1   Direct Thermal Recombinations

To begin, a graphic example of thermal transitions is shown in Fig. 20.1, where the edges of the conduction and valence bands are indicated with the same symbols used in Sect. 18.2; the transition marked with a is a recombination event, in which an electron belonging to an energy state of the conduction band transfers to an empty state of the valence band. The energy difference between the initial and final state is released to the lattice in the form of a phonon. The opposite transition, where the electron's energy increases due to phonon absorption, is an electron-hole generation and is marked with b in the figure. The transitions of type a and b are called *direct thermal recombination* and *direct thermal generation*, respectively. Let r_a be the number of direct thermal recombinations per unit volume and time, and r_b the analogue for the generations; considering the conduction band as a reference, the difference $r_a - r_b$ provides the contribution to the net thermal recombination rate U_n due to the direct thermal transitions. When the valence band is considered instead, the rates of electron transitions reverse; however, for the valence band the transitions of holes must be considered: as consequence, the contribution to U_p is again $r_a - r_p$. In conclusion,

$$U_{DT} = U_{DTn} = U_{DTp} = r_a - r_b, \tag{20.5}$$

where D stands for "direct" and T for "thermal". The expressions of r_a, r_b are determined by a reasoning similar to that used in Sect. 19.3.1 to express the collision term of the BTE; here, however, the analysis is carried out directly in the energy space instead of the $\mathbf{k}$ space.[1] Let $P(\mathbf{r}, E, t)$ be the occupation probability of a state at energy E; then, let C be the probability per unit time and volume (in $\mathbf{r}$) of an electron transition from a state of energy E to a state of energy E' belonging to a different band, induced by the interaction with a phonon.[2] Such a probability depends on the phonon energy $\hbar \omega$ (Sect. 12.5), and also on the position in $\mathbf{r}$ if the semiconductor is non uniform. Typically, the equilibrium distribution is assumed for the phonons, which makes C independent of time; as the collisions are point-like (Sect. 19.3.2), the spatial positions of the initial and final states coincide, whence $C = C(\mathbf{r}, \hbar \omega, E \to E')$.

Indicating with $g(E)$ the density of states of the band where the initial state belongs, the product $g \, dE \, P$ is the number of electrons within the elementary interval dE around the initial state; such a product is multiplied by C to find the number of unconditional $E \to E'$ transitions. On the other hand, the transitions take place only if the final states around E' are empty; as the fraction of empty states in that interval is $g' \, dE' \, (1 - P')$, the number of actual transitions from dE to dE' turns out to be $g \, dE \, P \, C \, g' \, dE' \, (1 - P')$. Now, to calculate the r_a or r_b rate it is necessary to add up all transitions: for r_a one lets E range over the conduction band and E' over the valence band; the converse is done to calculate r_b. As the calculation of the latter is

---

[1] A more detailed example of calculations is given below, with reference to collisions with ionized impurities.

[2] The units of C are $[C] = m^{-3} s^{-1}$. Remembering that the phonon energy equals the change in energy of the electron due to the transition (Sect. 14.8.2), it is $C = 0$ for $\hbar \omega < E_C - E_V = E_G$ (refer also to Fig. 20.1).

somewhat easier, it is shown first:

$$r_b = \int_{E_{VL}}^{E_V} g \, dE \, P \int_{E_C}^{E_{CU}} C \, g' \, dE' \left(1 - P'\right). \tag{20.6}$$

As in normal operating conditions the majority of the valence-band states are filled, while the majority of the conduction-band states are empty, one lets $P \simeq 1$ and $1 - P' \simeq 1$, whence, using symbol G_{DT} for r_b,

$$G_{DT}(\mathbf{r}, \hbar \omega) = \int_{E_{VL}}^{E_V} g \, dE \int_{E_C}^{E_{CU}} C \, g' \, dE'. \tag{20.7}$$

Thus, the generation rate is independent of the carrier concentrations. To proceed, one uses the relation $g = \Omega \gamma$, with γ the combined density of states in energy and volume, given by (15.65), and the definition (20.4) of the hole concentration. Thus, the recombination rate is found to be

$$r_a = \int_{E_C}^{E_{CU}} g \, dE \, P \int_{E_{VL}}^{E_V} C \, g' \, dE' \left(1 - P'\right) = p \int_{E_C}^{E_{CU}} K \, g \, P \, dE, \tag{20.8}$$

where $K(\mathbf{r}, \hbar \omega, E)$, whose units are $[K] = \text{s}^{-1}$, is the average of ΩC over the valence band, weighed by $g' (1 - P')$:

$$K = \frac{\int_{E_{VL}}^{E_V} \Omega C g' (1 - P') \, dE'}{\int_{E_{VL}}^{E_V} g' (1 - P') \, dE'}. \tag{20.9}$$

Strictly speaking, K is a functional of P'; however, the presence of P' in both numerator and denominator of (20.9) makes such a dependence smoother, so that one can approximate K using the equilibrium distribution instead of P'. By the same token one uses the definition of the electron concentration (20.3) to find

$$r_a = \alpha_{DT} \, n \, p, \qquad \alpha_{DT}(\mathbf{r}, \hbar \omega) = \frac{\int_{E_C}^{E_{CU}} \Omega K g P \, dE}{\int_{E_C}^{E_{CU}} g P \, dE}, \tag{20.10}$$

where the integrals are approximated using the equilibrium probability. In conclusion,

$$U_{DT} = \alpha_{DT} \, n \, p - G_{DT}, \tag{20.11}$$

where α_{DT} is the *transition coefficient* of the direct thermal transitions, with units $[\alpha_{DT}] = \text{m}^3 \, \text{s}^{-1}$, and G_{DT} their *generation rate* ($[G_{DT}] = \text{m}^{-3} \, \text{s}^{-1}$). As in the equilibrium case it is $r_a = r_b$, namely, $G_{DT} = \alpha_{DT} \, n^{\text{eq}} \, p^{\text{eq}}$, it follows $U_{DT} = \alpha_{DT} \, (np - n^{\text{eq}} \, p^{\text{eq}})$.

Fig. 20.2 Different types of
trap-assisted transitions

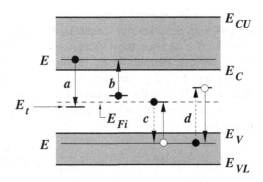

20.2.2   Trap-Assisted Thermal Recombinations

An important contribution to the thermal generation and recombination phenomena
is due to the so-called *trap-assisted transitions*. As mentioned in Sect. 19.3, among
the possible collisions undergone by electrons or holes are those with *lattice defects*.
The latter may originate from lattice irregularities (e.g., dislocations of the material's atoms occurring during the fabrication process, Sect. 24.1), or from impurities
that were not eliminated during the semiconductor's purification process, or were
inadvertently added during a fabrication step. Some defects may introduce energy
states localized in the gap; such states, called *traps*, may capture an electron from
the conduction band and release it towards the valence band, or vice versa. The
phenomena are illustrated in Fig. 20.2, where four traps located in the energy gap
are shown in order to distinguish among the different transition events, that are: *a*)
capture of a conduction-band electron by a trap, *b*) release of a trapped electron towards the conduction band, *c*) release of a trapped electron towards the valence band
(more suitably described as the capture of a valence-band hole by the trap), and *d*)
capture of a valence-band electron from the valence band (more suitably described
as the release of a hole towards the valence band). Each transition is accompanied
by the absorption or emission of a phonon. Thus, transitions of type *a* and *b* contribute to the net thermal recombination U_n of the conduction band, while those of
type *c* and *d* contribute to the net thermal recombination U_p of the valence band.
Also, a sequence of two transitions, one of type *a* involving a given trap, followed
by one of type *c* involving the same trap, produces an electron-hole recombination
and is therefore called *trap-assisted thermal recombination*; similarly, a sequence
of two transitions, one of type *d* involving a given trap, followed by one of type *b*
involving the same trap, produces an electron-hole generation and is therefore called
trap-assisted thermal generation.

   To calculate the contribution of the trap-assisted transitions to U_n and U_p it is
necessary to distinguish between two kinds of traps: those of *donor type*, that are
electrically neutral when the electron is present in the trap and become positively
charged when the electron is released, and those of *acceptor type*, that are electrically
neutral when the electron is absent from the trap and become negatively charged when

the electron is captured. In this respect, the traps are similar to the dopants' atoms. Instead, a strong difference is made by the position of the traps' energy within the gap. Consider, for instance, traps localized near the gap's midpoint (the latter is indicated by the intrinsic Fermi level E_{Fi} in Fig. 20.2); the phonon energy necessary for the transition is about $E_G/2$ in all cases, to be compared with the value E_G necessary for a direct transition. On the other hand, the equilibrium-phonon distribution (Sect. 16.6) is the Bose–Einstein statistics (15.55); it follows that the number dN_{ph} of phonons in the interval $d\omega$ is

$$dN_{ph} = \frac{g_{ph}(\omega)\, d\omega}{\exp\left[\hbar\,\omega/(k_B\, T)\right] - 1}, \tag{20.12}$$

with $\hbar\,\omega$ the energy and g_{ph} the density of states of the phonons. Due to (20.12), $dN_{ph}/d\omega$ rapidly decreases as the phonon energy increases, this making the probability of an electron-phonon interaction much larger at lower energies. For this reason, even in an electronic-grade semiconductor, where the concentration of defects is very small (Sect. 19.3.2), the traps are able to act as a sort of "preferred path" in energy for the inter-band transitions, to the extent that the contribution to U_n, U_p of the trap-assisted transitions is largely dominant over that of the direct transitions. Therefore, in the continuity Eqs. (20.13) below, and in the subsequent derivation of the trap-assisted, thermal-transition rates, symbols U_n, U_p refer only to the latter transitions, not any more to the sum of the trap-assisted and direct ones.

The net thermal-recombination terms U_n, U_p appear in (19.129) and (19.130) after replacing W_n, W_p with (19.132); this yields

$$\frac{\partial n}{\partial t} + U_n - \frac{1}{q}\,\mathrm{div}\mathbf{J}_n = G_n, \qquad \frac{\partial p}{\partial t} + U_p + \frac{1}{q}\,\mathrm{div}\mathbf{J}_p = G_p. \tag{20.13}$$

To introduce the trap-assisted transitions one formally duplicates (20.13) as if the acceptor and donor traps formed two additional bands; as the acceptor traps are either neutral or negatively charged, the charge and current densities of the band associated to them are thought of as due to electrons; instead, the charge and current densities of the band associated to the donor traps are thought of as due to holes. In summary, the two additional equations read

$$\frac{\partial n_a}{\partial t} + U_{na} - \frac{1}{q}\,\mathrm{div}\mathbf{J}_{na} = G_{na}, \qquad \frac{\partial p_d}{\partial t} + U_{pd} + \frac{1}{q}\,\mathrm{div}\mathbf{J}_{pd} = G_{pd} \tag{20.14}$$

with a and d standing for "acceptor" and "donor", respectively. To ease the calculation it is assumed that the non-thermal phenomena are absent, whence $G_n = G_p = G_{na} = G_{pd} = 0$. Combining (20.13) with (20.14), and observing that $\mathbf{J} = \mathbf{J}_p + \mathbf{J}_{pd} + \mathbf{J}_n + \mathbf{J}_{na}$ is the total current density of the semiconductor, yields

$$\frac{\partial[q\,(p + p_d - n - n_a)]}{\partial t} + \mathrm{div}\mathbf{J} = q\,(U_n + U_{na}) - q\,(U_p + U_{pd}). \tag{20.15}$$

As the net dopant concentration N is independent of time, it is $\partial[q\,(p + p_d - n - n_a)]/\partial t = \partial[q\,(p + p_d - n - n_a + N)]/\partial t = \partial\rho/\partial t$; thus, the left hand side of

(20.15) vanishes due to (4.23), and[3]

$$U_n + U_{na} = U_p + U_{pd}. \tag{20.16}$$

The two continuity Eqs. (20.14) are now simplified by observing that in crystalline semiconductors the current densities $\mathbf{J}_{pd}, \mathbf{J}_{na}$ of the traps are negligible. In fact, the trap concentration is so low that inter-trap tunneling is precluded by the large distance from a trap to another; the reasoning is the same as that used in Sect. 18.7.2 with respect to the impurity levels.[4] Letting $\mathbf{J}_{pd} = \mathbf{J}_{na} = 0$ makes the two Eqs. (20.14) local:

$$\frac{\partial n_a}{\partial t} = -U_{na}, \qquad \frac{\partial p_d}{\partial t} = -U_{pd}. \tag{20.17}$$

In steady-state conditions the traps' populations are constant, this yielding $U_{na} = U_{pd} = 0$ and, from (20.16), $U_n = U_p$. In equilibrium all continuity equations reduce to the identity $0 = 0$, whence the net-recombination terms vanish independently, $U_n^{eq} = U_{na}^{eq} = U_p^{eq} = U_{pd}^{eq} = 0$.

20.2.3 Shockley-Read-Hall Theory

The *Shockley-Read-Hall theory* describes the trap-assisted, net thermal-recombination term in a crystalline semiconductor based upon the steady-state relation $U_n = U_p$. In fact, the outcome of the theory is used also in dynamic conditions; this approximation is acceptable because, due to the smallness of the traps' concentration, the contribution of the charge density stored within the traps is negligible with respect to that of the band and dopant states; the contribution of the time variation of the traps' charge density is similarly negligible. The theory also assumes that only one trap level is present, of energy E_t; with reference to Fig. 20.2, the trap levels must be thought of aligned with each other. If more than one trap level is present, the contributions of the individual levels are added up at a later stage. In the theory it is not important to distinguish between acceptor-type or donor-type traps; however, one must account for the fact that a trap can accommodate one electron at most.

Still with reference to Fig. 20.2, let r_a be the number of type-a transitions per unit volume and time, and similarly for r_b, r_c, r_d. The derivation of these rates is

---

[3] The result expressed by (20.16) is intuitive if one thinks that adding up all continuity equations amounts to counting all transitions twice, the first time in the forward direction (e.g., using the electrons), the second time in the backward direction (using the holes). The reasoning is similar to that leading to the vanishing of the intra-band contribution in (19.63).

[4] In a polycrystalline semiconductor the traps' current densities are not negligible; in fact, the whole system of equations (20.13) and (20.14) must be used to correcly model the material [16, 17, 18]. The conduction phenomenon associated to these current densities is called *gap conduction*.

similar to that of the direct transitions and is shown in the complements; here the expressions of the net thermal-recombination terms are given, that read

$$U_n = r_a - r_b = \alpha_n \, n \, N_t \, (1 - P_t) - e_n \, N_t \, P_t, \tag{20.18}$$

$$U_p = r_c - r_d = \alpha_p \, p \, N_t \, P_t - e_p \, N_t \, (1 - P_t), \tag{20.19}$$

where N_t is the concentration of traps of energy E_t, P_t the trap-occupation probability, α_n, α_p the *electron- and hole-transition coefficients*, respectively, and e_n, e_p the *electron- and hole-emission coefficients*, respectively.[5] The ratios e_n/α_n, e_p/α_p are assumed to vary little from the equilibrium to the non-equilibrium case. From $U_n^{\text{eq}} = U_p^{\text{eq}} = 0$ one derives

$$\frac{e_n}{\alpha_n} = n^{\text{eq}} \left(\frac{1}{P_t^{\text{eq}}} - 1 \right), \qquad \frac{e_p}{\alpha_p} = p^{\text{eq}} \left(\frac{1}{P_t^{\text{eq}}} - 1 \right)^{-1}. \tag{20.20}$$

The occupation probability at equilibrium is the modified Fermi-Dirac statistics (compare with (18.21) or (18.36))

$$P_t^{\text{eq}} = \left[\frac{1}{d_t} \exp \left(\frac{E_t - E_F}{k_B T} \right) + 1 \right]^{-1}, \qquad \frac{1}{P_t^{\text{eq}}} - 1 = \frac{1}{d_t} \exp \left(\frac{E_t - E_F}{k_B T} \right), \tag{20.21}$$

with d_t the degeneracy coefficient of the trap. It follows, after introducing the shorthand notation $n_B = e_n/\alpha_n$, $p_B = e_p/\alpha_p$,

$$n_B = \frac{n^{\text{eq}}}{d_t} \exp \left(\frac{E_t - E_F}{k_B T} \right), \qquad p_B = p^{\text{eq}} \, d_t \, \exp \left(\frac{E_F - E_t}{k_B T} \right). \tag{20.22}$$

Note that $n_B \, p_B = n^{\text{eq}} \, p^{\text{eq}}$. Replacing (20.22) into (20.18), (20.19), and letting $U_n = U_p$ yields

$$\alpha_n \, n \, (1 - P_t) - \alpha_n \, n_B \, P_t = \alpha_p \, p \, P_t - \alpha_p \, p_B \, (1 - P_t), \tag{20.23}$$

whence

$$P_t = \frac{\alpha_n \, n + \alpha_p \, p_B}{\alpha_n \, (n + n_B) + \alpha_p \, (p + p_B)}, \qquad 1 - P_t = \frac{\alpha_n \, n_B + \alpha_p \, p}{\alpha_n \, (n + n_B) + \alpha_p \, (p + p_B)}. \tag{20.24}$$

In this way one expresses the trap-occupation probability as a function of two of the unknowns of the semiconductor-device model, namely, n and p, and of a few parameters. Among the latter, n_B and p_B are known (given the trap's energy) because

---

[5] It is $[\alpha_{n,p}] = \text{m}^3 \, \text{s}^{-1}$, $[e_{n,p}] = \text{s}^{-1}$.

they are calculated in the equilibrium condition. In conclusion, replacing (20.24) into
(20.18) or (20.19) yields, for the common value $U_{\text{SRH}} = U_n = U_p$,

$$U_{\text{SRH}} = \frac{n\,p - n^{\text{eq}}\,p^{\text{eq}}}{(n + n_B)/(N_t\,\alpha_p) + (p + p_B)/(N_t\,\alpha_n)}, \tag{20.25}$$

where the indices stand for "Shockley-Read-Hall". Eventually, the only unknown
parameters turn out to be the products $N_t\,\alpha_p$ and $N_t\,\alpha_n$ which, as shown in Sect. 25.2,
can be obtained from measurements.

The expression obtained so far, (20.25), has been derived considering a single
trap level E_t. Before adding up over the levels it is convenient to consider how
sensitive U_{SRH} is to variations of E_t; in fact, one notes that the numerator of (20.25)
is independent of E_t, whereas the denominator D has the form

$$D = c + 2\lambda\cosh\eta, \qquad \eta = \frac{E_t - E_F}{k_B T} + \frac{1}{2}\log\mu, \tag{20.26}$$

where

$$c = \frac{1}{N_t}\left(\frac{n}{\alpha_p} + \frac{p}{\alpha_n}\right), \qquad \lambda = \frac{1}{N_t}\sqrt{\frac{n^{\text{eq}}\,p^{\text{eq}}}{\alpha_p\,\alpha_n}}, \qquad \mu = \frac{1}{d_t^2}\frac{n^{\text{eq}}/\alpha_p}{p^{\text{eq}}/\alpha_n}. \tag{20.27}$$

The denominator has a minimum where $\eta = 0$; thus, U_{SRH} has a maximum there.
Moreover, the maximum is rather sharp due to the form of the hyperbolic cosine. It
follows that the trap level E_{tM} that most efficiently induces the trap-assisted transitions is found by letting $\eta = 0$. The other traps levels have a much smaller efficiency
and can be neglected; in conclusion, it is not necessary to add up over the trap levels.[6]
In conclusion, one finds

$$E_{tM} = E_F + \frac{k_B T}{2}\log\left(d_t^2\,\frac{p^{\text{eq}}/\alpha_n}{n^{\text{eq}}/\alpha_p}\right). \tag{20.28}$$

An estimate of E_{tM} is easily obtained by considering the non-degenerate condition,
whence $n^{\text{eq}} = N_C\exp\left[(E_F - E_C)/(k_B T)\right]$ and $p^{\text{eq}} = N_V\exp\left[(E_V - E_F)/(k_B T)\right]$
(compare with (18.28)). It follows

$$E_{tM} \simeq \frac{E_C + E_V}{2} + \frac{k_B T}{2}\log\left(d_t^2\,\frac{N_V\,\alpha_p}{N_C\,\alpha_n}\right). \tag{20.29}$$

Observing that the second term at the right hand side of (20.29) is small, this result
shows that the most efficient trap level is near the gap's midpoint which, in turn, is
near the intrinsic Fermi level E_{Fi}. In fact, combining (20.29) with (18.16) yields

$$E_{tM} \simeq E_{Fi} + \frac{k_B T}{2}\log\left(d_t^2\,\frac{\alpha_p}{\alpha_n}\right) \simeq E_{Fi}. \tag{20.30}$$

---

[6] This simplification is not applicable in a polycrystalline semiconductor.

Defining the *lifetimes*

$$\tau_{p0} = \frac{1}{N_t \, \alpha_p}, \qquad \tau_{n0} = \frac{1}{N_t \, \alpha_n}, \tag{20.31}$$

gives (20.25) the standard form

$$U_{\text{SRH}} = \frac{n\,p - n^{\text{eq}}\,p^{\text{eq}}}{\tau_{p0}\,(n + n_B) + \tau_{n0}\,(p + p_B)}, \tag{20.32}$$

which is also called *Shockley-Read-Hall recombination function*. In equilibrium it is $U_{\text{SRH}}^{\text{eq}} = 0$; in a non-equilibrium condition, a positive value of U_{SRH}, corresponding to an excess of the $n\,p$ product with respect to the equilibrium product $n^{\text{eq}}\,p^{\text{eq}}$, indicates that recombinations prevail over generations, and vice versa. In a non-equilibrium condition it may happen that $U_{\text{SRH}} = 0$; this occurs at the boundary between a region where recombinations prevail and another region where generations prevail.

In a non-degenerate semiconductor (20.22) become, letting $E_t = E_{tM} = E_{Fi}$ and using (18.12),

$$n_B = \frac{n_i}{d_t}, \qquad p_B = d_t \, n_i, \tag{20.33}$$

whence $n_B \, p_B = n_i^2$. This result is useful also in a degenerate semiconductor for discussing possible simplifications in the form of U_{SRH}.

20.2.3.1 Limiting Cases of the Shockley-Read-Hall Theory

The operating conditions of semiconductor devices are often such that the SRH recombination function (20.32) can be reduced to simpler forms. The first case is the so-called *full-depletion condition*, where both electron and hole concentrations are negligibly small with respect to n_B and p_B. Remembering that $n^{\text{eq}}\,p^{\text{eq}} = n_B\,p_B$ one finds

$$U_{\text{SRH}} \simeq -\frac{n_B\,p_B}{\tau_{p0}\,n_B + \tau_{n0}\,p_B} = -\frac{\sqrt{n_B\,p_B}}{\tau_g}, \qquad \tau_g = \sqrt{\frac{n_B}{p_B}}\,\tau_{p0} + \sqrt{\frac{n_B}{p_B}}\,\tau_{n0}. \tag{20.34}$$

In a non-degenerate condition n_B, p_B take the simplified form (20.33), whence $\sqrt{n_B\,p_B} = n_i$ and $\tau_g = \tau_{p0}/d_t + \tau_{n0}\,d_t$. In a full-depletion condition U_{SRH} is always negative, namely, generations prevail over recombinations; for this reason, τ_g is called *generation lifetime*.

The second limiting case of interest is the so-called *weak-injection condition*. This condition occurs when both inequalities below are fulfilled:

$$|n - n^{\text{eq}}| \ll c^{\text{eq}}, \qquad |p - p^{\text{eq}}| \ll c^{\text{eq}}, \tag{20.35}$$

where c^{eq} is the equilibrium concentration of the majority carriers in the spatial position under consideration. From the above definition it follows that the concept of weak injection is applicable only after specifying which carriers are the majority ones. Expanding the product $n\,p$ to first order in n and p around the equilibrium value yields $n\,p \simeq n^{eq}\,p^{eq} + n^{eq}\,(p - p^{eq}) + p^{eq}\,(n - n^{eq})$. As a consequence, the numerator of (20.32) becomes

$$n\,p - n^{eq}\,p^{eq} \simeq n^{eq}\,(p - p^{eq}) + p^{eq}\,(n - n^{eq}). \tag{20.36}$$

To proceed, it is necessary to distinguish between the n-type and p-type regions.

Weak-Injection Condition, n-Type Semiconductor
The weak-injection condition (20.35) reads $|n - n^{eq}| \ll n^{eq}$, $|p - p^{eq}| \ll n^{eq}$. As a consequence, one lets $n \simeq n^{eq}$ in the denominator of (20.32) and neglects n_B with respect to n^{eq}; in fact, in a non-degenerate condition it is $n_B \simeq n_i \ll n^{eq}$, and the same inequality is also applicable in a degenerate condition. As the lifetimes are similar to each other, the term $\tau_{n0}\,(p + p_B)$ in the denominator is negligible with respect to $\tau_{p0}\,n^{eq}$, because p is a concentration of minority carriers and p_B is similar to n_B. In conclusion, the denominator of (20.32) simplifies to $\tau_{p0}\,n^{eq}$, whence

$$U_{SRH} \simeq \frac{p - p^{eq}}{\tau_{p0}} + \frac{n - n^{eq}}{(n^{eq}/p^{eq})\,\tau_{p0}}. \tag{20.37}$$

The second term at the right hand side of (20.37) is negligible[7] because $n^{eq}/p^{eq} \gg 1$; letting $\tau_p = \tau_{p0}$ finally yields

$$U_{SRH} \simeq \frac{p - p^{eq}}{\tau_p}, \tag{20.38}$$

with τ_p the *minority-carrier lifetime* in an n-doped region.

Weak-Injection Condition, p-Type Semiconductor
The weak-injection condition (20.35) reads $|n - n^{eq}| \ll p^{eq}$, $|p - p^{eq}| \ll p^{eq}$. As a consequence, one lets $p \simeq p^{eq}$ in the denominator of (20.32) and neglects p_B with respect to p^{eq}; the other term in the denominator of (20.35) is neglected as above, this simplifying the denominator to $\tau_{n0}\,p^{eq}$. In conclusion,

$$U_{SRH} \simeq \frac{p - p^{eq}}{(p^{eq}/n^{eq})\,\tau_{n0}} + \frac{n - n^{eq}}{\tau_{n0}}. \tag{20.39}$$

The first term at the right hand side of (20.39) is negligible because $p^{eq}/n^{eq} \gg 1$; letting $\tau_n = \tau_{n0}$ finally yields

$$U_{SRH} \simeq \frac{n - n^{eq}}{\tau_n}, \tag{20.40}$$

with τ_n the *minority-carrier lifetime* in a p-doped region.

---

[7] Considering for instance the example given in Sect. 18.4.1, one has $n^{eq} \simeq 10^{15}$ cm^{-3}, $p^{eq} \simeq 10^5$ cm^{-3}, whence $n^{eq}/p^{eq} \simeq 10^{10}$.

The simplified expressions of U_{SRH} found here are particularly useful; in fact, in contrast to (20.32), the weak-injection limits (20.38) and (20.40) are linear with respect to p or n. Moreover, as (20.38) and (20.40) depend on one unknown only, they decouple the continuity equation of the minority carriers (the first one in (19.129) or in (19.130)) from the other equations of the semiconductor's model; thanks to this it is possible to separate the system of equations. The simplification introduced by the full-depletion condition is even stronger, because (20.34) is independent of the model's unknowns. On the other hand, all simplifications illustrated here are applicable only in the regions where the approximations hold; once the simplified model's equations have been solved locally, it is necessary to match the solutions at the boundaries between adjacent regions.

20.3 Auger Recombination and Impact Ionization

An important, non-thermal recombination mechanism is *Auger recombination*. The phenomenon is due to the electron-electron or hole-hole collision, and is illustrated in Fig. 20.3. With reference to case *a*, two electrons whose initial state is in the conduction band collide and exchange energy. The outcome of the collision is that one of the electrons suffers an energy loss equal or larger than the energy gap and makes a transition to an empty state of the valence band; the other electron absorbs the same amount of energy and makes a transition to a higher-energy state of the conduction band. The phenomenon is also indicated as an Auger recombination *initiated by electrons*. The analogue for holes is shown in case *c* of Fig. 20.3: two holes whose initial state is in the valence band collide and exchange energy. Remembering that hole energy increases in the opposite direction with respect to that of electrons (Sect. 19.2.3), the hole that suffers an energy loss equal or larger than the energy gap makes a transition to a filled state of the conduction band; the other hole absorbs the same amount of energy and makes a transition to a higher-energy state of the valence band. The phenomenon is indicated as an Auger recombination *initiated by holes*.

The phenomenon dual to Auger recombination is illustrated in Fig. 20.4 and is called *impact ionization*. With reference to case *b*, an electron whose initial state is in the conduction band at high energy collides and exchanges energy with an electron whose initial state is in the valence band. The initial energy E of the electron in the conduction band is such that $E - E_C$ is equal or larger than the energy gap, whereas the initial energy of the electron in the valence band is near E_V. The outcome of the collision is that, although the high-energy electron suffers an energy loss equal or larger than the energy gap, its final state is still in the conduction band; the other electron absorbs the same amount of energy and makes a transition to the conduction band. The phenomenon is in fact an electron-hole pair generation and is also indicated as an impact-ionization event *initiated by electrons*. The analogue for holes is shown in case *d* of Fig. 20.4: a hole whose initial state is in the valence band at high energy collides and exchanges energy with a hole whose initial state is in the conduction band. The initial energy E of the hole in the valence band is such that $|E - E_V|$

Fig. 20.3 Auger
recombinations initiated by
electrons (*a*) and holes (*c*)

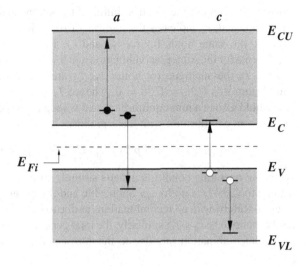

Fig. 20.4 Impact-ionization
transitions initiated by
electrons (*b*) and holes (*d*)

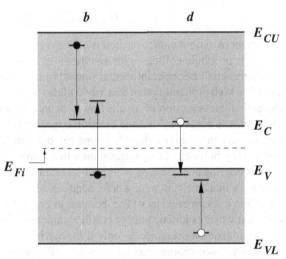

is equal or larger than the energy gap, whereas the initial energy of the hole in the
conduction band is near E_C. The outcome of the collision is that, although the high-
energy hole suffers an energy loss equal or larger than the energy gap, its final state
is still in the valence band; the other hole absorbs the same amount of energy and
makes a transition to the valence band. The phenomenon is in fact an electron-hole
pair generation and is also indicated as an impact-ionization event *initiated by holes*.

The derivation of the Auger and impact-ionization rates is shown in the complements; here the expressions of the net recombinations due to the Auger and
impact-ionization events are given, that read

$$U_n^{AI} = r_a - r_b = c_n\, n^2\, p - I_n\, n, \qquad U_p^{AI} = r_c - r_d = c_p\, p^2\, n - I_p\, p, \quad (20.41)$$

where U_n^{AI} refers to the electron-initiated transitions and U_p^{AI} to the hole-initiated ones. As usual, r_a indicates the number of transitions of type a per unit time and volume; the same holds for r_b, r_c, and r_d. In (20.41), c_n, I_n are the transition coefficients for the Auger recombination and impact ionization initiated by electrons, and c_p, I_p the analogue for holes; c_n, c_p are also called *Auger coefficients*.[8] In equilibrium it is $U_n^{AI} = U_p^{AI} = 0$, whence $I_n = c_n n^{eq} p^{eq}$, $I_p = c_p n^{eq} p^{eq}$. The above hold also in a non-equilibrium case as long as the operating conditions are not too far from equilibrium; with these premises it follows

$$U_n^{AI} = c_n n \, (n \, p - n^{eq} \, p^{eq}), \qquad U_p^{AI} = c_p \, p \, (np - n^{eq} \, p^{eq}), \qquad (20.42)$$

When the operating condition departs strongly from equilibrium, the simplification leading to (20.42) is no longer applicable and the general expressions (20.41) must be used. Referring to all recombinations as due to transitions of electrons, their rate is easily found to be $r_a + r_c$; similarly, the total generation rate is $r_b + r_d$. In conclusion, the net recombination rate due to the Auger and impact-ionization phenomena is given by

$$U_{AI} = U_n^{AI} + U_p^{AI}. \qquad (20.43)$$

For Auger recombination to occur it is necessary that an electron collides with another electron, or a hole collides with another hole. The probability of such an event is relatively small because in normal operating conditions and at room temperature there is a high probability that a carrier collides with a phonon; as a consequence, for the collisionless motion of an electron to be interrupted by a collision with another electron it is necessary that the electron concentration be very high. This situation occurs only in a heavily-doped, n-type region; similarly, an Auger recombination initiated by holes can be significant only in a heavily-doped, p-type region.[9]

Considering now the case of impact-ionization, for this phenomenon to occur it is necessary that an electron, or a hole, acquires a kinetic energy larger than the energy gap. This is a rare event as well,[10] because in general the carrier undergoes a phonon collision when its kinetic energy is still significantly lower than the energy gap. The impact-ionization event occurs only if the carrier acquires a substantial energy over a distance much shorter than the average collisionless path, which happens only in presence of a strong electric field.[11]

The qualitative reasoning outlined above explains why the conditions for a strong Auger recombination are incompatible with those that make impact-ionization

---

[8] The units are $[c_{n,p}] = cm^6 \, s^{-1}$ and $[I_{n,p}] = s^{-1}$.

[9] In fact, Auger recombination becomes significant in the source and drain regions of MOSFETs and in the emitter regions of BJTs, where the dopant concentration is the highest.

[10] In principle, high-energy electrons or hole exist also in the equilibrium condition; however, their number is negligible because of the exponentially-vanishing tail of the Fermi-Dirac statistics.

[11] The high-field conditions able to produce a significant impact ionization typically occur in the reverse-biased p-n juctions like, e.g., the drain junction in MOSFETs and the collector junction in BJTs.

dominant; in fact, a large charge density, like that imposed by a heavy dopant concentration, prevents the electric field from becoming strong. Vice versa, a strong electric field prevents a large carrier concentration from building up. It is therefore sensible to investigate situations where only one term dominates within U_{AI}.

20.3.1 Strong Impact Ionization

As indicated in Sect. 20.3, far from equilibrium the approximations $I_n = c_n \, n^{\mathrm{eq}} \, p^{\mathrm{eq}}$, $I_p = c_p \, n^{\mathrm{eq}} \, p^{\mathrm{eq}}$ are not valid, and the general expressions (20.41) must be used. Here the situation where impact ionization dominates over the other generation-recombination mechanisms is considered, using the steady-state case. If impact ionization is dominant, it is $U_n - G_n = U_p - G_p \simeq U_{AI} \simeq -I_n \, n - I_p \, p$. The continuity equations (the first ones in (19.129) and (19.130)) then become

$$\mathrm{div}\mathbf{J}_n = -q \, I_n \, n - q \, I_p \, p, \qquad \mathrm{div}\mathbf{J}_p = q \, I_n \, n + q \, I_p \, p. \tag{20.44}$$

As outlined in Sect. 20.3, impact-ionization dominates if the electric field is high. For this reason, the transport equations in (19.129) and (19.130) are simplified by keeping the ohmic term only, to yield $\mathbf{J}_n \simeq q \, \mu_n \, n \, \mathbf{E}$ and $\mathbf{J}_p \simeq q \, \mu_p \, p \, \mathbf{E}$. As a consequence, the electron and hole current densities are parallel to the electric field. Let $\mathbf{e}(\mathbf{r})$ be the unit vector of the electric field, oriented in the direction of increasing field, $\mathbf{E} = |\mathbf{E}| \, \mathbf{e}$; it follows $\mathbf{J}_n = J_n \, \mathbf{e}$ and $\mathbf{J}_p = J_p \, \mathbf{e}$, with J_n and J_p strictly positive. Extracting n, p from the above and replacing them into (20.44) yields

$$-\mathrm{div}\mathbf{J}_n = k_n \, J_n + k_p \, J_p, \qquad \mathrm{div}\mathbf{J}_p = k_n \, J_n + k_p \, J_p, \tag{20.45}$$

where the ratios

$$k_n = \frac{I_n}{\mu_n \, |\mathbf{E}|}, \qquad k_p = \frac{I_p}{\mu_p \, |\mathbf{E}|}, \tag{20.46}$$

whose units are $[k_{n,p}] = \mathrm{m}^{-1}$, are the *impact-ionization coefficients* for electrons and holes, respectively. Equations (20.45) form a system of differential equations of the first order, whose solution in the one-dimensional case is relatively simple if the dependence of the coefficients on position is given (Sect. 21.5).

20.4 Optical Transitions

The description of the optical transitions is similar to that of the direct thermal transitions given in Sect. 20.2.1; still with reference to Fig. 20.1, the transition marked with a can be thought of as an optical-recombination event if the energy difference between the initial and final state is released to the environment in the form of a photon. The opposite transition (b), where the electron's energy increases due to photon absorption from the environment, is an optical electron-hole generation. The expression of the net optical-recombination rate is similar to (20.11) and reads

$$U_O = \alpha_O \, n \, p - G_O, \tag{20.47}$$

whose coefficients are derived in the same manner as those of U_{DT} (Sect. 20.2.1).

In normal operating conditions the similarity between the direct-thermal and optical generation-recombination events extends also to the external agent that induces the transitions. In fact, the distribution of the phonon energies is typically the equilibrium one, given by the Bose–Einstein statistics (15.55) at the lattice temperature; as for the photons, the environment radiation in which the semiconductor is immersed can also be assimilated to the equilibrium one, again given by the Bose-Einstein statistics at the same temperature. The conditions of the optical generation-recombination events drastically change if the device is kept far from equilibrium. Consider for instance the case where the electron concentration of the conduction band is artificially increased with respect to the equilibrium value at the expense of the electron population of the valence band, so that both n and p in (20.47) increase. This brings about an excess of recombinations; if the probability of radiative-type generation-recombination events is high,[12] the emission of a large number of photons follows. The angular frequencies of the emitted photons are close to $(E_C - E_V)/\hbar$, because the majority of the electrons in the conduction band concentrate near E_C, and the final states of the radiative transitions concentrate near E_V. In this way, the energy spent to keep the artificially-high concentration of electron-hole pairs is transformed into that of a nearly-monochromatic optical emission. In essence, this is the description of the operating principle of a *laser*.[13] Another method for keeping the device far from equilibrium is that of artificially decreasing both the concentration of electrons of the conduction band and the concentration of holes of the valence band. The outcome is opposite with respect to that described earlier: the decrease of both n and p in (20.47) brings about an excess of generations, which in turn corresponds to the absorption of photons from the environment. The absorption may be exploited to accumulate energy (this leading to the concept of *solar cell*), or to provide an electrical signal whose amplitude depends on the number of absorbed photons (this leading to the concept of *optical sensor*).

In a non-equilibrium condition the amount of energy exchanged between the semiconductor and the electromagnetic field is not necessarily uniform in space. Consider, by way of example, the case of an optical sensor on which an external radiation impinges; as the non-equilibrium conditions are such that the absorption events prevail, the radiation intensity within the material progressively decreases at increasing distances from the sensor's surface. Therefore, it is important to determine the radiation intensity as a function of position.

---

[12] As indicated in Sect. 17.6.6, among semiconductors this is typical of the direct-gap ones.

[13] In fact, LASER is the acronym of Light Amplification by Stimulated Emission of Radiation.

Fig. 20.5 Sketch of photon
absorption in a material layer

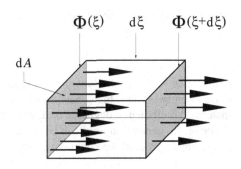

It is acceptable to assume that the absorption events are uncorrelated from each
other. Thus, one can limit the analysis to a monochromatic radiation; the effect of
the whole spectrum is recovered at a later stage by adding up over the frequencies.
When absorption prevails, (20.47) simplifies to $U_O \simeq -G_O$, where G_O is a function
of the radiation's frequency v and possibly of position. If the radiation's intensity
varies with time, G_O depends on time as well.[14] When the radiation interacts with
the external surface of the material, part of the energy is reflected; moreover, the
radiation is refracted at the boundary, so that the propagation direction outside the
material differs in general from that inside. Letting ξ be the propagation direction
inside the material, consider an elementary volume with a side $d\xi$ aligned with ξ
and a cross-section dA normal to it (Fig. 20.5). The monochromatic radiation can be
described as a flux of photons of equal energy $h\,v$, with h the Planck constant, and a
momentum's direction parallel to ξ. Let $\Phi(\xi)$ be the flux density of photons entering
the volume from the face corresponding to ξ, and $\Phi(\xi + d\xi)$ the flux density leaving
it at $\xi + d\xi$; the following holds, $\Phi = K\,u_f$, where $K(\xi)$ is the concentration of the
photons and u_f their constant phase velocity. Then,

$$\frac{\partial \Phi}{\partial \xi} = \frac{\partial K}{\partial (\xi/u_f)} = \frac{\partial K}{\partial t}. \tag{20.48}$$

The derivatives in (20.48) are negative because the photon concentration decreases
in time due to absorption; as the loss of each photon corresponds to the loss of an
energy quantum $h\,v$, the loss of electromagnetic energy per unit volume and time is
$-h\,v\,(\partial \Phi/\partial \xi)$. By a similar token one finds[15] that the energy absorbed by the optical-
generation events per unit time and volume is $h\,v\,G_O$. The latter is not necessarily
equal to $-h\,v\,(\partial \Phi/\partial \xi)$; in fact, some photons crossing the elementary volume may
be lost due to collisions with nuclei (this, however, is a rare event), or with electrons
that are already in the conduction band, so that no electron-hole pair generation

---

[14] In principle, a time-dependence of the intensity is incompatible with the hypothesis that the
radiation is monochromatic. However, the frequency with which the intensity may vary is extremely
small with respect to the optical frequencies.

[15] It is implied that $h\,v \geq E_C - E_V$, and that the two-particle collisions only are to be considered.

occurs. To account for these events one lets

$$G_O = -\eta \, \frac{\partial \Phi}{\partial \xi} > 0, \tag{20.49}$$

with $0 < \eta < 1$ the *quantum efficiency*. In moderately-doped semiconductors η is close to unity because the concentration of the conduction-band electrons is small; instead, the efficiency degrades in degenerate semiconductors. The spatial dependence of the generation term can be derived from (20.49) if that of the photon flux is known. To proceed, one defines the *absorption coefficient* as

$$k = -\frac{1}{\Phi} \, \frac{\partial \Phi}{\partial \xi} > 0, \tag{20.50}$$

with $[k] = \mathrm{m}^{-1}$. In general it is $k = k(\Phi, \xi, \nu)$; however, as the absorption effects are uncorrelated, the flux density lost per unit path $\mathrm{d}\xi$ is proportional to the flux density available at ξ. Then, k is independent of Φ; neglecting momentarily the dependence on ξ as well, one finds

$$\Phi(\xi) = \Phi_B \, \exp[-k(\nu)\xi], \tag{20.51}$$

with $\Phi_B = \Phi(\xi = 0^+)$ on account of the fact that, due to the reflection at the interface, the flux density on the inside edge of the boundary is different from that on the outside edge. When k is independent of position, its inverse $1/k$ is called *average penetration length of the radiation*. When k depends on position, (20.50) is still separable and yields

$$\Phi(\xi) = \Phi_B \, \exp(-k_m \, \xi), \qquad k_m = \frac{1}{\xi} \int_0^\xi k(\xi'; \nu) \, \mathrm{d}\xi'. \tag{20.52}$$

Combining (20.52) with (20.49), the optical-generation term is found to be

$$G_O = \eta \, \Phi_B \, k(\xi, \nu) \, \exp\left[-\int_0^\xi k(\xi', \nu) \, \mathrm{d}\xi'\right]. \tag{20.53}$$

20.5 Macroscopic Mobility Models

It has been shown in Sect. 19.5.2 that the carrier mobilities are defined in terms of the momentum-relaxation times. Specifically, in the parabolic-band approximation it is, for the electrons of the conduction band, $\mu_n = (\mu_l + 2\mu_t)/3$, with $\mu_l = q\,\tau_p/m_l$, $\mu_t = q\,\tau_p/m_t$, where τ_p is the electron momentum-relaxation time (19.87); similarly, for the holes of the valence band the carrier mobility is given by inserting (19.118) into the second relation of (19.121), namely, a linear combination of the heavy-hole and light-hole momentum-relaxation times. As, in turn, the inverse momentum-relaxation time is a suitable average of the inverse intra-band relaxation time, the

Matthiessen rule follows (Sect. 19.6.5); in conclusion, the electron and hole mobilities are calculated by combining the effects of the different types of collisions (e.g., phonons, impurities, and so on) suffered by the carrier.[16] In the case of electrons, the application of the Matthiessen rule is straightforward, leading to

$$\frac{1}{\mu_n} = \frac{m_n}{q} \left(\frac{1}{\tau_p^{\text{ph}}} + \frac{1}{\tau_p^{\text{imp}}} + \cdots \right), \tag{20.54}$$

where the index refers to the type of collision, and $1/m_n = (1/m_l + 2/m_t)/3$. For holes a little more algebra is necessary, which can be avoided if the approximation $\tau_{ph} \simeq \tau_{pl}$ is applicable.

In the typical operating conditions of semiconductor devices the most important types of collisions are those with phonons and ionized impurities. For devices like surface-channel MOSFETs, where the flow lines of the current density are near the interface between semiconductor and gate insulator, a third type is also very important, namely, the collisions with the interface. The *macroscopic mobility models* are closed-form expressions in which mobility is related to a set of macroscopic parameters (e.g., temperature) and to some of the unknowns of the semiconductor-device model; the concept is similar to that leading to the expressions of the generation-recombination terms shown in earlier sections.

20.5.1 Example of Phonon Collision

By way of example, a simplified analysis of the contribution to mobility of the electron-phonon collision is outlined below, starting from the definition of the ith component of the momentum-relaxation tensor τ_{pi} given by (19.87); the simplifications are such that the first-order expansion $f - f^{\text{eq}} \simeq (\mathrm{d}f/\mathrm{d}\lambda)^{\text{eq}} \lambda$ is not used here. Starting from the perturbative form (19.47) one considers the steady-state, uniform case and lets $\mathbf{B} = 0$, $\tau = \tau_v$, to find

$$\frac{q}{\hbar} \mathbf{E} \cdot \mathrm{grad}_{\mathbf{k}} f = \frac{f - f^{\text{eq}}}{\tau_v}. \tag{20.55}$$

Replacing f with f^{eq} at the left hand side of (20.55), and using the definition (17.52) of the group velocity, yields $\mathrm{grad}_{\mathbf{k}} f^{\text{eq}} = (\mathrm{d}f^{\text{eq}}/\mathrm{d}H)\hbar\,\mathbf{u}$, with H the Hamiltonian function defined in Sect. 19.2.2. Inserting into (19.87) yields

$$\tau_{pi} \iiint_{-\infty}^{+\infty} u_i\,\mathbf{E} \cdot \mathbf{u}\,(\mathrm{d}f^{\text{eq}}/\mathrm{d}H)\,\mathrm{d}^3k = \iiint_{-\infty}^{+\infty} u_i\,\mathbf{E} \cdot \mathbf{u}\,(\mathrm{d}f^{\text{eq}}/\mathrm{d}H)\,\tau_v\,\mathrm{d}^3k. \tag{20.56}$$

As the derivative $\mathrm{d}f^{\text{eq}}/\mathrm{d}H$ is even with respect to $\mathbf{k}$, the integrals involving velocity components different from u_i vanish because the corresponding integrand is odd;

---

[16] As mentioned in Sect. 19.6.5, it is assumed that the different types of collisions are incorrelated.

as a consequence, only the ith component of the electric field remains, and cancels out. A further simplification is obtained by replacing the Fermi-Dirac statistics with the Maxwell-Boltzmann distribution law, $f^{eq} \simeq Q \exp[(-E_e + q\varphi - E_C + E_F)/(k_B T)]$, to find

$$\tau_{pi} \iiint_{-\infty}^{+\infty} u_i^2 \exp[-E_e/(k_B T)]\, d^3k = \iiint_{-\infty}^{+\infty} u_i^2 \exp[-E_e/(k_B T)] \tau_v\, d^3k. \tag{20.57}$$

To proceed it is necessary to make an assumption about τ_v. Remembering the definition of the relaxation time given by the first relation in (19.43), it is reasonable to assume that the scattering probability S_0 increases with the kinetic energy E_e of the electron, so that the relaxation time decreases; a somewhat stronger hypothesis is that the relaxation time depends on E_e only, namely, the collision is isotropic.[17]

In this case, (20.57) is readily manipulated by a Herring-Vogt transformation. Following the same procedure as in Sect. 19.6.4, one finds that all numerical factors cancel out; as a consequence, one may replace the auxiliary coordinate η_i^2 with $\eta^2/3 = E_e/3$, this showing that $\tau_{pi} = \tau_p$ is isotropic as well. One eventually finds

$$\tau_p = \frac{\int_0^{+\infty} \tau_v(E_e)\, E_e^{3/2} \exp[-E_e/(k_B T)]\, dE_e}{\int_0^{+\infty} E_e^{3/2} \exp[-E_e/(k_B T)]\, dE_e}. \tag{20.58}$$

A simple approximation for the relaxation time is $\tau_v = \tau_{v0} (E_e/E_0)^{-\alpha}$, where τ_{v0}, E_0, and α are positive parameters independent of E_e. From (C.88) it follows

$$\tau_p = \tau_{v0} \frac{\Gamma(5/2 - \alpha)}{\Gamma(5/2)} \left(\frac{E_0}{k_B T}\right)^\alpha. \tag{20.59}$$

When the electron-phonon interaction is considered, $\tau_{v0} = \tau_{v0}^{ph}$ is found to be inversely proportional to $k_B T$ and to the concentration N_{sc} of semiconductor's atoms; moreover, for acoustic phonons[18] it is $\alpha = 1/2$ [62, Sects. 61, 62], whence

$$\tau_p^{ap} = \tau_{v0}(N_{sc}, T) \frac{4}{3\sqrt{\pi}} \left(\frac{E_0}{k_B T}\right)^{1/2}, \qquad \mu_n^{ap} \propto N_{sc}^{-1} (k_B T)^{-3/2}, \tag{20.60}$$

where "ap" stands for "acoustic phonon". More elaborate derivations, including also the contribution of optical phonons, still show that carrier-phonon collisions make mobility to decrease when temperature increases.

---

[17] The first-principle derivation of the scattering probabilities is carried out by applying Fermi's Golden Rule (Sect. 14.8.3) to each type of perturbation, using the Bloch functions for the unperturbed states [57]. An example is given later in the case of ionized-impurity scattering.

[18] Acoustic phonons are those whose momentum and energy belong to the acoustic branch of the lattice-dispersion relation (Sect. 17.8.5); a similar definition applies to optical phonons (Sect. 17.8.6).

20.5.2 Example of Ionized-Impurity Collision

As a second example one considers the collisions with ionized impurities. The interaction with a single ionized impurity is a perturbation of the Coulomb type; due to the presence of the crystal, the more suitable approach is the screened Coulomb perturbation, an example of which is shown in Sect. 14.7, leading to the perturbation-matrix element (14.34):

$$h_{\mathbf{kg}}^{(0)} = \frac{A/(2\pi)^3}{q_c^2 + q^2}, \qquad A = \frac{\kappa Z e^2}{\varepsilon_0}. \tag{20.61}$$

In (20.61), $e > 0$ is the elementary electric charge, Z a positive integer, ε_0 the vacuum permittivity, $q_c > 0$ the inverse screening length,[19] $q = |\mathbf{q}| = |\mathbf{k} - \mathbf{g}|$ and, finally, $\kappa = 1(-1)$ in the repulsive (attractive) case. The wave vectors $\mathbf{k}$ and $\mathbf{g}$ correspond to the initial and final state of the transition, respectively. In principle, (20.61) should not be used as is because it holds *in vacuo*; in fact, the eigenfunctions of the unperturbed Hamiltonian operator used to derive (20.61) are plane waves. Inside a crystal, instead, one should define the perturbation matrix $h_{\mathbf{kg}}(t)$ using the Bloch functions $w_{\mathbf{k}} = u_{\mathbf{k}} \exp(\mathrm{i}\,\mathbf{k} \cdot \mathbf{r})$ in an integral of the form (14.24). However, it can be shown that the contribution of the periodic part $u_{\mathbf{k}}$ can suitably be averaged and extracted from the integral, in the form of a dimensionless coefficient, whose square modulus G is called *overlap factor*. For this reason, the collisions with ionized impurities is treated starting from the definition (20.61) to calculate the perturbation matrix, with the provision that the result is to be multiplied by G and the permittivity $\varepsilon_{\mathrm{sc}}$ of the semiconductor replaces ε_0 in the second relation of (20.61).

Like in Sect. 14.6, a Gaussian wave packet (14.27) centered on some wave vector $\mathbf{b} \neq \mathbf{g}$ is used as initial condition. In this case the perturbation is independent of time, $h_{\mathbf{bg}} = h_{\mathbf{bg}}^{(0)} = \mathrm{const} \neq 0$; as a consequence, the infinitesimal probability $\mathrm{d}P_{\mathbf{b}}$ that such a perturbation induces a transition, from the initial condition (14.27), to a final state whose energy belongs to the range $\mathrm{d}E_{\mathbf{g}}$, is given by (14.32). In turn, the integral (14.32) providing $H_{\mathbf{b}}^{(0)}(E_{\mathbf{g}})$ is calculated in Problem 14.1. Assuming that the duration t_P of the interaction is large enough to make Fermi's Golden Rule (14.44) applicable, and inserting the overlap factor, one finally obtains

$$\mathrm{d}P_{\mathbf{b}} \approx G \left(\frac{2\pi m}{\hbar^2}\right)^{3/2} \frac{8\pi t_P \, \delta(E_{\mathbf{b}} - E_{\mathbf{g}}) \, A^2}{\lambda^3 \hbar \, (2\pi)^5 \, q_c^2 \, (q_c^2 + 8 m E_{\mathbf{g}}/\hbar^2)} \sqrt{E_{\mathbf{g}}} \, \mathrm{d}E_{\mathbf{g}}, \tag{20.62}$$

where the relation $E_{\mathbf{g}} = \hbar^2 g^2/(2m)$ has been used. Integrating over $E_{\mathbf{g}}$ and dividing by t_P provides the probability per unit time of a transition from the initial energy $E_{\mathbf{b}}$ to any final energy; letting $E_c = \hbar^2 q_c^2/(2m)$, one finds

$$\dot{P}(E_{\mathbf{b}}) = \frac{1}{\tau_{vc}} \frac{\sqrt{4 E_{\mathbf{b}}/E_c}}{1 + 4 E_{\mathbf{b}}/E_c}, \qquad \frac{1}{\tau_{vc}} = \frac{G A^2/\sqrt{2\pi m}}{8\pi^2 (\lambda^2 E_c)^{3/2}}. \tag{20.63}$$

---

[19] An example of derivation of the screening length is given in Sect. 20.6.4.

The above expression provides the contribution to the intra-band relaxation time of the scattering due to a single impurity. One notes that, since A is squared, the effect onto (20.63) of a positive impurity is the same as that of a negative one. If the effect of each impurity is uncorrelated with that of the others,[20] the probabilities add up; letting $N_I = N_D^+ + N_A^-$ be the total concentration of ionized impurities, the product $N_I \, d^3 r$ is the total number ionized impurities in the elementary volume $d^3 r$; it follows that the probability per unit time and volume is given by $\dot{P}(E_b) \, N_I \, d^3 r$. Considering that N_I depends on position only, mobility inherits the inverse proportionality with N_I; letting "ii" indicate "ionized impurity", one finds $\mu_n^{ii} \propto 1/N_I$.

The derivation of the dependence on N_I shown above is in fact oversimplified, and the resulting model does not reproduce the experimental results with sufficient precision. One of the reasons for this discrepancy is that the inverse screening length q_c depends on the dopant concentration as well. In order to improve the model, while still keeping an analytical form, the model is modified by letting $1/\mu_n^{ii} \propto N_I^\alpha$, with α a dimensionless parameter to be extracted from the comparison with experiments. One then lets

$$\frac{1}{\mu_n^{ii}(N_I)} = \frac{1}{\mu_n^{ii}(N_R)} \left(\frac{N_I}{N_R} \right)^\alpha , \tag{20.64}$$

with N_R a reference concentration.

20.5.3   Bulk and Surface Mobilities

Combining the phonon and ionized-impurity contributions using the Matthiessen rule yields $1/\mu_n^B(T, N_I) = 1/\mu_n^{ph}(T) + 1/\mu_n^{ii}(N_I)$, namely,

$$\mu_n^B(T, N_I) = \frac{\mu_n^{ph}(T)}{1 + c(T)(N_I/N_R)^\alpha}, \tag{20.65}$$

with $c(T) = \mu_n^{ph}(T)/\mu_n^{ii}(N_R)$. In practical cases the doping concentration ranges over many orders of magnitude; for this reason, (20.65) is usually represented in a semilogarithmic scale: letting $r = \log_{10}(N_I/N_R)$, $b = \alpha \log_e 10$, and $b_0 = \log_e c$, (20.65) becomes

$$\mu_n^B(T, N_I) = \frac{\mu_n^{ap}(T)}{1 + \exp(br + b_0)}. \tag{20.66}$$

The curves corresponding to $b = 1$, 1.5, 3 and $b_0 = 0$ are drawn in Fig. 20.6, using r as independent variable at a fixed T. Index "B" in the mobility defined in (20.65)

---

[20] In silicon, this assumption is fulfilled for values of the concentration up to about 10^{19} cm^{-3} [64, 84].

Fig. 20.6 Graph of the theoretical mobility curve (20.66), normalized to its maximum, for different values of b, with $b_0 = 0$. Each curve has a flex at $r = r_{\text{flex}} = -b_0/b$ and takes the value 0.5 there. The slope at the flex is $-b/4$

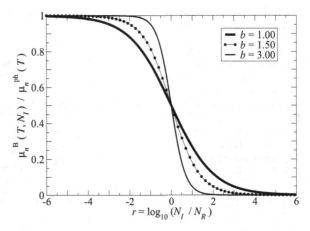

or (20.66) stands for "bulk". More generally, the term *bulk mobility* is ascribed to the combination of all contributions to mobility different from surface collisions.

As mentioned at the beginning of this section, in surface-channel devices the degradation of mobility produced by the interaction of the carriers with the interface between channel and gate insulator is also very important. The macroscopic models of this effect are built up by considering that the carrier-surface interaction is more likely to occur if the flow lines of the current density are closer to the interface itself; such a closeness is in turn controlled by the intensity of the electric field's component normal to the interface, $E_\perp$. In conclusion, the model describes the contribution to mobility due to surface scattering as a decreasing function of $E_\perp$, e.g., for electrons,

$$\frac{1}{\mu_n^s(E_\perp)} = \frac{1}{\mu_n^s(E_R)} \left(\frac{E_\perp}{E_R}\right)^\beta, \qquad (20.67)$$

with E_R a reference field and β a dimensionless parameter to be extracted from experiments. Combining the bulk and surface contributions using the Matthiessen rule yields $1/\mu_n(T, N_I, E_\perp) = 1/\mu_n^B(T, N_I) + 1/\mu_n^s(E_\perp)$, namely,

$$\mu_n(T, N_I, E_\perp) = \frac{\mu_n^B(T, N_I)}{1 + d(T, N_I)(E_\perp/E_R)^\beta}, \qquad (20.68)$$

with $d(T, N_I) = \mu_n^B(T, N_I)/\mu_n^s(E_R)$.

20.5.4 Beyond Analytical Modeling of Mobility

In general the analytical approaches outlined above do not attain the precision necessary for applications to realistic devices. For this reason, one must often resort to numerical-simulation methods; in this way, the main scattering mechanisms are

Fig. 20.7 Electron mobility in silicon calculated with the spherical-harmonics expansion method (HARM) as a function of the total ionized-dopant concentration N_I, using the lattice temperature T as parameter. The calculations are compared with measurements by Lombardi [73], Klaassen [64], and Arora [1] (courtesy of S. Reggiani)

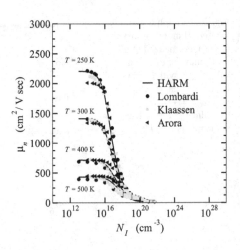

incorporated into the analysis (e.g., for silicon: acoustic phonons, optical phonons, ionized impurities, and impact ionization), along with the full-band structure of the semiconductor, which is included in the simulation through the density of states and group velocity defined in the energy space. The latter, in turn, are obtained directly from the corresponding functions in the momentum space by integrating the full-band system over the angles. The energy range considered to date allows for the description of carrier dynamics up to 5 eV.

As mentioned above, the ionized-impurity collisions can be treated as interaction between the carrier and a single impurity as long as the impurity concentration is below some limit. When the limit is exceeded, impurity clustering becomes relevant and must be accounted for [64]. In fact, at high doping densities the carrier scatters with a cluster of K ions, where K is a function of the impurity concentration. Finally, different outcomes are found for majority- or minority-mobility calculations: e.g., minority-hole mobility is found to be about a factor 2 higher than the majority-hole mobility for identical doping levels.

Figures 20.7 and 20.8 show the outcome of electron- and hole-mobility calculations for bulk silicon, obtained from the spherical-harmonics method illustrated in [112]. The method incorporates the models for the scattering mechanisms listed above. The electron and hole mobility have been calculated as a function of the total ionized-dopant concentration N_I, using the lattice temperature T as a parameter; in the figures, they are compared with measurements taken from the literature.

To include the surface effects in the analysis it is necessary to account for the fact that in modern devices the thickness of the charge layer at the interface with the gate insulator is so small that quantum confinement and formation of subbands must be considered. The typical collisions mechanisms to be accounted for at the semiconductor-insulator interface are surface roughness, scattering with ionized impurities trapped at the interface, and surface phonons. Figures 20.9 and 20.10 show the outcome of electron and hole surface-mobility calculations in silicon, also obtained from the spherical-harmonics method [84]. The electron and hole mobility

Fig. 20.8 Hole mobility in silicon calculated with the spherical-harmonics expansion method (HARM) as a function of the total ionized-dopant concentration N_I, using the lattice temperature T as parameter. The calculations are compared with measurements by Lombardi [73], Klaassen [64], and Arora [1] (courtesy of S. Reggiani)

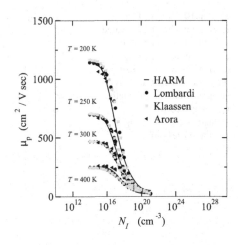

Fig. 20.9 Electron surface mobility in silicon calculated with the spherical-harmonics expansion method (HARM) method at room temperature, using the acceptor concentration N_A as parameter. The calculations are compared with measurements by Takagi [106] (courtesy of S. Reggiani)

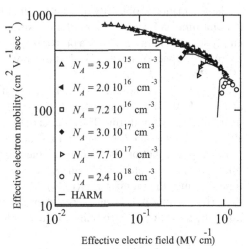

have been calculated as functions of the dopant concentration (N_A and N_D, respectively), at room temperature; in the figures, they are compared with measurements taken from the literature.

20.6 Complements

20.6.1 Transition Rates in the SRH Recombination Function

The expressions of the transition rates r_a, r_b, r_c, r_d to be used in the calculation of the Shockley-Read-Hall recombination function (20.32) are determined by the same reasoning as that used in Sect. 20.2.1 for the direct thermal transitions. Let $P(\mathbf{r}, E, t)$

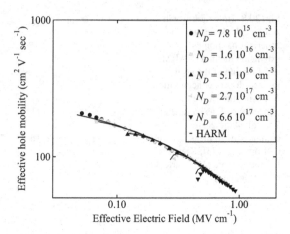

Fig. 20.10 Hole surface mobility in silicon calculated with the spherical-harmonics expansion method (HARM) at room temperature, using the donor concentration N_D as parameter. The calculations are compared with measurements by Takagi [106] (courtesy of S. Reggiani)

be the occupation probability of a state at energy E, and $C(E \rightarrow E')$ the probability per unit time and volume (in $\mathbf{r}$) of a transition from a filled state of energy E to an empty state of energy E'. Such a probability is independent of time; it depends on the energy of the phonon involved in the transition, and possibly on position. Then, define $P' = P(\mathbf{r}, E = E', t)$, $P_t = P(\mathbf{r}, E = E_t, t)$, where E_t is the energy of the trap. Finally, let $\gamma(E)$ be the combined density of states in energy and volume of the bands, and $\gamma_t(\mathbf{r}, E)$ the same quantity for the traps (the latter depends on position if the traps' distribution is non uniform). The number of transitions per unit volume and time, from states in the interval dE belonging to a band, to states in the interval dE' belonging to the trap distribution, is obtained as the product of the number $\Omega \gamma(E) \, dE \, P$ of filled states in the interval dE, times the transition probability per unit volume and time C, times the number $\Omega \gamma_t(\mathbf{r}, E') \, dE' \, (1 - P')$ of empty states in the interval dE'. Thus, letting ΔE_t be an energy interval belonging to the gap and containing the traps, the transition rate from the conduction band to the traps is given by

$$r_a = \int_{E_C}^{E_{CU}} \int_{\Delta E_t} \Omega \gamma(E) \, dE \, P \, C(E \rightarrow E') \Omega \gamma_t(\mathbf{r}, E') \, dE' \, (1 - P'). \quad (20.69)$$

By the same token, the transition rate from the valence band to the traps is

$$r_d = \int_{E_{VL}}^{E_V} \int_{\Delta E_t} \Omega \gamma(E) \, dE \, P \, C(E \rightarrow E') \Omega \gamma_t(\mathbf{r}, E') \, dE' \, (1 - P'). \quad (20.70)$$

In turn, the number of transitions per unit volume and time, from states in the interval dE' belonging the trap distribution, to states in the interval dE belonging to a band, is obtained as the product of the number $\Omega \gamma_t(\mathbf{r}, E') \, dE' \, P'$ of filled states in the interval dE', times $C(\mathbf{r}, E' \rightarrow E)$, times the number $\Omega \gamma(E) \, dE \, (1 - P)$ of empty states in the interval dE. Thus, the transition rates from the traps to conduction or valence band are respectively given by

$$r_b = \int_{E_C}^{E_{CU}} \int_{\Delta E_t} \Omega \gamma_t(\mathbf{r}, E') \, dE' \, P' \, C(\mathbf{r}, E' \rightarrow E) \Omega \gamma(E) \, dE \, (1 - P), \quad (20.71)$$

$$r_c = \int_{E_{VL}}^{E_V} \int_{\Delta E_t} \Omega \, \gamma_t(\mathbf{r}, E') \, dE' \, P' \, C(\mathbf{r}, E' \to E) \, \Omega \, \gamma(E) \, dE \, (1 - P). \quad (20.72)$$

The combined density of states of the traps is treated in the same manner as that of the dopant atoms (compare with (18.20) and (18.35)) by letting

$$\gamma_t(\mathbf{r}, E') = N_t(\mathbf{r}) \, \delta(E' - E_t), \quad (20.73)$$

where $N_t(\mathbf{r})$ is the trap concentration. Thanks to this, the integrals over ΔE_t are easily evaluated, to yield

$$r_a = N_t \, (1 - P_t) \, \Omega^2 \int_{E_C}^{E_{CU}} \gamma \, P \, C(\mathbf{r}, E \to E_t) \, dE = N_t \, (1 - P_t) \, \alpha_n \, n, \quad (20.74)$$

$$r_c = N_t \, P_t \, \Omega^2 \int_{E_{VL}}^{E_V} \gamma \, (1 - P) \, C(\mathbf{r}, E_t \to E) \, dE = N_t \, P_t \, \alpha_p \, p, \quad (20.75)$$

where the definitions (20.3), (20.4) of the electron and hole concentrations are used, and the transition coefficients for electrons and holes are defined as the weighed averages

$$\alpha_n = \Omega^2 \, \frac{\int_{E_C}^{E_{CU}} \gamma \, P \, C \, dE}{\int_{E_C}^{E_{CU}} \gamma \, P \, dE}, \qquad \alpha_p = \Omega^2 \, \frac{\int_{E_{VL}}^{E_V} \gamma \, (1 - P) \, C \, dE}{\int_{E_{VL}}^{E_V} \gamma \, (1 - P) \, dE}. \quad (20.76)$$

Like in the case of (20.10), the integrals in (20.76) are approximated using the equilibrium probability. The remaining transition rates r_b, r_d are determined in a similar manner, using also the approximation $1 - P \simeq 1$ in (20.71) and $P \simeq 1$ in (20.70). Like in Sect. 20.2.1, the approximation is justified by the fact that in normal operating conditions the majority of the valence-band states are filled, while the majority of the conduction-band states are empty. In conclusion,

$$r_b = N_t \, P_t \, \Omega^2 \int_{E_C}^{E_{CU}} \gamma \, (1 - P) \, C(\mathbf{r}, E_t \to E) \, dE \simeq N_t \, P_t \, e_n, \quad (20.77)$$

$$r_d = N_t \, (1 - P_t) \, \Omega^2 \int_{E_{VL}}^{E_V} \gamma \, P \, C(\mathbf{r}, E \to E_t) \, dE \simeq N_t \, (1 - P_t) \, e_p, \quad (20.78)$$

with the emission coefficients defined by

$$e_n = \Omega^2 \int_{E_C}^{E_{CU}} \gamma \, C \, dE, \qquad e_p = \Omega^2 \int_{E_{VL}}^{E_V} \gamma \, C \, dE. \quad (20.79)$$

20.6.2 Coefficients of the Auger and Impact-Ionization Events

The expression of the coefficients c_n, c_p and I_n, I_p, to be used in the calculation of the net recombination rates (20.41) due to the Auger and impact-ionization phenomena, are found in the same way as the transition rates of the SRH recombination function (Sect. 20.6.1) or the direct thermal recombinations (Sect. 20.2.1). Let $P(\mathbf{r}, E, t)$ be the occupation probability of a state of energy E, and $C_n(E_1, E_2 \rightarrow E_1', E_2')$ the combined probability per unit time and volume (in $\mathbf{r}$) of an electron transition from a filled state of energy E_1 in the conduction band to an empty state of energy E_1' in the conduction band, and of another electron from a filled state of energy E_2 to an empty state of energy E_2', where E_2 and E_2' belong to different bands.

Auger Coefficients
In an Auger recombination it is $E_1' > E_1$; also, E_2 belongs to the conduction band while E_2' belongs to the valence band. Due to energy conservation it is[21]

$$C_n = C_{n0}\, \delta \left[(E_1 - E_1') + (E_2 - E_2') \right], \tag{20.80}$$

where $E_2 - E_2' \simeq E_G$; it follows $E_1' \simeq E_1 + E_G$. Then, define $P_i = P(\mathbf{r}, E = E_i, t)$, $P_i' = P(\mathbf{r}, E = E_i', t)$, with $i = 1, 2$, and let $\gamma(E)$ be the combined density of states in energy and volume for the bands; in particular, let $g_i = \Omega\,\gamma(E_i)$ and $g_i' = \Omega\,\gamma(E_i')$. From the above definitions one finds, for the rate r_a of the Auger recombinations initiated by electrons,

$$r_a = \int g_1\, dE_1\, P_1\, g_2\, dE_2\, P_2\, C_n\, g_1'\, dE_1'\, (1 - P_1')\, g_2'\, dE_2'\, (1 - P_2'), \tag{20.81}$$

where $\int$ indicates a fourfold integral that extends thrice over the conduction band and once over the valence band. Observing that $P_1' \ll 1$ and integrating over E_1' with $C_n = C_{n0}\, \delta(E_1 + E_G - E_1')$ yields

$$r_a = \int_{E_C}^{E_{CU}} g_1\, dE_1\, P_1\, C_{n0}\, g_G \int_{E_C}^{E_{CU}} g_2\, dE_2\, P_2 \int_{E_{VL}}^{E_V} g_2'\, dE_2'\, (1 - P_2'), \tag{20.82}$$

where $g_G = g(E_1 + E_G)$ and $[C_{n0}\, g_G] = \mathrm{s}^{-1}\,\mathrm{m}^{-3}$. Thanks to (20.3) and (20.4), the second integral in (20.82) equals $\Omega\, n$ and the third one equals $\Omega\, p$. Letting

$$c_n = \Omega^3\, \frac{\int_{E_C}^{E_{CU}} C_{n0}\, g_G\, g_1\, P_1\, dE_1}{\int_{E_C}^{E_{CU}} g_1\, P_1\, dE_1}, \tag{20.83}$$

finally yields $r_a = c_n\, n^2\, p$. The derivation of $r_c = c_p\, p^2\, n$ is similar.

---

[21] The units of C_{n0} are $[C_{n0} = \mathrm{J}\,\mathrm{s}^{-1}\,\mathrm{m}^{-3}]$.

Impact Ionization's Transition Coefficients
Using the same symbols introduced at the beginning of Sect. 20.6.2, for an impact-ionization event initiated by an electron it is $E_1 > E_1'$; in turn, E_2 belongs to the valence band and E_2' belongs to the conduction band. It follows

$$r_b = \int g_1 \, dE_1 \, P_1 \, g_2 \, dE_2 \, P_2 \, C_n \, g_1' \, dE_1' \, (1 - P_1') \, g_2' \, dE_2' \, (1 - P_2'), \qquad (20.84)$$

where the fourfold integral extends thrice over the conduction band and once over the valence band. From the energy-conservation relation $E_1 + E_2 = E_1' + E_2'$ and from $E_2' - E_2 \simeq E_G$ it follows $E_1' \simeq E_1 - E_G$. Observing that $P_2 \simeq 1$, $P_1' \ll 1$, $P_2' \ll 1$, and integrating over E_1' with $C_n = C_{n0} \, \delta(E_1 - E_G - E_1')$ yields

$$r_b = \int_{E_C}^{E_{CU}} C_{n0} \, g_G \, g_1 \, P_1 \, dE_1 \int_{E_{VL}}^{E_V} g_2 \, dE_2 \int_{E_C}^{E_{CU}} g_2' \, dE_2', \qquad (20.85)$$

where $g_G = g(E_1 - E_G)$, and the product of the second and third integral is a dimensionless quantity that depends only on the semiconductor's structure. Indicating such a quantity with ν_n, and letting

$$I_n = \nu_n \, \frac{\int_{E_C}^{E_{CU}} C_{n0} \, g_G \, g_1 \, P_1 \, dE_1}{\int_{E_C}^{E_{CU}} g_1 \, P_1 \, dE_1}, \qquad (20.86)$$

finally yields $r_b = I_n \, n$. The derivation of $r_d = I_p \, p$ is similar.

20.6.3 Total Recombination-Generation Rate

The expressions for the most important generation-recombination terms have been worked out in this chapter. Only one of them, the SRH recombination function U_{SRH}, involves energy states different from those of the conduction and valence bands; in principle, such states would require additional continuity equations to be added to the semiconductor-device model. However, as discussed in Sect. 20.2.3, this is not necessary in crystalline semiconductors. The other mechanisms (direct thermal recombination-generation U_{DT}, Auger recombination and impact ionization U_{AI}, and optical recombination-generation U_O) do not involve intermediate states. As a consequence, with reference to (20.13) the generation-recombination terms of the electron-continuity equation are equal to those of the hole continuity equation. Finally, assuming that the different generation-recombination phenomena are uncorrelated, and neglecting U_{DT} with respect to U_{SRH} (Sect. 20.2.2), yields

$$U_n - G_n = U_p - G_p \simeq U_{SRH} + U_{AI} + U_{DO}. \qquad (20.87)$$

20.6.4   Screened Coulomb Potential

In the context of physics, the general meaning of *screening* is the attenuation in the electric field intensity due to the presence of mobile charges; the effect is treated here using the Debye-Hückel theory [29], which is applicable to a non-degenerate semiconductor where the dopants are completely ionized. For a medium of permittivity ε, with charge density ρ, the electric potential in the equilibrium condition is found by solving Poisson's equation

$$-\varepsilon \, \nabla^2 \varphi = \rho. \qquad (20.88)$$

One starts by considering a locally-neutral material, to which a perturbation is added due, for instance, to the introduction of a fixed charge $Z_c \, e_c$ placed in the origin; this, in turn, induces a variation in ρ. The corresponding perturbation of φ is calculated to first order by replacing φ with $\varphi + \delta\varphi$ and ρ with $\rho + (\partial\rho/\partial\varphi)\,\delta\varphi$, where the derivative is calculated at $\delta\varphi = 0$; the perturbed form of Poisson's equation reads:

$$-\varepsilon \, \nabla^2 \varphi - \varepsilon \, \nabla^2 \delta\varphi = \rho + \frac{\partial\rho}{\partial\varphi}\,\delta\varphi. \qquad (20.89)$$

As the unperturbed terms cancel out due to (20.88), a Poisson equation in the perturbation is obtained,

$$\nabla^2 \delta\varphi = q_c^2 \, \delta\varphi, \qquad q_c^2 = -\frac{\partial\rho/\partial\varphi}{\varepsilon}, \qquad (20.90)$$

where $1/q_c$ is the *screening length* or *Debye length*. The definition implies that $\partial\rho/\partial\varphi < 0$; this is in fact true, as shown below with reference to a non-degenerate semiconductor with completely-ionized dopants. Letting $N_D^+ = N_D$, $N_A^- = N_A$ in (19.125), and using the non-degenerate expressions (18.60), (18.61) of the equilibrium concentrations, one finds that $N = N_D - N_A$ is left unaffected by the perturbation, while the electron concentration[22] n transforms into $n \, \exp\left[e \, \delta\varphi/(k_B \, T)\right]$ and the hole concentration p transforms into $p \, \exp\left[-e \, \delta\varphi/(k_B \, T)\right]$. From $\rho = e\,(p - n + N)$ one obtains, to first order,

$$\frac{\partial\rho}{\partial\varphi} = -\frac{e^2}{k_B \, T}\,(n + p), \qquad q_c^2 = \frac{e^2\,(n + p)}{\varepsilon \, k_B \, T} > 0. \qquad (20.91)$$

The left hand side of the Poisson equation in (20.90) is conveniently recast using a set of spherical coordinates r, θ, ϕ whose origin coincides with the center of symmetry of the perturbation; using (B.25) one finds

$$\nabla^2 \delta\varphi = \frac{1}{r}\,\frac{\partial^2}{\partial r^2}(r \, \delta\varphi) + \frac{r^{-2}}{\sin\theta}\,\frac{\partial}{\partial\theta}\left(\sin\theta \, \frac{\partial\delta\varphi}{\partial\theta}\right) + \frac{r^{-2}}{\sin^2\theta}\,\frac{\partial^2\delta\varphi}{\partial\phi^2}. \qquad (20.92)$$

---

[22] The electron charge is indicated here with e to avoid confusion with q_c.

Considering a perturbation with a spherical symmetry, only the first term at the right hand side of (20.92) is left, whence (20.90) becomes an equation in the unknown $r\,\delta\varphi$:

$$\frac{d^2}{dr^2}(r\,\delta\varphi) = q_c^2\,(r\,\delta\varphi). \qquad (20.93)$$

The general solution of (20.93) is $r\,\delta\varphi = A_1\,\exp(-q_c\,r) + A_2\,\exp(q_c\,r)$, where it must be set $A_2 = 0$ to prevent the solution from diverging as r becomes large. In conclusion,

$$\delta\varphi = \frac{A_1}{r}\,\exp(-q_c\,r). \qquad (20.94)$$

The remaining constant is found by observing that for very small r the pure Coulomb case $\delta\varphi \simeq A_1/r$ is recovered, whence $A_1 = Z_c e_c e/(4\pi\,\varepsilon)$. This makes (20.94) to coincide with (14.33).

Part VII
Basic Semiconductor Devices

Chapter 21
Bipolar Devices

21.1 Introduction

The mathematical model of semiconductor devices, derived in Chap. 19, is applied here to the description of the fundamental bipolar device, the p–n junction. The term *bipolar* indicates that both electrons and holes contribute to the current. The analysis is carried out using the simple example of a one-dimensional abrupt junction in steady state, with the hypotheses of non-degeneracy and complete ionization, that lend themselves to an analytical treatment. The equilibrium condition is considered first, and the solution of Poisson's equation is tackled, showing that the structure can be partitioned into space-charge and quasi-neutral regions. Then, the Shockley theory is illustrated, leading to the derivation of the ideal $I(V)$ characteristic. The semiconductor model is then applied to illustrating two features of the reverse-bias condition, namely, the depletion capacitance and the avalanche due to impact ionization. The complements justify the simplification of considering only the diffusive transport for the minority carriers in a quasi-neutral region, and provide the derivation of the Shockley boundary conditions. Finally, the expression of the depletion capacitance is worked out for the case of an arbitrary charge-density profile.

21.2 *P–N* Junction in Equilibrium

A very simple, yet fundamental, semiconductor device is the *p–n junction*, whose one-dimensional version is sketched in Fig. 21.1. The device is fabricated by thermally diffusing (Chap. 23), or ion implanting *p*-type dopant atoms into an *n*-type substrate, or vice versa. As a consequence, the diffused or implanted profile is not spatially uniform. The substrate profile may in turn result from a similar process, so that in general it is not uniform either. The locus of points where the ionized dopant concentrations are equal to each other, $N_D^+ = N_A^-$, is a surface called *metallurgical*

© Springer Science+Business Media New York 2015
M. Rudan, *Physics of Semiconductor Devices,*
DOI 10.1007/978-1-4939-1151-6_21

Fig. 21.1 Schematic example
of a one-dimensional *p–n*
junction

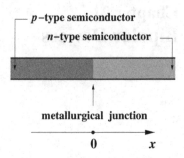

junction.[1] The theory of the *p–n* junction is carried out with reference to a simplified structure, where the device is one dimensional and aligned with the *x* axis; in this case the metallurgical junction is a plane normal to *x* and, as shown in Fig. 21.1, its position is made to coincide with the reference's origin. Also, the non-uniform dopant concentrations $N_A(x)$ and $N_D(x)$ are approximated by piecewise-constant functions, $N_A = \text{const}$ for $x < 0$ and $N_D = \text{const}$ for $x > 0$. The device obtained from this approximation is called *abrupt p–n junction*. Considering the actual form of the dopant distribution, the approximation is not realistic; however, the much simpler model based on it is still able to capture the essential features of the device characteristics. Moreover, the model assumes that the conditions of non-degeneracy and complete ionization hold; this assumption makes the analytical approach possible.

Within an integrated circuit the *p–n* junction is supplemented with contacts that connect it to the rest of the circuit. Such contacts are typically made of metals, metal silicides, or heavily-doped polycrystalline semiconductors; as a consequence, two more junctions are present: the first one is between the contact and the *p*-doped semiconductor, the other one between the contact and the *n*-doped semiconductor. It is implied that the contacts are made of the same material; if it is not so, more junctions must be considered as shown below.

21.2.1   Built-In Potential

A qualitative description of the device in the equilibrium condition starts from the assumption that the extension of the *p*-doped and *n*-doped regions along the *x* axis is large, so that, far away from the junction, the semiconductor can be considered as uniformly doped of the *p* or *n* type, respectively. This fixes the boundary conditions for the electron and hole concentrations:[2] in fact, remembering that in the non-degeneracy and complete-ionization conditions the equilibrium concentrations in a uniform semiconductor are given by (18.42) for the *p* type and by (18.30) for the *n* type, one finds

---

[1] The metallurgical junction is often indicated with the same term used for the whole device, namely, *p–n junction* or simply *junction*.

[2] The use of asymptotic conditions is not applicable to shallow junctions like, e.g., those used for the fabrication of solar cells. In this case, the theory is slightly more involved.

$$P_{p0} = p(-\infty) \simeq N_A, \qquad n_{p0} = n(-\infty) \simeq \frac{n_i^2}{N_A}, \tag{21.1}$$

$$n_{n0} = n(+\infty) \simeq N_D, \qquad p_{n0} = p(+\infty) \simeq \frac{n_i^2}{N_D}. \tag{21.2}$$

The above concentrations are also called *asymptotic concentrations*; the last approximations are derived from the assumption $N_A, N_D \gg n_i$ which, as outlined in Sect. 18.4.1, has a vast range of validity. The distance between the conduction-band edge and the Fermi level is found from $n = N_C \exp[-(E_C - E_F)/(k_B T)]$ (compare with (18.28)); combining with (21.1) and (21.2) yields

$$\frac{n_i^2}{N_A} \simeq N_C \exp\left[-\frac{E_C(-\infty) - E_F}{k_B T}\right], \quad N_D \simeq N_C \exp\left[-\frac{E_C(+\infty) - E_F}{k_B T}\right], \tag{21.3}$$

whence

$$E_C(-\infty) - E_C(+\infty) = k_B T \log\left(\frac{N_A N_D}{k_B T}\right). \tag{21.4}$$

An identical expression is found for $E_V(-\infty) - E_V(+\infty)$. These findings show that E_C, E_V are functions of position; their explicit form is determined below. Alternatively, one may use (18.60) and (18.61), to find

$$\psi_0 = k_B T \log\left(\frac{N_A N_D}{k_B T}\right), \qquad \psi_0 = \varphi(+\infty) - \varphi(-\infty), \tag{21.5}$$

where ψ_0 is called *built-in potential*.[3] One notes that so far the values of the constants $n^{(0)}$, $p^{(0)}$ in (18.60) and (18.61) have been left unspecified; remembering that $n^{(0)}$ is the value of n in the position(s) where $\varphi = 0$, and the same for $p^{(0)}$, the numerical values sought are determined by specifying the zero point of φ. Here such a point is fixed by letting $\varphi(+\infty) = 0$ whence, using (21.3) and (21.4), one finds

$$p^{(0)} = p(+\infty) = p_{n0}, \qquad n^{(0)} = n(+\infty) = n_{n0}. \tag{21.6}$$

The expressions of the carrier concentrations in terms of the band energies E_C, E_V must be coherent with those expressed in terms of φ. In fact, the relation between $E_C(x)$ and $\varphi(x)$ is found by combining $n = n^{(0)} \exp[q\,\varphi/(k_B T)]$ with $n = N_C \exp[(E_F - E_C)/(k_B T)]$ and using $n^{(0)} = N_C \exp\{[E_F - E_C(+\infty)]/(k_B T)\}$; a similar procedure is applied to E_V, to eventually find

$$E_C(x) = E_C(+\infty) - q\,\varphi(x), \qquad E_V(x) = E_V(+\infty) - q\,\varphi(x). \tag{21.7}$$

---

[3] The same quantity is also called *barrier potential* and is sometimes indicated with ψ_B.

Letting $N(x) = -N_A$ for $x < 0$ and $N(x) = +N_D$ for $x > 0$, the electric potential is found by solving the Poisson equation

$$\frac{d^2\varphi}{dx^2} = \frac{q}{\varepsilon_{sc}}\left[n_{n0}\exp\left(\frac{q\varphi}{k_B T}\right) - p_{n0}\exp\left(\frac{-q\varphi}{k_B T}\right) - N(x)\right] \qquad (21.8)$$

with boundary conditions $\varphi(-\infty) = -\psi_0$ and $\varphi(+\infty) = 0$. One notes that within each half domain the charge density in (21.8) has the form $\rho = \rho(\varphi)$, which makes the theory of Sect. 19.5.8 applicable. Therefore, it is convenient to separately solve (21.8) in each half space, and apply suitable matching conditions at $x = 0$ afterwards. When the regional-solution method is used, the boundary conditions must be modified with respect to $\varphi(-\infty) = -\psi_0$ and $\varphi(+\infty) = 0$; the new conditions are shown below. In the n-doped region the charge density reads $\rho = q(p-n+N_D)$; when $x \to +\infty$ the latter becomes $0 = p_{n0} - n_{n0} - N_D$: in fact, as at large distances from the origin the material behaves like a uniformly-doped semiconductor, local charge neutrality is fulfilled at infinity. Using the dimensionless potential $u = q\,\varphi/(k_B T)$, and indicating the derivatives with primes, gives the equation the form

$$u'' = \frac{1}{L_D^2} A_D(u), \quad A_D = \exp(u) - 1 + \frac{p_{n0}}{n_{n0}}[1 - \exp(-u)], \quad L_D^2 = \frac{\varepsilon_{sc} k_B T}{q^2 n_{n0}},$$
$$(21.9)$$

with L_D the *Debye length for the electrons*. The normalized charge density A_D vanishes for $u = 0$, and is positive (negative) when u is positive (negative). Note that, by letting $x \to +\infty$ in $n = n_i\exp[(\varphi-\varphi_F)/(k_B T)]$, $p = n_i\exp[(\varphi_F-\varphi)/(k_B T)]$, and using the normalized Fermi potential $u_F = q\,\varphi_F/(k_B T)$, one finds $p_{n0}/n_{n0} = (n_i/N_D)^2 = \exp(2\,u_F)$, where $u_F < 0$ on account of the fact that here an n-doped region is considered. Following the method illustrated in Sect. 19.5.8 transforms the left hand side of (21.8) into $u''u' = (1/2)[(u')^2]'$. This term is then integrated from $x = 0$ to $x = +\infty$; the result is simplified by observing that the region far from the junction is substantially uniform, whence the electric potential is constant there. As u' is proportional to the electric field, it follows that $u'(+\infty) = 0$: this is the second boundary condition to be added to $u(+\infty) = 0$. In conclusion, the integration of (21.9) from $x \ge 0$ to $+\infty$ yields

$$(u')^2 = \frac{2}{L_D^2} B_D(u), \quad B_D = \exp(u) - 1 - u + \frac{n_i^2}{N_D^2}[u + \exp(-u) - 1]. \quad (21.10)$$

It is easily found that B_D is non negative; in fact, from (21.10) one derives $B_D = 0$ for $u = 0$, and at the same time $dB_D/du = A_D(u)$ is positive for $u > 0$, negative for $u < 0$; as a consequence, B_D grows from zero in either direction when u departs from the origin. Letting $F_D^2 = B_D$ one finds $|u'| = \sqrt{2}\,F_D/L_D$; as the condition $u = 0$ holds only asymptotically, the modulus of u' always grows as u departs from the origin, showing that u is monotonic. Considering that u must fulfill the other boundary condition $u(-\infty) = -q\,\psi_0/(k_B T) < 0$, one concludes that u is a

Fig. 21.2 Solution of the one-dimensional Poisson equation 21.8 in an abrupt p–n junction at equilibrium, with $N_A = 10^{16}$ cm^{-3}, $N_D = 10^{15}$ cm^{-3}. The *continuous vertical line* marks the position of the metallurgical junction, the *dashed vertical lines* mark the edges of the space-charge region

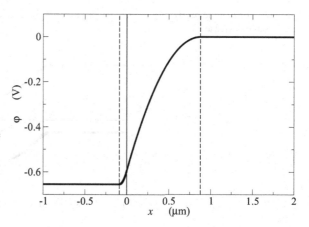

monotonically-growing function, whence, choosing the positive sign and separating the variables, one finds

$$\frac{du}{F_D(u)} = \frac{\sqrt{2}}{L_D} \, dx \qquad (21.11)$$

for $x \geq 0$. The above must be tackled numerically because it has no analytical solution.[4] Once the normalized potential is found from (21.11), its value in the origin, $u(x = 0)$, along with that of the derivative $u'(x = 0) = \sqrt{2} \, F_D[u(x = 0)]/L_D$, provide the boundary conditions for the solution in the p-doped region.

The solution for $x < 0$ follows the same pattern, where the asymptotic neutrality condition reads $n_{n0} \exp(-\chi_0) - p_{n0} \exp(\chi_0) + N_A = 0$, with $\chi_0 = q \, \psi_0/(k_B \, T)$. Letting $v = u + \chi_0$, $n_{p0}/p_{p0} = (n_i/N_A)^2 = \exp(-2 \, u_F) \ll 1$, and using (21.5) provides

$$v'' = \frac{1}{L_A^2} A_A(v), \quad A_A = \frac{n_i^2}{N_A^2} \left[\exp(v) - 1 \right] + 1 - \exp(-v), \quad L_A^2 = \frac{\varepsilon_{sc} \, k_B \, T}{q^2 \, p_{p0}}, \qquad (21.12)$$

where L_A is the *Debye length for the holes*. The rest of the procedure is similar to that used in the n-doped region.

21.2.2 Space-Charge and Quasi-Neutral Regions

The form of the electric potential φ is shown in Fig. 21.2 for a p–n junction at equilibrium with $N_A = 10^{16}$ cm^{-3}, $N_D = 10^{15}$ cm^{-3}. The form of the bands is

---

[4] The numerical procedure is outlined in the note of Sect. 22.2.1.

Fig. 21.3 Form of the bands
for the same device as in
Fig. 21.2

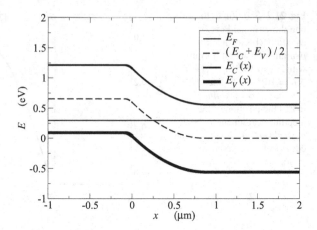

shown in Fig. 21.3 for the same device. It is interesting to note that the device can be
thought of as made of three regions: in the intermediate region, whose boundaries are
marked by dashed vertical lines in Fig. 21.2, the electric potential has a non-negligible
curvature, this showing that the charge density is large. The region is called *space-
charge region*, and contains the metallurgical junction, marked by the continuous
vertical line. In the two regions on the sides of the space-charge region, the electric
potential is nearly constant,[5] whence the electric field $-d\varphi/dx$ is negligibly small.
As a consequence, the charge density $\rho = -\varepsilon_{sc}\, d^2\varphi/dx^2$ is negligible as well; for
this reason, the two regions under consideration are called *quasi-neutral regions*.[6]

The transition from the space-charge region and one or the other quasi-neutral
region is sharp. Thanks to this, it is possible to identify the width l of the space-
charge region and correlate it with other parameters of the device; such a correlation
is worked out in Sect. 21.4 in a specific operating regime. For convenience, the width
of the space-charge region is expressed as $l = l_p + l_n$, where l_p is the extension of the
space-charge region on the p side of the metallurgical junction, and l_n the analogue
on the n side. As shown in Fig. 21.2, it is $l_n > l_p$; as explained in Sect. 21.4, this is
due to the global charge neutrality and to the fact that $N_D < N_A$.

If the equilibrium carrier concentrations corresponding to the electric potential of
Fig. 21.2 are drawn in a logarithmic scale, the curves look similar to that of Fig. 21.2,
apart from scaling factors and from the inversion due to the negative sign in $p =
p_{n0}\exp[-q\,\varphi/(k_B\,T)]$. This is shown in Fig. 21.4. A more realistic representation,
shown in Fig. 21.5, uses a linear scale. The hole concentration p ranges from $p_{p0} \simeq
N_A = 10^{16}$ cm^{-3} in the p-type quasi-neutral region to $p_{n0} \simeq n_i^2/N_D = 10^5$ cm^{-3} in

---

[5] The electric potential can not be exactly constant, because the solution of (21.8) is an analytical
function; as a consequence, if φ were constant in a finite interval, it would be constant everywhere.

[6] The inverse reasoning would not be correct: in fact, $\rho = 0$ may yield $d\varphi/dx = $ const $\neq 0$, which
makes φ a linear function of x; deducing $\varphi = $ const from $\rho = 0$ is correct only if the additional
condition of spatial uniformity holds.

Fig. 21.4 Electron and hole
concentrations in a
one-dimensional, abrupt *p–n*
junction at equilibrium, with
$N_A = 10^{16}$ cm^{-3},
$N_D = 10^{15}$ cm^{-3}. The figure
is drawn in a logarithmic
scale. The *continuous vertical
line* marks the position of the
metallurgical junction, the
dashed vertical lines mark the
edges of the space-charge
region

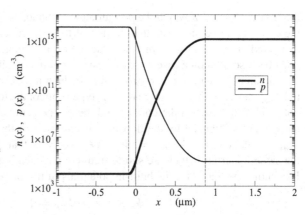

Fig. 21.5 The same
concentrations as in Fig. 21.4,
drawn in a linear scale

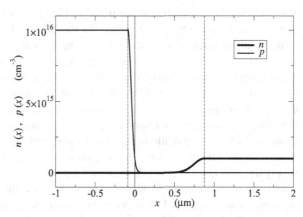

the *n*-type quasi-neutral region; similarly, the electron concentration *n* ranges from
$n_{p0} \simeq n_i^2/N_A = 10^4$ cm^{-3} in the *p*-type quasi-neutral region to $n_{n0} \simeq N_D = 10^{15}$
cm^{-3} in the *n*-type quasi-neutral region. This shows that the two concentrations
vary by several orders of magnitude over the space-charge region, whose length
is about 1 μm; for this reason, if these gradients existed alone, they would make
holes (electrons) to diffuse in the positive (negative) direction of the *x* axis. Such
diffusions in fact do not occur, because in the equilibrium condition they are balanced
by the electric field. The latter is negative: in fact, the electric potential increases
with *x*, so that the force associated to the electric field *E* is negative (positive) for
holes (electrons); the equations for the current densities of the semiconductor device
model in (19.129) and (19.130) yield in this case

$$-q\mu_p\, pE = -qD_p\, \frac{dp}{dx} > 0, \qquad -q\mu_n\, nE = +qD_n\, \frac{dn}{dx} > 0, \qquad (21.13)$$

whence $J_p = J_n = 0$. The description of the *p–n* junction in the equilibrium condition is completed by adding the contacts; as indicated above, this amounts to
introducing two more junctions. The contacts are made of materials different from

the semiconductor of which the p–n junction is made, hence the atomic structure of the contact's material must adapt to that of the semiconductor when the contact is deposited on it; for this reason, the structure of the contact-semiconductor junction must be described on a case-by-case basis. From the qualitative standpoint, one can use the analogy with the p–n junction to deduce the existence of a built-in potential Φ_{mp} between the contact and the p-type semiconductor, and of another built-in potential Φ_{mn} between the contact and the n-type semiconductor. The built-in potentials are influenced by the dopant concentration of the semiconductor, namely, $\Phi_{mp} = \Phi_{mp}(N_A)$ and $\Phi_{mn} = \Phi_{mn}(N_D)$. Assume that the contacts are made of the same material; if they are short-circuited, a closed loop is formed where, from Kirchhoff's voltage law, the built-in potentials fulfill the relation

$$\psi_0 + \Phi_{mn} - \Phi_{mp} = 0. \tag{21.14}$$

This situation is schematically illustrated in Fig. 21.6, where it is assumed that the material of the contacts is the same.[7] In the figure, the built-in potentials at the contacts are represented by discontinuities; in the practical cases, in fact, to prevent the contact-semiconductor junction from behaving like a rectifying device, a heavy dose of dopant is preliminarily introduced into the semiconductor region onto which the contact is to be deposited. For this reason, the spatial extension where Φ_{mp} or Φ_{mn} occurs is negligibly small with respect to the typical scale length of the device. Regardless of this, another important outcome of the fabrication process mentioned above is that the concentration of carriers available in the materials forming a contact is very large; for this reason, as mentioned in Sect. 19.5.6, a contact is able to supply the amount of charge necessary to keep the equilibrium and charge-neutrality conditions in the semiconductor layer adjacent to it. This also implies that, within some limits to be specified later, in a non-equilibrium condition the built-in potential is practically the same as in equilibrium. In circuit theory, a contact whose built-in potential is independent of the current that crosses it is called *ideal Ohmic contact*; this condition is equivalent to that of a vanishing differential resistivity of the contact.

21.3 Shockley Theory of the *P–N* Junction

The analytical derivation of the current-voltage characteristic of the p–n junction is based on the hypothesis that the device is not too far from equilibrium, so that the weak-injection condition (20.35) is fulfilled in the quasi-neutral regions. Within this limit, the approximation that the contacts are ideal is acceptable. A non-equilibrium condition is obtained by applying, e.g., a bias voltage V between the contacts; if V is such that the electric potential at the contact of the p region is higher than that of the n region, the condition is called *forward bias*; when it is lower, the condition is called *reverse bias*. Fig. 21.7 shows the symbol of the p–n junction used in circuit

---

[7] If it is not so, one must add to the left hand side of (21.14) the barrier between the two materials.

Fig. 21.6 Electric potential
for the same device as in
Fig. 21.2, including the
built-in potentials of the
contacts

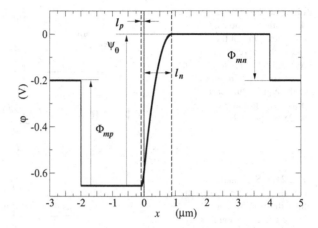

Fig. 21.7 Symbol and
typical *I*, *V* reference for the
p–n junction

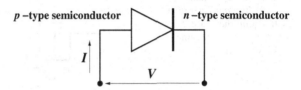

theory, along with the standard references for the applied voltage V and current I. In a one-dimensional case the current is given by $I = A_e J$, with A_e the device cross-sectional area and J the total current density.

Numerical solutions of the semiconductor-device model show that in the weak-injection condition the partitioning of the device into space-charge and quasi-neutral regions still holds; this implies that, when a bias voltage is applied between the contacts, such a voltage adds algebraically to the built-in potential. In fact, the discontinuities at the contacts are the same as in the equilibrium condition due to the contacts' ideality, and the electric potential in the quasi-neutral regions is nearly constant; the extension of the space-charge region, instead, changes due to the application of the external voltage, $l_p = l_p(V)$, $l_n = l_n(V)$. When a forward bias is applied, l_p and l_n slightly decrease, as qualitatively shown in Fig. 21.8 (the drawing is not in the same scale as Fig. 21.6, and is meant only to show the change in l; the solution in the space-charge region is omitted). The same applies to Fig. 21.9, that refers to the reverse bias and shows that in this case l_p and l_n increase. Using (21.14), the application of Kirchhoff's voltage law to either case yields, for the voltage drop ψ across the space-charge region, the expression

$$\psi + \Phi_{mn} + V - \Phi_{mp} = 0, \qquad \psi + V - \psi_0 = 0. \qquad (21.15)$$

In reverse bias ($V < 0$) it is always $\psi > \psi_0 > 0$; in forward bias ($V > 0$) a sufficiently large value of V in (21.15) could make ψ to become negative. However, when V becomes large the weak-injection condition does not hold any more, and

Fig. 21.8 Schematic
description of the change in
the extension l of the
space-charge region in a
forward-biased p–n junction
($V > 0$). The *thin lines* refer
to the equilibrium case. The
drawing is not in the same
scale as Fig. 21.6

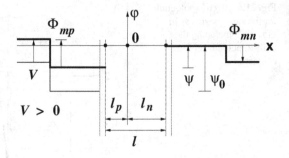

Fig. 21.9 Schematic
description of the change in
the extension l of the
space-charge region in a
reverse-biased p–n junction
($V < 0$). The *thin lines* refer
to the equilibrium case. The
drawing is not in the same
scale as Fig. 21.6

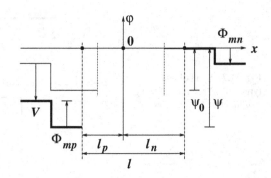

(21.15) does not apply; in conclusion, the range of forward biases to be considered
here is such that the condition $\psi_0 > \psi > 0$ is always fulfilled.

When a forward bias is applied, due to $\psi < \psi_0$ the electric field within the space-
charge region decreases with respect to the equilibrium case;[8] thus, the drift term in
the drift-diffusion equations of (19.129) and (19.130) becomes weaker. The diffusion
term, in contrast, becomes slightly stronger, because the values of the electron concentrations in the quasi-neutral regions are fixed by the asymptotic conditions, and the
width of the space-charge region slightly decreases with respect to the equilibrium
case. In conclusion, the diffusion term prevails and the current-density equations
yield

$$-q D_p \frac{\mathrm{d}p}{\mathrm{d}x} > -q \mu_p \, p \, E > 0, \qquad q D_n \frac{\mathrm{d}n}{\mathrm{d}x} > -q \mu_n \, n \, E > 0, \qquad (21.16)$$

so that $\mathbf{J}_p \cdot \mathbf{i} = q p \, v_p > 0$ and $\mathbf{J}_n \cdot \mathbf{i} = -q n \, v_n > 0$. The total current density
$(\mathbf{J}_p + \mathbf{J}_n) \cdot \mathbf{i}$ is positive as well.

When a reverse bias is applied, due to $\psi > \psi_0$ the voltage drop across the
space-charge region increases with respect to the equilibrium case. The region's
width increases as well; however, the increase in l is relatively weak, whence the

---

[8] The width of the space-charge region decreases as well (Fig. 21.8); such a decrease, however, is
small, and does not compensate for the decrease in the potential drop.

electric field within the space-charge region increases and the drift term in the drift-diffusion equations of (19.129) and (19.130) becomes stronger. The diffusion term, in contrast, becomes weaker, because the values of the electron concentrations in the quasi-neutral regions are fixed by the asymptotic conditions. In conclusion, the drift term prevails and the current-density equations yield

$$-q\mu_p\, pE > -qD_p\frac{\mathrm{d}p}{\mathrm{d}x} > 0, \qquad -q\mu_n\, nE > qD_n\frac{\mathrm{d}n}{\mathrm{d}x} > 0, \qquad (21.17)$$

so that $\mathbf{J}_p \cdot \mathbf{i} = qp\, v_p < 0$, $\mathbf{J}_n \cdot \mathbf{i} = -qn\, v_n < 0$. The total current density is negative as well.

The $I(V)$ relation of the p–n junction is worked out here in the one-dimensional and steady-state case, this leading to the *Shockley equations* [97, 98]. The steady-state form $\mathrm{div}\mathbf{J} = 0$ of the continuity equation (4.23) reduces in one dimension to $\mathrm{d}J/\mathrm{d}x = 0$, whence

$$J = J_p(x) + J_n(x) = \text{const.} \qquad (21.18)$$

The hole and electron current densities depend on position and fulfill the steady-state continuity equations of (19.129) and (19.130); in the latter, only the net thermal recombination term is considered, to find

$$\frac{\mathrm{d}J_p}{\mathrm{d}x} = -qU_{\mathrm{SRH}}, \qquad \frac{\mathrm{d}J_n}{\mathrm{d}x} = qU_{\mathrm{SRH}}. \qquad (21.19)$$

Once J_p, J_n are determined from (21.19), they are specified at a suitable position and added up. As shown below, such positions are the boundaries $-l_p$ and l_n between the space-charge and quasi-neutral regions, for instance, $J = J_p(-l_p) + J_n(-l_p)$. Observing that $-l_p$ is the boundary of the p-type region, $J_p(-l_p)$ is a majority-carrier current density, whereas $J_n(-l_p)$ is a minority-carrier current density. The opposite happens if the other boundary is chosen, to yield $J = J_p(l_n) + J_n(l_n)$. To proceed, it is convenient to seek for an expression of J where both current densities refer to minority carriers; this is achieved by integrating (21.19) over the space-charge region, to define the *recombination current density*

$$J_U = \int_{-l_p}^{l_n} qU_{\mathrm{SRH}}\ \mathrm{d}x = J_p(-l_p) - J_p(l_n) = J_n(l_n) - J_n(-l_p). \qquad (21.20)$$

Combining (21.20) with the expression of the total current density at, e.g., l_n provides

$$J = J_p(l_n) + J_n(l_n) = J_p(l_n) + J_n(-l_p) + J_U, \qquad (21.21)$$

which has the desired form.

21.3.1 Derivation of the $I(V)$ Characteristic

Remembering the simplified form (20.38) or (20.40) of the net thermal-recombination term in the weak-injection condition, and using $p^{eq} = p_{n0}$ in the n-type region and $n^{eq} = n_{p0}$ in the p-type region, transforms (21.19) into, respectively,

$$\frac{dJ_p}{dx} = q \frac{p - p_{n0}}{\tau_p}, \qquad x > l_n, \tag{21.22}$$

$$\frac{dJ_n}{dx} = q \frac{n - n_{p0}}{\tau_n}, \qquad x < -l_p. \tag{21.23}$$

In this way, the hole- and electron-continuity equations are decoupled from each other; also, they are to be solved over disjoint intervals. To the current-continuity equations one associates the corresponding drift-diffusion equation taken from (19.129) or (19.130); it follows that in each quasi-neutral region the drift-diffusion equation is that of the minority carriers. It can be shown that in a quasi-neutral region, when the weak-injection condition holds, the diffusion term of the minority carries dominates over the drift term (the details are worked out in Sect. 21.6.1). Thus, for $x < -l_p$ (p-type region), $J_n = q \mu_n n E + q D_n dp/dx \simeq q D_n dp/dx$. Inserting the latter into (21.23) yields $D_n d^2n/dx^2 = (n - n_{p0})/\tau_n$. In this derivation the diffusion coefficient $D_n = k_B T \mu_n/q$ is not subjected to the derivative; in fact, the two parameters that influence bulk mobility, T and N_A (Sect. 20.5), are independent of position. The equation then reads,

$$\frac{d^2(n - n_{p0})}{dx^2} = \frac{n - n_{p0}}{L_n}, \qquad L_n = \sqrt{\tau_n D_n}, \tag{21.24}$$

with L_n the *diffusion length* of the minority carriers in the p-type region. It is a second-order, linear equation in the unknown n; it is decoupled from the rest of the semiconductor-device model: in fact, the simplified form of the net-recombination term contains only the electron concentration, and the neglect of the drift term eliminates the coupling with the Poisson equation. The boundary conditions must be fixed at $x \to -\infty$ and $x = -l_p$; the former is $n(-\infty) = n_{p0}$, whereas the latter needs a more elaborate derivation, given in Sect. 21.6.2, whose outcome (also called *Shockley's boundary condition*) is $n(-l_p) = n_{p0} \exp[q V/(k_B T)]$. The general solution of (21.24) is

$$n = n_{p0} + A_n \exp(x/L_n) + B_n \exp(-x/L_n), \tag{21.25}$$

whence the asymptotic boundary condition yields $B_n = 0$. The other boundary condition provides $n(-l_p) = n_{p0} + A_n^-$, with $A_n^- = A_n \exp(-l_p/L_n)$. The electron current density is then found from

$$J_n = q D_n \frac{dn}{dx} = q \frac{D_n}{L_n} A_n \exp(x/L_n) = J_n(-l_p) \exp[(x + l_p)/L_n], \tag{21.26}$$

where, using $n(-l_p) = n_{p0} + A_n^-$ and the boundary condition at $x = -l_p$,

$$J_n(-l_p) = \frac{q D_n A_n^-}{L_n} = \frac{q D_n n_{p0}}{L_n} F, \qquad F(V) = \exp[q V/(k_B T)] - 1. \quad (21.27)$$

In the same manner one finds

$$J_p(l_n) = \frac{q D_p p_{n0}}{L_p} F, \qquad L_p = \sqrt{\tau_p D_p}, \quad (21.28)$$

where F is the same as in (21.27), and L_p is the diffusion length of the minority carriers in the n-type region. Inserting (21.27) and (21.28) into (21.21) yields the total current density,

$$J = J_p(l_n) + J_n(-l_p) + J_U = q \left(\frac{D_p p_{n0}}{L_p} + \frac{D_n n_{p0}}{L_n} \right) F + J_U. \quad (21.29)$$

Multiplying (21.29) by the cross-sectional area A_e and defining the *saturation current density*

$$J_s = q \left(\frac{D_p p_{n0}}{L_p} + \frac{D_n n_{p0}}{L_n} \right) = q n_i^2 \left(\frac{\sqrt{D_p/\tau_p}}{N_D} + \frac{\sqrt{D_n/\tau_n}}{N_A} \right), \quad (21.30)$$

yields the expression of the $I(V)$ characteristic of the p–n junction:

$$I = I_s \left[\exp\left(\frac{q V}{k_B T} \right) - 1 \right] + I_U, \qquad I_s = A_e J_s, \quad I_U = A_e J_U. \quad (21.31)$$

The characteristic fulfills the equilibrium condition $I(0) = 0$; in fact, at equilibrium it is $U_{SRH} = 0$. When $q V/k_B T \gg 1$, the exponential term in (21.31) prevails over the other terms and the characteristic becomes $I \simeq I_s \exp[q V/(k_B T)]$, namely, the well-known exponential form of the forward-bias case. Finally, when $q V./k_B T \ll -1$, the current becomes $I \simeq -I_s + I_U$. As the order of magnitude of I_s may be similar to that of I_U, it is necessary to calculate the latter explicitly. The analysis is made easier by the observation that the electric field (which in the reverse-bias condition prevails over diffusion) drains the holes from the space-charge region to the p-type quasi-neutral region; similarly, the electrons of the space-charge region are drained towards the n-type quasi-neutral region. As a consequence, the carrier concentrations in the space-charge region are negligible, and the full-depletion condition (20.34) applies there; using the non-degenerate expression one finds, for the reverse-bias current,

$$I \simeq -I_s + I_U \simeq -I_s + A_e \int_{-l_p}^{l_n} q \frac{-n_i}{\tau_g}\, dx = -I_s - q A_e \frac{n_i}{\tau_g} l(V) < 0, \quad (21.32)$$

where the expression (21.20) of the recombination current density has been used, and $l = l(V)$ is the width of the space-charge region. As shown in Sect. 21.4, in

Fig. 21.10 Charge density in
a reverse-biased p–n junction
using the ASCE
approximation, in arbitrary
units. The ratio N_A/N_D is the
same as in Fig. 21.6

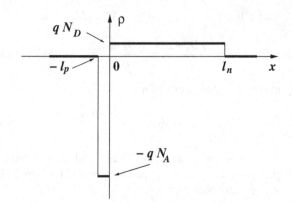

the reverse-bias condition and for an abrupt junction it is $l \propto \sqrt{\psi_0 + |V|}$; as a consequence, I_U increases with $|V|$.

The approximations that have been introduced to derive the $I(V)$ characteristic are many; in fact, (21.31) is referred to as the *ideal characteristic*. However, it captures quite well the general behavior of the device as long as the applied voltage is within the limit of the weak-injection approximation. When the forward bias exceeds such a limit, the drift term is not negligible anymore and the effect of the electric field in the quasi-neutral regions must be accounted for. Considering in turn the reverse-bias condition, at large values of $|V|$ the electric field in the space-charge region becomes sufficiently strong to induce impact ionization (Sect. 20.3) and, possibly, the junction breakdown due to avalanche (Sect. 21.5).

The dependence of the $I(V)$ characteristic on temperature is due, besides that of $q\,V/(k_B\,T)$, to the coefficients of J_s and J_U. To this purpose, the second form of (21.30) is more useful because it shows explicitly the term n_i^2, whose temperature dependence is exponential (18.14); in fact, the temperature dependence of D_p/τ_p, D_n/τ_n is much weaker. The same considerations apply to J_U, whose main dependence on temperature is due to factor n_i. The dependence on n_i of the reverse current (at constant temperature) has been used in the considerations about the parasitic currents in integrated circuits made in Sect. 18.7.

21.4   Depletion Capacitance of the Abrupt *P–N* Junction

It has been anticipated in Sect. 21.3.1 that, when a reverse bias is applied to the junction, the full-depletion condition holds in the space-charge region; as a consequence, the charge density ρ in the latter is essentially due to the dopant atoms. In the abrupt junction considered here, the dopants' concentration is piecewise constant; a simplified description of the charge density in the situation in hand is obtained from the *abrupt space-charge edge* (ASCE) approximation, which describes ρ with the form sketched in Fig. 21.10. In other terms, the approximation consists in replacing with a discontinuity the smooth change of ρ at $x = -l_p$, $x = 0$, and $x = l_n$; thus, the space charge is given by

$$\rho = -qN_A, \qquad -l_p < x < 0, \tag{21.33}$$

$$\rho = qN_D, \qquad 0 < x < l_n, \tag{21.34}$$

and $\rho = 0$ elsewhere. The electric field and the electric potential are continuous because there are no charge layers or double layers; letting $E_0 = E(0)$, from $dE/dx = \rho/\varepsilon_{sc}$ and (21.33), (21.34) one draws

$$\frac{E_0 - E(-l_p)}{l_p} = -\frac{qN_A}{\varepsilon_{sc}}, \qquad \frac{E(l_n) - E_0}{l_n} = \frac{qN_D}{\varepsilon_{sc}}, \tag{21.35}$$

the first of which holds for $-l_p \le x \le 0$, the second one for $0 \le x \le l_n$. In the quasi-neutral regions the field is negligible; in the order of approximation used here one lets $E = 0$ in such regions, whence $E(-l_p) = E(l_n) = 0$ and, from (21.35),

$$E_0 = -\frac{qN_D}{\varepsilon_{sc}}l_n = -\frac{qN_A}{\varepsilon_{sc}}l_p < 0, \qquad N_D l_n = N_A l_p. \tag{21.36}$$

In conclusion, the electric field is a piecewise-linear function whose form is shown in Fig. 21.11. Due to $d\varphi/dx = -E$, the integral of $-E$ over the space-charge region equals the potential drop ψ:

$$\psi = \varphi(l_n) - \varphi(-l_p) = -\int_{-l_p}^{+l_n} E \, dx = -\frac{1}{2} E_0 (l_n + l_p). \tag{21.37}$$

Inserting into (21.37) one or the other form of E_0 from (21.36), one obtains two equivalent expressions for $l_p + l_n$:

$$l_n + l_p = l_n + \frac{N_D}{N_A}l_n = N_D l_n \left(\frac{1}{N_D} + \frac{1}{N_A}\right) = N_A l_p \left(\frac{1}{N_D} + \frac{1}{N_A}\right). \tag{21.38}$$

Then, combining (21.38) with (21.37) one finds

$$\psi = \frac{q}{2\varepsilon_{sc}} \left(\frac{1}{N_D} + \frac{1}{N_A}\right) (N_D l_n)^2 = \frac{q}{2\varepsilon_{sc}} \left(\frac{1}{N_D} + \frac{1}{N_A}\right) (N_A l_p)^2, \tag{21.39}$$

whence

$$l_n = \frac{1}{N_D} \left(\frac{2\varepsilon_{sc}\psi/q}{1/N_D + 1/N_A}\right)^{1/2}, \qquad l_p = \frac{1}{N_A} \left(\frac{2\varepsilon_{sc}\psi/q}{1/N_D + 1/N_A}\right)^{1/2}, \tag{21.40}$$

and

$$l = l_n + l_p = \left[\frac{2\varepsilon_{sc}}{q}\left(\frac{1}{N_D} + \frac{1}{N_A}\right)\psi\right]^{1/2}, \qquad \psi = \psi_0 - V. \tag{21.41}$$

Fig. 21.11 Electric field
consistent with the charge
density of Fig. 21.10, in
arbitrary units

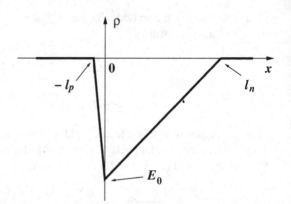

Multiplying by $q A_e$ both sides of the second relation in (21.36) yields $q N_D A_e l_n = q N_A A_e l_p$, that represents the global charge conservation in the device. Such a conservation is implied by the assumption that $E = 0$ in the quasi-neutral regions, as is found by integrating $\mathrm{div}\mathbf{D} = \rho$ over the space-charge region. In the charge-conservation relation the widths l_p, l_n depend on V through (21.40); it follows that in the reverse-bias condition the device can be assimilated to a non-linear capacitor where the charge per unit area of the two oppositely-charged sides is, respectively, $Q_p = -q N_A l_p$ and $Q_n = q N_D l_n$. The differential capacitance per unit area is defined as $C = \mathrm{d}Q_p/\mathrm{d}V = -\mathrm{d}Q_n/\mathrm{d}V$; from the definition,[9] two equivalent expressions follow,

$$C = -q N_A \frac{\mathrm{d}l_p}{\mathrm{d}V} = q \frac{\mathrm{d}(N_A l_p)}{\mathrm{d}\psi}, \qquad C = -q N_D \frac{\mathrm{d}l_n}{\mathrm{d}V} = q \frac{\mathrm{d}(N_D l_n)}{\mathrm{d}\psi}. \qquad (21.42)$$

Using (21.40), the differential capacitance per unit area of the abrupt p–n junction is found to be

$$C = \left[\frac{q \varepsilon_{sc}/(2 \psi)}{1/N_D + 1/N_A} \right]^{1/2} = \frac{[(q \varepsilon_{sc}/2)/(1/N_D + 1/N_A)]^{1/2}}{[\psi_0 (1 - V/\psi_0)]^{1/2}}, \qquad (21.43)$$

which is given the more compact form[10]

$$C = C_0 \left(1 - \frac{V}{\psi_0} \right)^{-1/2}, \qquad C_0 = C(V = 0) = \left[\frac{q \varepsilon_{sc}/(2 \psi_0)}{1/N_D + 1/N_A} \right]^{1/2}. \qquad (21.44)$$

---

[9] Definition $C = \mathrm{d}Q_p/\mathrm{d}V$ is coherent with the choice of the reference in Fig. 21.7. The units of C are $[C] = \mathrm{F\,cm}^{-2}$. Compare with the calculation of the MOS capacitance in Sect. 22.3.

[10] It is worth reminding that the result holds only in the reverse-bias condition. In the forward-bias condition the injection of carriers from the quasi-neutral regions into the space-charge region prevents one from neglecting the contribution of the carrier concentrations to the charge density and makes the use of (21.44) erroneous.

Combining (21.44) with (21.41) one derives the interesting relation

$$\frac{1}{C^2} = \frac{2}{q\varepsilon_{sc}} \left(\frac{1}{N_D} + \frac{1}{N_A} \right) \psi = \frac{l^2}{\varepsilon_{sc}^2}, \qquad C = \frac{\varepsilon_{sc}}{l}, \qquad (21.45)$$

namely, the standard expression for the capacitance per unit area of the parallel-plate capacitor. Such an expression is not limited to the case where the dopant concentration is piecewise constant; as shown in Sect. 21.6.3, it applies in fact to all cases.

From the standpoint of circuit design, the capacitance associated to a p–n junction is a parasitic effect that hampers the circuit's speed. However, the effect is also exploited to manufacture voltage-controlled capacitors, called *variable capacitors* or *varactors*. In these devices, that are operated in reverse bias, the geometry is designed to maximize the capacitance; they are used, e.g., in voltage-controlled oscillators, parametric amplifiers, and frequency modulation.[11] The bias range of these devices is such that the reverse current is negligibly small; if the modulus of the reverse bias is made to increase, and is eventually brought outside this range, carrier multiplication due to impact ionization (Sect. 20.3) takes place; this, as shown in Sect. 21.5, leads to a strong increase of the reverse current.

21.5 Avalanche Due to Impact Ionization

The situation where impact ionization dominates over the other generation-recombination mechanisms has been illustrated in Sect. 20.3.1, showing that in the steady-state case the continuity equations for electrons and holes reduce to (20.45). Such equations are applicable, for instance, to the space-charge region of a reverse-biased p–n junction; when the value of $|V|$ becomes large, the increase in the number of carriers due to impact ionization may give rise to an avalanche phenomenon that eventually leads to the junction's *avalanche breakdown*, namely, a strong increase in the current due to carrier multiplication.[12] The absolute value V_B of the voltage at which the phenomenon occurs is called *breakdown voltage*. To illustrate avalanche, the one-dimensional case is considered, so that (20.45) become

$$\frac{dJ_n}{dx} = k_n J_n + k_p J_p, \qquad \frac{dJ_p}{dx} = -k_n J_n - k_p J_p, \qquad (21.46)$$

---

[11] Varactors are also manufactured using technologies other than the bipolar one; e.g., with MOS capacitors or metal-semiconductor junctions.

[12] If the breakdown is accompanied by current crowding, the junction may be destroyed due to excessive heating. Special p–n junctions, called *avalanche diodes*, are designed to have breakdown uniformly spread over the surface of the metallurgical junction, to avoid current crowding. Such devices are able to indefinitely sustain the breakdown condition; they are used as voltage reference and for protecting electronic circuits against excessively-high voltages.

where the impact-ionization coefficients k_n, k_p depend on x through the electric field E and are determined experimentally.[13] To ease the notation the boundaries of the space-charge region are indicated with a, b; also, considering the reference's orientation (Fig. 21.7), it is $E, J_n, J_p < 0$, where the electric field is significant only for $a \le x \le b$. As in the one-dimensional and steady-state case it is $J = J_n(x) + J_p(x) = \text{const}$, eliminating J_p from the first equation in (21.46) yields

$$\frac{dJ_n}{dx} = k_n J_n + k_p (J - J_n) = (k_n - k_p) J_n + k_p J, \qquad (21.47)$$

namely, a first-order equation in J_n containing the yet undetermined parameter J. The equation is recast as

$$\frac{dJ_n}{dx} - \frac{dm}{dx} J_n = k_p J = k_n J - \frac{dm}{dx} J, \qquad m = \int_a^x (k_n - k_p) \, dx' \qquad (21.48)$$

where $m(a) = 0$, $dm/dx = k_n - k_p$. Multiplying by the integrating factor $\exp(-m)$ and dividing by J transforms (21.48) into

$$\frac{1}{J} \frac{d}{dx} [J_n \exp(-m)] = k_n \exp(-m) - \frac{dm}{dx} \exp(-m), \qquad (21.49)$$

where $-(dm/dx) \exp(-m) = d \exp(-m)/dx$. Integrating (21.49) from a to b and using $m(a) = 0$ yields

$$\frac{J_n(b)}{J} \exp[-m(b)] - \frac{J_n(a)}{J} = Y_n + \exp[-m(b)] - 1, \qquad (21.50)$$

where the *electron-ionization integral* is defined as

$$Y_n = \int_a^b k_n \exp(-m) \, dx. \qquad (21.51)$$

The above result is somewhat simplified by observing that, due to impact ionization, the concentration of electrons in the conduction band increases from a to b, whereas that of holes increases from b to a; the concept is rendered in Fig. 21.12: consider a portion $x_1 < x < x_2$ of the space-charge region, where it is assumed for simplicity that the electric potential is linear. As the electric field is oriented to the left, electrons (indicated by the black dots) are accelerated to the right, holes (the white dots) are accelerated to the left. The vertical lines indicate the exchange of energies involved in impact-ionization events initiated by electrons or holes. An electron transited from the valence to the conduction band is accelerated by the field and may acquire a kinetic energy sufficient for initiating an impact-ionization event itself; the same applies to holes. As a consequence, the number of conduction-band electrons is

---

[13] An example of model for k_n, k_p is that proposed by Chynoweth [14]: $k_n = k_{ns} \exp(-|E_{cn}/E|^{\beta_n})$, $k_p = k_{ps} \exp(-|E_{cp}/E|^{\beta_p})$, where the parameters depend on temperature [83, 111].

Fig. 21.12 Schematic
description of the avalanche
phenomenon. The details are
given in the text

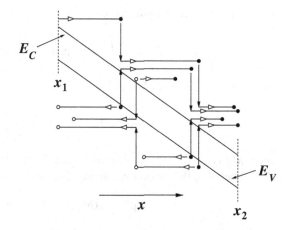

multiplied from left to right, namely, the number of those exiting at x_2 is larger than
the number entering at x_1; similarly, the number of valence-band holes is multiplied
from right to left. Due to the multiplication mechanism, taking $x = b$ by way of
example, the major contribution to $J = J_n(b) + J_p(b)$ is given by $J_n(b)$. Then, letting
$J \simeq J_n(b)$ in (21.50) and canceling out some terms provides

$$1 - \frac{1}{M_n} = Y_n, \qquad M_n = \frac{J_n(b)}{J_n(a)} \geq 1, \qquad (21.52)$$

where M_n is the *electron-multiplication factor*. The latter is a measure of the impact
ionization's level. As long as $Y_n < 1$, corresponding to a finite value of M_n, the
avalanche condition does not occur; when $Y_n \to 1$, then $M_n \to \infty$: in this case, the
injection of a negligibly-small number of electrons at a produces a large electron
current density at b. The operating conditions where M_n is large must be avoided
because an excessive current may damage the device. Note that in the design stage
of the device one first calculates Y_n from (21.51), then obtains M_n from (21.52).
Thus, it may well happen that $Y_n > 1$, corresponding to $M_n < 0$; this outcome is not
physically sound, and simply indicates that the parameters used in the calculation of
the ionization integral (21.51) are not consistent.[14]

The analysis of the impact-ionization condition can also be carried out starting with the elimination of J_n from the second equation in (21.46). The equation
corresponding to (21.48) reads

$$\frac{\mathrm{d}J_p}{\mathrm{d}x} = -k_p J_p - k_n (J - J_p) = \frac{\mathrm{d}m}{\mathrm{d}x} J_p - k_n J ; \qquad (21.53)$$

in turn, the equation corresponding to (21.50) is

$$\frac{J_p(b)}{J} \exp[-m(b)] - \frac{J_p(a)}{J} = -\frac{Y_p}{\exp[m(b)]} + \exp[-m(b)] - 1, \qquad (21.54)$$

---

[14] This happens, for instance, if a value of $|V|$ larger than the breakdown voltage is used in (21.51).

with the *hole-ionization integral* given by

$$Y_p = \int_a^b k_p \exp\left[m(b) - m\right] \mathrm{d}x. \tag{21.55}$$

Letting $J \simeq J_p(a)$ in (21.54) yields

$$1 - \frac{1}{M_p} = Y_p, \qquad M_p = \frac{J_p(a)}{J_p(b)} \geq 1, \tag{21.56}$$

where M_p is the *hole-multiplication factor*. Using the definition (21.48) of m, the relation between the ionization integrals is found to be

$$Y_n = \exp\left[-m(b)\right] Y_p + 1 - \exp\left[-m(b)\right]. \tag{21.57}$$

The above shows that $Y_p = 1$ corresponds to $Y_n = 1$, namely, the avalanche condition $Y_p = 1$ for the holes coincides with that of the electrons, as should be.

21.6 Complements

21.6.1 Weak-Injection Limit of the Drift-Diffusion Equations

In the calculation of the $I(V)$ characteristic of the p–n junction carried out in Sect. 21.3.1 it has been stated that in a quasi-neutral region, when the weak-injection condition holds, the diffusion term of the minority carries dominates over the drift term. To better discuss this issue, the case of a p-doped region is considered, so that the majority-carrier concentration is $c^{\mathrm{eq}} = p^{\mathrm{eq}} = p_{p0}$ and (20.35) become $|p - p_{p0}| \ll p_{p0}, |n - n_{p0}| \ll p_{p0}$. The latter may be recast as

$$|p - p_{p0}| \leq \alpha\, p_{p0}, \qquad |n - n_{p0}| \leq \alpha\, p_{p0}, \tag{21.58}$$

with $\alpha \ll 1$. Indicating with p_m, p_M the minimum and maximum values of p imposed by (21.58), one finds $p_M - p_{p0} = \alpha\, p_{p0}$, $p_{p0} - p_m = \alpha\, p_{p0}$, whence

$$p_M = (1 + \alpha)\, p_{p0}, \qquad p_m = (1 - \alpha)\, p_{p0}. \tag{21.59}$$

Similarly, for the minority-carrier concentration one finds $n_M - n_{p0} = \alpha\, p_{p0}$, $n_{p0} - n_m = \alpha\, p_{p0}$, whence

$$n_M = n_{p0} + \alpha\, p_{p0}, \qquad n_m = n_{p0} - \alpha\, p_{p0}. \tag{21.60}$$

The maximum absolute variation of p turns out to be:

$$p_M - p_m = 2\alpha\, p_{p0}. \tag{21.61}$$

Instead, the maximum variation of n must be treated with some care. In fact, using the non-degenerate case one finds

$$n_M = \frac{n_i^2}{p_{p0}} + \alpha \, p_{p0} = p_{p0} \left(\frac{n_i^2}{p_{p0}^2} + \alpha \right), \tag{21.62}$$

$$n_m = \frac{n_i^2}{p_{p0}} - \alpha \, p_{p0} = p_{p0} \left(\frac{n_i^2}{p_{p0}^2} - \alpha \right). \tag{21.63}$$

Even for a relatively low dopant concentration, say, $N_A \simeq p_{p0} = 10^{16} \, \text{cm}^{-3}$, at room temperature one has $n_i^2/p_{p0}^2 \simeq 10^{-12}$, which is much smaller than the reasonable values of α. It follows $n_M \simeq \alpha \, p_{p0}$, $n_m \simeq 0$, where the limit of n_m must be chosen as such because n is positive definite. In conclusion, the maximum relative variations of p and n with respect to the equilibrium values are given by

$$\frac{p_M - p_m}{p_{p0}} = 2\,\alpha, \qquad \frac{n_M - n_m}{n_{p0}} \simeq \alpha \, \frac{p_{p0}}{n_{p0}} = \alpha \, \frac{p_{p0}^2}{n_i^2} \gg 2\,\alpha. \tag{21.64}$$

By way of example, one may let $\alpha = 10^{-3}$, still with $N_A = 10^{16} \, \text{cm}^{-3}$. While the maximum relative variation of p is 2×10^{-3}, that of n is 10^9; it follows that the constraint imposed onto the derivative is strong in the case of p, much weaker for n. Within the same example, the maximum absolute variation is $2 \times 10^{13} \, \text{cm}^{-3}$ for p and $10^{13} \, \text{cm}^{-3}$ for n, in both cases much smaller than the majority-carrier concentration ($10^{16} \, \text{cm}^{-3}$). The conclusion is that in a quasi-neutral region, under the weak-injection conditions, the diffusive transport prevails for the minority carriers, whereas the transport of the majority carriers is dominated by drift. With reference to the p-doped region considered here, one has $J_p \simeq q \mu_p \, p \, E$ and $J_n \simeq q D_n \, \text{grad} \, n$, respectively.

21.6.2 Shockley's Boundary Conditions

The derivation of the analytical model of the p–n junction's $I(V)$ characteristic, worked out in Sect. 21.3.1, requires the boundary conditions for the minority-carrier concentrations at the boundaries of the space-charge region; specifically, one needs to determine $n(-l_p)$ and $p(l_n)$. The derivation is based on calculating approximate expressions for the ratios $n(-l_p)/n(l_n)$, $p(l_n)/p(-l_p)$, where the denominators are majority-carrier concentrations that are in turn approximated with $n(l_n) \simeq n_{n0} \simeq N_D$ and $p(-l_p) \simeq p_{p0} \simeq N_A$.

To proceed, one considers the electron drift-diffusion equation $J_n = q \mu_n \, n \, E + q D_n \, dn/dx$, and observes that in the space-charge region the drift and diffusion terms have opposite signs; also, their moduli are much larger than that of the current density. In fact, the latter is small due to the weak-injection condition, whereas the terms at the right hand side of the equation are large because the electric potential and the electron concentration have non-negligible variations over the space-charge region. It follows that the moduli of the drift and diffusion terms are comparable to each other: $-q\mu_n n E \simeq q D_n \, dn/dx \gg |J_n|$ and, similarly, $-q\mu_p p E \simeq -q D_p \, dp/dx \gg |J_p|$ for holes. Now, the approximation is introduced, that consists in neglecting J_n and J_p; this yields equilibrium-like expressions for the concentrations, $n \simeq n^{(0)} \exp[q\varphi/(k_B T)]$, $p \simeq n^{(0)} \exp[-q\varphi/(k_B T)]$, which are used to calculate the ratios sought:

$$\frac{n(-l_p)}{n(l_n)} \simeq \exp\left[\frac{q(V-\psi_0)}{k_B T}\right] = \frac{n_i^2}{N_A N_D} \exp\left(\frac{qV}{k_B T}\right), \tag{21.65}$$

$$\frac{p(l_n)}{p(-l_p)} \simeq \exp\left[\frac{q(V-\psi_0)}{k_B T}\right] = \frac{n_i^2}{N_A N_D} \exp\left(\frac{qV}{k_B T}\right). \tag{21.66}$$

The last form of (21.65), (21.66) is obtained from the definition (21.5) of the built-in potential. Using $n(l_n) \simeq n_{n0} \simeq N_D$ and $p(-l_p) \simeq p_{p0} \simeq N_A$ along with $n_{p0} = n_i^2/N_A$ and $p_{n0} = n_i^2/N_D$ (compare with (21.1,21.2)), finally yields the Shockley boundary conditions

$$n(-l_p) \simeq n_{p0} \exp\left(\frac{qV}{k_B T}\right), \qquad p(l_n) \simeq p_{n0} \exp\left(\frac{qV}{k_B T}\right). \tag{21.67}$$

21.6.3   Depletion Capacitance—Arbitrary Doping Profile

The expression of the depletion capacitance worked out in Sect. 21.4 for an abrupt p–n junction is extended here to an arbitrary doping profile, still in one dimension. Let $a < x < b$ be the region where the charge density ρ differs from zero, and assume for the electric potential that $\varphi = \varphi(a) = $ const for $x < a$ and $\varphi = \varphi(b) = $ const for $x > b$. Also, it is assumed that there are no single layers or double layers of charge, whence the electric field E and φ are continuous. The constancy of φ in the outside regions implies the global charge neutrality, as is found by integrating $\varepsilon_{sc} \, dE/dx = \rho$ from a to b and using the continuity of E:

$$\int_a^b \rho \, dx = 0. \tag{21.68}$$

Thanks to (21.68) one finds, for any x,

$$\int_a^x \rho \, dx + \int_x^b \rho \, dx = 0 \qquad Q = -\int_a^x \rho \, dx = \int_x^b \rho \, dx, \tag{21.69}$$

which provides the definition of the charge per unit area Q. The definition holds also if x is outside the interval $[a, b]$; in this case, however, one finds $Q = 0$. In the following it is assumed that x is internal to the space-charge region. The solution of the Poisson equation is taken from Prob. 4.2; using $E(a) = 0$ and the global charge-neutrality condition after letting $\psi = \varphi(b) - \varphi(a)$, yields

$$\varepsilon_{sc} \psi = \int_a^b x\rho \, dx. \tag{21.70}$$

If the voltage drop changes by a small amount, $\psi \leftarrow \psi + d\psi$, the space-charge boundaries are modified, $a \leftarrow a + da$, $b \leftarrow b + db$, whence Q changes as well:[15]

$$dQ = \int_b^{b+db} \rho \, dx = \rho(b) \, db = \int_a^{a+da} \rho \, dx = \rho(a) \, da. \tag{21.71}$$

On the other hand, from (21.70) it follows

$$\varepsilon_{sc} \, d\psi = \int_{a+da}^{b+db} x \, \rho \, dx - \int_a^b x\rho \, dx = b \, \rho(b) \, db - a \, \rho(a) \, da = (b-a) \, dQ. \tag{21.72}$$

Thus, the capacitance per unit area of the space-charge region is

$$C = \frac{dQ}{d\psi} = \frac{\varepsilon_{sc}}{b-a}, \tag{21.73}$$

which is the expected generalization of (21.45). Note that the absence of charge layers makes the variation dQ to depend on the variations in a and b only. As a consequence it is $a = a(\psi)$, $b = b(\psi)$, whence $C = C(\psi)$. If ψ and $\rho(x)$ are prescribed, the values of a, b are determined by the system of Eqs. (21.68) and (21.70).

It is interesting to note that if the charge density has a power form, $\rho \sim x^n$, then ψ depends on the $(n+2)$th power of $b - a$. Consider by way of example a *diffused junction*, namely, a junction obtained, e.g., by diffusing a dopant of the p type into a uniform n-type substrate. Expanding ρ to first order around the metallurgical junction, and using the full-depletion and ASCE approximations, yields $\rho \simeq k \, x$ for $-l/2 < x < l/2$ and $\rho = 0$ elsewhere. Using (21.70) and (21.72) then yields

$$\varepsilon_{sc} \psi = \frac{1}{12} k \, l^3, \qquad C = C_0 \left(1 - \frac{V}{\psi_0}\right)^{-1/3}, \qquad C_0 = \left(\frac{k \, \varepsilon_{sc}^2}{12 \, \psi_0}\right)^{1/3}. \tag{21.74}$$

The general expression (21.72) of the capacitance per unit area of the space-charge region finds a useful application in a measuring technique for the doping profile (Sect. 25.5).

---

[15] After the change in the boundaries' positions, x in (21.69) is still internal to the space-charge region.

21.6.4   Order of Magnitude of Junction's Parameters

Still considering an abrupt, p–n silicon junction with $N_A = 10^{16}$ cm^{-3}, $N_D = 10^{15}$ cm^{-3}, the built-in potential at room temperature is

$$\psi_0 = \frac{k_B T}{q} \log \left(\frac{N_A N_D}{n_i^2} \right) \simeq 0.65 \text{ V} \tag{21.75}$$

(compare with (21.5)). The carrier mobilities have been estimated in Sect. 19.6.6; in fact, hole mobility is smaller than electron mobility and, as outlined in Sect. 20.5.3, the mobility degradation due to impurity scattering is expected to vary from one side of the junction to the other because the doping concentrations are different. The experimental minority-carrier mobilities for the doping concentrations and temperature considered here are $\mu_n \simeq 1000$ cm^2 V^{-1} s^{-1} in the p region and $\mu_p \simeq 500$ cm^2 V^{-1} s^{-1} in the n region [103, Sect. 1.5], whence

$$D_n = \frac{k_B T}{q} \mu_n \simeq 26 \text{ cm}^2 \text{ s}^{-1}, \qquad D_p = \frac{k_B T}{q} \mu_p \simeq 13 \text{ cm}^2 \text{ s}^{-1}. \tag{21.76}$$

The experimental values of the minority-carrier lifetimes are $\tau_n \simeq 5 \times 10^{-5}$ s and $\tau_p \simeq 2 \times 10^{-5}$ s. The corresponding diffusion lengths (21.24), (21.28) are

$$L_n = \sqrt{\tau_n D_n} \simeq 360 \text{ } \mu\text{m}, \qquad L_p = \sqrt{\tau_p D_p} \simeq 160 \text{ } \mu\text{m}. \tag{21.77}$$

The above values provide for the saturation current density (21.30)

$$J_s = q \left(\frac{D_p \, p_{n0}}{L_p} + \frac{D_n \, n_{p0}}{L_n} \right) \simeq 14 \text{ pA cm}^{-2}. \tag{21.78}$$

From (21.41), the width of the depletion region at zero bias is found to be

$$l(V = 0) = l_n + l_p = \left[\frac{2 \, \varepsilon_{sc}}{q} \left(\frac{1}{N_D} + \frac{1}{N_A} \right) \psi_0 \right]^{1/2} \simeq 1 \text{ } \mu\text{m}, \tag{21.79}$$

with $l_n / l_p = N_A / N_D = 10$. The permittivity of silicon $\varepsilon_{sc} = 11.7 \times \varepsilon_0$ has been used, with $\varepsilon_0 \simeq 8.854 \times 10^{-14}$ F cm^{-1} the vacuum permittivity. Finally, the value of the differential capacitance per unit area at zero bias (21.44) is

$$C_0 = \left[\frac{q \varepsilon_{sc} / (2 \, \psi_0)}{1/N_D + 1/N_A} \right]^{1/2} = \frac{\varepsilon_{sc}}{l(V = 0)} \simeq 11 \text{ nF cm}^{-2}. \tag{21.80}$$

Problems

21.1 Evaluate the built-in potential at room temperature in an abrupt p–n junction with $N_A = 10^{16}$ cm^{-3} and $N_D = 10^{15}$ cm^{-3}.

21.2 Show that avalanche due to impact ionization is possible only if both coefficients k_n and k_p are different from zero.

Chapter 22
MOS Devices

22.1 Introduction

The mathematical model of semiconductor devices, derived in Chap. 19, is applied here to the description of two fundamental devices of the insulated-gate type: the MIS capacitor, whose most important implementation is the MOS capacitor, and the IGFET, whose most important implementation is the MOSFET. Both devices can be realized starting from either a p-doped or an n-doped substrate; only the first realization is illustrated here: the extension of the theory to the other one is immediate. The analysis of the MOS capacitor is carried out using the simple example of a one-dimensional device in steady state, with the hypotheses of non-degeneracy and complete ionization, that lend themselves to an analytical treatment. Observing that in a steady-state condition the device is in equilibrium, the theory needs the solution of Poisson's equation only. From the solution of the latter, the device's capacitance is calculated, followed by a number of other important relations, that are useful in the subsequent treatment of the MOSFET. The theory of the MOSFET is then tackled in two dimensions and in steady-state conditions, first deriving a general expression for the channel current that holds in the case of a well-formed channel. The calculation is then completed by introducing the gradual-channel approximation: the differential conductances are derived first, followed by the expression of the drain current as a function of the applied voltages. A further simplification leads to the linear–parabolic model, which is widely used in the semiqualitative analyses of circuits. The complements address the solution of the Poisson equation in the channel when a non-equilibrium condition holds, to provide a formal proof of the relation between the surface and quasi-Fermi potentials used in the gradual-channel approximation; finally, a few phenomena that are not accounted for by the gradual-channel approximation are discussed, and the neglect of the dependence on position of the average carrier mobility is justified.

© Springer Science+Business Media New York 2015
M. Rudan, *Physics of Semiconductor Devices*,
DOI 10.1007/978-1-4939-1151-6_22

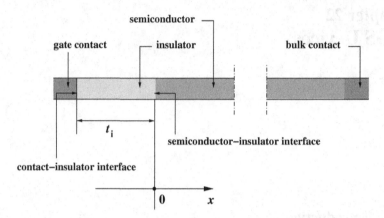

Fig. 22.1 Cross section of a metal–insulator–semiconductor capacitor. The thickness of the insulator layer is not realistic: in real devices the layer is much thinner than the contacts

22.2   Metal–Insulator–Semiconductor Capacitor

The Metal–Insulator–Semiconductor (MIS) capacitor is a fundamental device, that constitutes the basis for the field-effect transistors used in the fabrication of integrated circuits. The device has also extensively been used for studying the properties of semiconductor surfaces [103]. A one-dimensional version of it is sketched in Fig. 22.1: the structure is fabricated by depositing or thermally growing (Chap. 24) an insulator layer over a semiconductor substrate. The fabrication process must obtain an electrically clean interface; in fact, the number of localized electronic states at the interface must be kept to a minimum to avoid carrier trapping–detrapping processes. The contact deposited onto the insulator is called *gate contact*, the other one is called *bulk contact*.

In the standard silicon technology, the insulator is obtained by thermally growing silicon dioxide (Sect. 24.2). For this reason, the thickness of the insulator is indicated in the following with t_{ox} instead of the generic symbol t_i used in Fig. 22.1; by the same token, the insulator's permittivity is indicated with ε_{ox}, and the device is called MOS capacitor. In the last years, the progressive scaling down in the size of semiconductor devices has brought the insulator thickness to the range of nanometers. A smaller thickness has the advantage of providing a larger capacitance; however, it may eventually lead to dielectric breakdown and leakage by quantum tunneling. Silicon dioxide, which has been used as a gate insulator for decades, is being replaced in advanced devices with insulating layers made of materials having a larger permittivity (*high-k dielectrics*). Such layers are obtained by deposition (Sect. 24.5). Still with reference to the silicon technology, the conductive layers are made of metals, heavily-doped polycrystalline silicon, or metal silicides; here they will be indicated with the generic term "metal".

Like in the case of *p–n* junctions, the theory of the MOS capacitor is carried out with reference to a simplified structure, where the device is one dimensional

Fig. 22.2 The three materials forming the MOS capacitor shown separately. The symbols' meaning is illustrated in the text

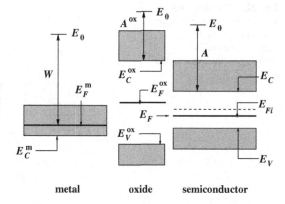

metal   oxide   semiconductor

and aligned with the x axis; in this case the semiconductor–insulator interface is a plane normal to x and, as shown in Fig. 22.1, its position is made to coincide with the reference's origin. A constant dopant concentration is present in the semiconductor region; to further simplify the analytical approach one also assumes that the conditions of non-degeneracy and complete ionization hold.

To describe the functioning of the device it is necessary to consider the fact that, in a region where the important electric phenomena occur, different materials are brought into an intimate contact. With reference to Fig. 22.2, the three materials (gate metal, oxide, and semiconductor) are initially considered separate from each other, and in the equilibrium condition. The left part of the figure shows the conduction band of the metal, with E_C^m the band's lower edge and E_F^m the metal's Fermi level. Due to the form of the Fermi-Dirac statistics, the probability that an electron's energy exceeds E_F^m is small; remembering the discussion of Sect. 7.2, the minimum energy necessary for an electron to transit from the metal into vacuum is the metal work function $W = E_0 - E_F^m$, with E_0 the vacuum level (left part of Fig. 22.2). For an insulator or a semiconductor, the electrons with maximum energy belong to states in the vicinity of the lower edge of the conduction band, E_C^{ox} (center) or E_C (right); in this case the minimum energy necessary for transiting into vacuum is the *electron affinity* $A^{ox} = E_0 - E_C^{ox}$ or $A = E_0 - E_C$, respectively. The semiconductor considered in the figure is uniformly doped of the p type, as shown by the fact that the Fermi level is below the intrinsic Fermi level E_{Fi} (Sect. 18.4.2). However, the analysis carried out here applies also to a semiconductor of the n type.

When the materials are brought into contact, they form a single, non-uniform system; as a consequence, in the equilibrium condition the Fermi levels must align with each other. On the other hand the vacuum levels must align as well, and the bands must adapt to compensate for a possible charge redistribution that occurs when the materials contact each other. The situation is similar to that represented in Fig. 21.3 for the p–n junction. The values of the parameters $W, A, \ldots$ selected for drawing Fig. 22.2 fulfill the relation

$$W - A = E_C - E_F. \tag{22.1}$$

Fig. 22.3 The three materials forming the MOS capacitor after being brought into contact. The symbols' meaning is illustrated in the text

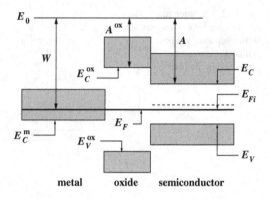

metal   oxide   semiconductor

As a consequence, there is no need for the bands to modify their shape; as shown in Fig. 22.3, which represents the situation after the materials have been brought into contact, the bands do not deform. The condition where the semiconductor's bands are everywhere parallel to the Fermi level is indicated with *flat-band condition*.[1] It is important to remark that condition (22.1) seldom occurs in realistic cases; however, as shown below, the case $W - A \neq E_C - E_F$ is easily incorporated into the analysis.

When the bulk contact is considered, Figs. 22.2 and 22.3 must be completed by adding the band structure of the contact's material to the right of that of the semiconductor. Assuming that the gate and bulk contacts are made of the same material, the structure to be added is identical to that already present on the left part of the figures. In the interior of each material, due to spatial uniformity, the electric potential is piecewise constant, thus the electric field is zero.

Consider now the case where a voltage V_G is applied between the gate and bulk contacts; the voltage reference is such that $V_G > 0$ when the electric potential of the gate contact is larger than that of the bulk contact, and vice versa. In steady-state conditions, the insulator prevents a current from flowing through the device; therefore, during the transient consequent to the application of V_G, the electric charge adjusts itself to the new boundary conditions. At the end of the transient the device is again in an equilibrium condition, while the form of the bands is different from that of Fig. 22.3. Similarly, the electric potential in the oxide and semiconductor is not constant any longer; its form is found by solving the Poisson equation in each region.

22.2.1 Surface Potential

The solution of the Poisson equation in the semiconductor region follows the same pattern as for the *p–n* junction (Sect. 21.2.1). Here a uniformly *p*-doped region

---

[1] The form of (22.1) is general enough to hold for both *p*- and *n*-type semiconductors.

is considered; its extension along the x axis is large, so that, far away from the semiconductor–insulator interface, the semiconductor behaves as if it were isolated. This fixes the carrier-equilibrium concentrations in the bulk; the asymptotic value of the electric potential is set to zero, $\varphi(+\infty) = 0$ whence, remembering that the non-degeneracy and complete-ionization conditions hold, it is

$$p^{(0)} = p(+\infty) = p_{p0} \simeq N_A, \qquad n^{(0)} = n(+\infty) = n_{p0} \simeq \frac{n_i^2}{N_A}. \qquad (22.2)$$

The Poisson equation in the semiconductor then reads

$$u'' = \frac{1}{L_A^2} A(u), \qquad A(u) = \frac{n_i^2}{N_A^2} \left[\exp(u) - 1 \right] + 1 - \exp(-u), \qquad (22.3)$$

with L_A the Debye length for the holes defined in (21.12). The normalized charge density $A(u)$ has the same sign as u (compare with Sect. 21.2.1). Multiplying by u' both sides of the first equation in (22.3), transforming its left hand side into $u'' u' = (1/2) \left[(u')^2 \right]'$, and integrating from $x \geq 0$ to $+\infty$, yields

$$\left(u' \right)^2 = \frac{2}{L_A^2} B(u), \qquad B(u) = \frac{n_i^2}{N_A^2} \left[\exp(u) - 1 - u \right] + u + \exp(-u) - 1. \qquad (22.4)$$

Following the same reasoning as for (21.10), one finds that B is non negative and u monotonic. However, in contrast with the case of the p–n junction, where the sign of u' is positive due to the boundary condition at $x \to -\infty$, here u may either increase or decrease monotonically; in fact, the sign of u' is fixed by the boundary condition V_G, which in turn may be either positive or negative. In conclusion, one finds

$$u' = \pm\frac{\sqrt{2}}{L_A} F(u), \qquad F(u) = \sqrt{\frac{n_i^2}{N_A^2} \left[\exp(u) - 1 - u \right] + u + \exp(-u) - 1}, \qquad (22.5)$$

where the sign must be found on a case-by-case basis. Separating (22.5), finally yields

$$\frac{du}{F(u)} = \pm\frac{\sqrt{2}}{L_A} dx, \qquad (22.6)$$

which must be solved numerically because it has no analytical solution.[2] Much information, however, is gained directly from (22.5), without the need of integrating

---

[2] The numerical evaluation from (22.6) of the inverse relation $x = x(u)$ is straightforward, though. Letting $\xi = \sqrt{2} x/L_A$, $\xi_0 = 0$, $u_0 = u_s$, $F_0 = F(u_0)$, it is $u_{k+1} = u_k \mp \delta u$, $F_{k+1} = F(u_{k+1})$, and $\xi_{k+1} = \xi_k + (1/F_k + 1/F_{k+1}) \delta u/2$, with $k = 0, 1, \ldots$.

Fig. 22.4 The cylinder used
to calculate the relation
between electric displacement
and charge per unit area
across an interface

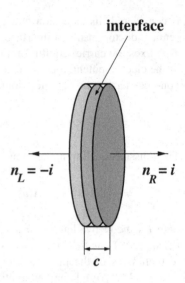

interface

$n_L = -i$ $n_R = i$

c

(22.6). To this purpose, one notes that the electric potential is continuous at the semiconductor–oxide interface; in fact, the normalized charge density in the semiconductor (22.3) has no charge layers in it, hence it can not contribute to a double charge layer at the interface. As a consequence, one can adopt the same symbol φ_s for the electric potential at $x = 0$, without distinguishing between the two sides of the interface; φ_s is called *surface potential*, whilst $u_s = q\varphi_s/(k_B T)$ is the *normalized surface potential*. In contrast, the electric field is discontinuous at the same interface; for this reason, one defines

$$u'_s = \lim_{x \to 0^+} \frac{du}{dx}, \qquad E_s = -\frac{k_B T}{q} u'_s, \qquad E_{ox} = -\lim_{x \to 0^-} \frac{d\varphi}{dx}. \qquad (22.7)$$

The relation between E_s and E_{ox} is found by considering a cylinder of thickness c placed across the semiconductor–oxide interface, such that the unit vector $\mathbf{n}_R$ normal to the right face is parallel to the unit vector $\mathbf{i}$ of the x axis, whereas the unit vector $\mathbf{n}_L$ normal to the left face is antiparallel to $\mathbf{i}$ (Fig. 22.4). Letting A_e be the common area of the two faces, the total charge within the cylinder is $A_e Q$, with Q the charge per unit area. Integrating $\mathrm{div}\mathbf{D} = \rho$ over the cylinder's volume and using (A.23) yields

$$A_e Q = \int_{A_e} \mathbf{D} \cdot \mathbf{n} \, dA_e = A_e [D_L \mathbf{i} \cdot (-\mathbf{i}) + D_R \mathbf{i} \cdot \mathbf{i}] = A_e (D_R - D_L), \quad (22.8)$$

with $D_R (D_L)$ the electric displacement on the right (left) face. From $\mathbf{D} = \varepsilon \mathbf{E}$ one then finds $Q = \varepsilon_{sc} E_R - \varepsilon_{ox} E_L$. It has been shown above that there are no charge layers on the semiconductor's side; as for the oxide layer, in principle it should be free of charges, although some contaminants may be present (Sect. 24.1). Here it is assumed that the oxide is free of charge; in conclusion, letting the cylinder's thickness c go to zero, one obtains $Q \to 0$, whence, using the limits (22.7),

$$\varepsilon_{sc} E_s = \varepsilon_{ox} E_{ox}. \qquad (22.9)$$

To find E_{ox} one observes that in a one-dimensional medium free of charge the electric potential is linear, whence E_{ox} is given by the negative potential drop across the oxide divided by the oxide thickness t_{ox}. To complete the analysis it is then necessary to consider the interface between oxide and gate metal.

In the interior of the metal the electric potential is uniform; its value with respect to the metal of the bulk contact is V_G. However, in the solution of the Poisson equation within the semiconductor, the asymptotic condition $\varphi(+\infty) = 0$ has been chosen, which holds inside the semiconductor region; the surface potential φ_s is referred to such a zero as well. It follows that V_G and φ_s are referred to two different zeros. Remembering the discussion carried out in Sect. 21.2.2, the difference between the external zero (namely, that within the bulk contact) and the internal zero (given by the asymptotic condition) is the built-in potential Φ_{mp} between the bulk contact and the p-type semiconductor; thus, the gate voltage referred to the internal zero is $V'_G = V_G - \Phi_{mp}$. Also, the electric potential is continuous across the interface between the oxide and the gate metal, because no double layer is present there. In contrast, as $E = 0$ within the metal while $E_{ox} \neq 0$, the electric field is generally discontinuous; in fact, a charge layer of density

$$\rho_m = Q_m \, \delta(x + t_{ox}^-), \tag{22.10}$$

with Q_m the charge per unit area of the metal, builds up at the gate–metal's surface. In conclusion, the electric field within the oxide reads

$$E_{ox} = \frac{V'_G - \varphi_s}{t_{ox}}. \tag{22.11}$$

22.2.2 Relation Between Surface Potential and Gate Voltage

Combining (22.5), (22.7), (22.9), and (22.11) one finds

$$C_{ox} \left(V'_G - \varphi_s \right) = \mp \varepsilon_{sc} \frac{k_B T}{q} \frac{\sqrt{2}}{L_A} F(\varphi_s), \qquad C_{ox} = \frac{\varepsilon_{ox}}{t_{ox}}, \tag{22.12}$$

where C_{ox} is the *oxide capacitance per unit area*, and $F(\varphi_s)$ is obtained by replacing u with $q\varphi_s/(k_B T)$ in the second relation of (22.5). The left hand side of (22.12) is the charge per unit area Q_m in the gate metal. This is easily found by considering the same cylinder as above, this time placed across the metal–oxide interface. Integrating $\text{div}\mathbf{D} = \rho$ over the cylinder's volume, using (22.10), and observing that $D_L = 0$ because the metal's interior is equipotential, yields

$$A_e Q_m = \int_{A_e} \mathbf{D} \cdot \mathbf{n} \, dA_e = A_e D_R = A_e C_{ox} \left(V'_G - \varphi_s \right). \tag{22.13}$$

Fig. 22.5 Normalized surface potential u_s in an MOS capacitor with a p-type substrate ($N_A = 10^{16}$ cm^{-3}), as a function of the normalized gate voltage u'_G

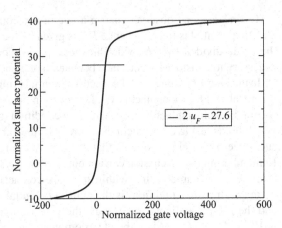

Due to the global charge neutrality, the following relation holds between the charge per unit area in the gate metal, Q_m, and that within the semiconductor, Q_{sc}:

$$Q_m + Q_{sc} = 0, \qquad Q_{sc} = \int_0^\infty q(p - n - N_A)\, dx = -C_{ox}(V'_G - \varphi_s).$$

$$(22.14)$$

In conclusion, (22.12) provides the relation between surface potential and gate voltage. When $\varphi_s = 0$, the electric potential vanishes everywhere in the semiconductor, namely, $V'_G = 0$ corresponds to the flat-band condition. When $V'_G > 0$, the charge in the gate metal is positive; as a consequence, the left hand side of (22.12) is positive as well, whence $V'_G > \varphi_s$ and the positive sign must be chosen at the right hand side. The opposite happens when $V'_G < 0$. An example of the $\varphi_s = \varphi_s(V'_G)$ relation is given in Fig. 22.5, showing the normalized surface potential u_s in an MOS capacitor as a function of the normalized gate voltage $u'_G = qV'_G/(k_B T)$. The semiconductor's doping is of the p type with $N_A = 10^{16}$ cm^{-3}, corresponding to $2\, u_F = 2\, q\varphi_F/(k_B T) = \log(p_{p0}/n_{p0}) \simeq 27.6$.

The $\varphi_s = \varphi_s(V'_G)$ relation lends itself to identifying different functioning regimes of the MOS capacitor. This identification can be carried out more accurately basing upon the values of the electron and hole concentrations at the semiconductor surface, $n_s = n(x = 0)$ and $p_s = p(x = 0)$. In the non-degenerate conditions considered here, the expressions of the surface concentrations read

$$n_s = n_{p0} \exp(u_s) = n_i \exp(u_s - u_F), \quad p_s = p_{p0} \exp(-u_s) = n_i \exp(u_F - u_s).$$

$$(22.15)$$

Depending on the value of u_s, several functioning regimes are identified, which are listed in Table 22.1. The regimes' designations are given by comparing the carrier concentrations at the surface with the intrinsic and asymptotic ones. When $u_s < 0$ the majority-carrier surface concentration (holes, in the example used here) exceeds the asymptotic one; the regime is called *accumulation*. When $u_s = 0$, both

Table 22.1 MOS capacitor, p substrate—functioning regimes

Norm. surface potential	Surface concentrations	Designation
$u_s < 0$	$n_s < n_{p0} < n_i < p_{p0} < p_s$	Accumulation
$u_s = 0$	$n_s = n_{p0} < n_i < p_{p0} = p_s$	Flat band
$0 < u_s < u_F$	$n_{p0} < n_s < n_i < p_s < p_{p0}$	Depletion
$u_s = u_F$	$n_{p0} < n_s = n_i = p_s < p_{p0}$	Mid gap
$u_F < u_s < 2u_F$	$n_{p0} < p_s < n_i < n_s < p_{p0}$	Weak inversion
$u_s = 2u_F$	$n_{p0} = p_s < n_i < n_s = p_{p0}$	Threshold
$2u_F < u_s$	$p_s < n_{p0} < n_i < p_{p0} < n_s$	Strong inversion

majority- and minority-carrier concentrations equal the corresponding asymptotic concentrations everywhere, and the already-mentioned *flat-band condition* holds. For $0 < u_s < u_F$, the majority-carrier concentration is smaller than the asymptotic one, while the minority-carrier concentration is smaller than the intrinsic one. By continuity, the majority-carrier concentrations is smaller than the asymptotic one not only at the semiconductor's surface, but also in a finite region of width x_d, which is therefore depleted from carriers; for this reason, the condition $0 < u_s < u_F$ is called *depletion regime*, and x_d is called *depletion width*.[3] When $u_s = u_F$, both majority- and minority-carrier concentrations at the surface equal the intrinsic concentration; remembering that in an intrinsic semiconductor the Fermi level practically coincides with the gap's midpoint (Sect. 18.3), this regime is called *mid gap*. When $u_F < u_s < 2u_F$, the minority-carrier concentration at the surface exceeds that of the majority carriers; however, it is still lower than the asymptotic concentration of the majority carriers: the regime is called *weak inversion*. When $u_s = 2u_F$, the surface concentration of the minority carriers equals the asymptotic concentration of the majority carriers, and vice versa; the regime is called *threshold of the strong inversion*, or simply *threshold*. Finally, when $u_s > 2u_F$, the minority-carrier concentration at the surface exceeds the asymptotic concentration of the majority carriers, and the regime is called *strong inversion*. In Fig. 22.5 the normalized surface potential at threshold, $2u_F$, is marked by the horizontal bar; one notes that in the strong-inversion regime the surface potential rapidly saturates as the gate voltage increases.

The form of the electric potential and charge density is shown in Fig. 22.6 for the accumulation regime. The upper part of the figure shows the charge density, which is schematically represented by a negative charge layer at the metal–oxide interface and by the thicker, positive layer at the semiconductor oxide interface. The lower part of the figure shows the electric potential along with the band structure of the semiconductor; note that two different vertical axes are used, in such a way that energy increases upwards and the electric potential increases downwards. The

---

[3] The depletion width x_d is conceptually the same thing as the extension l_p of the space-charge region on the p side of a metallurgical junction (Sect. 21.2.2). A different symbol is used to avoid confusion in the analysis of the MOSFET (Sect. 22.6).

Fig. 22.6 Schematic
representation of the charge
density and electric potential
in a p-substrate MOS
capacitor in the accumulation
regime

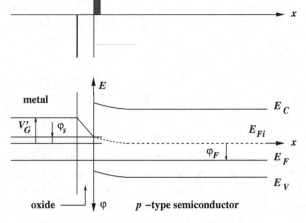

Fig. 22.7 Schematic
representation of the charge
density and electric potential
in a p-substrate MOS
capacitor in the mid-gap
condition

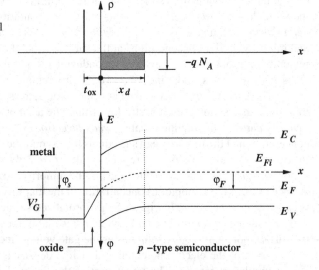

zero of the electric potential coincides with the horizontal part of the dashed line
(in fact, here it is $V_G' < \varphi_s < 0$). The mid-gap condition, $V_G' > 0$, $\varphi_s = \varphi_F$, is
illustrated in Fig. 22.7, whose general description is similar to that of Fig. 22.6; here
the charge layer on the gate metal is positive, and balances the negative charge of
the semiconductor. Due to the depletion that occurs in the region adjacent to the
semiconductor–oxide interface, the charge density is dominated by the contribution
from the negative acceptor ions, $\rho \simeq -qN_A$. In the figure, the charge density of
the semiconductor is schematically indicated by the shaded area, that corresponds
to a charge per unit area equal to $-qN_A x_d$. Finally, Fig. 22.8 shows the form of the

Fig. 22.8 Schematic
representation of the charge
density and electric potential
in a p-substrate MOS
capacitor at threshold

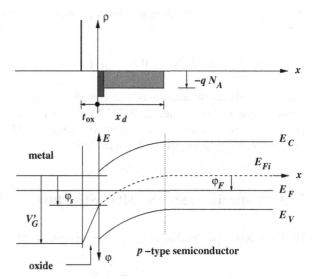

electric potential and charge density for the threshold condition, $V_G' > 0$, $\varphi_s = 2\varphi_F$.
Again, the general description is similar to that of Figs. 22.6 and 22.7; here, there
are two contributions to the negative charge of the semiconductor: the first one
comes from the contribution of the negative acceptor ions, whose charge density is
schematically indicated by the shaded area of width x_d (note that, due to the larger
value of V_G, the depletion width is larger than that of the mid-gap condition). The
second contribution to the semiconductor's charge is due to the electrons, whose
concentrations at the interface or near it is sufficiently large to be significant; they
form a negative layer, called *inversion layer*, whose width, albeit larger than that of
the positive layer located at the metal–oxide interface,[4] is much smaller than x_d.

Numerical solutions of the semiconductor-device model show that, with the exception of the accumulation regime, the semiconductor region of a uniformly-doped
MOS capacitor can be partitioned into a space-charge and a quasi-neutral region; the
quasi-neutral region behaves as an isolated, uniform semiconductor, whereas in the
volume of the space-charge[5] region the charge density is essentially dominated by
the ionized dopants. In the threshold and strong-inversion regimes, the layer of mobile charges near the semiconductor–oxide interface gives a significant contribution,
which must be accounted for; it can approximated as a charge layer at the interface.
Considering the range $\varphi_s > 0$ only, namely, excluding the accumulation regime for

---

[4] The width of the region where the charged layer at the metal–oxide interface is significant is of
the order of 1 nm. That of an inversion layer is of the order of 5 nm; an example is given below,
with reference to Fig. 22.13.

[5] With reference to MOS devices, the space-charge region is also called *depleted region*.

the p-substrate MOS capacitor, the charge per unit area in the semiconductor is found to be

$$Q_{sc} = \int_0^\infty \rho \, dx \simeq \int_0^{x_d} \rho \, dx \simeq -q \int_0^{x_d} (n + N_A) \, dx = Q_i + Q_b, \qquad (22.16)$$

where the first approximation is due to neglecting the charge of the quasi-neutral region, the second one to neglecting the holes in the space-charge region. Quantities $Q_i, Q_b < 0$ are, respectively, the integral of $-qn$ and $-qN_A$; they are called *inversion charge per unit area* and *bulk charge per unit area*.

22.3   Capacitance of the MOS Structure

The capacitance per unit area of the MOS structure is given by[6]

$$C = \frac{dQ_m}{dV_G} = \frac{dQ_m}{dV_G'}. \qquad (22.17)$$

Combining (22.17) with (22.12) and (22.14) yields

$$\frac{1}{C} = \frac{dV_G'}{dQ_m} = \frac{d\left(V_G' - \varphi_s\right) + d\varphi_s}{dQ_m} = \frac{1}{C_{ox}} + \frac{d\varphi_s}{d\left(-Q_{sc}\right)}. \qquad (22.18)$$

The above is recast in a more compact form by defining the *semiconductor capacitance per unit area*

$$C_{sc} = -\frac{dQ_{sc}}{d\varphi_s} = -\frac{q}{k_B T} \frac{dQ_{sc}}{du_s} = \pm \frac{\sqrt{2}\,\varepsilon_{sc}}{L_A} \frac{dF}{du_s} > 0, \qquad (22.19)$$

where the positive (negative) sign holds for $u_s > 0$ $\left(u_s < 0\right)$. In conclusion, the capacitance is the series of the oxide and semiconductor capacitances:

$$\frac{1}{C} = \frac{1}{C_{ox}} + \frac{1}{C_{sc}}. \qquad (22.20)$$

In (22.20) it is $C_{ox} = \text{const}$ while C_{sc} has a rather complicate dependence on u_s. However, basing on the second equation in (22.5), one may investigate the limiting cases of (22.20). For this, using $\exp\left(-2\,u_F\right) = n_i^2/N_A^2$, one finds for the asymptotic behavior of F in a p-substrate device,

$$F \simeq \exp\left(u_s/2 - u_F\right), \quad u_s \gg 1; \qquad F \simeq \exp\left(-u_s/2\right), \quad u_s \ll -1. \quad (22.21)$$

---

[6] Like in the case of the depletion capacitance of the p–n junction (Sect. 21.4), definition $C = dQ_m/dV_G$ is coherent with the choice of the voltage reference described in Sect. 22.2. The units of C are $[C] = \text{F m}^{-2}$.

Fig. 22.9 Normalized capacitance C/C_{ox} as a function of the normalized gate voltage u'_G, in a p-substrate MOS capacitor with $N_A = 10^{16}$ cm^{-3}, for different values of $r = \varepsilon_{sc} t_{ox}/(\varepsilon_{ox} \sqrt{2} L_A)$. The details of the calculations are in Problem 22.7.2

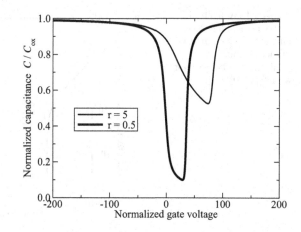

When, instead, it is $|u_s| \ll 1$, expanding the exponentials yields $\exp(\pm u_s) \simeq 1 \pm u_s + u_s^2/2$, whence, observing that $\exp(-2 u_F) \ll 1$,

$$F^2 \simeq \frac{1}{2}\left[1 + \exp(-2 u_F)\right] u_s^2 \simeq \frac{1}{2} u_s^2, \qquad F \simeq \pm \frac{u_s}{\sqrt{2}}. \tag{22.22}$$

Then, from (22.19) the asymptotic values of C_{sc} are found to be

$$C_{sc} \simeq \frac{1}{\sqrt{2}} \frac{\varepsilon_{sc}}{L_A} \exp(u_s/2 - u_F), \qquad u_s \gg 1, \tag{22.23}$$

$$C_{sc} \simeq \frac{1}{\sqrt{2}} \frac{\varepsilon_{sc}}{L_A} \exp(-u_s/2), \qquad u_s \ll -1. \tag{22.24}$$

Both limits correspond to the same asymptotic value of the capacitance per unit area,

$$C = \frac{C_{ox} C_{sc}}{C_{ox} + C_{sc}} \simeq C_{ox}, \qquad |u_s| \gg 1. \tag{22.25}$$

Near the origin, instead, one finds

$$C_{sc} \simeq \frac{\varepsilon_{sc}}{L_A}, \qquad |u_s| \ll 1. \tag{22.26}$$

The limit of C for $u_s \to 0$ is called *flat-band capacitance per unit area*; from (22.26) one finds

$$C \simeq C_{FB} = \frac{C_{ox}}{1 + C_{ox} L_A/\varepsilon_{sc}} < C_{ox}, \qquad |u_s| \ll 1. \tag{22.27}$$

Examples of capacitance's calculations are shown in Fig. 22.9.

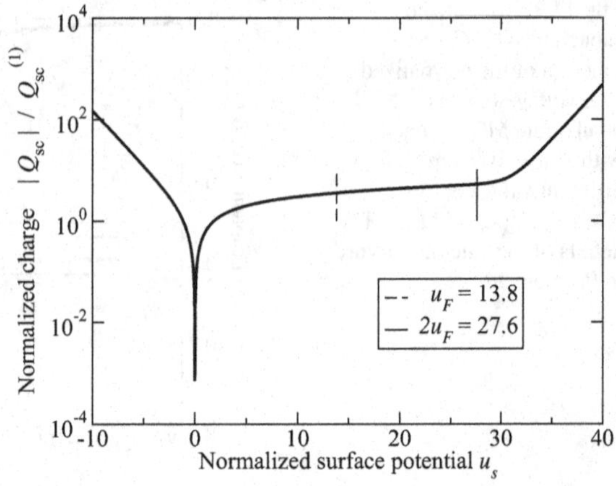

Fig. 22.10 Normalized charge per unit area as a function of the normalized surface potential, in a p-substrate MOS capacitor with $N_A = 10^{16}$ cm^{-3}

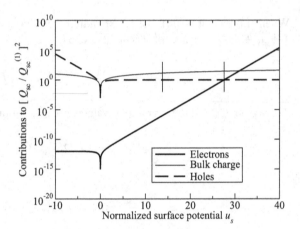

Fig. 22.11 Individual contributions of electrons, holes, and bulk charge to $F^2 = [Q_{sc}/Q_{sc}^{(1)}]^2$, as a function of the normalized surface potential u_s, in a p-substrate MOS capacitor with $N_A = 10^{16}$ cm^{-3}

22.4 Simplified Expression of the Inversion Charge

To the purpose of applying some results of the MOS capacitor's theory to the analysis of MOSFETs, it is convenient to determine a simplified form of the inversion layer's charge, that holds in all the functioning regimes with the exception of accumulation. For this, one starts from the expression of the semiconductor charge per unit area which, combining (22.12) and (22.13), reads

$$Q_{sc} = \pm Q_{sc}^{(1)}\, F(u_s), \qquad Q_{sc}^{(1)} = \varepsilon_{sc}\frac{k_B\, T}{q}\,\frac{\sqrt{2}}{L_A}, \tag{22.28}$$

where the negative (positive) sign must be chosen when $u_s > 0$ ($u_s < 0$), and $Q_{sc}^{(1)}$ is the value of Q_{sc} corresponding to $F = 1$. The relation $Q_{sc} = Q_{sc}(u_s)$ is shown in normalized units in Fig. 22.10. In turn, Fig. 22.11 shows, still in normalized form, the individual contributions of electrons, holes, and bulk charge to $F^2 = \left[Q_{sc}/Q_{sc}^{(1)}\right]^2$;

such contributions are, respectively, $\exp\left(-2\,u_F\right)\left[\exp\left(u_s\right)-1\right]$, $\exp\left(-u_s\right)-1$, and $\left[1-\exp\left(-2\,u_F\right)\right]u_s$. The contribution of holes dominates for $u_s < 0$, that of the bulk charge dominates for $0 < u_s < 2u_F$ and, finally, that of the electrons dominates for $u_s > 2u_F$.

When accumulation is excluded, in a p-substrate capacitor one must take $\varphi_s > 0$. The approximate dependence of F on the normalized potential is easily worked out from (22.5), whose limiting case in the depletion and weak-inversion regimes is

$$F \simeq \sqrt{u_s}, \qquad 0 < u_s < 2u_F. \tag{22.29}$$

Introducing (22.29) into (22.28) yields, for $0 < u_s < 2\,u_F$,

$$Q_{sc} \simeq -\frac{2\,\varepsilon_{sc}\,k_B\,T}{q L_A}\sqrt{\frac{q\varphi_s}{k_B\,T}} = -C_{ox}\,\gamma\,\sqrt{\varphi_s}, \qquad \gamma = \frac{\sqrt{2\,\varepsilon_{sc}\,q p_{p0}}}{C_{ox}}, \tag{22.30}$$

where the expression (21.12) of the Debye length L_A has been used.[7] It is interesting to note that a relation identical to (22.30) is obtained using the full-depletion and ASCE approximations (Sect. 21.4); in fact, letting $\rho = -qN_A$ for $0 < x < x_d$ and $\rho = 0$ for $x > x_d$ (compare, e.g., with Figs. 22.7 and 22.8) yields a simplified form of the Poisson equation,

$$\varphi'' \simeq \frac{qN_A}{\varepsilon_{sc}}, \qquad 0 < x < x_d. \tag{22.31}$$

The boundary conditions of (22.31) are obtained in the same manner as in the p–n junction, and read $\varphi(x_d) = 0$, $\varphi'(x_d) = 0$; the solution of (22.31) fulfilling the boundary conditions is $\varphi(x) = qN_A\left(x - x_d\right)^2/\left(2\,\varepsilon_{sc}\right)$. Letting $x = 0$ in the above, yields a relation between the surface potential and the depletion width; in turn, in the full-depletion and ASCE approximations the bulk charge per unit area is $Q_b = -qN_A\,x_d$. In summary,

$$\varphi_s = \frac{qN_A}{2\,\varepsilon_{sc}}\,x_d^2, \qquad Q_b = -qN_A\,x_d = -\sqrt{2\,\varepsilon_{sc}\,qN_A\,\varphi_s}. \tag{22.32}$$

Observing that for $0 < u_s < 2\,u_F$ it is $Q_{sc} \simeq Q_b$, and that $N_A \simeq p_{p0}$, one finds that the second relation in (22.32) coincides with the first one in (22.30). Combining $Q_{sc} \simeq Q_b$ with (22.30) and with the general expression (22.14) yields $V_G' - \varphi_s = \gamma\,\sqrt{\varphi_s}$; from it, one finds a simplified $\varphi_s = \varphi_s(V_G')$ relation, that holds in the depletion and weak-inversion conditions:[8]

$$\sqrt{\varphi_s} = \sqrt{V_G' + \left(\gamma/2\right)^2} - \gamma/2. \tag{22.33}$$

The contribution of electrons to the semiconductor charge per unit area, $Q_i = -q\int_0^{x_d} n\,dx$, becomes relevant from the threshold condition on. Remembering the

---

[7] The units of γ are $[\gamma] = V^{1/2}$.

[8] The negative sign in front of the square root in (22.33) must be discarded.

discussion of Sect. 22.2.2, the electron charge is approximated as a charge layer at the interface, $-qn \simeq Q_i \, \delta(x^+)$; as a consequence, the space charge can be considered as entirely due to the ionized dopant atoms also when $\varphi_s > 2\varphi_F$, so that (22.31) and (22.32) still hold. From (22.16) one then finds the result sought, that is, a simplified form of the inversion layer's charge, that holds in the depletion, weak-inversion, and strong-inversion regimes:

$$Q_i = Q_{sc} - Q_b = -C_{ox}\left[(V'_G - \varphi_s) - \gamma\sqrt{\varphi_s}\right] < 0. \qquad (22.34)$$

The theory worked out so far is based on the assumption that relation (22.1) holds between the gate metal's work function, the semiconductor's affinity, and the position of the semiconductor's Fermi level with respect to the band edges. Moreover, the gate oxide has been assumed free of charge. In fact, as the above conditions seldom occur, one must account for the effects of the difference $\Delta W = W - A - (E_C - E_F)$ and for the oxide charge. It can be shown that, if the oxide charge is fixed, the two additional effects produce a shift in V'_G with respect to the value $V'_G = V_G - \Phi_{mp}$ that holds for the ideal MOS capacitor with a p-type substrate [103, Sect. 9.4]. In fact, the combination of Φ_{mp}, ΔW, and of the oxide charge per unit area Q_{ox}, provides an expression of the form $V'_G = V_G - V_{FB}$, where the constant $V_{FB} = V_{FB}(\Phi_{mp}, Q_{ox}, \Delta W)$ is called *flat-band voltage*.[9]

22.4.1 Quantitative Relations in the MOS Capacitor

In the p-type silicon substrate considered so far it is $p_{p0} \simeq N_A = 10^{16}$ cm^{-3} and, at room temperature, $n_i \simeq 10^{10}$ cm^{-3}. In turn, the asymptotic minority-carrier concentration is $n_{p0} = n_i^2/N_A \simeq 10^4$ cm^{-3}; it follows $\exp(-2u_F) = n_{p0}/p_{p0} \simeq 10^{-12}$ and, as shown, e.g., in Fig. 22.5, $2u_F \simeq 27.6$. Using $k_B T/q \simeq 26$ mV then yields $2\varphi_F \simeq 0.72$ V. As $\varepsilon_{sc} \simeq 11.7 \times 8.854 \times 10^{-14} = 1.036 \times 10^{-12}$ F cm^{-1}, one finds

$$L_A = \sqrt{\frac{2\varepsilon_{sc} k_B T}{q^2 p_{p0}}} \simeq 5.8 \times 10^{-2} \ \mu\text{m}, \qquad (22.35)$$

$$Q_{sc}^{(1)} = \frac{2\varepsilon_{sc} k_B T}{q L_A} = \sqrt{2\varepsilon_{sc} k_B T \, p_{p0}} \simeq 9.3 \times 10^{-9} \ \text{C cm}^{-2}. \qquad (22.36)$$

The relation $Q_{sc}(V'_G)$ is found from $Q_{sc} = -C_{ox}\left[V'_G - \varphi(V'_G)\right]$, where $\varphi(V'_G)$ is obtained from (22.12). The result is shown in normalized form in Fig. 22.12.

---

[9] The designation is due to the fact that $V_G = V_{FB}$ establishes the flat-band condition in the semiconductor.

Fig. 22.12 Normalized semiconductor charge $Q_{sc}/Q_{sc}^{(1)}$ as a function of the normalized gate voltage u'_G, for a p-substrate MOS capacitor with $N_A = 10^{16}$ cm^{-3}

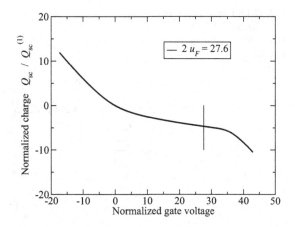

Fig. 22.13 Normalized concentrations n/p_{p0} and $(N_A - p)/p_{p0}$ as a function of position x/L_A, for a p-substrate MOS capacitor with $N_A = 10^{16}$ cm^{-3} in strong inversion $(u_s = 2.5\, u_F)$

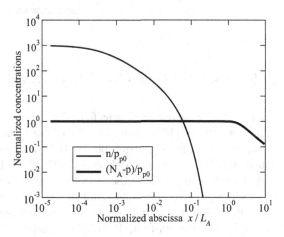

Note that the results illustrated so far have been obtained without the need of integrating (22.6). The result of a numerical integration of (22.6) is shown in Fig. 22.13, where the dependence on position of n and $N_A - p$ is drawn for a p-substrate MOS capacitor with $N_A = 10^{16}$ cm^{-3} in the strong-inversion regime $(u_s = 2.5\, u_F)$. The term $(N_A - p)/p_{p0}$ is significant in a surface region of the semiconductor, whose thickness is several units of x/L_A. The term n/p_{p0} is much larger, but only in a much thinner region near the surface. If the width of the inversion layer is conventionally taken at the intersection between the two curves of Fig. 22.13, that occurs at $x/L_A \simeq 0.1$, one finds from (22.35) a width of about 5 nm.

Fig. 22.14 Cross-section of
an n-channel MOSFET. The
black areas are the metal
contacts

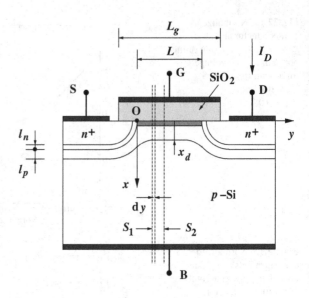

22.5   Insulated-Gate Field-Effect Transistor—MOSFET

The principle of the insulated-gate, field-effect transistor (IGFET) was demonstrated
in the early 1930s [103, Chap. 10]. The first structures using a thermally-oxidized
silicon layer were fabricated in 1960. The IGFET architecture using silicon dioxide as
gate dielectric is more commonly called MOSFET. This device architecture, jointly
with the continuous improvements in the silicon technology, made it possible the
tremendous progress in the fabrication of integrated circuits during the last decades,
and is the most common component in digital and analog circuits.

The electric current in a MOSFET is transported by one type of carriers only,
electrons or holes; for this reason the device is called *unipolar*. In a p-substrate
device, the carriers are the electrons that form the charge layer at the semiconductor–
insulator interface; therefore, this type of transistor is called *n-channel* MOSFET.
Conversely, in an n-substrate device the carriers are holes, and the transistor is called
p-channel MOSFET. The schematic cross-section of an *n-channel* MOSFET is shown
in Fig. 22.14. The starting point is a p-type silicon substrate, with an $N_A = $ const
dopant concentration, onto which a layer of silicon dioxide is thermally grown and
patterned; then, the gate contact (G) is deposited. The extension L_g of the gate metal
in the horizontal direction is called *geometrical length* of the gate. The next step is
the introduction of a heavy dose of an n-type dopant on the two sides of the gate. As
shown in the figure, lateral diffusion (Sect. 23.7.3) makes the gate oxide to partially
overlap the n-doped regions.[10]

---

[10] The n-type regions are typically obtained by ion implantation, whose lateral penetration is limited.
However, ion implantation is followed by a thermal process (*annealing*), during which thermal
diffusion takes place.

The metallizations of the n^+ regions provide two more contacts, called *source* (S) and *drain* (D); the bottom metal layer contacting the *p*-type substrate is indicated with *bulk* (B), and the term *channel* denotes the interfacial semiconductor region between the two junctions. To distinguish the applied voltages from one another, two letters are used; considering the bulk metallization as the reference contact, in an *n*-channel MOSFET a typical choice of the three independent voltages is $V_{GB} = V_G - V_B$, $V_{SB} = V_S - V_B$, and $V_{DB} = V_D - V_B$. As the standard MOSFET architecture is structurally symmetric, it is not possible to distinguish the source contact from the drain contact basing on geometry or dopant distribution: the distinction is to be based on the applied voltages; in fact, in the typical operating regime of the device the source–bulk and drain–bulk junctions are never forward biased, whence in an *n*-channel MOSFET it is $V_{SB} \geq 0$ and $V_{DB} \geq 0$. The drain contact is identified[11] by the condition $V_{DB} - V_{SB} = V_{DS} > 0$.

22.6 *N*-Channel MOSFET—Current-Voltage Characteristics

To work out the theory of the MOSFET, one introduces a reference whose x axis is normal to the semiconductor–insulator interface, while the y axis is parallel to it. The origin (O) is placed at the intersection of the source *p–n* junction and the interface (Fig. 22.14). The y coordinate corresponding to the intersection of the drain *p–n* junction and the interface is indicated with L; the latter is called *electric length* of the gate, or *channel length*. The device is considered uniform in the z direction; its width along such a direction is indicated with W.

Purpose of the analysis is to derive the steady-state characteristics, namely, the relations between the currents at the contacts and the applied voltages. To proceed, one assumes that the gate voltage V_{GB} is such that at all positions y along the channel the strong-inversion condition holds. Thus, a layer of electrons is present, indicated in Fig. 22.14 with the shaded area underneath the gate oxide; the term *well-formed channel* is used to denote this situation. The minimum gate voltage necessary for obtaining a well-formed channel will be identified later. In a steady-state condition there is no current through the gate contact because the gate insulator is in series to it; also, the current flowing through the bulk contact is negligibly small because the two junctions are reverse biased. Since the channel layer connects two heavily-doped regions of the n type, the application of a drain-source voltage $V_{DS} > 0$ gives rise to a current I_D that flows from the drain to the source contact (its reference is shown in Fig. 22.14); for a given V_{DS}, the drain current I_D is controlled by the amount of charge available in the channel, which is in turn controlled by the gate voltage V_{GB}. In other terms, the device is an electronic valve in which the gate–bulk port controls the current flowing into the drain–source port; moreover, in the steady-state

---

[11] In fact, in some types of logic circuits the source and drain contact may exchange their roles depending on the applied voltages.

condition the control port does not expend energy, because the gate current is zero. This, among other things, explains the success of the MOSFET concept.

Due to the uniformity in the z direction, the electron and hole current densities have the form

$$\mathbf{J}_n = J_{nx}\,\mathbf{i} + J_{ny}\,\mathbf{j}, \qquad \mathbf{J}_p = J_{px}\,\mathbf{i} + J_{py}\,\mathbf{j}, \qquad (22.37)$$

with $\mathbf{i}$, $\mathbf{j}$ the unit vectors of the x and y axes, respectively. On the other hand it is $J_{nx} = J_{px} = 0$ because no current can flow through the insulator, so only the y components J_{ny}, J_{py} are left in (22.37). In turn, the condition of a well-formed channel implies that the concentration of holes is negligibly small with respect to that of the electrons,[12] whence $|J_{ny}| \gg |J_{py}|$. It follows $\mathbf{J} = \mathbf{J}_n + \mathbf{J}_p \simeq \mathbf{J}_n = J_{ny}(x,y)\mathbf{j}$. The equality $\mathbf{J} = \mathbf{J}_n$ is the mathematical form of the MOSFET's property of being unipolar. It also entails $\mathrm{div}\mathbf{J} = \mathrm{div}\mathbf{J}_n$; therefore, remembering that in a steady-state condition it is $\mathrm{div}\mathbf{J} = 0$, it follows that $\mathrm{div}\mathbf{J}_n = 0$ as well.

Consider now two planes parallel to the x, z plane, placed at different positions y_1 and y_2 in the channel; their intersections with the x, y plane are respectively marked with S_1 and S_2 in Fig. 22.14. From the divergence theorem (A.23) and the property $\mathrm{div}\mathbf{J}_n = 0$ it follows[13]

$$\iint_2 \mathbf{J}_n \cdot \mathbf{j}\ \mathrm{d}x\ \mathrm{d}z - \iint_1 \mathbf{J}_n \cdot \mathbf{j}\ \mathrm{d}x\ \mathrm{d}z = 0; \qquad (22.38)$$

as a consequence, the channel current

$$I = \int_{z=0}^{W} \int_{x=0}^{x_d} \mathbf{J}_n \cdot \mathbf{j}\ \mathrm{d}x\ \mathrm{d}z = W \int_0^{x_d} J_{ny}\ \mathrm{d}x \qquad (22.39)$$

is independent of y. The last form of (22.39) derives from the uniformity in z. To express J_{ny} in (22.39) it is convenient to adopt the monomial form (19.141) of the electron current density, $\mathbf{J}_n = -q\mu_n\,n\,\mathrm{grad}\varphi_n$, with φ_n the electron quasi-Fermi potential. The components of $\mathbf{J}_n$ in monomial form read

$$J_{nx} = -q\mu_n\,n\,\frac{\partial\varphi_n}{\partial x}, \qquad J_{ny} = -q\mu_n\,n\,\frac{\partial\varphi_n}{\partial y}, \qquad (22.40)$$

where $J_{nx} = 0$ as found above. In the channel it is $n \neq 0$, whence for J_{nx} to vanish it must be $\partial\varphi_n/\partial x = 0$; as a consequence, φ_n in the channel[14] depends on y only. In conclusion,

$$I = W\,\frac{\mathrm{d}\varphi_n}{\mathrm{d}y}\,Q_i(y), \qquad Q_i = \int_0^{x_d} -q\mu_n\,n\,\mathrm{d}x < 0, \qquad (22.41)$$

---

[12] Compare with Fig. 22.13; the latter describes an equilibrium case, however the situation is similar to the one depicted here.

[13] In the integrals of (22.38) the upper limit of x is given by the depletion width $x_d(y)$ shown in Fig. 22.14. Compare also with Sect. 22.7.1.

[14] Far from the channel the semiconductor is practically in the equilibrium condition, whence $\varphi_n \to \varphi_F$ as x increases. However, in the bulk region where the dependence of φ_n on x is significant, the electron concentration is negligible; as a consequence, the integral in (22.39) is not affected.

where Q_i is the inversion-layer charge per unit area at position y in the channel. In the integral of (22.41) it is $n = n(x, y)$ and $\mu_n = \mu_n(x, y)$; defining the *effective electron mobility* as the average

$$\mu_e(y) = \frac{\int_0^{x_d} -q\mu_n n \, dx}{\int_0^{x_d} -qn \, dx} > 0, \qquad (22.42)$$

yields

$$I = W \frac{d\varphi_n}{dy} \mu_e(y) Q_i(y). \qquad (22.43)$$

In (22.43), μ_e and Q_i are positive- and negative-definite, respectively, and I, W are constant; it follows that $d\varphi_n/dy$ has always the same sign and, as a consequence, $\varphi_n(y)$ is invertible. Using the inverse function $y = y(\varphi_n)$ within μ_e and Q_i makes (22.43) separable; integrating the latter over the channel yields

$$\int_0^L I \, dy = L I = W \int_{\varphi_n(0)}^{\varphi_n(L)} \mu_e(\varphi_n) Q_i(\varphi_n) \, d\varphi_n. \qquad (22.44)$$

In turn, the dependence of μ_e on y is weak,[15] whence

$$I = \frac{W}{L} \mu_e \int_{\varphi_n(0)}^{\varphi_n(L)} Q_i(\varphi_n) \, d\varphi_n. \qquad (22.45)$$

22.6.1 Gradual-Channel Approximation

In the derivation of (22.45) the condition of a well-formed channel has not been exploited yet; this condition makes a number of approximations possible, which are collectively indicated with the term *gradual-channel approximation*;[16] they lead to an expression of (22.45) in closed form. First, one uses the definition of surface potential which, in the two-dimensional analysis considered here, is given by $\varphi_s(y) = \varphi(x = 0, y)$; it is shown in Sect. 22.7.1 that the condition of a well-formed channel entails the relation

$$\varphi_s = \varphi_n + \varphi_F. \qquad (22.46)$$

---

[15] This issue is discussed in Sect. 22.7.1.

[16] The gradual-channel approximation is not limited to the analysis of the MOSFET shown here. Indeed, it is a widely-used method to treat the Poisson equation in devices in which the geometrical configuration and applied voltages are such that the variation of the electric field in one direction is much weaker than those in the other two directions. Typically, the former direction is the longitudinal one (that is, along the channel), the other two the transversal ones. From the mathematical standpoint, the approximation amounts to eliminating a part of the Laplacian operator, so that the dependence on all variables but one becomes purely algebraic.

It follows that $d\varphi_n/dy$ in (22.43) can be replaced with $d\varphi_s/dy$, this showing that the transport in a well-formed channel is dominated by the drift term, $J_{ny} = -q\mu_n\, n\, d\varphi_s/dy = q\mu_n\, n\, E_{sy}$, with E_{sy} the y-component of the electric field at $x = 0$; using (22.46) one changes the variable from φ_n to φ_s in the integral of (22.45). The integration limits in terms of φ_s are found by the same reasoning leading to (22.46), and read (Sect. 22.7.1)

$$\varphi_s(0) = V_{SB} + 2\,\varphi_F, \qquad \varphi_s(L) = V_{DB} + 2\,\varphi_F. \qquad (22.47)$$

In conclusion, (22.45) becomes

$$I = \frac{W}{L}\,\mu_e \int_{V_{SB}+2\varphi_F}^{V_{DB}+2\varphi_F} Q_i(\varphi_s)\ d\varphi_s. \qquad (22.48)$$

The next step of the gradual-channel approximation consists in determining the relation $Q_i(\varphi_s)$, for which the solution of the Poisson equation in two dimensions is necessary. As shown in Sect. 22.7.1, one can exploit the strong difference between the strengths of the electric-field components in the x and y directions, to give the equation a one-dimensional form in which the y coordinate acts as a parameter. This is equivalent to assimilating each elementary portion of the channel, like that marked with dy in Fig. 22.14, to a one-dimensional MOS capacitor whose surface potential has the local value $\varphi_s(y)$. The final step of the gradual-channel approximation is the adoption of the full-depletion and ASCE approximations (Sect. 21.4), so that the inversion-layer charge per unit area at position y in the channel is given by (22.34), namely, $Q_i = -C_{ox}\left[(V'_{GB} - \varphi_s) - \gamma\sqrt{\varphi_s}\right] < 0$. Observing that $V_{DB} + 2\varphi_F > V_{SB} + 2\varphi_F$ while the integrand in (22.48) is negative, it follows that $I < 0$, whence $I_D = -I$ due to the reference chosen for the drain current (Fig. 22.14). In conclusion, (22.48) transforms into[17]

$$I_D = \beta \int_{V_{SB}+2\varphi_F}^{V_{DB}+2\varphi_F} \left[(V'_{GB} - \varphi_s) - \gamma\sqrt{\varphi_s}\right] d\varphi_s, \qquad \beta = \frac{W}{L}\,\mu_e\, C_{ox}. \qquad (22.49)$$

22.6.2 Differential Conductances and Drain Current

The drain current's expression (22.49) of the n-channel MOSFET provides a relation of the form $I_D = I_D(V_{GB}, V_{DB}, V_{SB})$, where $V_{GB} = V'_{GB} + V_{FB}$. In the integrated-circuit operation an important role is played by the *differential conductances* of the device, each of them defined as the partial derivative of I_D with respect to one of the applied voltages. In some cases the differential conductances can be found without the need of actually calculating the integral in (22.49); for this reason, here such conductances are calculated first. Prior to that, it is worth noting that in circuit

---

[17] The units of β are $[\beta] = A\ V^{-2}$.

applications it is often preferred to use the source contact, instead of the bulk contact, as a voltage reference. The transformation from one reference to the other is easily obtained from

$$V_{DS} = V_{DB} - V_{SB} > 0, \qquad V_{GS} = V_{GB} - V_{SB}, \qquad V_{BS} = -V_{SB} \le 0. \quad (22.50)$$

Then, the *drain conductance*[18] is defined as the derivative of I_D with respect to V_{DB}, at constant V_{SB} and V_{GB}; or, equivalently, as the derivative with respect to V_{DS}, at constant V_{BS} and V_{GS}:

$$g_D = \left(\frac{\partial I_D}{\partial V_{DB}}\right)_{V_{SB},V_{GB}} = \left(\frac{\partial I_D}{\partial V_{DS}}\right)_{V_{BS},V_{GS}}. \quad (22.51)$$

Remembering that the derivative of an integral with respect to the upper limit is the integrand calculated at such limit, from (22.49) one finds

$$g_D = \beta \left[(V'_{GB} - V_{DB} - 2\varphi_F) - \gamma\sqrt{V_{DB} + 2\varphi_F} \right]. \quad (22.52)$$

Using (22.47) and (22.34) yields

$$g_D = \beta \left[(V'_{GB} - \varphi_s(L)) - \gamma\sqrt{\varphi_s(L)} \right] = \frac{W}{L} \mu_e \left[-Q_i(L) \right], \quad (22.53)$$

namely, the output conductance is proportional to the inversion charge per unit area at the drain end of the channel. The quantity in brackets in (22.53) is non-negative by construction; its zero corresponds to the value of $\varphi_s(L)$ obtained from (22.33). Such a zero is indicated with φ_s^{sat} and is termed *saturation surface potential*. From (22.50), the *saturation voltage* in the bulk and source references is found to be

$$V_{DB}^{\text{sat}} = \varphi_s^{\text{sat}} - 2\varphi_F, \qquad V_{DS}^{\text{sat}} = V_{DB}^{\text{sat}} - V_{SB}, \quad (22.54)$$

respectively. Similarly, the current $I_D^{\text{sat}} = I_D\left(V_{DS}^{\text{sat}}\right)$ (which depends on V_{GS}) is called *saturation current*. If a value of V_{DS} larger than V_{DS}^{sat} is used, g_D becomes negative; this result is not physically sound and indicates that the gradual-channel approximation is not applicable in that voltage range.[19]

Still considering the drain conductance, it is also important to determine its limit for $V_{DB} \to V_{SB}$, or $V_{DS} \to 0$. Again, there is no need to calculate the integral in (22.49) which, in this limiting case, is the product of the integration interval V_{DS} times the integrand calculated in the lower integration limit; in turn, the derivative eliminates V_{DS}, whence

$$g_D(V_{DS} \to 0) = \beta \left[(V'_{GB} - V_{SB} - 2\varphi_F) - \gamma\sqrt{V_{SB} + 2\varphi_F} \right]. \quad (22.55)$$

---

[18] The drain conductance is also called *output conductance*; in this case it is indicated with g_o.

[19] In fact, beyond the saturation voltage the Poisson equation near the drain end of the channel can not be reduced any more to a one-dimensional equation where y is treated as a parameter (compare with Sect. 22.7.1).

Replacing V'_{GB} with $V_{GB} - V_{FB}$ and using (22.50) yields

$$g_D(V_{DS} \to 0) = \beta \ (V_{GS} - V_T), \qquad V_T = V_{FB} + 2\,\varphi_F + \gamma\sqrt{2\,\varphi_F - V_{BS}} \tag{22.56}$$

where, remembering that the junctions are never forward biased, it is $V_{BS} \leq 0$. The $V_T = V_T(V_{BS})$ voltage defined in (22.56) is called *threshold voltage*, and its dependence on V_{BS} is called *body effect*. Near $V_{DS} = 0$ the relation between I_D and V_{DS} is $I_D = \beta \ (V_{GS} - V_T) \ V_{DS}$: there, the current-voltage characteristics are approximated by straight lines whose slope, for a given V_{BS}, is prescribed by V_{GS}. At larger values of V_{DS} the limiting case (22.56) does not hold any longer: the slope of the $I_D = I_D(V_{DS})$ curves decreases, to eventually vanish when V_{DS} reaches V_{DS}^{sat}. Considering that I_D is non-negative, the theory depicted above is applicable as long as $\beta \ (V_{GS} - V_T) \geq 0$; this observation allows one to better specify the condition of a well-formed channel, used at the beginning: from the formal standpoint the condition of a well-formed channel is $V_{GS} > V_T$.

The integration of (22.49) is straightforward and yields $I_D = I'_D - I''_D$, where I'_D is obtained by integrating $\beta \ (V'_{GB} - \varphi_s)$. In this calculation many terms cancel out, to yield the relatively simple expression

$$I'_D = \beta \left[(V_{GS} - V_{FB} - 2\,\varphi_F) \ V_{DS} - \frac{1}{2} \ V_{DS}^2 \right]. \tag{22.57}$$

In turn, I''_D is obtained by integrating $-\beta \, \gamma \, \sqrt{\varphi_s}$, and reads

$$I''_D = \beta \frac{2}{3} \gamma \left[(V_{DS} + 2\,\varphi_F - V_{BS})^{3/2} - (2\,\varphi_F - V_{BS})^{3/2} \right]. \tag{22.58}$$

Comparisons with experiments show that the model $I_D = I'_D - I''_D$, where the two contributions are given by (22.57) and (22.58), provides a fair description of the drain current up to the saturation voltage. Beyond saturation, the model is not correct any longer: in fact, the terms with a negative sign within the expression of I_D give rise to a negative slope g_D; instead, the experiments show that for $V_{DS} > V_{DS}^{sat}$ the current tends to saturate. For this reason, the analytical model is given a regional form: for a prescribed pair V_{GS}, V_{BS}, the regional model first separates the *on* condition $V_{GS} > V_T$ from the *off* condition $V_{GS} \leq V_T$. The *on* condition is further separated into the *linear region*[20] $0 < V_{DS} \leq V_{DS}^{sat}$, where the drain current is described by the $I_D = I'_D - I''_D$ model worked out above, and the *saturation region* $V_{DS} > V_{DS}^{sat}$, where the regional model lets $I_D = I_D^{sat}$. Finally, in the *off* condition the model lets $I_D = 0$.

For a given bulk-source voltage V_{BS}, the $I_D = I_D(V_{DS})$ curves corresponding to different values of V_{GS} are called *output characteristics*. Other types of characteristics

---

[20] The term *linear* originates from the behavior of the curves near the origin, shown by (22.56). The term is ascribed to the region up to V_{DS}^{sat} despite the fact that far away from the origin the curves are blatantly non linear.

Fig. 22.15 Low frequency, small-signal circuit of an *n*-channel MOSFET

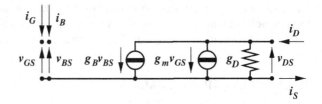

are also used to enrich the picture of the MOSFET's behavior: for instance, the *transfer characteristics* are the $I_D = I_D(V_{GS})$ curves drawn using V_{BS} as a parameter and letting $V_{DS} = \text{const}$, with a value of V_{DS} small enough to let the limiting case (22.56) hold.

Besides the drain conductance (22.51), two more differential conductances are defined in a MOSFET: the first one is the *transconductance* g_m, given by the derivative of I_D with respect to V_{GB}, at constant V_{DB} and V_{SB}; or, equivalently, as the derivative with respect to V_{GS}, at constant V_{DS} and V_{BS}. Observing that V_{GS} appears only in I_D' one finds

$$g_m = \left(\frac{\partial I_D}{\partial V_{GB}}\right)_{V_{SB}, V_{DB}} = \left(\frac{\partial I_D}{\partial V_{GS}}\right)_{V_{DS}, V_{BS}} = \beta\, V_{DS}. \qquad (22.59)$$

The second one is the *bulk transconductance* g_B, defined as the derivative of I_D with respect to V_{BS} at constant V_{DS} and V_{GS}:

$$g_B = \left(\frac{\partial I_D}{\partial V_{BS}}\right)_{V_{DS}, V_{GS}} = -\left(\frac{\partial I_D''}{\partial V_{BS}}\right)_{V_{DS}}. \qquad (22.60)$$

The small-signal circuit of an *n*-channel MOSFET is shown in Fig. 22.15. Since the circuit is derived from the steady-state transport model, it holds at low frequencies only. The small-signal voltages are indicated with v_{DS}, v_{GS}, and v_{BS}. The gate and bulk contacts are left open because the corresponding currents are zero; as a consequence, $i_D = i_S$ is the only non-zero small-signal current of the circuit. Observing that

$$i_D = g_D\, v_{DS} + g_m\, v_{GS} + g_B\, v_{BS}, \qquad (22.61)$$

the drain-source branch of the circuit is made of three parallel branches. One of them is represented as a resistor $1/g_D$ because the current flowing in it is controlled by the voltage v_{DS} applied to the same port; the other two branches are voltage-controlled generators because the current of each branch is controlled by the voltage applied to a different port.

It is worth adding that the body effect mentioned above is actually an inconvenience, because it introduces a complicate dependence on V_{BS} which must be accounted for during the circuit's design. The body effect is suppressed by letting

$V_{BS} = 0$: in a circuit's design, this is obtained by shorting the bulk and source contacts, which amounts to reducing the original four-contact device to a three-contact device.[21] This solution is adopted whenever the circuit's architecture allows for it.

22.6.2.1   Linear–Parabolic Model

In semiqualitative circuit analyses the whole term I_D'' is neglected, this leading to a simplified model $I_D \simeq I_D'$, called *linear–parabolic model*. As the neglect of I_D'' is equivalent to letting $\gamma \to 0$, it follows $g_B \simeq 0$ and, from the second relation in (22.56), the simplified threshold voltage reads

$$V_T \simeq V_{FB} + 2\,\varphi_F. \tag{22.62}$$

In turn, from $I_D \simeq I_D' = \beta\left[\left(V_{GS} - V_T\right) V_{DS} - V_{DS}^2/2\right]$ one finds for the drain conductance

$$g_D \simeq \beta\,(V_{GS} - V_{FB} - 2\,\varphi_F - V_{DS}) = \beta\,(V_{GS} - V_T - V_{DS}), \tag{22.63}$$

whence

$$V_{DS}^{\text{sat}} = V_{GS} - V_T. \tag{22.64}$$

The transconductance g_m is the same as in the general case. Note that the linear–parabolic expression of the drain current may be recast as $I_D \simeq \beta\left(V_{DS}^{\text{sat}}\,V_{DS} - V_{DS}^2/2\right)$, with V_{DS}^{sat} given by (22.64). As a consequence,

$$I_D^{\text{sat}} \simeq \beta\left[\left(V_{DS}^{\text{sat}}\right)^2 - \frac{1}{2}\left(V_{DS}^{\text{sat}}\right)^2\right] = \frac{1}{2}\,\beta\,(V_{GS} - V_T)^2. \tag{22.65}$$

The linear–parabolic model then yields for the saturation region

$$I_D = I_D^{\text{sat}}, \qquad g_D \simeq 0, \qquad g_m \simeq \beta\,(V_{GS} - V_T), \qquad g_B \simeq 0. \tag{22.66}$$

An example of the output characteristics of an n-type MOSFET obtained from the linear–parabolic model is given in Fig. 22.16, using $V_T = 1$ V, $\beta = 0.3$ A V^{-2}. The dashed curve represents (22.65).

22.7   Complements

22.7.1   *Poisson's Equation in the MOSFET Channel*

The derivation of the MOSFET's current carried out in Sect. 22.6 is based upon two integrals; the first one, (22.39), is calculated over a section of the channel at

---

[21] Note that letting $V_{BS} = 0$ also makes g_B to vanish.

Fig. 22.16 Output characteristics of an n-type MOSFET obtained from the linear–parabolic model, with $V_T = 1$ V, $\beta = 0.3$ A V^{-2}. The *dashed curve* represents (22.65)

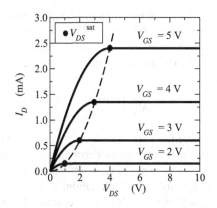

some position y, the second one, (22.44), is calculated along the channel from $y = 0$ to $y = L$. Apparently this procedure eliminates the need of solving the Poisson equation. In fact, the solution of the latter is deeply rooted in the relation (22.46), which is a fundamental point of the procedure itself, and in the choice of the integration limits (22.47), which are also related to (22.46).

The Poisson equation in a non-equilibrium condition is conveniently tackled by expressing the carrier concentrations in terms of the quasi-Fermi potentials φ_n and φ_p; the device considered here is the same n-channel MOSFET of Sect. 22.6. Using the normalized form $u_n = q\varphi_n/(k_B T)$ and $u_p = q\varphi_p/(k_B T)$, the concentrations read $n = n_i \exp(u - u_n)$ and $p = n_i \exp(u_p - u)$, respectively. Remembering that in the equilibrium limit it is $u_n, u_p \to u_F$, with u_F the normalized Fermi potential, it is useful to introduce the differences

$$\chi_n = u_n - u_F, \qquad \chi_p = u_p - u_F, \qquad (22.67)$$

by which the concentrations take the form

$$n = n_{p0} \exp(u - \chi_n), \qquad p = p_{p0} \exp(\chi_p - u). \qquad (22.68)$$

In the equilibrium limit it is $\chi_n, \chi_p \to 0$. Moreover, when a non-equilibrium condition holds, at any position y in the channel it is $\lim_{x \to \infty} \chi_n, \chi_p = 0$; in fact, as observed in Sect. 22.6, far from the channel the semiconductor is practically in the equilibrium condition, whence $\varphi_n \to \varphi_F$ as x increases. The same applies to φ_p. With these provisions, the charge density in the semiconductor reads $\rho = q\left[p_{p0} \exp(\chi_p - u) - n_{p0} \exp(u - \chi_n) - N_A\right]$, namely,

$$\rho = -qp_{p0}\, A, \qquad A = \frac{n_i^2}{N_A^2}\left[\exp(u - \chi_n) - 1\right] + 1 - \exp(\chi_p - u), \qquad (22.69)$$

and the Poisson equation takes the form

$$\frac{\partial^2 u}{\partial x^2} + \frac{\partial^2 u}{\partial y^2} = \frac{1}{L_A^2}\, A, \qquad (22.70)$$

with L_A the holes' Debye length defined in (21.12). One notes that (22.69), (22.70) are generalizations of (22.3). In the description of the MOSFET the accumulation condition is not considered, hence the holes' contribution $\exp(\chi_p - u)$ to A is negligible in the channel region; thus,

$$A \simeq \frac{n_i^2}{N_A^2} \left[\exp(u - \chi_n) - 1 \right] + 1 \simeq \frac{n_i^2}{N_A^2} \exp(u - \chi_n) + 1 > 0, \qquad (22.71)$$

where the term $n_i^2/N_A^2 = \exp(-2u_F)$ is negligible with respect to unity. Remembering that the quasi-Fermi potential in the channel does not depend on x, in (22.71) it is $u = u(x, y)$, $\chi_n = \chi_n(y)$, with $u(\infty, y) = 0$, $u'(\infty, y) = 0$ due to the charge neutrality of the bulk region; in fact, as shown by numerical solutions, both u and u' practically vanish when x reaches the value of the depletion width $x_d(y)$.

The Poisson equation is to be solved in two dimensions. If the condition of a well-formed channel holds, the components of the electric field along the x and y directions are quite different from each other. The x component at the semiconductor–oxide interface, E_{sx}, which is due to the voltage applied to the gate contact, is large because it maintains the strong-inversion condition of the surface. Moreover, the derivative $\partial E_x/\partial x = -\partial^2 u/\partial x^2$ is also large, because E_x changes from E_{sx} to zero in the short distance $x_d(y)$. In contrast, the y component of the electric field at the interface, E_{sy}, which is due to the voltage V_{DS} applied between the drain and source contacts, is small; in fact, V_{DS} is small in itself because the linear region only is considered, and the channel length L is larger than the insulator thickness. Moreover, numerical solutions show that for $0 < V_{DS} < V_{DS}^{sat}$ the dependence of both E_{sx} and E_{sy} on y is weak, which in particular makes $\partial E_y/\partial y = -\partial^2 u/\partial y^2$ also small at the semiconductor–insulator interface. In conclusion, one approximates (22.70) as

$$\frac{\partial^2 u}{\partial x^2} + \frac{\partial^2 u}{\partial y^2} \simeq \frac{d^2 u}{dx^2} = \frac{1}{L_A^2} A. \qquad (22.72)$$

The dependence of A on y remains, and y is treated as a parameter in the solution procedure. Due to the form of (22.72), the solution method is identical to that used in Sect. 22.2.1 to treat the equilibrium case; it yields

$$\left(\frac{qE_{sx}}{k_B T} \right)^2 = \frac{2}{L_A^2} F^2, \qquad F^2 = \exp\left(-\chi_n - 2u_F \right) \left[\exp(u_s) - 1 \right] + u_s. \quad (22.73)$$

Remembering that the accumulation condition is excluded, here it is $u_s(y) \geq 0$; the flat-band condition $u_s = 0$ corresponds to $F = 0$. In the strong-inversion condition the contribution of the electron charge (proportional to $\exp(u_s) - 1$ in (22.73)) is dominant; for this to happen it is necessary that the exponent $u_s - \chi_n - 2u_F$ in (22.73) be positive; it follows that the threshold condition is identified by $u_s = \chi_n + 2u_F$. Remembering the definition (22.67) of χ_n one then finds

$$u_s = u_n + u_F, \qquad 0 \leq y \leq L, \qquad (22.74)$$

that is, the normalized form of (22.46). Note that $u_s = \chi_n + 2\,u_F$ is coherent with the definition of the threshold condition at equilibrium (Table 22.1), which is obtained by letting $\chi_n = 0$. Also, specifying $u_s = \chi_n + 2\,u_F$ at the source and drain ends of the channel provides the integration limits (22.47) [103, Sect. 10.2].

The neglect of the variation in the y component of the electric field along the channel makes the general relation

$$Q_{sc} = -C_{ox}\,(V'_{GB} - \varphi_s) = -C_{ox}\,\frac{k_B\,T}{q}\,(u'_{GB} - u_s) \tag{22.75}$$

still valid at each position y along the channel, with $Q_{sc} < 0$ because $u_s > 0\,(Q_{sc} = 0$ in the flat-band condition $u_s = 0$). Also, when the inversion charge is approximated by a charge layer at $x = 0^+$, the volume charge is entirely due to the ionized dopants, whose contribution in the second relation of (22.73) is proportional to u_s like in the equilibrium case. Using the same relation $Q_i = Q_{sc} - Q_b$ as in Sect. 22.4 one finally finds that the theory worked out in this section makes (22.34) applicable also in a non-equilibrium condition in which the channel is well formed.

22.7.2 Inversion-Layer Charge and Mobility Degradation

In the model (22.49) for the drain current worked out in the previous sections, the modulus of the inversion-layer charge Q_i decreases from source to drain due to the increase in φ_s along the channel (compare with (22.34)). Considering that the MOSFET current is carried by the inversion-layer charge, the vanishing of the latter occurring at the drain end of the channel when $V_{DS} \to V_{DS}^{sat}$ may seem an oddity. However, it is important to remember that a number of approximations are necessary to reach the result expressed by (22.57) and (22.58); such approximations make the theory applicable only in the linear region and in the condition of a well-formed channel.

When $V_{DS} \to V_{DS}^{sat}$, the vertical component of the electric field at the interface, E_{sx}, is made weaker by the interplay between the voltages applied to the gate electrode and to the nearby drain electrode; for this reason, at the drain end of the channel the flow lines of $\mathbf{J}_n$ do not keep close to the interface any longer, but spread into the substrate, this decreasing the carrier density. The phenomenon is better understood with the aid of Fig. 22.17, where the linear-parabolic model is used with, e.g., $V_T = 0.5$, $V_S = 0$, $V_{GS} = 1.5$, $V_{DS} = 2\,\mathrm{V}$. As $V_{DS}^{sat} = 1\,\mathrm{V}$, the saturation condition $V_{DS} > V_{DS}^{sat} = 1$ holds. Along the dash-dotted line enclosed in the right oval, the direction of the electric field is that shown in the vector diagram in the upper-right part of the figure; in particular, the vertical component of the field at the position marked by the vertical dashed line points upwards. As a consequence, the channel electrons are repelled downwards and the flow lines of the current density detach from the interface. At the source end of the channel, instead, the vertical component of the field points downwards. By continuity, a position within the channel exists where, in saturation, the vertical component of the field vanishes; such a position (not shown in the figure) is called *inversion point*. Also, the large component of the electric field along the y direction,

Fig. 22.17 Illustration of the electric field's components at the channel ends in the saturation condition

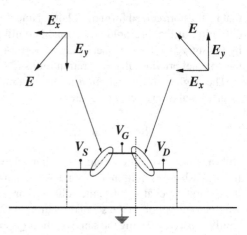

which exists within the space-charge region of the reverse-biased drain junction, makes the carriers' average velocity to increase at the drain end of the channel; as the total current is constant, a further decrease in the carrier concentration occurs. The two effects briefly illustrated above are not accounted for by the simplified model, and require a more elaborate approach in which the two-dimensional structure of the electric field is accounted for.

Another comment is useful, this time related to the *off* condition. When $V_{GS} \rightarrow V_T$, the current of a real device does not actually vanish; in fact, a current (called *subthreshold current*), due to the carriers present in the channel in the weak-inversion condition, flows for $V_{GS} < V_T$. Also in this case a more elaborate theory is necessary, showing that the subthreshold current vanishes exponentially as V_{GS} decreases [3].

It is worth concluding this section by commenting the simplification used in (22.45), where the effective electron mobility μ_e is assumed to be independent of the position y along the channel. The factors that affect mobility in a MOSFET are collisions with phonons, ionized impurities, and semiconductor–oxide interface; remembering the features of the macroscopic mobility models (Sect. 20.5), the electron mobility $\mu_n(x, y)$ is made to depend on the lattice temperature T, concentration of ionized impurities N_A, and x component of the electric field $E_x(x, y)$. The first two parameters, T and N_A, do not introduce a dependence on position because they are themselves constant. The x dependence of E_x is absorbed by the integral (22.42) that defines μ_e; it follows that the average mobility depends on y because E_x does. Such a dependence, in turn, is relatively weak in the strong-inversion condition as remarked in Sect. 22.7.1. Therefore, the dependence of μ_e on position is considered negligible.

Problems

22.1 Work out a method for drawing the curves of Fig. 22.9 without approximations.

Part VIII
Miscellany

Chapter 23
Thermal Diffusion

23.1 Introduction

The fabrication of integrated circuits requires the introduction into the semiconductor material of atoms belonging to specifically-selected chemical species. Such atoms are called *impurities* or *dopants*. As shown in Chap. 18, the inclusion of dopants into the semiconductor lattice attains the important goals of fixing the concentration of mobile charges in the material and making it practically independent of temperature.

Dopants are divided into two classes, termed *n-type* and *p-type*. With reference to silicon (Si), the typical *n*-type dopants are phosphorus (P), arsenic (As), and antimony (Sb), while the typical *p*-type dopants are boron (B), aluminum (Al), gallium (Ga), and Indium (In). When a dopant atom is introduced into the semiconductor lattice, in order to properly act as a dopant it must replace an atom of the semiconductor, namely, it must occupy a lattice position. When this happens, the dopant atom is also called *substitutional impurity*. An impurity atom that does not occupy a lattice position is called *interstitial*. Interstitials can not properly act as dopants, however, they degrade the conductivity and other electrical properties of the semiconductor.

The concentration of the dopant atoms that are introduced into a semiconductor is smaller by orders of magnitude than the concentration of the semiconductor atoms themselves. As a consequence, the average distance between dopant atoms within the lattice is much larger than that between the semiconductor atoms. Thus, the material resulting from a doping process is not a chemical compound: it is still the semiconductor in which some of the electrical properties are modified by the presence of the dopant atoms. In fact, while the presence and type of dopants are easily revealed by suitable electrical measurements, they may remain undetectable by chemical analyses.

As a high-temperature condition is necessary to let the dopant atoms occupy the lattice position, during the fabrication of the integrated circuit the semiconductor wafer undergoes several high-temperature processes. This, in turn, activates the dopant diffusion.

© Springer Science+Business Media New York 2015
M. Rudan, *Physics of Semiconductor Devices,*
DOI 10.1007/978-1-4939-1151-6_23

The chapter illustrates the diffusive transport with reference to the processes that are used for introducing impurities into a semiconductor in a controlled way. First, the expressions of the continuity equation and of the diffusive flux density are derived. These expressions are combined to yield the diffusion equation, whose form is reduced to a one-dimensional model problem. The model problem allows for an analytical solution, based on the Fourier-transform method, that expresses the diffused profile at each instant of time as the convolution of the initial condition and an auxiliary function.

Then, the solution of the model problem is used to calculate the impurity profiles resulting from two important processes of semiconductor technology, namely, the predeposition and the drive-in diffusion. In the last part of the chapter the solution of the model problem is extended to more general situations. Specific data about the parameters governing the diffusion processes in semiconductors are in [46, Chap. 3], [104, Chap. 10], [105, Chap. 7], [71, Chap. 12]. Many carefully-drawn illustrations of the diffusion process are found in [75, Sect. 1.5]. The properties of the Fourier transform are illustrated in [72, 118].

23.2   Continuity Equation

The continuity equation described in this section is a balance relation for the number of particles. Here it is not necessary to specify the type of particles that are being considered: they may be material particles, like molecules or electrons, particles associated to the electromagnetic field (photons), those associated to the vibrational modes of atoms (phonons), and so on. Although the type of particles is not specified, it is assumed that all particles considered in the calculations are of the same type.

The balance relation is obtained by considering the space where the particles belong and selecting an arbitrary volume V in it, whose boundary surface is denoted with S. The position of the volume is fixed. Let $\mathcal{N}(t)$ be the number of particles that are inside S at time t. Due to the motion of the particles, in a given time interval some of them move across S in the outward direction, namely, from the interior to the exterior of S. In the same interval of time, other particles move across S in the inward direction. Let $\mathcal{F}_{\text{out}}(t)$ and $\mathcal{F}_{\text{in}}(t)$ be the number of particles per unit time that cross S in the outward or inward direction, respectively, and let $\mathcal{F} = \mathcal{F}_{\text{out}} - \mathcal{F}_{\text{in}}$. The quantity $\mathcal{F}$, whose units are s^{-1}, is the *flux* of the particles across the surface S. If the only reason that makes $\mathcal{N}$ to change is the crossing of S by some particles, the balance relation takes the form of the first equation in (23.1). The minus sign at the right hand side is due to the definition of $\mathcal{F}$; in fact, $\mathcal{N}$ decreases with time when $\mathcal{F} > 0$, and vice versa.

Besides the crossing of the boundary S by some particles, there is another mechanism able to contribute to the time variation of $\mathcal{N}$, namely, the generation or destruction of particles inside the volume V. This possibility seems to violate some commonly-accepted conservation principle. However it is not so, as some examples given in Sect. 23.7.1 will show. As a consequence, the description of the particle generation or destruction must be included. This is accomplished by letting $\mathcal{W}_{ge}(t)$ and $\mathcal{W}_{de}(t)$ be the number of particles per unit time that are generated or, respectively, destroyed within the volume V. Defining $\mathcal{W} = \mathcal{W}_{ge} - \mathcal{W}_{de}$, the balance relation that holds when generation or destruction are present takes the form of the second equation in (23.1):

$$\frac{d\mathcal{N}}{dt} = -\mathcal{F}, \qquad \frac{d\mathcal{N}}{dt} = -\mathcal{F} + \mathcal{W}. \tag{23.1}$$

The quantity $\mathcal{W}$, whose units are s^{-1}, is the *net generation rate* within volume V.

It is convenient to recast (23.1) in local form. This is done basing on the second equation of (23.1), which is more general, and is accomplished by describing the motion of the particles as that of a continuous fluid. Such a description is legitimate if V can be partitioned into equal cells of volume $\Delta V_1, \Delta V_2, \ldots$ having the following properties: (i) the cells can be treated as infinitesimal quantities in the length scale of the problem that is being considered and, (ii) the number of particles within each cell is large enough to make their average properties significant. If the above conditions are fulfilled one lets $\Delta V_k \rightarrow dV$ and introduces the *concentration* $N(\mathbf{r}, t)$, such that $N \, dV$ is the number of particles that at time t belong to the volume dV centered at position $\mathbf{r}$. Similarly, one defines the *net generation rate per unit volume* $W(\mathbf{r}, t)$ such that $W \, dV$ is the net generation rate at time t within dV. The units of N and W are m^{-3} and $m^{-3} \, s^{-1}$, respectively. From the definitions of N, W it follows that the number $\mathcal{N}(t)$ of particles that are inside S at time t is found by integrating $N(\mathbf{r}, t)$ over V, and that the net generation rate $\mathcal{W}(t)$ is found by integrating W over V.

To recast in local form the part of (23.1) related to the flux $\mathcal{F}$, one associates a velocity $\mathbf{v}(\mathbf{r}, t)$ to the concentration $N(\mathbf{r}, t)$. In general, such a velocity is different from the velocity of each individual particle that contributes to the concentration N. In fact, $\mathbf{v}$ is a suitable average of the particles' velocities, whose definition (6.6) is given in Sect. 6.2. In the elementary time dt the concentration originally in $\mathbf{r}$ moves over a distance $\mathbf{v} \, dt$ in the direction of $\mathbf{v}$. As consequence, if $\mathbf{r}$ belongs to the boundary surface S, a crossing of S by the particles may occur, this contributing to the flux. To calculate the contribution to the flux at a point $\mathbf{r}$ belonging to S, one takes the plane tangent to S at $\mathbf{r}$ and considers an elementary area dS of this plane centered at $\mathbf{r}$ (Fig. 23.1). The construction implies that the surface S is smooth enough to allow for the definition of the tangent plane at each point of it.

Let $\mathbf{s}$ be the unit vector normal to dS, oriented in the outward direction with respect to S. If $\mathbf{v}$ is normal to $\mathbf{s}$, no crossing of S occurs and the contribution to the flux at point $\mathbf{r}$ is zero. If the scalar product $\mathbf{v} \cdot \mathbf{s}$ is positive, the crossing occurs in the outward direction and contributes to $\mathcal{F}_{out}$. Its contribution is found by observing that the elementary cylinder, whose base area and side are dS and, respectively, $\mathbf{v} \, dt$, has a volume equal to $\mathbf{v} \cdot \mathbf{s} \, dS \, dt$. Due to the sign of $\mathbf{v} \cdot \mathbf{s}$, the cylinder is outside the surface S. The number of particles in the cylinder is found by multiplying its volume by the concentration $N(\mathbf{r}, t)$. Letting $\mathbf{F} = N\mathbf{v}$, such a number reads $\mathbf{F} \cdot \mathbf{s} \, dS \, dt$. As the particles that are in the cylinder at time $t + dt$ were inside the surface S at time t, dividing the above expression by dt yields the elementary contribution of point $\mathbf{r}$

Fig. 23.1 Illustration of the
symbols used in the
calculation of the flux

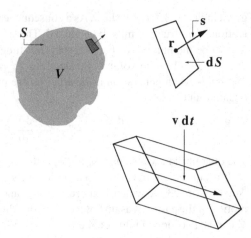

to the flux, $d\mathcal{F} = \mathbf{F} \cdot \mathbf{s}\,dS > 0$. The contribution from a point $\mathbf{r}$ where $\mathbf{v} \cdot \mathbf{s} < 0$ is
calculated in a similar way. The flux $\mathcal{F}$ is then found by integrating $\mathbf{F} \cdot \mathbf{s}$ over the
surface S. The quantity $\mathbf{F} \cdot \mathbf{s} = d\mathcal{F}/dS$, whose units are $m^{-2}s^{-1}$, is the *flux density*.

Introducing the relations found so far into the second form of (23.1), and
interchanging the derivative with respect to t with the integral over V, yields

$$\int_V \left(\frac{\partial N}{\partial t} - W \right) dV = -\int_S \mathbf{F} \cdot \mathbf{s}\,dS = -\int_V \mathrm{div}\mathbf{F}\,dV. \qquad (23.2)$$

The last equality in (23.2) is due to the divergence theorem (A.23), whereas the use
of the partial-derivative symbol is due to the fact the N, in contrast with $\mathcal{N}$, depends
also on $\mathbf{r}$. The procedure leading to (23.2) does not prescribe any constraint on the
choice of the volume V. As a consequence, the two integrals over V that appear in
(23.2) are equal to each other for any V. It follows that the corresponding integrands
must be equal to each other, this yielding the *continuity equation*

$$\frac{\partial N}{\partial t} + \mathrm{div}\mathbf{F} = W, \qquad \mathbf{F} = N\mathbf{v}. \qquad (23.3)$$

As mentioned above, (23.3) is the local form of the second equation of (23.1), which
in turn is a balance relation for the number of particles. In the steady-state condition
the quantities appearing in (23.3) do not depend explicitly on time, hence (23.3)
reduces to $\mathrm{div}\mathbf{F} = W$. In the equilibrium condition it is $\mathbf{v} = 0$ and (23.3) reduces to
the identity $0 = 0$. It is worth noting that in the equilibrium condition the velocity
of each particle may differ from zero; however, the distribution of the individual
velocities is such that the average velocity $\mathbf{v}$ vanishes. Similarly, in the equilibrium
condition the generation or destruction of particles still occurs; however, they balance
each other within any dV.

To proceed it is assumed that the net generation rate per unit volume W, besides
depending explicitly on $\mathbf{r}$ and t, may also depend on N and $\mathbf{F}$, but not on other
functions different from them.

23.3 Diffusive Transport

The continuity Eq. (23.3) provides a relation between the two quantities N and $\mathbf{F}$ (or, equivalently, N and $\mathbf{v}$). If both N and $\mathbf{F}$ are unknown it is impossible, even in the simple case $W = 0$, to calculate them from (23.3) alone. However, depending on the specific problem that is being considered, one can introduce additional relations that eventually provide a closed system of differential equations. The important case of the *diffusive transport* is considered in this section.

It is convenient to specify, first, that the term *transport* indicates the condition where an average motion of the particles exists, namely $\mathbf{F} \neq 0$ for some $\mathbf{r}$ and t. The type of transport in which the condition $\mathbf{F} \neq 0$ is caused only by the spatial nonuniformity of the particles' concentration N is called *diffusive*. Simple examples of diffusive transport are those of a liquid within another liquid, or of a gas within another gas. They show that in the diffusive motion of the particles, the flux density is oriented from the regions where the concentration is larger towards the regions where the concentration is smaller.

The analytical description of the diffusion process dates back to 1855 [36]. Here the relation between $\mathbf{F}$ and N in the diffusive case is determined heuristically, basing on the observation that $\mathrm{grad}\, N$ is a sensible indicator of the spatial nonuniformity of N. Specifically it is assumed, first, that $\mathbf{F}$ depends on N and $\mathrm{grad}\, N$, but not on higher-order derivatives of N. The dependence on $\mathrm{grad}\, N$ is taken linear, $\mathbf{F} = \mathbf{F}_0 - D\, \mathrm{grad}\, N$, with $\mathbf{F}_0 = 0$ because $\mathbf{F}$ must vanish when the concentration is uniform. Finally, one remembers that the particles' flux density is oriented in the direction of a decreasing concentration, namely, opposite to $\mathrm{grad}\, N$. It follows that $D > 0$, so that the relation takes the form

$$\mathbf{F} = -D\, \mathrm{grad}\, N, \qquad D > 0. \tag{23.4}$$

The above is called *transport equation of the diffusion type*, or *Fick's first law of diffusion*. Parameter D is the *diffusion coefficient*, whose units are $\mathrm{m}^2\ \mathrm{s}^{-1}$. From the derivation leading to (23.4) it follows that, if a dependence of $\mathbf{F}$ on N exists, it must be embedded in D. In the case $D = D(N)$ the relation (23.4) is linear with respect to $\mathrm{grad}\, N$, but not with respect to N. The diffusion coefficient may also depend explicitly on $\mathbf{r}$ and t. For instance, it depends on position when the medium where the diffusion occurs is nonuniform; it depends on time when an external condition that influences D, e.g., temperature, changes with time.

For the typical dopants used in the silicon technology, and in the temperature range of the thermal-diffusion processes, the experimentally-determined dependence on temperature of the diffusion coefficient can be approximated by the expression

$$D = D_0 \exp[\,-E_a/(k_B T)\,], \tag{23.5}$$

where k_B (J K^{-1}) is the Boltzmann constant and T (K) the process temperature. In turn, the *activation energy* E_a and D_0 are parameters whose values depend on the material involved in the diffusion process. The form of (23.5) makes it more convenient to draw it as an *Arrhenius plot*, that displays the logarithm of the function using the inverse temperature as a variable: $\log D = \log(D_0) - (E_a/k_B)(1/T)$. At the diffusion temperatures, E_a and D_0 can often be considered independent of temperature. In this case the Arrhenius plot is a straight line (examples of Arrhenius plots are given in Chap. 24). At room temperature the diffusion coefficient of dopants in silicon is too small to make diffusion significant. In order to activate the diffusion mechanism a high-temperature process is necessary, typically between 900 and 1100°C.

23.4   Diffusion Equation—Model Problem

Inserting (23.4) into (23.3) yields the *diffusion equation*

$$\frac{\partial N}{\partial t} = \text{div}(D\,\text{grad}N) + W, \tag{23.6}$$

where W depends on $\mathbf{r}$, t, N, and $\text{grad}N$ at most, while D depends on $\mathbf{r}$, t, and N at most. The above is a differential equation in the only unknown N. It must be supplemented with the initial condition $N_0(\mathbf{r}) = N(\mathbf{r}, t = 0)$ and suitable boundary conditions for $t > 0$. If the diffusion coefficient is constant, or depends on t at most, (23.6) becomes

$$\frac{\partial N}{\partial t} = D\nabla^2 N + W, \qquad D = D(t). \tag{23.7}$$

It is convenient to consider a simplified form of (23.7) to be used as a model problem. For this, one takes the one-dimensional case in the x direction and lets $W = 0$, this yielding

$$\frac{\partial N}{\partial t} = D\,\frac{\partial^2 N}{\partial x^2}. \tag{23.8}$$

Equation (23.8) is also called *Fick's second law of diffusion*. Thanks to the linearity of (23.8), the solution can be tackled by means of the Fourier-transform method, specifically, by transforming both sides of (23.8) with respect to x. Indicating[1] with $G(k, t) = \mathcal{F}_x N$ the transform of N with respect to x, and using some of the properties of the Fourier transform illustrated in Appendix C.2, one finds

$$\mathcal{F}_x \frac{\partial N}{\partial t} = \frac{dG}{dt}, \qquad \mathcal{F}_x D\,\frac{\partial^2 N}{\partial x^2} = D\mathcal{F}_x\,\frac{\partial^2 N}{\partial x^2} = -k^2 DG. \tag{23.9}$$

The symbol of total derivative is used at the right hand side of the first of (23.9) because k is considered as a parameter. The Fourier transform of the initial condition of N provides the initial condition for G, namely, $G_0 = G(k, t = 0) = \mathcal{F}_x N_0$.

---

[1] Symbol $\mathcal{F}_x$ indicating the Fourier transform should not be confused with the symbol $\mathcal{F}$ used for the particles' flux in Sect. 23.2.

Equating the right hand sides of (23.9) and rearranging yields $dG/G = -k^2 D(t) \, dt$. Integrating the latter from 0 to t,

$$\log(G/G_0) = -k^2 a(t), \qquad a(t) = \int_0^t D(t') \, dt', \tag{23.10}$$

with a an area. The concentration N is now found by antitransforming the expression of G extracted from the first of (23.10):

$$N(x,t) = \mathcal{F}_k^{-1} G = \frac{1}{\sqrt{2\pi}} \int_{-\infty}^{+\infty} G_0 \exp(ikx - ak^2) \, dk. \tag{23.11}$$

In turn, G_0 within the integral of (23.11) is expressed as the transform of N_0. After rearranging the integrals one finds

$$N(x,t) = \int_{-\infty}^{+\infty} N_0(\xi) \left\{ \int_{-\infty}^{+\infty} \frac{1}{2\pi} \exp\left[ik\,(x - \xi) - ak^2\right] dk \right\} d\xi. \tag{23.12}$$

As shown in Appendix C.7, the expression in braces in (23.12) is the integral form of the function $\Delta(x - \xi, t)$ defined by (C.75). As a consequence, the solution of the simplified form (23.8) of the diffusion equation is the convolution between Δ and the initial condition N_0, namely,

$$N(x,t) = \int_{-\infty}^{+\infty} N_0(\xi) \, \Delta(x - \xi, t) \, d\xi. \tag{23.13}$$

A straightforward calculation shows that Δ fulfills (23.8) for all ξ. As a consequence, (23.13) is a solution as well. In addition, due to (C.79), (23.13) also fulfills the initial condition N_0.

23.5 Predeposition and Drive-in Diffusion

Basing on the model problem worked out in Sect. 23.4 it is possible to describe the thermal diffusion of dopants in silicon. The modification induced in the electrical properties of the silicon lattice by the inclusion of atoms belonging to different chemical species (e.g., phosphorus or boron) are described elsewhere (Sects. 18.4.1 and 18.4.2). Here the analysis deals with the diffusion process in itself.

The formation of a diffused profile in silicon is typically obtained in a two-step process [46, 75, 105]. In the first step, called *predeposition*, a shallow layer of dopants is introduced into the semiconductor. The most common predeposition methods are the diffusion from a chemical source in a vapor form or the diffusion from a solid source (e.g., polycrystalline silicon) having a high concentration of dopants in it. In both methods the silicon wafers are placed in contact with the source of dopant within a furnace kept at a high temperature.

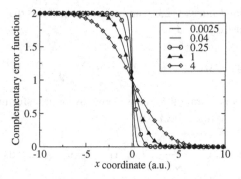

Fig. 23.2 Normalized profiles N/C produced at different instants by a predeposition, using the first of (23.15) as initial condition with arbitrary units for the x coordinate. The outcome is a set of complementary error functions whose expression is the first of (23.16). The legends show the value of $4\,a$ for each curve, also in arbitrary units, with $a = a(t)$ given by the second of (23.10)

During a predeposition step, new dopant atoms are continuously supplied by the source to the silicon region. As a consequence, the number of dopant atoms in the silicon region increases with time. When the desired amount of atoms is reached, the supply of dopants is blocked, whereas the diffusion process is continued. During this new step, called *drive-in diffusion*, the number of dopant atoms in the silicon region remains constant. The drive-in diffusion is continued until a suitable profile is reached.

Typically, the blocking of the flow of dopant atoms from the source to the silicon region is achieved by introducing oxidizing molecules into the furnace atmosphere, this resulting in the growth of a silicon-dioxide layer at the silicon surface (the details of the oxidation process are given in Chap. 24).

It is worth anticipating that in some processes the predeposition step is skipped, and the dopant atoms are introduced into the silicon wafers at low temperature by means of an ion-implantation process. The implanted wafers are then placed into the high-temperature furnace to activate the drive-in diffusion.

23.5.1 Predeposition

Figure 23.2 provides a schematic picture of the source–wafer structure during a predeposition step. The interface between wafer and source is assumed to coincide with the y, z plane, with the x axis oriented towards the wafer's bulk, and the initial condition N_0 is assumed constant in the source region. The diffusion coefficients in the source and wafer regions are provisionally taken equal to each other. Thanks to these assumptions the problem has no dependencies on the y, z variables, and the one-dimensional form (23.8) of the diffusion equation holds. In the practical cases the extent of the source region in the x direction is large and the concentration of the dopant atoms in it is high. As a consequence, the source is not depleted when the atoms diffuse into the wafer. The spatial form of the concentration N at a given time $t = t'$ is called *diffused profile*. Its integral over the semiconductor region,

$$Q(t') = \int_0^{+\infty} N(x, t = t') \, dx \qquad (\text{m}^{-2}), \qquad (23.14)$$

is called *dose*.

For convenience the constant value of the initial condition in the source region is indicated with $2C$ (m^{-3}). It follows that the initial condition of the predeposition step is given by the first of (23.15). In turn, the general expression (23.13) of the dopant concentration reduces to the second of (23.15):

$$N_0(\xi) = \begin{cases} 2C & \xi < 0 \\ 0 & \xi > 0 \end{cases} \quad ; \quad N(x, t) = 2C \int_{-\infty}^0 A(x - \xi, t) \, d\xi. \quad (23.15)$$

Using (C.78) and (C.71) one finds the following expressions for the diffused profile and dose of the predeposition step,

$$N(x, t) = C \operatorname{erfc}\left(\frac{x}{\sqrt{4a}}\right), \qquad Q(t) = C\sqrt{\frac{4a}{\pi}}, \qquad (23.16)$$

where the dependence on t derives from the second of (23.10). As parameter a increases with time, the dose increases with time as well, consistently with the qualitative description of predeposition given earlier in this section. In most cases the diffusion coefficient is independent of time, $a = Dt$, this yielding $Q \propto \sqrt{t}$.

Still from the second of (23.10) one finds $a(0) = 0$. Combining the latter with the properties (C.69) of the complementary error function shows that $\lim_{t \to 0^+} N(x, t)$ coincides with the initial condition given by the first of (23.15). Also, the solution (23.16) fulfills the boundary conditions $N(-\infty, t) = 2C$, $N(+\infty, t) = 0$ at any $t > 0$. Finally it is $N(0, t) = C$ at any $t > 0$. This explains the term *constant-source diffusion* that is also used to indicate this type of process. In fact, the concentration at the wafer's surface is constant in time. Figure 23.2 shows the normalized concentration N/C calculated from the first of (23.16) at different values of a.

The analysis of the diffusion process carried out so far was based on the assumption of a position-independent diffusion coefficient D. In the actual cases this assumption is not fulfilled because the dopant source and the wafer are made of different materials. As a consequence, the solution of (23.8) must be reworked. In the case of predeposition this is accomplished with little extra work, which is based on the first of (23.16) as shown below.

One assumes, first, that the diffusion coefficient in either region is independent of time, as is the standard condition of the typical processes. In each region the diffusion coefficient takes a spatially-constant value, say, D_S in the source and $D_W \neq D_S$ in the wafer. Now, observe that (23.8) is homogeneous and contains the derivatives of N, but not N itself. It follows that, if $C \operatorname{erfc}[x/(4a)^{1/2}]$ is the solution of (23.8) fulfilling some initial and boundary conditions, then $A \, \text{erfc}[x/(4a)^{1/2}] + B$ is also a solution of (23.8), fulfilling some other conditions that depend on the constants A and B. One then lets, with $t > 0$,

$$N_S = A_S \, \text{erfc}\left(\frac{x}{\sqrt{4D_S t}}\right) + B_S, \qquad x < 0, \tag{23.17}$$

$$N_W = A_W \, \text{erfc}\left(\frac{x}{\sqrt{4D_W t}}\right) + B_W, \qquad x > 0, \tag{23.18}$$

and fixes two relations among the constants in order to fulfill the initial conditions (23.15):

$$\lim_{t=0^+} N_S = 2A_S + B_S = 2C, \qquad \lim_{t=0^+} N_W = B_W = 0. \tag{23.19}$$

In order to fix the remaining constants one must consider the matching conditions of the two regional solutions (23.17, 23.18) at the source–wafer interface. The concentrations across an interface between two different media are related by the *segregation coefficient* k [105, Sect. 1.3.2]. Also, given that no generation or destruction of dopant atoms occurs at the interface, the flux density $-D \, \partial N/\partial x$ must be continuous there. In summary, the matching conditions at the source–wafer interface are

$$N_W(0^+, t) = k \, N_S(0^-, t), \qquad D_W \left(\frac{\partial N_W}{\partial x}\right)_{0^+} = D_S \left(\frac{\partial N_S}{\partial x}\right)_{0^-}. \tag{23.20}$$

Using (23.17, 23.18, 23.19) transforms (23.20) into $A_W = k \, (2C - A_S)$ and, respectively, $\sqrt{D_W} \, A_W = \sqrt{D_S} \, A_S$ whence, remembering the first of (23.19) and letting $\eta = D_W/D_S$,

$$A_S = \frac{k\sqrt{\eta}}{1 + k\sqrt{\eta}} \, 2C, \qquad B_S = \frac{1 - k\sqrt{\eta}}{1 + k\sqrt{\eta}} \, 2C, \qquad A_W = \frac{k}{1 + k\sqrt{\eta}} \, 2C. \tag{23.21}$$

Thanks to (23.21), the concentration of the dopant atoms in the source region at the source–wafer interface at $t > 0$ turns out to be $N_S(0^-, t) = A_S + B_S = 2C/(1 + k\sqrt{\eta})$. If, in particular, the source of dopant is in the gaseous phase, it is $\eta \ll 1$. As k is of order unity, one finds for the gaseous source $N_S(0^-, t) \simeq 2C$, namely, the interface concentration of the source region is practically equal to the asymptotic one. Figure 23.3 shows the diffused profile N calculated from (23.17, 23.18) at two different instants t_1 and $t_2 = 16\,t_1$, with $D_S = 400\,D_W$. The coefficients are, in arbitrary units, $A_S = 2$, $B_S = 78$, $A_W = 40$, $B_W = 0$. From the first of (23.19) it follows $C = 41$. Letting $(4 \, D_W \, t_1)^{1/2} = 1$ (a.u.) one has $(4 \, D_S \, t_1)^{1/2} = 20$, $(4 \, D_W \, t_2)^{1/2} = 4$, and $(4 \, D_W \, t_2)^{1/2} = 80$. These values are used to calculate the four curves shown in the figure.

Fig. 23.3 Diffused profiles
calculated at t_1 and $t_2 = 16\,t_1$
when two different materials
are involved. The calculation
is based on (23.17), (23.18) as
described at the end of
Sect. 23.5.1. The legends
show the $(4Dt)^{1/2}$ value for
each curve

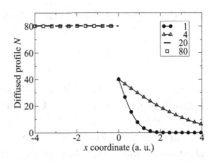

23.5.2   Drive-in Diffusion

As indicated at the beginning of this section, the drive-in diffusion is started when the
desired amount of atoms has been introduced into the silicon lattice, and is continued
until a suitable profile is reached.

In principle, the profile to be used as initial condition of a drive-in diffusion is
not exactly equal to the final profile of the predeposition step. In fact, the boundary
condition $(\partial N_W / \partial x)_{0^+}$ is different from zero during the predeposition step. Instead,
during the growth of the silicon-dioxide layer that blocks the supply of dopant atoms
from the source region, the boundary condition becomes equal to zero to adapt to
the situation of a vanishing flux density of dopants across the interface.

The calculation of the drive-in diffusion is tackled more easily by assuming that the
blocking of the supply of dopants atoms is instantaneous, so that the final profile of the
predeposition step is "frozen". Then, one considers the full domain $-\infty < x < +\infty$
instead of the wafer domain $0 \le x < +\infty$, with the same diffusion coefficient
$D = D_W$ everywhere. In this way one can still use the model problem (23.8). As for
the initial condition N_0, one mirrors the final profile of the predeposition step over
the negative axis, this making the initial condition even with respect to x. Letting
$x \leftarrow -x$ in (23.13) one easily proves that $N(-x, t) = N(x, t)$ if $N_0(-\xi) = N_0(\xi)$;
namely, if the initial condition is even, then the solution is even at all times. With
the provisions above one finds $(\partial N_W / \partial x)_{0^+} = -(\partial N_W / \partial x)_{0^-}$, which automatically
fulfills the condition of a vanishing flux density of dopants across the origin. Then,
the application of (23.13) provides the profile of the drive-in diffusion in the wafer
region $0 \le x < +\infty$.

The final profile (23.16) of the predeposition step, used as initial condition, does
not lend itself to an analytical calculation of the drive-in diffusion. Some examples
of calculation are given below, in which profiles of a simpler form than (23.16) are
used as approximations. Let Q be the dose present within the wafer region. As a first
example one lets

$$N_0(\xi) = 2Q\,\delta(\xi), \qquad N(x, t) = 2Q \int_{-\infty}^{+\infty} \delta(\xi)\,\Delta(x - \xi, t)\,d\xi. \qquad (23.22)$$

From the properties of the Dirac δ (Sect. C.4) it follows

$$N(x,t) = 2Q\,\Delta(x,t) = 2Q\,\frac{\exp[-x^2/(4\,a)]}{\sqrt{4\pi\,a}}, \qquad (23.23)$$

showing that, when the initial condition is a Dirac δ, the profile resulting from a diffusion process is Gaussian. Only the portion of (23.23) belonging to the wafer region, that is, $x \geq 0$, must in fact be considered. Integrating (23.23) from 0 to $+\infty$ and using (C.77) yields the expected value Q of the dose at all times. Although rather crude, the approximation of using a Dirac δ as initial condition is acceptable, because the profile obtained from a predeposition or an ion-implantation process is typically very thin.

As a second example one takes a Gaussian profile as the initial condition, specifically, the second of (23.23) where, to better distinguish the symbols, a is replaced with a_1. It is assumed that the drive-in diffusion to be calculated is characterized by another value of the parameter, say, a_2. The difference between a_2 and a_1 may be due to the duration of the diffusion process under investigation, to a temperature-induced difference in the diffusion coefficients, or both. As usual the instant $t = 0$ is set as the initial time of the diffusion process. Applying (23.13) yields

$$N(x,t) = 2Q \int_{-\infty}^{+\infty} \frac{\exp[-\xi^2/(4\,a_1)]}{\sqrt{4\pi\,a_1}}\,\frac{\exp[-(x-\xi)^2/(4\,a_2)]}{\sqrt{4\pi\,a_2}}\,d\xi. \qquad (23.24)$$

Using the auxiliary variable $\eta = \xi - a_1\,x/(a_1+a_2)$, whence $x-\xi = -\eta + a_2\,x/(a_1+a_2)$, transforms the exponent of (23.24) as

$$-\frac{\xi^2}{4\,a_1} - \frac{(x-\xi)^2}{4\,a_2} = -\frac{x^2}{4\,(a_1+a_2)} - \frac{a_1+a_2}{4\,a_1 a_2}\,\eta^2. \qquad (23.25)$$

Then, integrating with respect to $\sqrt{(a_1+a_2)/(4a_1 a_2)}\,\eta$ and using again (C.77) yields

$$N(x,t) = 2Q\,\frac{\exp[-x^2/(4\,a_1 + 4\,a_2)]}{\sqrt{4\pi\,(a_1+a_2)}}. \qquad (23.26)$$

As before, the integral of the profile from 0 to $+\infty$ yields the dose Q at all times. The result expressed by (23.26) is important because it shows that a diffusion process whose initial condition is a Gaussian profile yields another Gaussian profile. The parameter of the latter is found by simply adding the parameter $a_2 = \int_0^t D(t')\,dt'$ of the diffusion process in hand, whose duration is t, to the parameter a_1 of the initial condition. Clearly, the result is also applicable to a sequence of successive diffusion processes. In fact, it is used to calculate the final profiles after the wafers have undergone the several thermal processes that are necessary for the integrated-circuit fabrication.

23.6   Generalization of the Model Problem

The generalization of the model problem (23.8) to three dimensions, that is, Eq. (23.7) with $W = 0$ and initial condition $N_0(\mathbf{r}) = N(\mathbf{r}, t = 0)$, is still tackled by means of the Fourier transform. For this, it is necessary to define the vectors $\mathbf{r} = (r_1, r_2, r_3)$, $\mathbf{s} = (s_1, s_2, s_3)$, $\mathbf{k} = (k_1, k_2, k_3)$, and the elements $d^3 k = dk_1 \, dk_2 \, dk_3$, $d^3 s = ds_1 \, ds_2 \, ds_3$. Using (C.20) and following the procedure of Sect. 23.4, one finds again the relations (23.10). This time, however, it is $k^2 = k_1^2 + k_2^2 + k_3^2$. The solution $N(\mathbf{r}, t)$ is readily found as a generalization of (23.12), namely

$$N(\mathbf{r}, t) = \iiint_{-\infty}^{+\infty} N_0(\mathbf{s}) \left\{ \iiint_{-\infty}^{+\infty} \frac{1}{(2\pi)^3} \exp\left[i\mathbf{k} \cdot (\mathbf{r} - \mathbf{s}) - ak^2\right] d^3 k \right\} d^3 s. \tag{23.27}$$

The expression in braces in (23.27) is the product of three functions of the same form as (C.75). It follows

$$N(\mathbf{r}, t) = \iiint_{-\infty}^{+\infty} N_0(\mathbf{s}) \frac{\exp\left[-|\mathbf{r} - \mathbf{s}|^2/(4a)\right]}{(4\pi a)^{3/2}} d^3 s. \tag{23.28}$$

When the net generation rate per unit volume, W, is different from zero, it is in general impossible to find an analytical solution of (23.7). An important exception is the case where W is linear with respect to N and has no explicit dependence on $\mathbf{r}$ or t. In this case (23.7) reads

$$\frac{\partial N}{\partial t} = D\nabla^2 N - \frac{N - N_a}{\tau}, \qquad D = D(t), \tag{23.29}$$

where the two constants N_a (m^{-3}) and τ (s) are positive. This form of W is such that the particles are generated if $N(\mathbf{r}, t) < N_a$, while they are destroyed if $N(\mathbf{r}, t) > N_a$. Equation (23.29) is easily solved by introducing an auxiliary function N' such that $N = N_a + N' \exp(-t/\tau)$. In fact, N' turns out to be the solution of the three-dimensional model problem, so that, using (23.28), the solution of (23.29) reads

$$N(\mathbf{r}, t) = N_a + \exp(-t/\tau) \iiint_{-\infty}^{+\infty} N_0(\mathbf{s}) \frac{\exp\left[-|\mathbf{r} - \mathbf{s}|^2/(4a)\right]}{(4\pi a)^{3/2}} d^3 s. \tag{23.30}$$

23.7   Complements

23.7.1   Generation and Destruction of Particles

The discussion carried out in Sect. 23.2 about the continuity equation implies the possibility that particles may be generated or destroyed. To tackle this issue consider the problem "counting the time variation of students in a classroom". Assuming that the classroom has only one exit, to accomplish the task it suffices to count the students that cross the exit, say, every second. The students that enter (leave) the room are counted as a positive (negative) contribution.

Consider now a slightly modified problem: "counting the time variation of *non-sleeping* students in a classroom". To accomplish the task it does not suffice any more to count the students that cross the exit. In fact, a student that is initially awake inside the classroom may fall asleep where she sits (one assumes that sleeping students do not walk); this provides a negative contribution to the time variation sought, without the need of crossing the exit. Similarly, an initially-sleeping student may wake up, this providing a positive contribution. Falling asleep (waking up) is equivalent to destruction (creation) of a non-sleeping student.

In the two examples above the objects to be counted are the same, however, in the second example they have an extra property that is not considered in the first one. This shows that creation/destruction of a given type of objects may occur or not, depending on the properties that are considered. When particles instead of students are investigated, it is often of interest to set up a continuity equation for describing the time variation, in a given volume, of the particles *whose energy belongs to a specified range*. Due to their motion, the particles undergo collisions that change their energy. As a consequence a particle may enter, or leave, the specified energy range without leaving the spatial volume to which the calculation applies. In this example the origin of the net generation rate per unit volume W introduced in Sect. 23.2 is the extra property about the particles' energy.

23.7.2   Balance Relations

As indicated in Sect. 23.2, and with the provisions illustrated in Sect. 23.7.1, the continuity equation is a balance relation for the number of particles. Due to its intrinsic simplicity and generality, the concept of balance relation is readily extended to physical properties different from the number of particles; for instance, momentum, energy, energy flux, and so on. A detailed illustration of this issue is given in Chap. 19. It is also worth noting, in contrast, that the transport equation of the diffusion type (23.4), being based on a specific assumption about the transport mechanism, is less general than the continuity equation.

23.7.3   Lateral Diffusion

The treatment of predeposition and drive-in diffusion carried out in Sect. 23.5 is based on a one-dimensional model. This implies that the concentration of the dopant at the interface between the source and wafer regions is constant along the y and z directions. In the practical cases this is impossible to achieve, because the area over which the source is brought into contact with the wafer is finite. In fact, prior to the predeposition step the surface of the wafer is covered with a protective layer, called *mask*. As indicated in Sect. 24.1, in the current silicon technology the mask is typically made of thermally-grown silicon dioxide. Next, a portion of the mask is removed to expose the silicon surface over a specific area, called *window*, through which the predeposition step takes place.

From the description above it follows that the initial condition N_0 of the predeposition step is constant only within the window, while it is equal to zero in the other parts of the y, z plane. This makes the hypothesis of a one-dimensional phenomenon inappropriate, and calls for the use of the three-dimensional solution (23.28). The subsequent drive-in diffusions must be treated in three dimensions as well, due to the form of their initial conditions. An important effect is the diffusion of the dopant underneath the edges of the mask. This phenomenon, called *lateral diffusion*, makes the area where the doping profile is present larger than the original mask, and must be accounted for in the design of the integrated circuit.

23.7.4 Alternative Expression of the Dose

The definition of the dose Q deriving from the one-dimensional model problem is (23.14). Letting $W = 0$ in (23.3), using its one-dimensional form $\partial N/\partial t = -\partial F/\partial x$, and observing that it is $F(+\infty, t) = 0$ due to the initial condition, gives the following expression for the time derivative of the dose:

$$\frac{\mathrm{d}Q}{\mathrm{d}t} = -\int_0^{+\infty} \frac{\partial F(x,t)}{\partial x}\,\mathrm{d}x = F(0,t). \tag{23.31}$$

Integrating (23.31) and remembering that the dose at $t = 0$ is equal to zero yields

$$Q(t') = \int_0^{t'} F(0,t)\,\mathrm{d}t. \tag{23.32}$$

The procedure leading from the original definition (23.14) of the dose to its alternative expression (23.32) is based solely on (23.3), hence it does not depend on a specific transport model.

23.7.5 The Initial Condition of the Predeposition Step

The initial condition N_0 of the predeposition step is given by the first of (23.15). To carry out the solution of the diffusion equation it is necessary to recast N_0 in an integral representation of the Fourier type. However, (23.15) does not fulfill the condition (C.19) that is sufficient for the existence of such a representation.

Nevertheless the solution procedure leading to (23.13) is still applicable. In fact, remembering the definition (C.8) of the unit step function H, the initial condition can be recast as $N_0(\xi) = 2C\,[1 - H(\xi)]$. In turn, as shown in appendix C.4, H can be represented in the required form.

Problems

23.1 A Gaussian doping profile $N = 2Q\,\exp(-x^2/c_1)/\sqrt{\pi c_1}$ undergoes a thermal-diffusion process at a temperature such that $D = 10^{-11}$ cm^2/s. Assuming $c_1 = 1.6 \times 10^{-7}$ cm^2, calculate the time that is necessary to reduce the peak value of the profile to 2/3 of the initial value.

23.2 A Gaussian doping profile $N = 2Q\,\exp(-x^2/c_1)/\sqrt{\pi c_1}$, $c_1 = 9 \times 10^{-8}$ cm^2, undergoes a thermal-diffusion process with $c_2 = 16 \times 10^{-8}$ cm^2 yielding another Gaussian profile. Find the value $\bar{x}$ (in microns) where the two profiles cross each other.

23.3 A Gaussian doping profile $N = 2Q\,\exp(-x^2/c_1)/\sqrt{\pi c_1}$, $c_1 = 2.5 \times 10^{-6}$ cm^2, undergoes a 240 min-long thermal-diffusion process at a temperature such that the diffusion coefficient is $D = 2.5 \times 10^{-10}$ cm^2 s^{-1}. Determine the ratio between the peak value of the final profile and that of the initial one.

23.4 A Gaussian doping profile $N = 2Q\,\exp(-x^2/c_1)/\sqrt{\pi c_1}$, $c_1 = 10^{-6}$ cm^2, undergoes a thermal-diffusion process in which $c_2 = 3 \times 10^{-8}$ cm^2. Find the position $\bar{x}$ (in microns) where the value of the initial doping profile equals the value that the final profile has in $x = 0$.

23.5 A Gaussian doping profile $N = 2Q\,\exp(-x^2/c_1)/\sqrt{\pi c_1})$ undergoes a thermal-diffusion process in which $c_2 = 10^{-8}$ cm^2. The value of the final profile in the origin is equal to that of the initial profile at $x_0 = 1.1 \times \sqrt{c_1}$. Find the value of c_1 in cm^2.

23.6 A Gaussian doping profile $N = 2Q\,\exp(-x^2/c_1)/\sqrt{\pi c_1})$, $c_1 = 1.8 \times 10^{-8}$ cm^2, undergoes a thermal-diffusion process whose duration is $t = 10$ min, with $D = 10^{-11}$ cm^2 s^{-1}. At the end of the process the concentration at some point x_0 is $N_1 = 3 \times 10^{16}$ cm^{-3}. If the process duration were 20 min, the concentration at the same point would be $N_2 = 3 \times 10^{17}$. Find the value of x_0 in microns.

23.7 The doping profile resulting from a predeposition process with $D = 10^{-11}$ cm^2 s^{-1} is $N(x) = N_S\,\mathrm{erfc}(x/\sqrt{c})$. The ratio between the dose and surface concentration is $Q/N_S = \lambda/\sqrt{\pi}$, $\lambda = 1095$ nm. Find the duration t of the predeposition process, in minutes.

23.8 The initial condition of a drive-in diffusion is given by $N_0 = 2Q\,(h - x)/h^2$ for $0 \leq x \leq h$, and by $N_0 = 0$ elsewhere, where $Q > 0$ is the dose. Find the expression of the profile at $t > 0$.

Chapter 24
Thermal Oxidation—Layer Deposition

24.1 Introduction

High-quality oxide is essential in silicon technology. The most important applications of the oxide are the passivation of the wafer's surface, the isolation between metallizations, the formation of masks for, e.g., diffusion or implantation of dopants, the isolation between devices, and the formation of the gate insulator in MOS devices.

While the oxides used for passivation or isolation between metallization are typically obtained by chemical vapor deposition (Sect. 24.5), the oxide suitable for the other applications listed above is obtained by thermal oxidation. In fact, the extraordinary evolution of the VLSI technology in the last decades is due to a large extent to the excellent electrical properties of the thermally-grown layers of silicon dioxide and to the reliability and control of the growth process.

In crystalline silicon dioxide (quartz), one silicon atom forms chemical bonds with four oxygen atoms, creating a tetrahedral structure. In turn, one oxygen atom forms chemical bonds with two silicon atoms. The tetrahedra are thus connected to form a structure with a stoichiometric ratio 1 : 2 and density $\rho = 2.65$ g cm^{-3} (Fig. 24.1). In thermally-grown SiO$_2$ not all tetrahedra are connected, because at the silicon–oxide interface chemical bonds must be created with the pre-existing silicon crystal, whose interatomic distance is different from that of SiO$_2$. As a consequence, the oxide has a shorter-range order giving rise to a more open (amorphous) structure with density $\rho = 2.20$ g cm^{-3}. Because of this the diffusion of contaminants, in most cases Na and H$_2$O, is easier than in crystalline silicon dioxide.

Also, the need of adapting to the silicon crystal produces a mechanical stress in the oxide layer closer to the silicon surface, which in turn influences the concentration of substrate defects and the value of some electrical properties in MOS devices (typically, the threshold voltage, Sect. 22.6.2). The properties of the mechanically-stressed layer are influenced by the process temperature T. At relatively low process temperatures, $T < 950\,^\circ\text{C}$, the stressed layer is thinner and the mechanical stress in it is stronger; when $T > 950^\circ\text{C}$, the stress distributes over a thicker layer and becomes locally weaker.

The growth of a thermal oxide's layer is obtained by inducing a chemical reaction between the silicon atoms belonging to the wafer and an oxidant species that is

© Springer Science+Business Media New York 2015

M. Rudan, *Physics of Semiconductor Devices,*
DOI 10.1007/978-1-4939-1151-6_24

Fig. 24.1 Structure of
quartz. Silicon atoms are
represented in *gray*, oxygen
atoms in *white*. Within the
tetrahedron, the distance
between two oxygen atoms is
about 0.227 nm, that between
the silicon atom and an
oxygen atom is about
0.160 nm. The schematic
representation in two
dimensions is shown in the
lower-right part of the figure

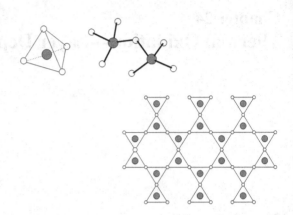

brought into contact with it. As mentioned above, another technique for obtaining
an oxide layer is deposition. The latter process has actually a broader scope, in fact
it is used for depositing several types of conducting or insulating materials that are
necessary in the fabrication of the integrated circuits. Deposition differs from the
thermal growth because the chemical reaction may be absent or, if present, it does
not involve the species that are in the solid phase. One special type of deposition is
epitaxy, that is used to grow a crystalline layer over another crystalline layer.

The chapter illustrates the oxidation of silicon, starting from the description of
the chemical reactions involved in it, and deriving the relation between the thickness
of the oxide layer and that of the silicon layer consumed in its growth. The kinetics
of the oxide growth is analyzed, the *linear–parabolic model* is worked out, and its
features are commented. Then, a brief description of the deposition processes is
given, followed by the description of the chemical reaction involved in the epitaxial
process and by the analysis of the epitaxial kinetics.

In the last part of the chapter a number of complementary issues are discussed.
Specific data about the parameters governing the thermal oxidation, deposition,
and epitaxial processes in semiconductors are in [46, Chap. 2], [104, Chap. 9],
[105, Chap. 3], [71, Chap. 3]. Many carefully-drawn illustrations are found in [75,
Sect. 1.2].

24.2 Silicon Oxidation

Silicon exposed to air at room temperature oxidizes spontaneously and forms a
shallow layer of SiO_2 of about 1 nm called *native oxide*. As soon as the native oxide
is formed, the oxygen molecules of the air can not reach the silicon surface any
more and the chemical reaction dies out. In fact, oxidation of silicon is caused by
the inward motion of the oxidant. To activate the reaction and grow a layer of the
desired thickness it is necessary to place the wafers at atmospheric pressure in a
furnace (Fig. 24.2) kept at a temperature in the range $800°C \leq T \leq 1200°C$. This

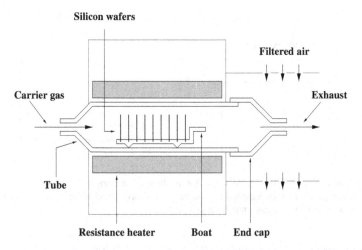

Fig. 24.2 Furnace for silicon oxidation. The intake of the carrier gas (O_2 or H_2O) is on the *left end* of the furnace, the exhaust on the *right end*. The tube, end cap, and boat, are made of fused quartz to avoid contamination

increases the diffusion coefficient of the oxidant. The latter penetrates the already-formed oxide layer and reaches the silicon surface, where new SiO_2 molecules are formed. The furnace is made of a quartz or polycrystalline-silicon tube heated by a resistance or by induction through a radiofrequency coil. To grow the oxide layer in a reproducible way it is necessary to control the temperature inside the furnace within $\pm 1\,°C$. The oxidant is introduced from one end of the furnace after being mixed with a carrier gas (typically, N_2 or Ar).

The chemical reactions involved in the growth of thermal oxide are different depending on the type of oxidant. The latter is either molecular oxygen (O_2) or steam (H_2O). The corresponding thermal growth is called, respectively, *dry oxidation* or *wet (steam) oxidation*. The reactions read

$$Si + O_2 \rightleftharpoons SiO_2, \qquad Si + 2H_2O \rightleftharpoons SiO_2 + 2H_2, \tag{24.1}$$

where the hydrogen molecules produced by the second reaction are eliminated by the carrier gas. The formation of SiO_2 molecules is accompanied by a change in volume. In fact, each newly-formed SiO_2 molecule uses up a silicon atom initially belonging to the silicon crystal. On the other hand, the concentration of the silicon atoms in a silicon crystal is about $N_1 = 5.0 \times 10^{22}$ cm^{-3}, while that of the silicon atoms in a thermally-grown SiO_2 layer is about $N_2 = 2.2 \times 10^{22}$ cm^{-3}. Thus the ratio between the volume V of SiO_2 and that of the silicon consumed for its formation is

$$\frac{V(SiO_2)}{V(Si)} = \frac{N_1}{N_2} \simeq 2.28. \tag{24.2}$$

As the oxide layer is free to expand only in the direction normal to the wafer, (24.2) is actually the ratio between the thickness s of the newly-formed SiO_2 layer

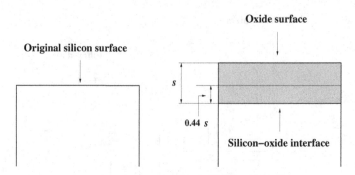

Fig. 24.3 The *left part* of the figure shows the position of the original silicon surface (prior to oxidation). The *right part* shows the position of the oxide's surface and of the silicon–oxide interface after an oxide layer of thickness s has been grown

and that of the silicon layer consumed in the process. It follows

$$\frac{s(\text{Si})}{s(\text{SiO}_2)} = \frac{V(\text{Si})/A}{V(\text{SiO}_2)/A} = \frac{N_2}{N_1} \simeq 0.44, \tag{24.3}$$

with A the area of the oxidized region. In other terms, when an oxide layer of thickness s is grown, the silicon–oxide interface shifts by $0.44\,s$ with respect to the original position (Fig. 24.3). If the oxidation takes place uniformly over the whole area of the wafer, the shift of the interface is uniform as well. However, in many cases the oxidation involves only portions of the wafer's area. This makes the silicon–oxide interface non planar, because the shift is different from one portion to another.

24.3 Oxide-Growth Kinetics

Growth kinetics is modeled after Deal and Grove [28]. The model describes the succession of steps by which the oxidant, initially in the gaseous phase, comes into contact with silicon and reacts with it. The steps are: the oxidant (i) diffuses from the source region into the already-formed oxide, (ii) crosses the oxide still by diffusion and reaches the silicon–oxide interface, (iii) produces a chemical reaction that forms a new SiO_2 molecule.

The motion of the gas parallel to the wafer surface is not considered. As a consequence, the only non-vanishing component of the oxidant average velocity has the direction x normal to the wafer surface. The corresponding flux density is $F = \mathbf{F} \cdot \mathbf{i}$, with $\mathbf{i}$ the unit vector parallel to x. The oxidant concentration N is assumed uniform over the wafer's surface, this making N and F to depend on x and t only.

The concentration of the oxidant in the bulk of the gaseous phase, N_G, is a known boundary condition because it is regulated by the microprocessors controlling the furnace. In the gaseous region, at the gas–oxide interface, and in the oxide region, no generation or destruction of oxidant molecules occurs. The flux density is given by

$$F = -D_S \frac{\partial N}{\partial x}, \qquad F = -D_O \frac{\partial N}{\partial x}, \qquad (24.4)$$

respectively in the source and oxide region. In (24.4), the symbol D_S (D_O) indicates the diffusion coefficient of the oxidant in the source (oxide) region. Each diffusion coefficient is taken independent of time and spatially constant in its own region. The matching conditions at the source–oxide interface are the same as in (23.20), namely,

$$N_O = kN_S, \qquad D_O \left(\frac{\partial N}{\partial x} \right)_O = D_S \left(\frac{\partial N}{\partial x} \right)_S, \qquad (24.5)$$

where k is the gas–oxide segregation coefficient, while the index S (O) attached to the concentration or its derivative indicates that the function is calculated at the source–oxide interface on the side of the source (oxide). As one of the two phases is gaseous, it is $D_S \gg D_O$ whence $|(\partial N/\partial x)_S| \ll |(\partial N/\partial x)_O|$. The situation here is similar to that illustrated in Fig. 23.3. It follows that the interface concentration of the source region, N_S, is practically equal to the boundary condition N_G. The first of (24.5) then yields $N_O = kN_G$.

To proceed one observes that, due to the thinness of the oxide layer, the oxidant concentration in it can be described by a linear approximation. Due to this, the flux density in the oxide layer (the second equation in (24.4)) becomes

$$F = -D_O \frac{N_I - N_O}{s} = D_O \frac{kN_G - N_I}{s}, \qquad (24.6)$$

where N_I is the oxidant concentration on the oxide side of the silicon–oxide interface, and s the oxide thickness. Note that the flux density in (24.6) is constant with respect to x, whereas it is time dependent because s increases with time.

When the oxidant reaches the silicon–oxide interface it reacts with silicon, so that there is no flux density of the oxidant on the semiconductor's side of this interface. In fact, the oxidant's molecules are destroyed at the interface to form molecules of SiO_2. The flux density F_I entering the silicon–oxide interface gives the number of oxidant molecules destroyed per unit area and time which, to a first approximation, is taken proportional to the concentration N_I. It follows

$$F_I = v_r N_I, \qquad (24.7)$$

where the constant v_r ($m\,s^{-1}$) is called *reaction velocity*. As F_I is just another symbol to denote the spatially-constant flux density within the oxide, one combines (24.7) with (24.6) to obtain

$$N_I = \frac{D_O kN_G}{v_r s + D_O}. \qquad (24.8)$$

At a given instant Eq. (24.8) expresses N_I in terms of the boundary condition N_G, the process parameters k, D_O, v_r, and the oxide thickness s.

24.4 Linear–Parabolic Model of the Oxide Growth

The relation between the oxidant's flux density F and the growth velocity ds/dt of the oxide layer is found as follows. Letting A be the area of the oxidized region, the product $A F$ provides the number of oxidant molecules reaching the silicon–oxide interface per unit time. Each molecule, in turn, makes the volume V of the oxide layer to increase by a finite amount w. As a consequence, the volume increase per unit time of the oxide layer is $dV/dt = wA F$. As shown in Sect. 24.2, the oxide layer is free to expand only in the direction normal to the wafer, so that $dV/dt = A\,ds/dt$. Combining the above relations and using (24.7, 24.8) yields a differential equation in the unknown s:

$$\frac{ds}{dt} = wF_I = wv_r N_I = v_r D_O \frac{w\,kN_G}{v_r\,s + D_O}. \tag{24.9}$$

The above is readily separated as $(s/D_O + 1/v_r)\,ds = wkN_G\,dt$ and integrated from $t = 0$, to yield

$$\frac{1}{c_p}\left(s^2 - s_i^2\right) + \frac{1}{c_l}\left(s - s_i\right) = t, \qquad \begin{cases} c_p = 2w\,kN_G D_O \\ c_l = w\,kN_G\,v_r \end{cases}, \tag{24.10}$$

with $s_i = s(t = 0)$. The relation between s and t given by (24.10) is called *linear–parabolic model* of the oxide growth. The quantities c_p (m^2 s^{-1}) and c_l (m s^{-1}) are the *parabolic coefficient* and *linear coefficient*, respectively. The model is recast as

$$\frac{1}{c_p}s^2 + \frac{1}{c_l}s = t + \tau, \qquad \tau = \frac{1}{c_p}s_i^2 + \frac{1}{c_l}s_i. \tag{24.11}$$

Using (24.11) and the definitions (24.10) of c_p and c_l one finds two limiting cases of the $s(t)$ relation. Specifically, it is $s \simeq c_l\,(t + \tau)$ when the oxide thickness is such that $v_r s \ll 2D$, while it is $s \simeq [c_p\,(t + \tau)]^{1/2}$ when the oxide thickness is such that $v_r s \gg 2D$. Due to the form of c_p and c_l, in the first limiting case the oxide growth does not depend on the diffusion coefficient D_O, whereas in the second limiting case it does not depend on the reaction velocity v_r. This is easily understood if one considers that the concentration $N_O = kN_G$ is prescribed. As a consequence, as long as the oxide thickness is small the derivative of the concentration (hence the flux density) is limited essentially by the flux density entering the silicon–oxide interface, (24.7); on the contrary, when the oxide thickness becomes large the flux density is limited essentially by the diffusion across the oxide because the value of the concentration N_I at the silicon–oxide interface becomes less important.

The differential form of (24.11),

$$\frac{dt}{ds} = \frac{2}{c_p}s + \frac{1}{c_l}, \tag{24.12}$$

is a linear relation between dt/ds and s. Such quantities can be measured independently from each other, this providing a method for measuring c_p and c_l. Repeating the measurement at different temperatures shows that the temperature dependence of the parabolic and linear coefficients is given by

$$c_p = c_{p0} \exp\left[-E_{ap}/(k_B T)\right], \qquad c_l = c_{l0} \exp\left[-E_{al}/(k_B T)\right]. \qquad (24.13)$$

The form of (24.13) is due to the temperature dependence of $D \propto \exp\left[-E_{ap}/(k_B T)\right]$ and, respectively, $v_r \propto \exp\left[-E_{al}/(k_B T)\right]$. In fact, the parameters w, N_G that appear in the definitions (24.10) are independent of temperature, whereas the temperature dependence of the segregation coefficient k, that can be measured independently, is shown to be relatively weak.

The measurement of c_p and c_l allows one to determine also other properties of the oxidation process; for instance, the effect of carrying out a steam or dry oxidation, and the influence of the substrate orientation. As for the first issue one finds

$$k(\text{Steam}) > k(\text{Dry}), \qquad D_O(\text{Steam}) > D_O(\text{Dry}). \qquad (24.14)$$

The Arrhenius plots of c_p and c_l are shown in Figs. 24.4 and 24.5, respectively. In each plot the upper (lower) continuous curve refers to the steam (dry) oxidation. As for the effect of the crystal orientation of the silicon wafer, one observes that the number of chemical reactions per unit time involved in the formation of SiO_2 molecules must depend on the surface density of silicon atoms at the silicon–oxide interface. Due to this, the reaction velocity is expected to depend on the orientation of the interface. The crystal planes that are typically used in the silicon technology are the (111) one, whose surface density is 11.8×10^{14} cm^{-2}, and those equivalent to the (100) one,[1] whose surface density is 6.8×10^{14} cm^{-2}. In fact the experiments show that

$$\frac{v_r[(111)]}{v_r[(100)]} = 1.68 \simeq \frac{11.8 \times 10^{14} \text{ cm}^{-2}}{6.8 \times 10^{14} \text{ cm}^{-2}}. \qquad (24.15)$$

The effect on c_l of the crystal orientation is shown by the dotted curves in the Fig. 24.5.

24.5 Layer Deposition and Selective Oxide Growth

The deposition of films of different materials is necessary at several steps of the integrated-circuit fabrication. Conducting materials provide the electrical connections among the individual devices of the integrated circuit, while insulating materials provide the electrical insulation between the metal layers, and the protection from the environment. The majority of the deposition processes take place in the vapor phase under reduced-pressure or vacuum conditions. One exception is the deposition of resist, which is carried out in the liquid phase.

---

[1] The definitions of the crystal planes are given in Sect. 17.8.1.

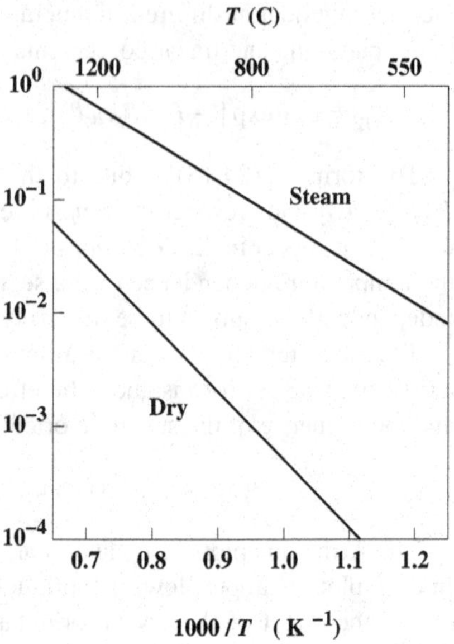

Fig. 24.4 Parabolic coefficient c_p as a function of $1,000/T$. The units are $\mu m^2\,h^{-1}$. The activation energy of the steam case is 0.71 eV, that of the dry case is 1.24 eV

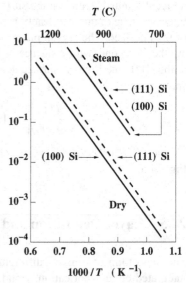

Fig. 24.5 Linear coefficient c_l as a function of $1,000/T$. The units are $\mu m\,h^{-1}$. The activation energy of the steam case is 2.05 eV, that of the dry case is 2.0 eV

When the material to be deposited does not react chemically with other substances, the process is called *physical vapor deposition* (PVD). An example of PVD is the deposition of a metal by evaporation *in vacuo*. When the material to be deposited is the product of a chemical reaction that takes place over the wafer surface or in its vicinity, the process is called *chemical vapor deposition* (CVD).

Table 24.1 Examples of CVD reactions

Product	Reaction[a]		Deposition temperature (°C)
Polysilicon	SiH_4	$\rightarrow Si + 2H_2$	575–650
Silicon dioxide	$SiH_4 + O_2$	$\rightarrow SiO_2 + 2H_2$	400–450
Silicon nitride	$3SiH_4 + 4NH_3$	$\rightarrow Si_3N_4 + 12H_2$	700–900

[a]SiH_4 and NH_3 are called *silane* and *ammonia*, respectively

The materials that are most widely used in CVD processes are polycrystalline silicon (also termed *polysilicon*), silicon dioxide (SiO_2), and silicon nitride (Si_3N_4). Examples of CVD reactions are given in Table 24.1. More examples are found in [105, Sect. 6.2].

The structure of the deposited layer depends on the substrate's properties and deposition conditions. In the manufacturing of integrated circuits the substrate is crystalline, that is, it has long-range order extending throughout the entire volume (Chap. 17). If the material to be deposited on a crystalline substrate is the same as that of the substrate, by means of a carefully-controlled process it is possible to obtain a deposited layer that replicates the substrate's structure. Such a process is called *epitaxy* and, with reference to silicon, is described in Sect. 24.6.

The structure of silicon deposited on a different material is polycrystalline, that is, it has a long-range order only within small volumes. Such volumes, called *grains*, have an average diameter of about 1 μm and are oriented randomly with respect to each other. Polycrystalline silicon is used for fabricating the gate electrodes in MOS devices, for obtaining ohmic contacts to shallow junctions, and for producing resistors. To increase the gate's conductivity, a layer of metal or metal silicide (like tungsten or tantalum silicide) may be deposited over the polycrystalline silicon.

The structure of deposited SiO_2 or Si_3N_4 is amorphous, that is, it has a short-range order only. The applications of SiO_2 have been illustrated in Sect. 24.1. Silicon nitride Si_3N_4 provides a strong barrier to the diffusion of water, that corrodes the metallizations, and of other contaminants, like sodium, that make the devices unstable by changing their threshold voltage. In addition, Si_3N_4 is resistant to high temperatures and oxidizes slowly. For these reasons it is used for passivating the wafer and for producing the masks that are necessary for the selective oxidation of silicon. The latter process, also called *local oxidation* (LOCOS), consists in depositing and patterning a Si_3N_4 layer over the areas where the substrate's oxidation must be prevented. As oxidation is isotropic, a lateral penetration of the oxidized region occurs under the edge of Si_3N_4. This produces a characteristic profile of the oxide layer called *bird's beak*. To compensate for the effect of the lateral penetration, the Si_3N_4 mask must be larger than the area whose oxidation is to be prevented.

A layer replicates the topography of the surface onto which it is deposited. For this reason it is important to avoid, or reduce, the formation of steps on the substrate. In fact, over a step the layer's thickness is smaller than on a flat surface which, in turn, may cause reliability problems in the final circuit. For instance, the non-uniform thickness of a metal line causes a non-uniform distribution of the current density. This may induce metal migration and the eventual breakdown of the metal connection.

24.6   Epitaxy

Epitaxy (from the Greek verb *epitàsso*, "to deploy side by side") is used to grow a monocrystalline layer over another monocrystalline layer. Most epitaxial processes use the CVD method. When the epitaxial layer is made of the same material as the substrate, e.g., silicon over silicon, the term *homoepitaxy* is also used, while the term *heteroepitaxy* is reserved to the case where the materials are different. Heteroepitaxy is possible when the difference between the lattice constants[2] of the two materials is small. An example of heteroepitaxy is the silicon-on-sapphire (SOS) process, that belongs to the silicon-on-insulator (SOI) technological family and consists in growing a thin layer of silicon (about $0.5\ \mu m$) on a wafer made of a sapphire crystal (Al_2O_3). Another application of heteroepitaxy is the fabrication of the heterojunctions that are necessary in optoelectronic devices.

In the silicon technology, epitaxy originated from the need of producing high-resistance layers in bipolar technology. This type of layers is necessary, e.g., for realizing the collector region of the bipolar junction transistor, whose dopant concentration must be substantially lower than that of the base region. Due to the high temperature of the CVD process (about $1200\,°C$), during an epitaxy a diffusion occurs of the substrate dopant into the epitaxial layer and of the epitaxial-layer's dopant into the substrate. This effect must be accounted for, and compensated, at the design stage of the process.

The fundamental reaction of epitaxy combines silicon tetrachloride $SiCl_4$ with molecular hydrogen in the vapor phase to obtain silicon in solid phase, while the hydrochloric acid HCl remains in the vapor phase and is eliminated:

$$SiCl_4 + 2H_2 \rightleftharpoons Si + 4HCl. \tag{24.16}$$

Reaction (24.16) is reversible: an excess of HCl removes silicon atoms from the wafer's surface and releases $SiCl_4$ and $2H_2$ in the vapor phases. This reaction is used in the first stages of the process to the purpose of cleaning the wafer's surface. Besides (24.16), a secondary reaction takes place as well, namely,

$$SiCl_4 + Si \rightarrow 2SiCl_2. \tag{24.17}$$

Reaction (24.17) removes silicon from the wafer's surface and releases silicon dichloride $SiCl_2$ in the vapor phase. For this reason, reactions (24.16) and (24.17) compete with each other. When the vapor concentration of $SiCl_4$ is sufficiently low the first reaction prevails and the thickness of the epitaxial layer increases with time. In contrast, at higher $SiCl_4$ concentrations the second reaction prevails and silicon is etched.

The epitaxial layer is doped by introducing hydrides of the dopants into the vapor phase. The hydride, e.g., *arsine* (AsH_3), *phosphine* (PH_3), or *diborane* (B_2H_6), is

---

[2] The definition of lattice constant is given in Sect. 17.6.4.

absorbed on the surface, decomposes, and is incorporated in the growing layer, e.g.,

$$2AsH_3 \rightarrow 2As + 3H_2. \tag{24.18}$$

24.7 Kinetics of Epitaxy

As in the case of the oxide-growth kinetics (Sect. 24.3), the motion of the vapor parallel to the wafer surface is not considered. As a consequence, the only non-vanishing component of the average velocity of the $SiCl_4$ molecules has the direction x normal to the wafer surface. The corresponding flux density is $F = \mathbf{F} \cdot \mathbf{i}$, with $\mathbf{i}$ the unit vector parallel to x. The $SiCl_4$ concentration N is assumed uniform over the wafer's surface, this making N and F to depend on x and t only.

The $SiCl_4$ concentration in the bulk of the vapor phase, N_G, is a known boundary condition because it is regulated by the microprocessors controlling the furnace. The flux density is given by

$$F = -D \frac{\partial N}{\partial x} \simeq v_G (N_G - N_I), \tag{24.19}$$

where D is the diffusion coefficient of $SiCl_4$ in the vapor phase and N_I the $SiCl_4$ concentration at the wafer's surface. The diffusion coefficient is taken independent of time and spatially constant. The form of the right hand side of (24.4), where the parameter v_G (m s^{-1}) is called *gas-phase mass-transfer coefficient*, is due to the observation that D is very large because the diffusion takes place in the vapor phase. As a consequence, the derivative $\partial N/\partial x$ is so small that a linear approximation for N is acceptable (the situation here is similar to that illustrated for the region on the left of the origin in Fig. 23.3). Note that the flux density in the vapor phase (24.19) is constant with respect to x. In principle it depends on time because the extension of the vapor phase decreases due to the growth of the epitaxial layer. However, this time dependence can be disregarded because the relative variation in the vapor-phase extension is negligible.

The flux density F_I entering the silicon surface gives the number of $SiCl_4$ molecules destroyed per unit area and time which, to a first approximation, is taken proportional to the concentration N_I. It follows

$$F_I = v_r N_I, \tag{24.20}$$

where the constant v_r (m s^{-1}), as in the case of the oxide-growth kinetics, is called *reaction velocity*. As F_I is just another symbol to denote the spatially-constant flux density, one combines (24.20) with (24.19) to obtain

$$N_I = \frac{v_G}{v_r + v_G} N_G. \tag{24.21}$$

At a given instant Eq. (24.21) expresses N_I in terms of the boundary condition N_G and process parameters v_G, v_r.

Fig. 24.6 Normalized growth velocity as a function of the normalized inverse temperature, as given by (24.23) and (24.24), at different values of the $r_v = v_G/v_{r0}$ ratio

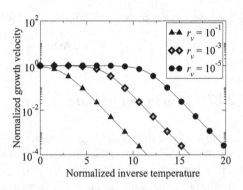

The relation between the flux density F of SiCl$_4$ and the growth velocity ds/dt of the epitaxial layer is found by the same reasoning as that used in Sect. 24.4 for the growth velocity of SiO$_2$. From (24.20, 24.21) it follows

$$\frac{ds}{dt} = wF_I = wv_r N_I = w\frac{v_r v_G}{v_r + v_G}N_G \tag{24.22}$$

whence, observing that $s(t = 0) = 0$,

$$s = c_l t, \qquad c_l = w\frac{v_r v_G}{v_r + v_G}N_G. \tag{24.23}$$

The $s(t)$ relation (24.23) is linear with respect to time. The growth velocity c_l of the epitaxial layer depends on the concentration N_G of SiCl$_4$ at the boundary and on the process parameters w, v_r, and v_G. The temperature dependence of the gas-phase mass-transfer coefficient v_G is weak. As w and N_G are independent of temperature, the temperature dependence of c_l is to be ascribed to v_r. It is found

$$v_r = v_{r0} \exp\left[-E_{al}/(k_B T)\right]. \tag{24.24}$$

When the temperature is such that $v_r \ll v_G$, which typically happens for $T <$ 1,150 °C, the second of (24.23) yields the limiting case $c_l \simeq wN_G v_r$, whence $c_l \propto \exp\left[-E_{al}/(k_B T)\right]$; when, instead, it is $v_r \gg v_G$, which typically happens for $T > 1,200$ °C, the limiting case is $c_l \simeq wN_G v_G =$ const. An Arrhenius plot of the normalized growth velocity $c_l/(wN_G v_G)$ as a function of the normalized inverse temperature $E_{al}/(k_B T)$ is shown in Fig. 24.6 for different values of the $r_v = v_G/v_{r0}$ ratio.

24.8 Complements

24.8.1 An Apparent Contradiction

In commenting (24.6) it was noted that the flux density F in the oxidation process is constant with respect to x, whereas it depends on time due to the time dependence of

Fig. 24.7 Oxidant concentration within the oxide at two different instants, t_1 and $t_2 > t_1$

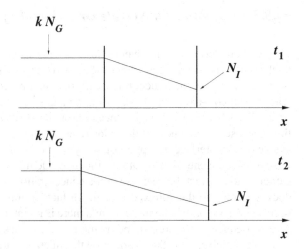

the oxide thickness s. This seems to bring about a contradiction. In fact, as the one-dimensional form of the continuity Eq. (23.3) with $W = 0$ yields $\partial N/\partial t + \partial F/\partial x = 0$, the constancy of F makes N independent of time. However, N does depend on time. This is demonstrated by Fig. 24.7, that shows the linear approximation of the oxidant concentration within the oxide at two different instants, t_1 and $t_2 > t_1$. The value $k\,N_G$ at the source–oxide interface is kept constant by the boundary condition as explained in Sect. 24.3, while the value N_I at the silicon–oxide interface changes with time due to (24.8), and the oxide's thickness changes as well.

The contradiction is eliminated by observing that the continuity Eq. (23.3) has been derived for the case where the boundary is fixed, whereas the growth of thermal oxide is a moving-boundary process. The motion of the boundary is not a rigid one (otherwise the problem could be fixed by moving the reference accordingly), because the oxide volume is actually expanding. In conclusion, Eq. (23.3) must not be used. In fact, the derivation of the linear–parabolic model of the oxide growth (24.10) is based solely on the definition of the flux density.

24.8.2 Elementary Contributions to the Layer's Volume

The relation $ds/dt = wF_I$ was used to connect, in Sect. 24.4, the growth velocity of the oxide layer to the flux density of the oxidant and, in Sect. 24.7, the growth velocity of the epitaxial layer to the flux density of $SiCl_4$. The coefficient w is the amount by which one oxidant or $SiCl_4$ molecule makes the volume of the layer to increase. To specify w for the oxidation process one must distinguish between the dry and steam cases. In the first one, each molecule of the oxidant produces one SiO_2 molecule. As a consequence, w is the volume of the SiO_2 molecule. In the steam case, two H_2O molecules are necessary for producing one SiO_2 molecule, hence w is half the volume of the latter. By the same token, in the epitaxial process w is the volume of a Si atom.

24.8.3   Features of the Oxide Growth and Epitaxial Growth

The quadratic term in the left hand side of (24.10) becomes dominant at larger oxide thicknesses. This in turn slows down the growth rate, as shown by (24.9). A qualitative explanation of the phenomenon is easily obtained by considering that, in order to reach the silicon–oxide interface, the oxidant must diffuse across the already-formed oxide. The slope of the oxidant concentration, hence its flux density, decreases with time because the thickness of the oxide region increases, while the value $k\,N_G$ at the source–oxide interface is kept constant by the boundary condition. The decrease in the oxidant concentration N_I at the silicon–oxide interface, shown by (24.8), is not sufficient to contrast the decrease in the concentration's slope. The reasoning above does not apply to the epitaxial growth; in fact, in this case the chemical reaction occurs at the vapor–silicon interface and there is no intermediate layer to be crossed. As a consequence, the corresponding model (24.23) has no quadratic term.

In the analysis of the oxide-growth kinetics carried out in Sect. 24.3 it is assumed that the interface concentration in the source region, N_S, is practically equal to the boundary condition N_G. The simplification is used in the expression (24.6) of the flux density in the oxide layer. The calculation then proceeds by considering only the oxidant diffusion across the already-formed layer and the chemical reaction at the silicon–oxide interface. In this respect, the assumption $N_S = N_G$ has the mere effect of introducing a negligible change in (24.6). A similar approximation would not be possible in the analysis of the epitaxial growth. In fact, letting $N_I = N_G$ in (24.19) would set the flux density to zero. The difference between the two cases is that in the epitaxial growth the flux density exists only in the vapor phase, while in the oxide growth it exists both in the gaseous and solid phases. However, as $D_S \gg D_O$, only the diffusion in the solid phase plays a significant role in determining the kinetics of the oxidation process.

24.8.4   Reaction Velocity

The reaction velocity v_r is among the parameters used in the analysis of the oxide-growth kinetics and epitaxial kinetics. This parameter controls the flux density through (24.7) or (24.20), and is found to depend also on the concentration N_{dop} of dopant atoms in the silicon lattice. The dependence is negligible as long as $N_{\text{dop}} \leq n_i(T)$, where n_i (called *intrinsic concentration*, Sect. 18.3) is calculated at the process temperature. When $N_{\text{dop}} > n_i(T)$, the reaction velocity increases with N_{dop}. It should be noted that $n_i \simeq 10^{18}$ cm^{-3} at $T = 1000\,°C$. As a consequence, the dependence of v_r on N_{dop} becomes important only at relatively high dopant concentrations.

Fig. 24.8 Typical growth velocity c_l of an epitaxial process, expressed in microns per minute, as a function of the mole fraction of tetrachloride. The *shaded area* shows the typical operating range

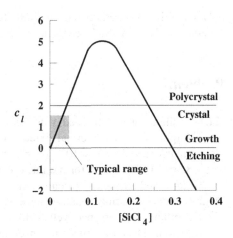

24.8.5 Molecular Beam Epitaxy

Epitaxy can also be obtained by a process different from CVD, that is called *molecular beam epitaxy* (MBE) and is based on evaporation. The main advantages of MBE are the low-temperature processing and the fine control of the dopant distribution throughout the epitaxial layer. On the other side, MBE has a low throughput and a higher cost. As a consequence, CVD is used is in the majority of cases for growing epitaxial layers in the silicon technology [105, Sect. 6.3].

24.8.6 Secondary Reaction in the Epitaxial Growth

The analysis of the epitaxial kinetics carried out in Sect. 24.7 is based on the hypothesis that only the fundamental reaction (24.16) is present. However, as mentioned in Sect. 24.7, the secondary reaction (24.17) also takes place, which removes silicon atoms from the wafer's surface and, therefore, competes with (24.16). At low concentrations of $SiCl_4$ the effect of the secondary reaction is negligible and the theory of Sect. 24.7 holds; in particular, as shown by (24.23), the growth velocity c_l is proportional to the concentration N_G of $SiCl_4$ in the bulk of the vapor phase. At larger tetrachloride concentrations the dependence of c_l on N_G becomes sublinear as shown in Fig. 24.8. Further increases in N_G make c_l to continuously decrease, and to eventually vanish when the secondary reaction balances the fundamental one. Further on, the secondary reaction prevails and c_l becomes negative, that is, silicon is etched.

The growth velocity also influences the structure of the epitaxial layer. When c_l is low the deposited silicon atoms match the preexisting crystalline structure, so that the newly-formed layer is crystalline as well. Higher values of c_l make the matching more and more difficult. Beyond a critical value of c_l the epitaxial layer is polycrystalline, as sketched in Fig. 24.8. To avoid the growth of a polycrystal it is necessary to keep the SiCl$_4$ concentration in the range shown in the figure.

Problems

24.1 A silicon wafer covered with an $s_i = 0.21$ μm-thick thermal oxide undergoes a second thermal oxidation whose duration is 136 min. Using the values $c_p = 4.43 \times 10^{-2}$ μm^2 h^{-1} for the parabolic coefficient and $c_l = 8.86 \times 10^{-1}$ μm h^{-1} for the linear coefficient, calculate the silicon thickness consumed during the second thermal oxidation.

24.2 Consider a thermal-oxidation process of silicon where the parabolic and linear coefficients are, respectively, 4.43×10^{-2} μm^2 h^{-1} and 8.86×10^{-1} μm h^{-1}. At some instant the oxidant concentration N_I at the silicon–oxide interface is 1×10^{12} cm^{-3}. Find the gradient of the oxidant concentration within the oxide layer at the same instant, expressed in 10^{17} cm^{-4} units.

24.3 A silicon wafer covered with an $s_i = 105$ nm-thick layer of thermal oxide undergoes a second thermal-oxidation process that consumes a 100 nm-thick layer of silicon. Letting the parabolic and linear coefficients be $c_p = 4.43 \times 10^{-2}$ μm^2 h^{-1} and $c_l = 8.86 \times 10^{-1}$ μm h^{-1}, respectively, determine the duration in minutes of the second oxidation process.

24.4 Consider a thermal-oxidation process of silicon where the parabolic and linear coefficients are, respectively, 0.12 μm^2 h^{-1} and 3 μm h^{-1}. At some instant the oxide thickness is 20 nm, and the oxidant concentration in the oxide at the source–oxide interface is $N_O = 3 \times 10^{12}$ cm^{-3}. Find the oxidant concentration at the silicon–oxide interface, expressed in 10^{11} cm^{-3} units.

24.5 A silicon wafer covered with an $s_i = 5$ nm-thick layer of thermal oxide undergoes a thermal-oxidation process that grows a $\Delta s_1 = 7$ nm-thick oxide layer and a successive thermal-oxidation process that grows a $\Delta s_2 = 20$ nm-thick oxide layer. In both processes the parabolic and linear coefficients are, respectively, 4.1×10^{-2} μm^2 h^{-1} and 8.5×10^{-1} μm h^{-1}. Determine the total duration in seconds of the two processes.

24.6 A silicon wafer covered with an $s_i = 80$ nm-thick layer of thermal oxide undergoes a thermal-oxidation process whose linear coefficient is 1 μm h^{-1}. The ratio between the diffusion coefficient of the oxidant within the oxide and the reaction velocity at the silicon–oxide interface is $r = D_O/v_r = 50$ nm. Find how many minutes are necessary to reach a final oxide thickness equal to 150 nm.

24.7 A silicon wafer covered with an $s_i = 40$ nm-thick layer of thermal oxide undergoes a thermal-oxidation process where the oxidant's diffusion coefficient in the oxide is $D_O = 4.5 \times 10^{-6}$ cm^2 s^{-1}, the reaction velocity is $v_r = 4$ cm s^{-1}, and the product of the parabolic coefficient and the process duration is $c_p t_P = 5 \times 10^{-11}$ cm^2. Calculate the final thickness of the oxide in nm.

24.8 A silicon wafer undergoes an epitaxial growth that produces a 12 μm-thick silicon layer. At the end of the process the wafer's weight has increased by 907 mg. Using $p_{Si} = 2.33$ g cm^{-3} for the specific weight of silicon, determine the wafer's diameter in inches (1 in. = 2.54 cm).

24.9 A 1.2 μm-thick epitaxial layer of silicon is grown by a 1 min-long process in which the reaction velocity is $v_r = 10$ cm s^{-1}. Find the concentration of SiCl$_4$ at the silicon surface expressed in 10^{16} cm^{-3} units.

24.10 The flux density of SiCl$_4$ in an epitaxial process in silicon is 8.33 $\times$ 10^{16} cm^{-2}s^{-1}. Remembering that the concentration of the silicon atoms in the crystal lattice is 5×10^{22} cm^{-3}, determine how many minutes are necessary to grow a 2 μm-thick epitaxial layer.

24.11 Determine the reaction velocity v_r (in cm min^{-1}) of an epitaxial process in silicon that in 5 min grows an $s = 2$ μm-thick layer. For the surface concentration of the silicon tetrachloride and the atomic volume of silicon use, respectively, the values 10^{16} cm^{-3} and $(5 \times 10^{22})^{-1}$ cm^3.

24.12 In an epitaxial process the ratio between the SiCl$_4$ concentration in the bulk of the vapor phase and at the wafer's surface is $a = N_G/N_I = 2$, while the ratio between the reaction velocity and growth velocity is $b = v_r/c_l = 4.87 \times 10^5$. Remembering that the concentration of the silicon atoms in the crystal lattice is 5×10^{22} cm^{-3}, determine the value of N_G in cm^{-3}.

Chapter 25
Measuring the Semiconductor Parameters

25.1 Introduction

A number of methods used for measuring the semiconductor parameters are illustrated here. Apart from the intrinsic usefulness, the methods are interesting because they show the connection with the theories worked out in other chapters. For example, the measurement of lifetimes exploits the features of the net thermal recombination and of optical generation, that are combined in a simplified form of the continuity equation for the minority carriers. Similarly, the measurement of mobility carried out with the Haynes-Shockley experiment is based on a clever use of the diffusion of optically generated carriers. The Hall effect, in turn, provides a powerful method to extract the information about the concentration and mobility of the majority carriers; the method exploits the effect of a magnetic field applied in the direction normal to that of the current density, and is widely used for determining, e.g., the dependence of carrier concentration and mobility on the concentration of dopants and on temperature. The chapter is completed by the illustration of a method for measuring the doping profile in an asymmetric, reverse-biased, one-dimensional junction; the procedure is based on the observation that, despite the fact that the relation between the applied voltage and the extension of the space-charge region is non linear, the differential capacitance of the junction has the same form as that of a parallel-plate, linear capacitor.

25.2 Lifetime Measurement

The lifetimes have been introduced in Sect. 20.2.3 with reference to the trap-assisted, thermal generation and recombination phenomena. A measurement method for the lifetimes is illustrated here with the aid of Fig. 25.1, that shows a uniformly-doped, thin layer of semiconductor of length L and cross-section A. The method is able to measure the minority-carrier lifetime; in the example shown in the figure, which refers to an n-doped material, the outcome is τ_p. The x axis is taken normal to the external surface of the semiconductor, with the origin placed on such a surface.

© Springer Science+Business Media New York 2015
M. Rudan, *Physics of Semiconductor Devices*,
DOI 10.1007/978-1-4939-1151-6_25

Fig. 25.1 Measurement scheme for the minority-carrier lifetime

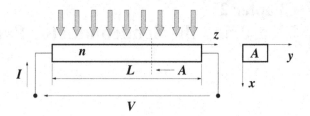

The latter, parallel to the y, z plane, is illuminated with a monochromatic radiation of frequency ν, whose intensity is uniform over the surface and constant in time. Remembering the expression (20.53) of the optical-generation term, and observing that due to uniformity the absorption coefficient does not depend on position, one finds

$$G_O = \eta\, \Phi_B\, k \, \exp(-kx), \qquad k = k(\nu). \tag{25.1}$$

At the same time, a constant voltage V is applied to the semiconductor, producing an electric field in the z direction. The device is thin in the x and y directions, and elongated in the z direction, to the extent that the flow lines of the current density are substantially parallel to the z axis; also, the small extension in the x direction makes it possible to neglect the x-dependence of G_O and let $G_O \simeq G_c = \eta\, \Phi_B k$. With these premises,[1] the material is spatially uniform also in the non-equilibrium case; it follows that the condition of local charge neutrality holds. For simplicity one assumes that the n-dopant concentration N_D is sufficiently low to insure that non-degeneracy and complete ionization hold, $N_D^+ = N_D$; thus, local neutrality reads

$$n = p + N_D. \tag{25.2}$$

The non-equilibrium condition prescribed by the combination of illumination and bias is adjusted in such a way that a weak-injection condition holds (Sect. 20.2.3.1); in this case the hole-continuity Eq. (19.124) reads

$$\frac{\partial p}{\partial t} + \frac{1}{q} \operatorname{div}\mathbf{J}_p = G - U \simeq G_c - \frac{p - p_{n0}}{\tau_p}. \tag{25.3}$$

In the steady-state, uniform condition considered here, (25.3) reduces to

$$\Delta p = p - p_{n0} = \tau_p\, G_c, \tag{25.4}$$

showing that optical generation is exactly balanced by thermal recombination. If, at time $t = 0$, the source of light is removed, G_c vanishes and recombination prevails; it follows that the semiconductor undergoes a transient to adapt to the new situation.

---

[1] The required thinness of the device can be achieved by growing an n-type epitaxial layer over a p-type substrate, and keeping the layer-substrate junction reverse biased.

Fig. 25.2 Measurement
scheme for the
minority-carrier lifetime

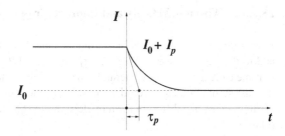

On the other hand, spatial uniformity still holds: during the transient the spatial
derivatives in (25.3) are still zero, and the equation reads

$$\frac{dp}{dt} = \frac{d(p - p_{n0})}{dt} = -\frac{p - p_{n0}}{\tau_p}, \qquad t > 0. \tag{25.5}$$

From (25.4), the initial condition of (25.5) is $\Delta p(t = 0) = \tau_p G_c$; thus, the solution
of (25.5) is found to be

$$\Delta p = \tau_p G_c \exp(-t/\tau_p). \tag{25.6}$$

To determine the conductivity of the device it is necessary to find the electron concentration n or, equivalently, the difference $\Delta n = n - n_{n0}$. This is easily accomplished
by observing that spatial uniformity holds also during the transient; as a consequence,
the condition of local charge neutrality (25.2) applies. At the end of the transient, the
latter becomes $n_{n0} = p_{n0} + N_D$ which, subtracted from (25.2), yields $\Delta n = \Delta p$ at
all times. The semiconductor conductivity (19.134) can then be written

$$\sigma = \sigma_0 + \sigma_p \exp(-t/\tau_p), \tag{25.7}$$

with $\sigma_0 = q(\mu_n n_{n0} + \mu_p p_{n0})$ and $\sigma_p = q(\mu_n + \mu_p)\tau_p G_c$. In the one-dimensional,
uniform case considered here, the current is the product of the current density times
the cross section A of the device, and the electric field is the ratio of the applied voltage
to the length L (Fig. 25.1). In conclusion, for $t > 0$ the current $I = V/R = V\sigma A/L$
is given by

$$I = I_0 + I_p \exp(-t/\tau_p), \tag{25.8}$$

with $I_0 = \sigma_0(A/L)V$, $I_p = \sigma_p(A/L)V$, namely, the current's decay time is
the minority-carrier lifetime. The quantity to be measured is the current through
the device which, for $t < 0$, is equal to the constant $I_0 + I_p$, while it decreases
exponentially towards I_0 for $t > 0$ (Fig. 25.2). The measurement of τ_p is easily
accomplished by observing that the tangent to the exponential branch drawn at the
origin intercepts the asymptotic value I_0 at $t = \tau_p$. Note that the hypothesis of a
monochromatic illumination is not essential; in fact, the analysis still holds if G_c in
(25.3) is replaced with the integral of G_c over the frequencies.

25.2.0.1 Thermal Velocity and Capture Cross-Section

In an n-doped material the hole lifetime is related to the hole-transition coefficient α_p and to the trap concentration N_t by $\tau_p = \tau_{p0} = 1/(\alpha_p N_t)$ (Sect. 20.2.3). It follows that measuring τ_p after deliberately introducing a known concentration N_t of traps into the semiconductor provides the value of the transition coefficient α_p. The latter is often recast in a different form by defining the carrier *thermal velocity* u_{th} with the relation

$$\frac{1}{2} m^* u_{\text{th}}^2 = \frac{3}{2} k_B T, \tag{25.9}$$

where $m^* = m_e$ for electrons and $m^* = m_h$ for holes, with m_e, m_h the average effective masses. Using the data of Table 18.1 one finds, for silicon, $m_e \simeq 2.98 \times 10^{-31}$ kg and $m_h \simeq 3.20 \times 10^{-31}$ kg. In turn, the *capture cross-sections* of the traps, for electrons and holes respectively, are defined by

$$\sigma_e = \frac{\alpha_n}{u_{\text{th},e}}, \qquad \sigma_h = \frac{\alpha_p}{u_{\text{th},h}}. \tag{25.10}$$

From the above definitions it follows that in an n-doped material

$$\tau_p = \tau_{p0} = \frac{1}{\alpha_p N_t} = \frac{1}{\sigma_h u_{\text{th},h} N_t}. \tag{25.11}$$

In particular, in silicon at $T_L = 300$ K it is from (25.9) $u_{\text{th},e} \simeq u_{\text{th},h} \sim 2 \times 10^7$ cm s^{-1}. The measure of α_p thus provides for the cross-section the value $\sigma_h \simeq 5 \times 10^{-15}$ cm^2. A qualitative picture of the cross-section as a circle centered at the trap yields the definition of a radius r_h such that $\sigma_h = \pi r_h^2$. It turns out $r_h \simeq 4 \times 10^{-8}$ cm, namely, r_h is of the order of the atomic radius. The measure of τ_n, α_n, and σ_e is carried out in a similar way, starting from a p-doped material.

25.3 Mobility Measurement—Haynes-Shockley Experiment

A measurement method for mobility is illustrated with the aid of Fig. 25.3, showing a uniformly-doped layer of semiconductor to which a constant positive voltage V is applied; this produces an electric field in the x direction, $\mathbf{E} = E \mathbf{i}_1$. The method is able to measure the minority-carrier mobility; in the example shown in the figure, which refers to an n-doped material, the outcome is μ_p.

Holes are generated at some position x by, e.g., illuminating the material with a laser pulse. The electric field makes the holes to drift to the right, where they are eventually collected after crossing a distance Δx_1 in the direction parallel to $\mathbf{E}$. The lower part of Fig. 25.3 shows three profiles of the hole distribution at successive instants of time, from left to right.[2] The leftmost, thin profile corresponds to the

---

[2] The generated electrons drift to the left and are absorbed by the left contact.

Fig. 25.3 Measurement scheme for mobility (Haynes-Shockley experiment)

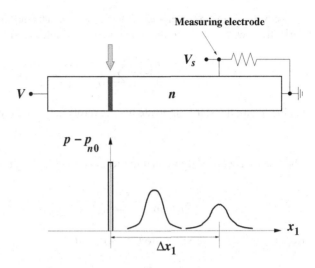

instant of the laser pulse; the other two profiles are shifted to the right because of the action of the field. The hole distribution becomes progressively wider and shorter because of diffusion; also, its area decreases with time due to recombination. When the profile crosses the section corresponding to the measuring electrode, the largest value of the measured voltage V_S corresponds to the profile's peak. This allows one to measure the time Δt necessary for the profile's peak to cover the distance from the section where the laser pulse is applied to that of the measuring electrode. Then, the average velocity of the peak is found from $\Delta x_1/\Delta t$.

The analysis of the experiment is carried out assuming that the perturbation is sufficiently small, so that a weak-injection condition holds. The continuity equation for the minority carriers reads

$$\frac{\partial p}{\partial t} + \frac{p - p_{n0}}{\tau_p} + \mathrm{div}\left(\mu_p \, p \, \mathbf{E} - D_p \, \mathrm{grad}\, p\right) = 0, \qquad (25.12)$$

with μ_p, D_p = const due to spatial uniformity. Remembering that $\mathrm{div}\,\mathbf{D} = \rho$, in (25.12) it is

$$\mathrm{div}\,(p\mathbf{E}) = \mathbf{E}\cdot\mathrm{grad}\, p + p\,\frac{\rho}{\varepsilon_{sc}} \simeq \mathbf{E}\cdot\mathrm{grad}\, p, \qquad (25.13)$$

on account of the fact that, due to the weak-injection condition, the perturbation with respect to the local charge neutrality is small, whereas $\mathrm{grad}\, p$ is large. Using the auxiliary function $f = (p - p_{n0})\exp(t/\tau_p)$ transforms (25.12) into

$$\frac{\partial f}{\partial t} - D_p\,\nabla^2 f + \mu_p\,\mathbf{E}\cdot\mathrm{grad}\, f = 0. \qquad (25.14)$$

The above equation is further simplified by applying a suitable change of the variables,

$$\mathbf{r}^\star = \mathbf{r}^\star(\mathbf{r}, t) = \mathbf{r} - \mathbf{v}\,t, \qquad t^\star = t^\star(\mathbf{r}, t) = t, \qquad (25.15)$$

where $\mathbf{v}$ is a constant velocity, yet undefined. The relations between the spatial derivatives with respect to the old and new variables read

$$\frac{\partial f}{\partial x_i} = \sum_{j=1}^{3} \frac{\partial f}{\partial x_j^{\star}} \frac{\partial x_j^{\star}}{\partial x_i} = \frac{\partial f}{\partial x_i^{\star}}, \tag{25.16}$$

so that, using the star to indicate the operators acting on the new spatial variables,

$$\mathrm{grad}_{\star} f = \mathrm{grad}\, f, \qquad \nabla_{\star}^2 f = \nabla^2 f. \tag{25.17}$$

The time derivatives are treated in the same manner, to find

$$\frac{\partial f}{\partial t} = \frac{\partial f}{\partial t^{\star}} \frac{\partial t^{\star}}{\partial t} + \sum_{j=1}^{3} \frac{\partial f}{\partial x_j^{\star}} \frac{\partial x_j^{\star}}{\partial t} = \frac{\partial f}{\partial t^{\star}} - \mathbf{v} \cdot \mathrm{grad}_{\star} f. \tag{25.18}$$

Replacing (25.17), (25.18) into (25.14) yields

$$\frac{\partial f}{\partial t^{\star}} = D_p \nabla_{\star}^2 f + (\mathbf{v} - \mu_p \mathbf{E}) \cdot \mathrm{grad}_{\star} f. \tag{25.19}$$

Exploiting the arbitrariness of $\mathbf{v}$ one lets $\mathbf{v} = \mu_p \mathbf{E}$, so that (25.19) simplifies to a diffusion equation, $\partial f / \partial t^{\star} = D_p \nabla_{\star}^2 f$. The solution of the latter is given by (23.28), namely,

$$f(\mathbf{r}^{\star}, t^{\star}) = \iiint_{-\infty}^{+\infty} f(\mathbf{s}, 0)\, \Delta(\mathbf{r}^{\star} - \mathbf{s}, t^{\star})\, \mathrm{d}^3 s, \tag{25.20}$$

with

$$\Delta(\mathbf{r}^{\star} - \mathbf{s}, t^{\star}) = \frac{1}{\left(4\pi D_p t^{\star}\right)^{3/2}} \exp\left(-\frac{|\mathbf{r}^{\star} - \mathbf{s}|^2}{4 D_p t^{\star}}\right), \tag{25.21}$$

and $f(\mathbf{s}, 0) = p(\mathbf{s}, 0) - p_{n0}$. Using again the old variables yields

$$p = p_{n0} + \exp(-t/\tau_p)\left(4\pi D_p t\right)^{3/2} \iiint_{-\infty}^{+\infty} f(\mathbf{s}, 0)\, \Delta(\mathbf{r} - \mathbf{v}t - \mathbf{s}, t)\, \mathrm{d}^3 s. \tag{25.22}$$

The input pulse can be selected in such a way that $f(\mathbf{s}, 0) \approx c\, \delta(\mathbf{s})$, where c is a dimensionless constant. In conclusion,

$$p = p_{n0} + \frac{c \exp(-t/\tau_p)}{\left(4\pi D_p t\right)^{3/2}} \exp\left[-\frac{(x_1 - \mu_p E t)^2 + x_2^2 + x_3^2}{4 D_p t}\right], \tag{25.23}$$

which, apart from the additive constant p_{n0}, is a Gaussian whose amplitude decreases in time and whose peak moves along the x_1 direction with the constant velocity $v = \mu_p E$. Measuring the time Δt_1 needed for the peak to cross the distance Δx_1 finally yields the mobility

$$\mu_p = \frac{1}{E} \frac{\Delta x_1}{\Delta t_1}. \tag{25.24}$$

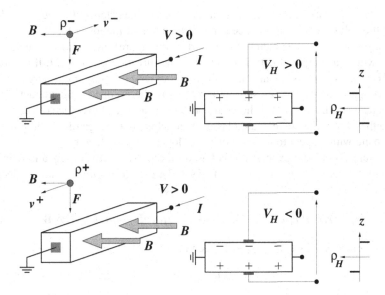

Fig. 25.4 Scheme of a Hall-voltage measurement

25.4   Hall-Voltage Measurement

The measurements based on the *Hall effect* are a powerful investigation tool that exploits the combined action of the electric and magnetic field. The Hall effect is the production of a voltage drop, transverse to the direction of the electric current, due to the application of a magnetic field. The qualitative features of the method are explained with the aid of Fig. 25.4. Consider a uniformly doped, prismatic block of semiconductor. The block is slender, and a constant voltage V is applied to it to produce an electric field $\mathbf{E}$ aligned with the longitudinal direction. As a consequence, one can assume that the flow lines of the current density be parallel to $\mathbf{E}$; due to spatial uniformity, such a current density is essentially due to the drift of majority carriers. At the same time, a constant magnetic-induction field $\mathbf{B}$ is applied, normal to one of the lateral faces. The upper part of the figure refers to an n-doped semiconductor; there, the majority carriers are electrons, whose average velocity is oriented opposite to the field; it follows that the Lorentz force $\mathbf{F} = \rho^- \mathbf{v}^- \wedge \mathbf{B}$ (Sect. 4.11) is oriented as shown in the figure. The negative indices in the expression of the Lorentz force remind one that the charge density and average velocity are those of negative charges. The mobile electrons are pushed by the Lorentz force towards the lower face of the device, where they form a negative charge layer. The flow lines of the current density are still parallel to the longitudinal direction; however, their density is not uniform any more. Due to the global charge neutrality, the negative charge layer is compensated by a positive charge layer that forms at the upper face. The two opposite layers, schematically indicated in the diagram in the upper-right part of Fig. 25.4, produce an electric field normal to the upper and lower faces; as a result, a measurable voltage drop (*Hall voltage*) between the two faces comes into existence: for the example in hand, the voltage of the upper face is larger than that of the lower face.

In a p-doped semiconductor (lower part of the figure), the majority carriers are holes, whose average velocity is oriented in the direction of the field; the Lorentz force $\mathbf{F} = \rho^+ \mathbf{v}^+ \wedge \mathbf{B}$ is oriented as in the previous case, because both charge density and average velocity change sign with respect to the n-doped semiconductor. The consequence is that the mobile holes are pushed towards the lower face of the device, where they form a positive charge layer. In conclusion, the sign of the Hall voltage is opposite with respect to the case of the n-doped semiconductor.

The analysis of the experiment is based on the drift-diffusion equations incorporating the magnetic terms, (19.107), (19.123); the diffusion terms are neglected, whence

$$\mathbf{J}_n = q\mu_n n\mathbf{E} - qa_n\mu_n^2 n\mathbf{E} \wedge \mathbf{B}, \qquad \mathbf{J}_p = q\mu_p p\mathbf{E} + qa_p\mu_p^2 p\mathbf{E} \wedge \mathbf{B}. \qquad (25.25)$$

The total current density $\mathbf{J} = \mathbf{J}_n + \mathbf{J}_p$ then reads

$$\mathbf{J} = \sigma\mathbf{E} + r\sigma^2\mathbf{E} \wedge \mathbf{B}, \qquad (25.26)$$

where $\sigma = q\,\mu_p\,p + q\,\mu_n\,n$ is the electric conductivity and

$$r = \frac{q}{\sigma^2}\left(a_p\,\mu_p^2\,p - a_n\,\mu_n^2\,n\right) = \frac{a_p\,\mu_p^2\,p - a_n\,\mu_n^2\,n}{q\left(\mu_p\,p + \mu_n\,n\right)^2} \qquad (25.27)$$

is the *Hall coefficient*. The two quantities σ and r can be measured independently as shown below; while σ is positive definite, r has a sign. In particular, the following limiting cases hold: for the p-type dopant it is $p \gg n$, whence $\sigma \simeq q\mu_p p$ and $r \simeq a_p/(qp) > 0$; thus,

$$p = \frac{a_p}{q\,r}, \qquad \mu_p = \frac{r}{a_p}\sigma \qquad (p \gg n). \qquad (25.28)$$

Similarly, for the n-type dopant it is $n \gg p$, whence $\sigma \simeq q\,\mu_n\,n$ and $r \simeq -a_n/(q\,n) < 0$; thus,

$$n = -\frac{a_n}{q\,r}, \qquad \mu_n = -\frac{r}{a_n}\sigma \qquad (n \gg p). \qquad (25.29)$$

From (25.28) and (25.29) it follows that the concentration and mobility of the majority carriers can be determined independently, provided σ and r are known. In turn, the measurement of σ and r is easily carried out by applying (25.26) to the prismatic sample of Fig. 25.4. Let $\mathbf{i}_L$ be the unit vector of the longitudinal direction, and $\mathbf{i}_W$ the unit vector parallel to $\mathbf{B}$, so that $\mathbf{B} = B\,\mathbf{i}_W$. Observing that $E_W = 0$, $J_W = 0$, and $\mathbf{E} \wedge \mathbf{B} = E_L B\mathbf{i}_H - E_H B\mathbf{i}_L$, it follows

$$J_L = \sigma E_L - r\sigma^2 E_H B, \qquad J_H = \sigma E_H + r\sigma^2 E_L B, \qquad (25.30)$$

with $J_H = 0$. In turn, B is small enough to make the following approximations possible:

$$J = J_L \simeq \sigma E_L, \qquad E_H = -r\sigma E_L B \simeq -r J B. \qquad (25.31)$$

On the other hand it is $E_L \simeq V_L/L$, $E_H \simeq V_H/H$, and $J = I/(W H)$, whence $V_H = -r B I/W$ and $I/(W H) = \sigma V_L/L$. In conclusion,

$$\sigma = \frac{LI}{WHV_L}, \qquad r = -\frac{WV_H}{BI}, \qquad (25.32)$$

namely, the two parameters are obtained by combining the Hall voltage with other known physical and geometrical parameters. Typical applications of the measurement scheme shown in this section are the measurements of majority-carrier concentrations and mobilities as functions of temperature and dopant concentration.

25.5 Measurement of Doping Profiles

The calculation of the depletion capacitance for an arbitrary doping profile, in the one-dimensional case, has been carried out in Sect. 21.6.3; the analysis has yielded, among others, the two relations (21.72) and (21.73), that are reported below:

$$dQ = \rho(b)\, db, \qquad C = \frac{dQ}{d\psi} = \frac{\varepsilon_{sc}}{b - a}, \qquad (25.33)$$

where C is the differential capacitance per unit area of the space–charge region, whose boundaries are a and b, and $\psi = \varphi(b) - \varphi(a)$. For a strongly-asymmetric junction it is, e.g., $b - a \simeq b$, and (25.33) become

$$C \simeq \frac{\varepsilon_{sc}}{b}, \qquad \varepsilon_{sc}\, d\psi \simeq b\, \rho(b)\, db = \frac{1}{2} \rho(b)\, db^2 = \frac{1}{2}\rho(b)\, d\left(\frac{\varepsilon_{sc}^2}{C^2}\right). \qquad (25.34)$$

As a consequence,

$$\rho(b) = \frac{2}{\varepsilon_{sc}}\left[\frac{d\left(1/C^2\right)}{d\psi}\right]^{-1}. \qquad (25.35)$$

Basing upon (25.35), a measurement scheme can be devised, which proceeds as follows:

1. C is measured at a given bias V, and b is determined from $b = \varepsilon_{sc}/C$.
2. C is measured again after slightly varying the bias from V to $V + dV = V + d\psi$.
3. A numerical calculation of the derivative yields $\rho(b)$.

An example of application of the above scheme is given with reference to an asymmetric p-n junction with, e.g., $a = -l_p$, $b = l_n$, and $l_n \gg l_p$. In the reverse-bias condition, and using the full-depletion approximation (20.34), it is $\rho \simeq q\,N$, whence

$$N(l_n) \simeq \frac{2}{q\varepsilon_{\text{sc}}} \left[\frac{\mathrm{d}\left(1/C^2\right)}{\mathrm{d}\psi} \right]^{-1}. \qquad (25.36)$$

This provides a method for measuring the dopant distribution.

Appendix A
Vector and Matrix Analysis

A.1 Scalar Product

Consider two complex, n-dimensional column vectors

$$
\mathbf{a} = \begin{bmatrix} a_1 \\ \vdots \\ a_n \end{bmatrix}, \quad
\mathbf{b} = \begin{bmatrix} b_1 \\ \vdots \\ b_n \end{bmatrix},
\tag{A.1}
$$

whose entries are the components in an n-dimensional Cartesian reference and may depend on position, time, and other parameters. The *scalar product* of the two vectors is indicated with $\mathbf{a} \cdot \mathbf{b}$ and is defined as

$$
\mathbf{a} \cdot \mathbf{b} = \sum_{i=1}^{n} a_i^* b_i.
\tag{A.2}
$$

with a_i^* the complex conjugate of a_i. Two non-vanishing vectors $\mathbf{a}$ and $\mathbf{b}$ are *orthogonal* if $\mathbf{a} \cdot \mathbf{b} = 0$. As $\mathbf{b} \cdot \mathbf{a} = (\mathbf{a} \cdot \mathbf{b})^*$, the order of the factors in the scalar product matters; in fact it becomes irrelevant only when the factors are real. The scalar product is distributive and bilinear; if, say, $\mathbf{a} = h_1 \, \mathbf{p}_1 + h_2 \, \mathbf{p}_2$, then

$$
\mathbf{a} \cdot (k_1 \, \mathbf{b}_1 + k_2 \, \mathbf{b}_2) = h_1^* k_1 \, \mathbf{p}_1 \cdot \mathbf{b}_1 + h_2^* k_1 \, \mathbf{p}_2 \cdot \mathbf{b}_1 + h_1^* k_2 \, \mathbf{p}_1 \cdot \mathbf{b}_2 + h_2^* k_2 \, \mathbf{p}_2 \cdot \mathbf{b}_2,
\tag{A.3}
$$

where h_1, h_2, k_1, k_2 are complex constants (in (A.3), the product $k_1 \, \mathbf{b}_1$ is the vector of components $k_1 \, b_{1i}$, and so on). The *modulus* of $\mathbf{a}$ is defined as

$$
a = |\mathbf{a}| = \sqrt{\mathbf{a} \cdot \mathbf{a}} = \left(\sum_{i=1}^{n} |a_i|^2 \right)^{1/2} \geq 0.
\tag{A.4}
$$

© Springer Science+Business Media New York 2015
M. Rudan, *Physics of Semiconductor Devices*,
DOI 10.1007/978-1-4939-1151-6

A.2   Schwarz Inequality and Generalizations

Using (A.2, A.3, A.4) one proves the *Schwarz inequality*

$$|\mathbf{a} \cdot \mathbf{b}| \leq a \, b. \tag{A.5}$$

The above is obvious if $\mathbf{a} = 0$ or $\mathbf{b} = 0$; let $\mathbf{b} \neq 0$ and define $\mathbf{c} = \mathbf{a} - (\mathbf{a} \cdot \mathbf{b}) \mathbf{b}/b^2$, whence $\mathbf{c} \cdot \mathbf{b} = 0$. It follows

$$a^2 = \left(\mathbf{c} + \frac{\mathbf{a} \cdot \mathbf{b}}{b^2} \mathbf{b}\right) \cdot \left(\mathbf{c} + \frac{\mathbf{a} \cdot \mathbf{b}}{b^2} \mathbf{b}\right) = c^2 + \frac{|\mathbf{a} \cdot \mathbf{b}|^2}{b^2} \geq \frac{|\mathbf{a} \cdot \mathbf{b}|^2}{b^2}, \tag{A.6}$$

which is equivalent to (A.5). The strict equality in (A.5) holds if and only if $\mathbf{b} = k \, \mathbf{a}$, with k any complex constant. Observing that $|\mathbf{a} \cdot \mathbf{b}|^2 = \Re^2(\mathbf{a} \cdot \mathbf{b}) + \Im^2(\mathbf{a} \cdot \mathbf{b})$, from (A.5) one also derives the inequalities $-ab \leq \Re(\mathbf{a} \cdot \mathbf{b}) \leq +ab$. Thanks to this, one defines the cosine of the angle ϑ between two non-vanishing vectors $\mathbf{a}$ and $\mathbf{b}$ as

$$\cos \vartheta = \frac{\Re(\mathbf{a} \cdot \mathbf{b})}{ab}. \tag{A.7}$$

Other types of products may be defined besides the scalar product, also involving higher-rank factors: for instance, $n \times n$ matrices of the second rank like

$$\mathbf{A} = \begin{bmatrix} A_{11} & A_{12} & \cdots & A_{1n} \\ A_{21} & A_{22} & \cdots & A_{2n} \\ \vdots & \vdots & \ddots & \vdots \\ A_{n1} & A_{n2} & \cdots & A_{nn} \end{bmatrix}, \quad \mathbf{B} = \begin{bmatrix} B_{11} & B_{12} & \cdots & B_{1n} \\ B_{21} & B_{22} & \cdots & B_{2n} \\ \vdots & \vdots & \ddots & \vdots \\ B_{n1} & B_{n2} & \cdots & B_{nn} \end{bmatrix}, \tag{A.8}$$

and so on. Given a second-rank matrix $\mathbf{A}$ of entries A_{ij}, its *transpose* $\mathbf{Q} = \mathbf{A}^T$ is the matrix of entries $Q_{ij} = A_{ji}$. Transposition applies also to vectors: the transpose of the column vector $\mathbf{a}$ defined in (A.1) is the row vector $\mathbf{a}^T = [a_1, \ldots, a_n]$. With these premises, given the column vectors $\mathbf{a}$, $\mathbf{b}$ and the matrices $\mathbf{A}$, $\mathbf{B}$, the products $\mathbf{A}\mathbf{B}$, $\mathbf{A}\mathbf{b}$, and $\mathbf{a}\mathbf{b}^T$ yield, respectively, an $n \times n$ matrix, an n-dimensional column vector, and an $n \times n$ matrix whose entries are

$$(\mathbf{A}\mathbf{B})_{ij} = \sum_{k=1}^{n} A_{ik} B_{kj}, \qquad (\mathbf{A}\mathbf{b})_i = \sum_{j=1}^{n} A_{ij} b_j, \qquad \left(\mathbf{a}\mathbf{b}^T\right)_{ij} = a_i \, b_j. \tag{A.9}$$

Applying definitions (A.9) one finds

$$(\mathbf{A}\mathbf{B})^T = \mathbf{B}^T \mathbf{A}^T, \qquad (\mathbf{A}\mathbf{b})^T = \mathbf{b}^T \mathbf{A}^T, \qquad \left(\mathbf{a}\mathbf{b}^T\right)^T = \mathbf{b}\mathbf{a}^T. \tag{A.10}$$

A.3 *Nabla* Operator

A further extension of the concepts introduced in this chapter consists in replacing one or more factors with an operator. An important example is that of the real, vector operator *nabla*,[1]

$$\nabla = \begin{bmatrix} \partial/\partial x_1 \\ \vdots \\ \partial/\partial x_n \end{bmatrix}, \tag{A.11}$$

where $x_1, \ldots, x_n$ are the coordinates of an n-dimensional Cartesian reference. The product of ∇ and a complex, scalar function $f(x_1, \ldots, x_n)$ is defined in the same manner as the product of a vector and a scalar quantity introduced above: ∇f is a vector of components $(\nabla)_i f$, namely,

$$\nabla f = \begin{bmatrix} \partial f/\partial x_1 \\ \vdots \\ \partial f/\partial x_n \end{bmatrix}. \tag{A.12}$$

In turn, the scalar product of ∇ and a complex vector $\mathbf{a}$ of the same dimension as ∇ yields

$$\nabla \cdot \mathbf{a} = \frac{\partial a_1}{\partial x_1} + \ldots + \frac{\partial a_n}{\partial x_n}. \tag{A.13}$$

The product defined by (A.12) is also called *gradient* of f, whereas the scalar product (A.13) is also called *divergence* of $\mathbf{a}$. The corresponding symbols are $\nabla f = \text{grad } f$ and $\nabla \cdot \mathbf{a} = \text{div } \mathbf{a}$, respectively. The scalar product of ∇ by itself is called *Laplacian operator*

$$\nabla^2 = \nabla \cdot \nabla = \frac{\partial^2}{\partial x_1^2} + \ldots + \frac{\partial^2}{\partial x_n^2}, \tag{A.14}$$

then,

$$\nabla^2 f = \frac{\partial^2 f}{\partial x_1^2} + \ldots + \frac{\partial^2 f}{\partial x_n^2}, \qquad \nabla^2 \mathbf{a} = \begin{bmatrix} \nabla^2 a_1 \\ \vdots \\ \nabla^2 a_n \end{bmatrix}. \tag{A.15}$$

Combining the above definitions yields the identities

$$\nabla^2 f = \nabla \cdot (\nabla f) = \text{div grad } f, \quad \nabla \cdot (f^* \mathbf{a}) = \text{div} (f^* \mathbf{a}) = f^* \text{ div } \mathbf{a} + \text{grad } f \cdot \mathbf{a}. \tag{A.16}$$

---

[1] Symbol ∇ is *not* a Greek letter. However, the term *nabla* is a Greek word, meaning "harp".

If, in turn, it is $\mathbf{a} = \text{grad}\, g$, the second relation of (A.16) with the aid of the first one yields the identity

$$\text{div}\,(f^* \,\text{grad}\, g) = f^* \,\nabla^2 g + \text{grad}\, f \cdot \text{grad}\, g. \tag{A.17}$$

A.4 Dyadic Products

Sometimes it is convenient to adopt a notation that uses the basis set of real, mutually-orthogonal unit vectors $\mathbf{i}_1, \ldots, \mathbf{i}_n$ associated with the axes of a Cartesian reference. By construction it is $\mathbf{i}_r \cdot \mathbf{i}_s = \delta_{rs}$, where the *Kronecker symbol* δ_{rs} is the entry of indices rs of a second-rank matrix defined as

$$\delta_{rs} = \begin{cases} 1 & s = r \\ 0 & s \neq r \end{cases} \tag{A.18}$$

The expression of vector $\mathbf{a}$ in terms of the basis vectors is $\mathbf{a} = a_1 \mathbf{i}_1 + \ldots + a_n \mathbf{i}_n$. The notation applies also to the higher-rank objects; for instance, in this notation the matrix $\mathbf{A}$ of (A.8) reads

$$\mathbf{A} = A_{11} \mathbf{i}_1 \mathbf{i}_1^T + A_{12} \mathbf{i}_1 \mathbf{i}_2^T + \ldots + A_{n,n-1} \mathbf{i}_n \mathbf{i}_{n-1}^T + A_{nn} \mathbf{i}_n \mathbf{i}_n^T, \tag{A.19}$$

A group like $\mathbf{i}_r \mathbf{i}_s^T$ is also called *dyadic product*. Observing that $\mathbf{i}_r$ is an n-dimensional column vector whose rth entry is equal to 1 while all the other entries are equal to 0, the application of the third equation in (A.9) shows that $\mathbf{i}_r \mathbf{i}_s^T$ is an $n \times n$ matrix whose entry of indices rs is equal to 1, while all the other entries are equal to zero. As a consequence, the form (A.19) expresses $\mathbf{A}$ as a sum of matrices, each associated to an individual entry. Using this notation, a product like $\mathbf{A}\,\mathbf{b}$ reads $\sum_{rs} A_{rs} \mathbf{i}_r \mathbf{i}_s^T \sum_k b_k \mathbf{i}_k$. On the other hand, due to the second equation in (A.9), the same product is equal to $\sum_{rs} A_{rs} b_s \mathbf{i}_r$. This shows that $\mathbf{i}_r \mathbf{i}_s^T \mathbf{i}_k = \mathbf{i}_r \delta_{sk}$, that is, the juxtaposition of the right unit vector of the dyadic product with the next unit vector must be treated as a scalar product.

The relation defined by the second equation in (A.9) applies also when $\mathbf{b}$ is replaced with a vector operator, with the provision that the operator is meant to act towards the left. For instance, replacing $\mathbf{b}$ with ∇ yields $(\mathbf{A}\,\nabla)_i = \sum_{j=1}^{n} \partial A_{ij}/\partial x_j$. It follows that the divergence of a second-rank matrix is a column vector of the form

$$\text{div}\,\mathbf{A} = \sum_{j=1}^{n} \frac{\partial A_{1j}}{\partial x_j} \mathbf{i}_1 + \ldots + \sum_{j=1}^{n} \frac{\partial A_{nj}}{\partial x_j} \mathbf{i}_n. \tag{A.20}$$

In turn, considering the product defined by the third equation in (A.9) and replacing $\mathbf{b}$ with ∇, still with the provision that the operator acts towards the left, yields $(\mathbf{a}\,\nabla^T)_{ij} = \partial a_i/\partial x_j$. It follows that the gradient of a column vector is a second-rank matrix of the form

$$\text{grad}\,\mathbf{a} = \frac{\partial a_1}{\partial x_1} \mathbf{i}_1 \mathbf{i}_1^T + \frac{\partial a_1}{\partial x_2} \mathbf{i}_1 \mathbf{i}_2^T + \ldots + \frac{\partial a_n}{\partial x_{n-1}} \mathbf{i}_1 \mathbf{i}_{n-1}^T + \frac{\partial a_n}{\partial x_n} \mathbf{i}_n \mathbf{i}_n^T \tag{A.21}$$

whence, from (A.20),

$$\text{div}\,(f\,\mathbf{A}) = f\,\text{div}\,\mathbf{A} + \mathbf{A}\,\text{grad}\,f, \qquad \text{div}\,(\mathbf{a}\,\mathbf{b}^T) = \mathbf{a}\,\text{div}\,\mathbf{b} + (\text{grad}\,\mathbf{a})\,\mathbf{b}. \quad \text{(A.22)}$$

A.5 Divergence Theorem

The *divergence theorem* (or *Gauss theorem*) states that

$$\int_V \text{div}\,\mathbf{v}\,dV = \int_S \mathbf{n}\cdot\mathbf{v}\,dS, \qquad\qquad \text{(A.23)}$$

where V is an n-dimensional volume, $dV = dx_1\ldots dx_n$, S the $(n-1)$-dimensional surface enclosing V, and $\mathbf{n}$ the unit vector normal to the surface element dS, oriented in the outward direction with respect to S. Letting $\mathbf{v} = f^*\,\text{grad}\,g$ and using (A.17) yields the *first Green theorem*

$$\int_S f^* \frac{\partial g}{\partial n}\,dS = \int_V \left(f^* \nabla^2 g + \text{grad}\,f \cdot \text{grad}\,g\right)\,dV, \qquad \text{(A.24)}$$

where $\partial g/\partial n = \mathbf{n}\cdot\text{grad}\,g$ is the derivative of g in the direction of $\mathbf{n}$. It is easily found that (A.24) is the generalization to n dimensions of the integration by parts. Rewriting (A.24) after letting $\mathbf{v} = g\,\text{grad}\,f^*$, and subtracting from (A.24), yields the *second Green theorem*

$$\int_S \left(f^* \frac{\partial g}{\partial n} - g \frac{\partial f^*}{\partial n}\right)\,dS = \int_V \left(f^* \nabla^2 g - g\nabla^2 f^*\right)\,dV. \qquad \text{(A.25)}$$

A special case of the first Green theorem occurs when vector $\mathbf{b} = \text{grad}\,g$ is constant; the relation (A.24) then reduces to

$$\int_S f^*\,\mathbf{n}\,dS \cdot \mathbf{b} = \int_V \text{grad}\,f\,dV \cdot \mathbf{b}, \qquad \mathbf{b} = \text{const.} \qquad \text{(A.26)}$$

As identity (A.26) holds for any choice of $\mathbf{b}$, the two integrals in it are equal to each other.

A.6 Vector Product

Another possible product between two vectors is the *vector product* $\mathbf{a} \wedge \mathbf{b}$, which yields a column vector. In contrast with the other products introduced in this section, the definition of the vector product will be limited to the three-dimensional case; it is given as the expansion of a determinant, namely,

$$\mathbf{a} \wedge \mathbf{b} = \begin{bmatrix} \mathbf{i}_1 & \mathbf{i}_2 & \mathbf{i}_3 \\ a_1 & a_2 & a_3 \\ b_1 & b_2 & b_3 \end{bmatrix} \quad\Rightarrow\quad \mathbf{a} \wedge \mathbf{b} = \begin{bmatrix} a_2 b_3 - a_3 b_2 \\ a_3 b_1 - a_1 b_3 \\ a_1 b_2 - a_2 b_1 \end{bmatrix}. \qquad \text{(A.27)}$$

From (A.27) it follows $\mathbf{b} \wedge \mathbf{a} = -\mathbf{a} \wedge \mathbf{b}$ and $\mathbf{a} \wedge \mathbf{a} = 0$. The latter also shows that, if two non-vanishing vectors are parallel to each other, say, $\mathbf{b} = k\,\mathbf{a} \neq 0$, then $\mathbf{a} \wedge \mathbf{b} = 0$. When the vector product involves the unit vectors associated to the axes of a right-handed Cartesian reference, the following relations are found:

$$\mathbf{i}_1 \wedge \mathbf{i}_2 = \mathbf{i}_3, \qquad \mathbf{i}_2 \wedge \mathbf{i}_3 = \mathbf{i}_1, \qquad \mathbf{i}_3 \wedge \mathbf{i}_1 = \mathbf{i}_2. \tag{A.28}$$

An intrinsic relation that provides the modulus of $\mathbf{a} \wedge \mathbf{b}$ is found by specifying (A.7) for the case of three-dimensional, real vectors, this yielding

$$\cos^2 \vartheta = 1 - \sin^2 \vartheta = \frac{\left(\sum_{i=1}^{3} a_i\, b_i\right)^2}{a^2\, b^2}. \tag{A.29}$$

As $\cos \vartheta = 1$ when the two vectors are parallel, $\mathbf{b} = k\,\mathbf{a}$, $k > 0$, while $\cos \vartheta = -1$ when they are antiparallel, $\mathbf{b} = k\,\mathbf{a}$, $k < 0$, the range of ϑ is $[0, \pi]$. Letting $r_{ij} = a_i\, b_j - a_j\, b_i$ and observing that $(\sum_{i=1}^{3} a_i^2)(\sum_{i=1}^{3} b_i^2) = (\sum_{i=1}^{3} a_i\, b_i)^2 + r_{23}^2 + r_{31}^2 + r_{12}^2$ provides

$$\sin^2 \vartheta = \frac{r_{23}^2 + r_{31}^2 + r_{12}^2}{a^2\, b^2} = \frac{|\mathbf{a} \wedge \mathbf{b}|^2}{a^2\, b^2}, \qquad |\mathbf{a} \wedge \mathbf{b}| = a\, b\, \sin \vartheta, \tag{A.30}$$

where $\sin \vartheta \geq 0$ due to the range of ϑ.

A.7 Mixed Product

The vector product $\mathbf{a} \wedge \mathbf{b}$ can in turn be scalarly multiplied by another vector $\mathbf{c}$, to yield a scalar quantity called *mixed product*. For the sake of simplicity, in the definition of the mixed product the three vectors will be considered real. From (A.2) one finds

$$\mathbf{a} \wedge \mathbf{b} \cdot \mathbf{c} = \sum_{i=1}^{3} (\mathbf{a} \wedge \mathbf{b})_i\, c_i = \begin{vmatrix} c_1 & c_2 & c_3 \\ a_1 & a_2 & a_3 \\ b_1 & b_2 & b_3 \end{vmatrix} = \begin{vmatrix} a_1 & a_2 & a_3 \\ b_1 & b_2 & b_3 \\ c_1 & c_2 & c_3 \end{vmatrix}. \tag{A.31}$$

The two determinants in (A.31) are equal because they transform into each other by interchanging rows an even number of times. On the other hand, from their equality it follows $\mathbf{a} \wedge \mathbf{b} \cdot \mathbf{c} = \mathbf{a} \cdot \mathbf{b} \wedge \mathbf{c}$, namely, the mixed product is invariant upon interchange of the "wedge" and "dot" symbols.

Considering three non-vanishing vectors $\mathbf{a}$, $\mathbf{b}$, $\mathbf{c}$, where $\mathbf{a}$ and $\mathbf{b}$ are not parallel to each other, and remembering the properties of determinants, one finds that the mixed product vanishes if $\mathbf{c}$ is parallel to $\mathbf{a}$ or parallel to $\mathbf{b}$. In fact,

$$\mathbf{a} \wedge \mathbf{b} \cdot \mathbf{a} = \mathbf{a} \wedge \mathbf{b} \cdot \mathbf{b} = 0. \tag{A.32}$$

It follows that the vector product $\mathbf{a} \wedge \mathbf{b}$ is normal to both $\mathbf{a}$ and $\mathbf{b}$, namely, is normal to the plane defined by the two non-parallel vectors $\mathbf{a}$ and $\mathbf{b}$. If one associates the plane of $\mathbf{a}$ and $\mathbf{b}$ with that of the unit vectors $\mathbf{i}_1$ and $\mathbf{i}_2$ then, using (A.28), the vector product simplifies to $\mathbf{a} \wedge \mathbf{b} = (a_1 b_2 - a_2 a_1) \mathbf{i}_3$, that provides the information about the direction of $\mathbf{a} \wedge \mathbf{b}$. Finally, using (A.27) twice provides the expression for the *double vector product*

$$\mathbf{a} \wedge (\mathbf{b} \wedge \mathbf{c}) = \mathbf{a} \cdot \mathbf{c}\,\mathbf{b} - \mathbf{a} \cdot \mathbf{b}\,\mathbf{c}, \qquad (\mathbf{a} \wedge \mathbf{b}) \wedge \mathbf{c} = \mathbf{a} \cdot \mathbf{c}\,\mathbf{b} - \mathbf{b} \cdot \mathbf{c}\,\mathbf{a}. \qquad (A.33)$$

A.8 Rotational of a Vector

The expressions involving the vector product can be extended to the case where one or two vectors are replaced with the nabla operator (A.11). The vector product

$$\nabla \wedge \mathbf{a} = \begin{bmatrix} \mathbf{i}_1 & \mathbf{i}_2 & \mathbf{i}_3 \\ \partial/\partial x_1 & \partial/\partial x_2 & \partial/\partial x_3 \\ a_1 & a_2 & a_3 \end{bmatrix} = \begin{bmatrix} \partial a_3/\partial x_2 - \partial a_2/\partial x_3 \\ \partial a_1/\partial x_3 - \partial a_3/\partial x_1 \\ \partial a_2/\partial x_1 - \partial a_1/\partial x_2 \end{bmatrix} \qquad (A.34)$$

is also called *rotational* of $\mathbf{a}$, the corresponding symbol being $\nabla \wedge \mathbf{a} = \mathrm{rot}\,\mathbf{a}$. Combining (A.34) with the three-dimensional case of (A.12) and (A.13) shows that the following identities hold:

$$\mathrm{rot}\,(f\,\mathbf{a}) = f\,\mathrm{rot}\,\mathbf{a} + \mathrm{grad}\,f \wedge \mathbf{a}, \qquad \mathrm{rot}\,\mathrm{grad}\,f = 0, \qquad \mathrm{div}\,\mathrm{rot}\,\mathbf{a} = 0, \quad (A.35)$$

$$\mathrm{rot}\,\mathrm{rot}\,\mathbf{a} = \mathrm{grad}\,\mathrm{div}\,\mathbf{a} - \nabla^2 \mathbf{a}, \qquad \mathrm{div}\,(\mathbf{a} \wedge \mathbf{b}) = \mathbf{b} \cdot \mathrm{rot}\,\mathbf{a} - \mathbf{a} \cdot \mathrm{rot}\,\mathbf{b}. \qquad (A.36)$$

Integrating the second equation in (A.36) over a three-dimensional volume V and using (A.23) yields the identity

$$\int_S \mathbf{n} \cdot \mathbf{a} \wedge \mathbf{b}\, dS = \int_V (\mathbf{b} \cdot \mathrm{rot}\,\mathbf{a} - \mathbf{a} \cdot \mathrm{rot}\,\mathbf{b})\, dV. \qquad (A.37)$$

A special case of (A.37) occurs when vector $\mathbf{a}$ is constant. In fact, noting that $\mathbf{n} \cdot \mathbf{a} \wedge \mathbf{b} = -\mathbf{n} \cdot \mathbf{b} \wedge \mathbf{a} = -\mathbf{n} \wedge \mathbf{b} \cdot \mathbf{a}$, (A.37) reduces to

$$\mathbf{a} \cdot \int_S \mathbf{n} \wedge \mathbf{b}\, dS = \mathbf{a} \cdot \int_V \mathrm{rot}\,\mathbf{b}\, dV, \qquad \mathbf{a} = \mathrm{const}. \qquad (A.38)$$

As identity (A.38) holds for any choice of $\mathbf{a}$, the two integrals in it are equal to each other.

Fig. A.1 Rotational theorem
(Sect. A.9): orientation of the
unit vectors

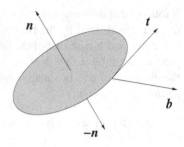

A.9 Rotational Theorem

The *rotational theorem* (or *Stokes theorem*) states that

$$\int_S \mathbf{n} \cdot \text{rot}\, \mathbf{v}\, dS = \int_C \mathbf{t} \cdot \mathbf{v}\, dC, \tag{A.39}$$

where C is the boundary curve of the open surface S, $\mathbf{t}$ the unit vector tangent to C, and $\mathbf{n}$ the unit vector normal to the surface element dS. The direction of the unit vectors is such that the orientation of $\mathbf{b} = \mathbf{t} \wedge \mathbf{n}$ is external with respect to the curve (Fig. A.1).

A.10 Helmholtz Theorem

A vector $\mathbf{u}$ such that $\text{rot}\, \mathbf{u} = 0$ is called *irrotational*. From the second identity in (A.35) one finds that, if $\mathbf{u} = \text{grad}\, f$, then $\mathbf{u}$ is irrotational. The inverse is not true in general; however, if the domain of $\mathbf{u}$ is simply connected, the condition $\text{rot}\, \mathbf{u} = 0$ implies that $\mathbf{u}$ can be expressed as a gradient: $\mathbf{u} = \text{grad}\, f$.

A vector $\mathbf{v}$ such that $\text{div}\, \mathbf{v} = 0$ is called *solenoidal*. From the third identity in (A.35) one finds that, if $\mathbf{v} = \text{rot}\, \mathbf{a}$, then $\mathbf{v}$ is solenoidal. The inverse is not true in general; however, if the domain of $\mathbf{v}$ is simply connected, the condition $\text{div}\, \mathbf{v} = 0$ implies that $\mathbf{v}$ can be expressed as a rotational: $\mathbf{v} = \text{rot}\, \mathbf{a}$.

The *Helmholtz theorem* states that a vector $\mathbf{w}$ defined in a simply-connected domain can be expressed in a unique manner as the sum of an irrotational and a solenoidal vector:

$$\mathbf{w} = \text{grad}\, f + \text{rot}\, \mathbf{a}. \tag{A.40}$$

Scalar f is found by taking the divergence of both sides of (A.40) and using the identities $\text{div}\,\text{grad}\, f = \nabla^2 f$, $\text{div}\,\text{rot}\, \mathbf{a} = 0$. In turn, vector $\mathbf{a}$ is found by taking the rotational of both sides of (A.40) and using the first identity in (A.36) along with the auxiliary condition $\text{div}\, \mathbf{a} = 0$. By this procedure it is found that f and $\mathbf{a}$ fulfill the relations

$$\nabla^2 f = \text{div}\, \mathbf{w}, \qquad \nabla^2 \mathbf{a} = -\text{rot}\, \mathbf{w}. \tag{A.41}$$

The right hand sides of (A.41) are known because $\mathbf{w}$ is prescribed. As a consequence, the problem of finding f and $\mathbf{a}$ is equivalent to solving a set of Poisson equations. The solution of (A.41) is unique provided that $\mathbf{w}$ vanishes at infinity faster than r^{-1} [65, Sect. XI.3]. Unless some additional prescriptions are imposed on f and $\mathbf{a}$, (A.40) still holds if one adds to f an arbitrary constant and, to $\mathbf{a}$, the gradient of an arbitrary scalar function.

A.11 Doubly-Stochastic Matrices

Consider a set of M square matrices of order M, $\mathbf{S}_1, \ldots, \mathbf{S}_M$, and a set of M real, non-negative numbers θ_k such that $\theta_1 + \ldots + \theta_M = 1$. The matrix

$$\mathbf{S} = \sum_{k=1}^{M} \theta_k \mathbf{S}_k \qquad (A.42)$$

is called *convex combination* of the $\mathbf{S}_k$ matrices.

The following theorem is easily proved: if the matrices $\mathbf{S}_k$ are doubly stochastic,[2] then $\mathbf{S}$ is doubly stochastic as well. In fact from the definition of $\mathbf{S}$ it is $(\mathbf{S})_{ij} = \sum_{k=1}^{M} \theta_k (\mathbf{S}_k)_{ij}$ whence, adding the terms row-wise,

$$\sum_{j=1}^{M} (\mathbf{S})_{ij} = \sum_{k=1}^{M} \theta_k \sum_{j=1}^{M} (\mathbf{S}_k)_{ij} = \sum_{k=1}^{M} \theta_k = 1. \qquad (A.43)$$

The same result is obtained when summing column-wise. As permutation matrices are doubly stochastic, from the above theorem the special case follows: a convex combination of permutation matrices is a doubly-stochastic matrix. The inverse property also holds: a doubly-stochastic matrix is a convex combination of permutation matrices [5].

A.12 Wronskian Determinant

The Wronskian determinant provides the condition of linear independence of functions [51, Sect. 5.2]. Although its properties hold for any number of functions, they will be discussed here for the case of two functions only, say, u and v defined on some interval of the independent variable x. It is convenient to seek for the condition of linear dependence first. If u, v are linearly dependent, then two non-vanishing constants c_1, c_2 exist such that

$$c_1 u + c_2 v = 0 \qquad (A.44)$$

---

[2] The definition of doubly-stochastic matrix is given in Sect. 7.6.1.

for all x in the interval. If (A.44) holds, it is easily found that both c_1 and c_2 must differ from zero. Also, as the function at the left hand side of (A.44) vanishes identically, its derivative vanishes as well. Such a derivative exists because u and v are supposed to be solutions of a second-order differential equation. Then,

$$c_1 u' + c_2 v' = 0 \qquad (A.45)$$

for all x in the interval. As (A.44, A.45) hold together, for all x the two constants c_1, c_2 are the non-trivial solution of a homogeneous algebraic system. Now, if the non-trivial solution of the algebraic system exists for all x, the determinant $W = u v' - u' v$ must vanish identically. That is, the condition $W = 0$ (identically) is necessary for the linear dependence of u, v. As a consequence, the condition $W \neq 0$ (identically) is sufficient for the linear independence of u, v.

Appendix B
Coordinates

B.1 Spherical Coordinates

When the problem in hand has a spherical symmetry it is convenient to describe the position of a particle by means of the *spherical coordinates*.

With reference to Fig. B.1, the transformation relations between the Cartesian (x, y, z) and spherical (r, ϑ, φ) coordinates are

$$
\begin{cases}
x &= r \sin \vartheta \cos \varphi \\
y &= r \sin \vartheta \sin \varphi \\
z &= r \cos \vartheta
\end{cases}
\qquad
\begin{cases}
r^2 &= x^2 + y^2 + z^2 \\
\cos \vartheta &= z/r \\
\tan \varphi &= y/x
\end{cases}
\tag{B.1}
$$

that are a special case of (1.26). The limits of the spherical coordinates are $0 \le r < \infty$, $0 \le \vartheta \le \pi$, $0 \le \varphi < 2\pi$. The 3×3 matrix of the partial derivatives of the Cartesian coordinates with respect to the spherical ones, expressed in terms of the latter (*Jacobian matrix*), is

$$
\frac{\partial(x, y, z)}{\partial(r, \vartheta, \varphi)} =
\begin{bmatrix}
\sin \vartheta \cos \varphi & r \cos \vartheta \cos \varphi & -r \sin \vartheta \sin \varphi \\
\sin \vartheta \sin \varphi & r \cos \vartheta \sin \varphi & r \sin \vartheta \cos \varphi \\
\cos \vartheta & -r \sin \vartheta & 0
\end{bmatrix},
\tag{B.2}
$$

where the left hand side is a short-hand notation for a matrix whose elements are $J_{11} = \partial x/\partial r$, $J_{12} = \partial x/\partial \vartheta$, and so on. The *Jacobian determinant* is

$$
J = \det \frac{\partial(x, y, z)}{\partial(r, \vartheta, \varphi)} = r^2 \sin \vartheta.
\tag{B.3}
$$

The matrix of the partial derivatives of the spherical coordinates with respect to the Cartesian ones, expressed in terms of the former, is

$$
\frac{\partial(r, \vartheta, \varphi)}{\partial(x, y, z)} =
\begin{bmatrix}
\sin \vartheta \cos \varphi & \sin \vartheta \sin \varphi & \cos \vartheta \\
(1/r) \cos \vartheta \cos \varphi & (1/r) \cos \vartheta \sin \varphi & -(1/r) \sin \vartheta \\
-(1/r) \sin \varphi / \sin \vartheta & (1/r) \cos \varphi / \sin \vartheta & 0
\end{bmatrix},
$$

$$\tag{B.4}$$

© Springer Science+Business Media New York 2015
M. Rudan, *Physics of Semiconductor Devices*,
DOI 10.1007/978-1-4939-1151-6

Fig. B.1 Cartesian (x, y, z)
and spherical (r, ϑ, φ)
coordinates

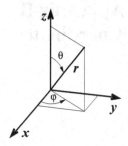

whence

$$\det \frac{\partial(r, \vartheta, \varphi)}{\partial(x, y, z)} = \frac{1}{r^2 \sin \vartheta} = \frac{1}{J}. \tag{B.5}$$

To calculate (B.4) consider, e.g., the last term of the second row, $(\partial \vartheta / \partial z)_{xy} = -(1/r) \sin \vartheta$. The second line of the second group of (B.1) yields $(\partial \cos \vartheta / \partial z)_{xy} = 1/r - z^2/r^3$, where $(\partial r / \partial z)_{xy} = z/r$ has been used, that in turn derives from the first line of the second group of (B.1). The relation $z = r \cos \vartheta$ then yields $(\partial \cos \vartheta / \partial z)_{xy} = (1/r) \sin^2 \vartheta$. On the other hand the same quantity can also be written as $(\partial \cos \vartheta / \partial z)_{xy} = - \sin \vartheta \, (\partial \vartheta / \partial z)_{xy}$. Comparing the two expressions above yields the result sought.

Differentiating with respect to time the first of (B.1) yields the relations

$$\begin{cases} \dot{x} &= \dot{r} \sin \vartheta \, \cos \varphi + r \dot{\vartheta} \, \cos \vartheta \, \cos \varphi - r \dot{\varphi} \sin \vartheta \, \sin \varphi \\ \dot{y} &= \dot{r} \sin \vartheta \, \sin \varphi + r \dot{\vartheta} \, \cos \vartheta \, \sin \varphi + r \dot{\varphi} \sin \vartheta \, \cos \varphi \\ \dot{z} &= \dot{r} \cos \vartheta - r \dot{\vartheta} \sin \vartheta \end{cases} \tag{B.6}$$

that express the components of the velocity in the Cartesian reference as functions of the generalized coordinates r, ϑ, φ and generalized velocities $\dot{r}, \dot{\vartheta}, \dot{\varphi}$ of the spherical reference. From (B.6) the expression of the kinetic energy in spherical coordinates follows:

$$T = \frac{1}{2} m \, (\dot{x}^2 + \dot{y}^2 + \dot{z}^2) = \frac{1}{2} m \, (\dot{r}^2 + r^2 \dot{\vartheta}^2 + r^2 \dot{\varphi}^2 \sin^2 \vartheta). \tag{B.7}$$

B.2 Polar Coordinates

To describe the motion of particles confined over a plane one may adopt, instead of the Cartesian coordinates x, y, the *polar coordinates* r, φ. The relations between the two groups of coordinates are

$$\begin{cases} x &= r\cos\varphi \\ y &= r\sin\varphi \end{cases} \qquad \begin{cases} r^2 &= x^2+y^2 \\ \tan\varphi &= y/x \end{cases} \tag{B.8}$$

The limits of the polar coordinates are $0 \le r < \infty$, $0 \le \varphi < 2\pi$. The Jacobian matrix and the Jacobian determinant are, respectively,

$$\frac{\partial(x,y)}{\partial(r,\varphi)} = \begin{bmatrix} \cos\varphi & -r\sin\varphi \\ \sin\varphi & r\cos\varphi \end{bmatrix}, \qquad J = \det\frac{\partial(x,y)}{\partial(r,\varphi)} = r. \tag{B.9}$$

Differentiating with respect to time the first of (B.8) yields the relations

$$\begin{cases} \dot{x} &= \dot{r}\cos\varphi - r\dot{\varphi}\sin\varphi \\ \dot{y} &= \dot{r}\sin\varphi + r\dot{\varphi}\cos\varphi \end{cases} \tag{B.10}$$

that express the components of the velocity in the Cartesian reference as functions of the generalized coordinates r, φ and generalized velocities $\dot{r}$, $\dot{\varphi}$ of the polar reference. From (B.10) the expression of the kinetic energy in polar coordinates follows:

$$T = \frac{1}{2}m\left(\dot{x}^2 + \dot{y}^2\right) = \frac{1}{2}m\left(\dot{r}^2 + r^2\dot{\varphi}^2\right). \tag{B.11}$$

B.3 Coordinate Rotation

Consider a coordinate transformation that consists in a rotation around the origin, bringing a right-handed system of coordinates $\mathbf{x} = (x_1, x_2, x_3)$ into another right-handed system $\mathbf{s} = (s_1, s_2, s_3)$. The transformation is described by the linear relations

$$\begin{cases} s_1 &= a_{11}x_1 + a_{12}x_2 + a_{13}x_3 \\ s_2 &= a_{21}x_1 + a_{22}x_2 + a_{23}x_3 \\ s_3 &= a_{31}x_1 + a_{32}x_2 + a_{33}x_3 \end{cases} \tag{B.12}$$

which can be recast in the matrix form $\mathbf{s} = \mathbf{A}\mathbf{x}$. It is known that a matrix describing this type of transformation is *orthogonal* [42, Sect. 4.2], namely,

$$\sum_{i=1}^{3} a_{ij}\,a_{ik} = \delta_{jk}, \qquad \det\mathbf{A} = 1, \qquad \mathbf{A}^{-1} = \mathbf{A}^T, \qquad j,k = 1,2,3, \tag{B.13}$$

where apex T indicates the transpose. From (B.13) it follows $\left(\mathbf{A}^{-1}\right)^T = \mathbf{A}$. As a consequence, the effect of the rotation onto the modulus of a particle's velocity is found from

$$\left(\dot{\mathbf{x}}^T, \dot{\mathbf{x}}\right) = \left[\left(\mathbf{A}^{-1}\dot{\mathbf{s}}\right)^T, \mathbf{A}^{-1}\dot{\mathbf{s}}\right] = \left(\dot{\mathbf{s}}^T\mathbf{A}, \mathbf{A}^{-1}\dot{\mathbf{s}}\right) = \left(\dot{\mathbf{s}}^T, \mathbf{A}\mathbf{A}^{-1}\dot{\mathbf{s}}\right) = \left(\dot{\mathbf{s}}^T, \dot{\mathbf{s}}\right). \tag{B.14}$$

In (B.14) the symbol $(\mathbf{a}^T, \mathbf{b})$ denotes the scalar product between the vectors $\mathbf{a}$ and $\mathbf{b}$, namely, it is equivalent to $\mathbf{a} \cdot \mathbf{b}$. The above calculation shows that $u^2 = (\dot{\mathbf{x}}^T, \dot{\mathbf{x}})$ is invariant under rotation of the coordinate system. The same reasoning applies to the modulus of position $r^2 = (\mathbf{x}^T, \mathbf{x}) = (\mathbf{s}^T, \mathbf{s})$.

B.4   Differential Operators Under Coordinate Transformations

Consider the coordinate transformation between the two sets $x_i, \xi_i, i = 1, 2, \ldots, n$:

$$\xi_i = \xi_i(x_1, \ldots, x_n), \qquad x_i = x_i(\xi_1, \ldots, \xi_n). \tag{B.15}$$

If a function f is transformed using the above, the following hold:

$$\frac{\partial f}{\partial x_i} = \sum_{j=1}^{n} \frac{\partial f}{\partial \xi_j} \frac{\partial \xi_j}{\partial x_i}, \quad \frac{\partial^2 f}{\partial x_i^2} = \sum_{j=1}^{n} \left(\frac{\partial f}{\partial \xi_j} \frac{\partial^2 \xi_j}{\partial x_i^2} + \sum_{k=1}^{n} \frac{\partial^2 f}{\partial \xi_j \partial \xi_k} \frac{\partial \xi_j}{\partial x_i} \frac{\partial \xi_k}{\partial x_i} \right). \tag{B.16}$$

Adding up over i in the second of (B.16) yields

$$\nabla^2 f = \sum_{j=1}^{n} \left(\frac{\partial f}{\partial \xi_j} \nabla^2 \xi_j + \sum_{k=1}^{n} \frac{\partial^2 f}{\partial \xi_j \partial \xi_k} \nabla \xi_j \cdot \nabla \xi_k \right), \tag{B.17}$$

where symbols ∇ and ∇^2 indicate, respectively, the gradient and the Laplacian operator with respect to the coordinates x_i. By way of example consider the transformation (B.1) from Cartesian to spherical coordinates. Remembering (B.4) one finds

$$\nabla r \cdot \nabla r = 1, \qquad \nabla \vartheta \cdot \nabla \vartheta = \frac{1}{r^2}, \qquad \nabla \varphi \cdot \nabla \varphi = \frac{1}{r^2 \sin^2 \vartheta}, \tag{B.18}$$

$$\nabla r \cdot \nabla \vartheta = 0, \qquad \nabla \vartheta \cdot \nabla \varphi = 0, \qquad \nabla \varphi \cdot \nabla r = 0, \tag{B.19}$$

whence

$$\nabla^2 f = \frac{\partial f}{\partial r} \nabla^2 r + \frac{\partial^2 f}{\partial r^2} + \frac{\partial f}{\partial \vartheta} \nabla^2 \vartheta + \frac{\partial^2 f}{\partial \vartheta^2} \frac{1}{r^2} + \frac{\partial f}{\partial \varphi} \nabla^2 \varphi + \frac{\partial^2 f}{\partial \varphi^2} \frac{1}{r^2 \sin^2 \vartheta}. \tag{B.20}$$

In turn, letting $\gamma \doteq \sin \vartheta \cos \vartheta / r$, $\zeta \doteq \sin \varphi \cos \varphi / r$, $\sigma \doteq \sin^2 \vartheta$, the terms $\nabla^2 r$, $\nabla^2 \vartheta$, $\nabla^2 \varphi$ are found from

$$\begin{bmatrix} \partial^2 r / \partial x^2 & \partial^2 r / \partial y^2 & \partial^2 r / \partial z^2 \\ \partial^2 \vartheta / \partial x^2 & \partial^2 \vartheta / \partial y^2 & \partial^2 \vartheta / \partial z^2 \\ \partial^2 \varphi / \partial x^2 & \partial^2 \varphi / \partial y^2 & \partial^2 \varphi / \partial z^2 \end{bmatrix} = \frac{1}{r} \times \tag{B.21}$$

$$\times \begin{bmatrix} 1 - \sigma \cos^2 \varphi & 1 - \sigma \sin^2 \varphi & \sigma \\ \gamma \left(\sin^2 \varphi / \sigma - 2 \cos^2 \varphi \right) & \gamma \left(\cos^2 \varphi / \sigma - 2 \sin^2 \varphi \right) & 2\gamma \\ 2\zeta / \sigma & -2\zeta / \sigma & 0 \end{bmatrix}, \qquad \text{(B.22)}$$

whence

$$\nabla^2 r = \frac{2}{r}, \qquad \nabla^2 \vartheta = \frac{1}{r^2} \frac{\cos \theta}{\sin \theta}, \qquad \nabla^2 \varphi = 0, \qquad \text{(B.23)}$$

$$\nabla^2 f = \frac{\partial f}{\partial r} \frac{2}{r} + \frac{\partial^2 f}{\partial r^2} + \frac{\partial f}{\partial \vartheta} \frac{1}{r^2} \frac{\cos \theta}{\sin \theta} + \frac{\partial^2 f}{\partial \vartheta^2} \frac{1}{r^2} + \frac{\partial^2 f}{\partial \varphi^2} \frac{1}{r^2 \sin^2 \vartheta} = \qquad \text{(B.24)}$$

$$= \frac{1}{r} \frac{\partial^2}{\partial r^2} (rf) + \frac{1}{r^2 \sin \vartheta} \frac{\partial}{\partial \vartheta} \left(\sin \vartheta \frac{\partial f}{\partial \vartheta} \right) + \frac{1}{r^2 \sin^2 \vartheta} \frac{\partial^2 f}{\partial \varphi^2}. \qquad \text{(B.25)}$$

B.5   Density of States

Consider a function s that depends on the coordinates u, v, w, and on one or more additional parameters that will collectively be indicated with σ. Let

$$S(\sigma) = \iiint s(u, v, w, \sigma) \, du \, dv \, dw, \qquad \text{(B.26)}$$

where the integration is carried out over the whole domain of u, v, w. Next, consider the transformation from the original variables to the new variables α, β, η,

$$\alpha = \alpha(u, v, w), \quad \beta = \beta(u, v, w), \quad \eta = \eta(u, v, w), \quad J = \frac{\partial(u, v, w)}{\partial(\alpha, \beta, \eta)}, \qquad \text{(B.27)}$$

which is assumed to be invertible, so that the Jacobian determinant J does not vanish. After the transformation is carried out, (B.26) takes the form

$$S(\sigma) = \iiint s(\alpha, \beta, \eta, \sigma) |J| \, d\alpha \, d\beta \, d\eta. \qquad \text{(B.28)}$$

It may happen that one is interested in the dependence of s on one of the new variables, say, η, rather than in the details about its dependence on the whole set of new variables. In this case one first carries out the integrals with respect to α and β in (B.28), to find

$$h(\eta, \sigma) = \iint s(\alpha, \beta, \eta, \sigma) |J| \, d\alpha \, d\beta. \qquad \text{(B.29)}$$

Then one defines

$$b(\eta) = \iint |J| \, d\alpha \, d\beta. \qquad \text{(B.30)}$$

A function like $b(\eta)$ plays an important role in many physical problems (e.g., Sects. 14.6, 15.8.1, 15.8.2). For this reason, although its derivation in this section is of a purely formal nature, $b(\eta)$ will be called *density of states in* η. Note that the density of states depends only on the structure of the variable transformation (B.27) and (at most) on η. The form of (B.29) and (B.30) shows that the ratio $\bar{s} = h/b$ is a weighed average of $s(\alpha, \beta, \eta, \sigma)$ over the two variables α and β, that uses $|J|$ as weight. Introducing the definition of $\bar{s}$ into (B.28) gives the latter the form

$$S(\sigma) = \int b(\eta)\,\bar{s}(\eta, \sigma)\,d\eta. \tag{B.31}$$

If s happens to be independent of α and β, definition (B.29) yields $h = s\,b$, whence $\bar{s}(\eta, \sigma) = s(\eta, \sigma)$. The derivation of b is not limited to a three-dimensional case; in fact it applies to any number of dimensions. In the following, a few examples in one, two, and three dimensions are given, in which one of the transformation relations (B.27), namely, $\eta = \eta(u, v, w)$, has a quadratic form; these examples are in fact particularly significant for the physical applications, where η stands for the energy and u, v, w stand for the generalized coordinates.

Considering a one-dimensional case with $\eta = u^2$, one finds[3]

$$u = \pm\eta^{1/2}, \qquad |J| = \left|\frac{\partial u}{\partial \eta}\right| = \frac{1}{2}\eta^{-1/2}, \qquad b(\eta) = 2\frac{1}{2}\eta^{-1/2} = \eta^{-1/2}. \tag{B.32}$$

This case is straightforward because there are no other variables involved in the transformation. Instead, in the two-dimensional case with $\eta = u^2 + v^2$, a convenient transformation involving the second variable is of the polar type (B.8), specifically, $u = \eta^{1/2}\cos\varphi, v = \eta^{1/2}\sin\varphi$. One finds

$$|J| = \frac{1}{2}, \qquad b(\eta) = \int_0^{2\pi} \frac{1}{2}\,d\varphi = \pi. \tag{B.33}$$

In the three-dimensional case with $\eta = u^2 + v^2 + w^2$, a convenient transformation involving the other two variables is of the spherical type (B.1), specifically, $u = \eta^{1/2}\sin\vartheta\,\cos\varphi, v = \eta^{1/2}\sin\vartheta\,\sin\varphi, w = \eta^{1/2}\cos\vartheta$. One finds

$$|J| = \frac{1}{2}\eta^{1/2}\sin\vartheta, \qquad b(\eta) = \int_0^{2\pi}\int_0^{\pi} \frac{1}{2}\eta^{1/2}\sin\vartheta\,d\vartheta\,d\varphi = 2\pi\,\eta^{1/2}. \tag{B.34}$$

The above examples show that, despite the fact that the $\eta = \eta(u, v, w)$ relation is quadratic in all cases, the form of $b(\eta)$ changes depending on the number of spatial dimensions.

Still considering the case where one of the transformation relations (B.27) has a quadratic form, the analysis can be extended to arbitrary values of the number

---

[3] Factor 2 in the last expression of (B.32) accounts for the fact that both positive and negative parts of the segment $[-\eta^{1/2}, +\eta^{1/2}]$ must be considered.

of spatial dimensions. As a starting point, and considering provisionally the three-dimensional case, one notes from (B.30) that the following equality holds:[4]

$$B = \int b(\eta)\,d\eta = \iiint |J|\,d\alpha\,d\beta\,d\eta = \iiint du\,dv\,dw. \qquad (B.35)$$

Remembering the definition of b, it follows that B is the number of states in the domain of u, v, w. Due to the last integral in (B.34), B is also equal to the volume of such a domain; in turn, due to the first integral, B can be thought of as the sum of the volumes of elementary shells of thickness $d\eta$, with $b(\eta)$ the area of each shell (that is, the area of the two-dimensional surface $\eta = $ const). These observations provide the key to extending the analysis to the case where η is a quadratic form in an arbitrary number of dimensions,

$$u_1^2 + u_2^2 + \ldots + u_n^2 = \eta, \qquad \eta = g^2. \qquad (B.36)$$

Letting $\eta = $ const, (B.36) is the equation of an $(n-1)$-dimensional sphere of radius $g \geq 0$ immersed into the n-dimensional space. The problem is thus reduced to expressing the area of the sphere in terms of η; although it can be solved by using a generalization of the spherical coordinates to n dimensions, a more elegant approach consists in finding a recursive expression involving also the sphere's volume.

To this purpose, let V_n indicate the volume of a sphere of an n-dimensional space, and let S_{n-1} indicate the surface area of the same sphere. When $n = 1$, the sphere is a segment whose volume is the length $V_1 = 2g$; for $n = 2$, the sphere is a circle whose volume is the area $V_2 = \pi g^2$; for $n = 3$ it is $V_3 = (4/3)\pi g^3$; for $n = 4$ it is $V_4 = \pi^2 g^4/2$, and so on; in turn, for $n = 2$ the surface is a circumference whose area is the length $S_1 = 2\pi g$; for $n = 3$ it is $S_2 = 4\pi g^2$; for $n = 4$ it is $S_3 = 2\pi^2 g^3$, and so on. Consistently with the expression of B as the integral of b given by (B.35), one finds from the above values the general relation

$$V_n = \frac{g}{n} S_{n-1}. \qquad (B.37)$$

Combining (B.37) with $V_1 = 2g$ also yields $S_0 = 2$, that is, the "surface" of the segment considered above; such a surface is made of the segment's endpoints $\{-1, +1\}$. From (B.37) it also follows that $V_n \propto g^n$ and $S_{n-1} \propto g^{n-1}$, whence $S_n \propto g\, V_{n-1}$ and $V_0 = $ const. From the values found above one finds $S_2/(g\, V_1) = S_3/(g\, V_2) = 2\pi$; it follows that $S_n = 2\pi g\, V_{n-1}$ and $V_0 = 1$. The latter is the "volume" of a sphere in a zero-dimensional space. The recursive relation involving the volumes then reads

$$V_n = \frac{g}{n} S_{n-1} = \frac{g}{n} 2\pi g\, V_{n-2} = \frac{2\pi g^2}{n} V_{n-2}, \qquad V_0 = 1, \qquad V_1 = 2g. \quad (B.38)$$

---

[4] In the practical applications of the concepts illustrated here, the integrands in (B.35) embed a constant factor Q_0, called *density of states in the u, v, w space* which, besides describing some properties of the physical problem under investigation, makes B dimensionless. Here, all variables involved are dimensionless, and Q_0 is set equal to unity.

The above can further by improved by observing that the sequence $V_0, V_1, \ldots$ embeds Euler's Gamma function of half-integer order; in fact, combining (B.37) and (B.38) with the definitions of Sect. C.10, yields

$$V_n = \frac{\pi^{n/2}}{\Gamma(n/2+1)} g^n, \qquad S_{n-1} = \frac{n\,\pi^{n/2}}{\Gamma(n/2+1)} g^{n-1}. \qquad \text{(B.39)}$$

The last step consists in expressing the result in terms of η. This is accomplished by noting that $b(\eta)\,d\eta = S_{n-1}(g)\,dg$, where $g = \sqrt{\eta}$ and $dg = d\sqrt{\eta} = d\eta/(2\sqrt{\eta})$; then, one finds

$$b(\eta)\,d\eta = \frac{n\,\pi^{n/2}\eta^{(n-1)/2}}{\Gamma(n/2+1)} \frac{d\eta}{2\eta^{1/2}}, \qquad b(\eta) = \frac{n\,\pi^{n/2}}{2\Gamma(n/2+1)} \eta^{n/2-1}. \qquad \text{(B.40)}$$

Letting $n = 1, 2, 3$ in the second expression of (B.40) renders (B.32), (B.33), (B.34), respectively.

Appendix C
Special Integrals

C.1 Sine Integral

Define the two functions

$$\text{si}(t) = -\frac{\pi}{2} + \int_0^t \frac{\sin x}{x}\,dx, \qquad N(a) = \int_0^\infty \frac{\sin(ax)}{x}\,dx. \qquad (C.1)$$

The first of them is called *sine integral* and fulfills the limit $\lim_{t \to \infty} \text{si} = 0$, whence $N(1) = \pi/2$. To demonstrate the above one starts from the functions

$$F(y) = \int_0^\infty \exp(-x)\,\frac{\sin(xy)}{x}\,dx, \quad G(y) = \int_0^\infty \exp(-xy)\,\frac{\sin x}{x}\,dx, \quad y \geq 0. \qquad (C.2)$$

The following hold true: $F(0) = 0$, $G(0) = N(1)$, $F(1) = G(1)$, and

$$\frac{dF}{dy} = \int_0^\infty \exp(-x)\,\cos(xy)\,dx, \qquad \frac{dG}{dy} = \int_0^\infty \exp(-xy)\,\sin x\,dx. \qquad (C.3)$$

Integrating (C.3) by parts twice yields $dF/dy = 1/(1+y^2)$, $dG/dy = -1/(1+y^2)$ whence

$$F(y) = \arctan y + F(0), \quad G(y) = -\arctan y + G(0), \quad 0 \leq y < \frac{\pi}{2}. \qquad (C.4)$$

It follows $F(1) = F(0)+\pi/4 = \pi/4$ and $F(1) = G(1) = G(0)-\pi/4 = N(1)-\pi/4$. Combining the above yields the result sought. This implicitly proves the convergence of the integrals in (C.1). The calculation of the second of (C.1) is now straightforward and yields

$$N(a) = \begin{cases} -\pi/2, & a < 0, \\ 0, & a = 0, \\ +\pi/2, & a > 0. \end{cases} \qquad (C.5)$$

© Springer Science+Business Media New York 2015
M. Rudan, *Physics of Semiconductor Devices*,
DOI 10.1007/978-1-4939-1151-6

The integrand in the second of (C.1) is even with respect to x. It follows that an integration carried out from $-\infty$ to $+\infty$ yields $2\,N(a)$. Basing on this one also finds

$$\int_{-\infty}^{+\infty} \frac{\exp\,(\mathrm{i}\,ax)}{\mathrm{i}\,x}\,\mathrm{d}x = 2\,N(a) + \int_{-\infty}^{+\infty} \frac{\cos\,(ax)}{\mathrm{i}\,x}\,\mathrm{d}x = 2\,N(a). \qquad (C.6)$$

When calculating the second integral in (C.6) one must let $z = \pm ax$, $\epsilon, Z > 0$ and use the principal part. In fact, observing that the integrand is odd one obtains

$$\int_{-\infty}^{+\infty} \frac{\cos\,(ax)}{\mathrm{i}\,x}\,\mathrm{d}x = \pm \mathrm{i}\,\lim_{\substack{\epsilon \to 0 \\ Z \to \infty}} \left(\int_{-Z}^{-\epsilon} \frac{\cos z}{z}\,\mathrm{d}z + \int_{+\epsilon}^{+Z} \frac{\cos z}{z}\,\mathrm{d}z \right) = 0. \qquad (C.7)$$

Combining (C.7) with (C.6) provides an integral representation of the Fourier type for the *step function*

$$H(a) = \begin{cases} 0 & a < 0 \\ 1/2 & a = 0 \\ 1 & a > 0 \end{cases} = \frac{1}{2} + \frac{1}{2\pi} \int_{-\infty}^{+\infty} \frac{\exp\,(\mathrm{i}ax)}{\mathrm{i}x}\,\mathrm{d}x. \qquad (C.8)$$

Still from (C.6), using the identity $2\,\mathrm{i}\,\sin x = \exp\,(\mathrm{i}\,x) - \exp\,(-\mathrm{i}\,x)$, one finds

$$\int_{-\infty}^{+\infty} \frac{\sin x}{x}\,\exp\,(-\mathrm{i}\,a\,x)\,\mathrm{d}x = N(-a+1) - N(-a-1) = \begin{cases} 0 & |a| > 1 \\ \pi/2 & a = \pm 1 \\ \pi & |a| < 1 \end{cases}$$
$$(C.9)$$

From (C.9) one derives integrals of a similar form, where $\sin x/x$ is replaced with $\sin^n x/x^n$, $n = 2, 3, \ldots$. The example with $n = 2$ is given below: one starts from

$$\frac{\mathrm{d}}{\mathrm{d}a} \int_{-\infty}^{+\infty} \frac{\sin^2 x}{x^2}\,\exp\,(-\mathrm{i}\,a\,x)\,\mathrm{d}x = \int_{-\infty}^{+\infty} \frac{\sin^2 x}{\mathrm{i}\,x}\,\exp\,(-\mathrm{i}\,a\,x)\,\mathrm{d}x, \qquad (C.10)$$

and uses the identity $2\,\sin^2 x = 1 - \cos\,(2\,x)$ to find

$$\int_{-\infty}^{+\infty} \frac{1 - \cos\,(2\,x)}{2\,\mathrm{i}\,x}\,\exp\,(-\mathrm{i}\,a\,x)\,\mathrm{d}x = N(-a) + \int_{-\infty}^{+\infty} \frac{\cos\,(2\,x)}{2\,x}\,\sin\,(a\,x)\,\mathrm{d}x,$$
$$(C.11)$$

where $N(-a)$ derives from (C.6) and the integral on the right hand side is obtained by eliminating the odd part of the integrand. From the identity $\sin\,[(a+2)\,x] + \sin\,[(a-2)\,x] = 2\,\sin\,(a\,x)\,\cos\,(2\,x)$ such an integral transforms into

$$\int_{-\infty}^{+\infty} \frac{\sin\,[(a+2)\,x] + \sin\,[(a-2)\,x]}{4\,x}\,\mathrm{d}x = \frac{1}{2}\,N(a+2) + \frac{1}{2}\,N(a-2), \quad (C.12)$$

where the second definition in (C.1) has been used. Combining (C.10), (C.11), and (C.12) yields

$$\frac{d}{da} \int_{-\infty}^{+\infty} \frac{\sin^2 x}{x^2} \exp(-iax)\,dx = \begin{cases} \pi/2 & -2 < a < 0 \\ -\pi/2 & 0 < a < 2 \\ 0 & |a| > 2 \end{cases} \qquad (C.13)$$

This result shows that the derivative with respect to a of the integral sought is piecewise constant in the interval $-2 < a < +2$, and vanishes elsewhere. The integral is also continuous with respect to a and does not diverge, because $|\sin^2 x/x^2| \le |\sin x/x|$ and (C.9) converges. This reasoning allows one to fix the integration constants, to finally obtain

$$\int_{-\infty}^{+\infty} \frac{\sin^2 x}{x^2} \exp(-iax)\,dx = \begin{cases} (\pi/2)(a+2) & -2 < a < 0 \\ -(\pi/2)(a-2) & 0 < a < 2 \\ 0 & |a| > 2 \end{cases} \qquad (C.14)$$

By a procedure similar to that used to prove (C.14) one finds

$$\frac{d}{da} \int_{-\infty}^{+\infty} \frac{\sin^2 (ax)}{x^2}\,dx = 2\,N(a), \qquad \int_{-\infty}^{+\infty} \frac{\sin^2 (ax)}{x^2}\,dx = \begin{cases} \pi a, & a > 0 \\ -\pi a, & a < 0 \end{cases}$$
$$(C.15)$$

C.2   Fourier Transform

Let $f(x)$ be a function defined over the entire x axis. Its Fourier transform is defined as the integral

$$G(k) = \mathcal{F}_x f = \frac{1}{\sqrt{2\pi}} \int_{-\infty}^{+\infty} f(x) \exp(-ikx)\,dx. \qquad (C.16)$$

In turn, the Fourier antitransform is defined as

$$f(x) = \mathcal{F}_x^{-1} G = \frac{1}{\sqrt{2\pi}} \int_{-\infty}^{+\infty} G(k) \exp(ikx)\,dk. \qquad (C.17)$$

Combining (C.16) and (C.17) provides a representation of f in the form

$$f(x) = \frac{1}{2\pi} \int_{-\infty}^{+\infty} \exp(ikx) \left[\int_{-\infty}^{+\infty} f(\xi) \exp(-ik\xi)\,d\xi \right] dk. \qquad (C.18)$$

A sufficient condition for the representation (C.18) is

$$\int_{-\infty}^{+\infty} |f(x)|\,dx < \infty. \qquad (C.19)$$

If f is discontinuous of the first kind at some point x_0, the left hand side of (C.18) must be replaced with $[f(x_0^+) + f(x_0^-)]/2$. As the condition (C.19) is sufficient, but not necessary, there are functions that admit an integral representation like (C.18) without fulfilling (C.19). An important example is the unit step function shown in Sect. C.1.

If f depends also on one or more parameters, $f = f(x, u, v, \dots)$, then it is $G = G(k, u, v, \dots)$. In an n-dimensional space, defining the vectors $\mathbf{x} = (x_1, \dots, x_n)$ and $\mathbf{k} = (k_1, \dots, k_n)$, the Fourier transform reads

$$G(\mathbf{k}) = \mathcal{F}_\mathbf{x} f = \frac{1}{(2\pi)^{n/2}} \int_{-\infty}^{+\infty} \dots \int_{-\infty}^{+\infty} f(\mathbf{x}) \exp(-i\mathbf{k} \cdot \mathbf{x}) \, dx_1 \dots dx_n. \quad \text{(C.20)}$$

A useful relation is found by differentiating both sides of (C.17). To this purpose, one must assume that the conditions for exchanging the derivative with the integral are fulfilled. It is found

$$\frac{df}{dx} = \frac{1}{\sqrt{2\pi}} \int_{-\infty}^{+\infty} ik \, G(k) \exp(ikx) \, dk. \quad \text{(C.21)}$$

Iterating the procedure yields

$$\frac{d^n f}{dx^n} = \frac{1}{\sqrt{2\pi}} \int_{-\infty}^{+\infty} (ik)^n \, G(k) \exp(ikx) \, dk, \quad \text{(C.22)}$$

showing that, if $G(k)$ is the Fourier transform of $f(x)$, then the Fourier transform of $d^n f/dx^x$ is $(ik)^n \, G(k)$. Relations like (C.21) and (C.22) are useful, for instance, in the solution of linear differential equations with constant coefficients, because they turn differential relations into polynomial relations (compare with the solution of the diffusion equation carried out in Sect. 23.4).

C.3   Gauss Integral

The relation

$$I_G = \int_0^{+\infty} \exp(-x^2) \, dx = \int_{-\infty}^0 \exp(-x^2) \, dx. \quad \text{(C.23)}$$

is called *Gauss integral* or *Poisson integral*. To calculate its value one may start from the double integral

$$F(R) = \iint_{\Sigma(R)} \exp[-(x^2 + y^2)] \, dx \, dy, \quad \text{(C.24)}$$

where $\Sigma(R)$ is a circle of radius R centered on the origin. Using the polar coordinates (B.8) yields

$$F(R) = \int_0^{2\pi} d\vartheta \int_0^R \exp(-\rho^2) \rho \, d\rho = \pi \, [1 - \exp(-R^2)], \quad \text{(C.25)}$$

whence $\lim_{R\to\infty} F(R) = \pi$. On the other hand, due to (C.24) it is also

$$\lim_{R\to\infty} F(R) = \iint_{-\infty}^{+\infty} \exp\left[-(x^2 + y^2)\right] dx\,dy = \lim_{a\to\infty} \left(\int_{-a}^{+a} \exp\left(-x^2\right) dx\right)^2.$$

$$(C.26)$$

Combining (C.25, C.26) with (C.23) provides

$$\int_{-\infty}^{+\infty} \exp\left(-x^2\right) dx = \sqrt{\pi}, \qquad I_G = \frac{\sqrt{\pi}}{2}. \tag{C.27}$$

From (C.27) it follows that for any $\lambda > 0$ it is

$$I_0(\lambda) = \int_0^\infty \exp\left(-\lambda x^2\right) dx = \frac{1}{2}\sqrt{\frac{\pi}{\lambda}}. \tag{C.28}$$

Another integral generated by $\exp\left(-\lambda x^2\right)$ is

$$I_1(\lambda) = \int_0^\infty x \exp\left(-\lambda x^2\right) dx = \frac{1}{2\lambda}. \tag{C.29}$$

Thanks to (C.28) and (C.29) it is possible to calculate all integrals of the form

$$I_n(\lambda) = \int_0^\infty x^n \exp\left(-\lambda x^2\right) dx, \qquad n \geq 0. \tag{C.30}$$

In fact, using the recursive relation

$$\frac{d}{d\lambda} I_n = \int_0^\infty \frac{\partial}{\partial\lambda} x^n \exp\left(-\lambda x^2\right) dx = -\int_0^\infty x^{n+2} \exp\left(-\lambda x^2\right) dx = -I_{n+2},$$

$$(C.31)$$

in combination with (C.29) yields all the integrals whose index is odd,

$$I_{2m+1} = \frac{m!}{2} \lambda^{-(m+1)}, \qquad m = 0, 1, 2, \ldots. \tag{C.32}$$

Similarly, combining (C.31) with (C.28) yields all the integrals whose index is even,

$$I_{2m}(\lambda) = \frac{(2m-1)!!}{2^{m+1}} \lambda^{-(m+1/2)} \sqrt{\pi}, \qquad m = 0, 1, 2, \ldots, \tag{C.33}$$

where

$$(2m - 1)!! = (2m - 1)(2m - 3)\ldots 3\cdot 1, \qquad (-1)!! = 1, \tag{C.34}$$

Finally, observing that the integrand of (C.30) is even (odd) if n is even (odd), one finds

$$\int_{-\infty}^{+\infty} x^{2m} \exp\left(-\lambda x^2\right) dx = 2\, I_{2m}(\lambda), \qquad \int_{-\infty}^{+\infty} x^{2m+1} \exp\left(-\lambda x^2\right) dx = 0.$$

$$(C.35)$$

The results of this section still hold for a complex λ with $\Re\lambda > 0$.

Fig. C.1 Generation of a
Dirac δ using a barrier-like
function. The width of the
peak is equal to a

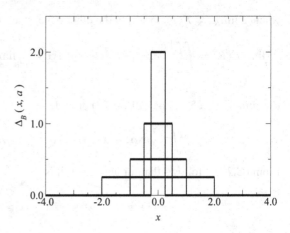

C.4   Dirac's δ

Consider a function $\Delta_B(x, a)$ defined as follows:

$$\Delta_B = \begin{cases} 1/a & -a/2 \leq x \leq +a/2 \\ 0 & x < -a/2, \quad x > a/2 \end{cases} \tag{C.36}$$

with $a > 0$.

The above definition yields

$$\lim_{a \to 0} \Delta_B = \begin{cases} 0 & x \neq 0 \\ +\infty & x = 0 \end{cases} \quad , \quad \int_{-\infty}^{+\infty} \Delta_B(x, a)\, dx = \frac{1}{a} \int_{-a/2}^{+a/2} dx = 1. \tag{C.37}$$

As the value of the integral in (C.37) is independent of a, the integral is equal to unity
also in the limit $a \to 0$. Figure C.1 shows how the form of Δ_B changes with a: the
width of the peak decreases as a decreases, while its height increases so that the area
subtending the function remains constant. Note that the procedure depicted above
gives a different result if one carries out the integration after calculating the limit.
In other terms, the integration and the limit are to be carried out in a specific order
(integration first). For a continuous function $f(x)$ the mean-value theorem provides

$$\int_{-\infty}^{+\infty} \Delta_B(x, a)\, f(x)\, dx = \frac{1}{a} \int_{-a/2}^{+a/2} f(x)\, dx = f(\bar{x}), \tag{C.38}$$

with $-a/2 < \bar{x} < +a/2$. As a consequence,

$$\lim_{a \to 0} \int_{-\infty}^{+\infty} \Delta_B(x, a)\, f(x)\, dx = f(0). \tag{C.39}$$

This result is expressed in a more compact form by defining a linear functional $\delta(x)$ (called *Dirac's symbol*) such that

$$\int_{-\infty}^{+\infty} \delta(x)\, f(x)\, dx = f(0). \tag{C.40}$$

The functional associates the number $f(0)$ to the function $f(x)$. If the reasoning leading to (C.40) is repeated after shifting Δ_B from the origin to another point x_0, one finds the generalization of (C.40)

$$\int_{-\infty}^{+\infty} \delta(x - x_0)\, f(x)\, dx = f(x_0). \tag{C.41}$$

From (C.41) and (C.16) one obtains

$$\int_{-\infty}^{+\infty} \delta(x - x_0)\, dx = 1, \qquad \mathcal{F}_x \delta(x - x_0) = \frac{1}{\sqrt{2\pi}} \exp(-i k x_0). \tag{C.42}$$

The antitransform (C.17) then reads

$$\delta(x - x_0) = \frac{1}{2\pi} \int_{-\infty}^{+\infty} \exp[i k (x - x_0)]\, dk, \tag{C.43}$$

that provides an integral representation of the Dirac δ. However, it is important to note that (C.43) has no meaning unless it is used within an integral like, e.g., (C.41). With this provision, one can consider the Dirac δ as the "derivative" of the step function; in fact, after a suitable change in the symbols, one finds that the integral at the right hand side of (C.43) is the derivative with respect to x of the integral at the right hand side of (C.8). More details about the integral representation of the Dirac δ are given in Sect. C.5.

The function $\Delta_B(x, a)$ defined above is an example of generating function of the Dirac δ. Several other examples may be given, as shown below. In all cases, if the generating function $\Delta(x, x_0, a)$ is centered at some point x_0, it is even with respect to x_0 and has the properties $\lim_{a \to 0} \Delta = 0$ if $x \neq x_0$ and $\lim_{a \to 0} \Delta = +\infty$ if $x = x_0$. Consider for instance the Lorentzian function (centered at $x_0 = 0$)

$$\Delta_L = \frac{a/\pi}{a^2 + x^2}, \qquad \int_{-\infty}^{+\infty} \Delta_L\, dx = \frac{1}{\pi} \int_{-\infty}^{+\infty} \frac{d}{dx} \arctan\left(\frac{x}{a}\right) dx = 1, \tag{C.44}$$

with $a > 0$. Apart from the limiting case $a \to 0$ the function has only one maximum that occurs at $x = 0$ and equals $1/(a\pi)$. For $x = \pm a$ the function's value halves with respect to the maximum, so $2a$ is conventionally taken as the width of Δ_L. The product $2/\pi$ of the maximum value by the conventional width is independent of a and is of order unity. Finally, for a continuous function $f(x)$ it is

$$\lim_{a \to 0} \int_{-\infty}^{+\infty} \Delta_L(x, a) f(x)\, dx = f(0). \tag{C.45}$$

Another example of a δ-generating function is the parameterized Gaussian function (centered at $x_0 = 0$)

$$\Delta_G = \frac{\exp{(-x^2/a^2)}}{a\sqrt{\pi}}, \quad a > 0, \quad \int_{-\infty}^{+\infty} \Delta_G(x,a)\,dx = 1 \tag{C.46}$$

(more details about this function and integrals related to it are given in Sects. C.3 and C.8). The function has only one maximum that occurs at $x = 0$ and equals $1/(a\sqrt{\pi})$. For $x = \pm a\sqrt{\log 2} \simeq \pm 0.833\,a$ the function's value halves with respect to the maximum, this yielding a conventional width of $2a\sqrt{\log 2}$. The product $2\sqrt{\log 2}/\sqrt{\pi}$ of the maximum value by the conventional width is independent of a and of order unity. For a continuous function $f(x)$ it is

$$\lim_{a \to 0} \int_{-\infty}^{+\infty} \Delta_G(x,a) f(x)\,dx = f(0). \tag{C.47}$$

A final example of a δ-generating function is the negative derivative of the Fermi-Dirac statistics (centered at $x_0 = 0$)

$$\Delta_F = -\frac{d}{dx}\frac{1}{\exp{(x/a)}+1} = \frac{\exp{(x/a)}}{a\left[\exp{(x/a)}+1\right]^2}, \quad a > 0, \tag{C.48}$$

$$\int_{-\infty}^{+\infty} \Delta_F(x,a)\,dx = \int_{+\infty}^{-\infty} \frac{d}{dx}\frac{1}{\exp{(x/a)}+1}\,dx = 1. \tag{C.49}$$

(more details about this function and integrals related to it are given in Sect. C.13). The function has only one maximum that occurs at $x = 0$ and equals $1/(4a)$. For $x = \pm a \log{(3+\sqrt{8})} \simeq \pm 1.76\,a$ the function's value halves with respect to the maximum, this yielding a conventional width of $2a \log{(3+\sqrt{8})}$. The product $(1/2) \log{(3+\sqrt{8})}$ of the maximum value by the conventional width is independent of a and of order unity. For a continuous function $f(x)$ it is

$$\lim_{a \to 0} \int_{-\infty}^{+\infty} \Delta_F(x,a) f(x)\,dx = f(0). \tag{C.50}$$

The δ-generating functions Δ vanish for $x \to \pm\infty$, otherwise they would not be integrable from $-\infty$ to $+\infty$. Assuming that Δ is differentiable with respect to x yields, after integrating by parts,

$$\int_{-\infty}^{+\infty} f(x)\frac{d\Delta(x,a)}{dx}\,dx = [\Delta(x,a)\,f(x)]_{-\infty}^{+\infty} - \int_{-\infty}^{+\infty} \Delta(x,a)\frac{df}{dx}\,dx, \tag{C.51}$$

with f a differentiable function. In (C.51) the integrated part is zero because Δ vanishes at infinity. Taking the limit $a \to 0$ at both sides of (C.51) and using (C.40) yields

$$\int_{-\infty}^{+\infty} f(x)\frac{d\delta(x)}{dx}\,dx = -\int_{-\infty}^{+\infty} \delta(x)\frac{df}{dx}\,dx = -f'(0), \tag{C.52}$$

Fig. C.2 Generation of a Dirac δ using a Lorentzian function. The width of the peak is proportional to a

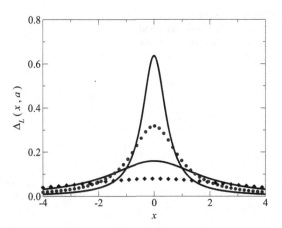

which is used as the definition of the derivative of δ. Such a definition generalizes to

$$\int_{-\infty}^{+\infty} f(x) \frac{d^n \delta(x)}{dx^n} \, dx = (-1)^n f^{(n)}(0). \tag{C.53}$$

C.5 Some Properties of Dirac's δ

An integral representation of δ is derived from (C.18) after rearranging it as

$$f(x) = \int_{-\infty}^{+\infty} \left[\int_{-\infty}^{+\infty} \frac{\exp[i k (x - \xi)]}{2\pi} \, dk \right] f(\xi) \, d\xi \tag{C.54}$$

and comparing with (C.41):

$$\delta(\xi - x) = \int_{-\infty}^{+\infty} \frac{\exp[i k (x - \xi)]}{2\pi} \, dk. \tag{C.55}$$

Replacing k with $-k$ in (C.55) shows that δ is even with respect to its argument, $\delta(x - \xi) = \delta(\xi - x)$. Also, comparing (C.55) with (C.16) shows that $\delta(\xi - x)$ is the Fourier transform of $\exp(i k x)/\sqrt{2\pi}$. The generalization of (C.55) to more than one dimension is immediate; e.g., the three-dimensional case reads

$$\delta(\mathbf{g} - \mathbf{x}) = \iiint_{-\infty}^{+\infty} \frac{\exp[i \mathbf{k} \cdot (\mathbf{x} - \mathbf{g})]}{(2\pi)^3} \, d^3k. \tag{C.56}$$

The discrete-case analogue of (C.56) is given by (C.117, C.121), where the generalization of the Kronecker symbol is given. Note that the latter is dimensionless, whereas the units of Dirac's δ depend on its argument: by way of example, the integral $\int_{-\infty}^{+\infty} \delta(\xi - x) \, d\xi = 1$ shows that the units of $\delta(\xi - x)$ are the inverse of those of

Fig. C.3 Generation of a
Dirac δ using a parameterized
Gaussian function. The width
of the peak is proportional to
a

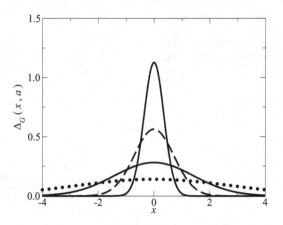

$\mathrm{d}\xi$; similarly, the integral $\iiint_{-\infty}^{+\infty} \delta(\mathbf{g} - \mathbf{x})\, \mathrm{d}^3 g = 1$ shows that the units of $\delta(\mathbf{g} - \mathbf{x})$
are the inverse of those of $\mathrm{d}^3 g$, and so on.

A generalization of Dirac's δ is found by replacing $\delta(x)$ with $\delta[q(x)]$, with $q(x)$ a
function having one or more zeros. Let x_1 be a simple zero of q, namely, $q'(x_1) \neq 0$,
and consider the contribution of it to the integral $\int_{-\infty}^{+\infty} \delta[q(x)]\, \mathrm{d}x$. Observing that in
a finite neighborhood I_1 of x_1 there are no other zeros, one can determine such a
contribution by replacing $q(x)$ with $q'(x_1)(x - x_1)$; in this way, to bring the calculation
back to the standard form one may provisionally scale the differential $\mathrm{d}x$ by $1/q'(x_1)$.
However, if the scaling factor were negative, the evenness of δ would be violated;
thus, the correct scaling factor is $|q'(x_1)|$, and

$$\int_{I_1} \delta[q(x)]\, f(x)\, \mathrm{d}x = \frac{1}{|q'(x_1)|}\, f(x_1). \tag{C.57}$$

If q has n simple zeros, from (C.57) it follows

$$\int_{-\infty}^{+\infty} \delta[q(x)]\, f(x)\, \mathrm{d}x = \frac{1}{|q'(x_1)|}\, f(x_1) + \ldots + \frac{1}{|q'(x_n)|}\, f(x_n). \tag{C.58}$$

C.6 Moments Expansion

For a given function $f(k)$ consider the integral

$$M_n = \int_{-\infty}^{+\infty} k^n f(k)\, \mathrm{d}k, \qquad n = 0, 1, \ldots \tag{C.59}$$

It is assumed that the integral converges for any n. This implies that f vanishes at
infinity with a strength larger than any power. As the present considerations apply to
a distribution function, the vanishing of f is typically of the exponential type. The

quantity M_n is called *moment of order n* of function f. Thanks to its properties, f can be Fourier transformed; let

$$g(y) = \mathcal{F}f = \frac{1}{\sqrt{2\pi}} \int_{-\infty}^{+\infty} f(k)\,\exp(-i\,y\,k)\,dk. \tag{C.60}$$

Using the Taylor expansion $\exp(-i\,y\,k) = \sum_{n=0}^{\infty} (-i\,y\,k)^n/n!$ yields

$$g(y) = \sum_{n=0}^{\infty} \frac{1}{n!} \frac{(-i)^n M_n}{\sqrt{2\pi}}\, y^n. \tag{C.61}$$

The latter is the Taylor expansion of g around the origin; it follows

$$\frac{(-i)^n M_n}{\sqrt{2\pi}} = \left(\frac{d^n g}{dy^n}\right)_0. \tag{C.62}$$

The above analysis shows that, if the moments M_n of $f(k)$ are known, from them one constructs the Fourier transform $g(y) = \mathcal{F}f$ by means of a Taylor series. Then, one recovers the original function from the inverse transform $f(k) = \mathcal{F}^{-1}g$. In conclusion, the knowledge of the set of moments of f is equivalent to the knowledge of f. The result holds true also in the multi-dimensional case $f = f(\mathbf{k})$, where

$$M_{l+m+n} = \iiint_{-\infty}^{+\infty} k_1^l\, k_2^m\, k_3^n\, f(\mathbf{k})\,d^3k, \qquad l,m,n = 0,1,\ldots \tag{C.63}$$

is the moment of order $l + m + n$ of f.

If only the lower-order moments are used, then the Taylor series for the Fourier transform is truncated and provides an approximation $\tilde{g}$ for g. As a consequence of this approximation, the inverse transform $\tilde{f} = \mathcal{F}^{-1}\tilde{g}$ provides an approximate form of the original function f.

An extension of the above concepts is obtained by replacing the monomial expression $k_1^l\, k_2^m\, k_3^n$ with a function $\alpha(\mathbf{k})$, that can be expressed by a polynomial interpolation. In this case, in fact, the integral of $\alpha(\mathbf{k})\, f(\mathbf{k})$ is a combination of moments of f. A further generalization consists in considering f, α, or both, as functions of other variables besides $\mathbf{k}$:

$$M_\alpha(\mathbf{r}, t) = \iiint_{-\infty}^{+\infty} \alpha(\mathbf{r}, \mathbf{k}, t)\, f(\mathbf{r}, \mathbf{k}, t)\,d^3k. \tag{C.64}$$

If $f(\mathbf{r}, \mathbf{k}, t)$ is the solution of a differential equation generated by an operator $\mathcal{A}$, say, $\mathcal{A}f = 0$, one can derive a set of moments from such an equation by selecting different forms of α:

$$\iiint_{-\infty}^{+\infty} \alpha\,\mathcal{A}f\,d^3k = 0. \tag{C.65}$$

Each moment depends on the other variables $\mathbf{r}, t$. If operator $\mathcal{A}$ contains the derivatives with respect to $\mathbf{r}$, t, or both, then the moment of $\mathcal{A}f = 0$ is a differential equation in $\mathbf{r}$, t, or both.

Fig. C.4 Generation of a
Dirac δ using a Fermi-Dirac
statistics. The width of the
peak is proportional to a

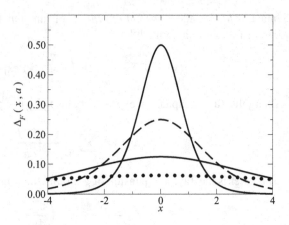

C.7 Error Function

The *error function* and the *complementary error function* are defined, respectively,
as

$$\text{erf}(x) = \frac{2}{\sqrt{\pi}} \int_0^x \exp(-\xi^2)\,d\xi, \qquad \text{erfc}(x) = 1 - \text{erf}(x). \qquad (C.66)$$

From the definitions (C.66) and from the Gauss integral (C.23) the following
properties are derived:

$$\frac{d}{dx}\text{erf}(x) = \frac{2}{\sqrt{\pi}}\exp(-x^2), \qquad \text{erf}(-x) = -\text{erf}(x), \qquad (C.67)$$

$$\text{erf}(-\infty) = -1, \qquad \text{erf}(0) = 0, \qquad \text{erf}(+\infty) = 1, \qquad (C.68)$$

$$\text{erfc}(-\infty) = 2, \qquad \text{erfc}(0) = 1, \qquad \text{erfc}(+\infty) = 0. \qquad (C.69)$$

Integrating by parts yields

$$\int_0^x \text{erfc}(\xi)\,d\xi = x\,\text{erfc}(x) + \frac{1}{\sqrt{\pi}}\left[1 - \exp(-x^2)\right]. \qquad (C.70)$$

Applying the de l'Hôpital rule shows that the first term at the right hand side of (C.70)
vanishes for $x \to +\infty$. It follows

$$\int_0^{+\infty} \text{erfc}(x)\,dx = \frac{1}{\sqrt{\pi}}. \qquad (C.71)$$

Still applying the de l'Hôpital rule shows that

$$\lim_{x \to 0}\frac{\text{erf}(x)}{x} = \frac{2}{\sqrt{\pi}}, \qquad \lim_{x \to +\infty}\frac{\text{erfc}(x)}{\exp(-x^2)} = \lim_{x \to +\infty}\frac{1/\sqrt{\pi}}{x}, \qquad (C.72)$$

whence

$$\text{erf}(x) \simeq \frac{2}{\sqrt{\pi}} x \quad \text{for} \quad |x| \ll 1, \qquad \text{erfc}(x) \simeq \frac{1}{\sqrt{\pi}} \frac{\exp(-x^2)}{x} \quad \text{for} \quad x \gg 1.$$

(C.73)

Other applications of the integration by parts yield

$$Y = \int_0^x \xi \, \text{erfc}(\xi) \, \mathrm{d}\xi = x^2 \, \text{erfc}(x) - Y - \frac{1}{\sqrt{\pi}} \int_0^x \xi \left[\frac{\mathrm{d}}{\mathrm{d}\xi} \exp(-\xi^2) \right] \mathrm{d}\xi =$$

$$= \frac{1}{2} x^2 \, \text{erfc}(x) + \frac{1}{4} \, \text{erf}(x) - \frac{1}{2\sqrt{\pi}} x \exp(-x^2).$$

(C.74)

C.8 Parametrized Gaussian Function

The relations introduced in Sects. C.3 and C.7 are useful for investigating the properties of function

$$\Delta(x - \xi, a) = \frac{\exp[-(x - \xi)^2/(4a)]}{\sqrt{4\pi a}}, \qquad a > 0.$$

(C.75)

The behavior of Δ in the limit $a \to 0$ depends on the argument $x - \xi$, namely

$$\lim_{a \to 0} \Delta(x - \xi, a) = \begin{cases} 0 & \xi \neq x \\ +\infty & \xi = x \end{cases}$$

(C.76)

In contrast, its integral over ξ is independent of x and a. In fact, using (C.23) after letting $\mu = (x - \xi)/\sqrt{4a}$ yields

$$\int_{-\infty}^{+\infty} \Delta(x - \xi, a) \, \mathrm{d}\xi = \frac{1}{\sqrt{\pi}} \int_{-\infty}^{+\infty} \exp(-\mu^2) \, \mathrm{d}\mu = 1.$$

(C.77)

Adopting the same variable change leading to (C.77) and using (C.23, C.66) yields

$$\int_{-\infty}^0 \Delta(x - \xi, a) \, \mathrm{d}\xi = \frac{1}{2} \, \text{erfc}\left(\frac{x}{\sqrt{4a}} \right).$$

(C.78)

The relations (C.77, C.78) hold also in the limit for $a \to 0$, provided the limit is calculated *after* the integration. This property is typical of the functions that generate the Dirac δ (Sect. C.4). In fact it can be shown that for a continuous function $g(x)$ the following holds:

$$\lim_{a \to 0} \int_{-\infty}^{+\infty} g(\xi) \, \Delta(x - \xi, a) \, \mathrm{d}\xi = g(x).$$

(C.79)

Other examples of δ-generating functions are given in Sect. C.4. This section is concluded by showing that $\Delta(x - \xi, a)$ admits an integral representation of the form

$$\Delta(x - \xi, a) = \frac{1}{2\pi} \int_{-\infty}^{+\infty} \exp\left[ik(x - \xi) - ak^2\right] dk. \tag{C.80}$$

To prove (C.80) one recasts the argument of the exponential by means of the identity

$$ik(x - \xi) - ak^2 = -\frac{(x - \xi)^2}{4a} - a\left[k - \frac{i(x - \xi)}{2a}\right]^2, \tag{C.81}$$

and uses (C.23) with $\sqrt{a}\,[k - i(x-\xi)/(2a)]$ as the integration variable. It is interesting to note in passing that letting $\xi = 0$, $a = \sigma^2/2$ in (C.80) yields

$$\exp\left[-x^2/(2\sigma^2)\right] = \frac{\sigma}{\sqrt{2\pi}} \int_{-\infty}^{+\infty} \exp\left(-\sigma^2 k^2/2\right) \exp\left(i k x\right) dk, \tag{C.82}$$

namely, the Gaussian function is the Fourier transform of itself.

C.9 Euler's Beta Function

The function defined by the integral

$$B(\lambda, \mu) = \int_0^1 x^{\lambda-1} (1 - x)^{\mu-1} dx, \tag{C.83}$$

with λ, μ complex numbers such that $\Re(\lambda) > 0$, $\Re(\mu) > 0$, is called *Euler's Beta function* or *Euler's integral of the first kind* [65]. Letting $x = y/(y+1)$ and replacing y with x gives (C.83) the equivalent form

$$B(\lambda, \mu) = \int_0^{+\infty} x^{\lambda-1} (1 + x)^{-(\lambda+\mu)} dx. \tag{C.84}$$

Limiting the variables' range to $0 < \Re(\lambda), \Re(\mu) < 1$ and letting

$$\mu = 1 - \lambda, \qquad T_0(\lambda) = B(\lambda, 1 - \lambda) \tag{C.85}$$

yields

$$T_0(\lambda) = \int_0^{+\infty} \frac{x^{\lambda-1}}{1 + x} dx = \frac{\pi}{\sin(\lambda\pi)}. \tag{C.86}$$

The last equality is demonstrated by applying Cauchy's residue theorem [114, Sect. 64] to the function $f(z) = z^{\lambda-1}/(1 + z)$, with z complex, that over the real axis reduces to the integrand of (C.86). The relation (C.86) can be exploited for calculating other integrals. For instance, for λ real one lets $\lambda = 1/(2\mu)$, $x = y^{2\mu}$ to find

$$\int_0^{+\infty} \frac{1}{1 + y^{2\mu}} dy = \frac{\pi/(2\mu)}{\sin[\pi/(2\mu)]}, \qquad \mu > \frac{1}{2}. \tag{C.87}$$

C.10 Euler's Gamma Function

The function defined by the integral

$$\Gamma(\lambda) = \int_0^{+\infty} x^{\lambda-1} \exp(-x) \, dx, \tag{C.88}$$

with λ a complex number such that $\Re(\lambda) > 0$, is called *Euler's Gamma function* or *Euler's integral of the second kind* [33, Sect. 1.3].[5] The negative of its derivative $\Gamma' = d\Gamma/d\lambda$ calculated for $\lambda = 1$ is called *Euler's constant*, $\gamma = -\Gamma'(1) = \int_0^{+\infty} \exp(-x) \log(x) \, dx \simeq 0.5772$. From (C.88) one finds $\Gamma(1) = 1$ and, after integrating by parts,

$$\Gamma(\lambda + 1) = \lambda \, \Gamma(\lambda). \tag{C.89}$$

If $\lambda = n = 1, 2, \ldots$ (C.89) yields

$$\Gamma(n + 1) = n \, \Gamma(n) = n(n - 1) \, \Gamma(n - 1) = \ldots = n!. \tag{C.90}$$

The definition of Γ is extended by analytic continuation to the complex plane with the exception of the points $\lambda = 0, -1, -2, \ldots, -n, \ldots$. At each negative integer $-n$, the function Γ has a simple pole with a residue equal to $(-1)^n/n!$ [65], namely,

$$\lim_{\lambda \to -n} (\lambda + n) \, \Gamma(\lambda) = \frac{(-1)^n}{n!}, \qquad n = 0, 1, 2 \ldots \tag{C.91}$$

A straightforward calculation shows that the beta and gamma functions are connected by the relation [65]

$$\Gamma(\lambda) \, \Gamma(\mu) = \Gamma(\lambda + \mu) \, B(\lambda, \mu). \tag{C.92}$$

Thanks to (C.92) one extends the definition of B to the complex plane with the exception of the points $\lambda, \mu, \lambda + \mu = 0, -1, -2, \ldots, -n, \ldots$. Moreover, limiting the variables' range to $0 < \Re(\lambda), \Re(\mu) < 1$ and letting $\mu = 1 - \lambda$ so that $\Gamma(\lambda + \mu) = \Gamma(1) = 1$, from (C.86) one finds

$$\Gamma(\lambda) \, \Gamma(1 - \lambda) = \int_0^{+\infty} \frac{x^{\lambda-1}}{1 + x} \, dx = T_0(\lambda), \qquad 0 < \Re(\lambda) < 1. \tag{C.93}$$

For $\lambda = 1/2$ (C.93) yields

$$\Gamma\left(\frac{1}{2}\right) = \sqrt{\pi} \tag{C.94}$$

---

[5] As remarked in [33], Legendre's notation $\Gamma(\lambda)$ is unfortunate because the argument that appears at the right hand side of the definition is $\lambda - 1$. Gauss used the notation $\Pi(\lambda - 1)$ for the left hand side of C.88.

whence, thanks to (C.89),

$$\Gamma\left(\frac{3}{2}\right) = \frac{1}{2}\Gamma\left(\frac{1}{2}\right) = \frac{1}{2}\sqrt{\pi}, \quad \Gamma\left(\frac{5}{2}\right) = \frac{3}{2}\Gamma\left(\frac{3}{2}\right) = \frac{3}{4}\sqrt{\pi}, \quad \dots \quad (C.95)$$

Iterating (C.95) and comparing with (C.33) shows that $\Gamma(m + 1/2) = 2\,I_{2m}(1)$, $m = 0, 1, 2, \dots$

C.11 Gamma Function's Asymptotic Behavior

Euler's Gamma function introduced in Sect. C.10, considered for real values of λ, lends itself to a significant application of the asymptotic analysis. Specifically, one seeks another function $f(\lambda)$, expressible through elementary functions, such that $\lim_{\lambda\to\infty}[\Gamma(\lambda + 1)/f(\lambda)] = 1$. The asymptotic analysis applied to the Γ function shows that [26]

$$\lim_{\lambda\to\infty} \frac{\Gamma(\lambda + 1)}{\lambda^{\lambda+1/2}\exp(-\lambda)} = \sqrt{2\pi}, \qquad (C.96)$$

namely, the function sought is $f(\lambda) = \sqrt{2\pi}\,\lambda^{\lambda+1/2}\exp(-\lambda)$. Equation (C.96) is called *Stirling's formula*. Remembering (C.90) one has $\Gamma(\lambda + 1) = \Gamma(n + 1) = n!$ when λ is a natural number. From (C.96) it follows

$$n! \simeq \sqrt{2\pi}\,n^{n+1/2}\exp(-n) = \sqrt{2\pi n}\,(n/e)^n, \qquad (C.97)$$

that provides an approximation to the factorial for $n \gg 1$. Letting by way of example $n = 10$, the rounded value of the right hand side of (C.97) turns out to be $3\,598\,696$, that differs from $10! = 3\,628\,800$ by less than 1%.

The asymptotic value of the derivative of $\log\Gamma$ is also of interest, for instance when determining the equilibrium distribution of particles in statistical mechanics (Sects. 6.4, 15.8.1, 15.8.2). Using (C.96) one finds

$$\frac{\mathrm{d}}{\mathrm{d}\lambda}\log\Gamma(\lambda + 1) \simeq \frac{1}{2\lambda} + \log\lambda \simeq \log\lambda, \qquad \lambda \gg 1. \qquad (C.98)$$

C.12 Integrals Related to the Harmonic Oscillator

Consider the integral

$$I(s) = \int_0^1 \frac{\mathrm{d}\xi}{\sqrt{1 - \xi^s}}, \qquad (C.99)$$

where s is a real parameter, $s > 0$. Letting $u = \xi^s$ one finds $1/\sqrt{1 - \xi^s} = (1 - u)^{1/2-1}$, $\mathrm{d}\xi = u^{1/s-1}\,\mathrm{d}u/s$ whence, using (C.83, C.92, C.94),

$$I(s) = \frac{1}{s}B(1/s, 1/2) = \frac{\sqrt{\pi}}{s}\frac{\Gamma(1/s)}{\Gamma(1/s + 1/2)}. \qquad (C.100)$$

By way of example $I(2) = \pi/2$, which can also be derived directly from (C.99). When $s \to \infty$ one can use (C.91) with $n = 0$. It follows

$$\lim_{s \to \infty} I(s) = 1. \tag{C.101}$$

Now consider the integral

$$J(s) = \int_0^1 \frac{d\xi}{\sqrt{1/\xi^s - 1}}, \tag{C.102}$$

still with $s > 0$. The same procedure used for calculating $I(s)$ yields

$$J(s) = \frac{1}{s} B(1/s + 1/2, 1/2) = \frac{\sqrt{\pi}}{s} \frac{\Gamma(1/s + 1/2)}{\Gamma(1/s + 1)} = \frac{\pi}{s} \frac{1}{I(s)}, \tag{C.103}$$

and $\lim_{s \to \infty} J(s) = 0$. By way of example $J(1) = \pi/2$, which can also be derived directly from (C.102). The integrals (C.100, C.103) appear in the theory of the harmonic oscillator (Sect. 3.3 and Problems 3.1, 3.2).

C.13 Fermi Integrals

The *Fermi integral* of order α is defined as

$$\Phi_\alpha(\xi) = \frac{1}{\Gamma(\alpha + 1)} \int_0^\infty \frac{x^\alpha}{1 + \exp(x - \xi)} \, dx, \qquad \alpha > -1, \tag{C.104}$$

where Γ is defined by (C.88) and α is a real parameter. The constraint $\alpha > -1$ guarantees the convergence of the integral. If $-\xi \gg 1$ one has $\exp(x - \xi) \geq \exp(-\xi) \gg 1$ and, from (C.88),

$$\Phi_\alpha(\xi) \simeq \frac{\exp(\xi)}{\Gamma(\alpha + 1)} \int_0^\infty x^\alpha \exp(-x) \, dx = \exp(\xi), \qquad \xi \gg -1. \tag{C.105}$$

A relation between Fermi integral of different order is found by considering, for some $\alpha > 0$, the integral of order $\alpha - 1$:

$$\frac{1}{\Gamma(\alpha)} \int_0^\infty \frac{x^{\alpha-1}}{1 + \exp(x - \xi)} \, dx = \frac{1}{\alpha \, \Gamma(\alpha)} \int_0^\infty \frac{x^\alpha \exp(x - \xi)}{[1 + \exp(x - \xi)]^2} \, dx, \tag{C.106}$$

where the right hand side is derived through an integration by parts. Observing that $\alpha \, \Gamma(\alpha) = \Gamma(\alpha + 1)$ and using again (C.104) shows that the right hand side of (C.106) is equal to $d\Phi_\alpha/d\xi$. Then,

$$\frac{d\Phi_\alpha}{d\xi} = \Phi_{\alpha-1}, \qquad \frac{d\log\Phi_\alpha}{d\xi} = \frac{\Phi_{\alpha-1}}{\Phi_\alpha}. \tag{C.107}$$

The Fermi integrals are positive by construction; from the first relation in (C.107) it then follows that the Fermi integrals are monotonically-increasing functions of the argument ξ. The Fermi integral of order 0 is expressed in terms of elementary functions,

$$\Phi_0 = \log\left[\exp\left(\xi\right) + 1\right]. \tag{C.108}$$

Approximations for the Fermi integrals are found, e.g., in [6, Appendix C]. In the applications to the semiconductor theory the Fermi integrals of small half-integer order (1/2, 3/2) are the most important ones (Sects. 18.2, 19.6.4). Remembering (C.94, C.95), they read

$$\Phi_{1/2}(\xi) = \int_0^\infty \frac{2\,x^{1/2}/\sqrt{\pi}}{1 + \exp\left(x - \xi\right)}\,\mathrm{d}x, \quad \Phi_{3/2}(\xi) = \int_0^\infty \frac{(4/3)\,x^{3/2}/\sqrt{\pi}}{1 + \exp\left(x - \xi\right)}\,\mathrm{d}x. \tag{C.109}$$

C.14 Hölder's Inequality

Hölder's inequality states that for any pair of real constants $b, c > 1$ such that $1/b + 1/c = 1$ it is

$$\int_\eta |F\,G|\,\mathrm{d}x \le \left(\int_\eta |F|^b\,\mathrm{d}x\right)^{1/b} \left(\int_\eta |G|^c\,\mathrm{d}x\right)^{1/c}, \tag{C.110}$$

where F, G are any complex functions defined over the real interval η and such that the integrals in (C.110) converge. The inequality is proven starting from the function $\varphi(r) = r^b - br + b - 1, r > 0, b > 1$, whose first derivative is $\varphi'(r) = b\,r^{b-1} - b$ and the second one $\varphi'' = b\,(b - 1)\,r^{b-2}$. As a consequence, for $r > 0$ the function has only one minimum, located at $r = 1$. The inequality $r^b + b \ge br + 1$ then holds, whence

$$\frac{r^{b-1}}{b} + \frac{1}{c\,r} \ge 1, \quad c = \frac{b}{b - 1} > 1. \tag{C.111}$$

Let $F_1(x)$ and $G_1(x)$ be any two complex functions defined over η and fulfilling the normalization condition

$$\int_\eta |F_1|^b\,\mathrm{d}x = \int_\eta |G_1|^c\,\mathrm{d}x = 1. \tag{C.112}$$

Letting $r^{b-1} = |F_1|^{b-1}/|G_1|$ and replacing in (C.111) yields

$$\frac{|F_1|^b}{b} + \frac{|G_1|^c}{c} - |F_1\,G_1| \ge 0, \quad \frac{1}{b} + \frac{1}{c} = 1. \tag{C.113}$$

Since the function at the left hand side of (C.113) is non negative, its integral is non negative as well. Integrating (C.113) over η and using the normalization condition (C.112) yields

$$\int_\eta |F_1 G_1|\, dx \le \frac{1}{b} + \frac{1}{c} = 1. \tag{C.114}$$

On the other hand the normalization condition also yields

$$\left(\int_\eta |F_1|^b\, dx\right)^{1/b} = \left(\int_\eta |G_1|^c\, dx\right)^{1/c} = 1, \tag{C.115}$$

whence

$$\int_\eta |F_1 G_1|\, dx \le \left(\int_\eta |F_1|^b\, dx\right)^{1/b} \left(\int_\eta |G_1|^c\, dx\right)^{1/c}. \tag{C.116}$$

As (C.116) is homogeneous, it still holds after replacing F_1, G_1 with $F = \lambda F_1$ and $G = \mu G_1$, where λ, μ are arbitrary positive real numbers. This proves Hölder's inequality (C.110).

C.15   Integrals Related to the Electromagnetic Modes

In several applications (e.g., calculations related to the modes of the electromagnetic field, Sect. 5.5) one must evaluate integrals of the form

$$Y = \int_V \exp\left[i\,(\mathbf{k} \pm \mathbf{k}') \cdot \mathbf{r}\right] d^3r, \tag{C.117}$$

where $\mathbf{k} = \mathbf{k}(n_1, n_2, n_3)$ is given by

$$\mathbf{k} = n_1 \frac{2\pi}{d_1}\, \mathbf{i}_1 + n_2 \frac{2\pi}{d_2}\, \mathbf{i}_2 + n_3 \frac{2\pi}{d_3}\, \mathbf{i}_3, \qquad n_i = 0, \pm 1, \pm 2, \ldots, \tag{C.118}$$

$\mathbf{i}_1, \mathbf{i}_2, \mathbf{i}_3$ being the unit vectors parallel to the coordinate axes. The integration domain in (C.117) is a box whose sides d_1, d_2, d_3 are aligned with the axes and start from the origin (Fig. 5.1). The volume of the box is $V = d_1 d_2 d_3$. As $(\mathbf{k} \pm \mathbf{k}') \cdot \mathbf{r} = (k_1 \pm k_1') x_1 + (k_2 \pm k_2') x_2 + (k_3 \pm k_3') x_3$, where the upper (lower) signs hold together, the integral becomes $Y = Y_1 Y_2 Y_3$, with

$$Y_i = \int_0^{d_i} \exp\left[i\,(k_i \pm k_i') x_i\right] dx_i = \frac{\exp\left[i\,(k_i \pm k_i') d_i\right] - 1}{i\,(k_i \pm k_i')}. \tag{C.119}$$

Letting $\theta_i = (k_i \pm k_i')\, d_i / 2 = \pi\,(n_i \pm n_i')$, (C.119) becomes

$$Y_i = d_i \exp(i\,\theta_i) \frac{\exp(i\,\theta_i) - \exp(-i\,\theta_i)}{2i\,\theta_i} = d_i \exp(i\,\theta_i) \frac{\sin\theta_i}{\theta_i}. \tag{C.120}$$

It follows that $Y_i = 0$ if $n_i \pm n'_i \neq 0$, while $Y_i = d_i$ if $n_i \pm n'_i = 0$. Combining the three integrals shows that it is $Y = 0$ if $\mathbf{k} \pm \mathbf{k}' \neq 0$, while it is $Y = V$ if $\mathbf{k} \pm \mathbf{k}' = 0$. The result is recast in a compact form by means of the three-dimensional extension of the Kronecker symbol (A.18):

$$Y = V \, \delta[\mathbf{k} \pm \mathbf{k}', 0] = V \, \delta[\mathbf{k} \pm \mathbf{k}'], \qquad (C.121)$$

where the last form is obtained by dropping the zero for the sake of conciseness. Compare (C.117, C.121) with (C.56) and the comments therein.

C.16   Riemann's Zeta Function

The function defined by

$$\zeta(\lambda, a) = \sum_{k=1}^{\infty} \frac{1}{(k+a)^{\lambda}}, \qquad (C.122)$$

where λ is a complex number with $\Re(\lambda) > 1$ and $a \geq 0$ is real, is called *Riemann's Zeta function*. It can be represented in integral form by combining it with the Gamma function (C.88): letting $x = (k + a) \, y$ in the latter, then replacing y back with x, yields

$$\Gamma(\lambda) = (k+a)^{\lambda} \int_{0}^{+\infty} x^{\lambda-1} \, \exp\left[-(k+a) \, x\right] dx. \qquad (C.123)$$

Dividing (C.123) by $(k + a)^{\lambda}$, letting $k = 1, 2, \ldots$, and adding over k provides

$$\Gamma(\lambda) \sum_{k=1}^{\infty} \frac{1}{(k+a)^{\lambda}} = \int_{0}^{+\infty} x^{\lambda-1} \, \exp(-ax) \left[\sum_{k=1}^{\infty} \exp(-kx) \right] dx, \qquad (C.124)$$

where $\sum_{k=1}^{\infty} \exp(-kx) = \exp(-x)[1 + \exp(-x) + \exp(-2x) + \ldots] = 1/[\exp(x) - 1]$, so that from (C.122),

$$\zeta(\lambda, a) = \frac{1}{\Gamma(\lambda)} \int_{0}^{+\infty} \frac{x^{\lambda-1}}{\exp(x) - 1} \, \exp(-a \, x) \, dx, \qquad \Re(\lambda) > 1. \qquad (C.125)$$

Remembering (C.89) one finds that (C.125) fulfills the recursive relation

$$\frac{\partial}{\partial a} \zeta(\lambda, a) = -\lambda \, \zeta(\lambda + 1, a). \qquad (C.126)$$

Also, letting $a = 0$ and $\lambda = 2m$, with $m = 1, 2, \ldots$ transforms (C.125) into

$$\int_{0}^{+\infty} \frac{x^{2m-1}}{\exp(x) - 1} \, dx = \Gamma(2m) \, \zeta(2m, 0) = \frac{(2\pi)^{2m}}{4m} \, |B_{2m}|, \qquad (C.127)$$

with $B_{2m} = (-1)^{m+1} |B_{2m}|$, $m \geq 1$ the *Bernoulli number*[6] of order $2m$ [44]. Thanks to (C.127) one calculates integrals used in different applications. For instance, letting $m = 2$ and using $B_4 = -1/30$

$$\int_0^{+\infty} \frac{x^3}{\exp(x) - 1} \, dx = \frac{1}{15} \pi^4, \tag{C.128}$$

that is used in (15.78) to calculate the Lagrangian multiplier in the equilibrium statistics for photons. From (C.125) one derives another important class of integrals; in fact, replacing x with $2x$ in the denominator of (C.125) yields

$$\int_0^{+\infty} \frac{x^{\lambda-1}}{\exp(2x) - 1} \exp(-ax) \, dx = 2^{-\lambda} \, \Gamma(\lambda) \, \zeta(\lambda, a/2), \qquad \Re(\lambda) > 1 \quad (C.129)$$

whence, using the identity $2/[\exp(2x) - 1] = 1/[\exp(x) - 1] - 1/[\exp(x) + 1]$ within (C.125), (C.129) provides

$$\int_0^{+\infty} \frac{x^{\lambda-1}}{\exp(x) + 1} \exp(-ax) \, dx = \Gamma(\lambda) \left[\zeta(\lambda, a) - 2^{1-\lambda} \, \zeta(\lambda, a/2) \right]. \tag{C.130}$$

Letting $a = 0$ and $\lambda = 2m$, $m = 1, 2, \ldots$ in the latter, and using (C.127), transforms (C.130) into

$$\int_0^{+\infty} \frac{x^{2m-1}}{\exp(x) + 1} \, dx = \frac{\pi^{2m}}{2m} \left(2^{2m-1} - 1 \right) |B_{2m}|. \tag{C.131}$$

For instance, for $m = 1$ and $m = 2$, (C.131) provides

$$\int_0^{+\infty} \frac{x}{\exp(x) + 1} \, dx = \frac{1}{12} \pi^2, \qquad \int_0^{+\infty} \frac{x^3}{\exp(x) + 1} \, dx = \frac{7}{120} \pi^4. \tag{C.132}$$

---

[6] The Bernoulli numbers are defined by the series expansion $x/[\exp(x) - 1] = \sum_0^\infty B_n x^n / n!$, with $|x| < 2\pi$. It is $B_0 = 1$, $B_1 = -1/2$, $B_2 = 1/6$, $B_4 = -1/30$. Apart from B_1, all Bernoulli numbers of odd index vanish.

Appendix D
Tables

Table D.1 Fundamental constants

Quantity	Symbol	Value[a]	Units
Vacuum permittivity	ε_0	8.85419×10^{-12}	$F\,m^{-1}$
Speed of light	c	2.99792×10^8	$m\,s^{-1}$
Electron charge	q	1.60219×10^{-19}	C
Electron rest mass	m_0	9.10953×10^{-31}	kg
Proton rest mass	M_0	1.67265×10^{-27}	kg
Boltzmann constant	k_B	1.38066×10^{-23}	$J\,K^{-1}$
Planck constant	h	6.62616×10^{-34}	J s
Reduced Planck c.	$\hbar$	1.05459×10^{-34}	J s
Atomic radius	r_a	$\sim 10^{-10}$	m
Electron radius	r_e	2.81794×10^{-15}	m

[a]The ratio between the proton and electron rest masses is $M_0/m_0 \simeq 1836$. The vacuum permeability is found from $\mu_0 = 1/(c^2 \varepsilon_0)$

Table D.2 Greek alphabet

Small	Capital[a]	Name	Small	Capital	Name
α	A	Alpha	ν	N	Nu, ni
β	B	Beta	ξ	$\varXi$	Xi
γ	$\varGamma$	Gamma	o	O	Omicron
δ	$\varDelta$	Delta	π	$\varPi$	Pi
ε	E	Epsilon	ϱ	P	Rho
ζ	Z	Zeta	σ	$\varSigma$	Sigma
η	H	Eta	τ	T	Tau
θ, ϑ	$\varTheta$	Theta	υ	$\varUpsilon$	Upsilon
ι	I	Iota	ϕ, φ	$\varPhi$	Phi
κ	K	Kappa	χ	X	Chi
λ	$\varLambda$	Lambda	ψ	$\varPsi$	Psi
μ	M	Mu, mi	ω	$\varOmega$	Omega

[a]Symbol ∇ is not a Greek letter. However, its name *nabla* is a Greek word, meaning "harp"

© Springer Science+Business Media New York 2015
M. Rudan, *Physics of Semiconductor Devices*,
DOI 10.1007/978-1-4939-1151-6

Solutions

Problems of Chap. 1

1.1 The distance between A and B is a functional of y:

$$G[y] = \int_{AB} \sqrt{dx^2 + dy^2} = \int_a^b \sqrt{1 + \dot{y}^2}\, dx.$$

As $g(y, \dot{y}, x) = \sqrt{1 + \dot{y}^2}$ it is $\partial g/\partial y = 0$, whence the Euler–Lagrange equation reads

$$0 = \frac{d}{dx}\frac{\partial g}{\partial \dot{y}} = \frac{d}{dx}\frac{2\dot{y}}{2\,g} = \frac{\ddot{y}\,g - \dot{y}(2\dot{y}\ddot{y}/2\,g)}{g^2} = \frac{\ddot{y}}{g^3}(g^2 - \dot{y}^2) = \frac{\ddot{y}}{g^3},$$

that is, $\ddot{y} = 0$, $y = c_1 x + c_2$. The two constants are found from $c_1 a + c_2 = y_a$, $c_1 b + c_2 = y_b$.

1.2 Letting $H = E$ one finds

$$\frac{x^2}{a^2} + \frac{p^2}{b^2} = 1, \qquad a = \sqrt{2E/c}, \quad b = \sqrt{2mE}.$$

The curves are ellipses whose axes are proportional to $\sqrt{E}$. The area of each ellipse is $\pi\,a\,b = 2\,\pi\,E/\omega$, with $\omega = \sqrt{c/m}$. As shown in Sect. 3.3, ω is the angular frequency of the oscillator, so that the area becomes $E\,T$, with $T = 2\,\pi/\omega$ the period. As time evolves, the phase point follows the curve in the clockwise direction; in fact, as the phase point reaches the maximum elongation $x_M > 0$ from the left, the momentum at x_M changes from positive to negative.

1.3 Letting $H = E$ one finds for the maximum elongation $x_M = (s\,E/c)^{1/s}$. Note that the units of c depend on the value of s. The form of the constant-energy curves becomes more and more rectangular as s increases. As in the previous exercise, the phase point follows the curve in the clockwise direction.

© Springer Science+Business Media New York 2015
M. Rudan, *Physics of Semiconductor Devices*,
DOI 10.1007/978-1-4939-1151-6

Problems of Chap. 2

2.1 From (2.49) one finds

$$J(E) = \sqrt{m\,c} \oint \sqrt{2E/c - x^2}\, dx,$$

where the integration path is the ellipse described in Problem 1.2. Letting $x = \sqrt{2\,E/c}\,\sin\varphi$ transforms the above into

$$J(E) = 2\sqrt{\frac{m}{c}}\, E \int_0^{2\pi} \cos^2\varphi\, d\varphi = \frac{2\pi}{\omega}\, E, \qquad \omega = \sqrt{\frac{c}{m}}.$$

The first of (2.51) then yields $v = \dot{w} = \partial H/\partial J = \partial E/\partial J = \omega/(2\pi)$.

Problems of Chap. 3

3.1 Like in problem 1.3, letting $H = E > 0$ one finds for the maximum elongation $x_M = (s\,E/c)^{1/s}$, where the units of c depend on the value of s. The motion is limited to the interval $[-x_M, +x_M]$ and the potential energy is symmetric with respect to the origin. Using (2.47) and exploiting the symmetry yields

$$T = 4\sqrt{\frac{m}{2}} \int_0^{x_M} \frac{dx}{\sqrt{E - V(x)}} = \sqrt{\frac{8\,m}{E}} \int_0^{x_M} \left[1 - (x/x_M)^s\right]^{-1/2} dx.$$

Letting $\xi = x/x_M$ and using (C.99, C.100) yields

$$T = \sqrt{\frac{8\,m}{E}}\, x_M \int_0^1 \frac{d\xi}{\sqrt{1 - \xi^s}} = \sqrt{8\,\pi\,m}\, \frac{(1/s)\,\Gamma(1/s)}{\Gamma(1/s + 1/2)} \left(\frac{s}{c}\right)^{1/s} E^{1/s - 1/2}.$$

The result shows that the case $s = 2$, namely, that of the linear harmonic oscillator, is special. In fact, the period does not depend on the total energy, whereas for $s \neq 2$ it does. Still in the case $s = 2$ one finds $T = 2\pi/\omega$, $\omega = \sqrt{c/m}$, as should be (compare with the results of Sect. 3.3). In turn, the case $s \to \infty$ yields $s^{1/s} \to 1$, $c^{1/s} \to 1$ whence, using (C.101), $\lim_{s \to \infty} T = \sqrt{8\,m/E}$. The above is the period in a square well of length 2 (compare with the description of Sect. 3.2). In fact, as $s \to \infty$, the potential energy $c|x|^s/s$ transforms into a square well with $x_M = 1$. The potential energy is shown in Fig. D.1 for some values of s. Thanks to the result of this problem one may tune the form of the potential energy to make the period proportional to a chosen power $h = 1/s - 1/2 \geq -1/2$ of the energy. For instance, letting $s = 2/3$ makes T proportional to E, namely, $T = \sqrt{m/(3c^3)}\,2\pi\,E$.

3.2 The solution is similar to that of Problem 3.1. Letting $H = E < 0$ one finds for the maximum elongation $x_M = [k/(s\,|E|)]^{1/s}$, where the units of k depend on the value of s. The motion is limited to the interval $[-x_M, +x_M]$ and the potential energy is symmetric with respect to the origin. Using (2.47) and exploiting the symmetry yields

Fig. D.1 Form of the potential energy $c|x|^s/s$ for $c = 1$ and different values of s (Problem 3.1)

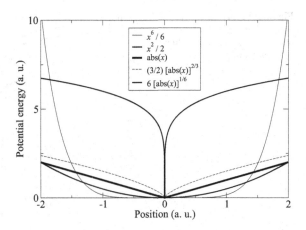

$$T = 4\sqrt{\frac{m}{2}} \int_0^{x_M} \frac{dx}{\sqrt{E - V(x)}} = \sqrt{\frac{8\,m}{|E|}} \int_0^{x_M} \left[(x_M/x)^s - 1\right]^{-1/2} dx.$$

Letting $\xi = x/x_M$ and using (C.102, C.103) yields

$$T = \sqrt{\frac{8\,m}{|E|}}\, x_M \int_0^1 \frac{d\xi}{\sqrt{1/\xi^s - 1}} = \sqrt{8\,\pi\,m}\, \frac{\Gamma(1/s + 1/2)}{s\,\Gamma(1/s + 1)} \left(\frac{k}{s}\right)^{1/s} |E|^{-1/s - 1/2}.$$

The Coulomb case $s = 1$ yields $T = \sqrt{2\,m}\,\pi\,k\,|E|^{-3/2}$ (in fact, in the Coulomb case and for a closed trajectory the period is always proportional to $|E|^{-3/2}$, compare with (3.81)). Note that in the case considered here the particle crosses the origin because the initial conditions are such that its trajectory is aligned with the x axis. The limit $s \to \infty$ yields $s^{1/s} \to 1$, $c^{1/s} \to 1$ whence, using (C.101, C.103), $\lim_{s \to \infty} T = 0$. The potential energy is shown in Fig. D.2 for some values of s.

3.3 The O reference is chosen as in Sect. 3.13.5, whence $T_{1a} = E = (m_1/m)\,E_B$. From (3.36) one extracts $\mu/s_0 = \tan[(\pi - \chi)/4]$, to find

$$\frac{2\,\mu/s_0}{1 - (\mu/s_0)^2} = \frac{2\,\mu\,s_0}{s_0^2 - \mu^2} = \tan\left(\frac{\pi - \chi}{2}\right) = \frac{1}{\tan(\chi/2)},$$

where s_0 is given by the second of (3.33). It follows that $s_0^2 - \mu^2 = 2\,\lambda\,s_0$ and $\tan(\chi/2) = \lambda/\mu$. Then, noting that (3.23) contains $\sin^2(\chi/2) = \tan^2(\chi/2)/[1 + \tan^2(\chi/2)]$, and using (3.73), one finds $\sin^2(\chi/2) = 1/(1 + c^2/\lambda^2)$. The expression of λ is taken from the first of (3.32), with α given by (3.75). Inserting the result into (3.23) yields

$$T_{1b} = \frac{\alpha^2(1 - m_1/m_2)^2 + c^2\,T_{1a}^2}{\alpha^2(1 + m_1/m_2)^2 + c^2\,T_{1a}^2}\,T_{1a}, \quad T_{1a} - T_{1b} = \frac{4\,(m_1/m_2)\,T_{1a}}{(1 + m_1/m_2)^2 + (c/\alpha)^2\,T_{1a}^2}.$$

Fig. D.2 Form of the
potential energy $-k|x|^{-s}/s$
for $k = 1$ and different values
of s (Problem 3.2)

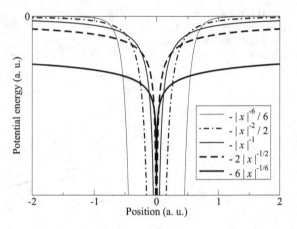

Fig. D.3 Normalized loss of
energy $c\,(T_{1a} - T_{1b})/\alpha$ as a
function of the normalized
initial energy $c\,T_{1a}/\alpha$
(Problem 3.3), for different
values of the ratio m_1/m_2

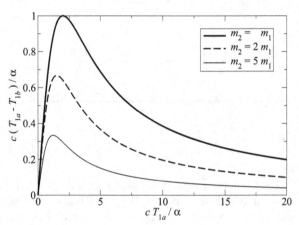

Obviously it is $T_{1b} < T_{1a}$. It follows that $T_{1a} - T_{1b}$ is the loss of energy due to the collision. It is also interesting to note that using the normalized energies $c\,T_{1a}/\alpha$ and $c\,T_{1b}/\alpha$ makes the expressions above to depend on the m_1/m_2 ratio only. The loss of energy is drawn in normalized form in Fig. D.3 for different values of m_1/m_2.

Problems of Chap. 4

4.1 Using primes to indicate derivatives, a first integration yields

$$\varphi' = \varphi'(c) - H, \qquad H(x) = \int_c^x \frac{\varrho(\xi)}{\varepsilon_0}\, d\xi,$$

where H is integrated by parts:

$$\int_a^x H(\xi)\, d\xi = x\, H(x) - a\, H(a) - \int_a^x \xi\, \frac{\varrho(\xi)}{\varepsilon_0}\, d\xi.$$

Integrating φ' and using the expression of $\int_a^x H(\xi)\,d\xi$ yields the solution

$$\varphi = \varphi(a) + \varphi'(c)(x-a) - x \int_c^x \frac{\varrho(\xi)}{\varepsilon_0}\,d\xi + a \int_c^a \frac{\varrho(\xi)}{\varepsilon_0}\,d\xi + \int_a^x \xi \frac{\varrho(\xi)}{\varepsilon_0}\,d\xi.$$

4.2 Letting $c = a$ in the solution to Problem 4.1 yields at any point x within $[a,b]$ the expression

$$\varphi(x) = \varphi(a) + \varphi'(a)(x-a) - x \int_a^x \frac{\varrho(\xi)}{\varepsilon_0}\,d\xi + \int_a^x \xi \frac{\varrho(\xi)}{\varepsilon_0}\,d\xi.$$

For $x > b$ it is $\varrho = 0$ so that the solution of $\varphi'' = 0$ is linear and has the form $\varphi(x) = \varphi(b) + \varphi'(b)(x-b)$. The term $\varphi(b)$ in the latter is obtained by letting $x = b$ in the above expression of $\varphi(x)$. One finds $\varphi(b) = \varphi(a) + \varphi'(a)(b-a) - b\,M_0 + M_1$, with

$$M_0 = \int_a^b \frac{\varrho(\xi)}{\varepsilon_0}\,d\xi, \qquad M_1 = \int_a^b \xi \frac{\varrho(\xi)}{\varepsilon_0}\,d\xi$$

the first two moments of ϱ/ε_0 (compare with Sect. C.6). The derivative φ' is found from Problem 4.1 with $c = a$, and reads

$$\varphi'(x) = \varphi'(a) - \int_a^x \frac{\varrho(\xi)}{\varepsilon_0}\,d\xi,$$

whence $\varphi'(b) = \varphi'(a) - M_0$. Using the expressions of $\varphi(b)$, $\varphi'(b)$ thus found yields

$$\varphi(x) = \varphi(a) + \varphi'(a)(x-a) - M_0\,x + M_1, \qquad x > b.$$

4.3 From the findings of Problem 4.2 one observes that the solution φ is invariant for any charge density $\tilde{\varrho}$ that leaves M_0 and M_1 unchanged. Due to this, if both M_0, M_1 differ from zero, the new charge density must contain two adjustable parameters in order to fit the values of M_0, M_1 through the expressions introduced in Problem 4.2. If only one moment differs from zero, one parameter suffices, while no parameter is necessary if both moments are equal to zero. Figure D.4 gives an example of charge density such that $M_0 = 0$ and $M_1 = 0$.

4.4 The starting point is the solution for $x > b$ found in Problem 4.2. When the charge density is removed, the new solution reads

$$\varphi(x) = \tilde{\varphi}(a) + \tilde{\varphi}'(a)(x-a).$$

For $x > b$ the two solutions become equal to each other by letting $\tilde{\varphi}(a) = \varphi(a) - M_0\,a + M_1$ and $\tilde{\varphi}'(a) = \varphi'(a) - M_0$.

Fig. D.4 Example of charge density such that $M_0 = 0$ and $M_1 = 0$

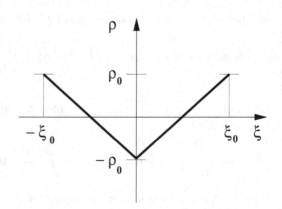

4.5 Considering that the value of h is unknown, the integrals that define the moments (Problem 4.2) must be extended from $-\infty$ to $+\infty$. One finds $\mu = M_0$, $h = M_1/M_0$. If $h \geq a$, the solution is given by $\varphi = \varphi(a) + \varphi'(a)(x-a) - M_0 x + M_1$ for $x \geq h$, while it is given by $\varphi = \varphi(a) + \varphi'(a)(x-a)$ for $x < h$. If $h < a$, the solution is given by $\varphi = \varphi(a) + \varphi'(a)(x-a)$ for $x \geq h$, while it is given by $\varphi = \varphi(a) + \varphi'(a)(x-a) - M_0 x + M_1$ for $x < h$. At h the electric potential is continuous, $\varphi(h^+) = \varphi(h^-) = \varphi(a) + \varphi'(a)(h-a)$, whereas the electric field is discontinuous, $\varphi'(h^-) - \varphi'(h^+) = M_0$. The case $M_0 \neq 0$, $M_1 = 0$ yields $\mu = M_0$, $h = 0$, while the case $M_0 = 0$, $M_1 \neq 0$ can not be fulfilled by $\mu\,\delta(x-h)$.

Problems of Chap. 5

5.1 From $n = n(x_1)$ one finds that grad$n = \mathbf{i}_1\, dn/dx_1$ is parallel to x_1 whereas $dn/dx_2 = dn/dx_3 = 0$. From the eikonal Eq. (5.57) it follows

$$\frac{d}{ds}\left(n\,\frac{dx_2}{ds}\right) = 0, \qquad \frac{d}{ds}\left(n\,\frac{dx_3}{ds}\right) = 0,$$

whence $n\,dx_2/ds = $ const, $n\,dx_3/ds = $ const. The ratio of the latter relations yields $dx_2/dx_3 = $ const, namely, $x_2 = ax_3 + b$, where a, b are constants. This expression is one of the two parametric equations $u(x_1, x_2, x_3) = 0$, $v(x_1, x_2, x_3) = 0$ describing the ray, and shows that the ray belongs to a plane parallel to x_1. By a suitable rotation around x_1, such that $x_2 \to x_2'$, $x_3 \to x_3'$, the plane of the ray is made parallel to the plane $x_1 x_2'$, so that the third coordinate x_3' is fixed. In the new reference, let ϑ be the angle between the direction of the ray and x_1 at some point P; it is $dx_1 = \cos\vartheta\,ds$, $dx_2' = \sin\vartheta\,ds$. The eikonal equation in the new reference then provides

$$n\,\frac{dx_2'}{ds} = \text{const}, \qquad n\sin\vartheta = \text{const}.$$

5.2 Like in Problem 5.1 one considers the case where the refraction index depends only on the x_1 coordinate. Let the medium be made of three regions separated by two planes parallel to each other.

The two external regions A and B have a constant refraction index, n_A and, respectively, $n_B \neq n_A$. The internal region I, whose thickness is s, has a refraction index that varies continuously from n_A to n_B as x_1 varies from the A-I interface to the I-B interface. Applying the solution of Problem 5.1 to this case shows that $n_A \sin \vartheta_A = $ const everywhere in region A, hence the ray is a straight line there; similarly it is $n_B \sin \vartheta_B = $ const everywhere in region B, with the same constant. It follows

$$n_B \sin \vartheta_B = n_A \sin \vartheta_A.$$

Unless $\vartheta_A = 0$, the position of the ray along the x_2' axis at the I-B interface is different from that at the A-I interface; if, however, s is made to vanish, the position becomes the same, this yielding the *Descartes law of refraction*: the ray crossing an interface between two media is continuous, whereas its slopes on the two sides of the interface fulfill (D). The result still holds in the cases where the interface between the two media is not planar, provided its curvature is small enough to make geometrical optics applicable.

Problems of Chap. 6

6.1 Letting $\vartheta = \beta h \nu$, with $\beta = 1/(k_B T)$, the Boltzmann distribution takes the form $N_n = N_0 \exp(-n \vartheta)$, whence

$$\sum_{n=0}^{\infty} N_n = N_0 \left[1 + \exp(-\vartheta) + \exp(-2 \vartheta) + \ldots \right] = \frac{N_0}{1 - \exp(-\vartheta)},$$

and

$$\sum_{n=0}^{\infty} n h \nu N_n = h \nu N_0 \left[\exp(-\vartheta) + 2 \exp(-2 \vartheta) + 3 \exp(-3 \vartheta) + \ldots \right].$$

Observing that $n \exp(-n \vartheta) = -d \exp(-n \vartheta)/d\vartheta$, one finds

$$\sum_{n=0}^{\infty} n h \nu N_n = -h \nu \frac{d}{d\vartheta} \left(\sum_{n=0}^{\infty} N_n - N_0 \right) = h \nu \frac{N_0 \exp(-\vartheta)}{\left[1 - \exp(-\vartheta)\right]^2},$$

whence

$$\overline{E}_n = \frac{\sum_{n=0}^{\infty} n h \nu N_n}{\sum_{n=0}^{\infty} N_n} = \frac{h \nu}{\exp(\vartheta) - 1} = \frac{h \nu}{\exp(\beta h \nu) - 1}.$$

Problems of Chap. 8

8.1 Consider the homogeneous equation associated to (8.76), $g'' + a g' + b g = 0$, and let $f = g h$ and $u = h'$; this splits (8.76) into the system

$$g'' + a g' + b g = 0, \qquad g u' + (2 g' + a g) u = c.$$

If g is known, then u is found by integrating a first-order equation, whence $f = g \int u \, d\xi$. To find g one lets $A(x) = \int a \, d\xi$, $g = \exp(-A/2) w$, this transforming the homogeneous equation for g into

$$w'' + q w = 0, \qquad q = b - a^2/4 - a'/2,$$

which is a time-independent Schrödinger equation.

Problems of Chap. 9

9.1 Inserting the expression of c_k into the one-dimensional form of (9.26) yields

$$A(x - u t; k_0) = \frac{\sqrt{\sigma/2}}{\pi^{3/4}} \int_{-\infty}^{+\infty} \exp\left[i (x - u t) (k - k_0) - \sigma^2 k^2/2\right] dk.$$

Following the same procedure as in Sect. C.8 one finds

$$i (x - u t) (k - k_0) - \frac{1}{2} \sigma^2 k^2 = -\frac{(x - u t)^2}{2 \sigma^2} - \frac{\sigma^2}{2}\left(k - j \frac{x - u t}{\sigma^2}\right)^2,$$

whence

$$A(x - u t; k_0) = \frac{1}{\pi^{1/4} \sqrt{\sigma}} \exp\left[-i k_0 (x - u t) - \frac{(x - u t)^2}{2 \sigma^2}\right].$$

The particle's localization is determined by

$$|A(x - u t)|^2 = \frac{1}{\sqrt{\pi} \sigma} \exp\left[-\frac{(x - u t)^2}{\sigma^2}\right].$$

Using again the results of Sect. C.8 yields $\|A\| = 1$.

9.2 Remembering that $|\psi|^2 = |A|^2$, the one-dimensional form of (9.23) reads

$$x_0(t) = \int_{-\infty}^{+\infty} x \, |A|^2 \, dx = \int_{-\infty}^{+\infty} (x - u t) |A|^2 \, dx + u t \int_{-\infty}^{+\infty} |A|^2 \, dx.$$

Letting $s = x - u t$ one finds that the integral of $s \, |A(s)|^2$ vanishes because the integrand is odd and the integration domain is symmetric with respect to the origin. Using the result $\|A\| = 1$ of Problem 9.1 then yields $x_0(t) = u t$.

Problems of Chap. 10

10.1 To determine the time evolution of the expectation value of the wavepacket for a free particle one starts from the general expression (9.5), with $w_k(\mathbf{r})$ and $E_k = \hbar\,\omega_k$ given by (9.22), and c_k given by the second relation in (9.6). The wave function is assumed normalized, $\int_{-\infty}^{+\infty} |\psi(\mathbf{r},t)|^2\,d^3r = \int_{-\infty}^{+\infty} |c(\mathbf{k})|^2\,d^3k = 1$. Using the first spatial coordinate x_1 and defining $m_k = c_k \exp(-i\,\omega_k\,t)$, the following are of use: $x_1\,w_k = -i\,\partial w_k/\partial k_1$, $x_1^2\,w_k = -\partial^2 w_k/\partial k_1^2$, and

$$-\int_{-\infty}^{+\infty} m_k \frac{\partial w_k}{\partial k_1}\,dk_1 = \int_{-\infty}^{+\infty} w_k \frac{\partial m_k}{\partial k_1}\,dk_1, \quad \int_{-\infty}^{+\infty} m_k \frac{\partial^2 w_k}{\partial k_1^2}\,dk_1 = \int_{-\infty}^{+\infty} w_k \frac{\partial^2 m_k}{\partial k_1^2}\,dk_1,$$

where the last two equalities are obtained by integrating by parts and observing that, due to the normalization condition, c_k and $\partial c_k/\partial k_1$ vanish at infinity. In turn,

$$i\frac{\partial m_k}{\partial k_1} = \left(u_1\,t\,c_k + i\frac{\partial c_k}{\partial k_1}\right)\exp(-i\,\omega_k\,t), \quad u_1 = \frac{\partial \omega_k}{\partial k_1} = \frac{\hbar\,k_1}{m},$$

$$-\frac{\partial^2 m_k}{\partial k_1^2} = \left[\left(u_1^2\,t^2 + i\frac{\hbar}{m}\,t\right)c_k + 2\,i\,u_1\,t\frac{\partial c_k}{\partial k_1} - \frac{\partial^2 c_k}{\partial k_1^2}\right]\exp(-i\,\omega_k\,t).$$

The expectation value $\langle x_1 \rangle = \langle \psi | x_1 | \psi \rangle$ involves an integration over $\mathbf{r}$ to calculate the scalar product, an integration over $\mathbf{k}$ to calculate the integral expression of ψ, and an integration over $\mathbf{k}'$ to calculate the integral expression of ψ^*. Performing the integration over $\mathbf{r}$ first, letting $c_k' = c(\mathbf{k}')$, $\omega_k' = \omega(\mathbf{k}')$, and using (C.56) yields

$$\langle x_1 \rangle = \iiint_{-\infty}^{+\infty} \left(u_1\,t\,|c_k|^2 + i\,c_k^* \frac{\partial c_k}{\partial k_1}\right)d^3k.$$

Letting $c_k = a_k + i\,b_k$, with a_k and b_k real, and using the asymptotic vanishing of c_k, one finds

$$\iiint_{-\infty}^{+\infty} i\,c_k^* \frac{\partial c_k}{\partial k_1}\,d^3k = x_{01}, \quad x_{01} = \iiint_{-\infty}^{+\infty} \left(\frac{\partial a_k}{\partial k_1} b_k - \frac{\partial b_k}{\partial k_1} a_k\right)d^3k,$$

where x_{01} is a real constant. Repeating the calculation for x_2 and x_3, and letting $\mathbf{u} = \mathrm{grad}_k\,\omega$, $\mathbf{r}_0 = (x_{01}, x_{02}, x_{03})$, finally yields

$$\langle \mathbf{r} \rangle = \mathbf{r}_0 + \iiint_{-\infty}^{+\infty} \mathbf{u}\,t\,|c_k|^2\,d^3k, \quad \frac{d}{dt}\langle \mathbf{r} \rangle = \iiint_{-\infty}^{+\infty} \mathbf{u}\,|c_k|^2\,d^3k = \text{const.}$$

If $|c_k|^2$ is even with respect to all components of $\mathbf{k}$, the expectation value of $\mathbf{r}$ does not change with respect to the initial value $\mathbf{r}_0$. Otherwise, it moves at constant speed.

10.2 The time evolution of the standard deviation of position is found following the same line and using the same symbols and relations as in Problem 10.1, starting with

$$\langle x_1^2 \rangle = \iiint_{-\infty}^{+\infty} \left[\left(u_1^2 \, t^2 + i \frac{\hbar}{m} t \right) |c_\mathbf{k}|^2 + 2i \, u_1 \, t \, c_\mathbf{k}^* \frac{\partial c_\mathbf{k}}{\partial k_1} - c_\mathbf{k}^* \frac{\partial^2 c_\mathbf{k}}{\partial k_1^2} \right] d^3k.$$

An integration by parts combined with the normalization condition for $c_\mathbf{k}$ shows that

$$\iiint_{-\infty}^{+\infty} 2i \, u_1 \, t \, c_\mathbf{k}^* \frac{\partial c_\mathbf{k}}{\partial k_1} \, d^3k = -i \frac{\hbar}{m} t + 2t \iiint_{-\infty}^{+\infty} u_1 \left(\frac{\partial a_\mathbf{k}}{\partial k_1} b_\mathbf{k} - \frac{\partial b_\mathbf{k}}{\partial k_1} a_\mathbf{k} \right) d^3k,$$

where the second term at the right hand side is real, whereas the first one cancels out in the expression of $\langle x_1^2 \rangle$. Finally, another integration by parts yields

$$-\iiint_{-\infty}^{+\infty} c_\mathbf{k}^* \frac{\partial^2 c_\mathbf{k}}{\partial k_1^2} \, d^3k = \iiint_{-\infty}^{+\infty} \left| \frac{\partial c_\mathbf{k}}{\partial k_1} \right|^2 d^3k.$$

In conclusion,

$$\langle x_1^2 \rangle = \iiint_{-\infty}^{+\infty} \left| u_1 \, t \, c_\mathbf{k} + i \frac{\partial c_\mathbf{k}}{\partial k_1} \right|^2 d^3k.$$

Repeating the calculation for x_2 and x_3 yields

$$\langle \mathbf{r} \cdot \mathbf{r} \rangle = \int_{-\infty}^{+\infty} \left| \mathbf{u} \, t \, c_\mathbf{k} + i \, \mathrm{grad}_\mathbf{k} c_\mathbf{k} \right|^2 d^3k,$$

where the definition of the squared length of a complex vector is found in (A.2) and (A.4). The standard deviation of the wave packet in the $\mathbf{r}$ space is the positive square root of $\langle \mathbf{r} \cdot \mathbf{r} \rangle - \langle \mathbf{r} \rangle \cdot \langle \mathbf{r} \rangle = \sum_{i=1}^{3} (\Delta x_i)^2$, where the expression of $\langle \mathbf{r} \rangle$ was derived in Prob. 10.1. It is easily shown that the standard deviation diverges with t. In fact, the leading term of $\langle x_1^2 \rangle$ and, respectively, $\langle x_1 \rangle^2$ is

$$\langle x_1^2 \rangle \sim t^2 \iiint_{-\infty}^{+\infty} u_1^2 \, |c_\mathbf{k}|^2 \, d^3k, \qquad \langle x_1 \rangle^2 \sim t^2 \left(\iiint_{-\infty}^{+\infty} u_1 \, |c_\mathbf{k}|^2 \, d^3k \right)^2,$$

the first of which is positive, whereas the second one is non negative. Letting $f = c_\mathbf{k}$, $g = u_1 c_\mathbf{k}$ in the Schwartz inequality (8.15) and using the normalization condition of $c_\mathbf{k}$ yields

$$\iiint_{-\infty}^{+\infty} u_1^2 |c_\mathbf{k}|^2 \, d^3k > \left(\iiint_{-\infty}^{+\infty} u_1 |c_\mathbf{k}|^2 \, d^3k \right)^2,$$

where the strict inequality holds because f and g are not proportional to each other. For the leading term it follows that $(\Delta x_1)^2 = \langle x_1^2 \rangle - \langle x_1 \rangle^2 \sim \mathrm{const} \times t^2$, where the constant is strictly positive. The same reasoning applies to x_2, x_3. In conclusion, the standard deviation Δx_i associated with the ith coordinate diverges in time with the first power of t.

10.3 Still with reference to the wave packet of a free particle used in Problems 10.1 and 10.2, the time evolution of the expectation value in the **p** space is found starting with the first component p_1 of momentum. The corresponding operator is $\hat{p}_1 = -i\hbar\,\partial/\partial x_1$, and the following relations are of use: $\hat{p}_1 w_{\mathbf{k}} = \hbar k_1 w_{\mathbf{k}}$, $\hat{p}_1^2 w_{\mathbf{k}} = \hbar^2 k_1^2 w_{\mathbf{k}}$. The expectation value $\langle p_1 \rangle = \langle \psi | p_1 | \psi \rangle$ involves an integration over **r** to calculate the scalar product, an integration over **k** to calculate the integral expression of ψ, and an integration over $\mathbf{k}'$ to calculate the integral expression of ψ^*. Performing the integration over **r** first, letting $c_{\mathbf{k}}' = c(\mathbf{k}')$, $\omega_{\mathbf{k}}' = \omega(\mathbf{k}')$, and using (C.56) yields

$$\langle p_1 \rangle = \iiint_{-\infty}^{+\infty} \hbar\, k_1\, |c_{\mathbf{k}}|^2\, d^3k = p_{01}.$$

The real constant p_{01} defined above is independent of time. In conclusion, repeating the calculation for p_2 and p_3, and letting $\mathbf{p}_0 = (p_{01}, p_{02}, p_{03})$, the following holds: $\langle \mathbf{p} \rangle = \mathbf{p}_0$. If $|c_{\mathbf{k}}|^2$ is even with respect to all components of **k**, the expectation value of **p** is zero.

10.4 The calculation of $\langle \psi | \hat{p}_1^2 | \psi \rangle$ is carried out following the same line as in Problem 10.3, leading to

$$\langle p_1^2 \rangle = \iiint_{-\infty}^{+\infty} \hbar^2\, k_1^2\, |c_{\mathbf{k}}|^2\, d^3k.$$

Repeating the calculation for x_2 and x_3 yields

$$\langle \mathbf{p} \cdot \mathbf{p} \rangle = \iiint_{-\infty}^{+\infty} \hbar\, \mathbf{k} \cdot \hbar\, \mathbf{k}\, |c_{\mathbf{k}}|^2\, d^3k.$$

In turn, the standard deviation of the wave packet in the **p** space is the positive square root of $\langle \mathbf{p} \cdot \mathbf{p} \rangle - \langle \mathbf{p} \rangle \cdot \langle \mathbf{p} \rangle = \sum_{i=1}^{3} (\Delta p_i)^2$. Letting $f = c_{\mathbf{k}}$, $g = \hbar k_1 c_{\mathbf{k}}$ in the Schwartz inequality (8.15) and using the normalization condition of $c_{\mathbf{k}}$ yields

$$\iiint_{-\infty}^{+\infty} \hbar^2\, k_1^2\, |c_{\mathbf{k}}|^2\, d^3k > \left(\iiint_{-\infty}^{+\infty} \hbar\, k_1\, |c_{\mathbf{k}}|^2\, d^3k \right)^2,$$

where the strict inequality holds because f and g are not proportional to each other. It follows that $(\Delta p_1)^2 = \langle p_1^2 \rangle - \langle p_1 \rangle^2$ is strictly positive and constant in time. The same reasoning applies to p_2, p_3. In conclusion, the standard deviation Δp_i associated with the ith component of momentum is constant in time.

10.5 One finds $\langle x \rangle = x_0$, $d\hbar\,\beta/dx = \hbar k_0$, $\langle p_e \rangle = \hbar k_0$,

$$\left\langle \frac{p_e^2}{2m} \right\rangle = \frac{\hbar^2 k_0^2}{2m}, \qquad \langle Q \rangle = \frac{\hbar^2}{8m\sigma^2}, \qquad \langle T \rangle = \frac{\hbar^2}{2m}\left(k_0^2 + \frac{1}{4\sigma^2} \right).$$

One notes that for a fixed $\langle T \rangle$ all non-negative values of the "convective" and "thermal" parts that add up to $\langle T \rangle$ are allowed. In the particular case of a free particle, where $\langle T \rangle = \langle E \rangle$, the above shows that different values of the average momentum and "dispersion" may combine to yield the same total energy.

Problems of Chap. 11

11.1 Letting $b^- = \pi a \sqrt{2m}(\sqrt{E} - \sqrt{E - V_0})/\hbar$, $b^+ = \pi a \sqrt{2m}(\sqrt{E} + \sqrt{E - V_0})/\hbar$ and remembering that $\sinh b \simeq b$ when $|b| \ll 1$ yields, with m fixed,

$$R(a \to 0) = \left(\frac{b^-}{b^+}\right)^2 = \frac{(\sqrt{E} - \sqrt{E - V_0})^2}{(\sqrt{E} + \sqrt{E - V_0})^2},$$

that coincides with the first relation in (11.11). Conversely, when $a > 0$ is fixed and m is let grow one finds

$$R \simeq \exp[2(b^- - b^+)] = \exp\left(-4\pi a \sqrt{2m}\sqrt{E - V_0}/\hbar\right),$$

namely, $\lim_{m \to \infty} R = 0$, this recovering the classical limit.

11.2 The maximum of the cotangent's argument $s\sqrt{2m(E - V_0)}/\hbar^2$ is found by letting $E = 0$. Thus,

$$\gamma = \frac{s}{\hbar}\sqrt{-2m V_0} \simeq 13.4, \qquad \frac{13.4}{\pi} \simeq 4.3.$$

As a consequence, the cotangent has four complete branches and one incomplete branch in the interval $V_0 < E < 0$, corresponding to five eigenvalues $E_1, \ldots, E_5$. Using the normalized parameter $0 < \eta = \sqrt{1 - E/V_0} < 1$, the equation to be solved reads

$$\frac{\eta^2 - 1/2}{\eta\sqrt{1 - \eta^2}} = \cot(\gamma \eta).$$

Over the η axis, the 5 branches belong to the intervals $(0, \pi/\gamma)$, $(\pi/\gamma, 2\pi/\gamma)$, $(2\pi/\gamma, 3\pi/\gamma)$, $(3\pi/\gamma, 4\pi/\gamma)$, $(4\pi/\gamma, 1)$.

Problems of Chap. 13

13.1 Letting $Z = 1$ one finds that the lowest total energy of the electron in the hydrogen atom has the value

$$E_1(Z = 1) = -\frac{m_0}{2\hbar^2}\left(\frac{q^2}{4\pi\varepsilon_0}\right)^2.$$

As noted in Sect. 13.5.2, the electron is bound as long as $E < 0$. As a consequence, the minimum energy for which it becomes free is $\lim_{n \to \infty} E_n = 0$. The hydrogen atom's ionization energy is thus found to be

$$E_{ion} = 0 - E_1(Z = 1) = |E_1(Z = 1)| = \frac{m_0}{2\hbar^2}\left(\frac{q^2}{4\pi\varepsilon_0}\right)^2.$$

Replacing the constants' values of Table D.1 yields $E_{ion} \simeq 2.18 \times 10^{-18}$ J $\simeq 13.6$ eV.

13.2 The time-dependent wave function is in this case $\psi = w(E_{min}) \exp(-i E_{min} t/\hbar)$, whence $|\psi|^2 = \exp(-2r/a)/(\pi a^3)$. Taking the Jacobian determinant $J = r^2 \sin \vartheta$ from (B.3) and using the definitions of Sect. 10.5 one finds

$$\langle r \rangle = \int_0^\infty \int_0^\pi \int_0^{2\pi} r \frac{\exp(-2r/a)}{\pi a^3} r^2 \sin \vartheta \, dr \, d\vartheta \, d\varphi = \frac{3}{2} a.$$

From (13.89) one finds $a_1 = a(Z = 1) = 4\pi \hbar^2 \varepsilon_0/(m_0 q^2) \simeq 5.3 \times 10^{-11}$ m $\simeq 0.53$ Å, where the constants' values are taken from Table D.1. Note that $a_1 = r_1/2$, with r_1 is the radius of the ground state derived from the Bohr hypothesis (Sect. 7.4.4). The expectation value of r turns out to be $\langle r \rangle \simeq 0.8$ Å.

Problems of Chap. 14

14.1 From $h_{bg}^{(0)} = [A/(2\pi)^3]/(q_c^2 + q^2)$ and $q = |\mathbf{b} - \mathbf{g}|$ one finds

$$H_b^{(0)}(E_g) = \frac{A^2}{(2\pi)^6} \int_0^\pi \int_0^{2\pi} \frac{1}{(q_c^2 + q^2)^2} \sin \vartheta \, d\vartheta \, d\varphi,$$

$A = \kappa Z e^2/\varepsilon_0$. Observing that $\mathbf{b}$ is a fixed vector one can use it as the reference for angle ϑ, so that $q^2 = (\mathbf{b} - \mathbf{g}) \cdot (\mathbf{b} - \mathbf{g}) = b^2 + g^2 - 2bg \cos \vartheta$. From $g = b$ it follows $q^2 = 4g^2 \sin^2(\vartheta/2)$. On the other hand it is $\sin \vartheta \, d\vartheta = d \sin^2(\vartheta/2)$ whence, integrating over φ and letting $\mu = \sin^2(\vartheta/2)$,

$$H_b^{(0)}(E_g) = \frac{A^2}{(2\pi)^5} \int_0^1 \frac{d\mu}{(q_c^2 + 4g^2 \mu)^2} = \frac{A^2/(2\pi)^5}{q_c^2(q_c^2 + 4g^2)}.$$

The dependence on E_g is found through the relation $E_g = \hbar^2 g^2/(2m)$.

Problems of Chap. 15

15.1 After selecting a number $0 < \eta < 1/2$, define E^+ and E^- such that $P(E^+) = \eta$, $P(E^-) = 1 - \eta$. It follows

$$E^+ - E^- = 2 k_B T \log \frac{1-\eta}{\eta}.$$

Letting, e.g., $\eta = 0.1$ one finds $E(P = 0.9) - E(P = 0.1) = 2 k_B T \log 9 \simeq 4.39 k_B T$. Similarly, letting $\eta = 0.01$ one finds $E(P = 0.99) - E(P = 0.01) = 2 k_B T \log 99 \simeq 9.19 k_B T$. At $T = 300$ K it is $k_B T \simeq 25.8$ meV. From the above results one finds $E(P = 0.9) - E(P = 0.1) \simeq 113$ meV and $E(P = 0.99) - E(P = 0.01) \simeq 237$ meV, respectively.

Problems of Chap. 19

19.1 Using (19.115) and adding up the two expressions in (19.118) one finds

$$\mu_p = q \left(\frac{m_{hh}^{1/2}}{m_{hh}^{3/2} + m_{hl}^{3/2}} + \frac{m_{hl}^{1/2}}{m_{hh}^{3/2} + m_{hl}^{3/2}} \right) \tau_p = \frac{q \, \tau_p}{\overline{m}_h}.$$

Using the values taken from Table 17.3 yields

$$\frac{m_0}{\overline{m}_h} = \frac{0.5^{1/2}}{0.5^{3/2} + 0.16^{3/2}} + \frac{0.16^{1/2}}{0.5^{3/2} + 0.16^{3/2}},$$

whence $\overline{m}_h \simeq 0.377 \, m_0$. As for a_p, using the common value of the relaxation time in (19.122) yields

$$a_p = \frac{q \, \tau_p}{\mu_p^2} \left(\frac{\mu_{ph}}{m_{hh}} + \frac{\mu_{pl}}{m_{hl}} \right).$$

Replacing the expressions (19.118) of μ_{ph}, μ_{pl},

$$a_p = \frac{m_{hh}^{3/2} + m_{hl}^{3/2}}{m_{hh}^{1/2} m_{hl}^{1/2} \left(m_{hh}^{1/2} + m_{hl}^{1/2} \right)} = \frac{0.5^{3/2} + 0.16^{3/2}}{0.5^{1/2} \, 0.16^{1/2} \left(0.5^{1/2} + 0.16^{1/2} \right)} \simeq 1.33.$$

Problems of Chap. 21

21.1 Using $k_B T/q \simeq 26$ mV (compare with (19.162)) and $n_i \simeq 10^{10}$ cm^{-3} (Table 18.2), one finds

$$\psi_0 = \frac{k_B T}{q} \log \left(\frac{N_A N_D}{n_i^2} \right) \simeq 0.65 \text{ V}. \tag{D.1}$$

21.2 The relations to be used are (21.51), (21.55), and (21.57). If $k_p = 0, k_n > 0$, one finds $Y_p = 0$, $Y_n = 1 - \exp[-m(b)]$. On the other hand it is in this case $m(b) = \int_a^b k_n \, dx > 0$, whence $Y_n < 1$. If, instead, $k_n = 0, k_p > 0$, one finds $Y_n = 0$, $Y_p = 1 - \exp[m(b)]$ with $m(b) = -\int_a^b k_p \, dx < 0$, whence $Y_p < 1$. In conclusion, the condition for avalanche never occurs.

Problems of Chap. 22

22.1 The differential capacitance of the MOS structure, extracted from (22.19) and (22.20), reads

$$\frac{C}{C_{ox}} = \frac{1}{1 + C_{ox}/C_{sc}}, \qquad C_{sc} = \pm \frac{\sqrt{2} \, \varepsilon_{sc}}{L_A} \frac{dF}{du_s} > 0,$$

where the plus (minus) sign holds for $u_s > 0$ ($u_s < 0$). From (22.3) and (22.5) one finds $dF^2/du_s = A(u_s)$; on the other hand it is $dF^2/du_s = 2F\, dF/du_s$, whence

$$C_{sc} = \pm\frac{\varepsilon_{sc}}{\sqrt{2}\, L_A}\frac{A}{F}, \qquad \frac{C_{ox}}{C_{sc}} = \pm\frac{F}{r\,A}, \qquad r = \frac{\varepsilon_{sc}\, t_{ox}}{\varepsilon_{ox}\sqrt{2}\, L_A}.$$

Then, the $C(V_G)$ relation is found by eliminating u_s from

$$u_G' = u_s \pm 2\,r\,F, \qquad \frac{C}{C_{ox}} = \frac{1}{1 \pm F/(r\,A)}.$$

In particular, from (22.26) one finds $C(V_G' = 0) = C_{ox}/[1 + 1/(\sqrt{2}\,r)]$.

Problems of Chap. 23

23.1 The maximum initial profile is $N(x = 0, t = 0) = 2Q/\sqrt{\pi c_1}$. Remembering (23.26), at the end of the diffusion process the profile has become $N(x, t = t_P) = 2Q\,[\pi(c_1+c_2)]^{-1/2} \times \exp[-x^2/(c_1 + c_2)]$, whence $N(x = 0, t = t_P) = 2Q\,[\pi(c_1+c_2)]^{-1/2}$, with t_P the process duration and $c_2 = 4D\,t_P$. From $N(x = 0, t = t_P) = (2/3)\,N(x = 0, t = 0)$ it follows $1/\sqrt{c_1 + c_2} = 2/(3\sqrt{c_1})$, $c_2 = (5/4)\,c_1$ and, finally, $t_P = (5/16)\,(c_1/D) = 5{,}000$ s.

23.2 The initial and final profiles are $N_i(x) = 2\,Q\,\exp(-x^2/c_1)/(\pi c_1)^{1/2}$ and, from (23.26), $N_f(x) = 2\,Q\,\exp[-x^2/(c_1 + c_2)]/[\pi(c_1 + c_2)]^{1/2}$. Letting $N_f = N_i$ and defining $r = [(c_1 + c_2)/c_1]^{1/2}$ and $a^2 = (c_1 + c_2)\,c_1/c_2$ yields $r = \exp(x^2/a^2)$, whence $\bar{x} = 10^4 \times a\,(\log r)^{1/2} \simeq 2.68\ \mu m$

23.3 Converting to seconds one finds $t_P = 14{,}400$ s, whence $c_2 = 4D\,t_P = 14.4 \times 10^{-6}\ cm^2$. Considering that $N(x, t = 0) = 2\,Q\,\exp(-x^2/c_1)/(\pi c_1)^{1/2}$ and, from (23.26), $N(x, t = t_P) = 2\,Q\,\exp[-x^2/(c_1 + c_2)]/[\pi\,(c_1 + c_2)]^{1/2}$, the ratio sought is $N(x = 0, t = t_P)/N(x = 0, t = 0) = [c_1/(c_1 + c_2)]^{1/2} = 0.385$.

23.4 Due to (23.26), the final profile is $N_f = 2\,Q\,\exp[-x^2/(c_1 + c_2)]/[\pi(c_1 + c_2)]^{1/2}$. From $N(\bar{x}) = N_f(0)$ one has $\exp(-\bar{x}^2/c_1) = [c_1/(c_1 + c_2)]^{1/2}$. As a consequence, the position sought is $\bar{x} = [(c_1/2)\,\log(1 + c_2/c_1)]^{1/2} = 0.83\ \mu m$.

23.5 Due to (23.26) the final profile is $N_f = 2\,Q\,\exp[-x^2/(c_1 + c_2)]/[\pi(c_1 + c_2)]^{1/2}$. Let $\alpha = 1.1$. From the condition $N_f(0) = N(x = \alpha\sqrt{c_1})$ one derives $1/[\pi(c_1+c_2)]^{1/2} = \exp(-\alpha^2)/(\pi c_1)^{1/2}$, whence $c_1 = c_2/[\exp(2\alpha^2)-1] = 0.976 \times 10^{-9}\ cm^2$.

23.6 Letting $c_2 = 4D\,t = 2.4 \times 10^{-8}\ cm^2$ and $r = c_2/(c_1 + c_2) = 4/7$, one eliminates Q at $x = x_0$ to find $N_1/N_2 = \sqrt{1+r}\,\exp[-r\,x_0^2/(c_1 + 2\,c_2)]$, whence $x_0 = 10^4 \times [(1/r)(c_1 + 2\,c_2)\,\log(\sqrt{1+r}\,N_2/N_1)]^{1/2} = 4.06\ \mu m$.

23.7 Remembering (C.71) one has $Q = \int_0^\infty N_S\,\mathrm{erfc}(x/\sqrt{c})\,dx = N_S\sqrt{c/\pi}$, whence $c = 4D\,t = \lambda^2$, $t = \lambda^2/(4\,D) = 2.99 \times 10^{16}\ nm^2\ s^{-1} \simeq 5$ min.

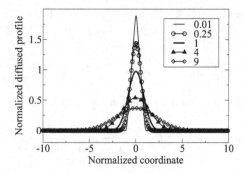

Fig. D.5 Normalized profiles $h\,N/Q$ resulting from the drive-in diffusion of Problem 23.8. The coordinate is $\mu = x/h$. Each profile corresponds to the value of $b(t)$ shown in the legend. The parameter is defined by $b = 4\,a/h^2$, while $a = a(t)$ is defined by the second of (23.10). As explained in Sect. 23.5.2, only the profile's portion on the right of the origin must be considered

23.8 As indicated in Sect. 23.5.2, the initial profile must preliminarily be mirrored onto the negative axis with $N_0 = 2\,Q\,(h + x)/h^2$ for $-h \le x \le 0$ and with $N_0 = 0$ for $x < -h$. Then, the profile is obtained from (23.13) as the portion of $N(x) = (2\,Q/h^2)(I^- + I^+)$ calculated for $x \ge 0$, where

$$I^- = \int_{-h}^{0} (h + \xi)\,\Delta(x - \xi, t)\,\mathrm{d}\xi, \qquad I^+ = \int_{0}^{+h} (h - \xi)\,\Delta(x - \xi, t)\,\mathrm{d}\xi$$

and $\Delta(x - \xi, t)$ is given by (C.75). Letting $\eta = (\xi - x)/(4\,a)^{1/2}$, $\mu = x/h$, $b = 4\,a/h^2$, and using (C.66) yields

$$N = (Q/h)\left[(\mu + 1)\,Y^- + (\mu - 1)\,Y^+ - (b/\pi)^{1/2}\,(Z^- + Z^+)\right],$$

with $Y^{\mp} = \mathrm{erfc}(\mu/\sqrt{b}) - \mathrm{erfc}[(\mu \pm 1)/\sqrt{b}]$ and $Z^{\mp} = \exp(-\mu^2/b) - \exp[-(\mu \pm 1)^2/b]$. When $t \to 0^+$ it is $b \to 0$. This makes the $Z^{\mp}$ terms to vanish, while the terms containing $Y^{\mp}$ render the initial condition N_0. When $t > 0$ the dose is found by integrating $h\,N$ over μ from 0 to $+\infty$. A somewhat lengthy calculation based on the expressions of Appendix C.7 shows that the integral of $(b/\pi)^{1/2}\,(Z^- + Z^+)$ vanishes whereas that of $(\mu+1)\,Y^- + (\mu-1)\,Y^+$ yields unity. As expected, the result of the dose calculation is Q. The normalized profile $h\,N/Q$ is shown in Fig. D.5 as a function of the normalized coordinate μ at different values of the parameter $b = b(t)$.

Problems of Chap. 24

24.1 The relation between time t and oxide thickness s is given by (24.11), $s^2/c_p + s/c_l = t'$ with $t' = t + s_i^2/c_p + s_i/c_l$. Solving for s and discarding the negative solution yields $s = [(c_p^2/c_l^2 + 4c_p t')^{1/2} - c_p/c_l]/2$, with $c_p/c_l = 0.05\,\mu\text{m}$. It follows $t' = t + s_i^2/c_p + s_i/c_l \simeq 136/60 + 0.995 + 0.237 = 3.50\,\text{h}$ and $4c_p t' = 0.620\,\mu\text{m}^2$.

The total oxide thickness and the thickness of silicon that is consumed in the second process are, respectively, $s = [(0.05^2 + 4c_p t')^{1/2} - 0.05]/2 \simeq 0.369$ μm and $h = 0.44\,(s - s_i) \simeq 70$ nm.

24.2 The gradient sought is found by remembering that, from (24.10), it is $c_p = 2\,w\,k_0\,N_G\,D_O$ and $c_l = w\,k_0\,N_G\,v_r$, whence $c_l/c_p = v_r/(2\,D_O) = 20$ μm^{-1}, $v_r/D_O = 4 \times 10^5$ cm^{-1}. On the other hand, from (24.7,24.6), $-D_O\,\mathrm{grad}N = -D_O\,dN/dx = v_r\,N_I$ whence $dN/dx = -N_O\,v_r/D_O = -4 \times 10^{17}$ cm^{-4}.

24.3 Converting the units one finds $c_p = 738$ nm^2 min^{-1}, $c_l = 14.8$ nm min^{-1}. Letting h be the thickness of silicon consumed one has $h = 0.44\,(s - s_i)$, $s = s_i + h/0.44$ whence, from (24.11), $t = (s^2 - s_i^2)/c_p + (s - s_i)/c_l = 150$ min.

24.4 Converting the units yields $c_p = 2000$ nm^2 min^{-1}, $c_l = 50$ nm min^{-1}. From (24.10) it follows $v_r/D_O = 2c_l/c_p = 0.05$, $1 + s\,v_r/D_O = 2$ whence, using (24.8), $N_I = N_O/(1 + s\,v_r/D_O) = 15 \times 10^{11}$ cm^{-3}.

24.5 Converting the units yields $c_p = 11.4$ nm^2 s^{-1}, $c_l = 0.237$ nm s^{-1}. From (24.10) the duration of the first process is found as $t_1 = (s_f^2 - s_i^2)/c_p + (s_f - s_i)/c_l$. Similarly, that of the second process is $t_2 = (s^2 - s_f^2)/c_p + (s - s_f)/c_l$. As the coefficients c_p, c_l are the same one adds the expressions of t_1 and t_2 to each other and lets $s = s_i + \Delta s_1 + \Delta s_2$. This yields $t_1 + t_2 = (s^2 - s_i^2)/c_p + (s - s_i)/c_l = 202$ s.

24.6 Converting the units yields $c_l = 16.67$ nm min^{-1}. Using the definitions (24.10) of c_p and c_l one finds $r = D_O/v_r = c_p/(2\,c_l)$, whence $t_P = [(s^2 - s_i^2)/(2r) + s - s_i]/c_l = 13.9$ min.

24.7 Letting $t = t_P$ and multiplying the first of (24.10) by c_p yields $s^2 + bs + c = 0$ with $b = c_p/c_l$, $c = -s_i^2 - (c_p/c_l)s_i - c_p t$. Here s_i and $c_p t_P$ are given while $c_p/c_l = 2D_O/v_r = 2.25 \times 10^{-6}$ cm. Solving for s and discarding the negative root provides the final thickness $s = [(b^2 - 4c)^{1/2} - b]/2 = 76.1$ nm.

24.8 From the relation $\Delta P = s\,\pi r^2\,\rho_{Si}$, where ΔP, s are the weight and thickness of the epitaxial layer, and r the wafer's radius, one finds $2\,r = 2\sqrt{\Delta P/(\pi s \rho_{Si})} \simeq 20.4$ cm $\simeq 8$ in.

24.9 The surface concentration N_S of SiCl$_4$ is found from the relations $s/t = c_l = w\,F_2 = w\,v_r\,N_S$, whence $N_S = s/(w\,v_r\,t) = 1 \times 10^{16}$ cm^{-3}.

24.10 Using $1/w = 5 \times 10^{22}$ cm^{-3} in the relations (24.22,24.23) yields $t = s/c_l = s/(w F_I) = 2$ min.

24.11 Letting t_P be the duration of the process one has, from (24.22,24.23), $c_l = s/t_P = w\,v_r\,N_I$, whence, using 1 μm $= 10^{-4}$ cm, $v_r = s/(w\,N_I\,t_P) = 200$ cm min^{-1}.

24.12 From (24.21) and the second of (24.23) one finds $b = (v_r + v_G)/(v_G\,w\,N_G) = (N_G/N_S)(w\,N_G)^{-1} = a/(w\,N_G)$, whence $N_G = a/(w\,b) = 2 \times (5 \times 10^{22}/4.87 \times 10^5) = 2.05 \times 10^{17}$ cm^{-3}.

Bibliography

References

1. N. D. Arora, J. R. Hauser, and D. J. Roulston. Electron and hole mobilities in silicon as a function of concentration and temperature. *IEEE Transactions on Electron Devices*, ED-29(2):292–295, Feb. 1982.
2. N. W. Ashcroft and N. D. Mermin. *Solid State Physics*. Saunders, 1976.
3. G. Baccarani, M. Rudan, and G. Spadini. Analytical IGFET model including drift and diffusion currents. *Solid-State and Electron Devices*, 2:62–68, 1978.
4. R. Becker. *Electromagnetic Fields and Interactions*. Dover, New York, 1982.
5. G. Birkhoff. Tres observaciones sobre el algebra lineal. *Univ. Nac. Tucuman Rev. Ser. A*, 5:147–150, 1946 (in Spanish).
6. J. S. Blakemore. *Semiconductor Statistics*. Dover, New York, 1987.
7. D. Bohm. *Quantum Theory*. Dover, New York, 1989.
8. D. Bohm and J. P. Vigier. Model of the causal interpretation of quantum theory in terms of a fluid with irregular fluctuations. *Phys. Rev.*, 96(1):208–216, 1954.
9. M. Born and E. Wolf. *Principles of Optics*. Pergamon Press, 6th edition, 1980.
10. A. Bravais. Mémoire sur les systèmes formés par les points distribués régulièrement sur un plan ou dans l'espace. *J. Ecole Polytech.*, 19(1):128, 1850 (in French. English version: Memoir 1, Crystallographic Society of America, 1949).
11. R. Brunetti, P. Golinelli, L. Reggiani, and M. Rudan. Hot-Carrier Thermal Conductivity for Hydrodynamic Analyses. In G. Baccarani and M. Rudan, editors, *Proc. of the 1996 ESSDERC Conference*, pages 829–832. Edition Frontiers, 1996.
12. R. Brunetti, M. C. Vecchi, and M. Rudan. Monte Carlo Analysis of Anisotropy in the Transport Relaxation Times for the Hydrodynamic Model. In C. Gardner, editor, *Fourth Int. Workshop on Computational Electronics (IWCE)*, Phoenix, 1995.
13. C. Y. Chang and S. M. Sze. *ULSI Technology*. McGraw-Hill, 1996.
14. A. G. Chynoweth. Ionization Rates for Electrons and Holes in Silicon. *Phys. Rev.*, 109(5):1537, March 1958.
15. C. Cohen-Tannoudji, B. Diu, and F. Laloë. *Quantum Mechanics*. John Wiley & Sons, New York, 1977.
16. L. Colalongo, M. Valdinoci, A. Pellegrini, and M. Rudan. Dynamic Modeling of Amorphous- and Polycrystalline-Silicon Devices. *IEEE Trans. El. Dev.*, ED-45:826–833, 1998.
17. L. Colalongo, M. Valdinoci, and M. Rudan. A Physically-Based Analytical Model for *a*-Si Devices Including Drift and Diffusion Currents. In K. Taniguchi and N. Nakayama, editors, *Simulation of Semiconductor Processes and Devices 1999 (SISPAD)*, pages 179–182, Kyoto, September 1999. IEEE.
18. L. Colalongo, M. Valdinoci, M. Rudan, and G. Baccarani. Charge-Sheet Analytical Model for Amorphous Silicon TFTs. In H. E. Maes, R. P. Mertens, G. Declerck, and H. Grünbacher,

editors, *Proc. of the 29th Solid State Device Research Conference (ESSDERC)*, pages 244–245, Leuven, September 1999. Edition Frontiers.

19. A. H. Compton. A Quantum Theory of the Scattering of X-Rays by Light Elements. *Phys. Rev.*, 21(5):483–502, May 1923.

20. L. N. Cooper. Bound Electron Pairs in a Degenerate Fermi Gas. *Phys. Rev.*, 104:1189–1190, November 1956.

21. C. R. Crowell. The Richardson constant for thermionic emission in Schottky barrier diodes. *Solid-State Electron.*, 8(4):395–399, 1965.

22. C. G. Darwin and R. H. Fowler. Fluctuations in an Assembly in Statistical Equilibrium. *Proc. Cambridge Phil. Soc.*, 21 (391):730, 1923.

23. S. Datta. *Quantum Transport: Atom to Transistor*. Cambridge University Press, Cambridge, 2006.

24. L. de Broglie. La mécanique ondulatoire et la structure atomique de la matière et du rayonnement. *J. Phys. Radium*, 8(5):225–241, 1927 (in French).

25. L. de Broglie. Interpretation of quantum mechanics by the double solution theory. In *Annales de la Fondation Louis de Broglie*, volume 12(4), pages 1–23. Fondation Louis de Broglie, 1987 (translated from the original 1972 version, in French).

26. N. G. de Bruijn. *Asymptotic Methods in Analysis*. Dover, New York, 1981.

27. E. De Castro. *Fondamenti di Elettronica—Fisica elettronica ed elementi di teoria dei dispositivi*. UTET, Torino, 1975 (in Italian).

28. B. E. Deal and A. S. Grove. General relationship for the thermal oxidation of silicon. *J. Appl. Phys.*, 36:3770, 1965.

29. P. Debye and E. Hückel. The theory of electrolytes. I. Lowering of freezing point and related phenomena. *Physikalische Zeitschrift*, 24:185–206, 1923.

30. P. A. M. Dirac. The Quantum Theory of the Emission and Absorption of Radiation. *Proc. R. Soc. Lond. A*, 114:243–265, 1927.

31. P. A. M. Dirac. The Quantum Theory of the Electron. *Proc. R. Soc. Lond. A*, 117:610–624, Feb. 1928.

32. P. A. M. Dirac. *The Principles of Quantum Mechanics*. Oxford University Press, Oxford, 4th edition, 1992.

33. H. M. Edwards. *Riemann's Zeta Function*. Dover, New York, 2001.

34. A. Einstein. Über einen über die Erzeugung und Verwandlung des Lichtes betreffenden heuristischen Gesichtspunkt. *Annalen der Physik*, 17(6):132–148, 1905 (in German). English translation: D. ter Haar, The Old Quantum Theory, Pergamon Press, 91–107, 1967.

35. A. Einstein. On the Movement of Small Particles Suspended in a Stationary Liquid Demanded by the Molecular-Kinetic Theory of Heat. In *Investigations on the theory of the Brownian movement*, chapter 1. Dover, New York, 1956.

36. A. Fick. Über Diffusion. *Ann. der Physik*, 94:59–86, 1855 (in German). Phil. Mag. 10, 30, 1855 (in English).

37. G. Floquet. Sur les équations différentielles linéaires à coefficients périodiques. *Annales Scientifiques de l'Ec. Norm. Sup.*, 12(2):47–88, 1883 (in French).

38. A. Forghieri, R. Guerrieri, P. Ciampolini, A. Gnudi, M. Rudan, and G. Baccarani. A new discretization strategy of the semiconductor equations comprising momentum and energy balance. *IEEE Trans. on CAD ICAS*, 7(2):231–242, 1988.

39. D. H. Frisch and L. Wilets. Development of the Maxwell-Lorentz Equations from Special Relativity and Gauss's Law. *Am. J. Phys.*, 24:574–579, 1956.

40. A. Gnudi, F. Odeh, and M. Rudan. Investigation of non-local transport phenomena in small semiconductor devices. *European Trans. on Telecommunications and Related Technologies*, 1(3):307–312 (77–82), 1990.

41. I. I. Gol'dman and V. D. Krivchenkov. *Problems in Quantum Mechanics*. Pergamon Press, London, 1961.

42. H. Goldstein, C. Poole, and J. Safko. *Classical Mechanics*. Addison Wesley, third edition, 2002.

43. P. Golinelli, R. Brunetti, L. Varani, L. Reggiani, and M. Rudan. Monte Carlo Calculation of Hot-Carrier Thermal Conductivity in Semiconductors. In K. Hess, J. P. Leburton, and U. Ravaioli, editors, *Proc. of the Ninth Intl. Conf. on Hot Carriers in Semiconductors (HCIS-IX)*, pages 405–408, Chicago, 1995. Plenum Press, New York.

44. I. S. Gradshteyn and I. M. Ryzhik. *Table of Integrals, Series, and Products*. Academic Press, New York, 1980.

45. T. Grasser, R. Korsik, C. Jungemann, H. Kosina, and S. Selberherr. "A non-parabolic six moments model for the simulation of sub-100nm semiconductor devices". *J. of Comp. Electronics*, 3:183–187, 2004.

46. A. S. Grove. *Physics and Technology of Semiconductor Devices*. J. Wiley & Sons, 1967.

47. G. Hardy, J. E. Littlewood, and G. Pólya. *Inequalities*. Cambridge University Press, Cambridge, second edition, 1952.

48. W. Heisenberg. Über den anschaulichen Inhalt der quantentheoretischen Kinematik und Mechanik. *Zeitschrift für Physik*, 43:172–198, 1927 (in German). English translation: J. A. Wheeler and H. Zurek, Quantum Theory and Measurement, Princeton Univ. Press, 62–84, 1983.

49. S.-M. Hong, A.-T. Pham, and C. Jungemann. *Deterministic Solvers for the Boltzmann Transport Equation*. Computational Microelectronics, S. Selberherr, Ed. Springer Verlag, Wien-New York, 2011.

50. K. Huang. *Statistical Mechanics*. Wiley, New York, second edition, 1987.

51. E. L. Ince. *Ordinary Differential Equations*. Dover, New York, 1956.

52. L. Infeld. On a New Treatment of Some Eingenvalue Problems. *Phys. Rev.*, 59:737–747, 1941.

53. J. D. Jackson. *Classical Electrodynamics*. John Wiley & Sons, New York, second edition, 1975.

54. C. Jacoboni. *Theory of Electron Transport in Semiconductors*. Springer, New York, first edition, 2010.

55. C. Jacoboni, R. Brunetti, and P. Bordone. *Theory of Transport Properties of Semiconductor Nanostructures*, volume 4 of *Electronics materials*, E. Schöll, Ed., chapter 3 "Monte Carlo simulation of semiconductor transport", pages 59–101. Chapman and Hall, first edition, 1998.

56. C. Jacoboni and P. Lugli. *The Monte Carlo Method for Semiconductor Device Simulation*. Computational Microelectronics, S. Selberherr, Ed. Springer Verlag, Wien-New York, 1989.

57. C. Jacoboni and L. Reggiani. The Monte Carlo method for the solution of charge transport in semiconductors with applications to covalent materials. *Rev. Mod. Phys.*, 55:645–705, 1983.

58. W. Jones and N. H. March. *Theoretical Solid State Physics*. Dover, 1973.

59. C. Jungemann, M. Bollhöfer, and B. Meinerzhagen. Convergence of the Legendre Polynomial Expansion of the Boltzmann Equation for Nanoscale Devices. In G. Ghibaudo, T. Skotnicki, S. Cristoloveanu, and M. Brillouët, editors, *Proc. of the 35th Solid State Device Research Conference (ESSDERC)*, pages 341–344, Grenoble, September 2005.

60. M. Kac. Some remarks on the use of probability in classical statistical mechanics. *Acad. Roy. Belg. Bull. Cl. Sci. (5)*, 42:356–361, 1956.

61. E. H. Kennard. Zur quantenmechanik einfacher bewegungstypen. *Zeitschrift für Physik*, 44:326, 1927 (in German).

62. P. Kiréev. *La Physique des Semiconducteurs*. MIR, Moscou, 1975 (in French).

63. C. Kittel. *Introduction to Solid State Physics*. J. Wiley & Sons, New York, seventh edition, 1953.

64. D. B. M. Klaassen. A unified mobility model for device simulation—I. Model equations and concentration dependence. *Solid-State Electr.*, 35(7):953–959, 1992.

65. C. Lanczos. *The Variational Principles in Mechanics*. Dover, New York, fourth edition, 1970.

66. L. Landau and E. Lifchitz. *Physique statistique*. MIR, Moscou, 1967 (in French).

67. L. Landau and E. Lifchitz. *Mécanique*. MIR, Moscou, 1969 (in French).

68. L. Landau and E. Lifchitz. *Théorie des Champs*. MIR, Moscou, 1970 (in French).

69. A. Landé. *New Foundations of Quantum Mechanics*. Cambridge University Press, 1965.

70. A. Landé. Solution of the Gibbs Entropy Paradox. *Philosophy of Science*, 32(2):192–193, April 1965.

71. ed. Levy, R. A. *Microelectronics Materials and Processes*, volume E-164 of *NATO ASI*. Kluwer, 1986.

72. M. J. Lighthill. *Fourier Analysis and Generalized Functions*. Cambridge University Press, Cambridge, 1962.

73. C. Lombardi, S. Manzini, A. Saporito, and M. Vanzi. A physically based mobility model for numerical simulation of nonplanar devices. *IEEE Transaction on CAD*, 7(11):1164–1171, novembre 1988.

74. A. M. Lyapounov. Problème Général de la Stabilité du Mouvement. *Ann. Fac. Sc. Univ. Toulouse*, 9(2):203–475, 1907 (in French).

75. W. Maly. *Atlas of IC Technologies: an Introduction to VLSI Processes*. The Benjamin/Cummings Publishing Co., 1987.

76. M. Marcus and H. Minc. *A Survey of Matrix Theory and Matrix Inequalities*. Dover, 1992.

77. E. Merzbacher. *Quantum Mechanics*. J. Wiley & Sons, New York, 1970.

78. A. Messiah. *Mécanique Quantique*. Dunod, Paris, 1969 (in French. English edition: Quantum Mechanics, Dover, New York, 1999).

79. M. Muskat and E. Hutchisson. Symmetry of the Transmission Coefficients for the Passage of Particles through Potential Barriers. *Proc. of the Nat. Academy of Sciences of the USA*, 23:197–201, April 15 1937.

80. D. A. Neamen. *Semiconductor Physics and Devices*. Irwin, 1992.

81. W. Pauli. The Connection Between Spin and Statistics. *Phys. Rev.*, 58:716–722, October 1940.

82. M. Planck. On an Improvement of Wien's Equation for the Spectrum. In *The Old Quantum Theory*, page 79. Pergamon Press, 1967.

83. S. Reggiani, M. Rudan, E. Gnani, and G. Baccarani. Investigation about the high-temperature impact-ionization coefficient in silicon. In R. P. Mertens and Cor L. Claeys, editors, *Proc. of the 34th Solid State Device Research Conference (ESSDERC)*, pages 245–248, Leuven, September 21–23 2004. IEEE.

84. S. Reggiani, M. C. Vecchi, A. Greiner, and M. Rudan. Modeling hole surface- and bulk-mobility in the frame of a spherical-harmonics solution of the BTE. In K. De Meyer and S. Biesemans, editors, *Simulation of Semiconductor Processes and Devices 1998 (SISPAD)*, pages 316–319. Springer-Verlag, Wien, Austria, 1998.

85. F. Reif. *Fundamentals of Statistical and Thermal Physics*. McGraw-Hill, New York, 1985.

86. M. Rudan and G. Baccarani. On the structure and closure condition of the hydrodynamic model. *VLSI Design* (Special Issue, J. Jerome, Ed.), 3(2):115–129, 1995.

87. M. Rudan, E. Gnani, S. Reggiani, and G. Baccarani. The density-gradient correction as a disguised pilot wave of de Broglie. In G. Wachutka and G. Schrag, editors, *Simulation of Semiconductor Processes and Devices 2004 (SISPAD)*, pages 13–16, Munich, September 2–4 2004. Springer.

88. M. Rudan, A. Gnudi, E. Gnani, S. Reggiani, and G. Baccarani. Improving the Accuracy of the Schrödinger-Poisson Solution in CNWs and CNTs. In G. Baccarani and M. Rudan, editors, *Simulation of Semiconductor Processes and Devices 2010 (SISPAD)*, pages 307–310, Bologna, September 6–8 2010. IEEE.

89. M. Rudan, A. Gnudi, and W. Quade. *Process and Device Modeling for Microelectronics*, chapter 2 "A Generalized Approach to the Hydrodynamic Model of Semiconductor Equations", pages 109–154. G. Baccarani, Ed. Elsevier, 1993.

90. M. Rudan, M. Lorenzini, and R. Brunetti. *Theory of Transport Properties of Semiconductor Nanostructures*, volume 4 of *Electronics materials*, E. Schöll, Ed., chapter 2 "Hydrodynamic simulation of semiconductor devices", pages 27–57. Chapman and Hall, first edition, 1998.

91. M. Rudan and F. Odeh. Multi-dimensional discretization scheme for the hydrodynamic model of semiconductor devices. *COMPEL*, 5(3):149–183, 1986.

92. M. Rudan, F. Odeh, and J. White. Numerical solution of the hydrodynamic model for a one-dimensional semiconductor device. *COMPEL*, 6(3):151–170, 1987.

93. M. Rudan, M. C. Vecchi, and D. Ventura. *Mathematical problems in semiconductor physics*, chapter "The Hydrodynamic Model in Semiconductors — Coefficient Calculation for the Conduction Band of Silicon", pages 186–214. Number 340 in Pitman Research Notes in Mathematical Series P. Marcati, P. A. Markowich, R. Natalini, Eds. Longman, 1995.

94. E. Schrödinger. Quantisierung als eigenwertproblem (erste mitteilung). *Annalen der Physik*, 384(4):361–376, 1926 (in German).

95. E. Schrödinger. *Statistical Thermodynamics*. Dover, New York, 1989.

96. M. Schwartz. *Principles of Electrodynamics*. Dover, New York, 1987.

97. W. Shockley. The Theory of *p-n* Junctions in Semiconductors and *p-n* Junction Transistors. *Bell Syst. Tech. J.*, 28:435, 1949.

98. W. Shockley. *Electrons and Holes in Semiconductors*. Van Nostrand Book Co., 1950.

99. J. C. Slater. *Quantum Theory of Matter*. McGraw-Hill, New York, 1968.

100. J. W. Slotboom and H. C. DeGraaff. Measurement of bandgap narrowing in silicon bipolar transistors. *Solid-State Electron.*, 19(2):857–862, 1976.

101. J. W. Slotboom and H. C. DeGraaff. Bandgap narrowing in silicon bipolar transistors. *IEEE Trans. El. Dev.*, 28(8):1123–1125, August 1977.

102. J. C. F. Sturm. Mémoire sur les équations différentielles linéaires du deuxième ordre. *Journal de Mathématiques Pures et Appliquées*, 1:106–186, 1836 (in French).

103. S. M. Sze. *Physics of Semiconductor Devices*. John Wiley & Sons, NewYork, 1981.

104. S. M. Sze. *Semiconductor Devices—Physics and Technology*. J. Wiley & Sons, 1985.

105. S. M. Sze. *VLSI Technology*. McGraw-Hill, 1988.

106. S. Takagi, A. Toriumi, M. Iwase, and H. Tango. On the Universality of Inversion Layer Mobility in Si MOSFET's: Part I—Effects of Substrate Impurity Concentration. *IEEE Trans. El. Dev.*, 41(12):2357–2362, December 1994.

107. D. ter Haar. On a Heuristic Point of View about the Creation and Conversion of Light. In *The Old Quantum Theory*, pages 91–107. Pergamon Press, 1967.

108. R. Thoma, A. Emunds, B. Meinerzhagen, H.-J. Peifer, and W. L. Engl. Hydrodynamic Equations for Semiconductors with Nonparabolic Band Structure. *IEEE Trans. El. Dev.*, 38(6):1343–1353, 1991.

109. R. C. Tolman. Note on the Derivation from the Principle of Relativity of the Fifth Fundamental Equation of the Maxwell-Lorentz Theory. *Phil. Mag. S. 6*, 21(123):296–301, 1911.

110. R. C. Tolman. *Statistical Mechanics*. Dover, NewYork, 1979.

111. M. Valdinoci, D. Ventura, M. C. Vecchi, M. Rudan, G. Baccarani, F. Illien, A. Stricker, and L. Zullino. Impact-ionization in silicon at large operating temperature. In K. Taniguchi and N. Nakayama, editors, *Simulation of Semiconductor Processes and Devices 1999 (SISPAD)*, pages 27–30, Kyoto, September 1999. IEEE.

112. M. C. Vecchi and M. Rudan. Modeling Electron and Hole Transport with Full-Band Structure Effects by Means of the Spherical-Harmonics Expansion ot the BTE. *IEEE Trans. El. Dev.*, 45(1):230–238, 1998.

113. G. H. Wannier. *Statistical Physics*. Dover, NewYork, 1996.

114. J. Ward Brown and R. V. Churchill. *Complex Variables and Applications*. McGraw-Hill, NewYork, 1996.

115. R. Weinstock. *Calculus of Variations*. Dover, NewYork, 1974.

116. H. Weyl. Quantenmechanik und gruppentheorie. *Zeitschrift für Physik*, 46(1–2):1–46, 1927 (in German).

117. J. A. Wheeler and W. H. Zurek. *Quantum Theory and Measurement*, pages 62–84. Princeton Univ. Press, 1983.

118. N. Wiener. *The Fourier Integral and Certain of Its Applications*. Dover, NewYork, 1958.